THE GREAT BASIN

The Great Basin

A NATURAL PREHISTORY
REVISED AND EXPANDED EDITION

DONALD K. GRAYSON

UNIVERSITY OF CALIFORNIA PRESS
Berkeley   Los Angeles   London

University of California Press, one of the most distinguished university presses in the United States, enriches lives around the world by advancing scholarship in the humanities, social sciences, and natural sciences. Its activities are supported by the UC Press Foundation and by philanthropic contributions from individuals and institutions. For more information, visit www.ucpress.edu.

For a digital version of this book, see the press website.

University of California Press
Berkeley and Los Angeles, California

University of California Press, Ltd.
London, England

© 2011 by the Regents of the University of California

Library of Congress Cataloging-in-Publication Data

Grayson, Donald K.
   The great basin : a natural prehistory / Donald K. Grayson. — Rev. and expanded ed.
      p.   cm.
   Rev. ed. of: The desert's past. c1993.
   Includes bibliographical references and index.
   ISBN 978-0-520-26747-3 (hardback)
   1. Geology, Stratigraphic—Pleistocene. 2. Geology, Stratigraphic—Holocene. 3. Geology—Great Basin. 4. Paleontology—Great Basin. 5. Indians of North America—Great Basin—Antiquities. 6. Paleo-Indians—Great Basin. 7. Great Basin—Antiquities. I. Grayson, Donald K. Desert's past. II. Title.
   QE697.G8   2011
   508.79—dc22   2010052100

The paper used in this publication meets the minimum requirements of ANSI/NISO Z39.48-1992 (R 1997)(Permanence of Paper).

Cover image: Black Rock Desert, Nevada. Photo by the author.

For Barbara, who was my life

CONTENTS

PREFACE ix
ACKNOWLEDGMENTS xiii

PART ONE
The Great Basins

1. Discovering a Great Basin / 3
   Chapter Notes / 9

2. Modern Definitions of the Great Basin / 11
   The Hydrographic Great Basin / 11
   The Physiographic Great Basin / 12
   The Floristic Great Basin / 17
   The Ethnographic Great Basin / 33
   Choosing a Great Basin / 40
   Chapter Notes / 40

PART TWO
Some Ice Age Background

3. Glaciers, Sea Levels, and the Peopling of the Americas / 45
   The Bering Land Bridge and the Human Arrival / 45
   Identifying the Earliest American Archaeology / 54
   The Earliest American Archaeology / 57
   Chapter Notes / 63

4. The End of the North American Pleistocene: Extinct Mammals and Early Peoples / 67
   The Mammals / 67
   A Note on Eurasian Extinctions / 73
   Back to North America / 74
   What Caused the Extinctions? / 79
   Chapter Notes / 84

PART THREE
The Late Ice Age Great Basin

5. The Late Pleistocene Physical Environment: Lakes and Glaciers / 87
   Modern Great Basin Lakes / 87
   Pleistocene Lakes in the Great Basin / 93
   Great Basin Glaciers / 120
   The Relationship between Pleistocene Glaciers and Lakes in the Great Basin / 124
   Pluvial Lakes, Glaciers, and Late Pleistocene Climates / 127
   Chapter Notes / 130

6. Late Pleistocene Vegetation of the Great Basin / 135
   Learning about Ancient Vegetation / 135
   Five Regional Pictures / 139
   A General Look at Late Pleistocene Great Basin Vegetation / 159
   Great Basin Conifers in Deeper Historical Perspective / 168
   Chapter Notes / 169

7. Late Pleistocene Vertebrates of the Great Basin / 173
   The Extinct Late Pleistocene Mammals of the Great Basin / 176
   Extinct Late Pleistocene Birds / 181
   Altered Late Pleistocene Distributions of Existing Great Basin Mammals / 186
   Why Are There Fishes in Devils Hole, but None in the Great Salt Lake? / 206
   Other Vertebrates / 210
   Chapter Notes / 211

PART FOUR
The Last 10,000 Years

8. The Great Basin during the Holocene / 217
   The Early Holocene (10,000 to 7,500 Years Ago) / 217
   The Middle Holocene (7,500 to 4,500 Years Ago) / 242
   The Late Holocene (The Last 4,500 Years) / 259
   Chapter Notes / 282

PART FIVE
Great Basin Archaeology

9. The Prehistoric Archaeology of the Great Basin / 289
   Pre-Clovis Sites in the Great Basin / 289
   The Latest Pleistocene and Early Holocene / 289
   The Middle Holocene / 302
   The Late Holocene / 313
   Chapter Notes / 333

PART SIX
Conclusions

10. The Great Basin Today and Tomorrow / 341
    Deer, Cougars, Porcupines, and Cattle / 341
    More Lessons from the Past / 342
    The Great Basin Today / 344
    Chapter Notes / 344

APPENDIX 1: RELATIONSHIP BETWEEN RADIOCARBON AND CALENDAR YEARS FOR THE PAST 25,000 RADIOCARBON YEARS / 347
APPENDIX 2: CONCORDANCE OF COMMON AND SCIENTIFIC PLANT NAMES / 351

REFERENCES / 355
INDEX / 409

PREFACE

Most North Americans grow up knowing that parts of our continent were once covered by glaciers, that now-extinct mammoths and sabertooth cats walked the same ground on which we now walk our dogs, that people discovered North America long before Columbus stumbled across it. Most of us acquired this knowledge so casually that, if we happen to be asked exactly when these things occurred, we have no real answer. We would probably know that mammoths looked like elephants, but not that they became extinct about 11,000 years ago in North America. We might know about Ice Age glaciers and still not know that the maximum expanse of the most recent glaciation of North America occurred about 18,000 years ago. We might just shrug if asked when people first got to the Americas.

Answers to these questions are easy to learn. It takes no great insight, and little effort, to register the fact that mammoths became extinct in North America about 11,000 years ago. It is, however, harder to grasp the nature and magnitude of change that has occurred throughout North America during and since the end of the Ice Age.

We tend to assume that landscapes and the life they support are relatively permanent affairs unless human activity modifies them. We are not surprised when the farmland that surrounds the town we grew up in gives way to subdivisions and shopping malls: that kind of change we are used to and have come to expect and perhaps regret. But it is very surprising to learn how ephemeral the assemblages of plants and animals that surround us today really are and, in most cases, how recently those assemblages came into being. The brevity of our lives easily misleads us into thinking that the way things are today is the way they have been for an immense amount of time. It is even easier to be misled into thinking that things are the way they are now because they have to be that way.

Until recently, life scientists attributed far greater stability, longevity, and predictability to biological communities than those communities actually possess. One of the great scientific gains of the past few decades is the recognition of the vital role that history has played in forming the plant and animal communities that now surround us, the recognition of how unpredictable changes in those communities can be, and the recognition of how fleeting their existence often is. Plant and animal communities appear stable and real to us only because we do not live long enough to observe differently. Bristlecone pines, which do live long enough, know better.

Today, most life scientists, especially those whose work has any significant time depth, also know better. Although we may not live as long as bristlecone pines, we do have techniques for extracting information about earth and life history that can tell us not only what specific landscapes were like in the past but also precisely when in the past they were like that. Even though we have made less progress toward understanding why they may have been that way, we have come a long way in this realm as well. We know now enough about at least the late Ice Age and the times that followed to be able to provide fairly detailed environmental histories for nearly all parts of North America.

In this book, I provide such a history for the Great Basin. I define the Great Basin in multiple ways in Chapter 2, but here, suffice it to say that the Great Basin centers on the state of Nevada, but also includes substantial parts of adjacent California, Oregon, and Utah. My goal is simple: to outline the history of Great Basin environments from about the time of the last maximum advance of glaciers in North America to the arrival of Europeans and their written records. In so doing, I hope to convey the dynamic nature of the landscapes and life of this region.

During the late Ice Age, camels lived near what is now Pyramid Lake in northwestern Nevada; massive glaciers existed in the high mountains of eastern Nevada; substantial lakes lay in settings as far-flung as Death Valley and the Great Salt Lake Desert; trees grew in the valleys of the Mojave Desert of southern Nevada. The camels, glaciers, lakes, and low-elevation trees are now gone. Today, pinyon-juniper woodlands drape across millions of mountain-flank acres in the Great Basin, and saltbush vegetation is common in many of the valleys that lie beneath the woodlands. To the south, the oddly attractive creosote bush is a dominant shrub in the valley bottoms. This is a remarkably recent state of affairs, and a prime goal of this book is to document these facts and to discuss why they are so.

It was not hard for me to decide what to cover here: glaciers and lakes, shrubs and trees, mammals and birds, the people. Others might have chosen a different set of topics, in some cases broader, in others narrower. The set I have chosen, however, not only strikes me as important but also reflects my background as a scientist trained in archaeology, vertebrate paleontology, and paleoecology. Someone who knows more about insects, leeches, and snails would have written a different book. In fact, I wish they would, since I would like to read it. The content of this book also reflects the fact that it has been my great fortune to know and to work with most of the people whose work is discussed here.

Deciding on the temporal coverage was more difficult. That this book would deal with the past 10,000 years was clear from the outset, as was the fact that it would also cover the waning years of the Ice Age or Pleistocene. These are, after all, years that were critical to the formation of the plant and animal communities of the Great Basin as we know it today. They also happen to be the years on which much of my own Great Basin work has focused. In the end, I decided to begin my coverage at about 25,000 years ago, though sometimes earlier and sometimes later. I made that decision both because 25,000 years ago allows me to discuss the Great Basin prior to and during the Last Glacial Maximum, and because our knowledge of Great Basin environments tails off sharply before that date.

Today, the term "natural history" is often used in a very general way to refer to the things that life and earth scientists study. That is the way I use it here, modifying it in the title to indicate that this book deals primarily with events that took place prior to the times for which written records are available. Thus, this book is very much a "natural prehistory," dealing with the landscapes of the Great Basin, and the life it supported, during the past 25,000 years or so.

In an important essay, Great Basin anthropologists Don and Kay Fowler discussed the fact that American Indians and other non-Western peoples became incorporated into Western notions of "natural history" not because all human life was so incorporated, but because non-Westerners were perceived as being more primitive, as "closer to nature" (Fowler and Fowler 1991:47), than western peoples. As usual, the Fowlers are completely right.

It is hard to shake the deeply embedded, pejorative implications of including non-Western peoples in a "natural history" of any place, but the shaking is needed. The fault lies not in the inclusion of non-Westerners in the examination of natural history, but in the exclusion of Western peoples. That we are all very much part of the natural world should be obvious, given such things as the impacts of Hurricane Katrina and the likely impacts of global warming. It is even possible that we are now "closer to nature" than are, say, the small-scale foraging and farming societies of Amazonia, given that we are far more vulnerable to massive losses due to environmental assaults. On the other hand, the prehistoric peoples of the Great Basin were "closer to nature" than the contemporary peoples of Reno or Salt Lake City are now, not in the nineteenth-century sense that they were further removed from God or closer to the beasts of the earth, but in the very real sense that they had to cope far more immediately, and on a daily basis, with the environmental challenges that nature dealt them. The very same is true for the early historic human occupants of the Great Basin, whether native or not. But because my emphasis is on the prehistoric Great Basin—the Great Basin prior to the time of written records—it is the prehistoric archaeological record that occupies me here.

I also note that there are a few places in this book where I repeat information given earlier. I have done this because I want people to dip into this book wherever they wish and have tried to make each chapter as independent as I could from earlier chapters. Doing that required some minor repetition, but, in the end, it does mean that the chapter on late Pleistocene vertebrates, for instance, can be read without having incorporated all that has come before. I hate flipping back and forth in lengthy books to remind myself of what came long before, and the sparing repetition is meant to help avoid that.

I would be pleased to discover that Great Basin archaeologists, ecologists, geologists, paleobotanists, and paleozoologists had read and learned something, no matter how minor, from this book. But the truth is that I did not write this book for my professional colleagues. Instead, I wrote it for those who know little if anything about the environmental history of the Great Basin, or even about the modern Great Basin. Although I have worked in this region for over forty years (I started young), I have yet to lose the excitement that comes from identifying the bones of an extinct horse or camel from Ice Age deposits. I continue to be awed by the Bonneville Basin, from its salt flats to the high terraces carved on its mountains, both products of Pleistocene Lake Bonneville. I will never forget the moment I discovered the remains of a 5,300-year-old heather vole in the deposits of central Nevada's Gatecliff Shelter or bushy-tailed woodrats living in the hot and dry Lakeside Mountains of Utah—both because these discoveries were unexpected and because of what they meant for our understanding of the histories of those animals in the Great Basin. I find the human prehistory of the Great Basin exciting, not because of the often-impressive nature of the artifacts people left behind but because of the varied and severe environmental challenges the people who made these things met successfully. I wrote this book because I wanted to share all of this with those who know little or nothing about it.

As a result, I have assumed that the readers of this book come to it with little knowledge of such things as radiocarbon dating, pollen analysis, packrat middens, equilibrium-line altitudes, and projectile point chronologies. I explain them here. I also assume that readers know little about the modern Great Basin (Chapter 2), about North America during the Ice Age, or about the initial peopling of the New World (Chapters 3 and 4). In the first two parts of this book, I have spent a good deal of time providing that essential background. Those already in the know might skip these parts, though I depend heavily on them in later sections of the book.

Some technical comments are needed. I provide both scientific and common names for plants and animals the first time I mention them in the book, then use one or the other (usually the common name) later on. An appendix provides a concordance of the common and scientific names of plant species. Concordances for the names of vertebrates used more than once are given in tables that accompany the text. As I discuss in Chapter 3, radiocarbon dates are not necessarily the same as calendar dates (see appendix 1); unless otherwise noted, the dates I provide here are the former.

It is standard in academic works to provide citations to the works of others in the text, as each work is called upon. With one exception, I have not done that here, because I do not want to interrupt the text with lengthy lists of the works on which I have depended so heavily. Instead, each chapter ends with a set of "Chapter Notes." Those notes provide the references I have used, often along with comments on those works. I have also used the chapter notes to discuss things that did not seem appropriate for the main text, including places that I think are worth visiting, from archaeological sites to local museums. The notes are an integral part of the book, but using them to provide the references has left the text far less cluttered than it otherwise would have been. The one exception I have made involves direct quotations: there, the source of the quotation is provided in the text itself.

If visitors to Great Basin National Park, Death Valley National Monument, Malheur National Wildlife Refuge, the Bonneville Salt Flats, Pyramid Lake, or the striking wilderness areas the Great Basin has to offer—Alta Toquima, Arc Dome, and Steens Mountain, for instance—have more meaningful trips for having read this book, I will be pleased. I have written it both for them and for those who live in the Great Basin today. If my scientific colleagues find it of value as well, I will be happier still.

This book represents a thoroughly updated version of an earlier work, *The Desert's Past: A Natural Prehistory of the Great Basin*, published in 1993 by the Smithsonian Institution and reissued in paperback in 1998. Most of the book is so different that it has earned a new title.

Donald K. Grayson
Seattle, Washington

ACKNOWLEDGMENTS

A list of the people who have taught me about the Great Basin may quite literally be found in the References section of this book. In addition, however, there are many people who discussed Great Basin issues with me, sometimes at great length, provided me with everything from references and manuscripts to reprints and maps, allowed me access to museum collections, and accompanied me on various field trips crucial to the preparation of this book. These friends and colleagues include Susan Abele, Ken Adams, Richard Adams, Mel Aikens, Scott Anderson, Bax Barton, Charlotte Beck, Larry Benson, Julio Betancourt, Jordon Bright, Paul Buck, Virginia Butler, Bill Cannon, Mike Cannon, Kim Carpenter, David Charlet, Chris and Chelise Crookshanks, Owen Davis, Don Fowler, Kay Fowler, Dorothy Friedel, Alan Gillespie, Stacy Goodman, Kelly Graf, Barbara Grayson, Margaret Helzer, Ann Haniball, Bob Hanover, Don Hardesty, Kim Harper, Gene Hattori, Vance Haynes, Jane Hill, Bryan Hockett, Richard Hughes, Joel Janetski, Angela Jayko, Dennis Jenkins, Stephanie Jolivette, Tom Jones, Darrell Kaufman, Chris Kiahtipes, Guy King, Lee Kreutzer, Ben Laabs, Sydney Lamb, Lisbeth Louderback, Karen Lupo, Patrick Lyons, David Madsen, Debby Mayer, Kelly McGuire, Jim Mead, Dave Meltzer, Jeff Meyers, Connie Millar, Cary Mock, Jan Nachlinger, Jim O'Connell, Jack Oviatt, Christy Parry, Bruce Pavlik, Lori Pendleton, Rachel Quist, Anan Raymond, Marith Reheis, Dave Rhode, Joe Rosenbaum, Dave Schmitt, Steve Simms, Denise Sims, Geoff Smith, Amanda Taylor, Dave Thomas, Bob Thompson, Claude Warren, Peter Wigand, and Jim Wilde.

Ken Adams, Charlotte Beck, Bryan Hockett, Dave Hunt, Lisbeth Louderback, Dave Meltzer, Connie Millar, David Madsen, Jim O'Connell, Jack Oviatt, and Dave Rhode read parts of this manuscript for me. Joel Janetski, Steve Simms, and Louis Warren read the entire thing. All provided me with many insightful suggestions about a broad variety of issues. The book is far better because of their kindness and generosity.

For help in obtaining permission to reproduce the illustrations that appear here as Figures 2-12, 4-3, 6-7, and 9-15, I thank Margaret Fisher Dalrymple of the University of Nevada Press, Paul and Suzanne Fish and Patrick Lyons of the Arizona State Museum, and Dave Thomas of the American Museum of Natural History. Thanks also to Peggy Corson, who prepared the projectile point illustrations that appear in chapter 9. I am deeply grateful to the National Science Foundation, the Legacy Program of the Department of Defense, and the University of Washington's Quaternary Research Center for supporting my own Great Basin research, much of which has been summarized in this book.

Were it not for the encouragement, friendship, and support of Charlotte Beck, Larry Benson, Joel Janetski, Tom Jones, David Madsen, Connie Millar, Dave Rhode, and Dave Thomas, this new version would not have happened.

My deepest debt, though, is to Barbara. This book would not have existed without her.

PART ONE

THE GREAT BASINS

CHAPTER ONE

Discovering a Great Basin

It was July 13, 1890, and the first Republican candidate for the presidency of the United States lay dying in a New York City boardinghouse, his son by his side, his celebrated wife, Jessie, in Los Angeles, a continent away. Seventy-seven years old, John C. Frémont had come to New York from Washington, where he had finally obtained a $6,000 yearly pension for his military service. That sum, Frémont hoped, would secure his family from the poverty that had marked their recent life, but he had not counted on dying so soon, and Congress had made no provision for continuing a pension in the absence of a pensioner.

The events that took place in Washington and New York that spring and summer echoed sequences that seemed to mark everything Frémont did: grand successes followed by remarkable failures. Born to loving parents but illegitimate at a time when that mattered; a hero to some in the Bear Flag Revolt of 1846 that led California to independence but court-martialed and convicted for what General Stephen Watts Kearny saw as mutiny; nominated for president but smeared as a "Frenchman's bastard" and defeated by James Buchanan; a millionaire in California but soon bankrupt; a Californian in the end but buried in New York because so many Californians opposed the use of public funds to bring him west for one last time. Ironies everywhere, but they especially surround his final resting place, overlooking a river named for the great explorer Henry Hudson—second place in death for one who desperately wanted first place in life but could never quite hold on to it. He was buried in New York, where not one significant place carries his name, and not in California, where he himself named so many significant things—Walker River, Owens Valley, and even the Golden Gate, above which he might have been buried. He was denied the final trip west by Californians, citizens of the very state his efforts had helped swing from Mexican to American control, citizens whose parents and perhaps even themselves had been spurred to come west by his *Report of the Exploring Expedition to Oregon and North California in the Years 1843–1844*. Buried not in California, where he had once been a hero, but in New York, the state to which he and Jessie had retreated in personal defeat after his twice-failed role as a Union general in the Civil War.

Of Frémont's successes, perhaps the grandest was his second expedition for the U.S. Bureau of Topographical Engineers. His first, in 1842, had gone from St. Louis to just beyond South Pass in the northern Rocky Mountains of Wyoming, an expedition that he made with Kit Carson—John and Jessie Frémont together turned Kit into a legend—as one of his guides. The second expedition was to go much farther.

Although following from, and funded as a result of, the expansionist dreams of Thomas Hart Benton, the powerful senator from Missouri and Frémont's father-in-law, it is not clear what unwritten goals Frémont carried with him on this second excursion deep into the American West. What is clear is that he went farther than his written orders allowed, wintering in Mexican California even though he was a representative of the American military. It is also clear that he had no written authorization to bring along a twelve-pound mountain howitzer, the famous Frémont cannon.

He left St. Louis in May 1843; three months later, he was back at South Pass, the terminus of his first expedition, but now simply the jumping-off point for the work that was to make him famous (figure 1-1). Accompanied once again by Kit Carson, Frémont made his way south to the Bear River, his description of which was to be crucial in guiding the Mormons to Salt Lake Valley in 1847. On September 6, the expedition reached the Great Salt Lake, Frémont's "Inland Sea, stretching in still and solitary grandeur far beyond the limit of our vision" (Frémont 1845:151).

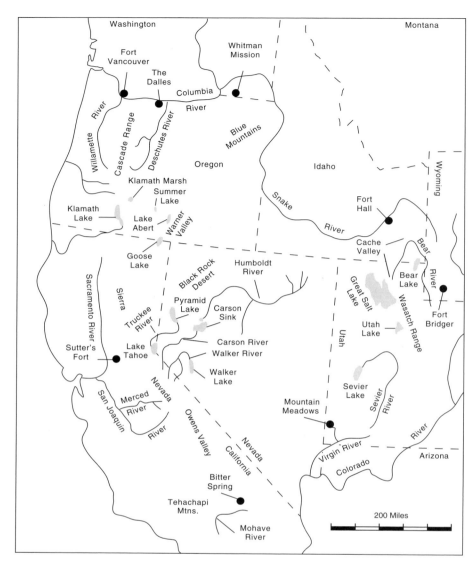

FIGURE 1-1 The American West.

The Great Salt Lake was a major target of the party's work and it spent nearly a week here, exploring the lake's shores by foot and its waters by boat. The expedition renewed its journey on September 12, heading north to Fort Hall on the Snake River, then down along the Snake River to Fort Boise in western Idaho. From here, the explorers cut across the Blue Mountains of northeastern Oregon, reaching Marcus Whitman's mission just east of the Columbia River near modern Walla Walla, in southeastern Washington, on October 24. They then traveled west along the Columbia River, arriving at Fort Vancouver on November 8.

Fort Vancouver moved several times during its history, but when Frémont arrived, it was on the north side of the Columbia River, just north of the mouth of the Willamette River. Today, it is within the city limits of Vancouver, Washington, and a reconstructed version exists as the Fort Vancouver National Historic Site. At 300 feet wide and 700 feet long, it was massive, the Hudson's Bay Company's prime redistributive and administrative center in this part of the world. Ships came up the Columbia River to supply the fort and to be supplied with furs (the British bark *Columbia* was there when Frémont arrived). The Fort also served as a stopping-off point for the growing number of American emigrants who were then entering Oregon's Willamette Valley, some 2,400 of them in 1843 and 1844.

Frémont was well treated at Fort Vancouver, as were all others who came here, but he stopped only long enough to stock up for the return home. His orders for that return were simple. "Return by the Oregon road," Colonel J.J. Abert had ordered, "and on again reaching the mountains, diverge a little and make a circuit of the Wind river chain" (Jackson and Spence 1970:160). But the "Oregon Road" was pretty much how Frémont had come to be where he was, and he was not about to return the way he had gotten there. When

Frémont stocked up, he did so for a far more rigorous journey, and by the time he had returned to The Dalles, on the eastern edge of the Columbia's passage through the Cascade Range, he had three month's worth of supplies for his twenty-five men, along with a herd of cattle and 104 mules and horses. And, rather than heading east from The Dalles, as Abert's orders indicated he should, he headed south.

Leaving The Dalles on November 25 in the midst of flurrying snow, Frémont moved south along the eastern flank of the Cascades, past the Metolius River, past the headwaters of the Deschutes, and south to Klamath Marsh. Arriving at the marsh on December 10, Frémont used his cannon for the first time, discharging it to impress the Indians whose fires were visible across the marsh. These were the Klamath, Frémont knew, but he was incorrect in thinking that this was Klamath Lake and that the river he had found here—the Williamson—was the Klamath River. In fact, Klamath Lake was still thirty miles to the south. But thinking he had found the lake, Frémont spent several days at this spot, resting his horses, exploring, and even buying a little dog that he named Tlamath. Once satisfied with what he had seen, he headed east, leaving on December 13. Three days later, after the explorers forced themselves, their animals, and the howitzer through deep and crusted snow, the woods suddenly ended:

> We found ourselves on the vertical and rocky wall of the mountain. At our feet—more than a thousand feet below—we looked into a green prairie country, in which a beautiful lake, some twenty miles in length, was spread along the foot of the mountains, its shores bordered with green grass.... Not a particle of ice was to be seen on the lake, or snow on its borders, and all was like summer or spring.... Shivering on snow three feet deep, and stiffening in a cold north wind, we exclaimed at once that the names of Summer Lake and Winter Ridge should be applied to these two proximate places of such sudden and violent contrast. (Frémont 1845:207)

These places still bear the names Frémont gave them, a highway marker on Oregon State Route 31 pointing out where Frémont and his men suffered their way down Winter Ridge on the evening of December 16, 1843, leaving the howitzer halfway up, to be retrieved the next day. Now, he said, they were "in a country where the scarcity of water and of grass makes traveling dangerous, and great caution was necessary" (Frémont 1845:208). Frémont had entered the Great Basin.

From here, the group continued south and east, farther into the Oregon desert. Lake Abert came next, so named by Frémont "in honor of the chief of the corps to which I belonged" (Frémont 1845:209), then farther south and east to Warner Valley, where Christmas Day was celebrated with a blast from the howitzer. Crossing the 42nd parallel, which today marks the boundary between Nevada and Oregon but which then marked his passage into Mexican territory, they moved deeper into northwestern Nevada: High Rock Creek, it seems; then Soldier Meadow; then, on New Year's Day, along the Black Rock Desert through what is now Gerlach, Nevada; and then, on January 11, to Pyramid Lake, "a sheet of green water, some twenty miles broad [that] broke upon our eyes like the ocean" (Frémont 1845:216). Here they rested, trading for cutthroat trout with the Northern Paiute who occupied the shores of the lake, allowing their horses to feed, killing the last of their cattle, and getting their howitzer unstuck from the steep shores of the Lake Range that forms the eastern edge of Pyramid Lake.

The lake itself they named from the "very remarkable rock" they saw jutting from it, a rock that to them "presented a pretty exact outline of the great pyramid of Cheops" (Frémont 1845:217). Once recuperated, they followed the Truckee River south, and then left it as it swung west toward the Sierra Nevada. Instead, they headed south to hit the Carson River, named by Frémont for the scout whose legend he had begun.

It was here, in the Carson Valley on January 18, that Frémont said he made his decision to cross the Sierra Nevada into California, though there are indications that the decision had been made well before. Faced with horses in poor condition and with no means of making shoes for them, Frémont "therefore determined to abandon my eastern course" (Frémont 1845:220) and to cross the Sierra Nevada into California.

The expedition's passage over the Sierra Nevada was one of remarkable hardship; it is to Frémont's great credit that the members all survived. Frémont himself became lost, knowing mainly that they had to go west and that they had to go up. They first tried going up the East Walker River, then gave up and followed the West Walker. They ended up in snow deep enough to bury their horses; the only way through was to build a road by stamping down the snow and covering it with pine boughs. On February 10, Frémont established what he called Long Camp, where, three days later, hunger forced them to eat their dog Tlamath. Thusly fueled, Frémont and his dyspeptic but talented cartographer and illustrator Charles Preuss then climbed nearby Red Lake Peak and became the first to record seeing Lake Tahoe. The party then forced its way over the top, through Carson Pass, eating horses and mules as it went, stumbling and crawling through the snow, emerging into the green California spring on February 24. On March 6, they finally reached the American River, only a mile from the Sacramento River and Sutter's Fort. They had left The Dalles with 104 horses and mules; they had begun their ascent of the Sierra Nevada with 67; they arrived at Sutter's Fort with 33 exhausted and nearly useless animals. Another, less animate loss was the howitzer: this they had abandoned on January 29, somewhere along the western flank of the Sweetwater Mountains, its whereabouts still a debated mystery. "If we had only left that ridiculous thing at home," Preuss had grumbled months earlier (1958:83), and now it was gone.

Three weeks at Sutter's Fort saw both men and animals revived. They left on March 24, this time with 130 horses and mules and some 30 cattle. Rather than moving north and out of Mexico, they went south and deeper into it, following the San Joaquin Valley to southern California's Tehachapi Mountains, crossing over them and into the Mojave Desert a few miles south of Tehachapi Pass. Back in the Great Basin again, they moved mostly east, hitting the Mojave River near what is now Victorville, California, and roughly following the Spanish Trail across southern California and southern Nevada into Utah. At Bitter Spring in southern California's Mojave Desert, Kit Carson and his companion Alexander Godey revenged the deaths of a party of Mexicans, and the stealing of their horses, by tracking down and scalping two of the Indians who had done the killing. "Butchery," Preuss disgustedly called it (1958:128), and it was a sign of the savage ferocity for which Carson was later to become infamous. Then, along the Virgin River near Littlefield, Arizona, one of Frémont's own—Jean Baptiste Tabeau—was killed by Indians on May 9, the first of his men to die (a second, François Badeau, was to die on May 23 from a gun-handling accident).

Leaving the scene of Tabeau's death, the expedition moved northward, reaching Mountain Meadows on May 12. Thirteen years later, this site, on the very fringe of the Great Basin in southwestern Utah, was to become the location of a Mormon-engineered massacre of some 120 emigrants from Arkansas and nearby states. For Frémont, however, it was simply a "noted place of rest and refreshment" (Frémont 1845:271). Equally important, as the expedition left Mountain Meadows, it was joined by Joseph Walker, one of the most famous of western backwoodsmen. It was Walker who guided the group north to the Sevier River and then to Utah Lake, south and east of the Great Salt Lake. Finally, on May 27, Frémont and his men headed east into the Wasatch Range and out of the Great Basin. On August 6, nearly fifteen months after his departure, Frémont was once again in St. Louis.

Frémont had thus struggled his way south from the Columbia River, deep along the eastern edge of the Great Basin, and then over the Sierra Nevada in the dead of winter. He had then moved even farther south in the interior valleys of Mexican California and then east across one of the most challenging deserts in North America, ultimately swinging north to nearly rejoin his original diversion into the Great Basin at Great Salt Lake.

His orders, however, directed him to return by the Oregon Trail, not by the Spanish Trail, some five hundred and more miles to the south, and those orders said nothing about California. He explained his entry into California by the situation in which he found himself in January 1844. Why, however, did he swing so far south from the Columbia River, and from the Oregon Trail, in the first place? Just as Frémont used his *Report* to justify his decision to enter California, he also used it to justify making this move.

There were, he said, three prime geographic reasons for making this "great circuit to the south and southeast." The first was to find Klamath Lake and explore the Klamath country, then poorly known. The second was to find and explore Mary's Lake, the sink into which the Humboldt River flows in western Nevada and that is now called Humboldt Lake. Third, he wished to locate, if it existed, the Buenaventura River, "which has had a place in so many maps, and countenanced the belief of the existence of a great river flowing from the Rocky mountains to the bay of San Francisco" (Frémont 1845:196).

Of these goals, Frémont approximated achieving the first, but failed at the second. He explored the Klamath country, but never found Klamath Lake, having mistaken it for Klamath Marsh, to the north and east. This was hardly his fault, since he also referred to the "imputed double character" of Klamath Lake as "lake, or meadow, according to the season of the year" (Frémont 1845:196), a description that applies not to the deep and permanent Upper Klamath Lake but that fits Klamath Marsh well.

After leaving the Klamath country, however, his movements south from Warner Valley and past Pyramid Lake brought him well west of Humboldt Lake, and his *Report* provides no clarification of the location and nature of the sink of the Humboldt except that it could not be found the way he went.

Ironically, even though Frémont himself was to name the river "Humboldt" during his next, 1845, expedition, he saw neither lake nor river until the summer of 1847. And, when he finally saw Humboldt Lake, he was heading eastward in the forced tow of Stephen Watts Kearny, who was to have him arrested and court-martialed for actions he had taken during the Bear Flag Revolution and the acquisition of California for the United States.

The Klamath country and Humboldt Lake existed, and in that sense were quite different from the Buenaventura, one of the most enduring myths that the geography of North American deserts was to provide, a myth that has its roots in the earliest entry of Europeans into the Intermountain West. In 1775, the Franciscan Father Francisco Garcés, a member of Juan Bautista de Anza's second expedition to forge overland routes linking the settlements of Sonora and New Mexico with those of coastal California, traveled up the Colorado River to somewhere near the current location of Needles, California. Garcés then headed west, reached and followed the Mojave River westward, and then crossed into the San Joaquin Valley. While there, he gained the impression that the Kern River cut through the Sierra Nevada; told of the San Joaquin River, he thought that this cut the Sierra Nevada as well. Indeed, Anza himself reached San Francisco Bay in 1776, and his diarist, Father Pedro Font, mistook the rivers that flowed into this bay for a large body of freshwater that reached east of the Sierra Nevada. So the myth began.

On July 29, 1776, Fathers Francisco Domínguez and Francisco Escalante left Santa Fe to find an acceptable overland route to Monterey, which had been established in 1770. They headed north through western Colorado and hit the Green River. This they named the San Buenaventura, after the biographer of St. Francis. From here, they moved west, crossing the mountains and reaching Utah Lake. While at

Utah Lake, they were told of the Great Salt Lake and may have assumed, from its salinity, that it had an outlet to the sea. Moving south, they crossed the Sevier River, which they thought was part of the San Buenaventura. Abandoning the idea of reaching Monterey on this trip—a wise decision, given the way they had gone—they continued south and returned to Santa Fe in January 1777.

Maps were soon produced that incorporated and compounded these errors. By the early 1800s, influential maps showed the Buenaventura River flowing from the far eastern Great Basin, and usually from Sevier or Great Salt Lake, to the Pacific Coast. John Melish's map in 1809, John Robinson's map in 1819, Alexander Finley's map in 1830, and even the map produced by the Society for the Diffusion of Useful Knowledge in 1842 provided for such a river.

A river that flowed from the Rockies to the Pacific Coast in this region would be of tremendous economic importance, since it would provide a means for the transportation of people and goods to and from California. Moving the former was important if the United States were to stretch from coast to coast. Moving the latter was important if the United States were to become a major player in trade with Asia. As Richard Francaviglia has noted in his important book on the mapping of the Great Basin, these early maps were "blueprints for American expansion" (2005:73), and the Buenaventura itself a river of empire. If the Buenaventura were real, it had to be discovered and charted.

But the Buenaventura was not real, and seasoned explorers of the West soon became aware of that. In 1826, Jedediah Smith traveled from Cache Valley to the Great Salt Lake, then south to the Colorado, turning west across the Mojave Desert to reach the San Bernardino Valley. Denied permission from the Mexican governor to travel north to San Francisco, he did so anyway, since he wanted to return by following "some considerable river heading up in the vicinity of the Great Salt Lake" (G.R. Brooks 1989:77–78).

Smith didn't find that river, and in May 1827, he left his men along the Stanislaus River and, accompanied by two others, became the first non-Indian known to have crossed the Sierra Nevada. Apparently moving through Ebbett's Pass south of Lake Tahoe, his return trip to the Great Salt Lake took him south of Walker Lake and through south-central Nevada. He headed back to his men almost immediately, taking the southern route via the Colorado River and Mojave Desert to San Bernardino Valley.

Late in 1827, he and his companions began the move north up the Sacramento Valley to the Trinity and Klamath rivers, then north up the Pacific Coast to the Umpqua River, where fifteen of his men were killed by Indians. Frémont later referred to this episode to explain why he fired his cannon at Klamath Marsh. By the time Smith reached Fort Vancouver in August 1828, he knew that no river south of the Columbia cut through either the Cascades or the Sierra Nevada.

Smith did not keep that information to himself, writing to William Clark (of Lewis and Clark, and then Superintendent of Indian Affairs) to tell him of his travels, thus informing him that the Buenaventura did not exist. "By Examination and frequent trials," Smith wrote, he "found it impossible to cross a range of mountains which lay to the East" (D.L. Morgan 1964:340).

Although Smith died before he could complete his projected book on his travels, William Clark and Thomas Hart Benton were good friends and fellow expansionists who routinely discussed what was known of the geography of western North America. Surely if by no other route than this, Frémont would have known that the Buenaventura did not exist. Indeed, in 1829 and 1830, Frémont's friend and guide, Kit Carson, had crossed the Mojave Desert from the east and traveled up the interior valleys of California. As Carson noted on seeing the Sacramento Valley with Frémont in 1844, "I knew the place well, had been there seventeen [sic] years before" (H.L. Carter 1968:90). There was even an important map, by David Burr, geographer for the U.S. House of Representatives, published in 1838, which incorporated data provided by Jedediah Smith and other explorers that depicted the Sierra Nevada as a massive barrier crossed by no river coming from the east.

Arrayed next to such information, however, were the many contemporary maps that did continue to show such a river. As Frémont noted, the Buenaventura formed "agreeably to the best maps in my possession, a connected water line from the Rocky mountains to the Pacific ocean" (Frémont 1845:205). If Frémont really did not know that the river was fictitious, then finding it could have been a legitimate and major goal of his intermountain explorations, and an excellent reason to move so far south from the Columbia River and the Oregon Trail. If he knew it did not exist, he did not let on, and the river became a prime justification for being where he had not been told to go, and it certainly became a major literary device in reporting the results of his explorations.

By the time Frémont reached southern California, he knew that there was no such river. In his *Report* entry for April 14, 1844, the day his party crossed over the Tehachapis, Frémont let his readers know as well:

> It had been constantly represented ... that the bay of San Francisco opened far into the interior, by some river coming down from the base of the Rocky mountains, and upon which supposed stream the name of Rio Buenaventura had been bestowed. Our observations of the Sierra Nevada ... show that this neither is nor can be the case. No river from the interior does, or can, cross the Sierra Nevada. (Frémont 1845:255)

The Columbia was the only river that led from the deep interior to the Pacific Ocean, and this was far to the north. Frémont had followed the mountains south from the Columbia to southern California, and he knew this to be the case. While other explorers had known for more than a decade that the Buenaventura did not exist, Frémont's *Report* brought the news to a wide audience and put the myth to a decided end.

But among scientists who work in the Desert West, Frémont is far better known for a very different pronouncement, since his *Report* carries an entry for October 13, 1843, that first introduced the term that is now used to characterize much of this region:

> "The Great Basin—a term which I apply to the intermediate region between the Rocky mountains and the next range [the Sierra Nevada], containing many lakes, with their own system of rivers and creeks, (of which the Great Salt is the principal,) and which have no connexion with the ocean, or the great rivers which flow into it" (Frémont 1845:175).

"The Great Basin": John C. Frémont's term for the area of internal drainage in the arid West. Between the crests of the Rocky Mountains on the east and the Sierra Nevada on the west, from the edge of the Columbia River drainage on the north to the edge of the Colorado River Plateau on the south, all waters that fall end up in enclosed basins that may, or may not, contain lakes; none of those waters reaches either the Atlantic or the Pacific oceans. "The Great Basin" is a great name for this area, and that is what it has been called ever since Frémont introduced the phrase in his 1845 *Report*.

As if to emphasize the importance of what historian Donald Jackson referred to as Frémont's "vital geographic discovery" (D. Jackson and Spence 1970:13), the message was repeated in a long arc of type running north–south through the heart of the American Desert West on the famous Preuss map that appeared in Frémont's *Report*: "THE GREAT BASIN: diameter 11° of latitude, 10° of longitude: elevation above the sea between 4 and 5000 feet: surrounded by lofty mountains: contents almost unknown, but believed to be filled with rivers and lakes that have no communication with the sea, deserts and oases which have never been explored, and savage tribes, which no traveler has seen or described."

Frémont is often credited not only with having named this region but also with having established the fact of internal drainage. He certainly did establish to his own satisfaction that no water flowed from the Great Basin across the Sierra Nevada and Cascades. However, his outbound route left a vast portion of the northern Great Basin and southern Columbia Basin unexplored by him. He could not have known by his own efforts that no waters flowed from this region into the Columbia system. In fact, Frémont never claimed to have established the fact of internal drainage. Indeed, he made it explicit that the information that allowed him to be sure that he was dealing with a "Great Basin" came from Joseph Walker.

Born in Tennessee in 1798, Walker began his explorations of the Great Basin in 1833, when he led a fur-trapping group down the Bear River to the Great Salt Lake. After exploring the western shore of the lake and discovering that it had no outlet, he headed west across the desert, reached the Humboldt River, and followed it down to the Humboldt Sink. Moving south from the sink, he passed Carson Lake, Carson River, and, it appears, Walker Lake. From here, Walker crossed the Sierra Nevada to the Merced River in California. On the way, his party became the first whites to see Yosemite Valley, on about October 30, 1833.

Walker wintered in California and began the trip back in February 1834. From the Tübatulabal Indians, he learned of a pass through the southern Sierra Nevada and took what is now called Walker Pass to reach Owens Valley. From here, the party generally followed the east flank of the Sierra Nevada northward, back to the Humboldt Sink. They then retraced their steps back along the Humboldt, leaving it to head north to the Snake and out of the Great Basin.

Although Walker himself left no detailed account of this trip, his clerk, Zenas Leonard, did, and it is primarily from Leonard's published travels that Walker's 1833–1834 crossings of the Great Basin are known. Leonard's *Narrative*, which appeared in book form in 1839, notes again and again that the lakes they saw—Great Salt Lake, Humboldt Lake, Carson Lake—had no outlets. By the time they had completed their journey, the general nature of the desert they had crossed had become completely clear, and Leonard himself came close to defining the Great Basin just the way Frémont was to do a few years later:

> This desert ... is bounded on the east by the Rocky mountains, on the west by the Calafornia [*sic*] mountain, on the North by the Columbia river, and on the south by the Red, or Colorado river.... There are numerous small rivers rising in either mountain, winding their way far towards the centre of the plain, where they are emptied into lakes or reservoirs, and the water sinks in the sand. Further to the North where the sand is not so deep and loose, the streams rising in the spurs of the Rocky and those descending from the Calafornia mountains, flow on until their waters at length mingle together in the same lakes. The Calafornia mountain extends from the Columbia to the Colorado river, running parallel with the coast about 150 miles distant, and 12 or 15 hundred miles in length with its peaks perpetually covered with eternal snows. There is a large number of water courses descending from this mountain on either side—those on the east side stretching out into the plain, and those on the west flow generally in a straight course until they empty into the Pacific; but in no place is there a water course through the mountain. (Quaife 1978:212–213)

Walker thus knew in 1834 not only that there was no Buenaventura cutting through to the Pacific but also that a vast amount of the intermountain area appeared to be of internal drainage. Unlike Frémont, he had traveled both the Humboldt and the area between the Humboldt and Snake rivers.

On May 12, 1844, while at Mountain Meadows, Frémont "had the gratification to be joined by the famous hunter and trapper, Mr. Joseph Walker" (Frémont 1845:271), who then became the group's guide. Frémont thus had the opportunity to discuss geographical and topographical matters with the older and more experienced explorer. We can be certain that they talked about the hydrology of the area Frémont called the Great Basin, since Frémont gives Walker full credit for confirming what Frémont may have suspected but certainly had not proved: that this vast area lacks any outlet to the sea.

The existence of the "Great interior Basin," Frémont observed,

> is vouched for by such of the American traders and hunters as have some knowledge of that region; the structure of the Sierra Nevada range of mountains requires it to be there; and my own observations confirm it. Mr. Joseph Walker, who is so well acquainted in those parts, informed me that, from the Great Salt lake west, there was a succession of lakes and rivers which have no outlet to the sea, nor any connexion with the Columbia, or with the Colorado of the Gulf of California. He described some of these lakes as being large, with numerous streams, and even considerable rivers, falling into them. In fact, all concur in the general report of these interior rivers and lakes.... The structure of the country would require this formation of interior lakes; for the waters which would collect between the Rocky mountains and the Sierra Nevada, not being able to cross this formidable barrier, nor to get to the Columbia or the Colorado, must necessarily collect into reservoirs, each of which would have its little system of streams and rivers to supply it. (Frémont 1845:275)

Frémont could not have been clearer in giving credit where credit was due, to Walker and the other "American traders and hunters" whose knowledge of the region far exceeded his own. Indeed, while Frémont argued that the small size of the streams that he saw coming into the Columbia from the south suggested that they did not reach very far south, and hence that the Columbia drainage must mark the northern edge of the Great Basin, even here he noted that "all accounts" concurred that the waters of the Great Basin were separated from those of the Columbia.

As extensive and impressive as they were, and as important as they would prove in bringing others west—in forwarding Thomas Hart Benton's dreams—Frémont's explorations in 1843 and 1844 were too limited to demonstrate on their own that no water flowed from this huge area. What he did do was to add his knowledge and insights to the travels and knowledge of others, most notably of Joseph Walker. He confirmed what they had found, coined a felicitous term, "the Great Basin," to encapsulate that confirmed knowledge, and presented all of this in one of the most readable, and most widely read, volumes of exploration ever published in and on North America, a book that made public both what he had learned and the private knowledge of those who had preceded him.

Chapter Notes

On Frémont, see D. Jackson and Spence (1970), Egan (1985), Herr (1987), Rolle (1991), Chaffin (2002), and Frémont himself (Frémont 1845, 1887). On Jessie Benton Frémont, see Herr (1987) and Herr and Spence (1993); this latter reference discusses her role in making Kit Carson famous. Rolle (1991) and Chaffin (2002) are the best places to start.

Although Frémont's death deprived his family of the $6,000 annual pension, Jessie Benton Frémont was soon awarded a pension of $2,000 per year as his widow. She died in 1902, at the age of seventy-eight. Frémont expert Bob Graham has identified Red Lake Peak as the spot from which members of the Frémont party saw Lake Tahoe. His website (www.longcamp.com/) includes a wide range of Frémont information, as well as detailed instructions for reaching the spot occupied by Frémont's Long Camp. B. Graham (2003) has also written a fascinating analytic guide to Frémont's 1844 winter crossing of the Sierra Nevada.

If you are interested in the history of the mapping of the Great Basin, Francaviglia (2005) is essential reading; I have relied heavily on it. Cline (1988) provides an excellent discussion of early explorations of the Great Basin in general and of the Buenaventura River myth in particular. She suggests that the saltiness of the Great Salt Lake led the Domínguez-Escalante expedition to assume that it had a connection to the sea. This may well be, but the Domínguez-Escalante journal itself does not say this (Chavez and Warner 1995).

For Kit Carson's memoirs, see Quaife (1966) and H.L. Carter (1968); for an analysis of Carson's relationships with American Indians, see Dunlay (2000). This latter author comes close to serving as Carson's apologist, but Carson's memoirs are hard to reconcile with that opinion. Preuss's description of the Carson-Godey attack as "butchery" matches Carson's own infamous description of his participation in an 1846 attack on Indians in the Sacramento Valley as "a perfect butchery" (H.L. Carter 1968:101; see and compare the discussions in H.L. Carter 1968:101–103, D. Jackson and Spence 1973:124–125, and Dunlay 2000:111–114). Carson claimed that at least some of his brutality was at Frémont's direct command (W.H. Davis 1929:345).

Charles Preuss's memoirs (Preuss 1958) are fascinating. Carson noted that "the old man was much respected by the party" (H.L. Carter 1968:90); Preuss was forty-one at the time, Carson thirty-four. Preuss hanged himself in Maryland in 1854.

On the route of the Oregon Trail, see W.E. Hill (1986). Among other places, Frémont's cannon is discussed in D. Jackson and Spence (1970), Townley (1984), Reveal and Reveal (1985), and Lewis (1992); see also Bob Graham's website, mentioned above.

On the Mountain Meadows massacre, see J. Brooks (1962), Bagley (2002), S. Denton (2003), Novak (2008), and R.W. Walker, Turley and Leonard (2008). The most recent of these important books take distinctly different approaches to the massacre. Novak (2008) emphasizes the emigrant group and its origins, while R.W. Walker, Turley, and Leonard (2008) emphasize the religious, political, and social context in which the massacre was embedded. Accidental discoveries of human bones at Mountain Meadows, and the ensuing controversy, are reviewed in a superb series of articles by Christopher Smith that appeared in the *Salt Lake City Tribune* on March 12, 13, and 14, 2000 (currently available at www.cesnur.org/testi/morm_01.htm or, for a fee, at www.sltrib.com/). Novak (2008) discusses these remains in some detail. The Mountain Meadows Monument Foundation (http://1857massacre.com/default01.htm) and the Mountain Meadows Association (www.mtn-meadows-assoc.com/) maintain superb websites that provide a wealth of historical and recent information on this horrid event.

On Walker, see B. Gilbert (1983) and, for Leonard's Narrative, Quaife (1978); on Smith, see D.L. Morgan (1964) and G.R. Brooks (1989). D. L. Morgan (1985) provides a fascinating discussion of the Humboldt River as a historic route across Nevada.

CHAPTER TWO

Modern Definitions of the Great Basin

The Hydrographic Great Basin

If scientists had just stuck with Frémont's fully appropriate definition, the Great Basin would be a well-defined hydrographic unit, the name applying only to that huge area of the arid West that drains internally. That definition, in fact, remains the common one, though few people working in the Great Basin follow it slavishly.

When it is so defined (figure 2-1), the Great Basin covers about 200,000 square miles, from the crest of the Sierra Nevada and southern Cascades to the western edge of the Uinta Mountains and Colorado Plateau, and from edge to edge of the Columbia and Colorado River drainages. In terms of modern political units, the hydrographic Great Basin centers on the state of Nevada, but also includes much of eastern California, western Utah, and south-central Oregon, as well as small portions of southeastern Idaho and adjacent Wyoming. Precipitation that falls in this vast region drains neither to the Pacific nor to the Atlantic, but instead generally flows into streams that empty into low, saline lakes—Frémont's "fetid salt lakes" (Frémont 1845:209)—or that simply disappear by evaporation and absorption into the ground—Jed Smith's "plains whose sands drank up the waters of the river and spring where our need was the greatest" (G. R. Brooks 1989:96). As Mark Twain put it in *Roughing It*: "There are several rivers in Nevada, and they all have this mysterious fate. They end in various lakes or 'sinks,' and that is the last of them . . . water is always flowing into them; none is ever seen to flow out." (Twain 1981 [1872]:171).

Those who are deeply familiar with the Great Basin may have seen a very different map of the hydrographic boundaries of this region. In that other version, the Great Basin has a long tail that extends south through southeastern California into the northern edge of Baja California. In following that pattern, these maps incorporate the Salton Basin and, within it, the Salton Sea. This latter, saline lake was created in 1905 when the Colorado River purged its way through irrigation canals and is today sustained largely by agricultural runoff. The connection with the Colorado River, however, is not recent. For at least the past 17,000 years, the Salton Basin has been connected with the Colorado far more often than it has been isolated from it. As a result, it is not properly considered to be part of the hydrographic Great Basin, and I do not consider it as such here.

Defined in this way, the hydrographic Great Basin does have many bodies of water that are called rivers (figure 2-2). The nature of Great Basin internal drainage can quickly be seen by following any of those rivers downstream. The Blitzen (or Donner und Blitzen) River, for instance, is a beautiful stream that rises in Oregon's Steens Mountain and flows into Malheur Lake, helping to create one of North America's finest wildlife refuges. When the waters of Malheur Lake rise, they flow into nearby Harney Lake, but these lakes have no outlet.

To the south and west, the Walker River rises in two branches—the East and West Walker—from the eastern flanks of the Sierra Nevada south of Lake Tahoe, and runs, through a curious course that I discuss in chapter 5, into Walker Lake. Walker Lake, however, also has no outlet.

The Carson River rises on the flanks of the Sierra Nevada closer to Lake Tahoe, and then flows north and east into Carson Lake and the Carson Sink. "Sink" says it all. Further north, the Truckee River rises in Lake Tahoe and, until recently, flowed north and east to fill both Pyramid and Winnemucca lakes. By 1939, the use of the Truckee's water for irrigation and the construction of a highway across the inlet of Winnemucca Lake had turned the lake into a playa. Today, a diminished Truckee feeds only Pyramid Lake, which, as Frémont learned, has no outlet. The Quinn

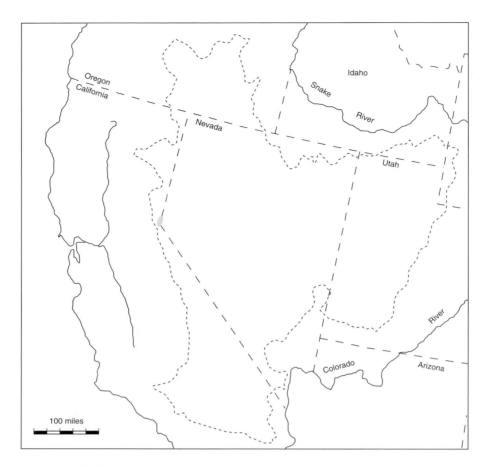

FIGURE 2-1 The hydrographic Great Basin (after R. B. Morrison 1991).

River rises in northern Nevada from the confluence of many small streams and ends in the Black Rock Desert: only in good years does it form even a shallow lake (see figure 2-3).

The Humboldt—the longest of the Great Basin rivers—begins in northeastern Nevada, then flows generally southwest across the northern part of the state. Because of its length and location, the Humboldt was extremely important in the Americanization of California, serving as the prime transportation route across the state during the nineteenth century. The Bidwell-Bartleson party came this way on its way to California in 1841. The Donner-Reed party followed the river dangerously late in 1846, on its way to disaster in the Sierra Nevada. Thousands more emigrants followed during the next decade, on their way, they hoped, to gold or land. But the Humboldt runs only as far as Humboldt Lake and Humboldt Sink—"sink" again—and beyond that point travelers were faced with a wicked stretch of desert.

The eastern Great Basin has its share of rivers as well. The Logan flows into the Bear River and the Bear into Great Salt Lake. The Ogden joins the Weber, which also feeds Great Salt Lake. The Provo flows into Utah Lake, and overflow from Utah Lake spills into Great Salt Lake via the Jordan River. Frémont was wrong in thinking that Utah Lake was part of the Great Salt Lake, but he certainly knew that neither had an outlet to the sea. To the south, the Sevier River runs into Sevier Lake. This lake would still have water in it were it not for irrigation demands on the Sevier River, but it, too, lacks an outlet.

In the southwestern Great Basin, the Amargosa River rises in western Nevada and flows—when it has any water in it—south along the east side of the Amargosa Range and then north into Death Valley, where any water it might have simply disappears. The Mojave River begins in the San Bernardino Mountains of Southern California and runs north and east until it, too, just disappears. The Owens River drains Owens Valley into Owens Lake. Its waters now reach the Pacific, but only because the city of Los Angeles has appropriated them. All Great Basin streams drain internally, and it is this fact that leads to the definition of a hydrographic Great Basin.

The Physiographic Great Basin

There are three other Great Basins to go along with the one defined by hydrographers, each created by a different set of scientists focusing on a different aspect of the Desert West. For those who like crisp regional boundaries, these other Great Basins can be frustrating, since only the hydrographic Great Basin allows any real certainty as to where the Great Basin starts and stops. Scientists defining one of the other

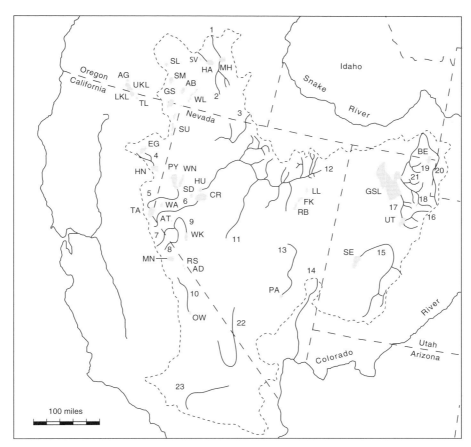

Rivers: 1. Silvies; 2. Donner und Blitzen; 3. Quinn; 4. Susan; 5. Truckee; 6. Carson; 7. West Walker; 8. East Walker; 9. Walker; 10. Owens; 11. Reese; 12. Humboldt; 13. White; 14. Muddy River/Meadow Valley Wash; 15. Sevier; 16. Provo; 17. Jordan; 18. Weber; 19. Logan; 20. Bear; 21. Ogden; 22. Amargosa; 23. Mojave

Lakes: AB. Abert; AD. Adobe; AG. Agency; AT. Artesia; BE. Bear; CR. Upper Carson; EG. Eagle; FK. Franklin; GS. Goose; GSL. Great Salt Lake; HA. Harney; HN. Honey; HU. Humboldt; LL. Little Lake; LKL. Lower Klamath Lake; MH. Malheur; MN. Mono; OW. Owens; PA. Lower and Upper Pahranagat; PY. Pyramid; RB. Ruby; RS. River Spring; SD. Soda Lakes; SL. Silver; SM. Summer; SE. Sevier; SU. Surprise Valley (Alkali); SV. Silver; TA. Tahoe; TL. Tule; UKL. Upper Klamath; UT. Utah; WA. Washoe; WK. Walker; WL. Warner Lakes

FIGURE 2-2 Rivers and lakes of the Great Basin. Agency, Lower Klamath, Tule, and Upper Klamath lakes are not in the Great Basin but are discussed elsewhere in the text.

Great Basins draw the edges of each of their Great Basins differently depending on the precise definitional criteria they use and on whether or not they think a given patch of the Desert West matches those criteria. Only for the most compulsive is this haziness of boundaries problematic, since in all cases the haziness reflects the continuum on the ground.

Physiographers, those who study the evolution and nature of the earth's surface, focus not on the drainages of the Intermontane West, but instead on the fact that much of this area has a distinctive topography: wide desert valleys flanked by often massive mountain ranges that run roughly north-south and that generally parallel one another. Topography of this sort covers an area much broader than the Great Basin itself, no matter how it is defined. This "Basin and Range Province," as it is called, runs from southeastern Oregon and southern Idaho deep into northern Mexico, including much of the southern Southwest on its way. But the northernmost part of this complex geological area is called the Great Basin Section of the Basin and Range Province and provides the second common definition of the Great Basin (figure 2-4).

So defined, three of the borders of the physiographic Great Basin are well marked: the Sierra Nevada and southern Cascades on the west, the Wasatch Range and Colorado Plateau on the east, and the Columbia Plateau on the north. In all three cases, the borders coincide well with the edges provided by the hydrographic definition. The same, however, is less true for the southern end, which is defined somewhat arbitrarily.

South of a line that runs roughly from Las Vegas on the east to just north of the Mojave River on the west, topographic relief becomes much less pronounced than it is in the north; the mountains become smaller and more irregular in outline. This line marks the southern edge of the

TABLE 2-1
Mountain Ranges in the Physiographic Great Basin with Summits above 10,000 Feet

Range	Summit	Elevation (feet)	Summit County
Antelope Range	Ninemile Peak	10,104	Nye, NV
Bodie Mountains	Potato Peak	10,235	Mono, CA
Carson Range	Freel Peak	10,881	Alpine, CA
Cherry Creek Range	Unnamed	10,458	White Pine, NV
Deep Creek Range	Ibapah Peak	12,087	Juab, UT
Diamond Mountains	Diamond Peak	10,614	Eureka/White Pine, NV
Duck Creek Range	Unnamed	10,328	White Pine, NV
East Humboldt Range	Hole-in-the-Mountain Peak	11,306	Elko, NV
Egan Range	South Ward Mountain	10,936	White Pine, NV
Grant Range	Troy Peak	11,298	Nye, NV
Hot Creek Range	Morey Peak	10,246	Nye, NV
Independence Mountains	McAfee Peak	10,439	Elko, NV
Inyo Mountains	Mount Inyo	11,107	Inyo, NV
Jarbidge Mountains	Matterhorn	10,839	Elko, NV
Monitor Range	Unnamed	10,888	Nye, NV
Oquirrh Mountains	Flat Tap Mountain	10,620	Utah, UT
Panamint Range	Telescope Peak	11,049	Inyo, CA
Pequop Mountains	Spruce Mountain	10,262	Elko, NV
Pilot Range	Pilot Peak	10,716	Elko, NV
Quinn Canyon Range	Unnamed	10,185	Nye, NV
Roberts Mountains	Roberts Creek Mountain	10,133	Eureka, NV
Ruby Mountains	Ruby Dome	11,387	Elko, NV
Schell Creek Range	North Schell Peak	11,883	White Pine, NV
Shoshone Mountains	North Shoshone Peak	10,313	Lander, NV
Snake Range	Wheeler Peak	13,063	White Pine, NV
Spring Mountains	Charleston Peak	11,918	Clark, NV
Stansbury Mountains	Deseret Peak	11,031	Tooele, UT
Sweetwater Mountains	Mount Patterson	11,673	Mono, CA
Toiyabe Range	Arc (Toiyabe) Dome	11,788	Nye, NV
Toquima Range	Mount Jefferson, South Summit	11,941	Nye, NV
Wassuk Range	Mount Grant	11,239	Mineral, NV
White Mountains	White Mountain Peak	14,246	Mono, CA
White Pine Range	Currant Mountain	11,513	White Pine, NV

FIGURE 2-3 Black Rock Desert, with the Jackson Mountains in the background.

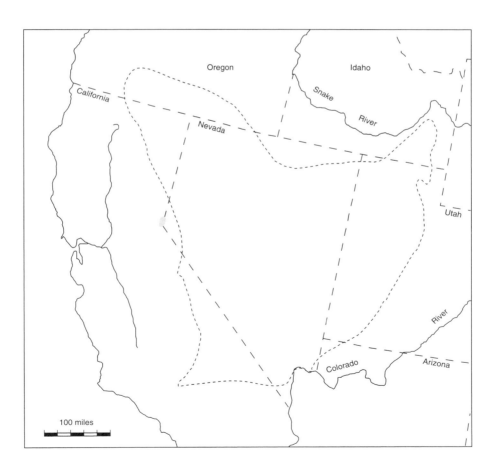

FIGURE 2-4 The physiographic Great Basin (after C. B. Hunt 1967).

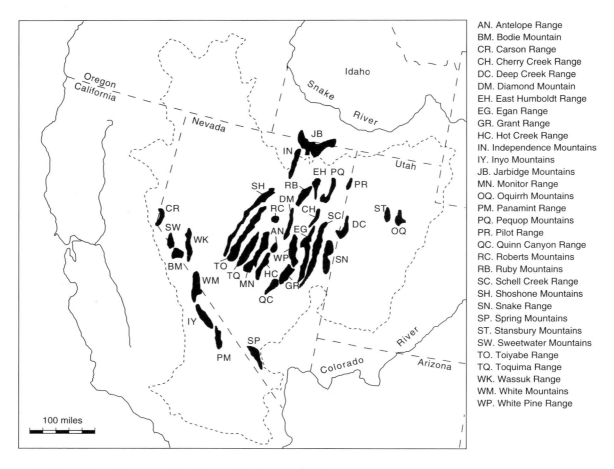

FIGURE 2-5 Mountain ranges in the physiographic Great Basin with summit elevations greater than 10,000 feet. The Duck Creek Range, which lies between the Schell Creek and Egan ranges, is not drawn.

physiographic Great Basin. While the northern, eastern, and western edges of this Great Basin match the drainage divide well, the southern boundary lops off a small part of the hydrographic Great Basin.

The mountains of this second Great Basin can be truly impressive. Their number depends on how a mountain is defined: Alvin McLane listed 314 of them in Nevada alone. Here, I simply point out that the physiographic Great Basin contains thirty-three ranges whose summits reach above 10,000 feet (figure 2-5). Of these thirty-three, fifteen have peaks that reach above 11,000 feet, while the summits of three—the Deep Creek Range, the Snake Range, and the White Mountains—rise above 12,000 feet. The highest point on the tallest of these mountains, White Mountain Peak in the White Mountains, reaches 14,246 feet (table 2-1).

In addition to being tall, these mountains are long and relatively narrow and are generally oriented north-south. For instance, a transect across Nevada at 39° N latitude (which passes through the southern edge of Lake Tahoe) cuts fifteen major Great Basin ranges. On the eastern edge of this transect, the Snake Range runs 68 miles from north to south and includes Wheeler Peak (13,063 feet), the second-highest point in the Great Basin. The Schell Creek Range, to the immediate west, runs 133 miles from north to south.

After it, the Egan Range, runs 110 miles in the same direction; and so on (table 2-2). In fact, even the smallest of the ranges encountered along this transect—the Pancake Range in central Nevada—is 24 miles long. The spines of all fifteen of these ranges run in the same general direction; hiking each of them would require one to walk 1,056 miles, an average of 70 miles per range.

Even though the mountains rise high above sea level, however, the elevations of the summits give no good indication of how far they rise above the surrounding valleys, because Great Basin valleys themselves tend to be quite high. The highest valleys are in the central Great Basin of central and eastern Nevada, where valley bottoms tend to lie between 5,300 and 6,000 feet, though some are higher still. From here, valley bottom elevations fall in all directions: to between 3,800 feet and 5,800 feet in northwestern Nevada, to between 3,800 feet and 5,000 feet in western Utah, to 2,500 feet and lower in the far southern Great Basin. In the far southwestern corner of the physiographic Great Basin, Death Valley's Badwater Basin falls to 282 feet below sea level, the lowest point in the Western Hemisphere not covered by water.

If the elevations of Great Basin valleys are plotted on graph paper, the diagram that results looks very much like an arch, with the peak of that arch lying in the central Great Basin. Figure 2-6 illustrates the archlike construction of the physiographic and hydrographic Great Basins. This illustration was built using the 1:250,000 scale U.S. Geological Survey topographic maps that cover this region, each of which displays two degrees of longitude and one degree of latitude. To create this figure, I chose the largest sizable intermountain valley nearest to the center of each map, and recorded the lowest point in that valley (the only exceptions being for maps along the edges of the Great Basin; here, I simply selected the largest valley still within the Great Basin). The figure shows the results, rounded to the nearest 100 feet. The high valleys of the central Great Basin are evident, as are the decreasing valley elevations found in all directions from here, with the steepest decline to the south.

A return to the 39° N latitude transect across Nevada provides an even better feel for the mountain and valley structure of the physiographic Great Basin and also helps place the high-elevation summits in better relief. Table 2-3 compares the lowest elevations in the valleys that intervene between the fifteen ranges along this transect with the highest points on the adjacent mountains. Valley bottom–to-mountaintop relief varies from 3,800 feet (between Little Smoky Valley and the top of the Antelope Range, and between Ione Valley and the top of the Shoshone Range) to 7,600 feet (from the bottom of Spring Valley to the summit of the Snake Range). Average maximum relief between valley floor and mountaintop along our transect is 5,800 feet—more than a mile.

The relief record for adjacent valley bottoms and mountain tops in the Great Basin is not set by any of those encountered along this transect. Instead, it is set by Death Valley and the adjacent Panamint Range: 11,331 feet of relief from the −282 feet depths of Badwater Basin to the 11,049 feet summit of nearby Telescope Peak. The elevational difference between the highest (White Mountain Peak at 14,246 feet) and the lowest (Badwater at -282 feet) points in the Great Basin is an impressive 14,528 feet (see figure 2-7).

Marked by often-massive north-south–trending mountain ranges with wide desert valleys between, the physiographic Great Basin coincides closely with the hydrographic one except in the south, where it excludes the southernmost extent of the area of internal drainage. A third approach to defining the Great Basin, however, excludes even more of the south.

The Floristic Great Basin

The third approach to defining the Great Basin is biological and draws its line around relatively distinctive assemblages of plants. However, biologists differ as to exactly what those distinctive assemblages of plants are. As a result, there are multiple possible floristic Great Basins. This is not necessarily bad, but it can make it a challenge to know what a particular botanist means when he or she refers to the botanical Great Basin, unless there is a map or a reference to a map included.

One of the most widely used definitions of the floristic Great Basin, that by Noel Holmgren and his colleagues (figure 2-8), provides a Great Basin whose lower elevations are characterized by plant communities in which saltbush and sagebrush are the dominant shrubs (figure 2-9) and in which the mountain flanks tend to be marked by some combination of pinyon and juniper woodland.

The western boundary of this floristic Great Basin is set by the appearance of the forests of the Cascades and Sierra Nevada, forests dominated by such trees as grand fir (*Abies grandis*), red fir (*Abies magnifica*), incense cedar (*Calocedrus decurrens*), and a variety of pines—jeffrey pine (*Pinus jeffreyi*), western white pine (*P. monticola*), sugar pine (*P. lambertiana*), and ponderosa pine (*P. ponderosa*). This boundary is well defined. As botanist Dwight Billings (1990:78) has noted, "The phytogeographic boundary between the Sierra Nevada and the Great Basin is about as sharp as one finds anywhere between two large biological regions."

The eastern edge is a little more problematic. Ecologist Forrest Shreve once included all of Utah except the state's highest mountains, along with parts of adjacent Colorado and southwestern Wyoming, in his easternmost Great Basin. Since Shreve's definition appeared in 1942, however, ecologists have tended to end the Great Basin in central Utah, along the Wasatch Range and the high Colorado Plateau. Here, Great Basin vegetation is interrupted by montane forests of white fir (*Abies concolor*), blue spruce (*Picea pungens*), Engelmann spruce (*P. engelmannii*), and Douglas-fir (*Pseudotsuga menziesii*)—all trees of the Rockies.

Holmgren and his colleagues draw the northern boundary at the base of the Ochoco and Blue mountains of eastern Oregon, and just north of the Snake River Plain in eastern

TABLE 2-2
A Transect across Nevada at 39° N Latitude
Mountains from East to West

Range	Summit Elevation (feet)	Length (miles)	Maximum Width (miles)
Snake	13,063	68	19
Schell Creek	11,883	133	16
Egan	10,936	110	15
White Pine	11,513	52	18
Pancake	9,240	24	15
Antelope	10,104	41	7
Monitor	10,888	106	14
Toquima	11,941	78	13
Toiyabe	11,788	126	18
Shoshone	10,313	106	13
Paradise	8,657	29	13
Wassuk	11,239	65	14
Singatse	6,778	28	6
Pine Nut	9,450	39	20
Carson	10,881	51	11

SOURCE: Mountain length and width figures primarily from McLane 1978.

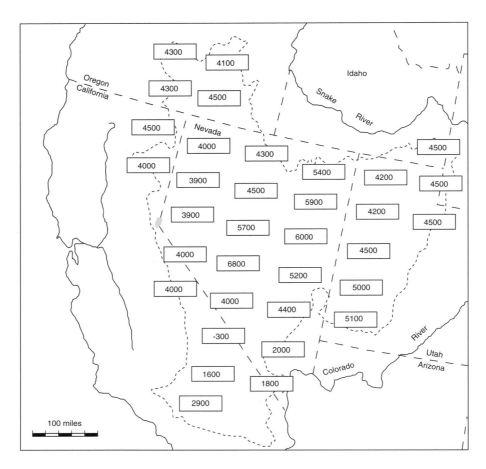

FIGURE 2-6 Valley bottom elevations in the Great Basin, according to the 1:250,000 scale U.S. Geological Survey map.

Idaho. At this point, the edge of the floristic Great Basin is marked by the appearance of ponderosa pine forests and by the occurrence of trees that do not occur to the immediate south, including grand fir and white pine. The match with the hydrographic Great Basin here is not perfect: much of southeastern Oregon and southern Idaho drains to the Snake River and ultimately to the Pacific.

What the floristic Great Basin gains in the north, however, it loses in the south. There are four biologically defined North American deserts—the Chihuahuan, Sonoran, Mojave, and Great Basin deserts (figure 2-10). The Great Basin is the coldest of these, the other three all having warmer winters and hot summers. Also, however much the three warm deserts differ from one another in other ways, they all have creosote bush (*Larrea tridentata*) in common.

The connection between North America's warm deserts and creosote bush is easy to explain. This shrub does not like prolonged periods of freezing temperatures or areas with annual precipitation above about seven inches. As a result, this species provides a very convenient dividing line between the colder botanical Great Basin to the north and the warmer Mojave Desert to the south. The boundary itself roughly follows the 4,000-foot contour through southern Nevada and adjacent California.

The transition is easy to see in any part of the southern Great Basin. U.S. Highway 395, for instance, cuts through Owens Valley in eastern California. Near Big Pine, the valley floor is covered by Great Basin shrubs, including shadscale (*Atriplex confertifolia*), fourwing saltbush (*Atriplex canescens*), and big sagebrush (*Artemisia tridentata*). Once near the dry bed of Owens Lake, however, scattered individuals of a tall, wispy, dark-green shrub begin to appear. A short distance further, just south of the Owens Lake playa and the town of Olanche, plants that look like oddly shaped, almost deformed, trees begin to appear. The spindly, dark-green shrub is creosote bush, the presence of which marks the southern boundary of the Great Basin as a botanical unit, and the northern edge of the Mojave Desert. The strangely shaped plant is the Joshua tree (*Yucca brevifolia*). This plant, one of the most famous of the North American deserts and one that Frémont found to be "the most repulsive tree in the vegetable kingdom" (Frémont 1845:256), is confined to the higher elevations of the Mojave Desert. In the space of a few miles—the trip from Big Pine to Little Lake—we move from the botanical Great Basin into the botanically defined Mojave Desert (see figure 2-11).

This well-accepted delineation of the Great Basin as a botanical unit reduces Shreve's 1942 definition by

TABLE 2-3
A Transect across Nevada at 39° N Latitude
Maximum Relief between Valleys and Adjacent Mountains

Range	Summit Elevation (feet)	Intervening Valley	Basin (nearest 100 feet)	Maximum Relief (nearest 100 feet)
Snake	13,063			
		Spring	5,500	7,600
Schell Creek	11,883			
		Steptoe	6,000	5,900
Egan	10,936			
		White River	5,400	6,100
White Pine	11,513			
		Railroad	5,100	6,400
Pancake	9,240			
		Little Smoky	6,300	3,800
Antelope	10,104			
		Antelope	6,300	4,600
Monitor	10,888			
		Monitor	6,500	5,500
Toquima	11,941			
		Big Smoky	5,500	6,500
Toiyabe	11,788			
		Reese River	6,200	5,600
Shoshone	10,313			
		Ione	6,500	3,800
Paradise	8,657			
		Walker River	4,100	7,100
Wassuk	11,239			
		Mason	4,300	6,900
Singatse	6,778			
		Smith	4,500	5,000
Pine Nut	9,450			
		Carson	4,700	6,200
Carson	10,881			

eliminating eastern Utah and adjacent Colorado and southwestern Wyoming, but includes, as Shreve did, the Owyhee Desert of southeastern Oregon and the Snake River Plain of southern Idaho. Many botanists, however, note that the Owyhee Desert and Snake River Plain do not fall within either the hydrographic or physiographic Great Basin: the water drains to the Snake River, and there are no massive north-south–trending mountains. These botanists also note the relative simplicity of the sagebrush-grass steppe vegetation here. Seeing all that, they eliminate these two areas from the floristic Great Basin.

If that is done, a Great Basin emerges that is well defined botanically and whose borders also match hydrography and physiography quite well, with the exception of the south. Here, the Mojave Desert is botanically quite distinct (creosote bush and Joshua trees are just two of the reasons), yet

FIGURE 2-7 Death Valley from the Panamint Range, with the Black Mountains, part of the Amargosa Range, in the distance.

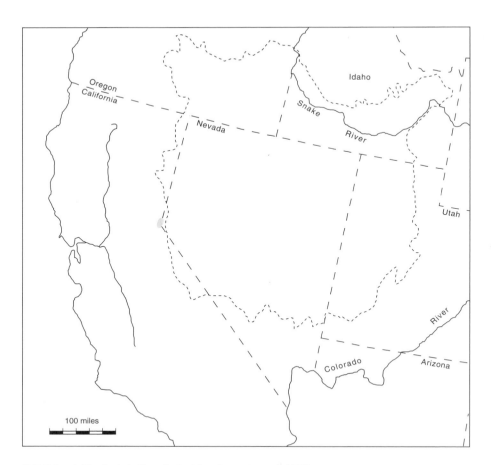

FIGURE 2-8 The floristic Great Basin (after Cronquist et al. 1972).

FIGURE 2-9 Sagebrush-covered lowlands in Oregon's Fort Rock Basin.

FIGURE 2-10 The four biologically defined North American deserts (after Spaulding, Leopold, and Van Devender 1983).

FIGURE 2-11 Joshua trees in Joshua Flats, eastern Inyo Mountains, California. Sagebrush and rabbitbrush dominate the understory.

incorporates large areas, including Death Valley and southern Owens Valley, that drain internally and that are within the physiographic Great Basin.

Noting that valleys within the floristic Great Basin are marked by either sagebrush or saltbush and that the lower elevations of the mountains are often covered with woodland composed of pinyon pine or juniper does not say very much about the vegetation of the Great Basin. A better feel for the botanical Great Basin can be gained by observing the nature of the vegetation from the bottom of a Great Basin valley to the top of an adjacent mountain.

A Botanical Walk from Monitor Valley to Mount Jefferson

Mount Jefferson (11,941 feet) is the highest point in central Nevada's Toquima Range and helps form the heart of the 38,000 acre Alta Toquima Wilderness area (figure 2-5). The bottom of Monitor Valley, due east of here, falls at about 6,830 feet, the elevation alone showing that we are in the high part of the Great Basin arch.

The vegetation of Monitor Valley beneath Mount Jefferson is dominated by two shrubs: little sagebrush (*Artemisia arbuscula*) and the yellow-flowered rabbitbrush (*Chrysothamnus viscidiflorus*). Although abundant at this spot, however, these shrubs do not dominate the entire valley bottom. A dozen miles to the north, for instance, the frequently dry lake bed or playa optimistically called Monitor Lake, which sits in the lowest part of the valley at an elevation of about 6,090 feet, is surrounded by a dense stand of bright-green big greasewood (*Sarcobatus vermiculatus*), a shrub highly tolerant of alkaline soils. Immediately adjacent to the playa, the greasewood is accompanied by saltgrass (*Distichlis spicata*), a plant whose name describes its habitat well. Moving away from the edge of the playa, little sagebrush and rabbitbrush become increasingly abundant and soon replace the greasewood almost entirely.

There are also two species of the plant genus *Atriplex* in Monitor Valley. This genus as a whole is amazingly widespread, found from Australia and Africa to China and Chile, generally in arid and saline contexts. Many of its species have the ability to concentrate salts in their leaves, much of which is then deposited near the leaf's surface, giving the plant a distinctive greenish-white color and making it more reflective to sunlight. This set of adaptations allows the plant access to water it could not otherwise obtain, keeps it cooler than it would otherwise be (because of its whiter shade of pale), and, since heavy doses of salts are not particularly palatable, provides a chemical defense against herbivores.

Both shadscale and fourwing saltbush are found in Monitor Valley. Where they appear, these shrubs are low, widely spaced, drab-gray affairs that typically cover only some 10 percent of the ground's surface. In other, lower-elevation Great Basin valleys, *Atriplex* often dominates the vegetation.

Shadscale grows in sediments high in salts and low in moisture. Up to 40 percent of the dry weight of its leaves may be salt, with most of that found near the surface of the leaves themselves. While this adaptation is impressive, equally impressive is the life that shadscale supports, including the kangaroo rats found here.

As a group, kangaroo rats are almost ideal desert mammals. Short front feet and legs combined with large rear ones and a long tail for balance make them excellent jumpers in open desert settings. An extremely large and efficient ear apparatus provides them with an effective means of detecting predators. Their ability to extract water from their food and to concentrate their urine allows desert-dwelling kangaroo rats the luxury of never having to find freestanding water to drink.

Of the five species of kangaroo rats found in the Great Basin, four are primarily seed-eaters. Only the chisel-toothed kangaroo rat (*Dipodomys microps*) habitually eats leaves, and the leaves it eats are from shadscale. The chisel-toothed kangaroo rat is well adapted to desert conditions, but shadscale leaves are so high in salt that living on them seems nearly impossible, even for such a well-adapted creature. How does it accomplish this?

Some years ago, biologist Jim Kenagy answered this question. The lower incisor teeth of chisel-toothed kangaroo rats are flat in cross-section with a chisel-like cutting surface (hence the name of the animal); all other kangaroo rats have rounded lower incisors with more awl-like cutting edges (figure 2-12). That fact had been known for a long time; Kenagy showed why it is only these kangaroo rats that have such teeth. They use them, he proved, to strip off the outer, salt-bearing portion of shadscale leaves. Holding the leaf in their forepaws, they strip first one side of the leaf, then the other. They leave behind a pile of leaf shavings but eat the starchy and moist leaf interior, whose salt content can be as low as 3 percent of that near the leaf's surface. As a result, they are able to survive on a diet that is essentially pure shadscale leaves, although they can also survive in settings that lack this plant.

Two species of kangaroo rats live in the Monitor Lake area: Ord's kangaroo rat (*Dipodomys ordii*) and the chisel-toothed kangaroo rat. They can both be found in the same set of shrubs, but they certainly don't compete intensely for food. Ord's kangaroo rat feeds primarily on seeds, while the chisel-toothed kangaroo rat eats mainly the leaves from Monitor Valley's shadscale.

If we move above the floor of Monitor Valley and onto the lower slopes of the Toquima Range, or into the wide canyon bottoms that emerge from the mountains, the dominant plants of the valley bottom give way to big sagebrush as the salt content of the soil decreases and as its moisture content rises. Slightly above 7,000 feet, the vegetation becomes dominated by big sage and other shrubs with which it is often found, including Nevada jointfir (*Ephedra nevadensis*) and two species of rabbitbrush. A variety of grasses is to be found here as well, among them cheatgrass (*Bromus tectorum*), an introduced Eurasian species that has unfortunately replaced many of our native grasses.

Nevada jointfir is one of several species of the genus *Ephedra* found in the Great Basin. The genus as a whole is widespread, found not only in the arid parts of western North America but also in parts of South America, North Africa, and much of southern Eurasia. The Great Basin forms as a group are routinely referred to as jointfir, with the names Nevada jointfir used for *Ephedra nevadensis* and Mormon tea for its close relative *Ephedra viridis*. These plants were, and are, commonly used by Great Basin Native Americans to

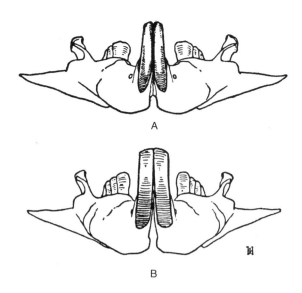

FIGURE 2-12 The lower incisors of (A) Ord's and (B) chisel-toothed kangaroo rats (from E. R. Hall 1946 [reissued by the University of Nevada Press, 1995]). Reproduced with the permission of the University of Nevada Press.

make tea for a wide variety of medical purposes. In fact, species of *Ephedra* are widely used outside North America for such purposes. The ancient Chinese herbal remedy Ma-huang comes from *Ephedra*, and the pharmaceutically valuable alkaloid known as ephedrine was first isolated from a Eurasian species of the genus in the nineteenth century. *Ephedra* can also contain a closely related compound, pseudoephedrine, which has the same chemical formula as ephedrine ($C_{10}H_{15}NO$) but differs in structure. It is this latter compound that is in the decongestant Sudafed, now hidden behind counters in pharmacies in the United States (and elsewhere) because both ephedrine and pseudoephedrine can be used to make methamphetamine. But manufacturers of illegal drugs can look elsewhere than to American species of *Ephedra* for their source of these compounds, since no North or South American member of this genus contains ephedrine, and only one (*Ephedra californica*) contains pseudoephedrine and even then only in minute amounts. *Ephedra viridis* and *Ephedra nevadensis* are, nonetheless, loaded with tannins, as well as with a compound known as 6-hydroxykynurenic acid. This latter compound may help protect the plant from the damaging effects of ultraviolet radiation and, along with the tannins, may help deter hungry herbivores.

Looking underfoot, or where you put your hands, is always a good idea in the Great Basin, where rattlesnakes (*Crotalus viridis*) are common. Fortunately, they want about as much to do with you as you do with them and will do their best to get away in most settings. The odd-looking material crunching beneath your feet as you walk might also escape if it could, since each step you take damages it. Those soil crusts go under a variety of names—cryptogamic, cryptobiotic, microbiotic, and microphytic—but there is now a strong tendency to refer to them as biological soil crusts. Found in an extremely wide variety of settings, though never in tropical rain forests, they are composed of a complex combination of organisms, including cyanobacteria ("blue-green algae"), green algae, fungi, bacteria, lichens, and bryophytes (liverworts and mosses), with the combination varying from place to place and with thicknesses varying from less than one-half inch to more than four inches. In the Great Basin, these living crusts tend to cover the otherwise bare interstices between plants and can be found from valley bottom to mountaintop, though they tend not to occur on highly saline soils.

Ecologist Jayne Belnap and her colleagues have shown that biological soil crusts are remarkably important components of the landscapes they cover. Among other things, they help fix nitrogen in the soil, provide nooks and crannies for the germination of plants with small seeds, and create nutrient-rich substrates for plant growth. By slowing the passage of water over the surface of the ground, they provide time for particles in the water to be deposited, and thus reduce erosion. They also reduce erosion by gluing soil particles together and by covering that soil in the first place. It is even possible that they inhibit the spread of one of the few organisms in the Great Basin that everyone wishes were not there—cheatgrass. Just as important, there is evidence that these crusts soak up carbon dioxide from the atmosphere and may thus help make arid lands a significant defense against global warming.

Unfortunately, biological soil crusts are easily damaged, especially when dry. Trampling by people and livestock and the use of off-road vehicles can destroy the crusts and thus increase soil erosion, reduce soil nutrients, perhaps increase atmospheric CO_2, and in general remove the benefits these crusts have. Full recovery from severe disturbance may take several centuries.

As we move higher in the Toquima Range, not only does the crust remain with us, but the sage gets thicker and taller, especially in and near the gentler canyons. While shadscale supports its share of life, the sagebrush community, with its greater plant cover, supports even more: more vertebrates, for example, may be found in the sage than in the shadscale.

A fair number of Utah junipers (*Juniperus osteosperma*) descend from the mountains onto the upper slopes of Monitor Valley. Junipers, in fact, are routinely the first conifers met along the flanks of Great Basin mountains, since these trees can exist in habitats drier and colder than those tolerated by singleleaf pinyon (*Pinus monophylla*).

The lowest pinyon pines are generally found at slightly higher elevations than the lowest junipers. At the mouth of Pine Creek Canyon, east of Mount Jefferson, the first, lone juniper is encountered at about 7,100 feet in elevation, but the pinyons begin at about 7,130 feet. From this point, the pinyon-juniper mix continues up the canyon to an elevation of about 8,200 feet. By that elevation, however, the junipers are stunted and rare, and the pinyons are thriving only on the canyon's south-facing slope. Just as the first juniper was encountered lower than the first pinyon, pinyon continues above the point where juniper meets its elevational match. From top to bottom here, the pinyon-juniper woodland is some 1,130 feet in elevational thickness, very close to its average elevational width of 1,150 feet.

The plants that accompany the pinyon-juniper in the Toquima Range vary from place to place, but sagebrush and grasses are common members of the understory, as are jointfir and rabbitbrush; in wetter areas, snowberry (*Symphoricarpos*) and silver buffaloberry (*Shepherdia argentea*) make their appearance. As in all pinyon-juniper woodland, the trees rarely get above twenty feet tall, and their branches rarely touch—this is woodland, not forest—and the most common shrub in the understory is big sagebrush, continuing up from below.

The pinyon in the Toquima Range pinyon-juniper belt is singleleaf, but there are two species of pinyon pine in the Great Basin. The other, two-needle or Colorado pinyon (*Pinus edulis*), has a wide distribution centering on the Four Corners area of the Southwest but entering the Great Basin primarily along its eastern edge. According to botanist David Charlet, the only reliable record for Colorado pinyon in Nevada is from the Snake Range, just west of the Utah-Nevada border, where it was discovered by fellow botanist Peter Wigand. Where the two pinyons meet, they may hybridize, but throughout most of the Great Basin, the pinyon in pinyon-juniper woodland is *Pinus monophylla*.

To say that there are two species of pinyon in the Great Basin—*Pinus monophylla* and *Pinus edulis*—glosses over some important complexities that these trees present. In fact, four different kinds of pinyon may be found in this region, and people have treated them in different ways. To begin with, there is the two-needled or Colorado pinyon, which everyone is happy to refer to scientifically as *Pinus edulis*. After this, things get tough, since there are three distinctly different kinds of single-needled pinyon. First, there is Great Basin singleleaf pinyon (*Pinus monophylla monophylla*), widespread in Great Basin uplands south of the Humboldt River. Second, there is Arizona singleleaf pinyon (subspecies *fallax*), found in southwestern Utah and adjacent Nevada and Arizona, with isolated populations in the Mojave Desert of southern Nevada and southeastern California. Finally, there is California singleleaf pinyon (*californiarum*), also found in the Mojave Desert of Nevada and California as well as deep into Baja California.

The needles of these trees are readily distinguishable, but different needle types can sometimes be found on the same tree, and the way scientists sort the trees into species depends on who is doing the sorting. Many experts accept only two species, *Pinus monophylla* and *P. edulis*, and treat the Arizona and California versions as subspecies of the former, giving us *Pinus monophylla monophylla*, *P. m. fallax*, and *P. m. californiarum*. Others treat *fallax* as a kind of Colorado pinyon, giving us *Pinus edulis fallax*. And still others treat the California version as a species on its own, *Pinus californiarum*. As botanist Ken Cole and his colleagues have observed, classifying the needles of these trees is often easier than classifying the species of the trees that produce them.

Having only single needles can be advantageous in times of drought, since trees with single needles have less surface area for evaporation than those with twice that number. In addition, however, the four kinds of pinyon are found in areas with very different precipitation patterns. Ken Cole and his colleagues have shown that both Great Basin (*Pinus monophylla monophylla*) and California (I will call it *Pinus monophylla californiarum*) singleleaf pinyon are found in areas that tend to be marked by dry summers and wet winters. Arizona singleleaf (I will call it *Pinus monophylla fallax*, but feel free to call it whatever you want) and Colorado (*Pinus edulis*) pinyon are found in areas with pronounced rainfall peaks during the summer. These distinctions, it turns out, may be helpful in understanding the history of Great Basin pinyon in the deeper past (chapter 8).

Both *Pinus monophylla* and *Pinus edulis* produce an extremely nutritious nut—10 percent protein, 25 percent fat, and 55 percent carbohydrate for singleleaf—that provided a crucial food for many Great Basin native peoples. Not for all such peoples, however, since pinyon barely makes it north of the Humboldt River in Nevada, as I relate later in this chapter. And, of course, without pinyon, pinyon-juniper cannot exist north of the Humboldt, though juniper does.

Rocky Mountain juniper (*Juniperus scopulorum*) can also be found in pinyon-juniper woodland. This tree is common in the eastern Great Basin, extending as far west as central Nevada, and both it and Utah juniper are found within the pinyon-juniper woodland in the Toquima Range. It never, however, forms an integral part of the pinyon-juniper. Seemingly less drought-resistant than Utah juniper, it is generally a tree of higher elevations than those in which pinyon-juniper woodland is found. When Rocky Mountain juniper is found in pinyon-juniper, it has often followed streams and washes down from above, and this is what it does in the Toquima Range.

Given that Colorado pinyon is found only along the eastern edge of the Great Basin and that Rocky Mountain juniper descends into the pinyon-juniper from above, it follows that almost all of the pinyon-juniper woodland in the Great Basin is assembled from singleleaf pinyon and Utah juniper. Nearly 18 million acres of the Great Basin are covered by this woodland, all on the lower and middle flanks of the mountains. Usually, pinyon-juniper woodland falls between 5,000 and 8,000 feet in elevation, and it generally occurs where mean annual precipitation is between twelve and eighteen inches.

It is not, however, precipitation alone that dictates where pinyon-juniper woodland occurs. Many years ago, Dwight Billings discovered that in the Virginia Range southeast of Reno, temperatures above and below the pinyon-juniper belt were colder than they were in the belt itself. Indeed, he found that valley bottom temperatures could be as much as 15°F or so cooler than those in the woodland and that these temperature differences lasted year-round. This belt of warmer temperatures is created by the downslope movement of colder, heavier air into the valley bottom. Such thermal belts are common—the rule rather than the exception—in the Great Basin, and Billings was cautiously impressed by the correlation he found between the placement of the thermal belt on the Virginia Range and the placement of the pinyon-juniper woodland here. Later work has confirmed Billings's observations.

Neil West and his colleagues have shown that while the average elevational thickness of the pinyon-juniper woodland is 1,150 feet, those widths decline to both the north and west in the Great Basin. Woodlands more than 1,300 feet in elevational thickness are confined to the southern half of the Great Basin. To the north in Nevada, the pinyon-juniper tends to cover narrower elevational bands and eventually disappears.

West and his coworkers suggest that the disappearance of the woodland is related to the disappearance of the inversion-caused thermal belts. The frontal systems that move into the Great Basin from the Pacific strengthen toward the northern reaches of this region (see chapter 5). These frontal storms break up thermal inversions, and it is this process, they suggest, that leads first to the diminishing elevational thickness, and then to the disappearance, of pinyon-juniper woodlands in the northern Great Basin. In fact, even in areas where pinyon-juniper generally does exist, this process seems to be at work. The west face of the northern Ruby Mountains of eastern Nevada, for instance, has no pinyon-juniper, apparently because there are no mountains west of here to break up inversion-destroying storm systems coming in from the west. No inversions, no woodlands.

As West and his colleagues also point out, the ecological amplitude of both pinyon and juniper is impressive. These trees occur alongside Joshua trees and pricklypear (*Opuntia*) in parts of the Mojave Desert and alongside high-elevation limber (*Pinus flexilis*) and bristlecone (*Pinus longaeva*) pines in the mountains to the north.

However, at least for now there is a limit, and virtually no pinyon lies to the north of the Humboldt River. Utah juniper makes it further north than pinyon, but it, too, finally gives way. So, in far northern Nevada, some ranges are devoid of pinyon, though still within the range of Utah juniper, and have Utah juniper woodlands—the Granite Range, for instance, in northwestern Nevada, and the Independence, Bull Run, and Jarbidge mountains to the east. Just a little farther north are ranges that have neither pinyon nor juniper—among them the Santa Rosa and Pine Forest ranges.

A little farther north of here, western juniper (*Juniperus occidentalis*) fills the role performed by Utah juniper to the south. Unlike Utah juniper, however, western juniper is essentially a Californian species, coming from the Sierra Nevada into the lower elevations of the mountains of far northeastern California and northwestern Nevada and spreading across southern Oregon into southwestern Idaho.

It is this juniper that is seen in the arid lands of south-central Oregon. Although it is sometimes said that western juniper never occurs with pinyon in the Great Basin, the two species can occasionally be found together, as in the Panamint Range of southeastern California.

Pinyon-juniper woodland and the western juniper woodlands to the north appear immutable today, as if they have always been there. This, however, is far from the case. In chapters 6 and 8, I talk about the deeper history of pinyon and juniper in the Great Basin. Here, I want to stress the surprising recent history of these trees.

Although pinyon-juniper covers some 18 million acres of the Great Basin today, that was certainly not true 150 years ago. A wide variety of researchers, including botanists Richard Miller, Robin Tausch, and Neil West, have shown that prior to the year 1850 or so, these trees were far less widespread in the Great Basin than they are today. Then, they, along with western juniper, began to spread across the Great Basin landscape. The rate of expansion seems to have peaked between about 1870 and 1920, but the process continues today.

There appear to be multiple reasons for this spread. First, because pinyon and juniper woodlands are susceptible to destruction by burning, anything that decreases fire frequency can lead to an increase in their abundance, as long as appropriate habitat is available for them. Two important factors seem to have led to just such an increase during the latter half of the nineteenth century. The introduction of livestock—cattle and sheep—into the Great Basin meant that plants that had once fed fires now fed sheep and cattle, causing a decline in fire frequency. That decline in turn allowed the expansion of both pinyon and juniper. Not only were trees, including seedlings, no longer removed by fire, but shrubby vegetation, which also would have been limited by fire, increased. Shrubs, in turn, can act as pinyon and juniper nurseries. Birds perching in shrubs drop seeds that can then germinate, and the shrubs themselves provide better growing conditions than does the space between them. Primarily for this reason, the expansion of pinyon and juniper began shortly after the introduction of domestic mammals into the Great Basin.

This advent, however, is not the only factor that would have led to a decrease in fire frequency at about this time. The alteration of Native American lifestyles that was well under way at about the same time would have had the same effect, both by reducing the number of purposefully set fires and by reducing the chances of accidental fires by people dispersed across the landscape.

Although the spread of pinyon and juniper across the landscape coincides with both of these things, it also coincides with milder and wetter climatic conditions. More recently, some combination of increased temperatures, increased atmospheric CO_2, and purposeful fire control have seen to it that this spread continues. The exact role that temperature is playing here is not at all clear, since recent work has shown the most significant expansion of pinyon-juniper, at least in central Nevada, has been into lower elevations and on south-facing slopes—the opposite of what might be expected were increasing temperatures driving this movement.

As pinyon-juniper continues to expand, large areas of sagebrush and grass are converted into woodland. Some have estimated that these woodlands today cover ten times as much territory as they covered prior to their expansion during the past 150 years or so, but this estimate is based on very limited data, and I doubt it is correct. Certainly, pinyon pine is not likely to have loomed as large in Native American uses of the landscape if it covered but 10 percent of the area it now covers. In fact, David Charlet has pointed out that the last half of the nineteenth century saw significant expanses of pinyon-juniper cut down for fuel and building material. Perhaps, he suggests, the expansion of pinyon-juniper we are seeing today simply reflects trees moving back into areas they occupied not long ago. On the other hand, the expansion of these woodlands is, without question, real; some of its possible consequences are shown in later chapters.

Climbing through the pinyon-juniper into higher elevations can be surprising, because once the trees begin to be left behind, the vegetation looks much like what was encountered before the pinyon-juniper woodland was entered in the first place. In the Toquima Range, the elevations at which this transition occurs vary, but along Pine Creek beneath Mt. Jefferson, the pinyon becomes confined to the canyon's south-facing slope by about 8,200 feet in elevation.

Here, the vegetation above the canyon bottoms is once again dominated by sagebrush, in some places much denser than beneath the woodland but accompanied by the same plants that occur with it before its dominance is interrupted by pinyon and juniper. There are additions, of course, perhaps the most notable of which is curl-leaf mountain mahogany (*Cercocarpus ledifolius*), which occurs individually and in small clusters in and just above the pinyon-juniper woodland. The first mountain mahogany along the Pine Creek trail makes its appearance, in shrubby form, at an elevation of about 7,660 feet, but soon takes on its more common, treelike stature. Not a true mahogany at all but a member of the rose family, this tree is widespread in the Great Basin. Curl-leaf mountain mahogany can reach great heights—well over thirty feet—but it was so heavily used as a source of charcoal during the mining booms of the nineteenth century that such large examples are now hard to come by.

Although plants like mountain mahogany can be found in the sagebrush above, but not beneath, the pinyon-juniper, the upper and lower sagebrush assemblages are quite similar to each other, both on the Toquima Range and elsewhere. Seeing these similarities, Dwight Billings observed in 1951 that it almost looks as if the pinyon-juniper woodland has simply been superimposed on a continuous sagebrush-grass zone, and he looked to the area north of the Humboldt Range to support this notion. Here, he noted, in places like the Santa Rosa Range where neither pinyon nor juniper exists, the sagebrush-grass vegetation begins in the valleys and continues virtually uninterrupted to elevations of nearly 10,000 feet. Subsequent research has shown not only that the pinyon-juniper does appear to have been superimposed on a more continuous sagebrush and grass plant community but also when this happened (chapter 8).

In the Toquima Range, the pinyon becomes confined to the south-facing slopes of Pine Creek Canyon starting at an elevation of about 8,200 feet. As this happens, another conifer begins to appear on the uppermost north-facing slopes and along drainages cutting down through those slopes. This conifer is limber pine, one of a number of subalpine conifers characteristic of the upper elevations of many of the highest Great Basin mountain ranges. As higher elevations are reached, the limber pines become taller and move lower down the slopes. By 9,100 feet, these trees cover north-facing slopes and have begun to cover the upper reaches of the south-facing wall of the canyon as well. Another 700 feet higher, and the limber pines have descended both slopes almost to the borders of the stream. We are now in subalpine conifer woodland.

On the Toquima Range, limber pine is the only subalpine conifer, growing here in number from about 9,600 feet to just over 11,000 feet in elevation (figure 2-13). There are much lower limber pines in the Toquima Range, since they follow cold-air drainages thousands of feet downward, and I have seen limber pine as low as 7,800 feet in this range, in the heart of pinyon-juniper woodland. But special situations are needed for them to do that, and most of the limber pine on the Toquima Range is restricted to 9,600 feet or so and above—nearly the uppermost parts of the mountain.

The higher elevations of other Great Basin mountains may have other subalpine conifers; many, in fact, have several of them. In ranges some distance from both the Sierra Nevada and the Wasatch Range, these additional subalpine conifers include bristlecone pine, whitebark pine (*Pinus albicaulis*), Engelmann spruce, and subalpine fir (*Abies lasiocarpa*). Any of these trees can be found at surprisingly low levels. The official Great Basin low elevation record for bristlecone pine—the species famous for providing the world's oldest living tree, which has survived on the White Mountains for about 4,800 years—is provided by an individual discovered by Travis Smith of Great Basin National Park, at 6,695 feet in Silver Creek Canyon at the southern end of the northern Snake Range. (An "official" record is one with supporting evidence that can be confirmed by others.) The lowest unofficial record, reported by biologist Chris Crookshanks, falls at 6,400 feet in Hendrys Creek Canyon, just to the north of Silver Creek. Until very recently, there was an even lower bristlecone, at 5,770 feet, in Smith Creek Canyon a few miles north of Hendrys Creek, but that tree has died. Even though bristlecone pine can occasionally descend to low elevations and limber pine can get well into the pinyon-juniper in the Toquima Range, these trees do not occur in any abundance beneath an elevation of 9,500 feet. Above that point, they can form anything up to and including true subalpine forests.

These five subalpine conifers do not occur only in the Great Basin, and the other places where they occur have

FIGURE 2-13 Treeline and alpine tundra in the Toquima Range, central Nevada, at approximately 11,000 feet. The trees are limber pine; the archaeological site of Alta Toquima Village, discussed in chapter 9, lies just above and adjacent to the trees in the left-center portion of the picture. The Toiyabe Range is in the background.

been used to suggest a general truth about the affinities of the montane floras of the Great Basin as a whole. The Sierra Nevada rises to the west of the Great Basin; the Rockies rise to the east. Both highlands would seem to provide an excellent potential source of plants for the mountains that fall in between. In fact, however, Sierran and Rocky Mountain contributions to the Great Basin flora differ greatly from each other. For instance, the five subalpine conifers I have mentioned are found in both the Rocky Mountains and the Great Basin, but only two of them also occur in the Sierra Nevada. That is, the affinities of the subalpine trees of the Great Basin appear to be to the east, not to the west—to the Rockies, and not to the Sierra Nevada.

We can take this proposition one step further and look not just at subalpine conifers, but at all conifers that occur well within the Great Basin and with or above the pinyon-juniper. "Well within the Great Basin" can mean different things to different people, so I have defined it to mean conifers that reach as far east or as far west as 116° W longitude, a line that runs north-south past Boise, Idaho, and Eureka and Pahrump, Nevada. There are ten of these species of conifer, listed in table 2-4. As this table shows, eight of them are found in the Rockies, but only two are found in the Sierra Nevada alone. These relationships are impressively unequal.

TABLE 2-4
Montane and Subalpine Conifer Species Whose Distributions Reach as Far as 116° W Longitude from either the East (Sierra Nevada) or West (Rocky Mountains)

Species (and subspecies)	Affiliation
Abies concolor concolor, Rocky Mountain white fir	Rocky Mountains
Abies lasiocarpa, subalpine fir	Rocky Mountains
Juniperus communis, common juniper	Sierra Nevada/ Rocky Mountains
Juniperus occidentalis, western juniper	Sierra Nevada
Juniperus scopulorum, Rocky Mountain juniper	Rocky Mountains
Pinus albicaulis, whitebark pine	Sierra Nevada/ Rocky Mountains
Pinus flexilis, limber pine	Sierra Nevada/ Rocky Mountains
Pinus longaeva, Great Basin bristlecone pine	Rocky Mountains
Pinus ponderosa ponderosa, Pacific ponderosa pine	Sierra Nevada
Pinus ponderosa scopulorum, Rocky Mountain ponderosa pine	Rocky Mountains

SOURCE: After Charlet 2007.

This is not to say that few Sierran conifers make it into the Great Basin at all; the table lists only those that extend some distance into the area. There are a number that penetrate a short distance into the western Great Basin from the Sierra Nevada. Incense cedar and Washoe pine (*Pinus washoensis*) make it to the Warner Mountains and Carson Range. Western white pine makes it into the Virginia Range, the Pine Nut Range, and the Warners. Jeffrey pine is in the White Mountains, the Wassuk Range, the Virginia Range, the Bodie Hills, and elsewhere. Sierran lodgepole pine (*Pinus contorta*) is also found in the mountains of far western Nevada, not far from the Sierra Nevada (and its Rocky Mountain cousin is in the far eastern and far northern Great Basin). But few Sierran conifers get very far east into the Great Basin, while Rocky mountain conifers make it all the way across.

This pattern applies to Great Basin montane floras as a whole. From grasses to sages to buttercups and pines, the affinities of the floras of Great Basin mountains are closer to the Rockies than to the Sierra Nevada. Kim Harper and his colleagues have shown that plant species from the Rockies appear to have been about four times more successful in reaching Great Basin mountains than have species from the Sierra Nevada. Indeed, even the floras of the mountains closest to the Sierra Nevada have a greater affinity to the Rocky Mountain flora than to that of the Sierra Nevada. In the sample amassed by Harper and his coworkers, only the White Mountains, just across the Owens Valley from the Sierra Nevada, violate this rule. In fact, the Rocky Mountain affinities of Great Basin montane floras remain even when plants that disperse in different ways are looked at separately: wind-dispersed plants, plants that disperse by sticking to animals, those with fleshy parts that are eaten—all behave the same way.

Why should this be? Why should mountain ranges a few dozen miles downwind from the massive Sierra Nevada and hundreds of miles away from the Rocky Mountains bear affinities to the latter and not to the former?

A number of factors seem to account for this seemingly odd relationship. First, Sierran species as a group are at a climatic disadvantage in any attempt to colonize the Great Basin. Some of these species are adapted to wetter, more maritime conditions than their Rocky Mountain counterparts, which are adapted to drier, more continental conditions far more like those encountered in the Great Basin. In addition, the soils of the Sierra Nevada are often acidic, developed as the result of the weathering of igneous rocks. The soils of the Rockies are often basic, developed from the weathering of calcareous rocks. Since the soils of Great Basin mountains are often basic, Rocky Mountain plants, adapted to the conditions such soils present, do much better here than do Sierran plants.

In a classic study of the distribution of Sierran ponderosa and jeffrey pines in the ranges just east of the Sierra Nevada, Dwight Billings showed that in every instance, these trees grow on acidic soils developed from volcanic rocks affected by heated waters charged with sulfur. Acidic soils, Sierran plants. More recent work, done by E. H. DeLucia, W. H. Schlesinger, and Billings, has shown that these altered sites are deficient in nutrients, particularly phosphorus, and that this deficiency prevents colonization by typical Great Basin vegetation, including sagebrush. The Sierran conifers, on the other hand, are excluded from adjacent, unaltered soil because their seedlings cannot compete with Great Basin grasses and shrubs for moisture.

Beyond considerations of climate and soil, time and geographic barriers have probably also played a role in determining the unequal affinities of the plants of the Sierra Nevada and Rocky Mountains to those of the Great Basin mountains. The Rocky Mountains are older than the mountains of the Great Basin, while the Sierra Nevada is younger. As a result, plants from the east have had more time to colonize this area than plants from the west. Rocky Mountain species may have been well established in the Great Basin by the time Sierran plants were able to make the attempt.

And, in making that attempt, they would have had to deal with a structural barrier that did not exist to the same extent to the east. With rare exceptions, the entire western edge of the floristic Great Basin, from its boundary with the Mojave Desert on the south to its junction with the Columbia Plateau on the north, lies below 6,600 feet. This relatively low area divides the Sierra Nevada and its immediate mountainous outliers—the White Mountains and the Wassuk Range, for instance—from the massive ranges of central Nevada. Not only does that low elevation provide less appropriate habitat for these plants today, but it also means that if the plants were able to make it to those mountains during cooler and wetter times in the past, their chances of survival during warmer, drier times would be diminished compared to their chances on higher, more massive ranges.

James Reveal has called this low-lying region the "Lahontan Trough." Although this trough seems to have acted as a conduit for some Mojavean species to move north, it may also have functioned as a barrier to at least reduce the rate at which propagules of Sierran plants reached the mountains to the east. Reduced rates of dispersal translates into reduced chances of success. Rocky Mountain species had no such low-elevation barrier. Although the huge Bonneville Basin, in which the Great Salt Lake sits, occupies the northern half of much of western Utah, and although much of the southern half of the Great Basin in Utah is below 6,600 feet in elevation, stepping stones in the form of major mountain masses lie at both the northern and southern ends of this eastern trough. These mountainous stepping stones provide high-elevation areas of access to the Great Basin from the Rockies. Comparable montane bridges do not, and did not, exist for Sierran plants.

For all of these reasons, and undoubtedly for others, Great Basin mountains are stocked with plants whose affinities are to the east rather than to the west. In fact, this situation pertains not just to the montane floras I have been discussing here, but to birds and butterflies as well.

Ornithologist William Behle has pointed out the evidence is greater for Rocky Mountain birds having penetrated deep into the Great Basin than for Sierran birds having done so, but the best data are available for butterflies. Entomologists George Austin and Dennis Murphy have shown that among butterflies that are widespread in the Great Basin, as opposed to occurring just along its edges, three times as many forms are shared with the Rocky Mountains as are shared with the Sierra Nevada. As with the conifers, Rocky Mountain and closely related Great Basin butterflies make it all the way across the Great Basin to the eastern slope of the Sierra Nevada, while Sierran forms rarely penetrate far to the east. The numbers for individual mountain ranges are impressive. Even in the White Mountains of eastern California, just across Owens Valley from the Sierra Nevada, twice as many butterfly species have Rocky Mountain affinities than have Sierran ones. By the time we reach the Snake Range on the Utah-Nevada border, the ratio is eighteen to one. While some aspects of this phenomenon might be explained by the distribution of the plants that play host to these butterflies, that can't be the whole story, since suitable habitat appears to exist for many of the missing species.

To judge from the current distributions of plants, birds, and butterflies, the ebb and flow of life in Great Basin mountains appears to have been east to west, not west to east. This, in turn, suggests that factors more general than those that affect plants alone must be at play here.

No matter where the trees came from, as we move upward through the subalpine conifers on a Great Basin mountain—for instance, through the limber pine on the Toquima Range—a number of things may happen depending largely, though not entirely, on the elevation and mass of the mountain. The trees may go all the way to the top. This, in fact, is what happens on parts of the Toquima Range due west of Monitor Lake. The mountain is simply not tall enough at this point to provide conditions that exceed the tolerance limits of limber pine. On truly massive mountains, however, as higher elevations are reached, the subalpine conifers ultimately give way to treeless vegetation—true alpine tundra and alpine desert above a true treeline (figure 2-13).

Treelines are never as sharp and are rarely as horizontal as they appear to be from a distance. The upper edge of tree distribution is generally caused by decreased summertime temperatures and, in particular, by a growing season that has decreased beyond the point that successful reproduction can occur. Indeed, both arctic and alpine treelines (one latitudinal, the other elevational) tend to fall at the 50°F July isotherm. Aspects of topography that increase summer temperatures on a given mountain also increase local treeline elevation, and those that decrease summer temperatures decrease that elevation. Thus, on Northern Hemisphere mountains, trees rise higher on southern and western exposures than they do on northern and eastern ones, since it is the southern and western slopes that gain the full benefit of the summer afternoon sun. Similarly, trees rise higher on ridges, and tundra descends lower in hollows, because of cold air drainage and also because snow accumulation can be so great in canyon heads as to prevent tree reproduction during the warmer months.

As a result, treelines are generally ragged, swinging upslope and down depending on local topography. The nature of the trees themselves may vary as we approach treeline, often becoming dwarfed and stunted, at times taking on the peculiar form known as krummholz (figure 2-14). Formed as the result of intense wind, krummholz trees are often low and matted; they are frequently single-tree thickets up to about 3 feet tall, sometimes taller, that can cover a patch of ground 50 feet or more wide. Small krummholz trees can also exist in the lee of rocks and in other sheltered places, but it is generally the shrubby thickets that catch the eye at upper treeline.

Not all trees form krummholz, but in the Great Basin, all five subalpine conifers—limber pine, bristlecone pine, Engelmann spruce, whitebark pine, and subalpine fir—do. Engelmann spruce krummholz, for instance, is on Mount Moriah in the Snake Range; subalpine fir krummholz on Willard Peak on the eastern flank of the Great Basin; limber pine krummholz on the Toiyabe Range; whitebark pine krummholz on the Carson Range; and bristlecone pine krummholz on Mt. Washington in the southern Snake Range (the only place in the Great Basin where this is known to happen). There is even pinyon pine krummholz on the Panamint Range.

Mechanical abrasion by winter winds and their pelting loads of ice and snow can be important in dictating the form of krummholz. However, those winds play an even more important role in krummholz formation by desiccating exposed needles at a time when stems and roots are frozen and so cannot replace lost water. Hence, krummholz height mirrors winter snow cover, since the covered needles are protected from the wind.

Botanist John Marr has even shown that winds can cause a single krummholz tree to walk across the landscape, as the upwind edge is battered and destroyed by wind desiccation while the downwind end roots its branches and reproduces vegetatively. If downwind growth exceeds upwind destruction, the plant can move into an area where it could not have begun life in the first place, leaving behind a trail of dead wood as it advances. In his study of Engelmann spruce and subalpine fir krummholz in the Front Range of the Colorado Rockies, the largest such trails Marr found were more than 50 feet long; the nearest sheltered site at which one of these might have gotten its start was some 165 feet away. Given that the most rapid growth downwind that Marr could document was on the order of 23 inches per century, this krummholz tree may have taken seven hundred years to get where Marr found it.

On the Toquima Range, there is only limber pine, and the interface between it and alpine tundra begins at about 11,000 feet. Limber pine krummholz, however, extends up to some 11,500 feet on Mount Jefferson in the central Toquima Range, while alpine tundra extends beneath 11,000 feet in topographic concavities that act as cold air traps and that accumulate massive quantities of snow in the winter. Indeed, just as the first limber pines we encountered on the way up the Toquima Range were on the north-facing slope of Pine Creek Canyon, the last ones found here are on the south-facing slope. The Toquima Range treeline is jagged, as treelines usually are.

FIGURE 2-14 Krummholz limber pine at 11,000 feet on the Toiyabe Range. Arc Dome, the summit of the range, is in the background.

Once above the trees, the vegetational landscape is distinctly different. Not only are the trees gone, but so are the shrubs that have accompanied us since we entered the foothills of the mountains. These above-treeline environments are marked by the dominance of low, perennial herbaceous vegetation and by dwarf shrubs. Not surprisingly, these are among the poorest known of Great Basin environments, not so much because they can be inhospitable, even in the summer, as because they are so inaccessible.

Just as with plants from lower elevations in Great Basin mountains, Rocky Mountain alpine tundra species are found farther west in the Great Basin than Sierran alpine tundra plants are found east. Work done by Dwight Billings suggests that strong similarities between the alpine floras of the Sierra Nevada and the Great Basin exist only within the closest Great Basin ranges, such as the White Mountains. Rocky Mountain similarities, on the other hand, extend deep into the west, to include the Deep Creek and Ruby mountains.

Other affinities extend to the true arctic tundra of the far north, the similarities here reflecting the opportunities for north-south movement along high-elevation routes. In the Beartooth Mountains of the Rockies, Billings has shown, nearly half of the plant species in the alpine vegetation are found in the Arctic as well, but that number drops dramatically as one enters the Great Basin. In the Deep Creek Range, 30 percent of the alpine species are shared with the Arctic; in the Ruby Mountain, 25 percent; and in the Toiyabe Range, 23 percent. The numbers reflect the poor long-distance colonizing abilities of tundra plants, as well as the route the Rockies provide to the direct north. Indeed, many Great Basin alpine tundra plants seem derived not from the west, east, or north, but instead from lower-elevation desert forms.

On the Toquima Range, alpine tundra vegetation is found from the upper treeline to the top of Mount Jefferson's south summit (11,941 feet), the highest point on the mountain. From the edge of the Mount Jefferson plateau, the Monitor Lake playa, some 5,000 feet below, is clearly visible. Also clearly visible is the vegetation encountered on the way up, vegetation that appears layered in zones (figure 2-15).

The Vegetation Zones of the Floristic Great Basin

These zones were named by Billings in 1951. Even though a much more detailed set of names now exists to describe the plant associations of the Great Basin, the names Billings provided remain appropriate, and I will use them here. In the valley bottom is Billings's Shadscale Zone, although poorly developed in high-elevation Monitor Valley. Slightly higher, the Sagebrush-Grass Zone is visible as a grayish-green apron on the upper valley and lower mountain flanks; then above it, the first stripe of timber, the Pinyon-Juniper Zone sits in the thermal inversion belt, and looks far more ragged from above than it looked from the valley bottom. The upper edge of the Pinyon-Juniper Zone is marked by another grayish-green interruption in the trees, the Upper Sagebrush-Grass Zone. Then comes the second stripe of timber, here composed only of limber pine, but generally called the Limber Pine–Bristlecone Pine Zone, reflecting the fact that limber pine is not always the only conifer in this subalpine belt (in fact, the Monitor Range, directly to the east, has both limber and bristlecone pines).

FIGURE 2-15 The central Toquima Range, as seen from the Monitor Range. Pinyon-juniper woodland forms the lower band of trees, while limber pine woodland forms the upper band.

The upper edge of the limber pine provides the second Toquima Range treeline, the first created by pinyon pine some 2,500 feet down ("double timberlines," Billings called them). From above, we can again see how ragged the interface between subalpine conifers and alpine herbaceous vegetation really is. Then, finally, we arrive at the vegetation that characterizes the uppermost reaches of the Toquima Range, Billings's Alpine Tundra Zone. In a little more than 5,000 feet in elevation, we have passed through the six classic vegetation zones to be found in the heart of the Great Basin.

While these zones exist in the central part of the Great Basin, however, they do not exist everywhere in the region. This holds true in a number of ways. Some mountains, for example, are not tall enough to reach above the Upper Sagebrush-Grass Zone, or even above the Pinyon-Juniper Zone. Pinyon-Juniper does not make it north of the Humboldt River, and there are even massive, tall mountains, like Steens in south-central Oregon, that have no subalpine conifers on them.

More important, as either the Rocky Mountains to the east or the Sierra Nevada to the west are approached, the montane vegetational zones become distinctly different. On the east, a chaparral of gambel oak (*Quercus gambelii*) and bigtooth maple (*Acer grandidentatum*) lies above the sagebrush-grass on the flanks of the Wasatch Range overlooking Great Salt and Utah lakes. This chaparral is, in turn, surmounted by forests of Douglas fir, white fir, and blue spruce. Above 10,000 feet, subalpine conifers—Engelmann spruce, subalpine fir, and on drier sites, limber pine—extend to treeline; above treeline, alpine tundra is once again encountered.

I have discussed the weak penetration of Sierran conifers east into the Great Basin. Sierran montane plant zones per se, as opposed to individual species of trees that have moved eastward, are well represented only on the Carson Range, between Lake Tahoe and Reno, and, of course, on the eastern flank of the Sierra Nevada itself. On the Carson Range, yellow pines—ponderosa and jeffrey—and white firs begin above the pinyon-juniper and extend upward to about 7,500 feet in elevation. Then come red firs, often alone, but often with lodgepole pine, western white pine, and mountain hemlock (*Tsuga mertensiana*). Above about 8,300 feet, the lodgepole pines and mountain hemlocks become dominant, but these give way to, or actually join, the subalpine conifers—whitebark pine and limber pine. Treeline falls at about 10,300 feet on the Carson Range; above this, alpine vegetation takes over.

Ecosystem-Based Definitions

The definition of the botanical Great Basin that I have described here is widely accepted and used, but as I mentioned earlier, it is not the only biologically oriented definition possible. In fact, in recent years, a tendency has emerged to adopt a narrower definition of a biologically based Great Basin, one that takes into account the fact that the flora of northern Nevada and of adjacent Oregon and Idaho are in many ways distinctive.

One of the best-known of these stems from ecologist James Omernik's definition of eco-regions for the continental United States, a system that has been adopted by the Environmental Protection Agency and the World Wildlife Fund. His eco-regions are based on a wide range of factors—soils, land use, wildlife, vegetation, hydrology, and so on. However, vegetation plays the major role in their description and they are, in that sense, floristic regions as well. His Central Basin and Range eco-region or, as the World Wildlife Fund terms it, Great Basin Shrub Steppe, coincides well with all but the northern boundaries of the more traditional floristic Great Basin that I discussed in the previous section. The parts of the latter that are excluded—southeastern Oregon and adjacent Idaho and parts of northwestern California and northern Nevada—are included in Omernik's Northern Basin and Range, or the World Wildlife Fund's Snake-Columbia Shrub Steppe. Unfortunately, there are no detailed descriptions of either of these eco-regions. In distinguishing these two eco-regions, for instance, the World Wildlife Fund's discussion simply mentions that "the Great Basin is hotter and drier than the Snake-Columbia, lacks distinct major watersheds, and exhibits vegetation associations indicative of its proximity to true desert regions like the Mojave" (Ricketts et al. 1999:326). This does not tell us very much at all.

Perhaps as a result, the biotically based definition of a Great Basin that has gained most traction in recent years is a version of one forwarded by Robert Bailey and modified by The Nature Conservancy. This differs from Omernik's version primarily by lopping off all of southern Idaho's Snake River drainage, thus producing a Great Basin that, in the words of The Nature Conservancy's *Conservation Blueprint* for this region, is "characterized by salt desert scrub and sagebrush shrublands in the valleys and on lower slopes, and by pinyon-juniper woodlands, mountain sagebrush, open conifer forests, and alpine areas in the mountain ranges" (Nachlinger et al. 2001:3). This is a reasonably coherent ecological unit, which is about as much as we can hope for given the complexities of nature on the ground. This coherency accounts for the wide use this definition of a Great Basin is now getting.

In what follows, I use the term "floristic" or "botanical Great Basin" to refer to the more traditional definition, the Great Basin that extends from the edge of the Mojave Desert into southeastern Oregon and adjacent Idaho.

The Vertebrates of the Floristic Great Basin

While the hydrographic Great Basin is an abstraction, seen only by circling it and following its waters and ultimately realizing that none flow out, the botanical Great Basin is the one that seems to impinge most directly on the senses. That Great Basin waters flow inward is certainly interesting, as well as both geologically and economically important, but it is the plants and the things that live in, on, and around them that form the immediacy of the Great Basin when you are in it.

Although the floristic Great Basin is defined on the basis of its plants, the vertebrates that live here are equally remarkable. There are, for instance, about 100 species of mammals known from the Great Basin. The Malheur National Wildlife Refuge lists 257 species of native birds recorded in this part of south-central Oregon; the Ruby Lake National Wildlife Refuge lists 245. Gordon Alcorn lists 456 species of birds recorded from, or probably to be found in, the state of Nevada alone. Add the native snakes, lizards, frogs, turtles, and fishes (44 species), and there are well over 600 species of vertebrates that live in the Great Basin.

How most of these animals reached, or reach, the Great Basin poses no mystery. The greater scaup (*Aythya marila*) breeds in the far north and winters on both Atlantic and Pacific coasts; it is occasionally found in the Great Basin as it makes its way from wintering to breeding grounds or vice versa. The northern shrike (*Lanius excubitor*) and snowy owl (*Nyctea scandiaca*) both breed north of the Canadian border; during winter, the shrikes frequently, and the owls occasionally, move as far south as the Great Basin. The western patch-nosed snake (*Salvadora hexalepis*) and the western ground snake (*Sonora semiannulata*) apparently moved north into Nevada (and for the western ground snake, into Oregon) by following the Lahontan trough out of the Mojave Desert. Merriam's kangaroo rat (*Dipodomys merriami*) follows shadscale up the same trough, reaching far northwestern Nevada.

Other vertebrates appear to pose real mysteries. The Devil's Hole pupfish (*Cyprinodon diabolis*) exists in only one place on earth: in a pool in a rock-bound hole only some twenty-three feet across in Ash Meadows, southern Nevada. How did it get here? The Salt Creek pupfish (*Cyprinodon salinus*) is a Death Valley specialty: how did it come to be in one of the hottest and driest places on earth?

Mammals provide equally intriguing questions. Pikas (*Ochotona princeps*) are small, short-eared mammals closely related to jackrabbits and cottontails. Though generally restricted to higher latitudes, they are also found at high elevations in western North America. Desert environments are absolutely inimical to pikas: they cannot take the summer heat of lower elevations and die in a few hours if exposed to it. So it is no surprise that there are no pikas in Great Basin valleys. However, pikas are found in a number of Great Basin mountains, each of which is surrounded by valleys in which no pika can or does live. In southeastern California, there are large populations of pikas in the White Mountains; in central Nevada, they live in the Desatoya, Toiyabe, Toquima, Monitor, and Shoshone ranges. In eastern Nevada, they are in the Rubies. How could this have happened? And equally intriguing, given that it did happen, why aren't there pikas on apparently similar mountains, like the Snake, Egan, and Schell Creek ranges? Just as important, if these animals are isolated on the tops of Great Basin mountains, what might the future hold for them?

Other mammals show similar distributional oddities, including the pygmy rabbit (*Brachylagus idahoensis*). This small animal—weighing less than one pound, it is the smallest member of the biological family that includes the rabbits and hares (the Leporidae)—can be distinguished from other rabbits not only by its size but also by the underside of its tail, which is gray, not white. Pygmy rabbits also make their own burrows. These are located in tall, dense stands of sagebrush, the same plant on which it is heavily dependent for food. Up to 99 percent of a pygmy rabbit's winter diet may be drawn from this plant, with the sage providing cover from predators as well. Because they dig their own burrows, these animals are also confined to areas with friable substrates that are easy to excavate but not so easy that they are prone to collapse. Pygmy rabbits are discontinuously distributed throughout the floristic Great Basin, but there is also an isolated, and endangered, population in southeastern Washington. A pygmy rabbit–less gap of about 150 miles separates the southernmost Washington record from the northernmost Oregon one. How, it is impossible not to wonder, did this happen? And, given that the Washington pygmy rabbits are on the federal endangered list, can understanding how this happened help save the animal? These are issues to which I return in later chapters.

The Ethnographic Great Basin

The hydrographic, physiographic, and botanic Great Basins look to drainage, landforms, and plants, respectively, to define different Great Basins. These definitions are known to one degree or another to almost every geologist or life scientist who works in the area. The fourth and last Great Basin, however, is defined on the basis of the native peoples who lived in the area at the time Europeans first arrived and is well known primarily to anthropologists.

For over a century, anthropologists have divided the peoples of the world into groups based on the similarity of the cultures possessed by those peoples. These "culture areas" cluster peoples that are behaviorally more similar to one another than they are to peoples outside that unit. Culture areas do not take into account how the peoples included perceive one another or how they get along with one another. Nor are the borders drawn around single sociopolitical entities like "tribes." Instead, culture areas group together sets of people on the basis of similarities in such things as the nature of their subsistence pursuits, their sociopolitical organization, their material manufactures, their religion, and so on.

The Great Basin Culture Area of anthropologists provides yet another Great Basin, a spatial grouping of peoples far more similar to one another than they were to peoples in adjacent culture areas. Although large-scale groupings of peoples based on cultural similarities do not often coincide well with groupings independently derived from considerations of linguistic relationships, the Great Basin is one place where the match between language and culture is nearly, though not quite, perfect.

When linguists classify languages into groups that reflect their historical relatedness, they do so hierarchically. A language—generally composed of a set of mutually intelligible dialects—is lumped with other, similar languages into a higher-order unit, generally called a language family. Similar language families are then grouped into an even higher order unit, generally called a phylum (though historical linguists, who tend to be hard-nosed scientists, are wildly inconsistent in their use of these terms). Among European languages, for instance, Italian, French, Portuguese, Provençal, and Romanian are grouped into a Romance family, all the members of which are derived from Latin. English, German, Yiddish, Dutch, Swedish, Danish, and others are grouped into a Germanic family. Ukrainian, Polish, Russian, Serbo-Croatian, and others are grouped into Slavic. Although the languages within each of these families show strong similarities to one another, the families themselves are also similar to one another and must share a common ancestor. As a result, Germanic, Slavic, and the Romance languages are grouped into a higher-order category called Indo-European. Other language families in this group include Celtic (which in turn includes Gaelic and Welsh), Baltic (Latvian and Lithuanian), and, in its own family, Greek. Such groupings directly reflect history because they directly reflect commonality and closeness of descent—linguistic evolution in action.

The number of languages spoken in North America at the time of European contact is not known, but there were hundreds of them; linguist Ives Goddard lists 329, of which 120 were extinct in the mid-1990s with many more spoken only by elders. The ties of historical relatedness indicated by analyses of those languages are fascinating. The speakers of languages belonging to the Uto-Aztecan language family, for instance, lived from southeastern Oregon to Panama. Nahuatl, the language of the Aztec and their modern descendants, is Uto-Aztecan, as is the language of the Hopi of Arizona. If the languages are related, the people are related, and so it is with the Hopi and the Aztec, just as it is with the French and the Germans. And so it is as well with nearly all of the native peoples of the Great Basin: with one major exception, they were and are Uto-Aztecan speakers and thus are closely related to both the Aztec and Hopi.

There are six Uto-Aztecan languages in the Great Basin, all very similar to one another, and all grouped into a single branch of Uto-Aztecan called Numic (figure 2-16). The similarities and differences among Numic languages show them to fall into three natural groups, each consisting of two languages. These three subdivisions are often simply called Western, Central, and Southern Numic, though linguists also call them by the names of the languages of which they are composed: Mono-Northern Paiute (or Paviotso), Panamint-Shoshone, and Kawaiisu-Ute.

There is a seventh Numic language, Comanche, spoken in the southern plains. In fact, Comanche speakers are descended from Shoshone speakers who moved from the Great Basin onto the plains sometime before 1700 A.D. and then moved southward where, during the 1700s, they displaced the Apache. The Apache, in turn, speak a language that belongs to a group called Na-Dene, which includes Navajo as well as a wide variety of languages found along the northern Northwest Coast (for instance, Tlingit) and subarctic western North America (for example, Dogrib and Chipewyan). Linguistic relationships show that the ancestors of the Navajo and Apache migrated to the Southwest from the subarctic, displacing the people who were there before them. When they arrived is not known with any certainty, but it may have been during the early 1600s or somewhat earlier (see chapter 9). Those who moved onto the southern plains were later displaced by the Comanche, whose ancestral homeland was the Great Basin.

Of the three pairs of Numic languages found in the Great Basin, one language of each pair occupied a very small part of the far southwestern Great Basin, while the second in the pair spread out over vast distances to the north and east. Mono speakers, for instance, occupied the Owens River Valley in southeastern California; Northern Paiute, or Paviotso, its match in the pair, spread from about Mono Lake well into southeastern Oregon. The speakers of Kawaiisu lived in an area running from the Mojave River to the mountains bordering the San Joaquin Valley. The speakers of Ute spread out from their border with the Kawaiisu to the east and north across southern Nevada, much of Utah, and ultimately into Colorado. The area between the Mono and Kawaiisu, focusing on Death Valley, was occupied by the Panamint; the speakers of the tandem in this language pair, Shoshone, spread out from here through central Nevada and across northern Utah into southern Idaho and adjacent Wyoming.

Languages are often composed of several mutually intelligible dialects, and the Numic languages are no exception. The pattern of Numic dialectical variation, however, is intriguing, since most of it occurs in the far southwestern corner of the Great Basin. Although Kawaiisu (which now sadly has fewer than ten speakers) has not been shown to have any dialects, dialectical variation within both Mono and Panamint appears to be greater than that within the far more widespread Paviotso, Shoshone, and Ute. In addition, dialectical variation within these three widespread Numic languages is greatest where they approach the borders of Mono, Paviotso, and Kawaiisu.

Given that dialectical variation develops as a result of diminished opportunities for contact between peoples, and that such divergence often reflects distance in both time and space, one might guess that those languages that are most widespread would also have the greatest dialectical diversity. This is not the case in the Great Basin. The fact that most of the linguistic diversity in the Great Basin is found in its southwestern corner has led some to argue that this corner—roughly centered on Death Valley—represents the homeland of Numic speakers, from which they have spread out across the Great Basin, perhaps in relatively recent times. This notion of relatively recent Numic expansion across the Great Basin, if accurate, could account for the geographic distribution of linguistic variation in the

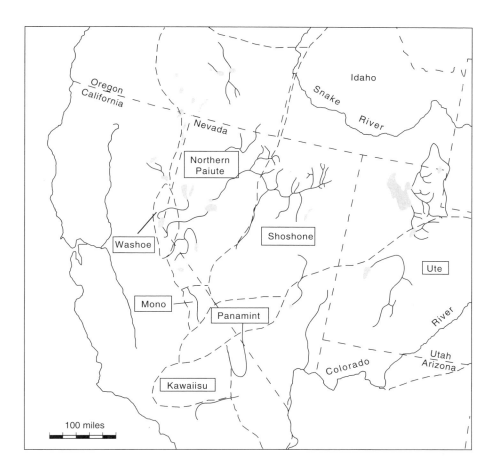

FIGURE 2-16 The Great Basin as a culture area (after D'Azevedo 1986).

Great Basin, since the longer residence times in the southwestern part of the area might explain why linguistic divergence is greater here. Whether or not such relatively recent expansion actually occurred, however, is a different issue, as I discuss in chapter 9.

In addition to Numic speakers, the anthropological Great Basin includes another group of people, the Washoe, whose territory roughly centered on Lake Tahoe, but that also extended along a one-hundred-mile wide strip on and adjacent to the eastern flanks of the Sierra Nevada. Today, both Reno and Carson City sit in territory once occupied only by the Washoe. Although the Washoe are culturally very similar to the Numic speakers of the Great Basin, their language is quite distinct. Along with a variety of primarily California and southwestern languages, it is often placed in a larger group called Hokan, though Hokan itself is something of a controversial linguistic category.

Living on the very edge of the Great Basin, the Maidu belong to a third substantial linguistic group, Penutian. Penutian speakers were and are found from southeastern Alaska deep into California, but Maidu speakers seem to be linguistically closest to the Penutian peoples of Oregon, including the Klamath and Modoc of northeastern California. During early historic times, Maidu territory apparently included Honey Lake and perhaps Eagle Lake as well, though other groups have also claimed that they used these areas. Most Maidu lands, however, were on the western side of the Sierra Nevada.

The classic picture of Native American lifeways in the Great Basin was drawn by the ethnographer Julian Steward. Steward's description and analysis of Great Basin peoples depended not only on the observations of other ethnographers but also on fieldwork he conducted among these peoples during the late 1920s and mid-1930s. Although he published a variety of papers and monographs dealing with the Great Basin, Steward's most influential work appeared in 1938, his *Basin-Plateau Aboriginal Sociopolitical Groups*. Now some seventy years old, this volume remains a key reference to the aboriginal lifeways of Great Basin native peoples.

Steward argued that the sparse, scattered, and unpredictable nature of resources in many parts of the Great Basin dictated that the basic economic unit in this area was the nuclear family—parents, children, and sometimes grandparents—averaging about six people in size. Throughout much of the Great Basin, plants in general and seeds in particular were of extreme dietary importance, and a highly sophisticated technology, from stone grinding tools to basketry seed-beaters and winnowing trays, developed to harvest and process plant foods. The gathering and preparation of plant foods was the realm of women, and their economic contribution was critical to the well-being of all families.

While plant foods formed the core of the diet, however, animal products were important not only for the nutrition they provided but also for such things as skins for clothing, bone for tools, and sinew for a wide variety of purposes. While both men and women took smaller game, the larger mammals—pronghorn (*Antilocapra americana*), deer (*Odocoileus*), mountain sheep (*Ovis canadensis*), and in some areas, bison (*Bison bison*)—were hunted by the men. Thus, the economic efforts of both men and women were essential to the success of both sexes. Accordingly, Steward argued, groups smaller than the nuclear family could not exist for long.

Steward also argued that, with rare exceptions, the food resources of the Great Basin were too meager to support large groups on a routine basis. In much of the area, Steward maintained, the largest single settlement was the winter village, where families gathered when new stores of plant foods could no longer be obtained. The location of a winter village was determined by ease of access to a series of critical resources. Perhaps most important, the villages were located near sites where foods gathered earlier in the year had been stored. Since pinyon nuts usually provided the prime over-wintering resource, winter villages were usually located in an area where substantial amounts of these nuts had been cached. In addition to stored foods, however, access to water and wood, for construction and fuel, were also critical. Within the list of places that met these needs, the winter village would also be located in such a way as to avoid the coldest temperatures winter might offer.

Given these requirements, it is probably no surprise that the winter settlements were often placed at the lower edge of the pinyon-juniper zone. Such placement minimized the costs incurred in transporting large quantities of pine nuts elsewhere. In addition, because of the thermal inversions that develop in the Great Basin—the same inversions that seem so important to the development of the pinyon-juniper zone in the first place—locating the winter villages at the base of the pinyon-juniper provided a setting as much as 15°F warmer than the valley bottoms. Those inversions, however, are destroyed by storm systems coming in from the west, especially on mountains not protected to the west by other ranges. The temperature advantage, then, would be best on the east flank of a range, and Steward's 1938 maps of winter-village locations show that most of those located in the mountains were on the east sides of those mountains.

Winter villages, however, could also be located in other places—in the mouths of canyons, for instance, or along valley bottom streams that could provide fish. In Owens Valley, to take one example, the decision as to whether to winter in the valley or in the pinyon-juniper depended heavily on the nature of the fall pinyon crop. Were that crop large, and the transportation costs involved in moving it down to the valley correspondingly high, people might winter in the mountains. Were the crop poor, and the transportation costs correspondingly low, greater advantages were to be had by remaining in the valley.

The decisions made in Owens Valley concerning winter-village location were similar to those made elsewhere in the Great Basin in that the location of these villages depended heavily on resource availability and resource productivity in a given year. Since those things changed from year to year, so did winter-village locations. With rare exceptions, Steward argued, people did not know during any given winter exactly where they would be spending the next one.

How many people a winter village held in a particular year depended on the abundance of the resources available near that village. If stored resources were meager, a single family might winter alone, but if it had been a good year and resources were truly abundant, twenty or more families—more than a hundred people—might pass the winter months together. While many kinds of foods could be stored—from roots and seeds to dried fish, bird, and mammal meat—in most parts of the Great Basin, pinyon nuts provided the key over-wintering resource. More than any other single item, these nuts approached the status of a true staple in those parts of the Great Basin that had pinyon trees in abundance. For this reason, I doubt modern estimates that pinyon now covers ten times the territory it covered just prior to the European arrival.

As I have mentioned, pinyon nuts are extremely nutritious. Singleleaf pinyon, the species found in most of the Great Basin, provides a nut that is 10 percent protein, 25 percent fat, and 55 percent carbohydrate; Colorado pinyon has a nut that is about 15 percent protein, 65 percent fat, and 20 percent carbohydrate. In both cases, the protein includes all twenty amino acids.

A given stand of pinyon does not produce every year: a good crop from a particular stand is likely to be followed by several lean years in that stand, presumably because the trees need to replenish the resources used to produce both cones and nuts. Fortunately for the people who depended on them, the general nature of a pinyon crop could be predicted well in advance, since the cones that bear the nuts are visible well more than a year before the crop matures. Because many things could intervene to destroy a developing crop, one couldn't predict a year in advance where exactly the crops would be best. However, one could predict, as archaeologist Dave Thomas has pointed out, where they would be bad.

Steward estimated that a family of four people working a good crop for four weeks could gather enough nuts to last them about four months, roughly the whole winter. Hence the importance of pinyon: this resource was critical to surviving the harshest and leanest time of the year. The larger the pinyon crop, the easier it would be to survive the winter, and the larger the winter village became.

Those villages remained occupied from the end of the pinyon season to the onset of spring, when new foods became available. As spring approached, the winter village would begin to break up, with individual families or, most often, small groups of families going their own way. The small groups of people that traveled together tended to be interrelated by marriage. According to Steward, it was rare that more than four or five such families—twenty-five or thirty people—would stay together for any length of time. His data suggest that the usual cluster was on the order of two or three families, or fifteen to twenty people. After leaving the winter village, those small groups spent the months from spring to fall moving from resource to resource, the length of time spent at any one place dictated by the abundance of the resources at that place.

Communal activities involving sets of related families did take place outside the winter village, many of which involved hunting. A wide variety of animals, from coots and ducks to jackrabbits and pronghorn, could be taken far more efficiently in cooperative hunts than by single individuals or by small groups of people. Sage grouse (*Centrocercus urophasianus*), for instance, congregate in large flocks from late fall to spring and could be taken en masse, sometimes in the hundreds, by hunters cooperating to drive them into nets or other surrounds. Indeed, Dave Thomas has described a complex series of low, intersecting rock walls extending for more than 1,300 feet on central Nevada's Monitor Range that may have been a communal sage grouse hunting facility.

While a number of different animals were taken cooperatively, it was the communal hunting of pronghorn and jackrabbits (*Lepus californicus*) that seems to have been of greatest importance. While these hunts could be conducted in different ways, the general picture of the way most were performed remains the same. Communal pronghorn hunts, for example, were often conducted under the leadership of a shaman who held spiritual power over the animals. Groups of people worked the animals in a given valley, forcing them between huge V-shaped wings that led into corrals, where they would be killed by waiting hunters, men and women alike. At times, the facilities that funneled and corralled the animals were made of fairly substantial materials—rock and wood—but, more commonly, they were of sagebrush or other vegetation. The V-shaped wings leading to the corral might be long—more than a mile in some cases—but they were not particularly high, since pronghorn rarely jump significant heights. Steward observed that communal pronghorn hunts of this sort were often so efficient that a dozen years might pass before a given valley would see another one.

Jackrabbit drives were conducted in much the same fashion, though here the animals were driven into huge nets, each of which was some three feet high and several hundred feet long. Robert Lowie described one such drive among the Washoe, in which two hundred people took between four hundred and five hundred jackrabbits a day.

Both jackrabbit and pronghorn drives could be held at various times of the year. If they were held during the winter months, people from different winter villages would join together, returning to their villages after the hunt had ended. For hunts held at other times of the year, different families or family groups would band together for the duration of the hunt and then continue on their separate ways.

At times, communal hunts coincided with another communal aspect of Great Basin life—the festival or "fandango." These gatherings took place two or three times a year. How long they lasted depended on how long the food lasted, food that was routinely provided by a jackrabbit or pronghorn drive or by the fall pine nut harvest. Indeed, a fall jackrabbit drive was often conducted by people who had gathered to harvest pine nuts, making both sets of resources available to support a fairly substantial fall festival. The purpose of these gatherings, lasting about a week, was strictly social: dancing, games, courting, gossip, and no doubt, the sharing of information about local resources all took place.

While communal activities of this sort did happen, however, Steward's analysis suggests that most time outside the winter months was spent in small groups, a few families moving across the landscape together in search of food. As fall approached, these small groups would move toward areas likely to produce a substantial pinyon crop, and a significant fraction of autumn would be occupied by gathering, processing, and storing these nuts. Steward noted that because the precise area in which a good pinyon crop would appear could not be tightly predicted a year in advance, people did not know during one winter where they would be spending the next one. And, because small groups of intermarried families left the winter villages separately, people also did not know with which other such groups they would be spending the next winter. Nuclear families, Steward emphasized, were the basic sociopolitical unit in the Great Basin.

In Steward's words, Great Basin native culture "was stamped with a remarkable practicality . . . starvation was so common that all activities had to be organized toward the food quest, which was carried on mostly by independent families" (1938:46). But to this basic description must be added one important caveat and a major modifier. Steward would have agreed with the modifier since he himself suggested it. However, the caveat he emphatically rejected.

The caveat first. Steward conducted his fieldwork during the 1920s and 1930s. The people he talked to described conditions and lifeways that had existed during the last half of the nineteenth century and during the first few decades of the twentieth. In addition to describing things they themselves had done, the people he interviewed also recalled what they had been taught by their elders. But none of these people could know with any certainty what life had been like prior to the monumental disruptions caused, directly or indirectly, by the arrival of Europeans in North America.

Pronghorn drives provide a simple example. Steward believed that a communal pronghorn hunt "so reduced their number that years might be required to restore the herd" (1938:33). However, archaeologist Brooke Arkush has observed that the information that led Steward to this conclusion came from a time well after the European arrival in the Great Basin. As a result, it also pertained to a time when, for reasons ranging from the introduction of livestock to a decrease in fire frequency (both leading to a reduction in preferred pronghorn food), pronghorn numbers must have been dramatically reduced. Arkush convincingly argues that in those parts of the Great Basin that were originally rich in the plant foods on which pronghorn depend, pronghorn drives are likely to have been possible on an annual basis.

In 1962, echoing concerns that had been raised twenty years earlier by ethnographer Omer Stewart, anthropologist Elman R. Service launched an attack on Steward's description of Great Basin lifeways on just these grounds. He argued that what Steward was describing was a system in disarray as a result of European contact. On all but the western edge of the Great Basin, Service argued, many Great Basin peoples had had their lifeways completely disrupted by Indians who had adopted the horse. Elsewhere, he suggested, Europeans themselves directly destroyed Great Basin cultures by usurping the resources on which they depended: cutting down pinyon for fuel, occupying critical springs and canyons mouths, disrupting the animals they hunted, and so on. Great Basin natives quickly became peripheral hangers-on in American towns. Indeed, the very people Steward interviewed generally occupied just such a position.

For Service, Steward's ethnographic Great Basin—a Great Basin founded on more or less independent nuclear families moving across the landscape, now coming together into larger groups, then fissioning once again—became a lifeway that reflected disruption caused by the European arrival, not a lifeway that had existed before that event.

The European arrival was disastrous to Native American lifeways throughout North America. Archaeological work has shown that, with rare exceptions, massive cultural change generally preceded actual face-to-face contact with Europeans by many years, with that change due to population declines caused by such diseases as whooping cough, smallpox, and measles. In some areas, population losses were so huge that when Europeans made their physical arrival, there were no human occupants to be found. Given our increasing understanding of these issues, Service's concerns over the meaning of Steward's descriptions are even more appropriate now than they were when Service first raised the issue nearly five decades ago. However, even those who tend to be very critical of some aspects of Steward's work tend to accept at least the general aspects of his descriptions of late prehistoric human lifeways in the Great Basin. In addition, mid-nineteenth century firsthand accounts of these lifeways routinely match Steward's descriptions.

The modifier needed to Steward's general description of Great Basin lifeways is not controversial. In fact, that modifier gave Steward's work on Great Basin native lifeways tremendous theoretical import.

Steward observed that the complexity of the societies he studied varied across the Great Basin, and that this variability was keyed to the nature of the resources to which people had access. Peoples who lived in settings in which diverse, predictable, and abundant resources were packed into relatively small areas were marked by a sociopolitical organization far more complex than that of peoples living in areas in which resources were sparse, widely scattered, and unpredictable. People living in the former areas tended to have, for instance, larger populations, more stable villages, private ownership of certain resources, territories they defended, and headmen whose positions might be inherited. These things were lacking in resource-poor areas.

At times, Steward almost seems to have thought of Great Basin peoples as having been divided into those living in resource-rich areas and possessing a relatively complex sociopolitical organization, and those living in areas of lesser abundance and thus having less sociopolitical complexity. But as Kay Fowler and others have pointed out, and certainly Steward recognized this as well, organizational complexity formed a continuum in the Great Basin keyed to resource distribution, predictability, and abundance.

Steward used the Owens Valley Paiute to exemplify the complex end of his continuum. These people lived, and live, in a well-watered, environmentally complex territory that included not only the Owens Valley itself but also the Sierra Nevada to the west and the Inyo and White mountains to the east. The resources of this area were extremely rich. The mountain flanks to the east held good stands of pinyon, those to the west provided abundant supplies of Pandora moth (*Coloradia pandora*) caterpillars (an important food item; see figure 2-17), and mountains in both directions were well stocked with a wide range of mammals and birds. The valley provided fish, birds, rabbits, pronghorn, and rich seed and root crops. While other parts of the Great Basin might provide the same set of resources, only rarely were they so tightly packed. As a result, Owens Valley held permanent villages—perhaps some three dozen of them, some of which may have held as many as two hundred people. Indeed, at one person per 2.1 square miles, Owens Valley had one of the highest estimated population densities known for the Great Basin. (The average may have been on the order of one person per 16 square miles, and the lowest at one person per 35 square miles, this for the Goshute of the Bonneville Basin in Utah. We have, however, no way of knowing what pre-European population numbers were like).

Owens Valley itself was divided into districts, with each district having one or more villages, and the people of the district communally owning the rights to hunt, fish, and gather plant foods therein. The people of each such district cooperated with one another in communal hunting and fishing, in holding ceremonies, and in some areas, in irrigating plots of wild seeds. These communal activities were routinely directed by one individual, a headman, chosen largely for his leadership abilities; the sons of headmen might inherit that position if they showed the same leadership qualities.

FIGURE 2-17 A jeffrey pine surrounded by a trench for trapping Pandora moth caterpillars as they descend from the tree. Piüga Park, Inyo National Forest, California.

All this was very different from the sociopolitical organization seen in many other parts of the Great Basin, where permanent villages did not exist, where land ownership or at least land control did not exist, where there were no leadership positions that might be inherited, and where resources were not routinely thought of as owned.

Although the Owens Valley Paiute provided Steward with an example of the sociocultural complexity that developed as resources increased in density, abundance, and predictability, it is evident that this relationship existed elsewhere in the Great Basin. This is the case even though we have no detailed, written descriptions of this relationship from other areas that can be compared with the one that Steward provided for the Owens Valley Paiute.

The Timpanogot—Ute speakers who lived in the area surrounding Utah Lake—provide a good example Their way of life was dramatically altered first by the arrival of the Spanish in the 1770s—the Domínguez-Escalante expedition mentioned in chapter 1—and then by other Europeans in the next century, most importantly, the Mormon settlers. As a result, there are no substantial, firsthand accounts of their lifeways prior to these disruptions. However, anthropologist Joel Janetski has scoured all the early accounts that do exist and has built as thorough a description of those lifeways as we are likely to have.

During early historic times, the Timpanogot depended heavily on the extraordinarily productive fishery then provided by Utah Lake. To this they added resources offered by the marshes that fringed the lake and the streams that fed it, as well as those available in the surrounding uplands. They defended their territory against incursions by the neighboring Shoshone, had a hierarchy of chiefly leaders, lived for a substantial part of the year in villages located near the lake, and along with the Owens Valley Paiute, had one of the highest population densities to be found in the Great Basin in early historic times.

In line with all of this, substantial late prehistoric villages are known from the Carson and Humboldt basins of western Nevada, the Snake River drainage of southern Idaho, and elsewhere. These villages suggest that comparable complexity existed in all of these places, in all instances associated with rich riverine and marsh resources.

Clearly, the ethnographic Great Basin was characterized by a resource-driven continuum in organizational complexity. Indeed, perhaps nowhere in North America south of the Arctic is the relationship between the organizational complexity of Native American lifeways and resource availability quite so clear. This will remain true even if it turns out that aspects of Great Basin cultures as described by Steward reflect the negative impacts of European contact. The late prehistoric archaeological record for the Great Basin strongly suggests, if it does not already show, that had Steward been able to talk to late prehistoric peoples, the continuum of organizational complexity would have been more, not less, evident.

The cultural Great Basin is larger than the Great Basin defined in other ways, since the peoples that make up this area extended well into Idaho (the Northern Shoshone and Bannock), Wyoming (the Eastern Shoshone), and Colorado (the Ute). Although the largest of the Great Basins, the cultural Great Basin is, in two distinct senses, also the most poorly known. Scientifically, it is the most poorly known because what characterized it—Native American lifeways in latest prehistoric times—no longer exists to be directly studied. It is also the most poorly known in the sense that what is known—and there is quite a bit—is primarily known to anthropologists and archaeologists, and not to others. This is unfortunate since Native American lifeways in the Great Basin were extraordinarily impressive, as even a few minutes' reflection on surviving on your own in Death Valley might suggest.

Choosing a Great Basin

Since this is a book about the natural prehistory of the Great Basin, it might seem that I should choose just one of these multiple Great Basins to examine. There are, however, some simple truths that constrain this choice and that place it in its proper perspective. The ethnographic Great Basin as anthropologists have so carefully defined it may not have existed 3,000 years ago. Indeed, as I discuss in chapter 9, some believe that Numic speakers may not even have been in much of the Great Basin at that time. Today's floristic Great Basin did not exist before 4,500 years ago or so, and one can legitimately argue that it is far younger than that, as I explore later in this book. The hydrographic Great Basin as it now exists has greater time depth, but it sprang leaks in a number of places between 25,000 and 10,000 years ago, spilling its waters into the Pacific, as I discuss in chapter 5. In fact, the Great Basin's hydrographic boundaries have been breached since that time as well (see chapter 5), but such violations were far less significant than the major leaks that occurred during the late Ice Age.

The only Great Basin that has actually continuously been here for the past 25,000 years is the physiographic Great Basin. The Great Basin has not stood still over these years, as anyone who was there for the 1872 Owens Valley earthquake (magnitude 7.6), the 1954 Fairview Peak (magnitude 7.2), or Dixie Valley (magnitude 6.8) earthquakes in central Nevada, or the 1915 quake in Pleasant Valley (magnitude 7.6) to the northeast, could attest. However, the general boundaries of the physiographic Great Basin have remained much the same for the entire period of time I will be dealing with. Adopting this Great Basin, though, would exclude much of the Mojave Desert, and doing that would make it far harder to fathom some major issues involved in understanding the development of the botanical Great Basin.

The thrust of this book is to watch the Great Basins develop and change over the past 25,000 years or so. Since that is the case, there is nothing wrong with using all four definitions at the same time, which is what I have done. The boundary of the Great Basin that you see drawn on many of the maps in this book, however, is the boundary of the hydrographic Great Basin. It is that boundary that I used, by far and large, in deciding what to include, and what to exclude, in what follows.

Chapter Notes

Li et al. (2008a, 2008b) provide detailed discussions of the drainage history of the Salton Basin. The Great Basin as a physiographic province is discussed in detail by C. B. Hunt (1967). I follow Hunt in excluding the Pavant Range (maximum elevation, 10,215 feet), Tushar Mountains (12,169 feet), and Wasatch Range (11,928 feet) from this province. McLane (1978) provides an excellent, historically oriented discussion of Nevada's mountain ranges. Badwater, at 282 feet below sea level, is often referred to as "the lowest point in the Western Hemisphere," to quote the Death Valley National Park 2008–9 *Visitor's Guide* (http://www.nps.gov/deva/upload/Visitor%20Guide%202008.pdf). A little thought, though, suggests that this is very unlikely. Are there no lakes whose bottoms fall beneath the elevation of Badwater? Of course there are. The lowest part of the basin in which Lake Superior sits, for instance, falls at about 700 feet below sea level, far lower than Badwater. As a result, Badwater Basin contains not the lowest point in the Western Hemisphere, but the lowest point not covered by water. Picky, but it is true.

The best general overview of the natural history of the modern Great Basin remains Trimble (1989). Pavlik (2008) overflows with knowledgeable discussions of the Great Basin, Mojave, and Sonoran deserts of California. For general reading on the plants of the Great Basin, see Gleason and Cronquist (1964), Lanner (1981, 1983b, 1996, 2007), and Mozingo (1987). Lanner (2007) discusses the ages of the oldest bristlecones. David Charlet's superb *Atlas of Nevada Conifers* (1996) contains almost everything one might want to know about the distribution of those trees in Nevada. What the book does not contain will be included in the forthcoming second edition, which, thanks to the author, I have used here. See Lanner and Van Devender (1998) and Cole et al. (2008) for discussions of the taxonomic placement of the various kinds of northern pinyons (on the many species of pinyon in addition to these, see Perry 1991, Lanner 1996, and Richardson and Rundel 1998). Among others, Lanner and Van Devender (1998) and R. A. Price, Liston, and Strauss (1998) treat all the single-needled pinyons as subspecies of *Pinus monophylla*. The distribution of the various kinds of pinyon in the Great Basin can be seen in the maps produced under the direction of Ken Cole and his colleagues (Cole et al. 2008), at http://sbsc.wr.usgs.gov/cprs/research/projects/global%5Fchange/RangeMaps.asp. My comments on these distributions follow Cole et al. (2008) and the associated maps. Shreve (1942) provides his classic botanical look at the deserts of North America. For a more detailed discussion of the plants of the Intermountain West as a whole (and the basis of much of my discussion), see Cronquist et al. (1972), and especially Noel Holmgren's lengthy article in that volume on the plant geography of the intermountain region. The superb collection of papers in Osmond, Pitelka, and Hidy (1990) focuses on physiological adaptations of Basin and Range plants, and thus supplements the essentially distributional approach taken in Cronquist et al. (1972). Dwight Billings won the Nevada Medal for scientific contributions to our understanding of what that state contains; those of his contributions that I have used in this chapter are Billings (1950, 1951, 1954, 1978, 1988,

1990). N. E. West (2000) provides an excellent discussion of the plant communities of the Intermountain West as a whole. A brief but excellent discussion of the plants that characterize the Mojave Desert is presented by L. Benson and Darrow (1981). MacMahon (2000) provides a more detailed discussion and places the Mojave Desert in the more general context of the North American warm deserts. DeDecker (1984) presents a detailed discussion of the distribution of the plants of the northern Mojave Desert.

The best reference on the mammals of Nevada remains E. R. Hall (1946). Zeveloff and Collett (1988) provide a general discussion of Great Basin mammals as a whole; for Great Basin birds, see Ryser (1985). Alcorn (1988) and T. Floyd et al. (2007) do for Nevada birds what Hall (1946) does for mammals; for fish, see W. F. Sigler and J. W. Sigler (1987). Information on the vegetation of Monitor Valley and the Toquima range is taken both from my own work, much of it done in association with Dave Rhode, and from R. S. Thompson (1983); on the alpine tundra of this area, see Charlet (2009b). Most of the elevations I give for the Pine Creek botanical walk are GPS-based. On salt accumulation in shadscale, see S. D. Smith and Nowak (1990); on chisel-toothed kangaroo rats, see Kenagy (1972, 1973) and Hayssen (1991). My discussion of Great Basin biogeography draws on Behle (1978), Harper et al. (1978), Tanner (1978), N. E. West et al. (1978), Reveal (1979), and P. V. Wells (1983). N. E. West et al. (1978) expand on the role of thermal inversions in the development of pinyon-juniper woodland. Tueller et al. (1979) and N. E. West, Tausch, and Tueller (1998) provide important discussions of Great Basin pinyon-juniper woodland as a whole; I have relied heavily on these (see also Hidy and Klieforth 1990). A huge literature deals with the recent expansion of pinyon and juniper; here, I have emphasized the contributions by Tueller et al. (1979); Tausch, West, and Nabi (1981); R. F. Miller and Rose (1995, 1999); West, Tausch, and Tueller (1998); R. F. Miller, Svejcar, and Rose (2000); R. F. Miller and Tausch (2001); R. F. Miller et al. (2005); Bradley and Fleishman (2008); and Charlet (2008). The estimate that pinyon-juniper woodlands now cover ten times the territory they covered some 150 years ago is repeated in many places: see, for instance, Suring et al. (2005). Janetski (1999b) provides an excellent summary of the use of pinyon-juniper woodlands by the native peoples of the Great Basin. Treelines in general are discussed in an excellent book by Arno and Hammersly (1984); see also Stevens and Fox (1991). For more on krummholz, see LaMarche and Mooney (1972), Marr (1977), and W. K. Smith and Knapp (1990).

Caveney et al. (2001) provide an important analysis of the chemical contents of *Ephedra*. Details on ephedrine and pseudoephedrine are widely available, but Salocks and Kaley (2003) provide a convenient synthesis. Many ethnographers have discussed the use of *Ephedra* by Native Americans in the Great Basin; Zigmond (1981), C. S. Fowler (1992), and Rhode (2002) are good places to start. My discussion of biological soil crusts draws heavily on Belnap (2003, 2006), Belnap and Eldredge (2003), Belnap et al. (2001), and Belnap, Phillips, and Troxler (2006). If you want a general introduction to these crusts, go to www.soilcrust.org (see "Crust 101"); Rosentreter, Bowker, and Belknap (2007) provide a brief introduction to the topic, along with a superb field guide to the organisms of which they are composed. Wohlfahrt, Fenstermaker, and Arnone (2008) argue that soil crusts may act as a carbon dioxide sink, but this suggestion is not accepted by all, and even soil crust expert Belnap questions this possibility (Stone 2008).

My discussion of Great Basin montane and subalpine conifers has relied not only on the references provided above but also on David Charlet's research, which has dramatically altered our understanding of the distribution of conifers in the Great Basin (Charlet 2007, 2009a). DeLucia and Schlesinger (1990) present an excellent review of the research that has been done on the disjunct distribution of Sierran conifers on altered soils in the western Great Basin. The full details of that research are in DeLucia, Schlesinger, and Billings (1988, 1989); Schlesinger, DeLucia, and Billings (1989); and DeLucia and Schlesinger (1991). Billings (1950) was the first to observe and interpret these Sierra "tree islands." R. S. Thompson (1984) first reported low-elevation bristlecone in Smith Creek Canyon. Although it was still alive in 1989 (I have a photo of it), we were unable to find it in 2009, and Connie Millar soon verified that it has died. Chris Crookshanks told me about the Hendrys Creek bristlecone, seen there in August 2009; Connie Millar confirmed the presence of bristlecone in Silver Creek Canyon in 2009 and provided the elevation I have used here.

The butterfly biogeography of the Great Basin has been the focus of exquisite research by a number of scientists. The key references I draw on here are Wilcox et al. (1986) and G. T. Austin and Murphy (1987). For sophisticated analyses of the butterflies of the Toquima Range, see Fleishman, Murphy, and Austin (1999, 2001).

My discussion of eco-region definitions of the Great Basin is based on Omernik (1995, 1999) and Ricketts et al. (1999) for the EPA eco-regions and, for Bailey's approach, on Bailey (1994, 1995, 1998), Nachlinger et al. (2001), and Wisdom, Rowland, and Suring (2005). A discussion of Bailey's eco-regions, including his map, is available at www.fs.fed.us/rm/analytics/publications/ecoregions_information.html; for the Nature Conservancy's blueprint for Great Basin conservation (Nachlinger et al. 2001), see http://conserveonline.org/browse_by_type?type=Ecoregional%20Plans.

For more on the current distribution of pikas in the Great Basin, see Rickart (2001); Beever, Brussard, and Berger (2003); and Beever et al. (2008). My discussion of pygmy rabbits draws heavily on Hall (1991), Green and Flinders (1980), Weiss and Verts (1984), Dobler and Dixon (1990), Verts and Carraway (1998), and Larrucea and Brussard (2008a, 2008b). Washington state pygmy rabbits have their own web site: http://wdfw.wa.gov/wlm/diversty/soc/pygmy_rabbit/index.htm.

Native American languages are discussed in Greenberg (1987), a book that linguists find highly controversial (see the lengthy multiple-authored review in *Current Anthropology* 28 [1987]:647–667); Ruhlen (1991b); and a wide variety of important papers in Goddard (1996) and L. Campbell (1997), among many other places. On Great Basin languages, see Jacobsen (1986) for Washoe and W. R. Miller (1986) for Numic; W. R. Miller (1983, 1984) provides a more general discussion of Uto-Aztecan languages; further references are provided in chapter 9. I have used the name Kawaiisu-Ute for the southern branch of Numic, but Kawaiisu–Southern Paiute would do just as well, since these are dialects of the same language. My discussion of the history of Comanche speakers is based largely on Sutton (1986), Bamforth (1988), and Kavanagh (2001); on the Comanche language, see W. R. Miller (1983) and, for an interesting discussion of the origin of the name "Comanche," Goss (1999). Towner (1996) provides a wide array of papers dealing with the timing of the arrival of Navajo and Apache speakers in the Southwest. W. R. Miller, Tanner, and Foley (1971) discuss dialectical variation within

Panamint. On the distribution of Great Basin languages and the implications it may have for the history of Great Basin native peoples, see chapter 9. On the Kawaiisu, see Zigmond (1986) and the Kawaiisu website (http://home.att.net/~write2kate/artbyhorseindex.html); for the current number of speakers, see Gordon (2005). My description of early historic Maidu territory is based on Kroeber (1953) and Simmons et al. (1997); see Golla (2007) for a discussion of the native languages of California.

Julian Steward's classic statement of Great Basin lifeways is in Steward (1938); see also Steward (1937b, 1970). Steward (1933) presents his ethnography of the Owens Valley Paiute. On Pandora moths in general, see the great book by Tuskes, Tuttle, and Collins (1996); as they show, the caterpillars of many members of the silk moth family, to which Pandora moths belong, were used as food. For this usage in the Great Basin, see C. S. Fowler and Walter (1985) and Sutton (1988). Virginia Kerns's (2003) superb biography of Steward should be read by all who are interested in his life and work. Kerns (1999) focuses on Steward's early experiences at Deep Springs school (now Deep Springs College), just southeast of the White Mountains, arguing that his experiences here were pivotal to his later anthropological work in the Great Basin. Clemmer, Myers, and Rudden (1999) present a reassessment of aspects of Steward's Great Basin work. Steward's archaeological work is reviewed by Janetski (1999a).

Dave Thomas's analysis of the predictability of pinyon yields is in Thomas (1972b); see also Lanner (1981, 1983b). The Monitor Range sage grouse communal hunting facility is discussed and illustrated in Thomas (1988). Adovasio, Andrews, and Illingworth (2009) provide a helpful review of net-based hunting in the Great Basin (and elsewhere). Kay Fowler's discussions of Great Basin subsistence and settlement patterns are found (among many other places) in C. S. Fowler (1982, 1986, 2000). C. S. Fowler (1994a) discusses her career, one that has become even more remarkable and influential since that paper was written. Thomas (1983a), Janetski (1991), and R. L. Kelly (2001) provide important discussions of the relationship between resource richness and sociocultural organization in the Great Basin. Lowie (1939) discusses the Washoe. Excellent syntheses of communal pronghorn hunts are provided by Arkush (1986, 1995, 1999a), Lubinski (1999), and Hockett and Murphy (2009). For other descriptions of the drive systems used to capture these animals, see L. S. A. Pendleton and Thomas (1983) and the references in these works.

Service's challenge to Steward appears in Service (1962). Thomas (1986) provides an extremely handy compilation of some of the most valuable sources that exist on Great Basin ethnography. Firsthand accounts of Great Basin Native American lifeways that match those provided by Steward are to be found in many places, including Frémont (1845) and Wilke and Lawton (1976); the latter contains the 1859 journal of J. W. Davidson's travels in Owens Valley. Wilke and Lawton (1976) also note the concordance between Davidson's descriptions and later ethnographic accounts. On the impact of disease on Native Americans, see Ramenofsky (1987) and the references therein, especially those to the work of S. F. Cook and H. Dobyns. On the Timpanogots, see Janetski (1986, 1990, 1991, 2007). Substantial late prehistoric Great Basin sites in areas marked by rich riverine and marsh resources are known from the Carson Sink (S. T. Brooks, Haldeman, and Brooks 1988; Raven and Elston 1988, 1989, 1990; Raven 1990; R. L. Kelly 2001), the Humboldt Sink (Livingston 1986, 1988b), the Lake Abert and Chewaucan Marsh basins (Pettigrew 1985; Oetting 1989, 1990), Utah Valley (Janetski and Smith 2007), and elsewhere (see the papers in Janetski and Madsen 1990 and in Hemphill and Larsen 1999). Crum (1999) criticizes Steward's disinterest in many nonmaterial aspects of Shoshone culture, as well as the very negative comments he made about the Western Shoshone in a report to the Bureau of Indian Affairs in 1936 (see also Pinkoski 2008). Nonetheless, he still sees great value in Steward's reconstructions of late prehistoric Western Shoshone lifestyles (Crum 1994).

On major historic earthquakes in the Great Basin, see R. E. Wallace (1984), Beanland and Clark (1994), Caskey et al. (1996), Caskey and Wesnousky (1997), dePolo and dePolo (1999), Caskey et al. (2000, 2004), W. R. Lund (2005), and Machette (2005).

The Pyramid Lake Visitor's Center and Museum, run by the Pyramid Lake Paiute tribe, is located just south of Pyramid Lake in Nixon, Nevada, an easy drive to the east from Reno. The collections are small but interesting; if you go, don't miss the public presentations, which are worth the visit by themselves.

Both Badwater and Telescope Peak are in Death Valley National Park. Badwater Basin can be reached by car. Reaching the top of Telescope Peak requires a seven-mile hike during which 2,916 feet of elevation are gained, but the trail is remarkably good—at least, it was when I was there—and the hike a relatively easy one (see Digonnet 2007 for a description of the walk). The magnificent view from the top includes Badwater, 11,331 feet below. Check with national park headquarters about the condition of the road to the Telescope Peak trailhead.

PART TWO

SOME ICE AGE BACKGROUND

CHAPTER THREE

Glaciers, Sea Levels, and the Peopling of the Americas

Archaeology deals with the entire time span of our existence, from our earliest tool-using ancestors in sub-Saharan Africa some 2.5 million years ago, to contemporary peoples from Tanzania to Tucson. Much archaeological research is fairly routine and, while important, causes little heated debate. But there are certain issues, as in all sciences, that become the focal point of loud argumentation and of lasting disagreement.

The archaeology of the Americas was once marked by many such debates. Did contact across the Pacific Ocean stimulate the development of the high civilizations of South and Middle America? Were the prehistoric pueblos of the American Southwest founded by immigrants from Mexico, or in some other way directly caused by cultural developments to the south? Both of these questions have now been answered in the negative, as archaeological research has shown ancient and local antecedents for all of the American peoples involved.

These debates have been replaced by others, two of which I address in this and the following chapter. When did people first arrive in North America? And what impact did they have on the native fauna when they got here?

The Bering Land Bridge and the Human Arrival

Pleistocene Glaciation

The human species evolved in Africa, skeletal evidence suggesting that this occurred between 200,000 and 150,000 years ago. Our species spread outward from Africa, with the biggest push beginning perhaps 50,000 years ago. Ultimately, we colonized the globe. As a result, there is no question that people reached the Americas from somewhere else. The problem, of course, is to figure out where they came from and when they came from there, all the while keeping in mind the possibility that there may have been, and probably were, multiple arrivals.

There is a chance that people first reached the Americas by some lucky accident that saw them drift across the Atlantic or Pacific oceans. In fact, there have been computer simulations meant to determine exactly how likely this might have been. The most recent of these was reported by Álvaro Montenegro and his colleagues, who focused on unintentional crossings in boats without sails. (I talk about the surprisingly deep history of sea voyaging later in this chapter.) Their interesting results suggest that, under present-day conditions, one could drift from Japan to North America in eighty-three days without paddling, or seventy days with, or from Europe to northern North America in about seventy-five days with or without paddling. Some crossings were even faster under the kinds of conditions that would have existed 18,000 years ago or so.

Some important strictures in the analyses of Montenegro and his colleagues impact the relevance of what they found. The most important of these is probably that the boats couldn't sink. Given that they were estimating crossing times for people who ended up on their way across the ocean without ever having intended to do so, their simulations actually tell us how unlikely it would be to survive such a crossing. It would be even more unlikely to survive with a group large enough to be reproductively viable.

Because of these long odds, the lucky voyage has not gotten much attention from archaeologists. Three other routes, however, have gotten more attention: via boat across the Pacific, via boat across the northern Atlantic, and by boat or foot across the area today occupied by the Bering Strait between Siberia and Alaska. Of these, only one has much support, and I will begin with it.

Virtually all scientists accept the proposition that the earliest Americans came from northeastern Siberia, crossing from Old World to New in the vicinity of what is now the Bering Strait (figure 3-1). Here, mainland Alaska and mainland Siberia are slightly less than sixty miles apart. When

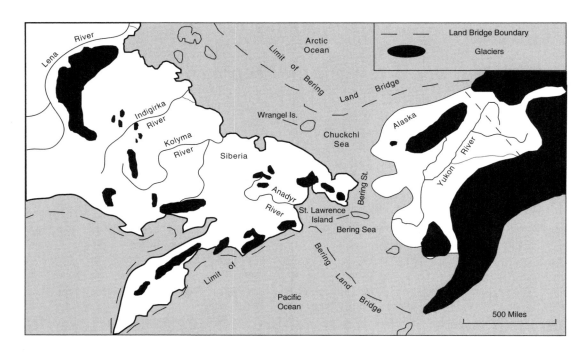

FIGURE 3-1 Beringia during the Last Glacial Maximum (modified from Morlan 1987).

the fog lifts long enough, one side can be seen from the other. Eskimos found crossing Bering Strait in the large, open boats called umiaks to be easy enough that a fairly routine trade developed between groups on both sides, hampered only by the fact that contacts were sufficiently close that these trading partners also became enemies.

The Eskimo, however, are a modern people with a sophisticated technology for dealing with their environment, and there is no reason to suppose that the first movement of people into the Americas had to be made by boat. Instead, the surmise is that the crossing was made when sea levels were lower, the Bering Strait did not exist, and people, along with a wide variety of other mammals, could have walked across. To understand how that could have happened requires some geology.

"Ice Age" is a popular name for that time in earth history that saw the alternate expansion and retraction of glacial ice over vast portions of the earth's surface. There were more than twenty of these episodes, with the earliest of them beginning some 2.6 million years ago. The official name for this epoch is the Pleistocene; by definition, the Pleistocene extends from 2.6 million to 10,000 years ago. Everything since then is referred to as the Holocene: it is the epoch we are in now. The Pleistocene and Holocene together are known as the Quaternary period.

The effects of Ice Age climate, and of glaciation itself, were huge. Much of the landscape of North America, including vast expanses that were never beneath glacial ice, cannot be understood without knowledge of those effects. Some of these I discuss in the following chapters. For now, I want to explore how it is that people could have walked from Siberia to Alaska on dry ground.

The last North American glaciation, named the Wisconsin because of the location of the deposits that helped define it, dates to between 80,000 and 10,000 years ago. During this glaciation, three massive ice sheets formed in the north, ultimately covering most of Canada and parts of the northern United States. Of least concern to us is the Innuitian Ice Sheet, which focused on the Arctic islands of Canada (north of 75° N latitude) and ultimately merged with the Greenland Ice Sheet to the east and the Laurentide Ice Sheet to the south. The massive Laurentide Ice Sheet in turn extended from Labrador to the eastern foothills of the Rocky Mountains and reached a maximum thickness of about 11,500 feet—a pile of ice over two miles high. To the west, the Cordilleran Ice Sheet grew mainly from the coalescence of glaciers that formed in the high mountains of British Columbia and the southern Yukon. Ultimately it covered all but the tallest pinnacles of the Canadian Rockies and Coast Range and sent its ice south beyond where Seattle now sits, reaching a thickness of at least 6,500 feet and probably much more. To the south of these huge ice sheets, many mountains harbored their own, alpine glaciers. The distinctive sawtooth ridges that mark the skylines of many parts of both the Sierra Nevada and the Rocky Mountains were largely formed by glacial action, while isolated mountains as far south as south-central Arizona and New Mexico were glaciated as well.

The timing of the earliest advances and retreats of North American glacial ice is not particularly well known, though glacial advances in the Sierra Nevada occurred some 2 million years ago or earlier and the same is true for northwestern Canada and eastern Alaska. We know far more about the latest Pleistocene advances and retreats of glacial ice. Our most precise knowledge is available for the very latest of them, and that is my focus here.

First, though, I need to discuss some terminology. Geologists are compulsive about the terms they use to describe the major divisions of earth history and the dates assigned to those divisions. There is an International Commission on Stratigraphy whose job it is establish, maintain, and publish a standard geological stratigraphic nomenclature and time table (the date that I used above for the Pleistocene comes from the most recent version of that timetable, published in 2009). However, as admirably compulsive as geologists are in this realm, they are less meticulous in other ways. They carefully distinguish between units that are named formally—for instance, those found on the commission's time scale—and those that are not. For instance, there are formal definitions for four stages within the Pleistocene, beginning at 2,588,000, 1,806,000, 781,000, and 126,000 years ago, the last of which is simply called the Upper Pleistocene.

Unfortunately, things like the subdivisions of the Wisconsin glaciation have no formal definitions. There is, however, widespread informal agreement that the Wisconsin glaciation is to be divided into three parts, aptly called the Early, Middle, and Late Wisconsin. Agreement is also widespread that these three divisions correlate well with reconstructions of glacial and temperature history derived from sea cores and glaciers.

The sea core–based reconstructions have been derived from analyzing ratios of two oxygen isotopes, oxygen-16 (^{16}O) and oxygen-18 (^{18}O). Both of these isotopes have eight protons, but ^{16}O has eight neutrons while ^{18}O has ten. Because of that, ^{16}O is the lighter of the two and, as a result, water molecules that contain it evaporate more easily than water that contains the heavier ^{18}O. In addition, since ^{18}O is heavier, water containing it is the first to precipitate; as storm systems get farther and farther from the ocean, the water they contain is composed of less and less ^{18}O and more and more ^{16}O. During times of glacial expansion in the north, much of that precipitation can be locked up on land in the form of glacial ice. All of this means that during times of glaciation, sea water comes to contain higher and higher concentrations of ^{18}O. At the same time, the water locked up in glacial ice comes to contain higher and higher concentrations of ^{16}O. When ice sheets melt, all that ^{16}O-rich water returns to the ocean, driving the concentration of ^{18}O in ocean water back down again. If we analyze the ratios of ^{18}O to ^{16}O in cores taken from both the deep sea and from glaciers that still exist (most notably in Greenland and Antarctica, but elsewhere as well), the ratios of these two isotopes can tell us when the land was heavily covered with glaciers and when it was not. The results of this work have told us that there have been at least twenty glacial episodes during the past 2.6 million years or so. Although these ratios vary continuously in time, they can be, and are, divided into separate pieces, called marine isotope stages (MIS; also called oxygen isotope stages, OIS), the most recent of which, and the one we are in now, is called MIS 1. In general, odd-numbered stages are warm and even-numbered ones cold. Those stages can then be dated in a variety of ways.

Given all this, and given that MIS 1 correlates with the Holocene, the stages that immediately predate MIS 1 must correlate with what was happening with Wisconsin ice on land. In fact, the Late Wisconsin correlates with MIS 2, the warmer Middle Wisconsin with MIS 3, and the colder Early Wisconsin with MIS 4.

If that is the case, then the dates for MIS 2 through 4 can be used to provide the dates for the Late, Middle, and Early Wisconsin, and this is what is normally done. That procedure gives dates of *about* 75,000 to 60,000 years ago for the Early Wisconsin, 60,000 to 30,000 for the Middle Wisconsin, and 30,000 to 10,000 for the Late Wisconsin. I stress "about," since different people use somewhat different dates for the MIS stages involved, but they are all roughly similar, and these are the dates I use here.

Another possible way of defining an Early Wisconsin is based on evidence for glaciation that started before this time, during MIS 5. Those who follow this approach date the beginning of the Wisconsin glaciation to as early as 122,000 years ago (the base of MIS 5d), but that is not the approach I take here.

We do not know when people first arrived in the Americas, but whenever they got here, it was long after the Early Wisconsin had ended. It is also true that our knowledge of Wisconsin-age glaciation picks up dramatically during the Middle Wisconsin. As a result, it is here that I start.

Since odd-numbered marine isotope stages are relatively warm and even-numbered ones cold, and since the Middle Wisconsin is equated with MIS 3, it follows that this was a relatively warm period of time. Even though this was the case, substantial parts of eastern Canada were glaciated at this time, with significant glacial advances sometime prior to 45,000 and at about 35,000 years ago. On the other hand, even though glaciers were more substantial in western North America during the Middle Wisconsin than they are now, most of the central and western North American landscape was ice free. By 30,000 years ago, however, glaciers had begun to expand in the west at the same time as the Laurentide Ice Sheet had started on its inexorable way south and west. By 18,000 years ago, the Cordilleran and Laurentide ice sheets had met just east of the Canadian Rocky Mountains, creating a continuous mass of ice across from the Pacific to the Atlantic across Canada and, via the Innuitian Ice Sheet, into Greenland.

On a global basis, the Last Glacial Maximum occurred between about 22,000 and 18,000 years, but it actually occurred at different times at different places. For instance, the maps prepared by glacial geologist Arthur Dyke and his colleagues show ice reaching its greatest southern extent in the eastern United States at about 18,000 years ago (on New York State's Long Island), but not until about 14,500 years ago south of the Great Lakes and in the Puget Sound area of western Washington State.

No matter when the northern ice sheets reached their maximum extent, however, they were in retreat by about 14,000 years ago. Sometime after 13,000 years ago, the eastern edge of the Cordilleran Ice Sheet parted company with the western edge of its Laurentide counterpart; by 11,500 years ago, a wide gap had opened between the two. This parting may have been an event of great archaeological and paleontological significance, as I discuss later.

As Arthur Dyke has also pointed out, the great North American ice sheets had lost less than 10 percent of their area by 14,000 years ago. By 7,000 years ago, only 10 percent of the north was more glaciated than it is today. Sometime between these two dates—by convention, it is 10,000 years ago—the Ice Age ended. This, though, was but one of many momentous events that seem to have occurred during this interval.

The Bering Land Bridge

The total amount of water on the earth is virtually constant. Today, about 97.5 percent of the earth's water resides in the oceans, with only a trace present in the atmosphere. The remainder is found on land. Of that remainder, about 75 percent is locked up in the form of ice, most of it—97 percent—in the Greenland and Antarctic ice sheets. Were the glaciers to expand again, the water they converted into ice would come from the ocean. As the glaciers expanded, ocean levels would fall; as they retreated, the oceans would rise. Indeed, there is enough water frozen in the ice sheets of Greenland and the Antarctic that if they were to melt, sea levels would rise some 210 feet.

Melting glaciers cause sea levels to rise; expanding glaciers cause them to fall. Estimates for the drop in sea level during the Last Glacial Maximum, between about 22,000 and 18,000 years ago, run between about 380 feet and 445 feet, with many estimates converging at 425 feet. Either end of this range of estimates would create a wide expanse of dry land between Alaska and Siberia.

Obviously, as sea level drops, more and more land is exposed along coastlines. Dozens of teeth from extinct North American elephants, for instance, have been dredged up from the Atlantic Coast. These teeth did not get washed offshore from contemporary land surfaces. Instead, they are the remains of now-extinct mammoths that at one time trudged along the exposed continental shelf along our East Coast, then exposed because the water that now covers this area was, prior to 11,000 years ago (by which time mammoths were becoming extinct), dry land. In fact, the shoreline receded sixty miles or more along the East Coast. Along the West Coast, where the continental shelf tends to be far narrower, the shoreline receded some thirty miles.

The Bering Strait connects the Bering Sea to the south with the Chuckchi Sea to the north (figure 3-1). The floor of the northeastern Bering Sea, Bering Strait, and Chukchi Sea forms a flat shelf running from the Siberian to the Alaskan coasts. Drop sea level by some 160 feet, and Siberia and Alaska begin to be connected by land. Drop sea level by 400 feet, and the two continents are joined by a broad land mass more than 1,000 miles wide from north to south. This figure—400 feet—is just about at the middle of the estimates for maximum sea level decline during the Last Glacial Maximum, and data from the Bering Strait itself suggest that this is precisely where things stood 18,000 years ago.

"Beringia" is the name given to the vast region connected by this land bridge—northeastern Siberia, Alaska, and the adjacent Yukon. Geographically, the boundaries are often taken to be the Lena River on the west and the western Yukon Territory of Canada. The Bering Land Bridge itself, along with the immediately adjacent parts of Alaska and Siberia, is often referred to as central Beringia.

The land bridge as a whole is so shallow that nothing approaching the volume of ice that covered the earth's surface during the Last Glacial Maximum is needed to expose huge parts of it. A sea level decrease of 180 feet, for instance, would create a land bridge several hundred miles wide. The shallowness of the land bridge, combined with the volume of water locked up in ice during much of the Wisconsin glaciation, means that a land bridge may have existed to one degree or another during much of the Wisconsin.

There is little agreement among scientists as to the history of the land bridge during Middle Wisconsin times (60,000–30,000 years ago). General sea level reconstructions suggest it may have been exposed for much of this interval. Others suggest exactly the opposite: that the only vertebrates intimately acquainted with the surface of the land bridge during much of the Middle Wisconsin were fish.

Nonetheless, there is no question that as Late Wisconsin glaciers advanced, the land bridge was exposed and remained exposed until near the very end of the Pleistocene. As the glaciers began to wane, the bridge narrowed. The Bering Strait itself seems to have been flooded for the last time between 12,000 and 11,000 year ago, leaving only a small corridor just north of St. Lawrence Island. Soon after 11,000 years ago, even this went under and the land bridge was gone.

Precisely what the vegetation of this land bridge was like is difficult to know, for the simple reason that it is now covered by water. It is, in fact, remarkable that we know anything about the vegetation at all, but, thanks to cores taken from the bottom of the Chuckchi and Bering seas, from lakes on what are now islands but were at one time land bridge uplands, and from the immediately adjoining eastern and western edges of the land bridge, we do have some understanding of what the landscape would have looked like during the Last Glacial Maximum and late glacial times. The identification of the insect and plant remains preserved in those cores, coupled with high-precision radiocarbon dating, has provided a paleoenvironmental record for the land bridge that extends back some 40,000 years, though it is most detailed for the past 20,000 years or so.

There is not much question that the northern edge of the land bridge would have been extremely cold, with vegetation similar to that in modern polar deserts. This was so because the very existence of the land bridge prevented relatively warm Pacific Ocean waters from flowing northwards. Deprived of warm waters, the Arctic Ocean froze, and the northern edge of the land bridge would have been ice-bound. To the south, different scientists investigating land bridge environments have obtained different results. Scott Elias and his colleagues, for instance, find evidence for a moist tundra in the central and eastern part of the land bridge, with shrub birch and graminoids (grasses and grasslike plants, including such things as sedges and rushes) very common and heaths and willows probably common in favorable locations. On the other hand, paleobotanist Tom Ager argues that the land bridge was both cold and arid and that it was largely covered by herb-dominated tundra down to about 14,000 years ago, after which a shrub-birch tundra developed.

These reconstructions are quite different, but that does not necessarily mean that one or the other is wrong, since Beringia as a whole seems to have been covered by a complex mosaic of different vegetational types. From a human perspective, it is unlikely that people who were adapted to the Siberian side of the land bridge would have had any particular difficulties coping with the challenges presented by the bridge itself. If you could survive in Siberia during the Pleistocene, you certainly could have crossed the Bering Land Bridge.

Although archaeologists widely assume that people crossed from Eurasia to the Americas when the Bering Land Bridge existed, it is also true that a crossing could have been made on sea ice, as perilous as that might have been. And while umiaks, used by Eskimos to cross the Bering Strait in recent times, may not have existed when people first entered the Americas (no later than about 12,000 years ago), boats sufficient to do the job existed in some parts of the world very early on.

What we know about the archaeology of Australia shows this very clearly. The earliest securely dated archaeological sites in Australia are about 45,000 years old, so people were certainly here by then. There is a healthy debate over when they first arrived, but it certainly wasn't long before this date. Although the evidence doesn't support it, let us assume that they arrived around 53,000 years ago, when sea levels in this part of the world were at the lowest point they have reached during the last 60,000 years. Even if they crossed at this time of minimum water levels, an open sea voyage of at least fifty-five miles would have been necessary, a distance comparable to that between Alaska and Siberia today. This might not even be the earliest evidence of a long-distance voyage by a member of our own genus. Flores Island, in eastern Indonesia, has artifacts that have been dated to about 840,000 years ago. If this is correct, then the makers of these artifacts would have been *Homo erectus,* and they would have required two separate sea crossings of at least fifteen miles to get there. (This part of the world seems to provide almost nothing but surprises for archaeologists: the famous "hobbit"—our 3.5-feet-tall, very controversial, latest Pleistocene relative—also comes from Flores.)

The archaeological record thus shows that our ancestral relatives had the ability to make substantial sea crossings, and there is no reason to think they could not have done so to get from what is now Siberia to what is now Alaska. Nonetheless, the assumption that people crossed when the land bridge was exposed appears to be a reasonable one, since our species was only one of many that made the crossing in one direction or another during the Pleistocene. For instance, North America gained the mammoth and caribou, among other things, from the Eurasian side. Eurasia gained camels and horses, among other things, from the American one. Crossings in the two directions were so common that what raises questions is not the fact of these crossings, but why some animals that might have crossed did not. The extinct wooly rhinoceros (*Coelodonta antiquitatus*) is known from the Siberian side, but not from the Alaskan one; the extinct giant short-faced bear (*Arctodus simus*) is known from the Alaskan side, but not the Siberian one. Why did they not cross when so many of their contemporaries did?

It seems most reasonable to think that the movement of our own kind across the Bering Land Bridge was part of a general faunal interchange between Eurasia and the Americas. What people accomplished in making the crossing wasn't anything special, and it is far more parsimonious to assume that they came across as part of this interchange than it is to think that they walked across sea ice or came by boat.

The fact that the crossing most likely took place when glaciers were sufficiently advanced to drop sea levels at least 160 feet suggests a simple question. Wouldn't the first Americans have been in for a rude shock, crossing the land bridge only to be confronted by glacial ice?

In fact, the geography of Alaska is such that that is not what happened (figure 3-1). Glaciers, of course, need precipitation to form. In Alaska, precipitation was borne by air masses moving north and east from the northern Pacific Ocean onto the southern Alaskan mainland. However, the entire southern front of Alaska, from the Aleutians to the British Columbia border, is marked by high mountain ranges, reaching their 20,300-foot pinnacle at Mount McKinley. As wet air moving out of the North Pacific encountered these mountains, it was forced upward, cooled, and condensed. The resultant precipitation allowed the formation of massive glaciers on the mountains themselves, glaciers that ultimately coalesced and formed the northwestern arm of the Cordilleran Ice Sheet.

By the time these storm systems got much beyond the crests of the mountains, they had been wrung nearly dry. It did not help that a massive area that could have added to the atmospheric moisture level had now been turned into land—the Bering Land Bridge—that helped block storm systems from penetrating deep into Beringia as a whole. As a result, central Alaska, low in elevation, was never glaciated, nor was the adjacent Yukon. Farther north, the high Brooks Range, whose peaks reach beyond 8,000 feet in elevation, saw the development of a massive glacial complex, but the low plains north of the Brooks Range also remained unglaciated. Central and far northern Alaska—some two-thirds of the state—was thus never covered by ice. No matter when people first entered the Americas via the Bering Land Bridge, they may have passed by Pleistocene glaciers, but they would not have had to deal with them. Central Alaska was ice free.

Ice free, however, does not necessarily mean livable. Environments cold enough to contain glaciers may be cold enough to be miserable for people, and glaciers themselves can modify surrounding environments tremendously. For one thing, when air is forced over glacial ice, it rapidly cools and, because of this cooling, becomes heavier as well. The resultant cold, heavy air rushes downslope, creating drainage winds that can chill areas far beyond the glacier itself. A similar, though far less pronounced, phenomenon is experienced by people who camp in the mouths of desert mountain canyons in the summer months. As the air in the high elevations of the mountain is cooled, it rushes downslope, funneled through the canyon itself, creating strong and relatively chilly winds that last until the temperature imbalance between the higher, colder and lower, warmer air has lessened. This phenomenon is one reason why subalpine conifers can follow canyons to lower elevations in places like the Toquima and Snake ranges.

In addition to creating their own winds, glaciers also shatter and grind the rock over which they pass. Some of the finely ground material remains beneath the ice until the glacier finally melts and exposes it to the surface. Some is washed out from beneath the ice by glacial waters, to be deposited in front of the glacial mass as "outwash alluvium." No matter which of these mechanisms exposes the finely ground rock on the surface, once it is there, it can be picked up by winds to form rock-dust clouds. Such material, called loess, blankets huge portions of the American Midwest, the ultimate source having been the outwash deposits from Laurentide ice. Loess deposits over 5 feet thick are common in such states as Illinois and Kansas.

It is thus fully possible that while the eastern terminus of the Bering Land Bridge was ice free, the environments here were inhospitable for people. In fact, there has been a lengthy debate over exactly what these environments would have been like, a debate that now seems to be over but that needs to be briefly recounted.

The debate was largely spurred by a series of arguments by paleontologist R. D. Guthrie. Guthrie's ideas were largely derived from his observation that Pleistocene-aged faunas in interior Alaska were dominated by large grazing animals, including mammoth, bison, and horse (see chapter 4). From this and other aspects of the fauna, Guthrie inferred that during the Wisconsin glaciation, interior Alaska supported an arid, productive grassland whose climate was marked by cold but short winters separated by a long growing season. Guthrie observed that similar faunas were found across Eurasia into southwestern Europe, and argued that in all places, these faunas showed the existence of a productive Ice Age grassland, the "Mammoth Steppe."

The only problem with this insightful set of arguments was that it conflicted with virtually all of the evidence paleobotanists were then developing on the nature of Beringian vegetation. In eastern Beringia, our direct knowledge of that vegetation comes primarily from the study of pollen grains that have accumulated in the sediments of the lakes and bogs that abound in the area.

Most wind-pollinated plants, and even some that depend on animals to do the job, produce pollen abundantly. Those tiny particles—a single pollen grain from an alder or a willow weighs only some twenty-billionths of a gram—can be carried far and wide by wind. Some pollen grains end up doing what they were meant to do, but most end up scattered across the landscape. When this "pollen rain" brings pollen to environments that protect it from oxidation and abrasion, as in lakes and bogs, it preserves beautifully. Samples of sediments from these protected environments can then provide a long sequence of vegetational history as represented by the changing pollen rain through time. Because the places that preserve pollen well are often good places for organic preservation in general, the deposits that provide the pollen can usually be dated by the carbon-14 method. Indeed, modern radiocarbon technology can even date samples of the pollen grains themselves. Pollen analysis is time-consuming, even tedious, but the results can tell us the nature of the vegetation on the landscape at a given known time in the past, and can also tell us how, if not always why, that vegetation changed through time.

The results of such exacting work have produced reasonable accord among pollen workers, virtually none of whom see anything resembling Guthrie's mammoth steppe. For instance, even though there is strong disagreement as to the nature of the vegetation that covered the Bering Land Bridge during the past 30,000 years or so, that disagreement involves the kind of tundra that was there. This is because no data suggest that this area was covered by a grassy steppe, mammoth or not. Indeed, as Scott Elias and his coworkers have pointed out, while their Bering Land Bridge cores are rich in beetle remains, none of those beetles were of the sort one finds in steppe environments. Instead, beetles typical of moist shrub-tundra were common.

Much the same is true on what is now the mainland east of the Bering Land Bridge. As paleobotanists Patricia Anderson and Linda Brubaker and others have pointed out, during the Middle Wisconsin, eastern Beringia seems to have been covered largely by various forms of tundra, especially tundra dominated by herbaceous plants and dwarf willows. At favorable times—most notably between 38,000 and 30,000 years ago—forests expanded, though never reaching anything like the area they now cover in Alaska, and even then, herb-willow tundra remained abundant. During Last Glacial Maximum times (25,000 to 14,000 years ago) eastern Beringia was covered by a mosaic of vegetation types, some of which included spruce, birch, and poplar trees lurking in restricted refugia. Most widespread, however, was a productive grass-forb tundra that developed under extremely arid and cold conditions. While this vegetation could support large mammals, including mammoths, there was no steppe involved.

Most of these vegetation types began to give way shortly after 14,000 years ago as the climate warmed and shrub tundra expanded across the landscape. Deciduous trees began to increase in number as well, especially as the end of the Pleistocene approached; the spruce-dominated forests for which interior Alaska is now so well known are a Holocene phenomenon.

Although the plant communities of eastern and western Beringia were in many ways quite different, they did share important similarities. Most obviously, during Late Wisconsin times, it was graminoid-dominated tundra, not steppe, that marked western Beringia, though some Russian scientists argue that there may have been enough steppe plants in this tundra to merit the name steppe tundra. Either way, it was certainly productive enough to support numbers of large grazing mammals. However we may describe these plant assemblages, they, too, were replaced by shrub tundra and then other vegetation types, including larch forest in southwestern Beringia, as the Pleistocene came to an end.

We do not know when people first crossed the Bering Land Bridge into the Americas, but the archaeological records on both sides of the land bridge suggest that it must have been some time during the Wisconsin glaciation. The chances are good that whenever they got here, the colonists would have found themselves in vegetation not that different from what they had left behind: some form of tundra with a thin, but perhaps locally rich, scattering of large mammals.

An Ice-Free Corridor?

During the Last Glacial Maximum, the western or Siberian end of Beringia was open to traffic into and out of the rest of Eurasia. But as alpine glaciers grew and then coalesced to form the Cordilleran Ice Sheet in northwestern North America, and as the Laurentide Ice Sheet grew to its maximum to the east, eastern Beringia became increasingly isolated from the rest of North America. During a significant part of the Wisconsin glaciation, Beringia is more accurately thought of as part of Asia than as part of the Americas. This is true for the simple reason that while movement into western Beringia was always possible, whether by people or by beasts, movement out of eastern Beringia into more southerly North America would have been blocked, to one degree or another, by glacial ice and attendant inhospitable conditions during a significant part of the Wisconsin. As a result, a human presence in eastern Beringia—the interior of Alaska, the Yukon, and the adjacent Northwest Territories—during the Wisconsin did not necessarily and quickly lead to a human presence farther south.

For decades, scientists considered the possibility that during the glacial maximum, an "ice-free corridor" ran between the Laurentide and Cordilleran ice sheets, providing a passageway from eastern Beringia to areas south of the glacial ice. Determining whether or not that was the case was not easy, because the glacial history of the area in which an ice-free corridor might have existed is extremely complex. Laurentide ice entered the area from the north and east, while both Cordilleran ice and ice contributed by smaller montane glaciers entered the area from the west. Although deposits laid down by these glaciers overlap in many areas, the fact that they overlap may mean little, since the expansion and retreat of the glaciers involved may not have been synchronous. Making matters worse is the tendency of later glaciation to obliterate deposits laid down during earlier glacial episodes. The result of all this is that the details of the glacial history of this area are very difficult to establish. Facts that seem securely established tend to become disestablished with surprising frequency.

Nonetheless, as it stands now, the geological evidence suggests that ice blocked the way south from interior Alaska beginning sometime after 30,000, and certainly by 21,000, years ago and continuing almost, but not quite, to the end of the Pleistocene. By about 12,000 years ago, deglaciation had reached the point that a corridor had emerged between the ice masses that may have been substantial enough to have allowed people to live in and walk through it. By 11,500 years ago, that corridor was substantial, and movement from north to south, or vice versa, would have been possible.

The Coastal Route

As we will see in the following chapter, the earliest secure, widespread North American archaeological record from south of glacial ice includes evidence of people who lived in the plains and southwestern United States around 11,000 years ago and who at least occasionally hunted large, now extinct mammals. Evidence of this sort was first discovered in New Mexico in 1927 and involved artifacts found tightly associated with extinct bison. Six years later, and once again in New Mexico, artifacts were discovered amongst the bones of mammoths. In that same year, 1933, Canadian geologist W. A. Johnston developed the idea of an ice-free corridor as a possible human migration route. All these pieces seemed to fit together well. The earliest Americans south of the ice were large-mammal hunters who lived in the heart of what is now the United States, and it made sense that such midcontinental hunters would have taken a midcontinental, ice-free route to get here.

While this might have been the route taken by some early peoples, however, archaeological evidence suggests it was unlikely to have included all of them. In addition, there is an alternative way whereby people could have moved south from eastern Beringia during times of glacial expansion. This route was suggested by paleobotanist Calvin Heusser in 1960 in a very brief postscript to a lengthy monograph on plant and environmental history in northwestern North America. Heusser noted that while unglaciated corridors may have existed in the interior of the continent at an appropriate time, it was also possible that people simply followed the Pacific Coast southward from Beringia. He even suggested doing archaeological surveys to discover whether or not evidence of appropriately ancient human occupation existed in this area. Subsequently, this reasonable idea was taken up by archaeologist Knut Fladmark and others, but only very recently has the idea been given the attention it deserves.

A coastal route would not have worked between about 23,000 and 16,000 years ago, the time that encompasses the Last Glacial Maximum. During that lengthy interval, massive glaciers along the Alaskan Peninsula flowed directly into the sea and would have cut off areas to the west from those to the east and south. However, everywhere along the coast, areas that had once been covered by ice had become ice free by 15,000 to 14,000 years ago. After 14,000 years ago or so, there would seem to be no reason why people could not have taken this route southward.

The most compelling way to document that the earliest Americans did in fact move south along a coastal route would be to find very early archaeological sites in that area. Unfortunately, the same rise in sea levels that submerged the Bering Land Bridge would also have submerged such sites in almost all areas. As Daryl Fedje and his colleagues have noted, almost all shorelines that would have existed along the western coast are now deeply submerged, and it would be prohibitively expensive to try to locate sites in this drowned context. However, as these scientists also note, some ancient shorelines in Alaska and British Columbia are still above water—stranded shorelines, they are called—and intensive archaeological survey of those have the potential of containing just the kinds of sites that could confirm an early coastal entry. In addition, early sites might be found in some drowned ancient shorelines that lie near the modern coast. There is, in other words, real hope for finding the ancient sites that the coastal route hypothesis suggests may be there.

All of this leaves us with two ways by which people could have gotten from eastern Beringia to the Americas south of glacial ice. They could have followed an ice-free corridor east of the Canadian Rockies before sometime between 21,000 and 30,000 years ago, and after about 12,000 years ago. They could have traveled down the Pacific Coast before about 23,000 years ago or after about 14,000 years ago. Or they could have done both.

The Human Occupation of Western Siberia

As I have mentioned, our species originated in Africa prior to 150,000 years ago, whence it spread throughout the world. At the moment, the earliest evidence for such peoples in Eurasia is from the Middle East, where they had arrived by about 90,000 years ago, but the big push seems to have begun around 50,000 years ago. Western Europe saw the arrival of morphologically modern peoples—"Cro-Magnon Man"—sometime after 40,000 years ago. In much, though not all, of the Old World, these spreading populations of morphologically modern peoples replaced earlier forms, such as Neanderthals, though precisely why and how that happened is anything but clear.

It is widely assumed that modern peoples were the first to colonize the Far North. The earliest well-dated archaeological site from that region is Mamontovaya Kurya, located near the very northern tip of Russia's Ural Mountains at the Arctic Circle (66°34' N). This site provided a small series of stone tools alongside a much larger number of mammoth bones, all dated to about 36,000 years ago. This site is well north of the most northerly known Neanderthal remains, but it is within the time range that makes it possible that either Neanderthals or modern peoples could be represented here. Either way, this site shows that people had the ability to live this far north at this time, though the occupation occurred during the Eurasian equivalent of the Middle Wisconsin, a time of relatively (but only relatively) mild climatic conditions.

Mamontovaya Kurya, however, is 1,800 miles west of the Lena River, and thus 1,800 miles west of the western edge of Beringia. As a result, it probably has little relevance to the peopling of the Americas. And as we move east, nothing whatsoever suggests that people who were not morphologically modern penetrated deeply into Siberia.

In fact, secure evidence for people in western Beringia prior to the end of the Last Glacial Maximum is very hard to come by. The only such site firmly within this region is the remarkable Yana RHS site. Located on the Yana River at 71° N, on the western edge of Beringia (but east of the Lena River), this site, well dated to about 27,000 years ago, provided a series of stone and bone artifacts (including a spear foreshaft made of rhino horn), along with the bones of a wide variety of animals including reindeer and mammoth. This site shows that people had the ability to live in far northern Beringia prior to the Last Glacial Maximum. It is still some 1,200 miles west of the land bridge, however, and no other sites of this age are known from western Beringia.

As skimpy as this evidence is, it is far better than the absolutely nonexistent evidence for occupation of western Beringia during the Last Glacial Maximum. This lack of evidence is not due to the fact that the area was beneath ice, since the region was only sparsely glaciated (figure 3-1). Instead, western Beringia appears to have been simply too cold for human occupation, given the available technology.

As conditions improved, people moved back, the archaeological record becomes far less skimpy, and we finally find archaeological sites getting closer to the Bering Land Bridge. The Berelekh site, on a tributary of the Indigirka River and about nine hundred miles from the land bridge, seems to date to between 11,000 and 14,000 years ago; the stone tools that it provided include one form very similar to those known from late Pleistocene Alaska. The earliest occupations at the sites of Ushki-1 and Ushki-5 date to 11,300 years ago; these are on the Kamchatka Peninsula, which juts into the Pacific south of the land bridge.

In short, if our knowledge of the archaeology of far northern Eurasia is accurate (it probably isn't), and if the earliest North Americans entered via the Bering Land Bridge (they probably did), then the earliest North American archaeological sites must postdate 40,000 years ago. Given the sparse pre–Last Glacial Maximum archaeological record for western Beringia, the total absence of known sites closer than about 1,000 miles to the land bridge at that time, and the complete absence of sites during the Last Glacial Maximum, it seems reasonable to guess that any colonization via this route probably postdated 16,000 years ago or so.

A Transpacific Route?

Although Hawai'i, Easter Island, and New Zealand are some of the most remote places on earth, people managed to occupy them prehistorically—Hawai'i around 1,000 years ago, Easter Island around 800 years ago, and New Zealand around 700 years ago. Accomplishing that feat required formidable seafaring skills, from building appropriate vessels and supplying the vessels with provisions to last for weeks on end, to navigating in open waters and being able to return home as needed. Polynesians not only occupied these islands but also seem to have reached South America. There seems to be no other way to explain the fact that sweet potatoes, an American plant, are found in Polynesia as early as 1,000 years ago. It is even possible that Polynesian sailors introduced chickens to the coast of Chile by 600 years ago—the famous Araucana variety, with its blue-green eggs—though better dating, and more detailed work on chicken genetics, will be needed before we can be certain of this.

Recently it has been suggested that purposeful voyaging brought far earlier peoples to South America as well. This argument does not assume that late Pleistocene sailors had the same nautical skills as late Holocene Polynesians. Instead, it observes that because of the global rise in sea level that I discussed earlier, many islands that once stuck their heads up above water might now be submerged. If this were the case—and as many as forty-three such islands have been suggested to have existed—then perhaps people made it across the Pacific by hopping from island to island.

There is nothing conceptually wrong with this argument, but there is no reason to think that things actually happened this way. Most important, the human prehistory of this part of the world is quite well known. Among other things, we know that the vast region north and east of the Solomon Islands—called Remote Oceania (for a good reason)—did not begin to be occupied until after about 3,100 years ago and that the most remote of these islands were not occupied, as I noted earlier, until about 700 years ago. Well stocked for these voyages, the people involved brought with them not only distinctive kinds of artifacts but also many species of plants, as well as pigs, chickens, dogs, and rats. There is no hint that anyone was on these islands before these people arrived—no early artifacts, no suggestion of any introduced organisms that lingered.

Of course, one could argue that there were people here before then but that the sites they left behind were drowned as water levels rose and the islands were abandoned. But this full set of considerations suggest, if they do not show, that the stepping-stone hypothesis for the transpacific colonization of South America during the late Pleistocene does not have much going for it.

A Transatlantic Route?

The possibility that people crossed the Atlantic to colonize North America from western Europe during the Pleistocene has been tossed around for well more than a century. The most recent version of the argument focuses on two archaeological phenomena: Clovis in North America, and the Solutrean in southwestern Europe.

I discuss the Clovis phenomenon in some detail in chapter 4. For now, I simply mention that Clovis is the name given to the cultural complex that represents the earliest securely dated, widespread archaeological material in North America, falling between about 11,200 and 10,800 years ago. It is marked by a distinctive set of artifacts, including beautifully made spear points and bone rods with beveled ends (chapter 4). The Solutrean, on the other hand, dates from about 20,500 to 17,000 years ago and is known from southern France and the Iberian Peninsula. It is also marked by a distinctive set of artifacts, including beautifully made spear points and bone rods with beveled ends.

The enigma about Clovis has long been that it appears full-blown south of glacial ice in North America. There is nothing like it in Beringia. The Solutrean, on the other hand, does show some similarities in artifact types and manufacturing techniques to things seen in Clovis. This has led archaeologists Bruce Bradley and Dennis Stanford, both experts in Clovis technology, to argue that the Solutrean represents the ancestor of Clovis. In particular, they argue that Solutrean peoples boated across the North Atlantic, skirting the sea ice, making use of the rich marine resources available along the way, and putting ashore somewhere along the Atlantic Coast of North America. Their descendants are, they argue, the people we call Clovis.

As interesting as this argument is, it comes associated with overwhelming difficulties. Most obviously, there is the 5,500-year gap between the end of the Solutrean and the beginning of Clovis. Bradley and Stanford attempt to fill this gap by pointing to a series of possible pre-Clovis sites in eastern North America, including Meadowcroft Rockshelter, which I discuss below. None of these sites has been fully published, however, and all are highly controversial. Even if they weren't controversial, however, none of them contains anything that looks like the Solutrean.

Archaeologists Lawrence Straus (Solutrean expert), David Meltzer (Clovis expert), and Ted Goebel (Beringian expert) also point out that the differences between Solutrean and Clovis tool kits far outweigh the similarities. There are other problems as well. Solutrean peoples made some of the cave art for which Paleolithic Europe is so justly famous, yet nothing like it has been found in the Americas. There is no evidence that Solutrean folk ever hunted sea mammals or took marine fish, even though these two possible food sources are supposed to have fueled the trip in the North Atlantic model. And, it is worth repeating, there is that 5,500 year gap between the two.

Finally, the last decade or so has seen an explosion of analyses of human DNA on a global basis, including studies of the native peoples of Asia and the Americas, along with the analysis of an admittedly small number of samples of ancient North American human DNA. These studies show conclusively that the closest relatives of Native Americans live in central Asia, not Europe. These studies also show that the immediate ancestors of the earliest Americans probably originated in southern Siberia.

As it stands now, everything we know points to the Bering Land Bridge as the route taken by people to colonize the Americas.

Some Caveats

It is important to recognize that no geological event of far northern North America or of the Old World can appropriately be used to set limits on the age of the earliest archaeological sites in the Americas. The timing of the initial entry or entries of people cannot be determined by when the Bering Land Bridge was open or closed, because there are other ways of passing from Siberia to Alaska and, as we have just seen, there are other possible avenues of entry, no matter how unlikely they may appear. The date or dates of entry south of glacial ice cannot be determined by examining when an ice-free corridor was available, because our understanding of that corridor can always change (though it was certainly open by 11,500 years ago) and because there were other ways of arriving south of the ice. The timing of the earliest entry cannot be determined even by the timing of the earliest entry of people into western Beringia, since we do not really know when that occurred and, again, there is some slim chance that the earliest entry occurred by some other route.

On the other hand, once we know when people first got here, we may be able to use our knowledge of such events as the first peopling of western Beringia, the opening and closing of the land bridge, and the chronology of an ice-free corridor or the nature of North Pacific coastal ice chronologies to explain the appearance of the archaeology. But the question of the timing of the entry of people into the Americas can be answered only by doing archaeology in the Americas.

Identifying the Earliest American Archaeology

So what are the earliest archaeological sites in the Americas, and what do they look like? I have already mentioned that the answer to this question is one of the most hotly debated issues in the archaeology of the Americas. This debate relates to the often slippery nature of simpler forms of archaeological data, as well as to the difficulties associated with the methods used to date archaeological sites. The debate also relates to the fact that so many American sites once thought to have been extremely old have been shown to be something else entirely. Because of all this, archaeologists use a straightforward set of criteria to judge whether or not a potentially ancient archaeological site really is ancient, and whether it is really archaeological.

Is the Site Really Archaeological?

Early in their career, archaeologists learn that the results of human handiwork called artifacts fall along a continuum of recognizability. On one end of that continuum are objects so complex that there can be no doubt that human hands played a role in their manufacture—a decorated bowl, for instance, or a finely made stone arrow point.

At the other end of the continuum are objects so simple that it can be extremely difficult, or even impossible, to know with certainty whether people had anything to do with producing them. People break stones to manufacture tools, but a wide variety of natural processes, from alternate freezing and thawing to tumbling in streambeds, breaks stones as well. The results of simple stone tool use or manufacture can be so similar to the results of natural stone fracture that the products of the two kinds of processes may be indistinguishable.

Most of the time such similarities cause no major problems, since the context of simple stone tools gives them away. If they are found among more complicated objects that obviously resulted from human behavior, there is usually no problem. If they are found outside such settings, however, problems arise, and if they are found in a setting known to produce stone mimics of simple artifacts—in a streambed, for example—caution is appropriate. Hence, the first criterion for any archaeological site to be accepted as such, including any early archaeological site, is that it must contain undoubted artifacts or, if not, other material that infallibly indicates the presence of people.

Is the Site Really That Old?

Dating can be a tricky matter in any archaeological setting. A wide variety of dating techniques can be used to place archaeological materials in time, some of which I discuss in other contexts. The prime technique that has been used to date the materials discussed in this book is radiocarbon dating.

RADIOCARBON DATING

Of the three isotopes of carbon that occur in nature, two (abbreviated ^{12}C and ^{13}C) are stable; one, ^{14}C, is radioactive. Carbon-14 is produced as a result of the bombardment of our atmosphere with cosmic rays. That bombardment produces neutrons, which in turn react with nitrogen-14 to produce radioactive carbon-14.

The radioactive carbon produced in this fashion combines with oxygen to form carbon dioxide, and in that form is distributed throughout the atmosphere, some ending up in plants and animals. During the lifetime of an organism, its carbon-14 content remains in equilibrium with that in the surrounding atmosphere, though there are some pernicious exceptions (the shells of both marine and, especially, terrestrial molluscs, for instance, can incorporate older carbon). When the organism dies, the balance between atmospheric and organismic carbon-14 is lost, and the carbon-14 begins to decay without being replaced.

The half-life of carbon-14 is about 5,730 years; every 5,730 years, half of it disappears, decaying into a beta particle and nitrogen-14. Because the half-life of carbon-14 is known, measuring the amount of it left in the remains of an ancient organism provides a means of measuring exactly when that organism died: after 5,730 years, half is gone; after 11,460 years, another half; and so on. The traditional technique for determining how much radiocarbon was left in a sample worked by counting the beta particles emitted as radiocarbon decayed in that sample. That required large samples and lengthy counting times. The modern method, called accelerator mass spectrometry dating (or AMS, or just accelerator dating), allows the actual number of radiocarbon atoms in a sample to be counted, making dating quicker and possible on much smaller samples. The limit of radiocarbon dating is set by the fact that after enough years pass—40,000 years or so—the amount of carbon-14 left is so minute that it is difficult to measure accurately.

Radiocarbon dates are expressed as "years before the present," or BP. The "present," however, doesn't mean "today." Instead, it means 1950, the year the technique came into being. This convention guarantees that dates "BP" determined at different times are comparable to one another. In addition, because there is always random error in making radiocarbon measurements, the dates come with an error term, or standard deviation. For instance, a date of 12,500 ± 50 ^{14}C years BP for a mammoth bone means that the best estimate of the date is 12,500 radiocarbon years ago, and that there is a 68 percent (one standard deviation) chance that the actual date lies between 12,450 and 12,550 radiocarbon years ago. If we want to be more certain than that, we can jump up to two standard deviations (in this case, 100 years on either side of the best estimate), which gives the 95 percent interval for the date. This error term is directly akin to the those given in modern polls. (For example: "Our poll shows 92 percent of nice people prefer dogs to cats, with a margin of error of ±3 percent"; if this were a Gallop poll, that figure would represent two standard deviations, or a 95 percent confidence interval).

When the radiocarbon dating technique was introduced, it was assumed that the amount of radiocarbon in the atmosphere remained constant through time. If that were the case, "radiocarbon years" would be the same as calendrical years. However, because radiocarbon is produced by bombardment of the atmosphere by cosmic radiation, variations in the amount of that radiation reaching the earth causes the amount of radiocarbon produced to vary through time. In addition, carbon dioxide dissolves in water, with significant amounts ending up in the oceans. If anything changes the rate at which radiocarbon is produced, or alters the nature of the exchange between atmosphere and oceans, the amount of radiocarbon in the atmosphere can change. Because this has, in fact, occurred, radiocarbon years and calendrical years are not identical.

As a result, scientists have devised a variety of ways to investigate the relationship between radiocarbon and calendrical years. Doing this requires the correlation of radiocarbon dates with other information relating to the ages of the dated material. Tree rings provide an obvious approach, currently allowing us to correlate radiocarbon and calendrical dates back to about 12,400 years ago. A variety of other means have been used to extend the chronology back to about 50,000 years ago.

By doing this, we can determine the relationship between a given radiocarbon date and a calendrical age, a process called calibration. We have also learned that the radiocarbon clock has at times either sped up (more radiocarbon produced per unit time) or slowed down (less radiocarbon produced). As a result, there are times when a single calendrical age may coincide with a range of radiocarbon dates. There are also times when a single radiocarbon date coincides with a range of calendrical ones, producing what is called a radiocarbon plateau.

The difference between radiocarbon and calendrical years can be quite significant. Take, for instance, some phenomenon that began 15,000 ^{14}C years ago and ended 10,000 ^{14}C years ago and so lasted 5,000 radiocarbon years. Once calibrated, these dates become 18,250 and 11,480 calendar years, respectively. Now, this phenomenon has lasted some 6,800 calendar years. Not only are the ages in years different, but the differences are different. Dating organic remains was easier, but much less accurate, before we knew all this.

All this, of course, raises an issue about the dates I use in this book, given that radiocarbon and calendrical years are not the same thing. For a number of reasons, I have opted to use radiocarbon years. First, all of the older, and a significant portion of the current, literature uses radiocarbon years. Second, the calibration curves have a strong tendency to change through time. Because of this, dates once calibrated have to be recalibrated, even though the underlying radiocarbon dates remain the same. As we get into the Holocene, the problem becomes less and less significant.

To make life easier, appendix 1 shows the current relationship between radiocarbon and calibrated years at 100-year intervals for the past 25,000 radiocarbon years.

GETTING GOOD DATES

Although radiocarbon dating works much the way it is supposed to, it is not always an easy matter to obtain valid radiocarbon dates. In the archaeological context, the older the site, the harder it is to get good dates, because the older the site, the poorer the preservation of organic materials tends to be, and the greater the chances that any organic material still present will be contaminated or will have moved from its original position. Indeed, some organic material is relatively easily contaminated, and great caution is required in using such materials as a source of radiocarbon dates. Bone is a prime example, since contaminants introduced by groundwater can easily make radiocarbon dates derived from it significantly older or younger than the animal's actual time of death.

Even with easily contaminated material, however, valid radiocarbon dates can often be obtained. Bone again provides a good example. About two-thirds of the composition of a fresh, dry bone is inorganic. The remaining, organic fraction primarily consists of small fibers called collagen. The organic content of bone may be quickly lost after an animal dies and its bones are scattered across the landscape. However, if the bone is deposited in an environment conducive to the preservation of organic material, as might be provided in a dry cave, the organic fraction of the bone may be preserved. If so, the collagen can be extracted, the individual amino acids that make up that collagen can be isolated, and those amino acids can be dated. Because the chances of contamination of certain amino acids in bone are quite small, the chances of getting an inaccurate date are correspondingly small. Indeed, to increase the odds of getting an accurate date, different amino acids from the same specimen can be dated and the results compared.

There are other ways of dating archaeological sites. Some of these—for instance, an approach called luminescence dating—are quite accurate. For early American sites, however, radiocarbon dating remains the standard unless the site is so old, or so bereft of organic material, that the radiocarbon technique cannot be used.

The second criterion, then, is that an archaeological site believed to be ancient must be provided with trustworthy dates. If those are compelling radiocarbon dates, we are quite happy, but we are also happy if the dates come from some other well-established technique (like luminescence dating). As it turns out, however, all the sites that we will have to deal with have had their ages provided by the radiocarbon method.

Is the Site Undisturbed?

In the best of all possible situations, archaeologists would be able to date the earliest American archaeological sites by dating wood or bone artifacts found through careful excavation. Unfortunately, such situations have proved difficult to come by. The oldest potential archaeological sites in the Americas tend to be marked by stone tools, not by organic ones, though there are some fortunate exceptions. Human bones will also do, but dating human bone in these contexts often entails a variety of appropriate social and political barriers. In addition, all truly early sites lack human bone anyway. As a result, most of these sites must be dated using organic material associated with the artifacts—bits of wood or seeds or animal bones so tightly associated with the stone tools that a strong argument can be made that they were laid down at the same time.

In many archaeological settings, it is not difficult to make that strong argument. A fireplace or hearth associated with living debris, for instance, can provide charcoal that, when dated, provides an excellent estimate of the age of that fireplace. But many, perhaps most, archaeological sites are disturbed in one way or another. Burrowing rodents, digging by people, the roots of trees and shrubs, alternating freezing and thawing, even the trampling of large animals (including people) can all move artifacts and other material horizontally across a site and vertically into sediments older and younger than those in which they were initially deposited. Some disturbances of this sort—Coke cans next to the bones of Pleistocene animals, for example—are easy to detect. Harder to detect are disturbances in which subtle processes have moved small numbers of objects up or down.

The depositional layering, or stratigraphy, of many sites can be so complex that modern excavations are routinely conducted by a team of scientists that includes geologists or geologically trained archaeologists whose job it is to unravel the depositional history of the debris that forms the site. The complex nature of most archaeological sites, no matter what their age, leads to a third criterion for evaluating a possibly early site. To be considered seriously, that site must have been excavated with exquisite care, the stratigraphy of the site carefully unraveled and possible means of disturbance eliminated.

Can the Results Be Evaluated?

To provide secure information on the earliest peoples of the New World, then, an archaeological site must contain infallible evidence indicating the past presence of people—that is, it must beyond doubt be an archaeological site. It must have radiocarbon or other dates that have been carefully extracted from either human bone or organic artifacts, or from material so tightly associated with the artifacts that contemporaneity cannot be doubted. And the deposits of the site must be free of significant disturbance and have been excavated in such a way that the integrity of the deposits cannot be questioned.

To these three prime criteria, a fourth must be added. The results of the work that led to the conclusion that a given site meets the first three criteria must be published in such a way that others can conduct the same evaluation. This last criterion, of course, exists in all sciences.

The Earliest American Archaeology

Given these criteria, what does the earliest American archaeology look like? As I have noted, there is no question that people were in North America around 11,200 years ago; this was the archaeological phenomenon I referred to as Clovis. While the dates for Clovis have recently become somewhat controversial, the controversy involves only a few hundred years (see chapter 4). The real question is whether there were people in the Americas prior to Clovis, and that is the issue I explore here.

I do this in four steps. First, I look at two notable misses—sites and specimens that some thought were truly ancient but that are now recognized as not being so at all. The story of these sites shows how easy it is to be wrong in this context. Second, I take a quick look at a site that may be early but that does not yet meet all four criteria. Third, I discuss the Monte Verde site and the Paisley Caves, the sites that provide the earliest secure evidence for people in the Americas. In the following chapter, I review the earliest widespread evidence for people in North America, as an essential backdrop for understanding what we know about the latest Pleistocene and early Holocene archaeology of the Great Basin.

Two Notable Misses

THE OLD CROW BASIN

The lead article in the journal *Science* for January 26, 1973, carried a riveting title: "Upper Pleistocene Radiocarbon-dated Artefacts from the Northern Yukon." The one-line synopsis beneath the title was even more surprising: "Man was in Beringia 27,000 years ago."

The contents of the story were straightforward, even though the site that had provided the date was not. Working in the Old Crow Basin in the northern Yukon—the eastern fringe of unglaciated Beringia—the paleontologist C. R. Harington had found a fossilized caribou tibia whose end had been carefully serrated to produce a tool identical to fleshing implements used by recent subarctic peoples to work hides. What was intriguing about the flesher was that it had been found amongst the bones of extinct Pleistocene mammals, including mammoths, suggesting by this placement that it might be extremely old.

Unfortunately, rather than having been found buried in an undisturbed context, the artifact had been found on the surface, leaving open the possibility that it was a recent tool lying among ancient bones. To Harington and his archaeological colleague W. N. Irving, this seemed unlikely, but the only way to be sure was to date the artifact itself. This they did, and *Science* reported the results: the tool was approximately 27,000 years old, showing that people had been in Beringia before the Last Glacial Maximum.

Or had they? Geologist C. Vance Haynes had challenged the presumed antiquity of the flesher even before the radiocarbon date appeared. The number of challenges grew after 1973, however, because the crucial date had been run on the radiocarbon content of the carbon dioxide gas that had been produced by treating the inorganic fraction of the bone with acid. After the flesher was dated, scientists became increasingly aware of the fact that this fraction of bone can be readily contaminated through contact with either older or younger carbonates in groundwater. In this case, the archaeological community suspected that the artifact was far younger than the radiocarbon date implied. At the time, nothing could be done about this, because all but the serrated end of the tool had been destroyed to produce the original date.

That situation changed with the development of accelerator dating and its ability to date milligram samples of carbon. In 1986, a tiny sample—0.3 grams—was removed from the flesher. The protein was extracted from that sample and then dated. The results were as stunning as the original date, and once again *Science* brought the news. The caribou flesher was only 1,350 years old. The flesher looked just like a modern tool because it was a modern tool.

In addition to the flesher, three other artifacts that had been made from caribou antlers and that had been found with the bones of extinct mammals were also dated: the oldest was only some 3,000 years old. The title of the *Science* paper said it all: "New Dates on Northern Yukon Artifacts: Holocene, not Upper Pleistocene."

CALICO HILLS

In the case of the Old Crow Basin, the sites were undoubtedly archaeological. The debates were over their age, and the outcome of those debates shows why archaeologists place so much stress on the dating criterion I discussed earlier. The Calico site, located near Barstow in the central Mojave Desert of southeastern California, and thus toward the southern edge of the hydrographic Great Basin, illustrates the problems that can result from a site that appears to be reasonably well dated but whose archaeological nature is murky.

In the early 1960s, famed archaeologist Louis Leakey became deeply impressed by a series of large, clumsy-looking artifacts from the Mojave Desert that had been shown to him by Ruth D. Simpson of the San Bernardino County Museum. Simpson thought that these materials were ancient, dating well into the Pleistocene, but she had been unable to demonstrate such an antiquity because all of her potentially ancient objects had been found on the surface and there was no way she could date them.

During a 1963 field trip to the area, Leakey and Simpson found what they thought was a buried archaeological site of tremendous antiquity exposed in a road cut. Leakey, flush from his epochal successes in Olduvai Gorge, was convinced that they had hit archaeological pay dirt. The following year, with funding from the National Geographic Society, and with Leakey as overall project director and Simpson leading the fieldwork, excavations began at the Calico site, so named for the nearby Calico Mountains. Although outside funding for the project ran out in 1970, and although Leakey himself died in 1972 (and Ruth Simpson in 2000), the work they began has continued thanks to a dedicated group of volunteers.

Four years after the work had begun, Leakey and his colleagues reported that they had discovered more than 170 artifacts at Calico, and they estimated that the deposits that provided these artifacts were between 50,000 and 80,000 years old. The true age of the deposits became clearer in 1981, when the results of the application of the uranium-thorium dating technique to materials in these deposits were published.

Like radiocarbon dating, uranium-thorium (U-Th) dating is a radiometric technique, relying on the breakdown of a radioactive isotope—in this case uranium-234—into a "daughter" isotope—in this case, thorium-230. While the half-life of radiocarbon is 5,730 years, limiting the current practical applicability of the approach to the past 40,000 years or so, the half-life of uranium-234 is 75,000 years, thus allowing much older material to be dated. And, while radiocarbon dating is applied to organic materials, U-Th dating can be applied to the calcium carbonate crusts that form on rocks as a result of groundwater activity.

The U-Th age for the deepest Calico deposits believed to have provided artifacts turned out to be 203,000 years. That is, the basal levels of this site predate the last interglacial. Leakey would have been pleased. He also would have been pleased at archaeologist Fred Budinger's report that over 11,000 objects thought to be artifacts have been collected from this site.

Budinger, as well as all other scholars who have worked on the Calico materials, knows full well what it means if he and his colleagues are right. As Budinger puts it, "The implications of a 200,000 year date for hominids in the New World are significant and far-reaching, and will certainly affect our understanding of both New and Old World populations during the Pleistocene Epoch. For one thing, the Calico evidence has made apparent the probability of pre-*Homo sapiens sapiens* man in the Western Hemisphere" (Budinger 1983:82). If Budinger and his colleagues are correct about Calico, that is exactly what the evidence means, even opening the possibility that modern peoples may have evolved from earlier stock in the New World as well as in the Old. There is no reason to question the Calico dates: the site does appear to be that old. But is it archaeological?

Many Calico specimens are fairly crude: cobbles of chert, jasper, limestone, and other kinds of rock that have had flakes removed from one or more edges and from one or more sides. Some, however, look very much like undoubted artifacts, and would likely be accepted as such if they came from an archaeological site in a very different context. These specimens include slender, parallel-sided stone flakes, as well as flakes that appear to have been worked along one edge.

Unfortunately, looking like an artifact and being an artifact can be two very different things and the context of the Calico site is one that is known to produce artifact mimics. The Calico specimens have been meticulously excavated from an alluvial fan, deposits that were brought by water, mud, or debris flows from the Calico Mountains up to four miles away. (Alluvial fans develop when water flowing along a steep gradient suddenly encounters a marked decrease in slope, as often occurs at the mouth of a canyon; the water, often but not always in the form of a stream, loses its ability to carry its load and deposits that material in a feature that is often shaped like a fan.) Archaeologists have long known that rocks tumbled as such fans are being formed can look much like artifacts. It is this fact that leads nearly all archaeologists to conclude that the specimens from Calico are far more likely to be what Vance Haynes has called "geofacts"—"artifact-like phenomena of geological origin" (C. V. Haynes 1973:305)—than they are to be artifacts.

Indeed, the archaeologist Louis Payen applied a time-worn test to eighty-three Calico specimens that Leakey selected as among the best produced by the site. The test Payen used measures the angles at the edges of stone tools and assumes that true stone tools will have edge angles more acute than those on stones that have been flaked by nature. Using a huge control sample of undoubted stone tools from sites other than Calico, Payen found their edge angles to average 72 degrees. He then measured the same angles on the Leakey sample of prime specimens, and found them to average 87 degrees. Finally, he measured edge angles on a sample of Calico specimens that Leakey had rejected as artifacts. They averaged 88 degrees. Payen concluded that the specimens selected by Leakey, and by extension all Calico specimens, were geofacts, not artifacts.

Scholars supporting the validity of the Calico site responded vociferously to Payen's analysis, as well as to Haynes's critique, but the context of the Calico site is such that natural rock breakage that produces specimens that look like true stone artifacts is to be expected. Even though we cannot demonstrate that none of these objects are artifacts, it is impossible to prove that any of them are. Calico has now entered the literature as another ancient archaeological site that wasn't, because it fails to meet one of the prime criteria: it does not contain undoubted artifacts.

A Possible Hit: Meadowcroft Rockshelter

The two sites I have discussed to this point are either archaeological but not old (Old Crow Basin), or old but not archaeological (Calico), and so illustrate the importance of the criteria with which I began this discussion. There are, however, a number of New World sites that may be both archaeological and old and are accepted as such by some archaeologists. However, in large part because work continues at most of them, none are fully published and so cannot be properly evaluated. There is, however, one on which quite a bit of research (but not enough) has been published.

Meadowcroft Rockshelter is located on a small tributary of the Ohio River some thirty miles southwest of Pittsburgh in far southwestern Pennsylvania. It is not huge by rockshelter standards—only some seven hundred square feet protected by the rock overhang—but the deposits within were some fifteen feet deep. It may be no accident that the archaeological potential of the site was tapped by an archaeologist, James Adovasio, with significant experience in the Great Basin. Great Basin archaeologists are always alert to the possibilities presented by cave deposits, while those working in northeastern North America rarely get to deal with such sites and, when they do, are often disappointed in the results.

Adovasio and his team began to work at Meadowcroft in 1973. Their excavations continued until 1978, with some additional work taking place in later years. By the end of their initial excavations, however, they had carefully revealed what may be one of the oldest known archaeological sites in the Americas. It is important to stress "carefully" here, since, as all who have worked with him know (and I have), Adovasio is a compulsively meticulous excavator.

The deepest Meadowcroft deposits appear to date to about 31,400 years ago, but those sediments, as well as the ones that immediately overlie them, contain no suggestion of human activities. Artifacts do not appear until just above the deepest part of the depositional unit called Stratum IIA. Those oldest artifacts were found alongside fireplaces or hearths that in turn provided organic materials that could be dated. Indeed, Meadowcroft is one of the best-dated archaeological sites in the world, Adovasio having obtained fifty-two radiocarbon dates for the site, ranging from 175 years ago at the top to 31,400 years ago at the bottom.

These dates produce a consistent picture of the accumulation of deposits in Meadowcroft, since with extremely rare exceptions, as the deposits get deeper, the dates get older, just as should happen. Lower Stratum IIA, which contains the earliest artifacts, has six radiocarbon dates said to be "unequivocally associated" with those artifacts. The oldest of these dates falls at 16,175 years ago, the youngest at 12,800 years ago. Averaged, they suggest to Adovasio and his colleague David Pedler that "humans were definitely present at this site (and by implication, throughout much and perhaps all of the Americas) sometime between 13,955 and 14,555 radiocarbon years ago" (Adovasio and Pedler 2005:26). That is, Meadowcroft seems to contain secure evidence that people were here at about 14,250 years ago.

The several hundred artifacts that come from deepest Meadowcroft have yet to be fully described in print, but in addition to the waste flakes that resulted from stone tool manufacture, the assemblage includes a number of small blades. Blades are stone flakes that are twice as long as they are wide and that, in this instance, were struck from carefully prepared cores of raw material (the cores themselves were not found at the site, though cores that would do the job are known from a nearby, undated site). In addition to these, a single projectile point, made from chert, was found in the uppermost part of lower Stratum IIA, bracketed between 11,300 and 12,800 years ago. Archaeologists use the term "projectile point" to refer to the points that tipped spears, darts, and arrows, since it is not always easy to tell the three kinds of points apart. The early Meadowcroft point, however, could not have been an arrow point, because it is too big (the bow and arrow did not appear in North America south of the Arctic until about 4,500 years ago and did not become widespread here until after 2,000 years ago).

Earlier in this chapter, I discuss the hypothesized Solutrean-Clovis connection and note that Bradley and Stanford helped fill the 5,500-year gap between the two with the Meadowcroft stone tool assemblage. Adovasio, however, sees no relationship between the early Meadowcroft material and what comes later, including Clovis. If the Meadowcroft dates are correct, this would remove yet another leg from the European hypothesis.

At one time, serious doubts surrounded the radiocarbon dates from the lower levels at Meadowcroft. In particular, Vance Haynes noted that this is coal country and that the lower levels of Meadowcroft may have been contaminated by ancient, soluble organic material carried into the shelter by groundwater. However, geoarchaeologists Paul Goldberg and Trina Arpin conducted a detailed analysis of the lower deposits at the site and showed to everyone's satisfaction that groundwater contamination had simply not occurred here. There is, as a result, no scientific reason not to trust the Meadowcroft dates.

There are, though, some troubling things about Meadowcroft. Carefully excavated archaeological sites routinely provide not only artifacts but also the remains of contemporary plants and animals. Meadowcroft, excavated with tremendous care, provided both animal and plant remains throughout its deposits. There are not many identified organic items from lower Stratum IIA, but those that exist could have come right out of the modern flora and fauna of the area.

The animals pose the least difficulties. The bones and teeth discovered in the site were identified and analyzed by zooarchaeologists John Guilday and Paul Parmalee, two of the best to ever ply the trade. While they were able to identify a very large sample of the Meadowcroft bones and teeth, only a tiny fraction of that sample came from deposits laid down between about 11,300 and sometime before 19,100 years ago. And of that tiny fraction, only eighteen specimens could be identified to the species level. Those identified specimens suggested that the animals living in the area included white-tailed deer (*Odocoileus virginianus*; five specimens), southern flying squirrel (*Glaucomys volans*; nine specimens), eastern chipmunk (*Tamias striatus*; three specimens), and passenger pigeon (*Ectopistes migratorius*; one specimen).

With the exception of the much-lamented passenger pigeon, every single one of these animals can be found around Meadowcroft today, and even the passenger pigeon could have been found there in the 1880s. That is, these are animals that do well in temperate woodlands. However, at the time of the Last Glacial Maximum, Meadowcroft was only about ninety miles south of the Laurentide ice front. As a result, Guilday and Parmalee, while noting that the animals they identified from these layers tend to have broad adaptive ranges, also thought that "the absence of boreal species is disturbing" (Guilday, Parmalee, and Wilson 1980:78). They concluded, though, that the identified sample was so meager that perhaps not much should be made of it.

That brings us to the plants, which present a bigger problem. Associated with radiocarbon dates between about 13,000 and 15,000 years ago, the plants represented in this sample include oak, hickory, and walnut, all components of the modern vegetation. Paleobotanical work, however, has shown that between 13,000 and 15,000 years ago, the vegetation surrounding Meadowcroft was most likely to have been, at best, a mixed parkland with both boreal (e.g., spruce) and deciduous species present. Oak certainly could have been nearby, but our understanding of the history of hickory in the eastern United States suggests that it did not arrive in the Meadowcroft area until 10,000 years ago.

There are two obvious ways to explain the nature of the Stratum IIA plant assemblage. First, it is possible that small enclaves of deciduous forest trees existed in sheltered areas in southwestern Pennsylvania during late Wisconsin times. That, in fact, is what Adovasio suggests. However, the proposition that this enclave included hickory is troublesome, since all we know about the history of this tree suggests that it was hanging out in the Southeast during full glacial times. If Adovasio is right, though, Meadowcroft will also have taught us something very significant about late Pleistocene plant history in this part of the world.

Second, it is possible that the tiny fragments of nutshells, seeds, and wood of deciduous trees from the depths of Meadowcroft represent contaminants from upper levels. This would not be surprising: the remains of these plants are not uncommon in higher levels of the site, and contamination of this sort may be virtually impossible to detect while a site is being excavated. The only way to be sure of the age of these remains is to date them directly. This appears not to have been done, but it is hard to tell because the published information does not say which species of plants have been dated.

In short, the plants and animals from Meadowcroft are fully consistent with having existed in a temperate environment, much like the one that exists there now; boreal species are absent. Many find this combination troublesome.

Most troublesome, though, is that Meadowcroft is not fully published. There are no detailed stratigraphic descriptions, no full descriptions of the relationship of the artifacts to the material that has been dated, and no complete descriptions of the artifact assemblage itself. As a result, one cannot evaluate the associations between dated material, artifacts, and the sediments that contain them. This is a real problem. Once that information is available, Meadowcroft may well become the earliest securely dated archaeological site in the Americas. Until that happens, however, it will remain in what archaeologist Dave Meltzer has appropriately called archaeological limbo.

The Oldest Securely Dated American Archaeological Sites

MONTE VERDE, CHILE

If people entered the Americas by crossing the Bering Land Bridge sometime prior to 12,500 years ago and then made their way southward, by foot or by boat or some combination of the two, how long would it take them to move 10,000 miles to the south?

That is the question raised by the remarkable Monte Verde site, which sat on two sides of small Chinchihuapi Creek in south-central Chile's Central Valley, some seven miles from the modern coastline. According to Tom Dillehay, who excavated the site between 1977 and 1985 and reported the results in two lengthy monographs, the area that was occupied by people covered some 8,600 square feet, of which about 4,600 were excavated. In addition to its age, which I will discuss shortly, what makes this site remarkable is the superb preservation of the organic remains it contains. The archaeological materials are capped by peat, and then by an impermeable layer of small volcanic gravel. This, plus the high water table, kept the cultural deposits continuously wet and led to superb preservation of the organic items in the archaeological layer.

Monte Verde contained a cluster of what appear to have been roughly rectangular wooden huts. The foundations of these structures were made from substantial logs up to twelve feet long. These in turn were held in place by wooden stakes with flattened heads, the flattening apparently having been produced when the stakes were hammered into the ground. The superstructures of these houses seem to have been poles draped with animal skins: fragments of skin still adhere to some of the poles. Inside the structures were plant remains, stone tools, and clay-lined pits that had once held fires. Wooden mortars and grinding stones were found near some of those hearths. There were also other structures on the site, one of which was a wishbone-shaped feature made of sand and gravel that had upright wooden posts located every few inches along both wings—apparently the remains of a pole frame. Adjacent to the structure, Dillehay found a concentration of stone tools, piles of wood, and the bones of a large extinct mammal.

In addition to the wood, bones, and hide, an extraordinary range of organic material was preserved at the site. Those organic materials include wooden artifacts, not just the stakes I mentioned but also such things as lances, fire-drills, and mortars. There are also quids—the spit-out remains of chewed plants, including the remains of seaweed and algae that must have been imported to the site. A rich collection of the bones of the extinct mastodon-like animal known as a gomphothere were found, as were chunks of animal meat (I'm not making this up—see Dillehay 1997, chapter 18), and the remains of a wide variety of economically useful plants, including pieces of wild potato

(*Solanum maglia*) and the fruits and seeds of a species of our old friend *Atriplex*, which were likely imported from the coast.

Dillehay obtained thirty-one radiocarbon dates for Monte Verde and for noncultural deposits in the site area. These are in good stratigraphic order and match the known geological history of the region quite well, including radiocarbon chronologies built by others. Of these dates, thirteen pertain to the cultural material beneath the peat, dates that fall between 11,800 and 12,800 years ago. These dates average 12,500 years ago.

Monte Verde has it all: good stratigraphy, undoubted artifacts, strong chronology, and a massive set of publications to back it all up. In addition, in 1997, the Dallas Museum of Natural History and the National Geographic Society funded a visit by a carefully selected group of experts (I was one of those invited) to examine the Monte Verde collections both in the United States and Chile and to visit the site itself. This we did. Although not everyone agreed about all the things we saw, we did agree on the basics: Monte Verde contains undoubted artifacts and Monte Verde is pre-Clovis old. That conclusion appeared in an article, signed by all of us, in the important American archaeological journal *American Antiquity* in 1997.

This is not to say that there are not still Monte Verde skeptics, for there are. Those skeptics will not be satisfied until there is another Monte Verde—replicability, they call it—or at least another equally compelling site of that age. My attitude is quite different. If I find a mammoth in my backyard and get thirteen dates that suggest it is 12,500 years old, I don't need to find another one to convince me that it had been there at that time.

PAISLEY CAVES, OREGON

But for those who do need a second site to be convinced that there were people in the Americas before Clovis, the Great Basin itself appears to have provided one. To fully appreciate this site for what it is, however, requires some history.

In 1932, Luther S. Cressman (1897–1994) of the University of Oregon initiated a major program of archaeological research in the northern Great Basin. He began with a survey of the rich store of rock art found in this region and then, in 1934, turned to an archaeological survey of Guano Valley in eastern Oregon. Between 1935 and 1938, he conducted the excavation of a series of caves and rockshelters in the Catlow, Fort Rock, and Summer Lake basins in the Great Basin section of eastern Oregon. This work was to put Oregon prehistory on the map.

Cressman was a remarkable scholar. He entered the graduate program in sociology at Columbia University in 1920 while still studying at General Theological Seminary (he was ordained as a priest in the Episcopal Church in 1923). Receiving his Ph.D. in 1925, he accepted a position at the University of Oregon in 1929, where he spent the rest of his career. Along the way, he founded the Department of Anthropology and the museums of anthropology and natural history at this institution.

What made Cressman's archaeological work so remarkable was not that he was self-trained: that was common at the time. Instead, it was that he recognized that archaeology should be an interdisciplinary enterprise. Almost from the beginning, his projects included a wide range of scientists drawn from other disciplines. Geologists Ernst Antevs and Ira Allison helped analyze the geological contexts of his sites; geologist Howel Williams studied the volcanic ashes contained within the caves he excavated; paleobotanist Henry Hansen analyzed pollen taken from nearby deposits; paleontologists Chester Stock, Alexander Wetmore, and Ida deMay identified the animal bones he excavated; and so on. This was some of the earliest interdisciplinary archaeology done in the Americas; what Cressman began in the 1930s is now standard procedure.

Cressman excavated Fort Rock Cave, in the Fort Rock Basin, in 1938, finding a large number of burned sagebrush-bark sandals and sandal fragments beneath a layer of volcanic ash that divided the deposits in the site in two. Some fifty miles to the south and east, Cressman excavated three caves at Fivemile Point, near the small town of Paisley (figure 3-2; see figure 5-15 for the location of Paisley). In all three of these sites, he again found artifacts beneath volcanic ash. There was no doubt in Cressman's mind that the artifacts lay beneath the ashes in these sites.

This relationship raised two important questions in his mind: what ashes were these, and when were they deposited? Howel Williams answered these questions for him: the ash in the Paisley Caves was from Mount Mazama; that in Fort Rock Cave, from Newberry Crater, to the north (he was right about the former, wrong about the latter, which was also Mazama ash).

At the time, Williams estimated that the cataclysmic eruption from Mt. Mazama occurred between 4,000 and 10,000 years ago; he soon focused on 5,000 years as his best guess. (We now know that the climactic eruption of Mount Mazama occurred 6,730 years ago, spewing twelve cubic miles of volcanic material into the atmosphere and creating the caldera in which Crater Lake now sits; ash from this eruption has even been found in Greenland ice cores.) Williams's estimate meant that Cressman's Paisley Caves artifacts were likely to be at least 5,000 years old. The Newberry Crater ash Williams identified within Fort Rock Cave was known to be younger, but how much younger was not known. Williams could estimate only that it might have been "several thousand years old" (Cressman, Williams, and Krieger 1940:78).

In addition to all of this, Cressman claimed that at one of the Paisley Caves he had found artifacts so closely associated with the remains of extinct Pleistocene-age mammals—horse and camel (see chapter 4)—that it was now clear "that man and an extinct fauna occupied this region of southern Oregon at some fairly remote time, probably . . . more than 10,000 years ago" (Cressman 1942:93). Few of Cressman's colleagues accepted his evidence at face value. The very first textbook of North American prehistory, published in 1947,

FIGURE 3-2 Fivemile Point, Summer Lake Basin, Oregon. The Paisley Caves are at the base of the basalt outcrop visible in the center distance.

ignored the careful work that Cressman had done. The oldest peoples of the northern Great Basin, this influential book asserted, were no more than two thousand years old.

Cressman was annoyed that while not all of his work had gone unnoticed, his conclusions about the age of what he had found had been ignored. Fortunately, he did not have to wait long to show that his assessment of the antiquity of southeastern Oregon's archaeological record was correct. One of the first radiocarbon dates ever produced was on a Fort Rock Cave sandal. That sandal, Cressman triumphantly announced in 1951, was 9,053 years old. Indeed, University of Utah archaeologist Jesse Jennings was soon to provide even more support for Cressman: in 1951, he received dates extending back nearly 11,000 years on organic materials associated with artifacts from Danger Cave, adjacent to the Bonneville Salt Flats in western Utah. From that time on, there was no doubt that the native peoples of the Great Basin have a very lengthy history.

The documented great antiquity of Fort Rock Cave made this site famous. It was to become even more famous, however, as a result of work done in 1967 by Cressman's student Stephen F. Bedwell. Bedwell's work showed that the Fort Rock Cave volcanic ash that Williams had identified as having come from Newberry Crater had actually come from Mount Mazama. Bedwell also provided a series of radiocarbon dates documenting that the human use of Fort Rock Cave had begun by at least 10,200 years ago. These dates, of course, provided even more confirmation for the arguments that Cressman had made nearly thirty years earlier. By now, however, these arguments were not controversial.

What was controversial was Bedwell's claim that the first human occupation in Fort Rock Cave dated to 13,200 years ago. He based that claim on a single radiocarbon date from organic material lying on top of the gravels found at the base of the cave. Nearby, and apparently lying on the same surface, he discovered a series of fourteen artifacts, including two projectile points and a fragmentary grinding stone. Bedwell argued that the 13,200-year date applied to these artifacts as well.

Bedwell's argument that Fort Rock Cave provides evidence for pre-Clovis human occupation in the Great Basin is not accepted by others, a result of the fact that he provided no detailed information on the nature of the association between the dated material and the artifacts. It is fully possible that materials of distinctly different ages were lying on the basal gravels at Fort Rock Cave, just as bullet shells and fluted points can now be found lying on the same surface in many parts of the Great Basin. Lacking the required contextual details, Fort Rock Cave must go down as just another site with inclusive evidence for pre-Clovis peoples in the Arid West.

Much the same was originally true for the Paisley Caves. Because the techniques Cressman used to excavate these sites were so crude, and his descriptions of the results of that work so vague, there was no reason to accept his claim that he had found evidence for a tight association between people and extinct Pleistocene mammals.

Cressman had, though, established that these sites might repay more detailed work, and another University of Oregon archaeologist, Dennis Jenkins, took up the challenge. Beginning in 2002, Jenkins conducted careful excavations at four of the Paisley Caves, all with the goal of testing Cressman's argument that artifacts and the remains of extinct mammals were to be found tightly associated with one another here. In fact, Jenkins found the remains of extinct mammals at three of these sites. However, it was at one of the sites that Cressman had not excavated—Paisley Cave 5—that Jenkins and his team discovered enough material to suggest that people really were in this area at the same time as extinct Pleistocene mammals, but also that they were here prior to Clovis times.

It would be nice to say that the evidence Jenkins uncovered consisted of beautifully made stone tools lying next to a carefully made fireplace filled with the charred remains of animals and plants that represented leftovers from the days' meals. It would be nice to say that, but it would be false. In fact, Jenkins did discover the remains of ancient human meals in the Paisley Caves, but they were in the form of what archaeologists refer to in print as coprolites.

In 1823, the famous British naturalist historian and Oxford geologist the Reverend William Buckland described the discovery of the "calcareous excrement of an animal that had fed on bones" (1823:20) in a British cave. He quickly showed that this calcareous excrement was, in fact, the fossilized dung of the ancient hyenas that had occupied the cave. He later coined the term "coprolite" (from the Greek for "dung" and "stone") to refer to such material and the name has stuck.

Human coprolites, then, are ancient human dung. The Paisley Caves provided six of them, which a series of analyses confirmed as being of human origin. These analyses also demonstrated that they contained human DNA with genetic markers characteristic of Native Americans. Because human DNA can be easily transferred from modern people to ancient items, the research team also analyzed the DNA of all sixty-seven people who might have come into contact with these items during or after their excavation. None came even close to providing a match.

This would not be such a big deal were it not for the ages of some of these specimens. After all, many Great Basin sites have provided human coprolites (see chapter 9). However, the high-precision dates obtained from these specimens show the earliest of them, all from Paisley Cave 5, to date between 12,260 ± 60 and 12,400 ± 60 ^{14}C years ago. This is well before Clovis and nearly as old as the average of the Monte Verde dates. Just as with Monte Verde, there appears to be no reason to think that this site—or more accurately, the human coprolites within it—are not just as old as the radiocarbon dates suggest.

A Pre-Clovis Conclusion

Other sites in the Americas may well be pre-Clovis in age, but Monte Verde and the Paisley Caves certainly are. These sites combine to show us that people were in the Americas by 12,500 years ago, long before Clovis. The long-held idea that Clovis people were the first to occupy the Americas is no longer tenable, but only future archaeological work will tell us how long before Clovis the first Americans got here.

Chapter Notes

The information on Eskimo movements across the Bering Strait comes from E. W. Nelson (1983). For those interested in the empirical record of human evolution, the book to read is Klein (2009). Recent, but highly technical, summaries focused on our own species are found in volume 106(38) of the *Proceedings of the National Academy of Sciences*, published in 2009. For more on the definition of the Pleistocene, Holocene, and Quaternary, see the websites of the Subcommission of Quaternary Stratigraphy (http://www.quaternary.stratigraphy.org.uk/) and the International Commission on Stratigraphy (http://www.stratigraphy.org/). For many of the dates I use here, see the "GeoWhen" section of the latter site. There is a strong tendency to equate the base of the Holocene with the end of the Younger Dryas cold interval, discussed in chapter 5. Doing so places the onset of the Holocene at about 11,700 calendar, or about 10,100 radiocarbon, years ago; see M. Walker et al. (2009).

On the Innuitian Ice Sheet, see Dyke et al. (2002) and England et al. (2006). The estimate of Laurentide Ice Sheet thickness is derived from Dyke et al. (2002) and Marshall, James, and Clarke (2002); the Cordilleran Ice Sheet, from Clague and James (2002); Marshall, James, and Clarke (2002); and Clague, Mathewes, and Ager (2004). On the earliest Ice Age glaciation in North America, see Barendregt and Duk-Rodkin (2004); on the Sierra Nevada, see A. R. Gillespie and Zehfuss (2004); on far northwestern North America, Duk-Rodkin et al. (2004). The dates used for MIS 3 and for the Middle Wisconsin are from Shackleton et al. (2004); see also Fullerton, Bush, and Pennell (2003). My discussion of Middle Wisconsin glaciation is based on Andrews (1987); Kirby and Andrews (1999); and Clague, Mathewes, and Ager (2004); for the Late Wisconsin, see Clague and James (2002); Dyke et al. (2002); Dyke, Moore, and Robertson (2003); Clague, Mathewes, and Ager (2004); and Dyke (2004). Excellent general reviews of Cordilleran and southern Laurentide ice history are provided by Booth et al. (2004) and Mickelson and Colgan (2004). The maps by Arthur Dyke and his colleagues are found in Dyke, Moore, and Robertson (2003) and Dyke, Giroux, and Robertson (2004). These can be downloaded for free from the Geological Survey of Canada's website (http://geopub.nrcan.gc.ca/); they provide a graphic chronological ice history of northern North America from 18,000 to 5,000 years ago. The best place to go for the chronology of the Last Glacial Maximum is P. U. Clark et al. (2009).

The estimates of Last Glacial Maximum sea level decrease (380 feet–440 feet) are taken from P. U. Clark and Mix (2002) and Milne, Mitrovica, and Schrag (2002); see P. U. Clark et al. (2009) on the 425-foot estimate. Data on modern global ice volumes and potential sea level change are from Lemke et al. (2007), available online at http://www.ipcc.ch/. On sea level changes during the Wisconsin Glaciation, see Lambeck, Yoyoyama, and Purcell (2002). For conflicting versions of Bering Land Bridge history during the Middle Wisconsin, see Clague, Mathewes, and Ager (2004); Hoffecker and Elias (2007); and the review by P. M. Anderson and Lozhkin (2001); on the closing of the Bering Land Bridge, see Elias et al. (1996), L. E. Jackson and Duk-Rodkin (1996), Manley (2002), and Keigwin et al. (2006). Manley (2002) provides a fascinating animation of Bering Land Bridge history from the Last Glacial Maximum to today (http://instaar.colorado.edu/QGISL/bering_land_bridge). The vegetation of the Bering Land Bridge is discussed by Elias et al. (1996); Elias, Short, and Birks (1997); Ager (2003); and Hoffecker and Elias (2007). There are multiple definitions of Beringia, but they differ little from one another; see, for instance, P. M. Anderson and Lozhkin (2001) and Edwards et al. (2005). In case you are thinking of swimming from Alaska to Siberia, read L. Cox (2004) first.

Fedje et al. (2004) provide an excellent review of the earliest archaeology of the northwestern North American coast, as well as a detailed analysis of the challenges and possibilities presented by searching for very early sites in this geologically complex region. That there is hope for finding ancient materials offshore in this area is shown by the fact that an apparent stone tool of unknown age has been collected from a depth of 174 feet off the coast of British Columbia (Fedje and Josenhans 2000). To my knowledge, the combined age and depth record for archaeological material collected from beneath the surface of the sea is provided by a collection of twenty-eight hand axes dredged from the floor of the North Sea eight miles east of the British coast. These came from beneath about 125 feet down—100 feet of water and 25 feet of sea-floor deposits. The artifacts are similar to those made by Neanderthals and so likely predate 40,000 years ago (Holden 2008).

Erlandson (2002) presents a concise, readable history of seafaring in the more distant human past. See O'Connell and Allen (2004); J. Allen and O'Connell (2008); and O'Connell, Allen, and Hawkes (2009) on the initial colonization of Australia; O'Sullivan et al. 2001 on *Homo erectus* on Flores; and Powledge (2006) on the Hobbit (which is what even archaeologists call this surprising human form).

Guthrie (2001), and the references it contains, discusses his approach to the Mammoth Steppe. My discussion of eastern Beringian biotas depends heavily on P. M. Anderson and Lozhkin (2001); P. M. Anderson, Edwards, and Brubaker (2004); Brubaker et al. (2005); and Edwards et al. (2005). For western Beringia, see P. M. Anderson and Lozhkin (2001); P. M. Anderson, Lozhkin, and Brubaker (2002); and Brigham-Grette et al. (2004).

The chronology of the ice-free corridor is discussed by Mandryk et al. (2001); Clague, Mathewes, and Ager (2004); Dyke (2004); and L. E. Jackson and Wilson (2004). Johnston (1933) introduced the idea. Heusser (1960:209–210) introduced the notion of a coastal route; Fladmark (1979) did so far more forcefully. My discussion of that route depends heavily on Mandryk et al. (2001) and Clague, Mathewes, and Ager (2004).

For more on western Beringian archaeology, see Goebel (2004) Hoffecker and Elias (2007); Goebel, Waters, and O'Rourke (2008); and Graf (2009); on the Yana RHS site, see Pitulko et al. (2004); on Mamontovaya Kurya, see Pavlov, Svensen, and Indrelid (2001).

Wyatt (2004) develops the argument for Pleistocene voyaging across the Pacific. The possibility of pre-Columbian chickens in South America is argued for by Storey et al. (2007) and against by Gongora et al. (2008). For the prehistory of the Pacific in general, see Kirch (2000); for prehistoric human impact on islands in general, see Grayson (2001), and on Pacific islands, see Steadman (2006). The debate over a possible Solutrean-Clovis connection and the associated transatlantic voyage is found in Stanford and Bradley (2002); B. Bradley and Stanford (2004); and Straus, Meltzer, and Goebel (2005). For a recent review of the implications of both ancient and modern DNA for the origins of Native Americans, see Goebel, Waters, and O'Rourke (2008) and Meltzer (2009).

A nice introduction to radiocarbon dating is found in the *PAGES Newsletter* 14(3), 2006. PAGES is an international scientific organization supported by U.S. and Swiss scientific organizations and dedicated to studying past global changes (hence the name) on earth. They publish themed newsletters—for instance, the 2006 one centered on radiocarbon dating—that provide excellent discussions of the given theme. Membership and the newsletter are free to all (http://www.pages.unibe.ch/about/index.html). Calibrating radiocarbon dates is simple, given that there are two widely used, free programs for doing that available on the Web. The two are CALIB (http://calib.qub.ac.uk/calib/), based at Queens University in Belfast, Ireland (though initially developed, I have to say, at the University of Washington), and OxCal (https://c14.arch.ox.ac.uk/), based at Oxford University, in England. Both are in very widespread use. The calibrations I provide in appendix 1 were done with the most recent version of CALIB. In some places in the text, I made use of the calibration curve presented by A. Schramm, Stein, and Goldstein (2000) to translate calendar into radiocarbon years.

If you are interested in the peopling of the Americas, there is only one place to start: Meltzer (2009). The best recent, short review of this topic is provided by Goebel, Waters, and O'Rourke (2008). On the criteria used to evaluate archaeological sites for validity, I have followed C. V. Haynes (1969) closely. The Old Crow Basin debate began with Irving and Harington (1973) and ended with D. E. Nelson et al. (1986). The Calico site can be pursued in C. V. Haynes (1973), R. E. Taylor and Payen (1979), Bischoff et al. (1981), Budinger (1983), Patterson (1983), Simpson (1989, 1999), and Grayson (1986); this last reference takes a deep historical look at Calico and Calico-like enigmas and what they do to archaeologists. Anyone interested in Calico should visit www.calicodig.org, a superb resource maintained by those who accept the site as archaeological. The site, maintained by the Bureau of Land Management, is easily visited and is worth the effort if you are nearby. Where else can you see a well-excavated, 200,000-year-old alluvial fan? For information on how and when to visit, see www.blm.gov/ca/st/en/fo/barstow/calico.html.

For the most recent statements on Meadowcroft, see Adovasio and Page (2002) and Adovasio and Pedler (2005). The first of these references includes a complete list of Meadowcroft publications. My discussion of the Meadowcroft fauna is based on Guilday, Parmalee, and Wilson (1980) and Guilday and Parmalee (1982). The Guilday, Parmalee, and Wilson paper is an unpublished discussion of the Meadowcroft fauna, but a copy is available at the Mercyhurst Archaeological Institute, Mercyhurst College. Goldberg and Arpin (1999) demonstrated that the deeper deposits of Meadowcroft had not been contaminated by groundwater. My discussion of the plant remains from Meadowcroft depends on Cushman (1982); T. Webb et al. (1998); and J. W. Williams, Schuman, and Webb (2001). Adovasio and Pedler (2005:24) note that the final Meadowcroft report is in preparation; the "archaeological limbo" comment is from Meltzer (2009:113). A simple example of the kind of problem that can occur without a detailed synthesis of the site is provided by the radiocarbon date of 19,600 years ago on what Adovasio et al. (1984:355) described as "cut bark-like material, possible basketry fragment," but Adovasio has also suggested that material this old from the site is not unequivocally archaeological. These kinds of problems will surely be solved when the monograph appears.

Good discussions of the history of the bow and arrow in North America are provided by Nassanay and Pyle (1999) and Ames, Fuld, and Davis (2010); see also chapter 9.

The key references on Monte Verde are Dillehay (1989, 1997), Meltzer et al. (1997), and Dillehay et al. (2008). For discussions of the Monte Verde site visit and its possible impact, see Grayson (2004), Meltzer (2009), and Adovasio and Page (2002). The last two of these references provide an accurate discussion of the inner dynamics, and at times combativeness, of the Monte Verde site visit mentioned in the text.

I have depended heavily on Cressman's 1988 autobiography; read it not only for the discussion of his archaeological work and the founding of the Department of Anthropology and the Museum of Natural History at the University of Oregon, but also for a lengthy discussion of his first marriage to famed anthropologist Margaret Mead. On the early work at the Paisley Caves and Fort Rock Cave, see Cressman, Williams, and Krieger (1940) and Cressman (1942, 1977). P. Si. Martin, Quimby, and Collier (1947) rejected Cressman's arguments concerning the deep antiquity of these sites. The date I use for the climactic eruption of Mount Mazama is from Hallett, Hills, and Clague (1997) and Zdanowicz, Zielinski, and Germani (1999); this latter reference also provides details on the magnitude of the eruption. Cressman (1951) announced the 9,053-year radiocarbon date for a Fort Rock sandal. Bedwell (1973) discusses the results of his excavations in Fort Rock Cave; see also Bedwell and Cressman (1971). Cressman (1937) presents his early survey of Oregon petroglyphs; his Guano Valley work is discussed in Cressman (1936). C. V. Haynes (1971) was the first to question the 13,200-year date from Fort Rock Cave. J. D. Jennings (1957:285) discusses the support provided by the early Danger Cave dates for Cressman's arguments; Danger Cave is discussed in more detail in later chapters. Cressman repeated his claim for an association between people and extinct mammals at the Paisley Caves in many places, including Cressman (1964, 1966, 1977, 1986). Rejections have been provided by, among others, Baumhoff and Heizer (1965), Heizer and Baumhoff (1970), Grayson (1982, 2006a), J. D. Jennings (1986), and Huckleberry et al. (2001).

The prime recent references on the Paisley Caves are D. L. Jenkins (2007) and M. T. P. Gilbert et al. (2008). Goldberg, Berna, and Macphail (2009) and Poinar et al. (2009) have raised questions about the site, but these have been addressed by M. T. P. Gilbert et al. (2009) and Rasmussen et al. (2009).

CHAPTER FOUR

The End of the North American Pleistocene: Extinct Mammals and Early Peoples

The end of the Pleistocene in North America was a time of remarkable change. Throughout much of this region, plant communities were reorganized as separate species of plants responded in their own individual ways to new climatic conditions. Huge glacial lakes that had formed in the Great Basin desiccated, some apparently with great speed. A substantial set of large and, from our modern perspective, exotic, mammals became extinct. In addition, people spread throughout a landscape that had apparently been occupied by only scattered human groups at best. I explore the late Pleistocene lakes and plants of the Great Basin in subsequent chapters, but the mammals and people require a broader perspective.

The Mammals

The episode of extinction that occurred in North America as the Pleistocene came to an end was astonishing in its breadth and depth, encompassing the loss of thirty-six genera of mammals, either in the sense that they became globally extinct at that time (thirty genera) or in the sense that they became extinct in North America while living on elsewhere (six genera; see table 4-1; the Pleistocene mammals of the Great Basin are discussed in chapter 6). Some of the mammals are well known to most Americans: the mammoth and sabertooth cat, for instance. Others, however, remain known only to scientists who specialize in this time period, even though these animals are no less fascinating. Many of these mammals lived in the Great Basin toward the end of the Pleistocene, and there is little doubt that the earliest human occupants of the Great Basin encountered at least some them.

Cingulates (Pampatheres and Glyptodont)

Until fairly recently, the armadillos, anteaters, and sloths were all placed in an order of mammals called the Edentata, or edentates. Then, for technical reasons, this named was changed to the Xenarthra. Next, it was decided that the armadillos on the one hand, and the sloths and anteaters on the other, were different enough from each other that they each deserved to be in their own order. Hence, we now have the Cingulata for the armadillos and the Pilosa for the sloths and anteaters. While these two groups are closely related, one of the differences is that the cingulates have external body armor (think "armadillo") while members of the Pilosa do not (think "sloth"). All of them are South American in origin.

Today, North America supports only one species of cingulate—the nine-banded armadillo, *Dasypus novemcinctus,* and even it may not have arrived here until the mid-nineteenth century. During the Pleistocene, however, southeastern North America had not only a now-extinct species of large armadillo—called *Dasypus bellus,* the beautiful armadillo—but also two genera of cingulates of enormous size.

Pampatheres were armadillo-like in that they were enclosed in a flexible armor of bony scutes, but they differed sufficiently from armadillos to be placed in their own family. There are two genera of later Pleistocene pampatheres that may have occurred in North America, but only one is securely known from here. The northern pamapathere, *Holmesina septentrionalis,* was some 6 feet long and 3 feet high and has been found in sites ranging from Kansas and North Carolina in the north to Texas and Florida in the south, with the greatest number of sites known from Florida. The southern pampathere, *Pampatherium,* has been reported from only a single site in Texas but this material may, in fact, be from *Holmesina*.

Simpson's glyptodont, *Glypotherium floridanum,* is known primarily from near-coastal localities in Texas, Florida, and South Carolina. This animal had a turtle-like carapace, an armored tail and skull, massive limbs, and a pelvic girdle fused to its shell. At some 10 feet long and 5 feet tall, it was

TABLE 4-1
The Extinct Late Pleistocene Mammals of North America
Genera marked with an asterisk live on elsewhere; those in bold
are also known from Eurasia.

Order/Family	Genus	Common Name
Cingulata		
Pampatheriidae	*Pampatherium*[a]	Southern Pampathere
	Holmesina	Northern Pampathere
Glyptodontidae	*Glyptotherium*	Simpson's Glyptodont
Pilosa		
Megalonychidae	*Megalonyx*	Jefferson's Ground Sloth
Megatheriidae	*Eremotherium*	Rusconi's Ground Sloth
	Nothrotheriops	Shasta Ground Sloth
Mylodontidae	*Paramylodon*	Harlan's Ground Sloth
Carnivora		
Mustelidae	*Brachyprotoma*	Short-faced Skunk
Canidae	***Cuon*** *	Dhole
Ursidae	*Tremarctos**	Florida Cave Bear
	Arctodus	Giant Short-faced Bear
Felidae	*Smilodon*	Sabertooth
	Homotherium	Scimitar Cat
	Miracinonyx	American Cheetah
Rodentia		
Castoridae	*Castoroides*	Giant Beaver
Caviidae	*Hydrochoerus**	Holmes's Capybara
	Neochoerus	Pinckney's Capybara
Lagomorpha		
Leporidae	*Aztlanolagus*	Aztlan Rabbit
Perissodactyla		
Equidae	***Equus*** *	Horses
Tapiridae	***Tapirus*** *	Tapirs
Artiodactyla		
Tayassuidae	*Mylohyus*	Long-Nosed Peccary
	Platygonus	Flat-Headed Peccary
Camelidae	*Camelops*	Yesterday's Camel
	Hemiauchenia	Large-Headed Llama
	Palaeolama	Stout-Legged Llama
Cervidae	*Navahoceros*	Mountain Deer
	Cervalces	Stag-Moose
Antilocapridae	*Capromeryx*	Diminutive Pronghorn
	Tetrameryx	Shuler's Pronghorn
	Stockoceros	Pronghorns
Bovidae	***Saiga*** *	Saiga
	Euceratherium	Shrub Ox
	Bootherium	Harlan's Muskox
Proboscidea		
Gomphotheriidae	*Cuvieronius*	Cuvier's Gomphothere
Mammutidae	*Mammut*	American Mastodon
Elephantidae	***Mammuthus***	Mammoths

SOURCE: After Grayson 2007.

[a] The single record for this genus may actually pertain to *Holmesina* and should be reanalyzed (Grayson 2006d).

roughly the shape and height of a Volkswagen Beetle, though it was a few feet shorter than its metallic counterpart. From the settings in which its remains have been found, Simpson's glyptodont appears to have lived along lakes, streams, and marshes and may have been semiaquatic.

Pilosa (Sloths)

Modern ground sloths are arboreal; the fattest ones may weigh twenty pounds. The four genera of extinct ground sloths of North America were something quite different. As a group, these animals were found from Florida in the Southeast to Alaska in the Northwest, but not all of them were so widely distributed.

The largest, *Eremotherium laurillardi*, combined the height of a giraffe with the bulk of an elephant and is known from coastal plain environments from New Jersey south through Florida and west into Texas. The Shasta ground sloth, *Nothrotheriops shastensis*, which weighed about 330 pounds, was the smallest of the North American sloths and is known only from western North America. Aspects of the ecology of this sloth are well understood since its baseball-sized dung balls have been found in dry caves of the American Southwest. Analyses of that material have revealed the remains of parasites, DNA from the plants the animals ate, and bits and pieces of the plants themselves. As a result, we know that while the sloth has been gone for some 10,000 years, the plants that it consumed remain common in the general region. Jefferson's ground sloth, *Megalonyx jeffersoni*, was the most widespread of the North American ground sloths, distributed from Florida to Alaska (the far northwestern specimens appear to date to the last interglacial). This was the first of the North American sloths to be described, by Thomas Jefferson, who concluded from its massive claws ("megalonyx") that it was a carnivore, a conclusion that was soon corrected. Harlan's ground sloth, *Paramylodon harlani* (placed by some in the genus *Glossotherium*), was also widespread in North America, found from coast to coast. Distinguished in part by the small bones embedded in its skin, this was the most abundant sloth at Southern California's Rancho La Brea tar pits.

Carnivores

The dhole (*Cuon alpinus*) is a pack-hunting, highly carnivorous member of the dog family that is widespread (but dwindling) in Asia. During the Pleistocene, it was found from southwestern Europe to Alaska and the Yukon, with a single additional site known from northern Mexico. It must have lived somewhere in between, perhaps even in the Great Basin, but geographically intermediate specimens have yet to turn up.

Dire wolves (*Canis dirus*), on the other hand, were broadly distributed in North America. These large canids (they are estimated to have weighed 130 pounds or so—the size of a large gray wolf) seem to have been pack hunters capable of taking prey that weighed well in excess of 1,000 pounds. Dire wolves do not appear on table 4-1, because the genus to which they belong includes the wolves and coyotes (and domestic dogs), so the genus still exists in North America.

Since jaguars (*Panthera onca*) exist in North America as well, table 4-1 also excludes the giant American lion, *Panthera leo* (or *Panthera atrox*, depending on the author). Huge lions were not uncommon in North American open environments during the Pleistocene. Weighing some 950 pounds, they were the largest cat North America had to offer; their tracks, known from a Missouri cave, are significantly broader than those of the African lion, *Panthera leo*, which weighs about 400 pounds.

They were not, however, the only huge member of the cat family to be found here. The famous saber-toothed cat, *Smilodon fatalis*, weighed an estimated 850 pounds and ranged from coast to coast. The less well-known scimitar cat, *Homotherium serum* (found from Alaska to Texas and Florida), weighed in at 400 pounds, and the American "cheetah," *Miracinonyx trumani*, now known to be closely related to the cougar (*Puma concolor*), weighed an estimated 150 pounds.

The largest late Pleistocene carnivore, however, was the giant short-faced bear, *Arctodus simus*. Large males averaged 1,500 to 1,800 pounds, and the biggest of them may have reached 2,000 pounds. Found from Pennsylvania to California, south into Mexico, and north to Alaska and the Yukon, these long-limbed bears were highly carnivorous, with the meat they consumed perhaps obtained largely by scavenging. Some 40 percent of *Arctodus simus* sites in the United States are located in caves, and the individuals from those caves tend to be smaller than those found in open settings. This suggests that the smaller female bears used caves as den sites and that their remains represent animals that died while denning.

The second extinct genus of North American bear, *Tremarctos*, continues to exist today in the form of the spectacled bear, *T. ornatus*, which occupies the mountains of northwestern South America. The North American Pleistocene form, however, was more powerfully built and larger, with an estimated weight of 300 pounds. Known from Texas to South Carolina and Florida, it appears to have been omnivorous.

Not all of the now-extinct North American Pleistocene carnivores were huge. The short-faced skunk (*Brachyprotoma*), whose later Pleistocene distribution included eastern North America, the Yukon, and the Great Basin, was roughly equivalent in size to a spotted skunk (*Spilogale*), the genus to which it appears to be most closely related.

Rodents

North America did not lose many rodents toward the end of the Pleistocene, but the ones that were lost were impressive. The giant beaver, *Castoroides ohioensis*, ranged from New York to Florida in the East and to Alaska in the Far Northwest but appears to have been most common in the Great Lakes region. Giant beavers had incisors one inch across, but these teeth had rounded and blunt tips, and this and other aspects of their anatomy shows that they were not dam builders, even though their preferred habitat appears to have been lakes, ponds, marshes, and swamps. Recent analyses suggest that they weighed some 170 pounds, less giant than was once thought. However, since these animals were the size of black bears, this estimate may be incorrect, and others suggest the weight to have been closer to 350 pounds.

The two North American capybaras—southeastern in distribution—are both related to the world's largest living rodent, *Hydrochoerus hydrochaeris* of central and northern South America, which weighs about 130 pounds. The extinct *Hydrochoerus holmesi* was slightly larger than this, but *Neochoerus pinckneyi* was perhaps half again larger.

Lagomorphs

Only one genus of North American lagomorph (the order that includes the rabbits, hares, and pikas) was lost toward the end of the Pleistocene. This was the small Aztlan rabbit (*Aztlanologus agilis*), known from eastern Nevada through New Mexico and Texas south into central Mexico. The few reasonably secure radiocarbon dates available for it suggest that it may not have survived the Last Glacial Maximum, some 18,000 years ago.

Perissodactyls (Odd-Toed Ungulates)

After evolving in the New World and crossing into the Old via the Bering Land Bridge, horses became extinct in North America at the end of the Pleistocene; the wild horses of the West are recent introductions. To judge from the number of sites in which they have been found, horses were especially common in western North America, including the Far North. The number of species represented by these horses is unknown, and even experts routinely identify them only to the genus level. Tapirs, of at least two species, were also reasonably common, known from Pennsylvania to California, with the largest number of sites found in eastern North America.

Artiodactyls (Even-Toed Ungulates)

North America lost thirteen genera of artiodactyls, from peccaries and camels to muskoxen and pronghorn, a variety almost bewildering in scope.

Today, the collared peccary (*Tayassu tajacu*) ranges north into the southwestern United States and Texas, but it appears to be a very recent arrival, and peccaries were far more widespread in North America during the late Pleistocene. The long-nosed peccary, *Mylohyus nasutus,* apparently flourished in wooded environments and was primarily eastern in distribution, though specimens are known from as far west as western Texas. The flat-headed peccary, *Platygonus compressus*, on the other hand, was distributed from coast to coast and seemed to prefer more open settings; unlike *Mylohyus*, it also seems to have been gregarious, to judge from the discoveries of "fossil herds."

Like horses, camels are New World natives, but, unlike horses, some—the llamas (*Lama* and *Vicugna*)—survived here. Toward the end of the Pleistocene, North America supported three members of the camel family. *Camelops hesternus* looked something like a longer-legged, narrower-headed version of the dromedary (*Camelus dromedarius*) and was very common in the western half of North America. To judge from aspects of the skull, this camel fed on a diet derived from both grazing and browsing, a view that matches food remains plucked from the teeth of individuals at Rancho La Brea. The large-headed llama *Hemiauchenia macrocephala* was widespread in the southern half of North America, with sites known as far north as Iowa and Idaho. Also long-limbed, this camelid seems to have been a mixed feeder with a preference for browse. The stout-legged llama *Palaeolama mirifica* has been reported from South Carolina and Florida west to southern California, but most records are from the Southeast. As the common name suggests, it had relatively short and robust limbs, perhaps an evolutionary response to the need to escape predators in closed to semiclosed habitats. Analyses of *Palaeolama mirifica* dentition, and of stable isotopes from that dentition, suggest that it was a browser.

The large deer *Navahoceros fricki* had short, robust limbs and fairly simple antlers and was found from southeastern California to the plains. Detailed analyses of its skull suggest that its closest living relatives are reindeer and caribou, which belong to the genus *Rangifer*. The "elk-moose" or "stag-moose," *Cervalces scotti,* is far better known, with sites known from southern Canada and the eastern and central United States; specimens that may or may not belong to the same species are known from the Yukon and Alaska. The contexts in which it has been found suggest that this animal may have been similar to the moose (*Alces alces*) in habitat preference, to which species it was also similar in size.

There is only a single species of Antilocapridae in North America today, the pronghorn, *Antilocapra americana*. These speedy animals have two laterally compressed horn cores, each covered by a sheath with a forward-projecting prong. During the late Pleistocene, this animal existed alongside three other genera of the same family, all of which were characterized by the fact that they had four, rather than two, horns. The smallest of these was the aptly named diminutive pronghorn, *Capromeryx minor*, found from California to Texas. Some 1.5 feet tall at the shoulder, estimates for the weight of this long-limbed antilocaprid range from twenty to thirty-five pounds. The other two extinct antilocaprids were closer to the pronghorn in size; *Stockoceros* is known from late Pleistocene contexts in the Southwest and Texas, while *Tetrameryx* has been reported from Texas, New Mexico, and southern Nevada.

Three genera of mammals related to cattle (family Bovidae) became extinct in North America toward the end of the Pleistocene. Of these, one, the saiga, *Saiga tatarica*, today lives on in the arid steppes of Eurasia. During the later Pleistocene, however, it was to be found as far southwest as northern Spain and as far northeast as Alaska, the Yukon, and the Northwest Territories. The helmeted muskox, *Bootherium bombifrons*, is known from much of unglaciated North America; only the far Southeast and far Southwest appear to have been without it. Compared to the extant muskox, *Ovibos moschatus*, the helmeted muskox had longer legs and stood taller, but was shorter from head to tail and had shorter pelage (known from preserved material from the Far North). Dale Guthrie has noted that the longer legs of the extinct form imply that they were likely far more mobile than the species familiar to us today. The horn cores of male helmeted muskoxen fused in the midline, unlike those of *Ovibos*, and it is this feature that provides the common name. Those of the female differed so considerably that males and females were at one time assigned to separate genera. The shrub-ox, *Euceratherium collinum*, is characterized by horn cores that arise from near the rear edge of the frontal bones, sweep up and back, then out and forward, curling upward near their tips. Their skulls and neck vertebrae imply that, like mountain sheep (*Ovis canadensis*), they were head-butters. These animals seem to have occupied foothills or low mountainous terrain; their remains have been found from California to Iowa and south into Mexico.

The mountain goat *Oreamnos americanus* is doing well in northwestern North America, but Harrington's mountain goat, *Oreamnos harringtoni*, failed to survive the end of the Pleistocene. Sites that have provided the remains of this animal are known from the central Great Basin south through the Colorado Plateau into northern Mexico and as far east as southeastern New Mexico. The greatest amount of material, however, has come from sites in the Grand Canyon region. Harrington's mountain goat was about 30 percent smaller than its modern counterpart and had a narrower face with thinner and smaller horns, robust feet suitable for negotiating rough terrain, and a mixed diet, incorporating plants ranging from grasses to spruce (*Picea* sp.). It is even known that while the animal was smaller than its modern counterpart, its dung was larger (perhaps due to the nature of its diet), and that it had white hair.

Proboscideans

Gomphotheres are related to elephants and, among the extinct forms, to mastodons and mammoths, through they are far closer to the former than the latter. They differ from mastodons in the form of their tusks and in the greater complexity of the cusps of their molar teeth. Primarily central and south American in distribution during the late Pleistocene (their remains were found at Monte Verde), gomphotheres were also found, according to recent work in northwestern Mexico by geoarchaeologist Vance Holiday, on at least the southern edge of North America during the very late Pleistocene.

Along with the saber-toothed cat, mammoths and the American mastodon are the iconic large mammals of the North American Ice Age. The mastodon (*Mammut americanum*) was widespread in unglaciated North America, found from coast to coast and from Alaska into Mexico, but was particularly abundant in the woodlands and forests of eastern North America; on the Wasatch Plateau, they have been found at elevations as high as 9,800 feet. As paleontologist Jeff Saunders (1996:274) has noted, these were "low, long, and stocky" animals; they had shoulder heights of 6.5 feet to 10 feet and weighed an estimated 6,000 pounds. They also had distinctive cheek teeth with large cusps arranged in pairs to form ridges that ran at right angles to the main axis of the tooth—appropriate to a browsing diet, though at least some individuals ate significant amounts of grass. The fact that their remains are frequently found as single individuals has suggested to some that they were fairly solitary, but others argue that they did form social groups.

Disagreement continues over the number of species of full-sized mammoths that occupied North America during the late Pleistocene. Some experts recognize three (Columbian mammoth, *Mammuthus columbi*; Jefferson's mammoth, *Mammuthus jeffersoni*; and woolly mammoth, *Mammuthus primigenius*), while others treat Columbian and Jefferson's mammoths as belonging to *M. columbi*. I follow the latter approach. There is also a diminutive form, the pygmy mammoth, *M. exilis*, known from the Channel Islands off the coast of California. Derived from *M. columbi*, these animals had shoulder heights that varied from 4 feet to 8 feet, compared to shoulder heights that ranged from 7.5 feet (woolly mammoth females) to 13 feet (Columbian mammoth males) for the other species.

The famed woolly mammoth was the northern form, common in eastern Beringia and the upper Midwest and adjacent Canada and, to a lesser extent, the northeastern United States. The Columbian mammoth was found coast to coast, but was the common form west of the Mississippi River, known from elevations as high as 9,000 feet (figure 4-1). Mammoth cheek teeth were flat and high-crowned, each composed of a series of enamel-bordered plates running at right angles to the main axis of the tooth, forming an efficient grinding device. The analysis of dried Columbian mammoth dung in southwestern caves shows that mammoths' diet in this area was dominated by grasses, sedges, and reeds, though they ate woody plants as well. A similar combination of browse and graze is represented in the remains of well-preserved *Mammuthus primigenius* innards from Siberia.

Some Other Things

The mammals were not alone in suffering significant extinctions. Some twenty genera of birds were also lost during the North American late Pleistocene. While nine of

FIGURE 4-1 Columbian mammoth under excavation by the Desert Research Institute at the Delong Mammoth site, Black Rock Desert, Nevada.

these were predators or scavengers whose extinction may have been driven by the loss of large mammals, the others ranged from storks and flamingos to shelducks and jays. There was even a species of tree—the spruce *Picea critchfieldii*—that disappeared as the North American Pleistocene ended.

At the same time as these extinctions were occurring, other animals were undergoing often-massive distributional changes. Caribou (*Rangifer tarandus*), for instance, are known from as far south as northern Mississippi during the late Pleistocene, and muskoxen (*Ovibos moschatus*) from as far south as Tennessee. Very small mammals, which underwent no extinctions, also moved dramatically across space as the Pleistocene ended. For instance, late Pleistocene Tennessee was home to both the taiga vole, *Microtus xanthognathus*, and the heather vole, *Phenacomys intermedius*. Today, the taiga vole is found no farther south than central Alberta, and the heather vole no farther south than the Canadian border in eastern North America.

Although there is an obvious question here—what accounts for all of this change?—there is no obvious answer, in spite of nearly two centuries of searching for a satisfying explanation. Getting that answer is going to require that we know quite precisely when the animals were lost.

Timing of the North American Extinctions

We can be reasonably certain that the extinctions were nearly over by 10,000 years ago. Of the thousands of well-dated archaeological and paleontological sites known from throughout North America that are younger than 10,000 years old, none contain the bones or teeth of any of the extinct mammals, with rare and detectable exceptions caused by the incorporation of older bone into younger deposits. The lack of skeletal remains of these animals in deposits dating to the past 10,000 years shows that the extinctions had either occurred by then or that the animals had diminished in number to the point that their skeletal signatures had become archaeologically and paleontologically invisible. The latter possibility is a very real one, since detecting the last populations of any species that became extinct long ago poses a substantial problem, requiring a lot of looking and a lot of luck. In fact, the recent discovery of mammoth and horse DNA in frozen sediments from central Alaska that date to between 9,200 and 7,000 years ago strongly suggests that at least some species in some places made it well into the Holocene.

We do know that of the thirty-six genera that became extinct, populations belonging to at least sixteen of them survived in some parts of North America until between 12,000 and 10,000 years ago. We can be sure of that because we have radiocarbon dates that document this to have been the case (table 4-2).

If sixteen made it this late, what are the chances that most or all of the rest did as well? A statistical analysis of the radiocarbon chronology of the North American extinctions by archaeologists Tyler Faith and Todd Surovell shows that nearly all of the genera involved may have breathed their last just as the Pleistocene was coming to an end. Even if this were the case, however, it remains very possible that these populations had begun to dwindle long before this time.

This is an important issue, since a set of extinctions that occurred rapidly, with no forewarning, requires a different kind of explanation than one that took place over thousands of years. We are only beginning to have

TABLE 4-2
Extinct Late Pleistocene Mammalian Genera with Radiocarbon Dates of Less Than 12,000 Radiocarbon Years and Illustrative Sites

Genus	Illustrative Site	References
Arctodus	Sheridan Cave, OH	Tankersley 1997; Bills and McDonald 1998; Tankersley, Redmond, and Grove 2001
Bootherium	Wally's Beach, AB	Kooyman et al. 2001
Camelops	Casper, WY	Frison 2000
Castoroides	Dutchess Quarry Caves, NY	Steadman, Stafford, and Funk 1997
Cervalces	Kendallville, IN	Farlow and McClain 1996
Equus	Rancho La Brea, CA	Marcus and Berger 1984
Euceratherium	Falcon Hill, NV	Dansie and Jerrems 2005; Kropf, Mead, and Anderson 2007
Mammut	Pleasant Lake, MI	D. C. Fisher 1984
Mammuthus	Dent, CO	Stafford et al. 1991
Megalonyx	Little River Rapids, FL	Muniz 1998; S. D. Webb, Hemmings, and Muniz 1998
Mylohyus	Sheriden Cave, OH	Redmond and Tankersley 2005
Nothrotheriops	Muav Caves, AZ	A. Long and Martin 1974; J. I. Mead and Agenbroad 1992
Palaeolama	Woody Long, MO	R. W. Graham 1992, pers. comm.
Platygonus	Sheriden Cave, OH	Tankersley 1997; Bills and McDonald 1998; Tankersley, Redmond, and Grove 2001
Smilodon	Rancho La Brea, CA	Marcus and Berger 1984
Tapirus	Lehner, AZ	Haury, Sayles, and Wasley 1959; C. V. Haynes 1992

the tools to address this issue in a compelling way. The most important of these tools is currently provided by, of all things, the dung-loving genus of fungus known as *Sporormiella*. The spores of this organism are found almost (but not quite) exclusively on the dung of herbivores. Many years ago, paleobotanist Owen Davis showed that the greater the number of large herbivores on the landscape, the greater the number of *Sporormiella* spores end up in the sediments that accumulate in lakes, ponds, and other wetlands. He showed not only that their abundance in such sediments spiked after Europeans introduced livestock to North America but also that their abundance declined as the Pleistocene came to an end—clearly a marker of the late Pleistocene extinctions.

This means that a late Pleistocene history of *Sporormiella* abundance probably also tells us about the abundance of large herbivores on the landscape. Jacquelyn Gill and her colleagues applied this logic to a lake in Indiana and showed that the abundance of dung fungus spores began to decline in this lake sometime prior to 12,600 years ago, hitting bottom at around 10,500 years ago (though the chronology at this point is not very strong). Similar analyses from the Hudson River Valley have similar implications: the extinctions appear to have been less of an event than a process, one that might have been spread out over several thousand years.

In addition to the possibility that at least some of the extinctions were spread over thousands of years, it is also quite likely that different animals were lost at different times at different places. In fact, this is exactly what happened in Eurasia.

A Note on Eurasian Extinctions

Although fewer genera were lost in Europe and northern Asia (table 4-3), the end of the Eurasian Pleistocene was also a time of mammalian extinction and of massive changes in mammal distributions, as can be seen by looking at almost any area within this vast region. Table 4-4 lists the larger mammals known from the late Pleistocene of southwestern France, many of which were as large as those lost in North America. Of these twenty-one mammals, ten either are extinct or no longer exist in this part of the world. The closest reindeer are in Scandinavia, the closest saiga in central Eurasia, and the closest muskoxen in North America.

As in North America, range adjustments and extinctions among large mammals were accompanied by huge range changes among small mammals toward the end of the Pleistocene. The arctic fox, *Alopex lagopus*, is common in the late Pleistocene faunas of southwestern France, but today the closest populations are in Scandinavia. The narrow-headed vole *Microtus gregalis* is known from a substantial number of late Pleistocene sites in this region, but is now found no closer than Siberia and China. Much the same is true for the British Isles, Italy, and elsewhere.

At one time it was assumed that most or all of the Eurasian extinctions occurred synchronously at the end of the

TABLE 4-3
The Extinct Late Pleistocene Mammals of Northern Eurasia
Genera marked with an asterisk live on elsewhere; those in bold are also known from North America. *Bison priscus* is ancestral to *Bison bonasus*.

Order/Family	Genus	Common Name
Carnivora		
Hyaenidae	*Crocuta**	Spotted Hyena
Felidae	***Homotherium***	Scimitar Cat
Perissodactyla		
Rhinocerotidae	*Coelodonta*	Wooly Rhinoceros
	Stephanorhinus	Rhinoceros
Artiodactyla		
Hippopotamidae	*Hippopotamus**	Hippopotamus
Cervidae	*Megaloceros*	Giant Deer
	Cervalces	Stag-Moose
Bovidae	*Spirocerus*	Spiral-horned Antelope
	*Ovibos**	Muskox
Proboscidea		
Elephantidae	*Palaeoloxodon*	Straight-Tusked Elephant
	Mammuthus	Mammoths
Species level losses		
	Panthera leo (*spelaea*)	Lion
	Ursus spelaeus	Cave Bear
	*Bison priscus**	Extinct Bison

SOURCES: Koch and Barnosky 2006; Reumer et al. 2003; Stuart 1991, 1999; Vereshchagin and Baryshnikov 1984.

Pleistocene, but we now know that this was not the case. Mammoths seem to have been lost from significant parts of Europe at or soon after 12,000 years ago (and perhaps much earlier from Iberia), but to have lasted until 10,000 years ago in Estonia, until shortly after 10,000 years ago on mainland northern Siberia and on the northern Russian Plain, and until shortly after 4,000 years ago on Wrangel Island in the Arctic Ocean (figure 3-1). Similarly, the giant deer *Megaloceros giganteus* seems to have disappeared from southwestern France between 12,000 and 11,000 years ago, but from the Urals and western Siberia after 7,000 years ago. The muskox seems to have disappeared from southwestern France at around 19,000 years BP, but it survived in northern Siberia until at least 2,700 years ago and perhaps well beyond that.

That is, in Europe and northern Asia, different taxa were lost at different times at different places toward the end of the Pleistocene. The same phenomenon can be seen at single places as long as sufficient work has been done to construct a reliable chronology. In southwestern France, for instance, muskoxen were gone by 19,000 years ago, wooly rhinos by about 15,000, mammoth shortly after 12,000, and reindeer by about 11,000 (see figure 4-2). Everyone agrees that the Eurasian losses were staggered in time and space.

Back to North America

In short, all we can be sure of at the moment is that at least sixteen of the genera became extinct after 12,000 years ago and that the extinctions were over by 10,000 years ago. Even though this is the case, most scientists assume that all of the North American extinctions occurred within this narrow time frame. We will see why this is so shortly.

The People

As I discussed in Chapter 3, there is no question that people were in the Americas well prior to 12,000 years ago. This does not change the fact that you have to look far and wide to find archaeological sites that are, or might be, more than 11,200 years old. Beginning about 11,200 years ago, however, you have to look far and wide to find a significant patch of habitable ground that lacks traces of human occupation.

This change in archaeological visibility at 11,200 years ago or so may be real, or it may simply reflect tricks played on us by the archaeological record. Since the same change occurs in the Great Basin—one site documented prior to about 11,200 years ago (the Paisley Caves), but dozens soon after that time—understanding these sites

TABLE 4-4

The Late Pleistocene (<40,000 Years BP) Larger Mammals of Southwestern France

Extinct taxa indicated by †; late Pleistocene extirpations in bold. *Bison priscus* is ancestral to *B. bonasus*; *Capra ibex* and *C. pyrenaica* are not routinely distinguishable from fragmentary material.

Order/Family	Genus/Species	Common Name
Carnivora		
Canidae	*Canis lupus*	Wolf
Felidae	***Panthera spelaea*†**	Cave Lion
Hyaenidae	***Crocuta crocuta spelaea*†**	Spotted Hyena
Ursidae	*Ursus arctos*	Brown Bear
	***Ursus spelaeus*†**	Cave Bear
Perissodactyla		
Equidae	*Equus caballus*	Horse
	***Equus hydruntinus*†**	Wild Ass
Rhinocerotidae	***Coelodonta antiquitatis*†**	Woolly Rhinoceros
Artiodactyla		
Suidae	*Sus scrofa*	Wild Boar
Cervidae	*Alces* sp.	Moose
	Capreolus capreolus	Roe Deer
	Cervus elaphus	Red Deer
	***Megaloceros giganteus*†**	Giant Deer
Bovidae	***Bison priscus*†/*B. bonasus***	Bison
	Bos primigenius	Aurochs
	Capra ibex/*C. pyrenaica*	Ibex
	Ovibos moschatus	Muskox
	Rangifer tarandus	Reindeer
	Rupicapra rupicapra	Chamois
	Saiga tatarica	Saiga
Proboscidea		
Elephantidae	***Mammuthus primigenius*†**	Woolly Mammoth

SOURCES: Delpech 1983, 1999; Grayson and Delpech 2006; terminology follows Guérin and Patou-Mathis 1996.

requires that we understand, as well, something about the archaeology of western North America as a whole during this time.

Clovis and Some Clovis Sites

In 1952, Ed Lehner was thinking about buying a ranch in the San Pedro Valley in Cochise County, in far southeastern Arizona. Walking across property soon to be his, he noticed bone protruding from the base of the steep bank of a dry streambed, some eight feet beneath the surface. In the first of many moves that were to give him a place in the history of North American archaeology, Lehner removed some of the bones and brought them to the Arizona State Museum, where they were identified as mammoth. Archaeologist Emil Haury of the University of Arizona soon visited the site and realized that although no artifacts were visible in the stream bank, the age of the deposits was such that they might well be there. Haury and Lehner decided to keep a close eye on what was then simply an interesting paleontological find.

Three years later, Lehner got in touch with them again: heavy summer rains had exposed much more of what was there. Haury returned and decided it was worth a closer look.

Excavations began in late November 1955. Soon after the first dirt had been moved, the first artifact appeared. Indeed, the Lehner site, as it came to be called, proved to be so important that work here, work on the collections the site provided, and work on similar sites subsequently discovered in the area have continued ever since. Thanks ultimately to Ed Lehner, we have learned a tremendous amount about the late Pleistocene prehistory of the Southwest. Thanks to the generosity of both Ed and Lyn Lehner, the site itself is now part of the San Pedro Riparian National Conservation Area and is strictly monitored by the Bureau of Land Management.

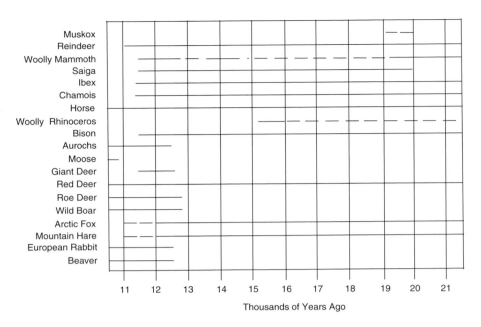

FIGURE 4-2 The staggered losses of late Pleistocene mammals in southwestern France (after Delpech 1999 and personal communication). Dashed lines indicate rare species.

The deposits at the Lehner Site are fairly simple. At the base of a deep set of Holocene sediments is a dark organic deposit called the black mat. Black mats are widespread in arid North America; work by geologist Jay Quade has shown that they are highly organic deposits that form during relatively wet times when spring discharge and water tables are high. The base of the black mat at Lehner has been dated to about 10,800 years ago, and the artifacts at the Lehner site lie directly beneath it. When Haury worked his way through the black deposits, he came on the decaying remains of nine mammoths and a single bison; later work at the site has brought the total number of mammoths here to thirteen.

Haury also discovered what he interpreted to be two hearths, along with the mandible of a mammoth and bones from both tapir and horse. Later work by Vance Haynes has suggested that these hearths may be natural stains and not archaeological at all, but that does not affect the main part of the site, since along with the mammoths, Haury found twenty-one stone artifacts. Of these, eight appeared to be simple cutting tools, perhaps used to butcher one or more of the mammoths. The other thirteen were projectile points, eleven of which were found in the immediate vicinity of mammoth or bison bone.

Because of the care with which Haury worked, and because of the research that has been done here since that time, the setting for what appears to represent a series of ancient mammoth kills is fairly well understood. At the time the site was forming, this area was a sand- and gravel-floored streambed bordered on the south by a nearly vertical clay wall about six feet high. The nature of the north wall is unknown, but it probably looked much the same.

Analysis of pollen taken from dirt samples along the arroyo has shown that the area surrounding the ancient stream—appropriately called Mammoth Kill Creek—was desert grassland, slightly wetter and cooler than today but not dramatically different. Haury suggested that mammoths probably walked up the streambed to reach a watering spot—the Lehner site—where they were attacked and killed by human hunters. Haury also argued that these animals had not been killed all at once but represented instead a series of kills that had extended over some fairly short period of time. Later work has done nothing to suggest otherwise. It is possible that a combination of killing and scavenging was involved in the formation of this site, but its age is well understood. Lehner dates to 10,940 years ago.

The projectile points from Lehner are truly distinctive. They range in size from about one to four inches long, but it is their form that makes them distinctive. They are all lanceolate, with slightly concave bases, and nearly all have thin flakes removed from both sides, producing a very characteristic fluting that begins at the base and extends less than halfway toward the tip. The fluting appears to have aided in hafting the point to a spear or to a foreshaft that fit into a spear. The edges just above the bases of the points are dulled by grinding; this was probably done either to prevent the sinew that held the point in place from being cut or to remove irregularities in the point that would otherwise have hastened breakage.

Lehner is not the only site in the San Pedro Valley to have provided fluted points of this distinctive style. Indeed, when Haury first visited Lehner, he was fresh from his work at the Naco site, twelve miles southeast of Lehner and just north of

FIGURE 4-3 Clovis points from the Naco site (the longest measures 4.6 inches). Photograph by E. B. Sayles, courtesy of the Arizona State Museum, Tucson.

the Mexican border. There he had found the remains of a single mammoth that had fallen on a sandbar adjoining a streambed. Among the bones of this mammoth were eight fluted points virtually identical to those from Lehner (figure 4-3). The Naco mammoth may have been one that got away, since there are no artifacts here other than the points.

Not far from these two sites, and still in San Pedro Valley, are Murray Springs and Escapule, both of which have mammoths and the same kind of fluted points. The better studied of these sites is Murray Springs, thanks to exacting excavations and a substantial monograph by Vance Haynes. Murray Springs provided the remains of a single mammoth that was likely killed by Clovis people (the associated artifacts are not distinctive enough to be sure who did the killing). In addition, it yielded the remains of eleven bison associated with a series of stone and bone tools, including Clovis points, all dated to almost exactly the same time as Lehner: 10,900 years ago. Amazingly enough, Murray Springs also includes well-preserved mammoth tracks in the area of the one secure mammoth kill. As at Lehner, all this was found immediately beneath the late Pleistocene San Pedro Valley black mat.

Farther afield, there is the Miami site in Texas, Dent in Colorado, Domebo in Oklahoma, Blackwater Draw in New Mexico, Colby in Wyoming, Lange/Ferguson in South Dakota, and a number of others, all of which have both mammoths and the points. As we saw at Murray Springs, however, not all sites with the points and large mammals have mammoth as that mammal. The Sheaman site in Wyoming substitutes bison; the Kimmswick site in Missouri has mastodon. Although the animals may change, however, the dates don't. The sites that have been well dated fall within a few hundred years of 11,000 years ago. The most expansive dating places Clovis between 11,570 and 10,800 years ago, while a recent, more conservative assessment places it between 11,050 and 10,800 years ago. From our perspective, it doesn't matter. Clovis was a fairly short-lived phenomenon that dates to about 11,000 years ago.

The points, and the cultural complex they represent, take their name from the town of Clovis, New Mexico, because it was near here, at the Blackwater Draw site, that they were first found and recognized to be ancient. There is, of course, more to Clovis than just the distinctive points and large mammals. Clovis sites have yielded a variety of stone tools made on both flakes and blades (long, narrow flakes struck from specially prepared cores of raw material), with some Clovis blades being almost as distinctive as the points themselves. These sites have also provided bone and ivory tools, including what may be spear foreshafts. There are also many known Clovis sites that are not large mammal kill sites. Most notably, a large series of what appear to be cache sites have been found, containing materials that seem to have been carefully stored for future use. These sites are known primarily from the Great Plains and Northwest—Drake in Colorado, Simon in Idaho, Anzick in Montana, and Richey-Roberts in eastern Washington are examples—but more than twenty have been found. Some, like Anzick, may have been associated with human burials, but others were not. These others, archaeologist Dave Meltzer suggests, may have been resupply depots.

I have mentioned that even with the hints now available concerning pre-Clovis occupations in the Americas, the origins of Clovis are not known. Clovis appears shortly after the ice-free corridor became available for human transit, but there are no fluted points known from the Asiatic side of the Bering Land Bridge. There are fluted points from Alaska and the Yukon and in the ice-free corridor, but all the ones that have been dated are younger than Clovis. Archaeologists uniformly believe that these points represent people, or ideas, moving into the North from the South, not the other way around. The best guess is that Clovis points originated exactly where they are found—south of glacial ice in North America.

On the other hand, Alaskan archaeological assemblages assigned to what is called the Nenana complex, which dates to around 11,300 years ago, do show many similarities to Clovis assemblages, as long as the projectile points are excluded (Nenana points are small, delicate, and unfluted). Based on detailed assessments of the stone tools in these two sets of assemblages, Ted Goebel and his colleagues argue that Nenana and Clovis may represent the northern and southern archaeological remnants of the same group of early entrants into the Americas.

Monte Verde and the Paisley Caves show us that people were in North and South America well before Clovis. Clovis, however, provides us with the earliest detailed and widespread evidence we have of people in North America. On the other hand, even though we know a reasonable amount about Clovis, it has some treacherous aspects that can be highly misleading.

The first of these involves big-game hunting. As soon as the first fluted point was found with the remains of an extinct large mammal, it began to be assumed that the people who made the points gained a significant part of their livelihood by hunting big game. Even though many have discussed the possibility that the hunting of large mammals may have been a relatively rare event in a Clovis life, sites such as Naco and Lehner provide such a compelling picture of cunning human hunters taking massive game that many have found it difficult to think of Clovis in any other way.

But the fact that Clovis peoples were capable of taking such mammals as mammoths does not mean that big-game hunting provided a critical part of their diet or that it was even an important part of their lives. The apparent importance of mammoths to Clovis people may result instead from the very biased way in which our sample of Clovis sites has accumulated. With very few exceptions, buried Clovis sites have been discovered because someone—often an astute rancher like Ed Lehner—saw large bones protruding from the ground. Subsequent excavations showed that artifacts were associated with these bones. Sites that cannot be found in this way have a much reduced chance of being discovered: eroding projectile points are much less visible than eroding mammoth bones, Clovis sites tend to have small numbers of artifacts anyway, and most people would not recognize a Clovis artifact as such to begin with.

As a result, if Clovis peoples in the West spent most of their time hunting rabbits and gathering berries, we probably would not know it. "Dr. Haury? This is Ed Lehner in the San Pedro Valley. There are rabbit bones eroding from an arroyo here. Would you like to take a look?" It doesn't sound right and it doesn't happen.

This is not to say that Clovis peoples were not big-game hunters. A thorough review of Clovis-aged sites in North America has shown that fourteen of them have the remains of either mammoths (twelve sites) or mastodons (two sites) so tightly associated with artifacts that it is reasonable to conclude that the animals involved were hunted by people. In addition, sites like Lehner suggest that Clovis peoples took elephants in situations that required considerable skill. There is no question that Clovis people killed very large mammals.

What I am saying instead is that because of the way we have learned about most of our Clovis sites—large bones found first—our sample may be so biased that we have gotten a mistaken impression of the role that large mammals played in their subsistence. Indeed, archaeologists Mike Cannon and Dave Meltzer reviewed all faunal evidence from sites that date to this time period and showed that these sites tend to have a wide range of small vertebrates, from turtles and rabbits to birds and deer.

Clovis may also mislead us into thinking that human populations densities in North America took a major jump upward around 11,200 years ago.

As I mentioned, there is a lot more to the Clovis tool kit than the Clovis points themselves. However, the points are usually what allow Clovis sites to be recognized as such, because the points are so distinctive and because, unlike the bone and ivory tools, they preserve in and on the ground. Because Clovis points are readily recognizable and well dated, when archaeologists find one, they know what it is and how old it is. In addition, because Clovis is both old and relatively rare, many, perhaps even most, discoveries of these points end up getting published, even if only as a dot on a map. Although nearly all later North American projectile point styles also have a fixed and known range in time, not every last discovery of these later points is published, because we know so much more about these later periods.

Clovis points provide the earliest recognizable artifact type in North America south of glacial ice. Monte Verde provides no artifacts that, by themselves, would be indicative of a great antiquity. Much the same is true of Meadowcroft, were that site to prove to be as old as the dates might suggest. Were the artifacts from either of these sites to be found lying on the surface, few would rush to the conclusion that they pre-dated Clovis. Because Clovis is so easily recognized, and because we know of nothing earlier that is, Clovis may mislead us into thinking that a major population increase occurred in North America around 11,200 years ago, when all that really happened was the invention of an easily recognized artifact type.

The difficulty is obvious. Even so, I suspect that a population increase really did occur, the real question being how big it was. That is the only way I can reconcile the large number of buried sites that begin to appear between 11,500 and 11,000 years ago with the vanishingly small number that we may know about before that time.

Finally, it is also important to realize that while nearly all Clovis points are fluted (there are rare unfluted Clovis points as well—Lehner had three), not all fluted points are Clovis. Clovis in the Great Plains, for instance, is followed by a cultural complex called Folsom. Folsom sites date from about 10,900 to 10,200 years ago and contain not mammoth but bison as the prime large mammal in kill sites. Folsom points are fluted, but unlike the Clovis version, the flutes run all the way from the base to the tip. Many eastern fluted points are superficially most similar to Clovis points from the plains and Southwest: they are lanceolate with concave bases and flutes that extend less than halfway up from the base. But while they are similar, these eastern points are often not identical to the Clovis version. Those that have been well dated are often between 10,600 and 10,200 years old. That is, they tend to be contemporary with Folsom, not Clovis.

Eastern archaeologists no longer automatically think, as they once did, that superficial similarities between their fluted points and Clovis points mean similarity in time, since we know they do not. Western archaeologists have not always been so careful: when they see fluted points superficially similar to Clovis points, they often call them Clovis, and thus assume that they are the same age as Clovis. They may in fact be the same age, but they may also be older or younger. I will return to this issue in chapter 9.

What Caused the Extinctions?

It is hard not to be struck by the fact that just as the large Pleistocene mammals were breathing their last, Clovis peoples were busy creating the sites marked by fluted points and mammoths (and, in two instances, mastodons). For some animals the correlation between the dates for Clovis and the dates for extinction may seem too much to be coincidental.

The dates for Shasta ground sloths in the Southwest provide an excellent example. Thanks largely to the painstaking efforts of paleoecologist Paul Martin, more than three dozen radiocarbon dates are available for Shasta ground sloth dung and soft tissue from southwestern sites. The youngest of these falls at 10,500 years ago.

Harrington's mountain goat provides a second example. The remains of this animal are fairly well known from caves within Arizona's Grand Canyon. Detailed work by paleontologist Jim Mead provided thirty-seven radiocarbon dates for the remains of these goats. Of those dates, thirty-five fall before 11,000 years ago, two fall between 11,000 and 10,000 years, and none are more recent. These animals seem to have been on their way to extinction as Clovis peoples and other fluted point makers were spreading across the landscape.

Even though this is the case, there is no consensus that people caused the extinction of the thirty-six genera of North American mammals discussed at the beginning of this chapter. There are two reasons for this. First, the end of the Pleistocene in North America saw not only the emergence of Clovis but also the massive climate changes that are discussed in the following chapters. That is, at least sixteen of the extinctions (those dated to after 12,000 years ago), Clovis, and climate change all coincided. Second, and just as important, there is very little evidence that people caused those extinctions. To understand this second point, we need to understand the arguments in favor of a human cause in the first place.

Pleistocene Overkill

The argument that people caused the extinction of Pleistocene mammals was first made during the second half of the nineteenth century, but the current version of this hypothesis began to be developed by Paul Martin some fifty years ago. Martin maintains that the North American herbivores were directly doomed by human hunting. The carnivores, in turn, were doomed because they lost their food supply. All the extinctions, he argues, happened with great speed. This idea has come to be known as "overkill," the name Martin himself gave to it.

Everyone encounters this argument someplace, since it is almost impossible not to: it is mentioned in newspapers, on television, in popular books on the environment, and in ecology textbooks. It is a special favorite of those writing about the modern human impact on the environment, who use it to make points about the proper management of natural resources. For instance, recent books by ecologist Jared Diamond and paleontologist Peter Ward blame Clovis peoples for the extinctions and use this blame to help establish how intrinsically dangerous our species is and how careful we must be to prevent debacles ranging from the extinction of salmon to a nuclear holocaust. Martin's important argument merits attention here both because it is scientifically important and because of its direct relevance to modern issues.

One of the great strengths of that argument is its simplicity, since it consists of four key assertions:

1. Archaeological and paleontological research has demonstrated that prehistoric human colonization of islands was followed by often-massive vertebrate extinctions;
2. Clovis most likely represents the first people to have entered North America south of glacial ice, and certainly represents the first people known to have hunted large mammals in this area;
3. Clovis people preyed on a diverse variety of now-extinct mammals; and,
4. The late Pleistocene North American mammal extinctions occurred at or near 11,000 years ago.

Martin concludes from these assertions that Clovis peoples quickly caused the extinction of North America's Pleistocene mammalian fauna, that "large mammals disappeared not because they lost their food supply but because they became one" (P. S. Martin 1963:70).

It may already be clear that not all of these premises are well supported by the archaeological and paleontological records, but I will examine each of them in turn to show the strengths and weaknesses of each.

ISLAND COLONIZATION IS FOLLOWED BY EXTINCTION

Martin relies heavily on comparing extinctions on islands to those on continents. It is this comparison that has allowed him to base a significant part of his argument on the chronological correlation between human colonization and extinction. Were it not for islands, that correlation would have to stand largely on the American case alone, and the argument would be far less compelling.

The evidence that extinction inexorably followed the prehistoric human colonization of the world's islands is incontrovertible. The first well-documented case was provided by New Zealand. Just prior to permanent human settlement around seven hundred years ago, New Zealand supported nine or more species of moas—large, flightless birds that ranged in weight from fifty to more than five hundred pounds. A few hundred years after human colonization, all were extinct. A role for human predation in moa extinction is indicated by the fact that there are some three hundred archaeological sites in New Zealand that have been interpreted as related to moa hunting and more than one hundred sites documented to contain moa remains.

However, it wasn't just moas that became extinct after people arrived. At least another twenty-five species of birds were lost, from wrens and ravens to ducks and geese (as well as Haast's eagle, the largest eagle known and a moa predator), as were such things as frogs and lizards. No one questions that people were behind all this, but it was not just hunting that did it. The human colonizers of these islands brought with them Pacific rats (*Rattus exulans*) and domestic dogs. One can debate whether rats affected moas—by eating their eggs or competing with them for food—but it is widely agreed that they played a significant role in driving the loss of lizards, frogs, wrens, and other small animals on New Zealand.

In addition, the human arrival here was quickly followed by massive, fire-caused deforestation. Within a few hundred years of human settlement, almost all of the lowland forest of both North and South islands had been destroyed by fire, with higher, wetter sites affected as well. No one doubts that this habitat disruption also played a role in driving prehistoric extinctions in New Zealand.

In short, the wave of extinctions that followed the human colonization of New Zealand was a result of the multiple consequences associated with this event: massive burning, the introduction of nonhuman competitors and predators, and direct predation by people themselves. It was the overwhelming sum of these processes that led to the losses. In no case can we decipher which cause or causes led to the extinction of any given animal.

In fact, no matter where you look in the world, prehistoric human colonization of islands appears to have been followed, sooner or later, by vertebrate extinction. And, in all of these places, those extinctions were not caused by hunting alone but by a wide variety of effects associated with that colonization. Superb ecologist that he is, Martin recognizes that this was the case, noting that "the side effects of the arrival of prehistoric colonizers were severe and involved fire, habitat destruction, and the introduction of an alien fauna" (P. S. Martin 1984:396).

"Sooner or later" is an important qualification of the speed with which extinction followed prehistoric island colonizations. While some have argued that the extinctions always followed rapidly and have used this assertion to bolster their argument that North American continental extinctions followed rapidly as well, there are many known instances in which the prehistoric colonization of islands was not followed by the rapid loss of animals that appear to have been vulnerable. In Cuba, people coexisted with ground sloths for more than 1,000 years. In Puerto Rico, they coexisted for some 2,000 years with the thirty-pound rodent known to scientists as *Elasmodontomys obliquus*. People and a wide variety of large vertebrates shared Madagascar for as many as 2,000 years before the latter, including the 1,000-pound elephant bird (the apparent source of the roc legend), became extinct. In the Mediterranean's Belearic Islands, including Mallorca, the small, tanklike goat *Myotragus balearicus* survived with people for more than 3,000, perhaps more than 4,000, years. And, in southeastern California, the flightless, ground-nesting sea duck *Chendytes lawi* managed to live alongside people for some 8,000 years. Island extinction does not always follow human colonization with the blink-of-an-eye rapidity assumed by those who think that the North American extinctions happened quickly and as a result of human predation.

Nonetheless, there is no question that prehistoric island faunas were very vulnerable to the human arrival. This is no surprise. As every island specialist knows, native island vertebrates are vulnerable. They have relatively small population sizes; they are confined to well-delineated areas of land that may undergo rapid environmental change; they have likely lost, and in some cases have clearly lost, the mechanisms needed to cope successfully with introduced predators, pathogens, and competitors; and they have reduced genetic diversity. In addition, the isolation of oceanic islands means that there is no ready source of individuals from the same species to replenish dwindling populations. Island faunas, simply put, are far more extinction-prone than are continental faunas. It is this fact that makes their post-human colonization extinction records so pronounced.

If it is agreed that island biotas are far more vulnerable to extinction than are continental ones and that prehistoric post-colonization extinctions were due to a complex set of interacting causes, then the relevance of island extinctions to continental ones is unclear since the overkill hypothesis stipulates that continental extinctions were driven by human predation alone. As a result, the fact that people caused extinctions on islands and that those extinctions came later than the continental ones—perhaps the key argument in Martin's position—becomes irrelevant. No one denies that the island extinctions were anthropogenic, and no one denies that islands were generally colonized later than mainlands. Given the known differences in the vulnerability of island versus continental faunas, and given the stark contrast in causes that have been invoked to explain the extinctions in these two very different contexts, there is no reason to assume that they are meaningfully comparable.

NORTH AMERICA WAS FIRST COLONIZED BY CLOVIS PEOPLES

The overkill hypothesis combines the Clovis phenomenon with the apparent magnitude and chronology of the North American extinctions, observes that the prehistoric human colonization of islands was routinely followed by extinction, and concludes that the Clovis colonization of North America caused extinction here as well. This argument has deep intuitive appeal. However, it is weakened by the existence of two sites that document that the presence of people in the Americas long before Clovis: Monte Verde in southern Chile and the Paisley Caves in eastern Oregon (chapter 3). On the other hand, it is true, as Martin has said, that "whether or not prehistoric people were in the Americas earlier, 11,000 B.P. is the time of unmistakable appearance of Paleo-Indian hunters using distinctive projectile points" (P. S. Martin 1984:363). From this perspective, it does not matter that people were here prior to Clovis times, as long as they were not significant predators of the animals that were lost.

CLOVIS PEOPLES HUNTED THE NOW-EXTINCT HERBIVORES

The overkill hypothesis requires that early Americans hunted a diverse variety of now-extinct mammals in substantial numbers. However, archaeologists have long observed that this would seem to require that we have some evidence of that hunting. As I have mentioned, however, such evidence exists only for mammoths and mastodons, a fact that overkill supporters do not dispute. There is no secure evidence that people hunted any of the other mammals involved.

Importantly, this lack of evidence is not the result of some sampling fluke (à la "Dewey Defeats Truman"). Horses and camels are extremely well represented in the late Pleistocene faunal record, but none of their remains are securely associated with artifacts in such a way as to suggest that people hunted them. Paul Martin is well aware of this, noting that "in the New World, archaeological remains are virtually unknown in secure association with extinct animal remains" (2005:138). As a result, he has long argued that the extinctions occurred so quickly that associations with people are not to be expected. Even if this were somehow true, however, it could not account for the fact that convincing associations are not simply rare but are available only for mammoths and mastodons. Nor does it match the New Zealand case, where moa remains are abundant in archaeological sites.

In short, the archaeological evidence that the overkill hypothesis would seem to require from North America is simply not there.

THE EXTINCTIONS OCCURRED DURING CLOVIS TIMES

Nearly all scientists have accepted that the North American late Pleistocene extinctions occurred around 11,000 radiocarbon years ago. This is the case even though the Aztlan rabbit cannot be shown to have survived the Last Glacial Maximum and the most recent date for the huge ground sloth *Eremotherium* falls at 39,000 years ago. Even if the analysis by Tyler Faith and Todd Surovell mentioned earlier is correct and the last populations of all or nearly all of these animals coincides closely with Clovis, it would remain entirely possible that the extinction process had begun earlier. Remember the lessons that the dung fungus

Sporormiella seems to be teaching us and that muskoxen were lost from southwestern France about 19,000 years ago but from northern Siberia only about 2,700 years ago. Likewise, an extinction chronology being developed by Dale Guthrie for Alaska suggests that many of the mammals were lost from parts of that region long before Clovis times, including giant beaver, camels, short-faced bears, and horses. In short, we cannot yet document that the North American extinctions were synchronous in the sense that the overkill hypothesis requires—that the extinctions began during Clovis times and ended then or soon after.

Even if it turns out that the great majority of these extinctions did occur at the very end of the Pleistocene, this would not make North America unique. In Ireland, the latest radiocarbon date for the giant deer (or "Irish elk") *Megaloceros giganteus* falls at 10,610 years ago; the latest date for reindeer (*Rangifer tarandus*), at 10,250 years ago. The same slice of time that saw the arrival of Clovis in North America saw the disappearance of reindeer, mammoths, saigas, and giant deer from southwestern France (figure 4-2). In the southern Jura and northern French Alps, reindeer disappeared shortly after 12,000 years ago. In the Taimyr Peninsula of northern Siberia, mammoths disappeared from the mainland shortly after 10,000 years ago (they persisted well into the Holocene on Wrangel Island). Human hunters cannot account for these Eurasian extinctions. People had hunted reindeer for tens of thousands of years in France, yet the animals persevered until the end of the Pleistocene. There were no Clovis hunters in northern Siberia, and no one can blame people for the Irish extinctions, because there were no people there at that time. While all of this was happening, Harrington's mountain goat and the Shasta ground sloth disappeared from the American Southwest, caribou (North American reindeer) retreated from their late Pleistocene ranges in the American Midwest and Southeast, and mammoths and mastodons (among others) were lost from the American landscape. Genetic data even suggest that cheetahs in Africa and cougars in North America may have undergone severe population declines as the Pleistocene ended. That is, the end of the Pleistocene was a traumatic time in the Northern Hemisphere for many mammals, not just those who might have come into contact with Clovis hunters.

SOME OTHER MATTERS

These are not the only problems with the North American version of the overkill hypothesis. The focus on large mammals that marks the debate over the North American extinctions may lead an unwary reader into thinking that other organisms were unaffected by whatever it was that caused those extinctions. However, as I mentioned above, the North American and Eurasian latest Pleistocene was also a time of dramatic alterations in the ranges of many small mammals. Some twenty species of birds were lost in North America as well, although we know relatively little about the chronology of these extinctions. Not until we know far more about the timing of these massive changes in the North American fauna toward the end of the Pleistocene will we have a realistic chance of understanding the cause or causes of them.

CLOSING WORDS ON PLEISTOCENE OVERKILL

I have gone through all of these arguments simply to convince you that when you read or hear about Clovis people driving animals to extinction at the end of the North American Ice Age, it might be good to keep in mind that there is very little evidence for it.

Since most archaeologists and paleontologists whose work focuses on the North American late Pleistocene reject Martin's argument—as he has acknowledged—one might think that we have compelling alternatives to make up for the failure of the overkill account. Regrettably, we do not.

Climate-Based Explanations

Scientists have built explanations for the extinctions that call on climate change as the cause, but none are convincing. For instance, paleontologists Russ Graham and Ernie Lundelius focused on evidence that changing seasonal swings in temperature were at the root of it all. Put simply, this argument maintains that while the late Pleistocene may have seen colder annual temperatures, it was also a time when temperature swings from winter to summer were dampened, in part because Arctic air masses were blocked from moving south into central North America by the massive ice sheets that covered Canada. As the Pleistocene ended, Graham and Lundelius suggest, average annual temperatures rose, but seasonal differences in temperature increased dramatically. These increased seasonal temperature swings caused massive changes in North American plant communities, leading to habitats that were structurally far less complex than those that supported the Pleistocene mammals. In this view, the reorganization of vegetational communities caused increased competition among herbivores and decreased the availability of the plant foods utilized by these animals. This severe ecological disruption caused tremendous range changes for many smaller mammals and extinction for many larger ones.

This intriguing argument, first made for southwestern Europe in the late 1800s, might account for certain aspects of biotic change in the North American plains. However, paleobotanical studies provide no evidence for the changes in seasonality required by the argument, and it is hard to see how it might apply to, for instance, California west of the Sierra Nevada. There are many other climatically oriented attempts to explain late Pleistocene mammal extinctions in North America, but none of them is convincing.

As Bruce Huckell and Vance Haynes have put it, something happened at this period of time that we have yet to fully understand. The way out of this explanatory difficulty is probably to be had by following the approach taken by scientists working in Eurasia. Here, detailed work has built exquisite archaeological, paleontological, and paleoenvironmental histories. As a result, for some parts of this region, we know when extinctions occurred; we know what environments were like before, during, and after those events; and we know the same about the people who lived in those areas. In nearly but not quite all cases, climate change, not people, was behind the losses. In a sophisticated analysis of Eurasian woolly mammoth extinction, for instance, David Nogués-Bravo and his colleagues have shown that climate change was likely the cause of a dramatic reduction in Eurasian mammoth habitat and population sizes as the Pleistocene came to an end. It would not take much, they observe, for human hunters to have delivered the final blow to these climatically decimated populations.

The kinds of detailed information available for parts of Eurasia are not yet available for North America, but at least the successful Eurasian approach provides a model for solving this problem—or more likely, these problems—in North America as well. Unless, of course, neither climate nor people were involved.

Extraterrestrial Impact?

R. B. Firestone and his colleagues have argued that the demise of both the North American Pleistocene mammals and the end of Clovis was due to an extraterrestrial impact event, of the same general sort that did in the dinosaurs. They suggest that a comet fragmented on its way toward the earth's surface and exploded in the atmosphere, with surviving parts perhaps ultimately crashing into the Laurentide Ice Sheet in northeastern North America. The airbursts caused continent-wide burning, the melting of the eastern North American ice sheet, the extreme cold snap known as the Younger Dryas (see chapter 5), the extinctions, and the demise of Clovis.

Firestone's team amassed multiple sorts of evidence in support of this position. They argue that deposits that date to the time of the hypothesized impact incorporate a variety of things that can only be explained by a major extraterrestrial impact event. These include, among others, magnetic microspherules, miniscule diamonds ("nanodiamonds"), carbon spherules, a particular kind of carbon molecule called a fullerene that contains extraterrestrial helium, and a spike in the abundance of iridium similar to that associated with the extinction of the dinosaurs. They also observe that some Clovis sites are covered with black mats of the sort found at Lehner and attribute the formation of these mats to the impact event.

Much of this evidence is not compelling, even to someone who is not an expert in the effects of extraterrestrial impact events (that is, me). A recent analysis of the history of North American wildfires at the end of the Pleistocene showed that there was no continent-wide burning at the hypothesized time of the impact, and did not even show that this was a time of particularly high fire frequency. Jay Quade's work on the formation of black mats, noted above and discussed in detail in later chapters, has demonstrated that such mats form in response to high water tables, not impact events. Indeed, geologist Vance Haynes, cited by Firestone and his colleagues as a black mat expert (he is), interprets the mats that formed immediately above Clovis layers as indicating "a rise of water tables to either emergence as ponds or saturation of lowland surfaces" (C. V. Haynes 2005:119). Black mats have formed off and on throughout the Holocene in the Arid West and continue to form today. They are, in fact, important indicators of climate change.

To date, all those who have independently examined the evidence presented by Firestone and his colleagues have rejected it. Two research teams, one led by Todd Surovell and the other by François Paquay, have examined a geographically widespread series of deposits of the proper age but were unable to find any evidence for the hypothesized impact—and the sites they examined include some of the same ones analyzed by Firestone and his colleagues. Detailed examination of the famous Murray Springs Clovis site in Arizona by Vance Haynes and his coworkers also failed to replicate Firestone and colleagues' findings. Andrew Scott and his co-researchers have shown that the carbon spherules, including those from two of the sites in the initial study, are parts of fungi, though some also seem to be termite fecal pellets. Just like the black mats, these are common in deposits that are both earlier and later than the time of the hypothesized impact and do not indicate intense burning. Vance Holliday and Dave Meltzer have shown that changes in the archaeological record from Clovis to what followed are inconsistent with the idea that Clovis ended catastrophically. Others have questioned nearly all aspects of the analyses presented by Firestone and his team, including the fullerenes containing extraterrestrial helium and the presence of nanodiamonds. Vance Haynes even reported that he found magnetic spherules in the dust on the roof of his house in Tucson, Arizona.

Extreme skepticism seems required here. This is what good science is like. Bold ideas, followed by intense criticism, followed, one hopes, by resolution of the issue. It is also worth noting that few believed Walter Alvarez when he suggested that the dinosaurs were done in by a similar mechanism. Likewise, few believed Harlan Bretz when he argued that massive late Pleistocene floods, now called the Missoula Floods, had carved out significant chunks of eastern Washington (see chapter 5). Both, however, were right. Nonetheless, results to date provide little, if any, reason to think that an impact event can explain either the extinctions or the disappearance of Clovis. Or that the event even occurred.

No matter what caused those extinctions, however, when the Pleistocene ended, the animals were gone, and people had already been here for quite some time.

Chapter Notes

On the Pleistocene mammals of North America, see Kurtén and Anderson (1980), still the best general introduction to the animals involved. Grayson (2006a) provides a more recent review of these mammals and associated references to the literature; I have drawn on that paper heavily (but see Anyonge and Roman [2006] on dire wolves, F. A. Smith et al. [2003] for giant beaver body mass, and H. G. McDonald and Lundelius [2009] on *Eremotherium*). My inclusion of *Cuvieronius* on the list of late Pleistocene North American extinct mammals is based on the results of Vance Holliday's ongoing excavations at the El Fin-del-Mundo site in northern Sonora, Mexico. For general discussions of the modern relatives of the animals discussed here (and elsewhere in the book), see R. M. Nowak (1999) and, for the surprising recent history of armadillos, Taulman and Robbins (1996). Unless otherwise indicated, distributional data for extinct North American genera is from FAUNMAP Working Group (1994). For the general literature on Pleistocene extinctions, see Grayson (2001, 2006a, 2006c, 2007), P. S. Martin (2005), Barnosky et al. (2004), and Koch and Barnosky (2006). The extinct Pleistocene birds of North America are discussed in Grayson (1977c), Emslie (1998), and Van Valkenburgh and Hertel (1998). Eurasian extinctions are reviewed in Stuart (1991), Koch and Barnosky (2006), and Grayson (2007); the data from southwestern France are discussed by Delpech (1999).

Haile et al. (2009) provide DNA evidence for the survival of mammoth and horse into the Holocene in Alaska; Faith and Surovell (2009) analyze the chronology of North American late Pleistocene mammalian extinctions. O. K. Davis (1987) pioneered the use of *Sporormiella* as a proxy measure of large herbivore abundance on the landscape; Davis and Shafer (2006) and Raper and Bush (2009) have added important methodological details. J. L. Gill et al. (2009) provide the Indiana dung spore history; Robinson, Burney, and Burney (2005) provide comparable results from the Hudson River Valley.

On the Lehner site, see Haury, Sayles, and Wasley (1959); Mehringer and Haynes (1965); N. Allison (1988); and C. V. Haynes (1991). Haury (1953) describes his discoveries at Naco. Black mats are discussed in Quade et al. (1998), C. V. Haynes and Huckell (2007), and in chapter 8. Murray Springs and other Clovis sites in the San Pedro Valley are discussed in C. V. Haynes and Huckell (2007). On the dates for Clovis, see Meltzer (2004, 2009), G. Haynes et al. (2007), and Waters and Stafford (2007a, 2007b). For Sheaman, see Frison and Stanford (1982); for Kimmswick, R. W. Graham et al. (1981). For a sampler of the remarkable artifacts that come from Clovis caches, see Frison and Bradley (1999). Goebel, Powers, and Bigelow (1991) and Meltzer (2009) discuss the similarities between Nenana and Clovis stone tool assemblages. On the northward movement of fluted and similar points, see Hoffecker and Elias (2007). Holliday (2009) provides a superb review of, among other things, the biases that afflict our understanding of Clovis. Grayson and Meltzer (2002) review North American Clovis-aged sites with associated extinct mammals; see G. Haynes (2002) for a similar conclusion. M. D. Cannon and Meltzer (2004) cover the same ground for non-extinct animals. Meltzer (2006) is the place to go if you want to know more about Folsom. The dates for ground sloth and Harington's mountain goat extinctions in the Southwest are in Martin, Thompson, and Long (1985) and J. I. Mead et al. (1986); see also the discussion in P. S. Martin (2005).

Diamond (1992) and Ward (2000), among many others, use assumed Clovis-caused extinctions to bolster arguments about contemporary human environmental impacts. On New Zealand extinctions, see Worthy and Holdaway (2002), and on the extinct birds in particular, Tennyson and Martinson (2006); both of these are superb books. The most recent examination of the number of species of moas is provided by A. J. Baker et al. (2005). Guthrie (2003, 2004, 2006) provides his chronological refinements of the Alaskan extinctions chronology. R. W. Graham and Lundelius (1984) present their climate-based argument for the extinctions; Guthrie (1984) presents an alternative to it. J. W. Williams, Schuman, and Webb (2001) find no evidence for the seasonality changes required by the R. W. Graham and Lundelius (1984) account. Grayson (1984a) puts all of these arguments in historical perspective. The Huckell and Haynes comment is in C. V. Haynes and Huckell (2007). Nogués-Bravo et al. (2008) analyze Eurasian mammoth extinction in a paper that should be read in conjunction with Kuzmin (2010).

On slow extinctions on islands—the process I have called Holocene underkill (Grayson 2008)—see Ramis and Bover (2001) for the goat-tank *Myotragus*; Burney et al. (2004) for Madagascar; MacPhee, Iturralde-Vinent, and Jiménez Vázquez (2007) for Cuba; Turvey et al. (2007) for Puerto Rico; and T. L. Jones et al. (2008) for the flightless sea duck.

Firestone et al. (2007) present the initial argument for a terminal Pleistocene impact event (for an earlier version, see Firestone and Topping 2001); Kennett et al. (2009a, 2009b) provide further support. Negative responses are in Buchanan, Collard, and Edinborough (2008); Kerr (2008, 2009, 2010); Pinter and Ishman (2008a); Paquay et al. (2009); Surovell et al. (2009); Daulton, Pinter, and Scott (2010); Haynes et al. (2010); Holliday and Meltzer (2010); and A. C. Scott et al. (2010); see also C. V. Haynes (2008). The telling history of North American wildfires is in Marlon et al. (2009). Firestone's team has provided a lengthy series of brief responses to some of these criticisms. Many are referenced in, and immediately precede, the rebuttals by Pinter and Ishman (2008b) and Collard, Buchanan, and Edinborough (2008); but see also Bunch et al. (2010) and Paquay et al. (2010).

PART THREE

THE LATE ICE AGE GREAT BASIN

CHAPTER FIVE

The Late Pleistocene Physical Environment: Lakes and Glaciers

The fastest anyone has ever officially gone on land is 763.035 miles per hour, a speed hit by Royal Air Force fighter pilot Andy Green in the jet-powered Thrust SSC II on October 15, 1997, on northwestern Nevada's Black Rock Desert. That accomplishment broke the record that he had set some three weeks earlier, 714.144 MPH, also on the Black Rock Desert. The latter speed broke the record of 633.468 MPH set by Richard Noble in the Thrust II, on October 4, 1983, again on the Black Rock, which in turn broke the record set by Gary Gabelich on October 23, 1970, when his Blue Flame hit 622.407 MPH on the Bonneville Salt Flats, in western Utah. Kitty O'Neil, who holds the land-speed record for women, clocked 524.016 MPH on December 6, 1976, driving her jet-powered Motivator on the Alvord Desert in eastern Oregon. The fastest bicycle ever pedaled was pedaled 152.284 miles an hour by ex-Olympian John Howard, drafting behind a car on the Bonneville Salt Flats.

Since 1933, when Sir Malcom Campbell hit 272.465 MPH at Daytona, all land-speed records have been set on the Black Rock Desert, the Bonneville Salt Flats, or other places like them. The attributes these places have in common are fairly obvious: they are big, flat, and salt-encrusted. While the shared attributes are obvious, however, the reason those attributes are shared is less so. All of these places were once at the bottom of lakes that formed during the Pleistocene, and it was deposition within, and the desiccation of, those lakes that created the almost perfectly level surfaces on which land-speed records are now set. The Alvord Desert was once beneath Pleistocene Lake Alvord. The Black Rock Desert was once beneath Pleistocene Lake Lahontan. Australia's Lake Eyre, the scene of Donald Campbell's 1964 record for wheel-driven turbines (429.311 MPH), was once beneath Pleistocene Lake Dieri. When Campbell's record fell in 2001, it was bested by Don Vesco, who hit 458.440 MPH on the Bonneville Salt Flats, which, along with the rest of the Great Salt Lake Desert, was once beneath Pleistocene Lake Bonneville.

Modern Great Basin Lakes

Lakes are rarely the first thing that comes to mind when traveling through the Great Basin today. While uplands like the Ruby and East Humboldt mountains have their share of alpine lakes, lakes, as opposed to ephemeral sheets of water, are harder to come by in the valleys. It can even be hard to know what to call a real lake, since bodies of water in the valley bottoms that appear to be truly substantial—Goose Lake, straddling the Oregon-California border, for instance, or Lake Abert in south-central Oregon, or Malheur Lake to the east—have all, at one time or another, been completely dry during historic times. One of the remarkable things about today's Great Basin is how variable lakes in the valleys are. In any given year, a list of lakes made in the spring and early summer will be long, recording as lakes what are actually playas temporarily filled with runoff from adjacent mountains. The same list made in late summer or early fall will be far shorter.

Any list of lakes varies not only by season but also across the years. During wet years, such as the middle 1980s, existing lakes expand and flood areas previously thought secure, and playas fill with water year-round. During dry spells, as occurred from 1928 to 1935, lakes that seem permanent shrink and dry. Outside of alpine settings, Great Basin lakes that have not disappeared completely during historic times are rare—Great Salt Lake, Utah Lake, Pyramid Lake, Mono Lake, Walker Lake. Not coincidentally, these lakes lie close to either the Wasatch Range or the Sierra Nevada, and all lie at the terminus of streams that flow from these mountains, streams whose catchments tend to be both high and large.

Added to all these natural variations are fluctuations that have occurred because of human intervention during historic times. Owens and Mono lakes provide good, if disturbing, examples.

Owens Lake is now essentially dry, a result not so much of the diversion for irrigation that began during the 1870s and 1880s as of withdrawal of water from Owens Valley for use in Los Angeles, beginning with the completion of the Los Angeles–Owens Valley Aqueduct in 1913. Just prior to the time these withdrawals began, the lake covered about 110 square miles with some 49 feet of water, reaching an elevation of about 3,596 feet. The lake was dry by 1924, and quickly became what may be the largest source of airborne dust particles less than ten microns (ten-millionths of a meter) in diameter in the United States. Fortunately, and as a result of lengthy legal battles, on December 6, 2006, water that had previously been diverted to Los Angeles was once again allowed to flow into the lower sixty-two miles of the Owens River, in turn allowing, by 2010, almost thirty-five square miles of the Owens Lake basin to be covered by water. This is not only helping to control airborne dust pollution but is also re-creating significant amounts of wildlife habitat.

Mono Lake's tributary streams began to be diverted by the Los Angeles Department of Water and Power in 1941. Prior to the diversion, the surface of the lake stood at 6,417 feet. As Lisa Cutting of the Mono Lake Committee has pointed out, the lake had declined 45 feet by 1990, doubling its salinity and destroying significant amounts of freshwater habitat. In addition, the decline opened land bridges to what had been islands, allowing predators to reach previously isolated colonies of breeding birds on those former islands. The good news is that here, too, legal battles have led to stream restoration and to an agreement to restore the lake level to a surface elevation of 6,391 feet—still 26 feet beneath what it was but far better than what it had become. As I write, the Mono Lake Committee's website tells me that the lake level is currently 6,383.3 feet, some 11 feet higher than it was in 1990, prior to the restoration efforts.

Winnemucca Lake used to receive its waters from the Truckee River when the river was high. Completion of the Derby Dam on the Truckee in 1905, diversion of Truckee River water into the Carson River basin, and other uses of Truckee water spelled the end of Winnemucca Lake. Today, the channel that used to carry water from the Truckee to Winnemucca Lake is closed by highway fill. Sevier Lake in central Utah was dry by 1880, a result of diversion of water from the Sevier River for irrigation; not until 1983 did the lake begin to fill again, the temporary result of extremely high precipitation along the Wasatch Front.

All of this variation—natural changes in lake levels from season to season and from year to year, and changes that we have caused—makes the compilation of a simple list of Great Basin valley-bottom lakes a difficult affair. I have done it anyway, and the results are in table 5-1 (see also figure 2-2). In compiling this table, I generally included all those valley-bottom lakes that are indicated as "permanent" on the U.S. Geological Survey 1:250,000 scale topographic maps, while eliminating those that are usually dry during the summer (for instance, Franklin Lake in Ruby Valley). The second column in the table, "Pleistocene System," provides the name of the Pleistocene lake that filled the basin in which the modern lake sits. The Pleistocene lakes themselves I discuss later in this chapter.

The third column claims to show the area of the lakes in acres, but it is important to remember that such areas change dramatically from season to season and from year to year. In some cases, I have provided figures that predate major modification of the drainage system—Owens, Mono, Winnemucca, and Sevier lakes are examples. In other cases, the areas represent historic average levels—Great Salt Lake and Utah Lake are examples. In still other cases, I simply measured lake areas from 1:24,000 or 1:100,000 U.S. Geological Survey topographic maps.

This table includes a few lakes—most notably Tahoe and Summit—that are not true valley-bottom lakes at all; I include these because their basins held significant Pleistocene lakes. In addition, there are four lakes on my list—Upper Klamath, Agency (which could just as well have been treated as part of Upper Klamath Lake), Lower Klamath, and Tule—that actually fall outside of the hydrographic Great Basin, since they drain to the Pacific via the Klamath River. These I include both because they all fall within the basin of a large Pleistocene lake (that likewise drained to the Pacific) and because the area in which they are located falls within the botanical Great Basin.

My list comprises forty-five lakes, with a total area of some 2,500,000 acres. Of that acreage, almost half is contained in Great Salt Lake, the rest differentially scattered across the remaining systems. A better idea as to how this water is distributed can be gained from figure 5-1, which shows the acreage of Great Basin lakes that fall within each of the 1:250,000 scale maps that cover the Great Basin. It would have been convenient if each lake fell squarely within a given map, but while most lakes do fall that way, some are divided among maps. Goose Lake, for instance, is shared between the Alturas and Klamath Falls sheets. In all such cases, I simply assigned the entire lake to the Great Basin map that contains most of it. Alturas thus got all 97,390 acres of Goose Lake, even though roughly one-third of it (and even more if measured in terms of drainage area) falls on the Klamath Falls map. This way of assigning lake areas does not affect the simple points I want to make from this figure. Great Basin lakes of any permanence are primarily found on the eastern, western, and northern fringes of this region. Lakes become rarer toward the heart of the Great Basin and toward the south. The reasons for this are simple, as I discuss in the next section.

Great Basin Climate and Modern Lakes

The distribution and amount of rainfall received by the Great Basin is, of course, determined by the air masses that carry moisture into, and within, the region. In an important

TABLE 5-1
Valley Bottom Lakes in the Great Basin
Lakes shown in figure 2-2.

Lake	Pleistocene System	Area (acres)	Reference
California			
Eagle	Acapsukati	24,100[a]	Sweet and McBeth 1942; B. Davis, pers. comm.
Adobe	Adobe	110	Planimetric measurement from USGS 1:100,000 series map
River Spring	Adobe	140	Planimetric measurement from USGS 1:100,000 series map
Honey	Lahontan	57,600[b]	K. N. Phillips and Van Den Burgh 1971
Lower Klamath	Modoc	30,000[c]	D. M. Johnson et al. 1985
Tule	Modoc	96,000[c]	R. H. Jackson and Stevens 1981
Owens	Owens	71,000[d]	Smales 1972
Mono	Russell	54,900[e]	Sweet and McBeth 1942; B. Davis, pers. comm.
Tahoe	Tahoe	121,000	Sweet and McBeth 1942
Nevada			
Ruby	Franklin	9,000	Sweet and McBeth 1942
Humboldt	Lahontan	4,200	Sweet and McBeth 1942
Pyramid	Lahontan	142,000[f]	Sweet and McBeth 1942; B. Davis, pers. comm.
Soda Lakes	Lahontan	260	Gale 1915
Upper Carson	Lahontan	25,600[g]	Planimetric measurement from USGS 1:100,000 series map
Walker	Lahontan	65,000[g]	Sweet and McBeth 1942; B. Davis, pers. comm.
Winnemucca	Lahontan	61,000[h]	Sweet and McBeth 1942; B. Davis, pers. comm.
Summit	Summit	530	Sweet and McBeth 1942
Washoe	Washoe	4,100	Sweet and McBeth 1942
Artesia	Wellington	1,000	Sweet and McBeth 1942
Lower Pahranagat	White River	585	Sweet and McBeth 1942
Upper Pahranagat	White River	370	Sweet and McBeth 1942
Oregon			
Abert	Chewaucan	36,540	Harding 1965; planimetric measurement from USGS 1:24,000 series map
Summer	Chewaucan	25,000	Harding 1965; planimetric measurement from USGS 1:24,000 series map
Goose	Goose	97,390	Harding 1965; planimetric measurement from USGS 1:24,000 series map
Harney	Malheur	26,400	Harding 1965; planimetric measurement from USGS 1:24,000 series map
Malheur	Malheur	49,700	Harding 1965; planimetric measurement from USGS 1:24,000 series map
Agency	Modoc	9,300	Harding 1965
Upper Klamath	Modoc	61,540	Harding 1965
Anderson	Warner	410	W. F. Sigler 1962
Bluejoint	Warner	6,920	W. F. Sigler 1962
Campbell	Warner	2,340	W. F. Sigler 1962
Crump	Warner	7,680	Harding 1965; planimetric measurement from USGS 1:24,000 series map
Fisher	Warner	300	W. F. Sigler 1962
Flagstaff	Warner	3,580	W. F. Sigler 1962

TABLE 5-1 (CONTINUED)

Lake	Pleistocene System	Area (acres)	Reference
Hart	Warner	7,230	Harding 1965; planimetric measurement from USGS 1:24,000 series map
Mugwump	Warner	210	W. F. Sigler 1962
Pelican	Warner	230	W. F. Sigler 1962
Stone Corral	Warner	1,000	W. F. Sigler 1962
Swamp	Warner	850	W. F. Sigler 1962
Turpin	Warner	300	W. F. Sigler 1962
Upper Campbell	Warner	980	W. F. Sigler 1962
Utah			
Bear	Bonneville	70,400	Currey, Atwood, and Mabey 1984
Great Salt	Bonneville	1,152,200	Russell 1885b
Sevier	Bonneville	120,320	G. K. Gilbert 1890; Oviatt 1989
Utah	Bonneville	94,100	Rush 1972

[a]1960 level. [d]1872. [g]1882.
[b]1867. [e]1930–1939 average. [h]1891–1902 average.
[c]ca. 1905. [f]1891–1902 average.

analysis of Great Basin rainfall patterns, meteorologist John Houghton has documented how three main sources of precipitation interact to establish those patterns.

Most important of all sources of precipitation here are the low-pressure storm systems brought into the Great Basin by westerly winds moving off the Pacific Ocean. During the warmer months of the year, those westerlies lie far to the north. The cold water on the surface of the Pacific Ocean along Oregon and California chills the air that comes into contact with it, slowing evaporation and creating a layer of cool, stable air over the water itself. As a result, moisture-bearing westerlies rarely penetrate into the Great Basin during summer.

During winter, however, the westerlies move southward and bring storm systems—Pacific fronts—onto the coasts of Oregon and California, where they soon encounter the Sierra Nevada and Cascade Range. Forced upward, these air masses are cooled and condensed and, as a result, drop often massive amounts of precipitation on the western (windward) sides of these ranges. How much of their moisture that the air masses lose in this way, and that is thus lost to the Great Basin, depends not only on their initial moisture load but also on the mass and elevation of the parts of the Sierra Nevada and Cascades they encounter.

While the Pacific contribution of Great Basin rainfall is the most important component of precipitation in this region, it plays the heaviest role in the western and northern Great Basin, since the former area is closest to the Pacific source and the latter receives the greatest number of Pacific frontal storms. In addition, Pacific moisture is an important source of rainfall wherever the frontal systems encounter massive mountain ranges. The Wasatch Range, for instance, receives its maximum precipitation in the winter from Pacific storm systems, although it also benefits from lying downwind of Great Salt Lake.

The Pacific component of Great Basin precipitation comes mainly between October and April. In the summer months—July and August—the westerlies are far weaker and precipitation from this source is at its minimum. During these warm months, monsoonal storms, driven by the temperature differential between air masses over the gulfs of California and Mexico on the one hand and those over the land on the other enter the Great Basin from the south. Although the southeastern Great Basin is the major recipient of precipitation from these storms, receiving some 35 percent of its annual rainfall in this way, much of the southern and eastern Great Basin benefits from these southerly rains. Indeed, some two-thirds of all summer rain in the Great Basin is derived from the Gulf component of Great Basin precipitation.

The spring and fall months see a third source of precipitation rise in importance. Between April and June on the one hand and October and November on the other, low-pressure systems develop over the Great Basin itself, often east of the most massive parts of the Sierra Nevada. These systems— the Great Basin or Tonopah Lows—draw their moisture from a number of sources. However, they generally drop more precipitation in the spring than in the fall, suggesting that a significant source of the moisture they carry lies in the snowpacks of Great Basin mountains and in the low elevation lakes of the area: both snowpacks and lake levels are at their seasonal highs during the spring and early summer. Rains from this Continental component are most important in an

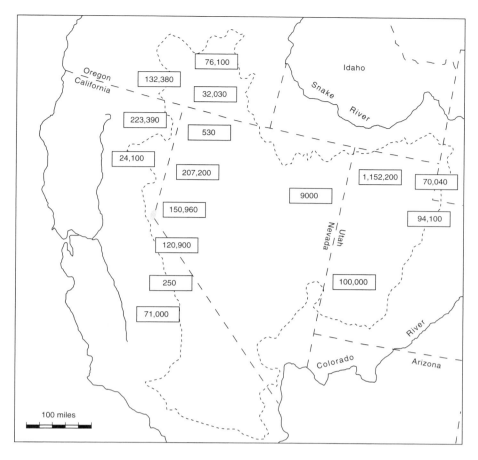

FIGURE 5-1 Total area, in acres, of Great Basin lakes, according to 1:250,000 scale U.S. Geological Survey maps.

area that roughly straddles the Utah-Nevada border. Here, about half the average annual precipitation comes from this source, with nearly all the rest contributed by the Pacific storms of winter.

Figure 5-2 shows the contributions of these three precipitation regimes—Pacific, Gulf, and Continental—to the annual total rainfall received in the Great Basin. The average annual rainfall from all these sources combined is illustrated in figure 5-3. These rainfall patterns alone go far to explain why lakes are found where they are in the Great Basin. All other things being equal, areas of decidedly low precipitation have no lakes.

However, it is the precipitation that falls on bordering mountains that accounts for many Great Basin lakes. The Sierra Nevada receives more than 50 inches of precipitation a year, while the Wasatch Range receives more than 40. On the west, Lake Tahoe lies on the eastern edge of the Sierra Nevada, while the sources for the major lakes flanking the western Great Basin—Pyramid, Winnemucca, Upper Carson, Walker, Mono, and Owens—all lie within the Sierra Nevada. On the east, Great Salt Lake, Utah Lake, and Sevier Lake have their sources in streams that flow from the Uinta Mountains, Wasatch Range, and nearby uplands. Ultimately, these lakes are fed primarily by precipitation brought by winter storms coming from the Pacific. These storms bring their moisture at a time when evaporation is at a minimum and when moisture is readily stored in the mountains in the form of snow. While precipitation directly on the lakes is important, it is the mountain-born rivers that sustain the sizable lakes that fringe the far eastern and far western edges of the Great Basin.

For instance, precipitation provides 31 percent of the input to Great Salt Lake, but surface streams provide 66 percent (the remaining 3 percent is provided by groundwater). Of this 66 percent, nearly all comes from rivers that flow directly or indirectly from the Uinta Mountains and Wasatch Range: 59 percent from the Bear River, 20 percent from the Weber, and 13 percent from the Jordan. On the other side of the Great Basin, Walker Lake receives 83 percent of its input from the Walker River, which in turn flows from the Sierra Nevada. Only 11 percent of Walker Lake inflow comes directly from precipitation, the remaining 6 percent coming from groundwater and local runoff.

Thus, lakes fringe the eastern and western edges of the Great Basin because they are fed by streams that originate in adjoining massive uplands. These uplands, in turn, receive their precipitation from winter storms moving off the Pacific.

THE LATE PLEISTOCENE PHYSICAL ENVIRONMENT 91

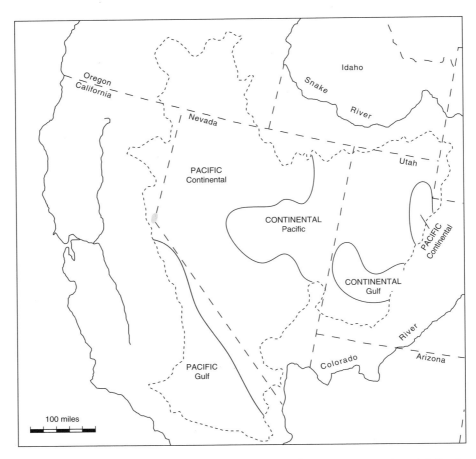

FIGURE 5-2 The contribution of Pacific, Gulf, and Continental air masses to annual precipitation in the Great Basin. The upper name in each pair is the primary, the lower name the secondary, precipitation source (after Houghton 1969).

The Great Basin is also fairly lake-rich in its northern reaches. The Ruby and Warner valleys and Harney Basin, for instance, all have substantial lakes, but the waters that feed them come neither from the Sierra Nevada–Cascades nor from the Wasatch Range and its neighbors. All do have major uplands nearby—the Ruby, Warner, and Steens mountains in these three cases—but even more substantial uplands are found to the south, where comparable lakes do not exist.

These northern lakes exist for two prime reasons. First, the Great Basin north of 40° N latitude—that is, north of a line that runs through Pyramid and Humboldt lakes, through the Ruby Valley south of Ruby Lake, and just south of Utah Lake—receives a greater share of Pacific frontal storms than do areas that lie south of that latitude. The northern Ruby Mountains, for example, lie directly in the path of these storms; the upper reaches of the Rubies may receive nearly 50 inches of precipitation a year as a result. Further south, a high-pressure system, the Great Basin High, often prevents Pacific frontal storms from entering; it is when this high-pressure system deteriorates that Pacific storms sweep across this area as well. The greater frequency of winter storms originating in the Pacific helps account for the greater number of lakes in the northern Great Basin.

Evaporation rates also play a key role, however. These rates are far higher in the southern Great Basin than they are to the north. In the north, mean annual lake evaporation rates are on the order of 40 to 45 inches a year. On the southern edge of the Great Basin, they are on the order of 70 inches a year or more (figure 5-4). These differences are largely a function of the increased temperatures encountered in moving from north to south. The mean annual temperature in Las Vegas, for instance, is 66°F; in Elko, in the northeastern part of the state, it is 45°F. Decreased evaporation, coupled with increased precipitation, accounts for the relative abundance of lakes in the north and for their absence in the south.

So the distribution of lakes in the Great Basin can be accounted for by modern Great Basin geography and climate. Pacific storms moving into the Great Basin between October and April provide much of the precipitation. On the western and eastern edges of the area, that water collects in the fringing high mountains and is ultimately deposited in the lakes. On the north, the water is also largely deposited

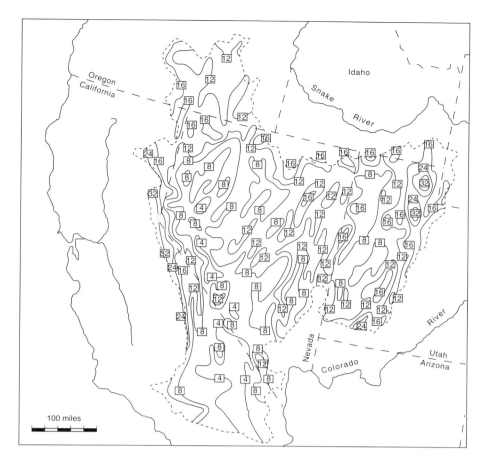

FIGURE 5-3 Distribution of average annual rainfall, in inches, in the Great Basin (after U.S. Department of Commerce 1983).

in the fringing mountains, but here the existence of lakes is assisted by the increased frequency of Pacific storms north of 40° N latitude and by evaporation rates far lower than those to the south. The distribution of lake areas shown in figure 5-1 makes good sense when seen in climatic and geographic perspective.

Pleistocene Lakes in the Great Basin

Geologists use the term "pluvial lake" to refer to Pleistocene lakes whose levels were higher because of altered ratios between precipitation and evaporation. Later, I discuss current estimates of late Pleistocene temperatures and precipitation in the Great Basin. For now, I just note that these estimates routinely suggest that late Ice Age temperatures were lower, and average annual precipitation higher, than those of today. Pluvial lakes were the result.

When lake basins that have outlets to the sea receive greater amounts of water or lose less to evaporation, or both, excess water can flow outward through swollen stream channels. Because the separate basins of which the Great Basin is composed have no outlets, they respond in a predictable way to increased amounts of water. Basins that have no lakes get them, just as ephemeral lakes appear today in many Great Basin playas when runoff is high. Basins that have lakes get bigger ones. Since lake levels cannot be controlled by simply increasing outflow through drainage channels, lakes continue to rise until one of a number of things happen. In most basins, as lakes rise, their surfaces areas increase. As that happens, more water is exposed to evaporation, so more is lost to this process. In some cases, the lakes can rise to the point that increased evaporation balances increased input and the lake stabilizes at this level.

If that does not happen, the increased flow leads inexorably to the point that the lake overflows—no different from what happens when you forget to turn off the water in your bathtub. Water then pours into an adjacent basin. If increased evaporation from that process does not lead to an equilibrium between water loss and water gain, then the adjacent basin—probably filling on its own anyway—can fill as well. Given enough water, these now-conjoint basins can overflow yet again. The process does not stop until evaporation balances or exceeds precipitation, or until the altered precipitation-to-evaporation ratios that led to the lake level

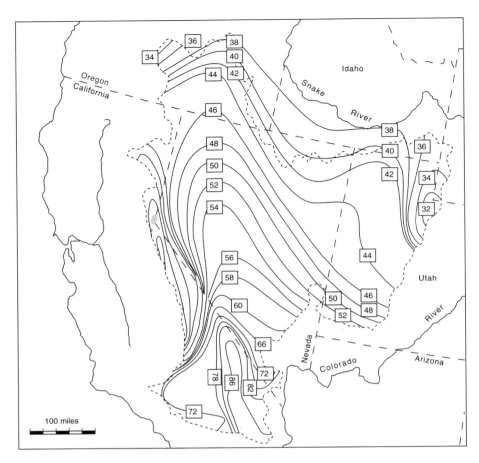

FIGURE 5-4 Lake evaporation rates, in inches per year, in the Great Basin (after U. S. Department of Commerce 1983).

rise in the first place change once again. Alternatively, the lake can rise until it finds a way to spill to the sea. Goose Lake, for instance, rose so high in 1868 that it overflowed into the Pit River system, thus stabilizing its level at 4,716 feet (in typical Great Basin fashion, Goose Lake also became completely dry in 1926).

Although it is easy to state the general principles involved in the rise and fall of Great Basin lakes, the details can be extremely complex. For instance, lake basins can lose water through underground flow to other basins. To give but one example, eastern Nevada's northern Butte Valley provides groundwater to Ruby Valley to the north. In addition, the increased evaporation that occurs as lakes expand their surface areas can be partially offset by the fact that expanded lake surfaces can now catch greater amounts of precipitation. As lakes rise and fall, their salinity changes, and changing salinities alter evaporation rates. Nonetheless, the general principles are fairly simple. Increased precipitation or decreased evaporation, or both, leads to higher lake levels. Since both things happened in the Great Basin during the Pleistocene, pluvial lakes were abundant, and in some cases huge.

Today's Great Basin has, or would have if waters were not diverted for other purposes, some 2,500,000 acres of lakes that at least tend to be there more often than they are not (table 5-1). During the late Pleistocene, the Great Basin held about 27,860,000 acres of lakes (table 5-2 and figure 5-5; by tradition, I have included Pleistocene Lakes Chemult and Modoc even though these are not in the hydrographic Great Basin). This figure may be conservative, since small ancient lakes are difficult to detect long after the fact. On the other hand, it is also possible that some of the smaller, unstudied lakes listed in table 5-2 did not actually exist. Assuming that these things balance one another out, approximately eleven times more of the Great Basin's surface was covered by water during parts of the late Pleistocene than is so covered today.

While the Great Basin held some eighty pluvial lakes during the later Pleistocene, the largest of these lakes, Lahontan and Bonneville, are also the best known, and I begin with those.

Pleistocene Lake Bonneville

In mid-April 1846, twenty-two members of the families of George and Jacob Donner and James Reed left Springfield,

TABLE 5-2
Pluvial Lakes of the Great Basin
Basins thought to have been filled with marshes are not included. Lakes shown in figure 5-5.

Lake	Location	Approximate Area (square miles)	References
California			
Acapsukati*	Eagle Lake Basin	62	T. R. Williams and Bedinger 1984
Adobe*	Black Lake	6	T. R. Williams and Bedinger 1984
China	China Lake Basin (area included with Searles Lake Basin)		
Cuddeback	Cuddeback Lake	35	T. R. Williams and Bedinger 1984
Deep Spring	Deep Springs Valley	17	T. R. Williams and Bedinger 1984
Harper*	Harper Lake Basin	85	Orme 2008b
Horse*	Horse Lake Basin	8	T. R. Williams and Bedinger 1984
Koehn	Fremont Valley	42	T. R. Williams and Bedinger 1984
Madeline*	Madeline Plain	300	Bowers and Burke 2005
Manix*	Troy Lake, Coyote Lake	83	Meek 1989; S. G. Wells et al. 2003
Manly	Death Valley	618	Orme 2008b
Mojave*	Soda Lake, Silver Lake	112	Wells et al. 2003
Owens*	Owens Lake	195	Orme and Orme 2008
Panamint*	Panamint Valley	297	Orme 2008b
Russell*	Mono Lake Basin	305	Orme 2008b
Searles*	Searles Lake Basin (area includes China Lake Basin)	351	Orme 2008b
Surprise	Surprise Valley	506	Orme 2008b
Tahoe*	Lake Tahoe Basin	212	Orme 2008b
Thompson*	Antelope Valley	367	Orme 2008a
Idaho			
Thatcher*	Bear River Basin	100	T. R. Williams and Bedinger 1984
Utaho	Pocatello Valley	42	T. R. Williams and Bedinger 1984
Nevada			
Antelope	Antelope Valley	48	T. R. Williams and Bedinger 1984
Bristol	Dry Lake Valley	35	T. R. Williams and Bedinger 1984
Buffalo*	Buffalo Valley	77	T. R. Williams and Bedinger 1984
Carpenter	Lake Valley	134	T. R. Williams and Bedinger 1984
Cave	Cave Valley	69	T. R. Williams and Bedinger 1984
Clover	Clover-Independence valleys	344	Orme 2008b
Coal	Coal Valley	69	T. R. Williams and Bedinger 1984
Columbus	Columbus Salt Marsh	54	Reheis 1999b
Corral	Little Smoky Valley	9	T. R. Williams and Bedinger 1984
Crooks*	New Year Valley	5	T. R. Williams and Bedinger 1984
Desatoya*	Smith Creek Valley	168	T. R. Williams and Bedinger 1984
Diamond*	Diamond Valley	293	Orme 2008b
Dixie	Dixie Valley	421	Orme 2008b
Edwards	Edwards Creek Valley	102	T. R. Williams and Bedinger 1984
Franklin	Ruby Valley	471	Orme 2008b
Gale	Butte Valley	159	T. R. Williams and Bedinger 1984
Garfield	Garfield Flat	3	T. R. Williams and Bedinger 1984

TABLE 5-2 (CONTINUED)

Lake	Location	Approximate Area (square miles)	References
Gilbert	Grass Valley	208	Orme 2008b
Gold Flat	Gold Flat	26	T. R. Williams and Bedinger 1984
Granite Springs	Granite Springs Valley	40	T. R. Williams and Bedinger 1984
Groom	Emigrant Valley	36	T. R. Williams and Bedinger 1984
Hawksy Walksy*	Hawksy Walksy Valley	12	T. R. Williams and Bedinger 1984
High Rock*	High Rock Basin	12	T. R. Williams and Bedinger 1984
Hubbs	Long Valley	205	Reheis 1999b
Jakes	Jakes Valley	63	Garcia and Stokes 2006 (after Mifflin and Wheat 1979)
Kawich	Kawich Valley	22	T. R. Williams and Bedinger 1984
Kumiva	Kumiva Valley	15	T. R. Williams and Bedinger 1984
Labou*	Fairview Valley	20	T. R. Williams and Bedinger 1984
Lahontan	Lahontan Basin	8,284	Reheis 1999b
Laughton	Cold Spring Valley	7	T. R. Williams and Bedinger 1984
Lemmon	Lemmon Valley	13	T. R. Williams and Bedinger 1984
Macy	Macy Flat	9	T. R. Williams and Bedinger 1984
Maxey	Spring Valley	81	T. R. Williams and Bedinger 1984
Meinzer	Long Valley	355	Orme 2008b
Mud	Ralston Valley	133	T. R. Williams and Bedinger 1984
Newark*	Newark Valley	307	Redwine 2003a:187, 2003b
Paiute*	Bawling Calf Basin	4	T. R. Williams and Bedinger 1984
Parman*	Summit Lake Basin	7	T. R. Williams and Bedinger 1984
Railroad	Railroad Valley	525	Orme 2008b
Reveille*	Reveille Valley	41	T. R. Williams and Bedinger 1984
Rhodes	Rhodes Salt Marsh	13	T. R. Williams and Bedinger 1984
Spring	Spring Valley	336	Orme 2008b
Toiyabe	Big Smoky Valley	251	Orme 2008b
Tonopah	Big Smoky Valley	90	T. R. Williams and Bedinger 1984
Waring	Goshute and Steptoe valleys	514	T. R. Williams and Bedinger 1984
Washoe*	Washoe Lake Basin	23	T. R. Williams and Bedinger 1984
Wellington*	Smith Valley	117	T. R. Williams and Bedinger 1984
Yahoo	Stevens Valley	2	T. R. Williams and Bedinger 1984
Oregon			
Alkali*	Alkali Lake Basin	228	Friedel 1993
Alvord	Alvord Desert	490	Orme 2008b
Catlow*	Catlow Basin	347	Orme 2008b
Chemult	Klamath Marsh	151	Conaway 2000
Chewaucan	Abert and Summer Lake basins	480	Licciardi 2001
Coyote	Coyote Basin	175	Hemphill-Haley, Lindberg, and Reheis 1999
Fort Rock	Fort Rock Basin	751	Friedel 1993
Goose*	Goose Lake Basin	367	Orme 2008b
Malheur*	Harney Basin	919	Orme 2008b
Modoc	Klamath and Tule Lake basins	1,096	Dicken 1980; Conaway 2000
Warner (Coleman)	Warner Valley	483	Orme 2008b
Utah			
Bear	Bear Lake	225	B. Laabs, pers. comm.
Bonneville*	Great Salt Lake; Great Salt Lake and Sevier deserts	19,800	Currey, Atwood, and Mabey 1984; Currey 1990
Pine (Wah Wah)	Pine Valley	41	T. R. Williams and Bedinger 1984†

*Lake overflowed into an adjacent basin at maximum levels.
†Currey (1983) notes that "Pine" has priority over "Wah Wah" for the name of this lake (which is also in Pine, not Wah Wah, Valley).

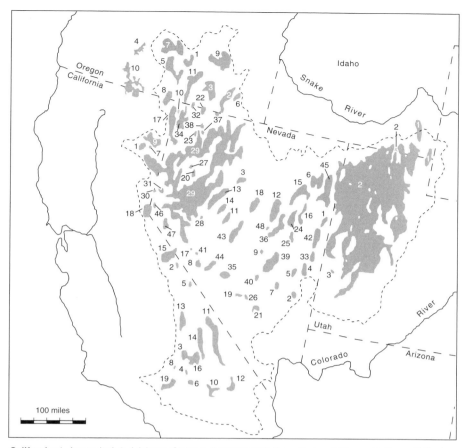

California: 1. Acapsukati; 2. Adobe; 3. China; 4. Cuddeback; 5. Deep Spring; 6. Harper; 7. Horse; 8. Koehn; 9. Madeline; 10. Manix; 11. Manly; 12. Mojave; 13. Owens; 14. Panamint; 15. Russell; 16. Searles; 17. Surprise; 18. Tahoe; 19. Thompson

Idaho: 1. Thatcher; 2. Utaho

Nevada: 1. Antelope; 2. Bristol; 3. Buffalo; 4. Carpenter; 5. Cave; 6. Clover; 7. Coal; 8. Columbus; 9. Corral; 10. Crooks; 11. Desatoya; 12. Diamond; 13. Dixie; 14. Edwards; 15. Franklin; 16. Gale; 17. Garfield; 18. Gilbert; 19. Gold Flat; 20. Granite Springs; 21. Groom; 22. Hawksy Walksy; 23. High Rock; 24. Hubbs; 25. Jakes; 26. Kawich; 27. Kumiva; 28. Labou; 29. Lahontan; 30. Laughton; 31. Lemmon; 32. Macy; 33. Maxey; 34. Meinzer; 35. Mud; 36. Newark; 37. Paiute; 38. Parman; 39. Railroad; 40. Reveille; 41. Rhodes; 42. Spring; 43. Toiyabe; 44. Tonopah; 45. Waring; 46. Washoe; 47. Wellington; 48. Yahoo

Oregon: 1. Alkali; 2. Alvord; 3. Catlow; 4. Chemult; 5. Chewaucan; 6. Coyote; 7. Fort Rock; 8. Goose; 9. Malheur; 10. Modoc; 11. Warner (Coleman)

Utah: 1. Bear; 2. Bonneville; 3. Pine (Wah Wah)

FIGURE 5-5 Late Pleistocene lakes in the Great Basin. Lake numbers begin anew for each state. Sources: C. T. Snyder, Hardman, and Zdenk (1964), G. I. Smith and Street-Perrott (1983), Reheis (1999a), Negrini et al. (2000), Negrini (2002), Pinson (2008), and those listed in table 5-2.

Illinois, headed for disaster in the snows of the Sierra Nevada. On August 22, they and sixty-five other members of what has come to be known as the Donner Party emerged from the mouth of Emigration Canyon in the Wasatch Range to enter the far eastern edge of the basin of Pleistocene Lake Bonneville (figure 5-6). Continuing west, they passed south of Great Salt Lake, a Lake Bonneville remnant. To the south lay Utah and Sevier lakes, two more "orphans of the Pleistocene" (Trimble 2008:38) and the only other substantial lakes that still occupy the Bonneville Basin. They had well more than one hundred miles to go before they would finally be out of the largest of all Great Basin Pleistocene lake basins. East of Floating Island, they mired their wagons and lost many of their cattle but still had some thirty miles to go before they reached the Pilot Range, whose Pilot Peak provided the visible target for those coming this way and whose Pilot Springs provided the water so desperately needed after the crossing (figure 5-7). The end of the Bonneville Basin lay a few miles beyond this range.

Had the Donner party attempted to cross the Bonneville Basin a few hundred years earlier, during the heart of what

FIGURE 5-6 The northwestern Bonneville Basin. The northern end of the Silver Island Range curves to the east in the near distance; beyond it lies Donner-Reed Pass, taken by the Donner Party to reach the Pilot Range on the far western edge of the Bonneville Basin.

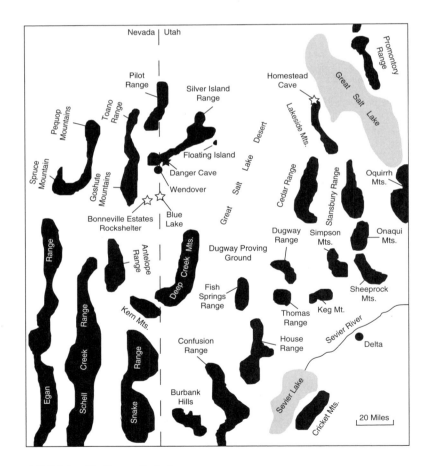

FIGURE 5-7 The Bonneville Basin.

is known as the Little Ice Age (chapter 8), they would have found much of the Great Salt Lake Desert, including the Bonneville Salt Flats, under shallow water. Floating Island—which from a distance appears to be a floating island—was then at the end of a small peninsula jutting into the waters of a much expanded Great Salt Lake.

Had they tried to make the same trip some 16,000 years earlier, however, they would not have even gotten out of the lower reaches of Emigration Canyon. Instead, they would have seen almost nothing but water facing them to the west, with some of today's mountains emerging from the blue as islands—the Cedar, Lakeside, and Newfoundland mountains, for instance, and the Silver Island and Pilot ranges.

The Great Salt Lake is one of the most saline permanent lakes on earth. Nearly 20 percent of the weight of the lake's contents consists of dissolved solids, about 5 billion tons of them, with sodium chloride—common table salt—leading the way and making the Great Salt Lake some seven times saltier than the ocean. A large proportion of these salts are there because they are derived from the waters of the much larger Lake Bonneville that preceded Great Salt Lake in this basin, but as new waters reach the basin today, new salts are added as well. Recently, dissolved solids have been added to the lake at a rate of some 2 million tons a year.

Exactly how saline the lake is varies from season to season and from year to year: the higher the lake level, the lower the salinity. In 1873, when the lake reached one of its historic elevational highs, at about 4,212 feet, the proportion of dissolved minerals dropped to slightly below 12 percent. In 1963, when the lake retreated to its historic low of 4,191 feet, dissolved solids formed nearly 30 percent of the its contents. It is always easy to float in the Great Salt Lake, but how easy it is depends on how high the lake is.

The level of Great Salt Lake depends, of course, on the balance between evaporation and inflow. Inflow, as I have mentioned, is provided by streams, precipitation, and groundwater, with streams derived from the Uinta Mountains and Wasatch Range being of greatest importance by far. The Great Salt Lake is in the northern latitudes of the Great Basin; Ted Arnow and Doyle Stephens have calculated that at an average elevation of 4,196 feet, evaporation from the surface of the lake is about 45 inches a year, roughly the same as it is for Pyramid Lake (48 inches a year) and Walker Lake (about 49 inches). During historic times, the average elevation of the surface of Great Salt Lake has been about 4,200 feet, at which point the lake has an area of about 1,700 square miles (official flood stage lies at 4,202 feet). At its historic low, 4,191.35 feet in 1963, the lake had an area of about 950 square miles and a maximum depth of about 25 feet. A high of 4,212 feet was reached in June 1986 (and again in March 1987), at least equaling its 1873 high. At this point, the lake had an area of approximately 2,300 square miles and a maximum depth of about 45 feet.

Although the historic low and high levels of Great Salt Lake are separated by only twenty-three years, the increase to 4,211.85 feet did not take nearly that long. In response to both increased precipitation and decreased evaporation, the lake rose 12.2 feet between September 18, 1982, and June 3, 1986. Indeed, the rise would have been some 2.5 feet greater were it not for the fact that water is withdrawn from feeding streams for human use. Even so, the economic loss from this unexpected rise in the level of Great Salt Lake approached $300 million (far greater losses were caused by the increased precipitation in higher-elevation areas). Rapid fluctuations in levels of this sort are fully typical of Great Basin lakes.

Great Salt Lake lies in the eastern part of the Bonneville Basin, the Great Salt Lake Desert in the western. The lowest thresholds between lake and desert lie just south and southwest of the Newfoundland Mountains between 4,214 feet and 4,216 feet in elevation. Were Great Salt Lake to rise above this level, the Great Salt Lake Desert would begin to be flooded, the magnitude of the flooding depending, of course, on the magnitude of the rise. This last happened between three hundred and four hundred years ago, when the lake increased to about 3,700 square miles and flooded a large part of the Great Salt Lake Desert.

The late prehistoric high, however, was nothing compared to Lake Bonneville during its late Pleistocene heyday. At its greatest extent, Pleistocene Lake Bonneville covered seven intermontane subbasins in Utah, Nevada, and Idaho. Four of these (the Tule, Rush, Cedar, and Puddle valley subbasins) were quite small, but the remaining three were substantial—the Great Salt Lake Desert and Great Salt Lake subbasins in the north and the Sevier subbasin in the south (figure 5-8). Lake Bonneville then had an area of about 19,800 square miles, roughly the size of Lake Michigan (22,400 square miles), and was about 1,220 feet deep. Indeed, the only reason the lake did not get any bigger than that is that it began to overflow into the Snake River drainage in southern Idaho. When it did this, it was no longer part of the hydrographic Great Basin, but instead belonged to the Columbia River system.

Driving through the Bonneville Basin today, one cannot but be impressed by the terraces that have been carved into the mountains on the edges of, and within, the basin (see figure 5-9). Standing on the salt-encrusted floor of the old lake and looking toward terraced hills, one does not need much imagination to picture the body of water that must have been here, even though it has taken tremendous imagination to work out the history of the lake that carved those terraces.

There are five named (and many unnamed) shorelines in the Bonneville Basin, each of which consists of a number of beach ridges at roughly the same elevation. The names and approximate elevations of these are shown in table 5-3, but these elevations are truly approximate. The weight of the water that made up Lake Bonneville

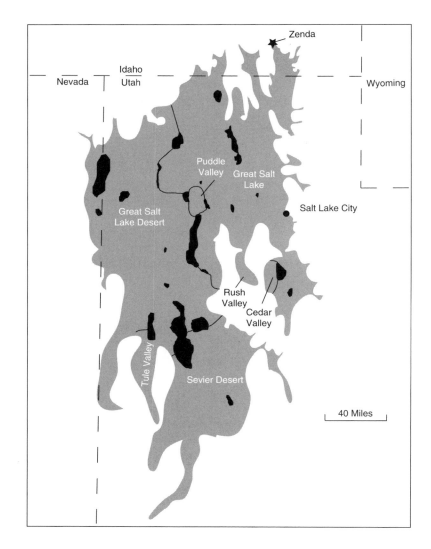

FIGURE 5-8 Pleistocene Lake Bonneville and its subbasins (after Sack 2002).

FIGURE 5-9 Lake Bonneville terraces on Table Mountain, adjacent to the Old River Bed connecting the Sevier and Great Salt Lake subbasins of Pleistocene Lake Bonneville.

TABLE 5-3
Lake Bonneville Shorelines

Shoreline	Age (years ago)	Elevational Range (feet above sea level)	Surface Area (square miles)
Stansbury (Lake Rise)	22,000–20,700	4,420–4,520	9,300
Sub-Provo (Lake Rise)	19,400	4,725	
Bonneville (Lake Decline)	15,000–14,500	5,090–5,335	19,800
Provo (Lake Decline and Rise)	14,500–12,500	4,740–4,930	14,400
Gilbert (Lake Decline)	10,500–10,000	4,230–4,300	6,600
Great Salt Lake	Modern	4,200	1,700

SOURCES: After Oviatt 1997; Sack 1999; Oviatt, Madsen, and Schmitt 2003; Godsey, Currey, and Chan 2005; Oviatt et al. 2005; Patrickson et al. 2010; Oviatt, pers. comm.

depressed the entire basin by as much as 240 feet. Once the water began to recede, the basin began to rebound, but it has done so differentially, so that the elevations of a given shoreline vary from place to place. In the vicinity of Great Salt Lake, for instance, the Provo shoreline varies between 4,790 feet and 4,925 feet in elevation; the Gilbert shoreline here varies from about 4,240 feet to about 4,300 feet in elevation.

The major shorelines, and the separate beach ridges of which they are composed, mark the rises and falls of Pleistocene Lake Bonneville toward the end of the Pleistocene. Although the precise dates of all these events are not fully clear, the general picture is quite clear, due in large part to major advances made during the past few decades in our understanding of the history of this lake. In recent years, geologists Jack Oviatt and Don Currey have played major roles in assembling this history, though, as we will see, many others have made significant contribution as well.

This work shows that just prior to 28,000 years ago, Lake Bonneville was at about the level of today's Great Salt Lake (4,200 feet). Soon thereafter, the lake began to rise and, by 26,500 years ago, had reached an elevation of about 4,400 feet (figures 5-10 and 5-11; for comparison, the elevation of the Salt Lake City Airport is 4,229 feet). The lake continued to rise and, by about 22,000 years ago, had reached an elevation of approximately 4,500 feet, at which point it had an area of some 9,300 square miles. For the next 1,300 years or so, the lake stayed at roughly this level, albeit rising and then falling some 150 feet during this interval (called the Stansbury Oscillation), and creating the complex of features known as the Stansbury shoreline.

After the low point of the Stansbury Oscillation was reached, some 20,700 years ago, the lake rose above the Stansbury terraces and then began an almost inexorable rise to its high point, reached sometime between 15,500 and 15,000 years ago. I say "almost inexorable" because there were several oscillations, watery second thoughts, between about 20,000 years ago, when the low point of the Stansbury Oscillation was reached, and about 15,000 years ago, by which time the lake had reached its peak. The first of these oscillations occurred after the lake had climbed to about 4,725 feet, creating a series of shorelines confusingly close to, but just beneath, another set of terraces, known as the Provo shoreline, that the lake was to form on its way down. Geologist Dorothy Sack has given the name "sub-Provo" to the ones formed by the rising lake; they date, she has shown, to about 19,400 years ago.

As figure 5-10 shows, the lake then fell rapidly, rose again, fell again, rose again, fell again, and then, after about 16,000 years ago, began working its way upward. Sometime between 15,500 and 15,000 years ago, it reached its peak at an elevation of 5,090 feet, cutting the Bonneville shoreline and covering some 19,800 square miles of Utah, Nevada, and Idaho with water. This was the highest the lake was to reach, not because there was insufficient water to drive it higher, but because it had become so high that it reached the threshold that separated it from the Columbia River drainage system and began to overflow into the Snake River drainage, controlled by a threshold near Zenda in southern Idaho (figure 5-8). This overflow continued until about 14,500 years ago.

Then, it did something that was to change the face of the western landscape from southern Idaho to the Pacific Ocean. It cut through the threshold at Zenda and unleashed one of the largest floods known to have occurred in earth history. The upper 350 feet of Lake Bonneville passed through Red Rock Pass in southern Idaho, rushing first into the Snake River—which was flooded in places to depths of over

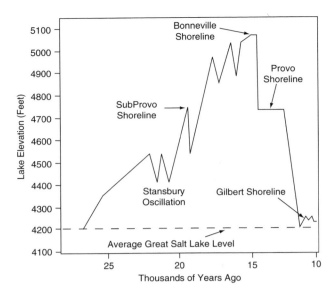

FIGURE 5-10 The changing levels of late Pleistocene Lake Bonneville. Sources: Sack 1999; Broughton, Madsen, and Quade 2000; Oviatt, Madsen, and Schmitt 2003; Oviatt et al. 2005; Godsey, Currey, and Chan 2005; Patrickson et al. 2010; and personal communications from C. G. Oviatt.

400 feet—and then into the Columbia River and, finally, into the Pacific. Although the size of Lake Bonneville is easy to imagine from within the Bonneville Basin, the magnitude of the Bonneville Flood is anything but easy to visualize.

Geologist Jim O'Connor has calculated that at its peak discharge, water tore through Red Rock Pass at the rate of about 35,300,000 cubic feet per second and perhaps even a bit more. The flooding stopped when the lake reached resistant rock at Red Rock Pass, stabilizing its levels at the Provo shoreline. During the flood, a volume of water almost equal to that contained within Lake Michigan (1,169 cubic miles, compared to the 1,140 cubic miles disgorged through Red Rock Pass) was shunted into the Snake River basin. By the time it was all over, Lake Bonneville had declined from the Bonneville shoreline, at 5,090 feet, to the Provo shoreline, at 4,740 feet, and the lake had decreased from some 19,800 square miles to some 14,000 square miles in area.

How long Lake Bonneville took to unleash all of this water is unknown. However, O'Connor calculates that the entire episode lasted about three hundred days, with discharge greater than about 18,000,000 cubic feet per second for about eighteen days. Given that the lake fell 350 feet during this time, the average rate of lake-level decline would have been 1.2 feet a day. Had people been there, they could have watched it fall.

Because it is extremely difficult to visualize what such a flood would have been like, a few comparisons might help provide some perspective. The average discharge rate for the Amazon River is 6,180,000 cubic feet per second—approximately one-fifth of the peak discharge rate of the Bonneville Flood. The largest known flood in historic times occurred in 1953, when the Amazon discharged 13,595,000 cubic feet per second—only 39 percent of the peak discharge of the Bonneville Flood. Nothing in historic times comes even remotely close to what happened at Red Rock Pass some 14,500 years ago. Indeed, the average discharge rates of all the world's rivers to the oceans is 42,376,000 cubic feet per second; the peak discharge rate for the Bonneville Flood is 83 percent of that figure.

As massive as the Bonneville flood was, it does not come close to being the largest flood known in earth history. That record may belong to the largest of the Missoula floods. These occurred when glacial ice dammed Clark Fork of the Columbia River drainage in southern Wyoming, creating a lake that contained as much as seven hundred cubic miles of water. The dam ultimately gave way, only to grow and fail again. This occurred many times between about 16,000 and 13,500 years ago, creating, among other things, the remarkable Channeled Scablands of eastern Washington. The largest of the floods discharged at least 600,000,000 cubic feet of water per second at its peak and perhaps nearly double that amount. A second contender is provided by another ice-dammed lake, this one along the Chuya River in the Altai Mountains of south-central Siberia. There were multiple floods here as well, beginning at about 23,000 years ago, all caused by failure of the ice dam. Victor Baker and his colleagues calculated that one of these failures unleashed a flood that, at its peak, discharged more than 636,000,000 cubic feet per second.

The Bonneville flood was big enough, though, and with the level of the lake now established by the new threshold at Red Rock Pass, some 1.5 miles south of the old threshold at Zenda, the lake stabilized at an elevation of about 4,740 feet. The details of what happened next are still being worked out, but according to recent work by Holly Godsey and her colleagues, it appears that the lake stood at this level until 12,500 years ago. Then, it fell rapidly. Work by Jack Oviatt, Holly Godsey, and their colleagues suggests that the lake reached modern—Great Salt Lake—levels between 11,500 and 11,000 years ago. If that estimate is correct, and much of what we know suggests that it is, overflow continued at Red Rock Pass for some 2,000 years.

In the northern Bonneville Basin, it does seem that Lake Bonneville had fallen to something close to modern levels between 11,500 and 11,200 years ago. That this is the case is suggested by the history of Lake Bonneville fishes.

There are no fish in Great Salt Lake today, because the lake is far too saline for them, but there were fish in Lake Bonneville. These I discuss in some detail in chapter 7. All I want to mention now is the fish history reconstructed by Jack Broughton and his colleagues from the paleontological site of Homestead Cave, Utah. Homestead Cave sits on a northern spur of the Lakeside Mountains, just west and south of the Great Salt Lake at an elevation of 4,612 feet; the barren playa of Lake Bonneville lies to the immediate

FIGURE 5-11 Lake Bonneville shorelines and the location of Homestead Cave, Lakeside Range, Utah (shorelines after Madsen 2000b).

west and northwest of the site (figures 5-7 and 5-11). Excavated by David Madsen, David Rhode, and myself, the site provided a huge sample of the remains of small vertebrates that spans the past 11,300 years.

The fish bones were analyzed by Broughton and his colleagues. They showed that the lowest stratum of the site, Stratum I and dating to between 11,300 and 10,200 years ago, contains more than 90 percent of the fish remains deposited during the entire 11,300-year regional history recorded in the site. In fact, there were exactly 13,536 fish bones that Broughton was able to identify from Stratum I.

Today, there is no water whatsoever near Homestead Cave, and while fish can be transported great distances by raptors and other wide-ranging fish-eating predators and scavengers, the fact that most of the fish bones in Homestead come from the lowest stratum certainly means that fish were then to be had nearby. Broughton and his colleagues showed that these bones were introduced into the site by owls who were scavenging local fish die-offs produced by severe declines in the levels of Lake Bonneville. Local fish die-offs mean, of course, that there was at least intermittent shallow water in the Homestead Cave area between 11,300 and 10,200 years ago.

Broughton showed that the deepest part of Stratum I contained fishes that do not tolerate saline waters or high temperatures, members of the whitefish genus *Prosopium* and the sculpin genus *Cottus*. The upper part of the stratum contained a somewhat different assemblage, marked by an increase in abundance of fishes tolerant of moderately saline and warm waters. He also showed that some of the fishes deposited in the lower part of Stratum I were larger than those deposited in the upper part, and inferred from this that at least two, and perhaps multiple, fish die-off events are represented at the site. The first of these events occurred around 11,200 years ago, marking the decline of the lake to something close to modern Great Salt Lake levels. This closely matches the reconstruction of late Pleistocene Lake Bonneville history provided by Oviatt and his colleagues. The lake rose again after this time, now supporting a new set of fishes, ones tolerant of saltier, warmer waters. These fishes, in turn, may have become extinct soon after 10,400 years ago. I return to this second, latest-Pleistocene interlude in Lake Bonneville history below.

So we seem to know when post-Provo Lake Bonneville fell to modern, fish-killing levels—around 11,200 years ago. We also have information from the southern Bonneville Basin that fits well with this sequence.

I mentioned earlier that there are seven separate subbasins within the Bonneville Basin as a whole. As figure 5-8 shows, the southern part of the Bonneville area is largely occupied by the Sevier subbasin. This drainage is separated from the more northerly parts of the Bonneville Basin by what is called the Old River Bed, an abandoned, broad river valley that sits at an elevation of 4,560 feet between Keg Mountain and the Simpson Mountains, some fifty-five miles north of today's Sevier Lake (figure 5-7). Whenever the waters of Lake Bonneville climbed above 4,560 feet, the Old River Bed area was flooded. This happened shortly after 20,000 years ago as the lake climbed toward its Bonneville shoreline high. Obviously, whenever Lake Bonneville dropped beneath this level, separate lakes formed to the north and south of the Old River Bed channel. The late Pleistocene lake that formed in the Sevier subbasin gets its own name, Lake Gunnison, so called after Lieutenant J. W. Gunnison, a U.S. government surveyor who lost his life in the Sevier River Valley in 1853.

When did Lake Bonneville fall so much that it was beneath the 4,560 feet Old River Bed threshold, creating separate lakes to the north and south? There are, at the moment, two possibilities here. The first is that this occurred around 12,500 years ago, when Lake Bonneville fell rapidly from the Provo level. Some radiocarbon dates from the Sevier Basin suggest that this may have been the case. On the other hand, radiocarbon dates for deposits left behind by Lake Gunnison itself range from 11,400 to 10,100 years ago. Lake Gunnison was certainly in existence by 11,400 years ago and may have existed some 1,100 years earlier.

Oviatt, David Madsen, and Dave Schmitt have shown that on the Dugway Proving Ground, not far to the northwest of the Old River Bed threshold, high-energy stream channels created as a result of water flowing northward through this area from Lake Gunnison had begun to form by at least 10,500 years ago (see chapter 8). The very earliest dates for these channels are not known, so this does not help us determine when the lake came into existence, but it does fit well with another fact, known for more than a century. After Lake Bonneville fell beneath the Old River Bed threshold, Lake Gunnison overflowed to the north, into what is now the Great Salt Lake subbasin. In fact, after Lake Bonneville declined to beneath the elevation of the Old River Bed, the water in the Sevier subbasin seems to have remained high enough to flow north until sometime after 10,000 years ago.

Why Lake Gunnison climbed high enough to flow northward is a mystery, though it is known that this is what created the Old River Bed in the first place. Over the years, Oviatt has suggested two very different possibilities. Some years ago, he wondered whether Lake Gunnison might have formed as a result of a strengthening of the southwestern monsoons—Houghton's Gulf Component of Great Basin precipitation. In this view, summer monsoonal rains, and accompanying clouds, may have intruded far enough north to have fed the lake that existed in the Sevier Basin at this time but not far enough north to have maintained Lake Bonneville itself. This hypothesis, however, does not appear to be supported by the history of plants and animals during the late Pleistocene in the eastern Great Basin (chapters 6 and 7), and Oviatt himself no longer thinks it is likely to be correct.

Instead, he suspects that the explanation may lie in the fact that the Sevier Basin, in which Lake Gunnison formed, is quite small compared to the massive basins to the north. As a result, he reasons, the streams feeding this small basin may have been able to fill it to higher levels than streams to the north were able to fill theirs. He couples this with the observation that groundwater today flows from the Sevier Desert northward, a process that would have been amplified during the wetter times that marked the latest Pleistocene and early Holocene. These two things together, he suggests, allowed water to flow to the north at this time.

No matter the explanation, the geology itself shows that Lake Gunnison overflowed into the northern basins of Pleistocene Lake Bonneville from sometime before 11,400 years to 10,100 years ago and apparently beyond that time (chapter 8). Indeed, as I discuss in Chapter 9, people were here then and took substantial advantage of the rich stream and marsh environments that this overflow created.

The fall of Lake Bonneville to essentially modern levels by about 11,200 years ago was not its last gasp. Oviatt and his colleagues have provided strong evidence, from radiocarbon dates on two very low shorelines—about 4,250 feet—that the lake rose once again sometime between about 10,500 and 10,000 years ago. The creation of two separate beaches suggests that the lake may have risen to this level twice during this interval, with the whole episode perhaps lasting only a few decades. It is also possible, but by no means certain, that this final, feeble episode in the history of Lake Bonneville saw the return of the fishes that succumbed to the second die-off represented at Homestead Cave. The lake terraces created by this rise are referred to as the Gilbert shoreline, in honor of the great Grove Karl Gilbert, the late-nineteenth-century geologist whose work on the history of Lake Bonneville was so astonishingly accurate that all that has happened since involves filling in the details he left for us to deal with.

The Gilbert episode in the history of Lake Bonneville occurs at almost exactly the same time as a significant climatic event first defined from northwestern Europe. This event, the Younger Dryas, takes its name from a herbaceous tundra plant, the mountain avens, genus *Dryas*. This plant is a marker for the rapid replacement of forest by arctic grasses, herbs, and shrubs in northwestern Europe shortly after 11,000 years ago. Glaciers expanded in many places in the Northern Hemisphere, from Norway to British Columbia. Temperatures fell dramatically, to as much as 27°F below modern in Greenland, though temperature decreases were less dramatic elsewhere. Methane levels declined from about 700 parts per billion by volume (ppbv) to about 475 ppbv. Since methane is produced by wetlands, the implication is that wetlands decreased dramatically in area, or that their biological activity declined dramatically because of the reduced temperature, or both. The Younger Dryas thus represents a cold snap of glacial intensity, one felt most strongly in the North Atlantic but affecting other areas as well. The Younger Dryas ended shortly before 10,000 years ago with astonishing rapidity. In Greenland, where our data are most precise, it began around 10,600 years ago and ended almost exactly 500 years later. When it ended, temperatures climbed at least

18°F within a span of about 60 years, and whatever the atmospheric event was that led to this change seems to have occurred in as little as three years. Along with climbing temperatures, glaciers and tundra vegetation retreated, and methane values climbed even higher than they had been before the Younger Dryas began.

The cause of this severe climatic episode is under debate. One widely discussed possibility is that it was caused by a massive flood of fresh, cold water that got dumped into the North Atlantic from a huge glacial lake (Lake Agassiz) that formed to the north and west of where the Great Lakes now sit. That influx of cold water then shut down the oceanic circulatory system that brings warm waters to the North Atlantic from the south. This system is known as the thermohaline circulation because it is driven by the heat and salinity of ocean water. Warm waters arriving in the far north from the south are cooled and, at times, frozen. Freezing the water squeezes the salts out of it, causing the nonfrozen water to become denser. This cold, dense water then sinks and flows southward, ultimately to rise and flow northward, repeating the cycle. Dramatically changing both heat and salinity can cause the system to break down, and a massive influx of cold, fresh water could do that. Another possibility is that this circulation was shut down by a massive flotilla of icebergs dumped into the North Atlantic from eastern North American glaciers, or that sea ice that formed in the Arctic Ocean played the same role. None of these explanations is as yet fully supported by the empirical record, and it is quite possible that multiple mechanisms were involved. In addition, some have argued that the Younger Dryas cold snap may be a "normal" event at the end of glaciations and that no exceptional mechanism may be needed to explain it. Since we do not fully understand why the Younger Dryas occurred, we obviously cannot fully understand the relationship between it and the contemporary rise and fall of Lake Bonneville. The correlation between the two, though, is clear.

No matter how this short episode in the life of Lake Bonneville is to be explained, when the lake retreated from the Gilbert shoreline, Lake Bonneville came to an end. The retreat of Lake Bonneville from the Gilbert shoreline saw the birth of the Great Salt Lake Desert, the Bonneville Salt Flats, and Great Salt Lake itself. Later fluctuations in water levels were miniscule compared to what happened during the Pleistocene.

A NOTE ON PRE–LAKE BONNEVILLE LAKES IN THE BONNEVILLE BASIN

Lake Bonneville wasn't the only deep lake to form in the Bonneville Basin. A large number of cores have been extracted for both commercial and scientific reasons from the Bonneville Basin, cores that incorporate up to 16,000,000 years of sedimentary history. Analyses by Owen Davis, Jack Oviatt, Bob Thompson, and other scientists have shown that these cores contain evidence for four major lake cycles during the past 800,000 years, but none before that time. In addition, each of these cycles occurs at, or just after, a major Northern Hemisphere glacial episode: around 620,000 years ago (the Lava Creek cycle), 420,000 years ago (tentatively named Pokes Point), 150,000 years ago (Little Valley), and, of course, 28,000–10,000 years ago (Bonneville). There was also a shallower cycle, called Cutler Dam, that dates to around 60,000 years ago. All the terraces seen in the Bonneville Basin today, however, were built by Lake Bonneville itself, making it difficult to know exactly how deep those earlier lakes were. Other shallow cycles probably remain to be discovered, but the cores suggests that major lakes began to form within the Bonneville Basin shortly after 800,000 years ago, during or just after periods of major glacial advance.

Today, the Bear River provides about 60 percent of the water that flows into the Bonneville Basin. As figure 2-2 shows, this river originates in the mountains of northern Utah, then flows north into southern Idaho, then quickly changes direction to flow back toward, and into, the Bonneville Basin. This dramatic change in direction was caused by lava flows that blocked the northward movement of the river. Before that barrier appeared, the upper portion of the Bear River flowed into the Snake River drainage. With the barrier in place, it flowed into the Bonneville Basin.

Much of the water—66 percent of it—that the Bear River provides to the Bonneville Basin comes from streams that feed the river from beyond the diversion point—the Cub, Logan, and Little Bear rivers, for instance. These tributaries seem to have been flowing into the Bonneville Basin all along. This suggests that the diversion of the upper Bear River would have increased Bear River inflow into the Bonneville Basin by some 33 percent, about the same proportion as the Weber and Jordan rivers provide today. This, of course, would have made a difference.

Perhaps, it might be argued, it was not until the Bear River was diverted in this fashion that lakes began to form in the Bonneville Basin. It might even be argued that the Bear River was diverted multiple times—the lava flowed, the river was diverted south until it cut through the barrier and flowed north again, the lava flowed again, and so on. Perhaps this process could help account for the history of deep lakes in the Bonneville Basin.

It turns out that it cannot, however. Owen Davis's work on Bonneville Basin sediment cores shows that a dramatic vegetation change occurred in the basin around 750,000 years ago. Prior to this time, plants that mark hot and dry environments, including *Atriplex*, dominate the terrestrial pollen record contained by the cores. After that time, plants that mark cooler and moister environments, including sagebrush and pine, dominate. Soon thereafter, the major lake cycles begin. There is no evidence in Davis's cores for an increased influx of Bear River waters until about 310,000 years ago, long after the Lava Creek

(620,000 years ago) and Pokes Point (420,000 years ago) lake cycles had ended.

This evidence matches that provided by William Hart, Jay Quade, and their colleagues. They took advantage of the fact that the waters that flow into lake basins, including the Bonneville Basin, often have distinctive chemical signatures derived from the rocks over which they flow. The chemical signature they examined was the ratio of two isotopes of strontium, ^{87}Sr and ^{86}Sr. Because the Bear River has a very distinctive isotopic signature, and because that signature is retained in the materials deposited in the lakes that occupied the basin, they were able to show that the Bear River was flowing into the Bonneville Basin during all the lake highstands that occurred in the basin back to and including the Little Valley cycle, about 150,000 years ago (their data extend no further back in time than this).

So, while the later lake highstands would not have been as high without the full contribution of the Bear River, highstands occurred both before and after the southerly diversion of this river. It now looks most likely that global cooling accounts for the onset of pluvial lake cycles in the Bonneville Basin, of which the Lake Bonneville cycle was the most recent.

Pleistocene Lake Lahontan

When land speed records are set on the Bonneville Salt Flats, they are set in the basin of Pleistocene Lake Bonneville. Those set on the Black Rock Desert are set in the basin of Pleistocene Lake Lahontan, the second largest of the Great Basin's pluvial lakes (see figure 5-12).

At its maximum, Lake Lahontan covered 8,284 square miles of western and northwestern Nevada, as well as a small part of adjacent California, reaching a maximum depth of about nine hundred feet at the site of today's Pyramid Lake. In historic times, lakes in the Lahontan Basin have had a maximum surface area of about six hundred square miles. Thus, the Pleistocene lake at its largest was nearly fifteen times larger in surface area than the modern lakes at their largest.

The Lahontan Basin is far more complicated than the Bonneville Basin, in large part because six major rivers drain into it from different directions. These six are the Humboldt, which terminates in Humboldt Lake and the Humboldt Sink; the Truckee, which flowed into Pyramid and Winnemucca lakes; the Carson, which flowed into the Carson lakes and Carson Sink; the Walker, ending in Walker Lake; the Susan, flowing into Honey Lake; and the Quinn, ending in the Black Rock Desert. Today, the Susan and Quinn rivers contribute about 4 percent of the waters flowing into the Lahontan Basin; the rest are contributed by the Carson (17 percent), Humboldt (38 percent), Truckee (27 percent), and Walker (14 percent) (table 5-4).

With the exception of the Quinn, each of these rivers flows into a more or less substantial lake. Given that the sinks of the Humboldt and Carson rivers can merge in times of high water today, and that the Truckee River flows into two separate basins (Pyramid and Winnemucca), these rivers provide six separate subbasins within the Lahontan Basin as a whole. In addition, there is a seventh, the Buena Vista subbasin, to the northeast of Carson Sink, which has no river at all (figure 5-12).

These seven separate drainages were once united beneath the waters of Lake Lahontan. However, since each of these basins is separated from every other by a threshold, or sill, at a different level, the rates of rise and fall of water within each is independent of those rates in every other, unless the water rises high enough to connect one or more of them. When that occurs, of course, the waters in the connected basins form a single lake. Table 5-5 shows the names and elevations of these thresholds, and figure 5-12 shows their locations. The highest threshold in the system connects the Walker Lake basin to the Carson Sink and lies at an elevation of 4,291 feet. Only above this level are all the separate subbasins connected into a single Lake Lahontan. As a result, as geologist Jonathan Davis has noted, there really is no single Lahontan chronology, but instead a family of chronologies and lake histories—fewer in number when the lake is high, greater in number when the lake is low.

To makes matters even more complex, not all of the rivers that empty into the Lahontan Basin have stayed put during the past 100,000 years or so. In the 1980s, Davis suggested that sometime during the later Pleistocene, the Humboldt River flowed not into Humboldt Lake as it does today, but instead ran well north of what is now Rye Patch Reservoir in north-central Nevada and into the Black Rock Desert. More recent research has shown him to be right. Using the same strontium isotopes that have been so effective in tracing the history of the Bear River, Larry Benson and his colleagues have shown that the late Pleistocene Humboldt River flowed into the Black Rock Desert until sometime between 12,500 and 11,000 years ago. Possible diversions of the Truckee River have been less well studied, but it appears that it may have briefly changed its channel during the relatively recent past. As Lake Lahontan fell from its late Pleistocene highstand sometime after 14,000 years ago (see below), the Truckee may have flowed not north toward Pyramid Lake but east toward Fernley, Nevada, and the nearby Fernley Sink.

Walker River has been even more fickle. Just as with the Bear River, the geography of the Walker suggests that odd things might happen here. The east and west branches of Walker River flow north and east out of the Sierra Nevada, to join in Mason Valley just south of Yerrington, Nevada (figure 2-2). The enlarged river then flows almost due north, to the end of Mason Valley, then turns in the opposite direction and flows almost due south, along the east side of the Wassuk Range, to end up in Walker Lake. The abrupt turn taken by the river is curious, not only because it is so abrupt but also because there is an old river channel

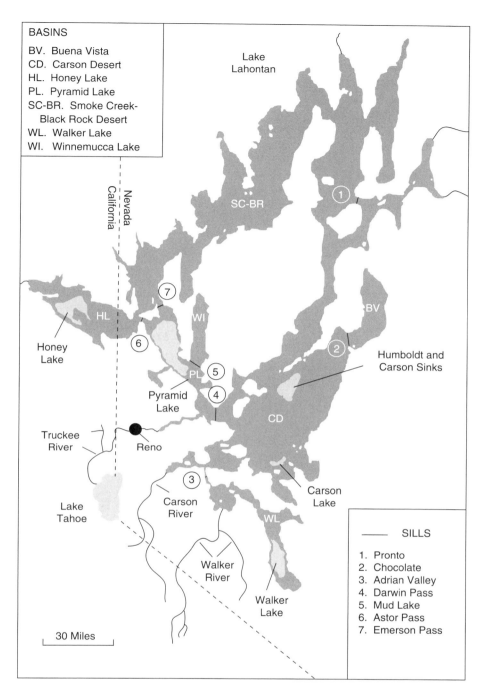

FIGURE 5-12 Lake Lahontan and its subbasins (after Benson 2004a).

heading off to the northwest, out of Mason Valley and, by a nearly level pass called Adrian Valley, into the Carson River drainage. Indeed, Adrian Valley contains the threshold between the Walker Lake and Carson Lake drainage systems. There is no doubt that in the past the Walker River occupied this channel, flowing north into the Carson drainage. Whenever that occurred, the effect on Walker Lake would have been devastating. As I mentioned, Walker River provides 83 percent of the inflow to Walker Lake. Without that source, Walker Lake would be a puddle. The challenge is not to figure out if this happened but rather to figure out when and why.

How complicated the Lake Lahontan system is should now be obvious. Six different rivers feed the system, of which four come from the Sierra Nevada and provide about 61 percent of the modern river inflow into the basin. The

THE LATE PLEISTOCENE PHYSICAL ENVIRONMENT 107

TABLE 5-4
River Discharge into the Lahontan Basin

River	% of Total Discharge	Source Area
Carson	17	Sierra Nevada
Humboldt	38	Northern and Eastern Nevada
Quinn	1	Northern and Eastern Nevada
Susan	3	Sierra Nevada
Truckee	27	Sierra Nevada
Walker	14	Sierra Nevada
Total	100	

SOURCE: After Benson and Paillet 1989.

two others come from the mountains of northern and eastern Nevada and provide the remaining 39 percent of river inflow (table 5-4). There are seven different subbasins within this area, each separated from the others by thresholds that fall at different elevations. Two rivers, the Walker and Humboldt, have even switched basin allegiance during the past, and the Truckee may have done so as well. Even with these complexities, however, tremendous progress has been made during the past few decades toward gaining a detailed understanding of the history of Lake Lahontan and its separate parts.

As with the Bonneville Basin, the Lahontan Basin contained substantial pluvial lakes long before late Pleistocene Lake Lahontan appeared. Work by Mareth Reheis and her colleagues has shown that at least six or seven lakes in this basin reached levels higher than those reached by Lake Lahontan, three of which date to between 2.6 and 1 million years ago. In separate work, Yong Lao and Larry Benson have provided rough dates of about 290,000 and 130,000 years ago for pre-Lahontan highstands, although these may or may not be related to the lakes detected by Reheis and her coworkers. None of these lakes is well dated, but it does seem that the older the lake, the higher the shoreline, a phenomenon known from some other Great Basin drainages, though not the Bonneville Basin. We do not know why this is the case, though there are some obvious possibilities. For instance, other basins may have drained into the Lahontan Basin in the past, only to become disconnected through time, thus decreasing water flow into the basin. The uplift of rainshadow-producing western mountains might have played a role, as might changing positions of the jet stream, a mechanism I discuss in more detail later in this chapter.

Significant parts of Lake Lahontan history remain murky into much later times. Larry Benson and his colleagues argue that there was no Lake Lahontan between about 27,500 and 23,200 years ago (figure 5-13). During that interval, they suggest, Pyramid Lake levels were beneath the elevation of Mud Lake Slough (3,862 feet), the threshold between Pyramid and Winnemucca lakes. Geologist Ken Adams, on the other hand, has evidence suggesting that, at 27,300 years ago, the western basins of Lake Lahontan held a lake that reached an elevation of 3,995 feet. By 23,200 years ago, he argues, the lake had climbed to an elevation of 4,115 feet and covered some 2,625 square miles, an idea first forwarded by Jonathan Davis many years ago. Because there is, as yet, no way to reconcile these two distinctly different views of Lake Lahontan history during this period, I have included them both on figure 5-13.

It does seem that Lake Lahontan had risen to or near the level of Darwin Pass (4,150 feet) by 21,000 years ago. At this level, all of the separate basins that comprise the Lahontan Basin would have been joined, except for the Walker Lake drainage (table 5-5), and Lake Lahontan would have had an area of 5,524 square miles.

TABLE 5-5
Lake Lahontan Subbasin Thresholds and Their Elevations
Sill locations are provided in figure 5-12.

Threshold	Elevation (feet)	Basins Joined
Adrian Valley	4,291	Pyramid/Winnemucca/Smoke Creek–Black Rock/Honey/Buena Vista/Carson Desert–Walker Lake
Pronto	4,239	Secondary joining of all basins listed below
Darwin Pass	4,150	Pyramid/Winnemucca/Smoke Creek–Black Rock/Honey/Buena Vista–Carson Desert
Chocolate	4,140	Pyramid/Winnemucca/Smoke Creek–Black Rock/Honey Lake–Buena Vista
Aster Pass	4,009	Pyramid/Winnemucca/Smoke Creek–Black Rock–Honey Lake
Emerson Pass	3,960	Pyramid/Winnemucca–Smoke Creek–Black Rock Desert
Mud Lake Slough	3,862	Pyramid Lake–Winnemucca Lake

SOURCES: Benson and Mifflin 1986. For slightly different sill elevations, see K. D. Adams et al. 2008.

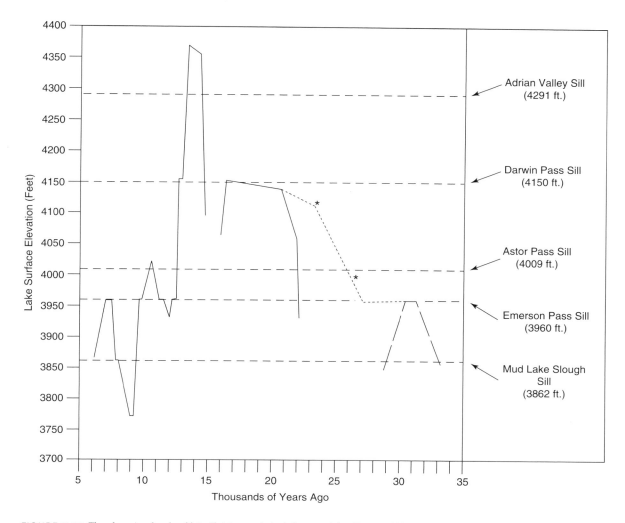

FIGURE 5-13 The changing levels of late Pleistocene Lake Lahontan (after Benson 2004a, 2004b, K. D. Adams et al. 2008, and K. D. Adams 2010). The asterisks at 27,300 and 23,200 years ago represent lake levels reconstructed for those times by K. D. Adams (2010). I have interpolated lake level histories through these points (dashed line).

The lake remained around this level until about 16,000 years ago, when it began an apparently rapid decline. By 15,500 years ago, it had fallen to somewhere beneath 4,110 feet, and perhaps down to or even beneath the Emerson Pass threshold, at 3,960 feet.

That decline did not last long, for the lake began a rapid rise around 15,000 years ago. Larry Benson and his colleagues estimate that by about 14,500 years ago, the lake had reached its maximum elevation (at 4,380 feet) and area (8,284 square miles). Here, they suggest, it stayed for perhaps 900 years, beginning its decline around 13,600 years ago. Work by Ken Adams and Steve Wesnousky, however, suggest a different possibility. They have radiocarbon dates of between 12,690 and 13,070 years ago from the bones of the extinct camel *Camelops hesternus* (see chapter 4) from sediments laid down after the lake had retreated from its high point. They argue that the lake must have fallen from its high point immediately before this camel met its unhappy fate. They also argue that the Lahontan highstand could not have lasted more than 150 years and that it might been of much shorter duration than that.

No matter how these interpretive differences are resolved, all agree that the lake was dropping rapidly by 13,000 years ago. Benson and his colleagues argue that by 12,400 years ago, the lake was well beneath the Mud Lake Slough threshold, at 3,862 feet, the point at which Pyramid and Winnemucca lakes join. If the lake was still at its highstand at 13,100 years ago, this suggests a retreat from that highstand of about nine inches a year. In the Carson Sink, Dave Rhode, Ken Adams, and Bob Elston have estimated the rate of retreat at between four and eight inches a year. Either way, it was substantial.

THE LATE PLEISTOCENE PHYSICAL ENVIRONMENT 109

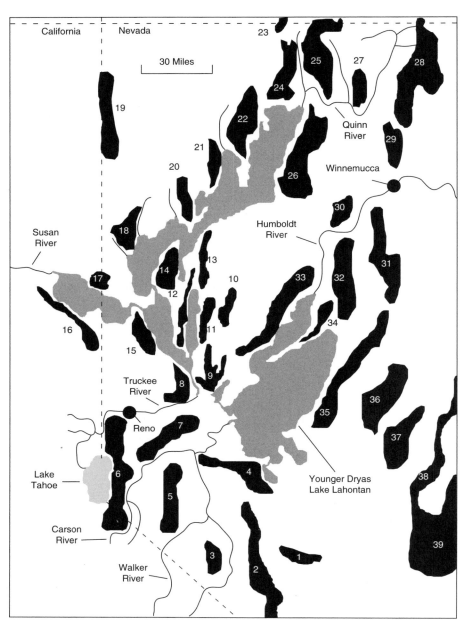

FIGURE 5-14 Younger Dryas–aged Lake Lahontan (after Adams et al. 2008).

Here it stayed until shortly after 11,000 years ago when, during the Younger Dryas, and just like Lake Bonneville, it rose for one last time, reaching an elevation as high as 4,035 feet—enough for a joined Pyramid and Winnemucca Lake to spill into the Smoke Creek–Black Rock Desert and Honey Lake basins one last time (figure 5-14). The chronological control over this period of the lake's history is not as good as one might like. Richard Briggs and his colleagues have shown that it stood at this level at 10,800 years ago, but how long it remained here is not known. By 10,000 years ago the lake had retreated to beneath Emerson Pass, and Lahontan's last gasp was over.

WALKER LAKE

One of the most-studied subbasins within the Lahontan Basin as a whole is the Walker Lake Basin. I have not said much about Walker Lake to this point, for two reasons. First, as I have mentioned, changing levels of this lake may reflect either regional climatic change or the diversion of the Walker River into the Carson Sink. Second, what we know about the history of Walker Lake doesn't make a lot of sense.

The Walker Lake Basin was the focus of intense investigation by a U.S. Geological Survey scientific team, in the hopes that a long-term climatic record could be developed that would help determine the suitability of Yucca Mountain, in the Mojave Desert of southwestern Nevada, as an appropriate long-term repository for nuclear waste. Given that "long-term" in this context came to mean hundreds of thousands of years, knowledge of future climatic change, and the potential that such change has for altering groundwater characteristics, was essential. Since understanding future climatic change requires an understanding of past climates, huge efforts were directed toward gaining that knowledge. Those efforts focused on virtually everything that might tell us what the past of this region was like, including lake histories and vegetational change. The Yucca Mountain project has been shelved, but the science done in association with it was of high importance.

Larry Benson led the interdisciplinary group that examined the Walker Lake Basin, emphasizing the extraction and dating of lake sediment cores and studying virtually everything in those cores that might provide information on the history of the lake. The results of their efforts provided some evidence that the Walker Lake basin held a deep lake between at least 32,000 and 25,000 years ago. There is even better evidence that the basin held not a lake but a saline marsh between about 22,000 and 14,000 years ago. During this interval, it appears that the Walker River was flowing not into Walker Lake, but instead north into the Carson Basin, where, as we have seen, a sizable lake existed for much of this time.

The most puzzling aspect of the Walker Lake picture involves the next 10,000 years, since the sediment cores extracted by the USGS team have supplied no convincing evidence that a lake existed in this basin between about 14,000 and 4,700 years ago. This, in turn, suggests that the Walker River must have flowed northward during this entire interval. The difficulty is that this could not have been the case if the reconstruction of Lake Lahontan that I have just presented—much of it built by the same people—is correct. That reconstruction has Lake Lahontan so high between 14,500 and 13,000 years ago that it incorporated the Walker Lake Basin. In fact, radiocarbon-dated shorelines within the Walker Lake Basin itself imply that a deep lake existed here at about that time. Given this situation, the Walker River could not have flowed north even if it had wanted to, but must instead have flowed directly into an expanded Lake Lahontan.

There are some obvious possibilities for resolving this discrepancy. As Benson has noted, it is possible that the lake sediments laid down during the 14,000-year-old Lake Lahontan highstand were subsequently eroded, presumably by wind action, from the places sampled by the USGS team. It is also possible that the lake that existed at the Lahontan maximum simply did not leave much of a sedimentary record, though this seems extremely unlikely. No matter what the explanation, there must be one, since it is impossible that Lake Lahontan rose so high that it entered the Walker Lake Basin between 14,500 years and 13,000 years ago at the same time as the Walker River flowed north and there was no lake here. Worse, if the 14,500-to-13,000-year-old highstand did exist in the Walker Lake basin (as surely must have been the case) and the USGS team missed it, then the arguments the team has made that there was no lake here anytime between 14,000 and 4,700 years ago cannot be given too much credence, either. If the highstand could be missed, certainly other, lower lake levels could be missed as well. I return to this matter in chapter 8, which deals with the last 10,000 years.

LAHONTAN AND BONNEVILLE COMPARED

Comparing what we know in general about the histories of Lake Lahontan and Lake Bonneville shows that those histories are roughly, though not perfectly, synchronous during the past 28,000 years ago or so. Both lakes seem to have been low around 28,000 years ago and to have remained so for several thousand years after that time. By 22,000 years ago, Bonneville had climbed to the Stansbury shoreline, at about the same time that Lahontan was rising to the level of Darwin Pass. Lahontan stayed at about this level until 16,000 years ago, then declined rapidly, then grew again, reaching its maximum extent at or soon after 14,500 years ago. During this same period, Bonneville had its ups and downs but trended steadily upward and, in this way, differed from what we now know about Lahontan at this time. However, both were very high—at their late Pleistocene maxima, in fact—between about 15,000 and 13,000 years ago, after which time both fell dramatically, though how dramatically remains to be worked out. Then, after dropping to very low levels, both climbed again during the Younger Dryas, between 11,000 and 10,000 years ago.

While the parallels between the two lakes' histories aren't perfect, they are too close to be accidental. It is important to realize that these similarities exist even though the water supply for Lake Bonneville lies entirely in the eastern Great Basin, and largely in the Uinta Mountains and Wasatch Range, while more than 60 percent of the river flow into the Lahontan Basin today comes from the Sierra Nevada. Given such different sources of inflow, the similarities must reflect broad, regional climatic control over the sizes of these lakes.

It is, however, one thing to conclude that climate did it and quite another to specify exactly what it was about late Pleistocene climates that caused not only the growth of pluvial lakes in the Great Basin but also the general synchroneity in the histories of its two largest lakes.

Lake Chewaucan

Because they have been most heavily studied, the histories of Lakes Bonneville and Lahontan are among the best understood of all Great Basin pluvial lakes. Although the histories of many other Great Basin lakes have hardly been studied at all and little can be said about them, some have been the focus of significant research efforts.

Lake Chewaucan in south-central Oregon is one of those reasonably well studied, if somewhat frustrating, lakes. At its maximum, this lake covered some 480 square miles to a maximum depth of about 375 feet. Today, two lakes, Summer and Abert, are found within its basin. Summer Lake tends to cover about seventy square miles to a maximum depth of about 3 feet. Lake Abert is deeper, about 16 feet at its deepest, but a little smaller, covering about sixty square miles (figure 5-15). As is typical for Great Basin lakes, however, these numbers vary greatly, and at times during the past one hundred years these lakes were virtually dry.

The Lake Chewaucan basin is simpler than the Lahontan basin, but it still has its complexities. The northern part of the basin, containing Summer Lake, is separated from the rest of the basin by a threshold near the town of Paisley—the same Paisley for which the Paisley Caves are named (chapter 3). This threshold falls at about 4,390 feet; a rise in lake level above this point would see the lakes forming to the north and south coalesce. Geologist Ira Allison called the Pleistocene lake that formed in the northern basin Winter Lake; the lake to the south he called ZX Lake, named for a ranch in the area. Lake Chewaucan forms when the two rise high enough to merge.

Today, the northern basin gets most of its water from springs and the spring-fed Ana River. These springs seem to get their water by underground flow from the Fort Rock Basin, which lies directly to the north (and is the home of Fort Rock Cave, discussed in chapter 3). As a result, when the Fort Rock Basin is very wet, the Summer Lake basin will also be wet. The southern, ZX Lake basin now gets much of its water from the Chewaucan River, which ultimately flows into Lake Abert.

A glance at figure 5-15 shows that the Chewaucan River has a course that looks suspiciously like that of both Walker and Bear rivers—it flows in one direction for a long way and then changes its mind and flows in another. The point at which it changes its mind represents the 4,390-foot threshold between the northern and southern lake basins. That threshold, in turn, is composed of deposits laid down by the river itself—a combination of what geologists call a fan

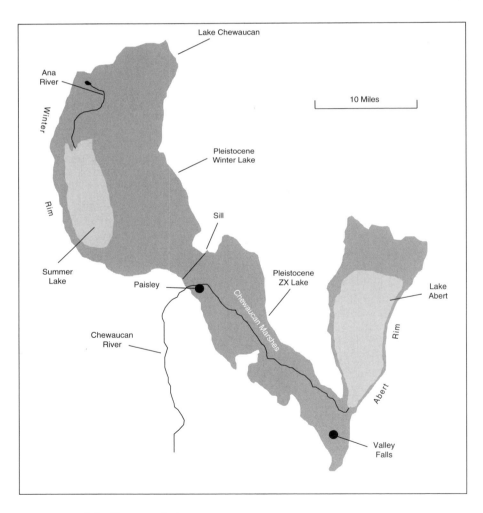

FIGURE 5-15 Lake Chewaucan Basin.

and a delta. Fans are deposited when streams lose force and deposit the sediments they carry onto a dry surface; deltas are formed when streams empty into lakes or seas and deposit those sediments. The fan-delta deposits that mark the boundary between northern and southern subbasins in the Lake Chewaucan Basin mark a time when the Chewaucan River flowed north, into the basin that now contains Summer Lake. Somehow, probably related to the buildup of this barrier, the river was diverted to the south, into the ZX Lake Basin. We don't know when that happened, but these factors do make understanding the history of the lakes in this area more difficult, just as it did in the case of the Walker and Bear rivers.

In 1982, Allison provided basic descriptions of Lake Chewaucan and its basin. Soon thereafter, Jonathan Davis demonstrated that sediments exposed along the Ana River, which feeds Summer Lake from the north, contains fifty-four separate layers of volcanic ash or tephra. Volcanic ashes from different sources, and often even those that originate from the same source at different times, have different chemical compositions. As a result, they can be identified wherever they are found. The ashes can also be dated in a variety of ways. Those deposited within the past 40,000 years, for instance, can be placed in time by obtaining radiocarbon dates for organic materials above, below, or within the ash. Older tephras can be dated using a method that depends on the decay of radioactive potassium to inert argon (K/Ar dating); still others can be dated using thermoluminescence (TL dating).

TL dating depends on the fact that electrons accumulate in substances that undergo natural irradiation. These electrons, along with light, are released when the material is heated. The rate at which electrons accumulate in a given setting can be calculated, and the amount of light released on heating is proportional to the amount of those electrons. As a result, materials that have been heated in the past and exposed to radiation can be dated either by measuring the amount of light given off on reheating or by measuring the number of accumulated electrons directly. The event dated is the time of last heating, which resets the TL clock. TL dating has wide application in archaeology. It can, for instance, be applied to ceramics or to stone tools that had been heated as part of the manufacturing process. It can also be used to date tephras. The combination of radiocarbon, K/Ar, and TL dating can allow tephras of a wide variety of ages to be dated.

Because tephras are scattered explosively and then quickly settle from the air, they mark a virtual instant in time. Because their chemical composition often allows them to be identified, far-flung deposits can be correlated with tremendous precision. Just as important, once a deposit has been dated, identifying a given ash in a depositional sequence also reveals exactly how old that part of the deposit is.

The detailed characterization and dating of tephras—tephrochronology—can provide such a powerful tool for dating and correlating deposits across vast expanses that a great deal of scientific effort has been dedicated to doing just that. Thanks to the work of many scientists, the ages and sources of many of the tephras represented in the Ana River exposure are now reasonably well known.

The youngest of the tephras in the Ana River area was provided by the eruption of Mount Mazama, some sixty-five miles to the west, about 6,730 years ago. The Mount St. Helens Mp tephra represented in these deposits fell some 20,500 years ago; this came from the same Mount St. Helens in southwestern Washington that erupted in 1980, 180 miles north and slightly west of Summer Lake. The next tephra down in the sequence, the Trego Hot Springs tephra, was given that name by Davis, after the hot springs by that name in Nevada's Black Rock Desert. Although we do not know exactly where it came from, other than the fact that it originated somewhere in the Cascade Mountains, Larry Benson and his colleagues have shown that it dates to 23,200 years ago. The Wono tephra (named by Davis after the Paiute word for Pyramid Island, where it is well represented) dates to 27,300 years ago; it probably comes from Newberry Crater, in the Cascades some fifty-five miles to the northwest. All the other tephras in the sequence here predate 35,000 years ago, with the oldest ones identified from the Summer Lake Basin falling at about 200,000 years ago.

In his initial investigation, Davis made no attempt to extract lake history from these deposits, since that was not the focus of his work. A few years later, however, he returned to the Summer Lake Basin with Robert Negrini. While Davis brought his skills in Pleistocene geology to the analysis of the deposits of Lake Chewaucan, Negrini brought his expertise in the study of variations in the declination, inclination, and intensity of the earth's magnetic field through time.

It is well-known that the declination (the difference between magnetic north and true north at any one point on the earth's surface) and inclination (the vertical component of the earth's magnetic field at any one point) change as time passes. These variations are often well recorded in fine-grained sediments that contain magnetic minerals, since those minerals align themselves with the magnetic field as they are being deposited. In many areas, curves have been built showing the precise nature and ages of these variations through time. One such curve, available from the Mono Lake Basin of California, covers the period from 36,000 to 12,000 years ago. Negrini and Davis hoped that they would be able to detect paleomagnetic variations in the fine-grained deposits of Lake Chewaucan and then correlate those variations with the well-dated Mono Lake sequence. Such correlations would allow them to infer precise dates for the Lake Chewaucan deposits, above and beyond those already available from the volcanic ashes. The sediments of Lake Chewaucan proved to contain a paleomagnetic record very similar to that of the Mono Lake Basin, albeit not nearly as continuous. The detailed similarities between these curves allowed Negrini and Davis to provide a much finer-grained chronology for the Lake Chewaucan deposits than could have been built from the tephras alone.

Armed with this knowledge, Negrini and his colleagues returned to Summer Lake with a vengeance. Among other things, they managed to take two cores from the Summer Lake Basin, one reaching a depth of 100 feet, the other 29 feet. This gave them three deep sequences to examine: these two, plus the original Ana River outcrop that had been the focus of earlier work. They correlated the sediments in these three sequences through the tephras they contain and by looking at the magnetic properties of those sediments. In the end, they were able to show that during the past 200,000 years, Lake Chewaucan had reached very high levels three times. The first of these falls between 190,000 and 165,000 years ago and corresponds closely to marine isotope stage (MIS) 6, a time of glacial expansion in the north (see chapter 3). The second dates to about 89,000 to 47,000 years ago, corresponding to MIS 4, another significant cold episode that saw glaciers expanding in the north.

But it is the third, late Pleistocene highstand that is of most concern here. The Mono Lake paleomagnetic sequence I describe above shows a significant but relatively short-lived change in the inclination and declination of the earth's magnetic field during the later-Pleistocene Mono Lake excursion, dated to about 28,600 years ago. Identifying this alteration in the deposits at Summer Lake allowed Negrini and his colleagues to date the section of the deposits in which it was registered. This, in turn, allowed them to show that Lake Chewaucan was fairly low at that time. Soon after that date, the lake dropped to extremely low levels, so low that although the bottom of the Summer Lake Basin still contained water, it was not high enough to reach the section exposed along the Ana River. By 23,000 years ago, however, it had risen to at least half of its maximum level, at about 4,305 feet. We know this because the 23,200-year-old Trego Hot Springs tephra was deposited in lake sediments at this elevation and because a radiocarbon date shows the lake was at this level 22,100 years ago. The lake then dropped significantly, only to rise again by 20,500 years ago (the age of the Mount St. Helens Mp tephra), reaching its highest late Pleistocene level at around 17,500 years ago.

For the next 5,000 years, though, Lake Chewaucan has disappointed us. All of the deposits that have been examined so far are truncated at around 17,000 years ago. At the Ana River exposure, for instance, the Mazama tephra (6,730 years old) directly overlies lake deposits dated to 16,700 years ago. The best guess is that erosion has removed the deposits that were once here and, with them, the environmental record they contained.

Thanks largely to work by geologists Joseph Licciardi and Dorothy Friedel, we do have information on the latest Pleistocene history of Lake Chewaucan. Licciardi focused his efforts on the southern section of the lake basin, the section that contains the Chewaucan Marshes and Lake Abert. Working widely throughout that region, he coupled detailed geological descriptions of exposed sediments with the dating of snail shells from past shoreline or near-shoreline contexts. Those dates showed that the lake stood at a level of at least 4,347 feet 11,930 years ago; Licciardi suggests that the lake may have risen high enough at this time to overflow the 4,390-foot threshold near Paisley and form a single, continuous Lake Chewaucan. His dates also suggest that the lake fell rapidly from this point and that by 11,750 years ago, it stood at about 4,300 feet, a drop in lake level of about 90 feet in 250 years.

This suggestion was supported by independent work done by Friedel. She obtained dates from both the Winter Lake and ZX Lake basins that show Winter Lake rising from 4,298 feet around 12,500 years ago, reaching a peak at 4,386 feet about 12,000 years ago, and then falling to 4,334 feet by 11,900 years ago.

This history provides us with a major mismatch between the latest Pleistocene history of lakes in the Chewauacan, Lahontan, and Bonneville basins. In both of those more southerly, and larger, basins, there is strong evidence that lake levels were low between 12,000 and 11,000 years ago and that they rose during the Younger Dryas, between about 11,000 and 10,000 years ago. In the Chewaucan Basin, however, the evidence seems to suggest higher lake levels between roughly 12,000 and 11,000 years ago and low ones after that.

The Owens Lake–Death Valley System

Death Valley not only has bragging rights to the lowest point in the Americas not beneath water but is also one of the hottest and driest places on earth. On July 13, 1913, the air temperature at Death Valley's Furnace Creek reached 134°F, the highest temperature ever recorded in the Western Hemisphere, and the second highest ever recorded on the planet (the highest, 136°F, was reached in 1922, in northwestern Libya). Even January is warm in Death Valley; the average January minimum temperature is 37°F, and temperatures as high as 85°F have been recorded here during this month. In Death Valley's hottest month, July, temperatures average 102°F, and the average maximum July temperature is 116°F. The difference between the lowest and highest temperature in a given year can easily exceed 100°F. Even these readings, however, are glacial compared to the heights that can be reached by ground temperatures: those have been measured as high as 190°F and may go beyond that.

Precipitation is as low as the temperature is high, averaging a meager 1.65 inches a year. On top of that, the evaporation rate is extraordinarily high: 155 inches of evaporation was measured at the National Park Service headquarters on Cow Creek between May 1, 1958, and May 1, 1959, the highest evaporation rate ever measured in the United States. Even the highest elevations of the lofty Panamint Range, which forms the western border of the valley and reaches 11,049 feet, receive only some 12 to 15 inches of precipitation a year.

All this may make it hard to believe that Death Valley once held substantial lakes, but it did, and the record of those lakes is, after many years of work, finally becoming clearer.

Those lakes cannot be understood without recognizing that Death Valley was at one time part of a complex system of lake basins that stretched all the way to Mono Lake in eastern California (figure 5-16). When this system was operating at its maximum, the Pleistocene lake in the Mono Basin overflowed into Lake Adobe in Adobe Valley. Lake Adobe, in turn, overflowed into the Owens River, where it joined more direct Sierran runoff to flow into Owens Lake. Once Owens Lake had reached a depth of about 200 feet, it overflowed into China Lake. When China Lake reached a depth of about 30 feet, it overflowed into Searles Lake. Once Searles Lake had reached a depth of about 650 feet, it both coalesced with China Lake and overflowed into Panamint Lake in the Panamint Valley. Once Panamint Lake filled to overflowing, it dumped into Death Valley's Lake Manly. Although it has been suggested that at some time in the deep past, Death Valley might have been connected somehow to the Colorado River, there is no compelling geological evidence to support this (see chapter 7). Certainly, during the Pleistocene, Death Valley was the end of the

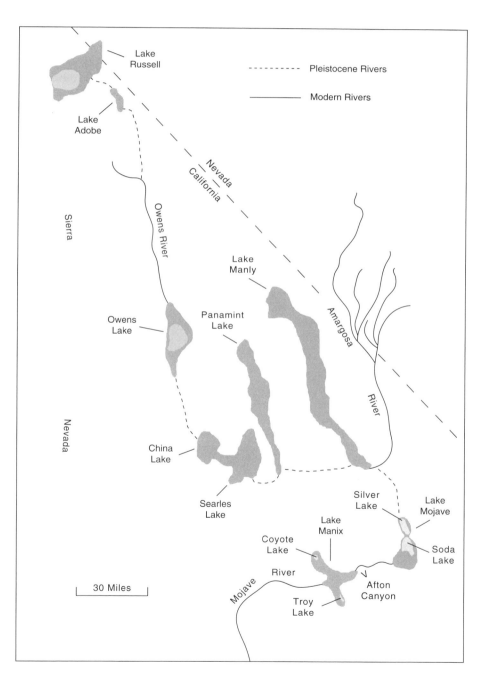

FIGURE 5-16 Mono Lake–Death Valley drainage system.

line in this interconnected system of rivers and lakes that ran some 350 miles. Death Valley, however, received water not only from the Owens River–Death Valley system but also from the Amargosa River, whose headwaters are to the northeast of the valley; from the Mojave River system to the south; and from groundwater coming from as far away as the Spring Mountains near Las Vegas.

I discuss this complex system in lake order, beginning with the Mono Basin and ending in Death Valley.

MONO LAKE

For those whose focus is on the past 30,000 years or so, life is made a little easier by the fact that Mono Lake has not participated in this system during that time. Historically, and before Los Angeles began diverting water from the Mono Lake watershed in 1941, the levels of Mono Lake varied between 6,404 feet and 6,427 feet. At the midpoint of these two figures, 6,416 feet, the lake is about 187 feet deep. The threshold that separates the lake's basin from Adobe Valley, and thus from Lake Adobe, falls at 7,185 feet. Obviously, the lake would have to rise an additional 769 feet to reach this point.

We know it did that at some time in the past because the spillway itself is full of features that can be explained only by significant amounts of water having moved through it—a well-defined terrace, for instance, and abundant beach cobbles. The difficulty lies in determining when this event last happened. We know that the last two highstands in the basin occurred at about 18,000 and about 13,000 years ago, with a significant lowstand, dated to roughly 17,000 to 14,000 years ago, in between. The elevation reached by the earliest of these highstands is not known, but the 13,000-year highstand reached 7,070 feet, at which point the lake had a depth of about 841 feet, 115 feet short of its Adobe Valley threshold. In addition, Mareth Reheis and her colleagues have obtained a radiocarbon date from the vicinity of the spillway that shows that the last overflow must have occurred prior to 29,000 years ago. How much prior is not known, but other dates suggest that it was probably before 50,000 years ago and perhaps well before that.

In short, while the Mono Lake basin contained a substantial lake during the Pleistocene—one that has come to be known as Lake Russell, in honor of geologist Israel Russell—it has not dumped its waters into Adobe Valley since long before 30,000 years ago.

OWENS LAKE

Things are different for Owens Lake. The bottom of the now-dry lake bed here falls at about 3,560 feet in elevation. The threshold that separates the lake basin from overflowing into Searles Lake, far downstream, falls at about 3,755 feet, meaning that whenever the depth of Owens Lake exceeded about 197 feet, the lake overflowed, with China and then Searles Lake the recipient of its waters. For comparison, the maximum known depth of the lake during historic times is about 50 feet, reached in 1872. Whether it might have gone higher than that during the last eighty-five years or so is not known, since Los Angeles began diverting water from the basin in 1913.

Our knowledge of the history of Owens Lake goes back some 800,000 years, thanks in large part to a 1,060-feet-deep core (core OL-92) drilled and analyzed by a team of U.S. Geological Survey scientists led by George Smith and James Bischoff. That work led to a remarkable understanding of the general history of the lake across this time period but has to date provided only very general information on the past 30,000 years of that history.

Our knowledge of this more recent period of Owens Lake history is derived from two very different kinds of studies. The first of these approaches, by Larry Benson and his colleagues, focuses on detailed analyses of the chemical and magnetic properties of the lake's sediments as revealed by a different USGS core (core OL-84B). Here, the emphasis has been on four different things that can be measured from those sediments: oxygen isotope ratios, the total amount of inorganic carbon, the total amount of organic carbon, and changes in magnetic susceptibility through time. I will focus on the oxygen isotope ratios, since the results from these different approaches are generally very consistent.

Oxygen isotope values can be gotten from lake sediments in a number of ways. One can, for instance, examine the oxygen in the shells of the tiny crustaceans known as ostracodes that are often abundant in the waters of Great Basin lakes. If ostracodes are not available, one can examine the oxygen locked up in the form of carbonates (for instance, calcium carbonate, $CaCO_3$, the same as limestone or chalk). When oxygen isotopes from both kinds of records are examined, the results are quite similar, meaning that either one will do. In chapter 3 I discuss the logic of the analysis of oxygen isotopes from sea cores—oxygen 18 (^{18}O) and oxygen 16 (^{16}O)—but the situation for Great Basin lakes is different enough that it needs to be discussed here.

As in the oceans, a prime determinant of the quantities of these two isotopes in lake water is the evaporation rate: the greater the evaporation rate, the more of the lighter isotope (^{16}O) is preferentially lost to the atmosphere, producing lake water that has a higher proportion of ^{18}O, and vice versa. The causes of the evaporation rate from the surface of a lake are complex, dependent not only on temperature but also on humidity, the speed of winds over the lake, and so on. But insofar as evaporation rate depends heavily on temperature, the ratio of the two isotopes in lake waters in Great Basin lakes should respond in part to that.

The behavior of oxygen-isotope ratios in closed-basin lakes is, though, far more complicated than this. As I explain in chapter 3, the farther a moisture-carrying air mass is from the ocean, the lighter the water in it becomes, since the water that is formed from the heavier ^{18}O precipitates first. As a result, the waters that create streams in the Great Basin tend to be rich in ^{16}O (I will refer to this as "light water").

Now picture what happens during wet periods in a closed-basin Great Basin lake. As evaporation occurs, light water is being preferentially evaporated from the lake at the same time as the streams that feed it are bringing in more light water. As the lake gets deeper as a result of this light-water inflow, the heavier water gets diluted by lighter water, and the ratio of ^{18}O to ^{16}O declines. The faster the water depth increases, the faster the dilution process occurs. During a dry period, things differ considerably. Now evaporation is reducing the amount of light water in the lake, but less light water is being brought in by streams to replace it. The faster the water depth decreases, the faster the ratio of ^{18}O to ^{16}O increases

Things get complicated when the lake gets so high that it begins to overflow, since the greater the volume of water that moves through the overflow channel, the less time it has spent in the lake itself, and thus the less time it has had to shed light water through evaporation. There are many other possible complications, and the proper interpretation of these ratios from a given lake requires substantial knowledge of the local context. But, all other things being equal, the ratio of ^{18}O to ^{16}O in the past should tell us about the alternation between wet and dry periods through time. Higher ratios of ^{18}O to ^{16}O suggest that times were dry and the lake relatively low; lower ratios suggest that the lake was deep and the climate wet. The speed with which the ratios changed tell us something about how fast these changes were taking place.

While changing isotope ratios from Great Basin lakes can tell us whether a lake was relatively deep or relatively shallow, and about the speed of lake level changes through time, they cannot tell us how shallow or how deep that lake was. That is best done by examining and dating shoreline and near-shoreline geological features formed by the lake, as we saw in the cases of Lakes Bonneville and Lahontan. This work can be augmented by conducting geological analyses of deposits left behind by the lake as it grew and shrank. The most important recent Owens Lake work in this realm has been done by Steven Bacon and his colleagues and by Antony and Amalie Orme. In the best of all possible worlds, chemical analyses, such as those provide by oxygen-isotopes, have the same implications as the more purely geological studies, and that is the case for Owens Lake.

The Owens Lake oxygen isotope record amassed by Benson and his coworkers extends back over 50,000 years, while the shoreline-based study goes back about 23,000 years ago. At this latter time, and continuing to about 16,500 years ago, the oxygen isotopes tell us that the lake was deep, and the shorelines studies tell us that it was in general so deep— about 245 feet—that it overflowed its threshold, allowing its waters to flow southward into the China Lake and Searles Lake basins. Between about 16,400 and 14,600 years ago, the lake fell dramatically. In fact, it may have even dried entirely during this interval: the core analyzed by Benson and his colleagues has no sediments from between 15,500 and 13,700 years ago, and the bottom of this depositional gap is marked by features that suggest the lake bed was dry at this time. After this, however, the lake rose again, overflowing between 13,000 and 12,000 years ago. That overflow episode may have been brief: within a few hundred years it had fallen beneath its threshold, perhaps never to overflow again. That doesn't mean that the lake has not gone up and down during the past 13,000 years or so; it has. In fact, it was relatively deep (almost 115 feet) just before the Younger Dryas began; the Ormes even suggest that the lake may even have risen so high at around 10,750 years ago that it once again transgressed its threshold. After the Younger Dryas ended, it deepened once again, reaching perhaps 50 feet—about the same as its brief historic maximum. During the Younger Dryas itself, however, the lake may have dried completely.

That is, when lakes Lahontan and Bonneville seem to have been rebounding, Owens Lake was doing the opposite, just as was the case with Lake Chewaucan. Why that might be we simply don't know.

SEARLES LAKE

The history of Searles Lake has been the target of a lengthy series of important studies by George Smith of the U.S. Geological Survey and his colleagues. Their work has focused on the analysis of deep cores taken from deposits within the Searles Lake Basin. The deepest of these cores not only penetrated 2,274 feet, and some 3.2 million years, to the bottom of lake-deposited sediments here, but then went an additional 728 feet through sands and gravels deposited by intermittent streams prior to the formation of Searles Lake, and then through nearly 50 feet of bedrock.

Smith has constructed his analysis of the latest Pleistocene history of Searles Lake by combining information gained from a large number of cores. A chronology for these sediments has been built by extracting and dating a wide variety of materials provided by these cores, while the nature of the lake at any given point in the past has been inferred from the nature of the sediments themselves. Smith has used the general nature of these sediments to build an outline of lake history, with alternating muds and salt beds, for instance, presenting a picture of alternating deeper and shallower lakes. In addition, he has conducted highly detailed analyses of the chemical and mineral content of the various depositional units revealed by the cores. He has thus been able to use the chemistry and mineralogy of the deposits to refine our understanding of the history of Searles Lake.

Unfortunately, different interpretations of the hydrological meaning of the chemistry and mineralogy of the late Pleistocene deposits of Searles Lake provide different estimates of lake depth. Things improve a bit for the past 30,000 years, but much remains to be learned. We do seem to know that the lake level underwent a series of six increases and decreases between about 33,000 and 23,000 years ago; during the increases, Owens Lake must have overflowed into the Searles Lake Basin. It would be nice to be able to pinpoint when the latest highstand of the lake was reached, but conflicting data make that difficult. Attempts to date those highstands by Jo Lin and his colleagues suggest that the lake may have been quite high between about 14,000 and 13,500 years, though uncertainties about the dating mean that the last highstand may have been as many as 1,000 years later than this. If the former dates are correct, Searles Lake would have been quite high when the Owens Lake Basin was dry or close to it. That situation is hard to imagine, since Owens Lake is the prime source for Searles Lake waters, though being hard to imagine doesn't make it wrong. If the more recent dates are correct, however, the histories of the two lakes would agree with one another, with Owens high when Searles was high, and vice versa, just as it should be.

PANAMINT VALLEY

We have known for more than a century that there was a Pleistocene-aged lake or lakes in Panamint Valley, but until very recently even the general history of these lakes has been poorly known. Thanks to geologist Angela Jayko and her colleagues, however, some aspects of this history are now far better understood.

Most important from our perspective, their work shows that during the past 200,000 years or so, this valley has held two substantial lakes, both of which were fueled largely by water coming downstream from Owens Valley. The earlier of the two is not particularly well dated, but it seems to have existed either between about 185,000 and 130,000 years ago (MIS 6) or, less likely, between 75,000 and 60,000 years ago (MIS 4). This earlier episode may have been deep enough to overflow into Death Valley via Wingate Pass (at an elevation of 1,975 feet). If, as appears likely, the early lake formed during MIS 6, it correlates well with a large lake that seems to have existed in Owens Valley at the same time.

The later of the two lakes, called Lake Gale, seems to have reached its maximum at two separate times, between about 26,000 and 22,000 years ago and again between 18,000 and 15,000 years ago, with a retreat separating the two highstands. Although Jayko and her colleagues stress that the dating needs work, this is a fairly good fit with the history of Owens Lake. Just as important, while latest Pleistocene Lake Gale reached a remarkable depth of over six hundred feet, this was still some three hundred feet too shallow to breach Wingate Pass and flow into Death Valley. The last time that happened was well more than 120,000 years ago.

DEATH VALLEY AND LAKE MOJAVE

Today, tourists who spend time in Death Valley are warned again and again to take the high temperatures, and attendant water loss through evaporation, very seriously. "Drink at least one gallon (four liters) per day" is the official advice provided by the National Park Service, yet people routinely ignore this suggestion and die as a result. But as arid as the valley now is, we have known for more than eighty years that it once supported substantial lakes. Thanks to work by geologist Roger LeB. Hooke, we have known for more than thirty years that some of those lakes existed between sometime before 26,000 years ago to about 10,000 years ago.

The general picture of Death Valley lake history was provided by analyses of a six-hundred-foot-deep core taken from the heart of the valley, near Badwater. This work, by Teh-Lung Ku and his colleagues, shows that the valley contained substantial lakes on two separate occasions during the past 200,000 years. The first of these occurred between about 186,000 and 120,000 years ago and may have reached a depth of six hundred feet. This occurred, as I mentioned above, at the same time that there were also large lakes in both Owens and Panamint valleys, with the latter apparently overflowing into Death Valley. The second, shallower lake existed between about 35,000 and 10,000 years ago, as Hooke suggested in the 1970s.

Hooke also noted that the lake seems to have existed continuously during that span, though work by Diana Anderson has shown it had its ups and downs. The first of these ups and downs is the least controlled chronologically, but it happened sometime before 26,000 years ago. The second dates to about 18,000 years ago, bracketed by dates of 19,100 and 17,500 years ago. The third and last dates to between about 12,500 and 11,500 years ago, though there is also a date of 13,450 years ago that suggests a deeper lake may have been in existence by then. We don't know how deep each of these highstands got, since it is not yet possible to correlate each with the shorelines it created, but the maximum depth seems to have been between about 260 and 295 feet. By 10,000 years ago, the lakes were gone, and the problem for tourists shifted from drowning to dying of thirst.

Geologists have long referred to Death Valley's Pleistocene lake as Lake Manly, after William Lewis Manly (see chapter 6). Diana Anderson argues that this name should be restricted to the lake that occurred here between 186,000 and 120,000 years ago, and she named the three highstands of the later lake DVPL-1 (prior to 26,000 years ago), DVPL-2 (about 18,000 years ago), and DVPL-3 (about 12,000 years ago), with PL standing for "perennial lake." Others have continued to use the name Lake Manly for both of them, which is what I do. To make things easy, I will refer to the early lake as Lake Manly I and the later one as Lake Manly II.

Given how arid Death Valley is today, one may well wonder where enough water to create these lakes came from. Remember that when the Mono Lake–Death Valley system was going full-bore, water from Mono Lake could end up in Death Valley, flowing via the lakes in the Owens, China, Searles, and Panamint basins. I have already noted that this may have happened when Lake Manly I was in existence, though this is not certain. There is, however, no evidence that significant amounts of water have flowed through Wingate Pass between Panamint Valley and Death Valley since that time. This, of course, means that Lake Manly II was not getting water from this source.

How about the Amargosa River? Recall that this river has its headwaters to the north and east of Death Valley and that it now flows into that valley. Work by a series of geologists, including Diana Anderson and Christopher Menges, have shown that the Amargosa has an enormously complex history. Until very recently, it was thought that up to about 185,000 years ago, the river ended in a substantial lake, called Lake Tecopa, just north of where the Amargosa now swings to the west toward Death Valley (figure 5-16). It was also thought that this lake breached the southern end of its basin, allowing the Amargosa to flow south from here. Recent work, however, suggests that most, or perhaps even all, of the sediments in the Tecopa Basin that were once thought to have reflected the existence of a deep lake are more likely to have been deposited as a result of groundwater activity. If so, a substantial Pleistocene Lake Tecopa, about which much has been written, may not have existed. However, there is no reason to question that the divide at the southern edge of this basin was cut between about 200,000 and 150,000 years ago, and that the cutting allowed the Amargosa to flow into Death Valley and thus help create Lake Manly I. It has been flowing into Death Valley ever since and certainly played a role in helping to create Lake Manly II.

This leaves the Mojave River, whose headwaters are in the San Bernardino Mountains of southern California and whose history has been the focus of work by a legion of geologists, most recently Stephen Wells and his many colleagues, by Marith Reheis and Joanna Redwine, and by Norman Meek. What we know of the complex history of the river shows that it could not have played a role in helping to form Lake Manly I, but it certainly did play that role during Lake Manly II times.

Two major lakes formed along the course of the Mojave River during the late Pleistocene (figure 5-16). The first of these, Lake Manix, filled three separate basins (the Coyote Lake, Troy Lake, and Afton basins) to cover, at its maximum, about ninety square miles with water. Reheis and Redwine suggest that this lake reached highstands at 34,000 to 29,000, 27,500 to 25,000, and 22,500 to 20,800 years ago.

The second lake, called Lake Mojave, grew in the area that is now the endpoint of the Mojave River: the Soda Lake and Silver Lake basins, straddling the small desert town of Baker. Winter floods can still lead to the formation of a lake in these basins today, including one that was twenty-two feet deep in 1938. These modern events, though, are short-lived.

This was not the case during the late Pleistocene, when long-lasting lakes covered some 110 square miles with up to forty feet of water. As work by Wells and his colleagues has shown, Pleistocene Lake Mojave began to come into existence around 22,600 years ago as the basin began to fill with shallow, sporadic lakes. By at least 18,400 years ago, a permanent lake, Lake Mojave I, had formed. This lake, fed by overflow from the Lake Manix basin, lasted until some 16,600 years ago. For the following 3,000 years—to 13,700 years ago—the Lake Mojave basin was once again the scene of a series of intermittent lakes, interrupted by an arid event so severe that, at about 15,500 years ago, the entire basin dried. At about 13,700 years ago, a second permanent lake, Lake Mojave II, flooded the Soda Lake and Silver Lake basins, lasting until around 11,400 years ago. This lake was then replaced by another bout of intermittent, shallow lakes; by 8,700 years ago or soon thereafter, the basin was dry (see chapter 8).

There are two different versions of how the waters of the Lake Manix basin came to feed Lake Mojave. In one version, forwarded by Norman Meek, Lake Manix cut through its outlet at the east end of the Afton Basin, causing its waters to pour downstream into the Lake Mojave Basin and quickly creating much of today's Afton Canyon. By "quickly," Meek means *very* quickly, at one time estimating that the upper four hundred feet of this canyon may have been removed in something on the order of ten hours, after which the cutting slowed. In the other version, forwarded by Wells and his colleagues, there was no catastrophic failure here. Instead, they argue, the creation of this canyon was a slow process, taking thousands of years. Reheis and Redwine's interpretation suggests that both views may in part be right. They suggest that the initial incision appears to have been relatively rapid, but then, once the bedrock floor had been reached, the cutting slowed considerably. This process, Reheis and her colleagues suggest, began sometime after 21,000 years ago.

No matter how long the process took, Mojave River waters ended up in the Lake Manix Basin, and those waters ended up in Lake Mojave. What all of this has to do with Death Valley is straightforward. Remember that during the late Pleistocene, Death Valley held substantial lakes sometime before 26,000 years ago (DVPL-1), around 18,000 years ago (DVPL-2), and around 12,000 years ago (DVPL-3). Lake Mojave was in full bloom between 18,400 and 16,600 years ago (Lake Mojave I), and between 13,700 and 11,400 years ago (Lake Mojave II). That is, lakes were high in both places at about 18,000 and at about 12,000 years ago. This doesn't necessarily mean that the two systems were connected, since both basins could have simply been responding to the same climatic events. However, if Lake Mojave did overflow at these times, it would have overflowed to the north, into the Silurian Valley. From there, it is just a short hop to the Amargosa River Valley. In fact, an intermittent stream, Salt Creek, runs from the Silver Lake basin through the Silurian Valley into the Amargosa River. If Lake Mojave got high enough, this is where its waters would have gone, and geological work in the Silurian Valley shows that this is exactly what happened. A lake, called Lake Dumont, formed in this valley, reaching its maximum between 19,400 and 18,000 years ago, during which time it overflowed to the north, dumping its waters into the Amargosa and thus into Death Valley (DVLP-2). Then, sometime after 18,000 years ago, the waters of Lake Dumont simply cut a new channel, allowing the waters of Lake Mojave to flow directly to the Amargosa whenever they got high enough.

This they did during Lake Mojave II times (13,700–11,400 years ago), helping to create the third of the late Pleistocene lakes in Death Valley (DVLP-3).

So that is the complicated story of the three lakes that formed in Death Valley during the past 30,000 years. Water that came from afar had three possible sources, other than groundwater. Of these three, the Amargosa was always involved, while the Mono Lake–Owens Valley–China Lake–Searles Lake–Panamint Valley chain played no role at all. The Mojave River system seems to have played no role in forming the earliest of the three Death Valley lakes (DVLP-1), which was fed by the Amargosa River and local sources alone. The following two lakes, though, were fed by both the Amargosa and the Mojave, the latter by overflow through the Silurian Valley.

Escaping the Great Basin

The Bonneville Flood provides the most impressive example of a Great Basin lake breaking its geographic ties and escaping to the Pacific Ocean, but it is not the only such instance. I have already mentioned that Goose Lake, in northeastern California, has overflowed into the Pit River system during historic times, thus contributing its waters to the Sacramento River drainage and, ultimately, San Francisco Bay. Malheur Lake, in south-central Oregon, has overflowed into the Malheur River, a tributary of the Snake River (which in turn flows into the Columbia). Geologist Dan Dugas has presented evidence suggesting that this event last happened around 1,000 years ago. Ira Allison once suggested that the Pleistocene lake in the Fort Rock Basin escaped into the Deschutes River drainage through an outlet on its western edge. There is no evidence for this, but it is possible that Fort Rock Basin water escapes underground to reach the Deschutes system anyway.

A far more dramatic escape, however, was managed by Pleistocene Lake Alvord, which formed in the Alvord and Pueblo basins east of the Steens and Pueblo mountains in southeastern Oregon. At its maximum, this lake was some eighty miles long and 280 feet deep; after it dried, it provided the salt flats on which Kitty O'Neil set the land speed record mentioned at the beginning of this chapter. Whenever Lake Alvord got high enough, it overflowed into Coyote Basin to the east. Whenever the lake in Coyote Basin got high enough, it overflowed into the Crooked Creek drainage, a tributary of the Owyhee River, which flows into the Snake. Relatively gentle overflow appears to have been common during the late Pleistocene, but geologist Deron Carter and his colleagues have shown that between about 13,000 and 14,000 years ago, when Lake Alvord was about 225 feet deep, the threshold at its eastern edge failed catastrophically, sending some 2.7 cubic miles of water into the Coyote Lake basin, and then into Crooked Creek, in the space of a few weeks. And even this massive flood may have been significantly smaller than one or more earlier floods that Lake Alvord unleashed to the east.

Great Basin Glaciers

The mountains that fringe the Great Basin on three sides—the Sierra Nevada to the west, the Wallowa Mountains to the north, and the Rocky Mountains to the east—support active glaciers today. Within the Great Basin, however, there is but a single glacier, stretching some 2,950 feet along the north-facing slope of Mount Wheeler in the Snake Range at an elevation of about 11,200 feet. Even this glacier is, in glaciologist Gerald Osborn's words, "nearly defunct" (G. Osborn 2004: 67). Given that glaciers formed as far south as the mountains of south-central Arizona and New Mexico during the Pleistocene (see chapter 3), and given that the Great Basin has thirty-three mountains ranges with peaks above 10,000 feet (see table 2-1 and figure 2-5), it will probably come as no surprise that substantial numbers of Great Basin mountains were glaciated during the Pleistocene.

The Pleistocene glaciers of the Great Basin have been the focus of relatively little work. Geologists interested in alpine glaciers have had richer peaks to climb in the American West, including those of the Sierra Nevada, the Cascades, and the Rocky Mountains. Compared to those settings, Great Basin glaciers must seem uninteresting. In addition, nearly all Great Basin geologists interested in the Pleistocene have, understandably, focused not on the mountains but on the valley bottoms, and on the more accessible, more tractable, and unique information they provide on Pleistocene lake histories. It is telling that a recent review of Pleistocene glaciation in the mountains of the western United States dedicates just two sentences, out of a total of forty-one pages, to Great Basin Pleistocene glaciers. That short shrift is a function of the relatively little work done on these glaciers and of the negligible excitement those glaciers have aroused in scientists working elsewhere.

Table 5-6 provides a list of all Great Basin highlands known or thought to have been glaciated during the Pleistocene. Following the lead of glaciologists Gerald Osborn and Ken Bevis, whose work has done much to clarify the extent of glaciation in the Great Basin during the Pleistocene, this list includes all glaciated areas whose waters drain into the Great Basin. If we eliminate places that also drain externally—the high plateaus of Utah and the Sierra Nevada, for instance—we are left with twenty-five mountains on the list. Two of these, Nevada's Wassuk Range and Shoshone Mountains, are not on the Osborn and Bevis list, but have been included because others have made the case. With only five exceptions, all of these mountains have peaks in excess of 10,000 feet, and all five exceptions—Hart Mountain (8,017 feet), Drake Peak (8,407 feet) and Steens Mountain (9,733 feet) in south-central Oregon, and the Pine Forest (9,397 feet) and Santa Rosa (9,732 feet) ranges in northern Nevada—are in the northerly reaches of the Great Basin, where temperatures and evaporation rates are lower and precipitation higher.

TABLE 5-6
Great Basin Highlands Glaciated during the Pleistocene

Range	Maximum Elevation (feet)	References
Great Basin Drainage Only		
Carson Range, NV	10,881	G. A. Thompson and White 1964
Deep Creek Range, UT	12,087	Ives 1946
Drake Peak, OR	8,407	G. Osborn and Bevis 2001; G. Osborn 2004
East Humboldt Range, NV	11,306	Wayne 1984
Grant Range, NV	11,298	G. Osborn and Bevis 2001; G. Osborn 2004
Hart Mountain, OR	8,017	G. Osborn and Bevis 2001; G. Osborn 2004
Monitor Range, NV	10,888	Dohrenwend 1984
Oquirrh Mountains, UT	10,620	Ives 1946
Pavant Range, UT	10,215	G. Osborn and Bevis 2001; G. Osborn 2004
Pine Forest Range, NV	9,428	Rennie 1987
Ruby Mountains, NV	11,387	Wayne 1984
Santa Rosa Range, NV	9,732	Dohrenwend 1984
Schell Creek Range, NV	11,883	Willden 1964
Shoshone Mountains, NV	10,313	Dohrenwend 1984
Snake Range, NV	13,063	Whitebread 1969
Stansbury Mountains, UT	11,031	Ives 1946
Steens Mountain, OR	9,733	E. H. Lund and Bentley 1976
Sweetwater Mountains, CA, NV	11,673	Halsey 1953
Toiyabe Range, NV	11,788	G. Osborn 1989
Toquima Range, NV	11,941	Dohrenwend 1984
Tushar Mountains, UT	12,169	G. Osborn and Bevis 2001; G. Osborn 2004
Wasatch Range, UT	11,928	G. Osborn and Bevis 2001; G. Osborn 2004
Wassuk Range, NV	11,239	Dohrenwend 1984
White Mountains, CA, NV	14,246	LaMarche 1965; Elliot-Fisk 1987
White Pine Range, NV	11,513	Piegat 1980
Great Basin and External Drainage		
Boulder Mountain	11,309	G. Osborn and Bevis 2001; G. Osborn 2004
Copper Mountains, NV	9,911	G. Osborn and Bevis 2001; G. Osborn 2004
Fish Lake Mountain, UT	11,633	G. Osborn and Bevis 2004
Gearhart Mountain, OR	8,370	G. Osborn and Bevis 2004
Independence Mountains, NV	10,439	Blackwelder 1934
Jarbidge Mountains, NV	10,839	Coats, Green, and Cress 1977; Coats 1964
Markagunt Plateau	11,306	G. Osborn and Bevis 2004
Paulina Peak, OR	7,984	G. Osborn and Bevis 2004
Raft River Mountains, UT	9,925	G. Osborn and Bevis 2001; G. Osborn 2004
Sevier Plateau, UT	11,227	G. Osborn and Bevis 2004
Sierra Nevada, CA	14,494	G. Osborn and Bevis 2004
Spring Mountains (?)	11,918	J. Osborn, Lachniet, and Saines 2008
Strawberry Mountains, OR	9,038	G. Osborn and Bevis 2004
Uinta Mountains, UT	13,528	G. Osborn and Bevis 2004
Warner Mountains, CA	9,892	G. Osborn and Bevis 2001; G. Osborn 2004
Wasatch Plateau, UT	11,285	G. Osborn and Bevis 2004
Yamsay Mountain, OR	8,196	G. Osborn and Bevis 2004

SOURCE: After G. Osborn and Bevis 2001.

FIGURE 5-17 Kiger Gorge, Steens Mountain, Oregon.

Of these twenty-five glaciated uplands, the Ruby and East Humboldt mountains have received by far the most attention. As early as 1931, geologist Eliot Blackwelder presented evidence that these ranges were the scene of two separate episodes of glacial expansion and retreat, episodes that he called the Angel Lake and Lamoille stages, named for Angel Lake in the East Humboldt Range and Lamoille Canyon in the northern Ruby Mountains. Not long thereafter, Robert Sharp confirmed Blackwelder's arguments, and more details on these two episodes have been provided by William Wayne and by Osborn and Bevis.

That the Ruby Mountains and East Humboldt Range were glaciated is obvious even to those with only general knowledge of the effects that glaciers have on the landscape. The existence of U-shaped valleys provides an excellent example. Unless something intervenes, stream-cut valleys generally have a V-shaped profile, produced by stream incision. Glaciers flowing down a valley, however, pluck and grind rocks from valley sides and bottoms, often exposing bedrock and typically producing a valley that, in cross-section, looks like a huge U. Kiger and Big Indian gorges in Steens Mountain are textbook examples (figure 5-17), but the northern Ruby Mountains and East Humboldt Range have superb instances as well, including Lamoille and Rattlesnake canyons on the west side of the northern Rubies (figure 5-18). The U-shaped profile of upper Lamoille Canyon is particularly easy to see, since a paved road follows the canyon nearly to its top.

The term "moraine" refers to the materials accumulated by glaciers, whether at their sides (lateral moraines), at their ends (terminal moraines), or beneath them (ground moraine). In the northern Ruby Mountains, both lateral and end moraines emerge from Lamoille Canyon and a number of valleys to the south; the outermost moraines fronting Lamoille Canyon sit a mile from the edge of the mountain itself. All of these moraines were deposited during the Lamoille glaciation, the earlier of the two known glacial episodes in this area. Although some moraines from the later Angel Lake glaciation descend far down the canyons, none get as far as the Lamoille moraines, and none emerge from the canyons onto the broad valley floor fronting the mountain itself. Indeed, Sharp estimated that Lamoille moraines reached an average altitude five hundred feet lower than that reached by Angel Lake moraines.

Osborn and Bevis estimate that the Lamoille-age glacier in Lamoille Canyon was about twelve miles long, making it the longest glacier known from within the Great Basin. The Angel Lake glacier that formed here was about half that length. To the south, in Rattlesnake Canyon, Sharp reported, the Lamoille-age glacier covered roughly eight miles of the valley, while the Angel Lake glacier covered less than four. He also estimated that in the northern Rubies and East Humboldts together, Angel Lake glaciers covered 80 percent of the area occupied by Lamoille-age glaciers. Even though this was the case, however, Angel Lake glaciation was impressive. By mapping the upper limit of the glacier-scoured wall in Lamoille Canyon, Sharp showed that the Angel Lake glacier here reached a maximum thickness of some nine hundred feet; the Lamoille glacier must have been thicker yet.

In addition to the end moraines, lateral moraines, and U-shaped valleys, both the northern Rubies and East Humboldts show a wide range of other glacial phenomena, from glacially polished bedrock to glacial outwash deposits that extend well into Huntington Valley, to the west of the northern Rubies. Perhaps the most obvious of these other

FIGURE 5-18 Lamoille Canyon, Ruby Mountains, Nevada.

remnants of glaciation, however, are the numerous high-elevation cirques, or ampitheater-shaped depressions carved by glacial ice. These features are found at the heads of many of the higher valleys in both ranges. Wayne mapped about fifty of them for that part of the northern Rubies centering on Lamoille Canyon.

Each of these cirques marks the head of a glacier. In the northern Rubies, the cirques lie at an average elevation of about 9,385 feet, but the average elevations decrease toward the north, and cirques in the northern reaches of the East Humboldt Range lie at an average elevation of about 8,890 feet. Many of these are readily visible to those who hike the Ruby Crest Trail here. In fact, many now contain lakes that form one of the scenic highlights of the trail. Both on and off the trail, these cirque lakes include Verdi Lake just north of Verdi Peak, Echo Lake at the head of Echo Canyon, Lamoille Lake at the head of Lamoille Canyon, and Liberty Lake just south of Liberty Pass.

Since these lakes most likely formed soon after deglaciation, the chances are good that the sediments contained in their basins can provide a partial environmental history of the area since the ice retreated. If nothing else, dating the earliest sediments in these lakes should provide an excellent indication of when the ice retreated. On Steens Mountain to the north, for instance, Peter Mehringer's analyses of the deepest sediments within Wildhorse Lake, a cirque lake lying at 8,480 feet, shows that this cirque was free of glacial ice by 9,400 years ago (see chapter 8); lower on Steens Mountain, Fish Lake (7,380 feet) was free of ice by about 13,000 years ago.

While detailed analyses of the sediments contained within the cirque lakes of the northern Ruby Mountains and the East Humboldt Range would undoubtedly repay the effort, that work has yet to be done. Paleobotanist Bob Thompson tried, but the sediments he encountered were too rocky to be penetrated by his coring equipment. He did extract a core from Upper Dollar Lake, a noncirque lake in the upper reaches of Lamoille Canyon, but he hit Mazama ash, deposited some 6,730 years ago, and was unable to go deeper.

In fact, there is only one radiocarbon date from either the northern Ruby or East Humboldt mountains that pertains to the chronology of glaciation and deglaciation here. Coming from a small bog that sits on a moraine at an elevation of 8,660 feet in Lamoille Canyon, the date indicates that this part of the canyon had become ice free by 13,000 years ago. Beyond that, estimates of the ages of the Lamoille and Angel Lake glacial episodes depend on more subjective criteria and on correlation of these glaciations with better-dated sequences in the Rocky Mountains and Sierra Nevada.

Since Blackwelder's work in 1931, it has been evident that the Lamoille glaciation was the older of the two. For instance, Angel Lake moraines sit atop areas that had already been glaciated during the Lamoille episode. In addition, Lamoille glacial deposits are far more heavily eroded than those of the Angel Lake episode. To take but one example, the end moraines of Angel Lake glaciers are cut by streams running through narrow passages, but those of the Lamoille episode are cut by far wider stream channels, as both Sharp and Wayne have observed.

Wayne's detailed consideration of the relative weathering of the deposits associated with these two episodes of glaciation, coupled with an analysis of glaciation in both the Sierra Nevada and Rocky Mountains, led him to argue that the Lamoille glaciation is not Wisconsin in age at all. Instead, he argues, it correlates with the preceding North American Pleistocene glaciation, the Illinoian, and occurred at about 150,000 years ago. He attributes the Angel Lake glaciation, on the other hand, to the Wisconsin glacial maximum, between about 22,000 and 13,000 years ago. More recent work, reported by Osborn and Bevis, supports these conclusions. As such, the Angel Lake glacial advance correlates with far better known glacial episodes in both the Rocky Mountains (termed the Pinedale glaciation) and the Sierra Nevada (the Tioga glaciation). In addition, it correlates well with what we know of glacial history on the eastern edge of the Bonneville Basin.

The White Mountains of eastern California were also the scene of significant glaciation, even though they are just east of the Sierra Nevada and thus in the rainshadow of that huge range. They were, in fact, the most southerly glaciated range within the Great Bain, with significant glacial advances known from a series of valleys on the range's east slope. Unfortunately, this is also an area whose glacial history is not very well known.

Late Pleistocene glacial activity here at least roughly correlates in time with the Angel Lake glacial advance in the Ruby Mountains. Following terminology introduced by Deborah Elliot-Fisk and her coworkers, Osborn and Bevis refer to this as the Perry Aiken glaciation (after Perry Aiken Creek). They present evidence that, unlike the situation in the Ruby Mountains, there were two significant glacial advances during this time. Attempts to date these advances, by Marek Zreda and Fred Phillips, suggests that they occurred around 16,000 radiocarbon years ago.

Little more can be said about the late Pleistocene glaciation of the White Mountains with any certainty. This is an area aching for additional work, the twin draws being that the glacial sequence here is in dramatic need of clarification, and that providing that clarification would involve working in one of the most beautiful areas the Great Basin has to offer.

Late Pleistocene Climates: Estimates from Glacial Phenomena

The net budget of a glacier is determined by calculating the difference between the amount of ice and snow added to it and the amount lost from it. In glaciers that have reached a steady state, the net budget is zero, with new material added at the same rate as the old is lost. In such steady state glaciers, the line that divides the area of snow and ice accumulation from the area of snow and ice loss is called the equilibrium line, and the altitude at which this line falls is called the equilibrium-line altitude, or ELA.

Geologists have spent a good deal of time calculating modern ELAs, studying why they fall where they do, and comparing modern ELAs with their Pleistocene counterparts. My colleague Stephen Porter, for instance, has taken a detailed look at the factors that determine the elevation at which glaciers form in the Cascade Range of Washington State, and while these elevations (called the glaciation threshold) are not identical to ELAs, they are very close to them. Porter found that 90 percent of the variation in these elevations is accounted for by two factors: mean annual temperature

TABLE 5-7
Equilibrium-Line Altitudes (ELAs) for Great Basin Late Pleistocene Glaciers as Reconstructed by Bevis (1995)

Region	ELA (feet)	Summit (feet)
Strawberry Mountains	6,855	9,038
Steens Mountain	7,445	9,733
Pine Forest Range	8,170	9,428
Ruby Mountains	9,020	11,387
Deep Creek Mountains	9,940	12,087
Tushar Mountains	10,465	12,173

and the amount of precipitation that falls between October and April (the "accumulation season," or the months when accumulation of these glaciers exceeds the losses they suffer).

Given that we appear to know why ELAs (and similar elevational parameters) are where they are, this understanding can be coupled with our knowledge of the position of late Pleistocene ELAs to infer various aspects of full-glacial climates. Porter used paleobotanical data from the Olympic Peninsula to infer that accumulation season precipitation was probably no more than 20 to 30 percent greater than that of today. Were that the case, he observed, a decrease in mean annual temperature of about 7.6°F is needed to account for the fact that full-glacial ELAs were some 2,950 feet lower than modern ELAs in the Washington Cascades.

The most detailed attempt to reconstruct ELAs in the Great Basin and immediately adjacent areas has been made by geologist Ken Bevis, who worked in a wide variety of mountains in and adjacent to the Great Basin. His Great Basin results are shown in table 5-7 (I cheated a bit by including the Strawberry Mountains, which lie just north of the boundary of the hydrographic Great Basin in central Oregon). Given that there is only one glacier in the Great Basin today, these cannot be compared to modern ELAs in this region. However, they can be used to estimate the magnitude of temperature and precipitation changes that may have coincided to produce late Pleistocene glaciers having these ELA values. The results of this work provide estimates of summer temperature decreases that range from 5.4°F to 16.2°F and of winter precipitation increases of 142 to 242 percent.

The Relationship between Pleistocene Glaciers and Lakes in the Great Basin

Given that both glaciers and lakes expanded in the Great Basin during the late Wisconsin, it is obvious to wonder about the temporal relationship between the two. Such wondering has, in fact, gone on since the nineteenth century, but far more detailed attention has been paid to this relationship during the past few decades.

Three prime ways are available for deciphering the relationship between lake and glacier histories in the Great

Basin. First, well-dated lake histories could be coupled with well-dated glacier histories; then, the relationship between the two could simply be read from time charts. Unfortunately, this cannot be done now, since the fine-scale glacial chronology needed to do it is not available, even though we do have some dates that can help in certain instances.

Second, if there are places where streams draining from glaciers ended up in Great Basin lakes, we might look at sediments from those lakes for indicators of the history of the glaciers themselves. If the indicators are there and we can figure out when they were deposited, perhaps that chronology could be put together with the lake's history to decipher the relationship between the two.

The third and conceptually simplest way is to look for situations in which glacier and lake deposits were laid down in the same place, and study the temporal relationship between the two on the ground.

I begin with the third of these approaches. Only two places are known in the Great Basin where lakes rose high enough, and glaciers descended low enough, for their deposits to overlap: the Mono Lake basin on the west, where deposits from Sierran glaciers come into contact with deposits laid down by Lake Russell, and the Bonneville basin on the east, where deposits from glaciers that emerged from the Wasatch Range come into contact with deposits laid down by Lake Bonneville.

Mono Basin

Geologist William Putnam, who made the first analysis of such deposits in the Mono Basin, concluded that the late Pleistocene highstand of Lake Russell (Putnam gave the lake this name) coincided with the peak advance of Sierran glaciers in the area. The most far-reaching attempt to correlate the two sets of deposits, however, was made by Ken Lajoie as part of a remarkably broad attempt to understand the geological history of this region. Lajoie found the key glacial and lake deposits to be so poorly exposed and so complex that he could not correlate them. However, what he did accomplish at that time, combined with later work he did here, led him to conclude that Lake Russell reached its last Pleistocene highstand after glaciers had receded from elevations within reach of the lake. Others have agreed, not so much on the basis of the physical relationship between glacial and lake deposits, which remains poorly understood, as on reliable evidence that the highest late Pleistocene shoreline in the Mono Basin, which falls at 7,070 feet, dates to about 13,000 years ago. If correct, that date puts this highstand after the retreat of local Sierran glaciers from their latest Pleistocene maximum extent. What remains unknown, however, is whether earlier highstands reached the 7,070 feet level at a time when nearby Sierran glaciers were reaching the same elevation.

Eastern Bonneville Basin

We seem to be better off as regards the relationship between glacial and lake deposits on the eastern edge of the Bonneville Basin. The work that has clarified this relationship has focused on Little Cottonwood and Bells canyons, which drain part of the western slope of the Wasatch Range some fifteen miles south and a little east of Salt Lake City. Little Cottonwood Canyon is a familiar place to Utah skiers, since State Highway 210 goes from the floor of Pleistocene Lake Bonneville, at an elevation of about 5,000 feet, to the ski lifts of Alta, some 4,000 feet higher. Like many other west-draining canyons in the Wasatch Range, Little Cottonwood Canyon was glaciated during the Pleistocene. But it was only here, and in Bells Canyon to the immediate south, that glaciers were of sufficient magnitude to reach beyond the mouths of the canyons and to come within the elevational range of the waters of Lake Bonneville.

The glaciers in both Little Cottonwood and Bells canyons were fed from north-facing cirques on the canyons' south sides, some ten of them in Little Cottonwood Canyon alone. Twice during the Pleistocene, these glaciers grew massive enough to reach nearly a mile beyond the front of the Wasatch Range, forming moraines that now have houses built on them (risky, since the Wasatch Fault Zone runs through here).

The deposits of the earlier of these two glacial advances have been called the Dry Creek till, so I will call this the Dry Creek glaciation. Although this episode is not well dated, the deposits that it left are very heavily weathered, and it is obviously quite old. The best guess is that it correlates with the Bull Lake glaciation of the Rocky Mountains, which occurred about 150,000 years ago. If so, the Dry Creek advance probably happened at much the same time as the Lamoille glaciation in the Ruby Mountains and East Humboldt Range.

The age of the second, far more recent glaciation in Little Cottonwood and Bells canyons is much better known. The deposits left by this episode of glacial advance are called the Bells Canyon till, so I will call this the Bells Canyon glaciation. Glaciers reached the mouths of both Little Cottonwood and Bells canyons several times during this episode. A number of years ago, David Madsen and Don Currey made a joint effort to understand the glacial and vegetational history of Little Cottonwood Canyon during latest Pleistocene and Holocene times (chapter 8) and, as part of that effort, obtained radiocarbon dates that bracket the Bells Canyon glacial episode. They obtained a date of 26,000 years on organic material from beneath Bells Canyon till at the mouth of Bells Canyon, showing that the advance did not occur until after that time. They were also able to show that by 12,300 years ago, glacial ice had retreated from an elevation of 8,100 feet, far up the canyon. The last glacial episode, then, must fall between these two dates.

More recent work by Elliott Lips and his colleagues has shown that Bells Canyon till both overlies and is overlain by deposits associated with the highstand of Lake Bonneville. The dates they obtained for the more recent of these glacial advances showed that it occurred at around 13,400 years ago, suggesting that this advance occurred when the lake was at the Provo shoreline. All of this evidence in turn suggests that glaciers on the western edge of the Wasatch Range reached their maximum limits at about the time that Lake Bonneville was doing the same thing. They were not the only glaciers in the region to do this. Much the same is true for the southwestern Uinta Mountains, where glaciers also reached their maximum during the Bonneville highstand.

Interestingly, these maxima were reached after glaciers in the Rocky Mountains had begun to retreat. If glaciers along the Wasatch Front and southwestern Uinta Mountains were expanding while glaciers in the Rockies were retreating, there must have been something special about this area. Fortunately, we don't have to guess what that special something was. As Steve Hostetler and his colleagues pointed out some time ago, Lake Bonneville helped create its own climate. The thermal inertia of this huge body of water made winter temperatures warmer, and summer temperatures cooler, than they would have been without the lake, at the same time as evaporation from the lake's surface provided significant amounts of precipitation that would not otherwise have existed here. While some of that precipitation would have fallen directly on the lake, some of it would also have been lofted to the east, ending up in the Wasatch Range and western Uinta Mountains, helping to form the glaciers that appear to have been contemporary with the highstands of Lake Bonneville.

So while the relationship between glacial advances and lake highstands in the Mono Lake Basin is not fully understood, beyond the fact that the very latest Pleistocene highstand of Lake Russell came after the glaciers retreated, the same is not true for the eastern edge of the Bonneville Basin. Here, glaciers expanded from the Wasatch Range into the Bonneville Basin itself at the same time as the lake was high.

Owens Lake

As I mention above, a second approach can be taken to understanding these relationships, one that depends on examining the contents of Pleistocene lake sediments. This approach is far more complicated than examining the direct physical relationships between geological features created by lakes and glaciers. For the lakes at issue here, this approach depends, in one way or another, on the fact that as glaciers expand, they scour the bedrock beneath them, producing fine-grained rock dust, much as filing your fingernails produces fine-grained fingernail dust. The larger the glacier, the more of this material, called rock flour, is produced. Streams issuing from the fronts of glaciers then carry it downward. If those streams flow into lakes, the rock flour becomes incorporated into the sediments that accumulate on the bottoms of those lakes. As long as the glacially produced rock flour is composed of minerals different from those that can enter the lake in other ways, it can be securely identified as such. The greater the amount of rock flour in those sediments, the more substantial the glaciation in the nearby mountains. In addition, if that rock flour contains magnetic minerals—for instance, the aptly named magnetite (a form of iron oxide, or rust)—then the greater the amount of flour, the greater the degree to which the sediments that contain it can be magnetized, a phenomenon known as magnetic susceptibility.

Rock flour also changes the carbon content of lakes, but to understand this, one needs to know that carbon in lakes comes in two different forms. The first of these comes from inorganic sources—the rocks, for instance, that help form the lake basin or that are scoured to produce the rock flour in the first place. The second form, organic carbon, comes from living things—plants that live in the water, ducks that poop and die in it, leaves that are washed into it from the surrounding land, bacteria that spend their lives there. Increasing amounts of rock flour in lakes dilute the total amount of organic carbon in those lakes. The greater the amount of that flour, the less light that penetrates the water and, as a result, the less biologically productive the lake becomes and the less the amount of organic carbon the lake contains. Organic carbon tends to be a less sensitive indicator of nearby glacier history than rock flour or magnetic susceptibility, because it can virtually disappear fairly early in a glacial cycle but it can still help decipher the relationship between lakes and glaciers.

Other things in lake sediments can be examined to figure out the degree to which surrounding mountains were glaciated at the times those sediments accumulated, but these are the biggies: the amount of rock flour and organic carbon in, and the degree of magnetic susceptibility of, those sediments.

The greatest effort in this realm in the Great Basin has been spent on Owens Lake. The waters flowing into Owens Lake come primarily from the Sierra Nevada, so the glacial history of this mountain range must be reflected in the sediments of that lake. Larry Benson and his colleagues have shown that during the last 55,000 years, peaks in magnetic susceptibility and minima in organic carbon values occurred between 25,500 and 15,470 years ago. This, they argue, represents the late Wisconsin Tioga glaciation in the Sierra Nevada. They have no data from 15,470 to 13,700 years ago; this is the time of the very low lake levels that I mentioned before, when Owens Lake may have dried entirely. When the record begins again, magnetic susceptibility is low and organic carbon high, suggesting the Tioga glaciation had ended by then.

The most detailed analysis of rock flour from the Owens Lake sediments has been provided by geologists James Bischoff and Kathleen Cummins. They showed that rock flour abundance in these sediments is tightly correlated with magnetic susceptibility and organic carbon: when rock flour is abundant, susceptibility is high and organic carbon low, and vice versa. Like Benson and his coworkers, Bischoff and Cummins concluded that the Tioga glaciation began around 25,500 years ago. However, they also concluded that this glaciation lasted until 12,700 years ago and that the lake overflowed continuously during this time.

This creates something of a problem. The sediments that Bischoff and Cummins analyzed are the same as those analyzed by Benson and his colleagues, but the latter team argued convincingly that Owens Lake was very low, and may have dried completely, between 15,470 and 13,700 years ago. If this were the case, it certainly couldn't have been overflowing at this time. Perhaps not coincidentally, Bischoff and Cummins have no rock flour data from this interval. When their record resumes, around 13,500 years ago, rock flour is abundant but then immediately drops to very low levels. This, of course, coincides with the episode of low magnetic susceptibility and high organic carbon noted by Benson and his colleagues. If we take this depositional hiatus into account, the Bischoff and Cummins rock flour sequence agrees extremely well with the picture drawn by Benson's team—and provides no evidence that the lake overflowed continuously through this time.

How does all this compare with the history of Owens Lake I discuss above? Stephen Bacon's analysis of that history shows that Owens Lake levels were so high between about 23,000 and 16,500 years ago that the lake overflowed, and that high levels, and perhaps overflow, characterized the lake between about 14,600 and 13,000 years ago. Insofar as this history is based on shoreline and near-shoreline features, it is independent of the sediment-based reconstructed history I have just discussed. It also means that when glaciers were advancing in the Sierra Nevada, Owens Lake was deep and getting deeper. In confirmation of this, the dates we have for glacial advances in the nearby Sierra Nevada, determined by Fred Phillips and his colleagues directly from glacial deposits, suggest that glacial advances occurred here at around 21,000, 17,500, and 14,000 years ago. These dates have significant margins of error—on the order of 15 percent—but they do match the highstands of Owens Lake quite well. Indeed, the highstand of Owens Lake that occurred during Marine Isotope Stage 6, somewhere between 185,000 and 130,000 years ago, may also correlate with glacial expansion in the Sierra Nevada, though the chronology is not sufficiently well controlled to know this with any certainty.

All of this might suggest that where we have the most detailed data, glaciers and lakes seem to have reached their peaks at about the same time. Mono Lake would then become a possible exception, perhaps explained by the fact that we know little about the levels of this lake between about 25,000 and 15,000 years ago.

Bear Lake

Things are not quite that simple, though. We know that because of recent, intense efforts to understand the late Pleistocene and Holocene history of Bear Lake.

Bear Lake is tucked into the far northeastern corner of the Great Basin, where it straddles the Utah-Idaho border (figure 2-2). Today, most of the water it receives comes from the Bear River, whose headwaters lie in the Uinta Mountains to the south and east. That, though, is only because water is now diverted from the river to the lake and has been since about 1912. Without that diversion, the river would now flow past the lake some eight miles to the north. This, in fact, has been the case for much, though not all, of the Holocene, with Bear Lake getting most of its water from streams that issue from the Bear River Mountains to the immediate west.

Recent work by geologist Joseph Rosenbaum and his colleagues has provided a detailed reconstruction of the late Pleistocene and Holocene history of this lake. Their results show that Bear Lake reached its late Pleistocene highstand between 13,500 and 12,700 years ago. During this episode, called the Raspberry Square phase, the lake reached a depth of approximately 236 feet, nearly 30 feet deeper than its modern average. Their work also shows that glaciers in the Bear River Valley reached their maximum extent much earlier, between some 16,850 and 15,800 years ago, at a time when the lake was near its modern level. Although the history of Bear Lake is determined by both climate and the wanderings of Bear River, the late Pleistocene highstand of Bear Lake came after, not during, the last great expansion of glaciers in this area, just as may be the case in the Mono Lake area.

Discordant Histories

So the situation is this. In the far western Great Basin, Owens Lake seems to have reached very high levels at the same time as extensive glaciation was occurring in the adjacent Sierra Nevada. To the north, Lake Russell seems to have reached its last highstand after local Sierran glaciers had already retreated. In the far eastern Great Basin, glacial expansion in the Wasatch Range and the southwestern Uinta Range apparently coincided with very high levels of Lake Bonneville. Not very far away, Bear Lake reached its last significant Pleistocene highstand long after local glaciers had begun their retreat. One might be tempted to conclude that perhaps the lakes in the Mono and Bear Lake basins reached their highstands because they were fed by melting glaciers. But that conclusion does not work for a number of reasons, including the fact that glacial retreat and lake highstands are separated by too much time for this process to account for what happened. Instead, the obvious conclusion is that we still have much to learn about the relationship between late Pleistocene glaciers and lakes in the Great Basin.

Pluvial Lakes, Glaciers, and Late Pleistocene Climates

During the mid-1980s, much of the northern Great Basin saw increased precipitation and a corresponding increase in the size of the lakes the region holds. The Great Salt Lake reached its historic high in June 1986; it would have risen even further were it not for the human activities that now divert so much water from its feeding streams. Malheur Lake flooded roads, ranches, and critical wildlife habitat on the Malheur National Wildlife Refuge. Humboldt Lake flooded roads and other human constructions and nearly overflowed into the Carson Sink.

Eventually, the high rate of precipitation that fed these increases ended, and the lakes began to wane. But what if the precipitation had not declined? In the Lahontan Basin in 1983, for instance, incoming rivers discharged at a maximum rate nearly 2.5 times their historic averages. If this discharge rate had continued while evaporation rates remained the same, would Lake Lahontan have reappeared?

The answer, Larry Benson and Fred Paillet have shown, is no. Had water input into the Lahontan Basin remained at 1983 values with the same evaporation rates, lakes in the Lahontan Basin would have grown substantially, ultimately covering 2,672 square miles, but this is only 30 percent of the area of Lake Lahontan at its maximum (8,284 square miles). The extremely high precipitation rates experienced in the Lahontan Basin in the mid-1980s would not have re-created anything like Lake Lahontan if evaporation rates had remained unchanged.

Today, evaporation rates in the Lahontan Basin are on the order of 48 inches a year. What if the 1983 water inflow figures were combined with the lowest monthly evaporation rates known to have occurred during historic times? Those lowest known rates yield 24.8 inches of evaporation a year, about half the average amount. Given the 1983 inflow rates (2.5 times the average) and the minimum monthly evaporation rates (about 0.5 times the average), Benson and Paillet showed that Lake Lahontan could, in fact, reappear.

So combining known precipitation maxima with known evaporation minima over a sufficient amount of time can recreate a Lake Lahontan at its peak. But even though such combinations could bring back the lake, there is no guarantee that it actually worked this way. Evaporation rates may have declined even more dramatically, or precipitation rates risen to even higher levels, to create the lake. What we do know something about is temperature.

Estimates of temperatures during the Last Glacial Maximum in the Lahontan and Bonneville areas have been derived from a broad variety of phenomena, from late Pleistocene vegetation (chapter 6) and glaciers to the chemical composition of invertebrates (which can change sensitively through time) and the rates at which sediment accumulated along lake edges. Computer simulations have also been used toward the same end. While the results vary, they do tend to focus on a mean annual temperature decrease of about 10.8 to 16.2°F compared to today (table 5-8).

This figure is satisfyingly similar to independently derived calculations of the global mean annual temperature decrease during this time. A recent estimate of that decrease places it at between about 7.5 and 13.7°F, with the magnitude greater toward the poles and less toward the equator. Indeed, when applied to the Great Basin, this work suggests that temperatures here would have been between 9 and 23.4°F colder than today.

Decreased temperatures would, of course, decrease the evaporation rate. If the temperature decreases were combined with an increase in precipitation and cloudiness, which would decrease evaporation beyond that expected from a temperature decline alone, little mystery might remain in the growth of Great Basin pluvial lakes and glaciers. In fact, Yo Matsubara and Alan Howard have suggested that Lake Lahontan and Lake Bonneville could be re-created with various combinations of temperature decreases of between 0.4 and 10.4°F, associated with increases in precipitation ranging from twice those of modern times to no change at all from those of today. The same study suggests that re-creating Lake Manly requires climate change ranging from a temperature decrease of 17.6°F with only a modest increase in modern precipitation, to precipitation nearly three times what the area now receives coupled with only a slight temperature increase.

But where would any increased precipitation and cloudiness come from? Here the answer seems to lie in the effects of the huge glacial mass that blanketed much of northern North America during the Wisconsin maximum.

This answer was, in fact, first suggested many years ago by geologist Ernst Antevs, about whom I say much more in chapter 8. In 1922, at the invitation of the Carnegie Institution of Washington, Antevs spent two months studying the Pleistocene deposits found in the Lahontan, Bonneville, and Mono lake basins. In his report, published in 1925, he noted that one possible effect of the formation of massive glaciers in northern North America during the Pleistocene might have been to produce low-pressure, moisture-bearing weather systems far south of the ice, impacting many areas, perhaps including the Great Basin.

In 1948, he dropped the "perhaps" and argued that one effect of the massive northern North America late Pleistocene glaciers had likely been to deflect storm tracks southward, bringing clouds and moisture to areas that otherwise would have been far more arid. "Hence," he concluded, "the West had a pluvial period" (Antevs 1948:170). According to this view, when the glaciers shrank, the storm tracks retreated northward, and the great Pleistocene lakes retreated as well. Many aspects of Antevs's brief 1948 paper quickly became enormously influential. However, his suggestion that the growth of pluvial lakes was caused by the displacement of storm tracks southward did not begin to gain real traction until the 1980s, when computer models showed that what Antevs had hypothesized did, in fact, likely occur.

Today, the winter jet stream, which marks the boundary between cold and warm air masses, passes over the West Coast of North America at about 50° N latitude. Climatic modeling of atmospheric circulation during the late Pleistocene by climatologist John Kutzbach and his colleagues showed that one effect of the vast northern ice mass was to split the jet stream in two, with one arm skirting the northern edge of that ice, the other swinging far south of it in western North America. The more southerly location of the jet stream would have increased the frequency of Pacific storms reaching the Great Basin, bringing both increased precipitation and increased cloud cover.

Therein, it now seems, lies the origin of pluvial lakes and the contemporary glaciers in the Great Basin: decreased temperatures coupled with increased precipitation and cloudiness—just as Antevs said. Decreased temperatures

TABLE 5-8
Selection of Temperature Estimates (Differences from Modern Values) for the Great Basin during the Last Glacial Maximum
MAT = Mean Annual Temperature

Area	Temperature	Source of Estimate	Reference
Great Basin	−9.9 to −12.6°F (MAT)	Glacier formation	Dohrenwend (1984)
Strawberry Mountains	−14.4°F (Summer)	Glacier formation	Bevis (1995)
Steens Mountain	−16.2°F (Summer)	Glacier formation	Bevis (1995)
Pine Forest Range	−10.8°F (Summer)	Glacier formation	Bevis (1995)
Lahontan region	−14.4 to −16.2°F (MAT)	Vegetation change	Wigand and Rhode (2002)
Lake Lahontan surface	−14.4°F (MAT)	Computer modeling	Hostetler and Benson (1990)
Bonneville and Lahontan lake surfaces	0 to +2.7°F (January)	Computer modeling	Hostetler et al. (1994)
Bonneville and Lahontan lake surfaces	−23.4°F (July)	Computer modeling	Hostetler et al. (1994)
Northern Uinta Mountains	−9.9 to −14.4°F (Summer)	Glacier formation	Munroe and Mickelson (2002)
Western Uinta Mountains	−9 to −12.6°F (MAT)	Glacier formation	Refsnider et al. (2008)
Northern Bonneville Basin	−5.4 to −16.2°F (MAT)	Vegetation change	O. K. Davis (2002)
Bonneville Basin	−10.8 to −12.6°F (MAT)	Computer modeling	Laabs, Plummer, and Mickelson 2006
Bonneville Basin	−12.6 to −23.4°F (MAT)	Ostracode amino acids	Kaufman (2003)
Bonneville Basin	−23.4°F (MAT)	Delta sedimentation rates	Lemons, Milligan, and Chan (1996)
Ruby Mountains	−12.6°F (Summer)	Glacier formation	Bevis (1995)
Deep Creek Mountains	−5.4°F (Summer)	Glacier formation	Bevis (1995)
Tushar Mountains	−7.2°F (Summer)	Glacier formation	Bevis (1995)
East-Central Nevada	−16.2 to −18°F (Summer)	Vegetation change	R. S. Thompson (1984, 1990)
Lake Newark, Newark Valley	−3°F (MAT)	Lake levels	Redwine (2003a, 2003b)
Southern Nevada	≥−10.1°F (MAT)	Ostracode oxygen isotopes	Quade, Forester, and Whelan (2003)
Southern Nevada (Yucca Mountain)	−14.4°F (MAT)	Vegetation change	Thompson, Anderson, and Bartlein (1999)
Southern Nevada (Yucca Mountain)	≥−15.3 to −17.1°F (MAT)	Vegetation change	Forester et al. (1999)
Southern Nevada (Nevada Test Site)	≥−10.8°F (Winter)	Vegetation change	Spaulding (1985)
Southern Nevada (Nevada Test Site)	−10.8 to −12.6°F (MAT)	Vegetation change	Spaulding (1985)
Southern Nevada (Nevada Test Site)	−12.6 to −14.4°F (Summer)	Vegetation change	Spaulding (1985)
Owens Valley, California	−18°F (MAT)	Plant hydrogen isotopes	S. A. Jennings and Elliot-Fisk (1993)
Death Valley, California	−14.4 to −25.2°F (Summer)	Vegetation change	Woodcock (1986)

NOTE: Bevis (1995) suggests that inaccurate estimates of modern temperatures in the Deep Creek and Tushar mountains may make these figures underestimates of the actual temperature decreases.

alone would decrease evaporation rates. In fact, by increasing ice cover on lakes, decreased temperatures would likely have decreased evaporation rates well beyond those expected from the temperature drop alone, as Larry Benson and others have pointed out. Increased cloudiness would have enhanced these effects, and increased precipitation would have provided the lakes and glaciers with the stuff of which they are made.

Many people think that science is largely built by carefully amassing facts, from which something approaching the truth ultimately emerges. In fact, science is a complex interplay between empirical knowledge and insightful speculation. Antevs couldn't have speculated about changing positions of storm tracks had he not known about glaciers, Great Basin pluvial lakes, and meteorology, but other scientists knew the same things without making the connections he did. The computer models suggested that Antevs was right, but computer modeling of this sort requires many assumptions, not all of which are likely to be correct (as modelers are fully aware). The important thing about the combination of Antevs's hypothesis and the results of the computer models is that it has implications for what the western North American world should have been like if all this is correct.

Multiple obvious, testable implications emerge from all of this work. First, as the northern glaciers grew and forced storm tracks southward, northern Great Basin lakes should have grown in response. Second, as those glaciers continued to grow, ultimately reaching their maximum extent, the storm tracks should have been forced even farther south. As this occurred, the northern Great Basin lakes should have fallen, and this should have happened at about the same time as lakes farther south were getting bigger. Third, as the northern glaciers retreated, the jet stream should have moved north once again, causing southern lakes to fall and the northern Great Basin lakes to grow yet again. Finally, as the glaciers waned, Great Basin pluvial lakes should have simply disappeared.

These predictions are in some important ways supported by our knowledge of late Pleistocene lake history in the Great Basin. Larry Benson and his colleagues suggest that the steep rise in the level of Lake Lahontan at about 22,000 years ago marks the initial pass of the southern jet stream over this lake basin, a function of the growth of the ice sheets in the Far North. While Ken Adams's reconstruction of lake level history for this general interval differs from that of Benson's team, it also allows for a steep rise in lake level by about 23,000 years ago (figure 5-13).

No matter what the timing of the rise, it was followed by a decline at about 16,000 years ago, marking the movement of the jet stream even farther south in response to maximum growth of the northern ice. Then, at about 15,000 years ago, as the ice retreated, the lake rose yet again, a result of the return of the jet stream over northern Nevada. As the ice dwindled even further, the jet stream moved even farther north, causing the lake to decline dramatically at around 13,000 years ago.

The correlations with Lake Bonneville are somewhat less impressive. The dramatic growth in the lake necessary to reach the Stansbury shoreline at around 21,000 years ago roughly matches the suggested first arrival of the jet stream; the rise to the Bonneville shoreline around 15,000 years ago matches the suggested second arrival of the jet; and the dramatic drop in lake level at around 13,000 years ago matches the hypothesized northward departure of the jet stream. What is missing, however, is the steep decline between the two jet stream visits that seems evident in the Lake Lahontan record (figures 5-10 and 5-13). The general match, though, is not bad. And the fact that Lake Alvord got so high that it broke through its eastern edge and flooded out of the Great Basin between 13,000 and 14,000 years ago matches this general picture as well.

If the northern lakes were high between about 13,000 and 15,000 years ago because the jet stream was stationed here, then we would expect lakes in the southern lake basins to be relatively shallow at that time. To some extent, that seems to have happened. Owens Lake was low, and might not even have existed, between 15,000 and 13,700 years ago. Lake Mojave was low between 16,600 and 13,700 years ago. And while the history of Lake Searles does not seem well enough understood to help us here, Death Valley's Lake Manly was low after 18,000 and before 13,700 years ago. This is exactly what we would expect if the jet stream argument is correct.

On the other hand, all of the lakes I have discussed here were relatively high at about 18,000 years ago, from Death Valley in the southeast to Lake Lahontan and Lake Chewaucan in the northwest, and that doesn't match the prediction that when the northern lakes are high, the southern ones should be low, and vice versa. In fact, even lakes in the southwestern United States had highstands at about this time. For instance, work by Bruce Allen and his colleagues in the Estancia basin of central New Mexico has shown that the lake that formed here was very high at 18,000 years ago. This matches a highpoint from an even more southerly lake, Lake Cloverdale, in far southwestern New Mexico, which had a significant peak between 20,000 and 18,000 years ago. This, in turn, comes close to matching the history of Lake San Agustin in western New Mexico, which reached peaks at 22,000, 19,000, and 17,000 years ago. Neither Lake San Agustin nor Lake Cloverdale were deep between 15,000 and 13,000 years ago, when the northern lakes were high, matching the predictions of the jet stream model. But Lake Estancia was quite high between 13,800 and 13,400 years ago, as was yet another Pleistocene lake, Lake Cochise in southeastern Arizona. These late highstands do not seem to meet the requirements of the jet stream model as it now stands.

None of these mismatches means that the jet stream argument is wrong. Not only does the history of glaciers in the far north require that something like this happened, but there is no other mechanism that can explain as much about pluvial lake history. On the other hand, the mismatches do mean that we have a lot to learn about Pleistocene lake history in the Arid West and about the position of the jet stream, including its sinuosity and width, as the Pleistocene came to an end.

Chapter Notes

The land speed records are taken from three places: the *Guinness Book of World Records*, the website of the Féderation Internationale de l'Automobile, and www.landracing.com. On the Bonneville Salt Flats, see Turk (1973).

The surface area, depth, and elevation figures I provide for Owens Lake are from Orme and Orme (2008). For an excellent account of the diversion of Owens Valley waters to Los Angeles, see Reisner (1993), which discusses the "development" of water resources in the western United States as a whole and is required reading for all interested in western resources in general. On the history of the Metropolitan Water District of Southern California, see Schwarz (1991) and perhaps the movie *Chinatown*. For a local account of the Owens Valley story, see Chalfant (1975).

The best way to familiarize yourself with the recent history of Mono Lake is by visiting the Mono Lake Committee's website (www.monolake.org). Since 1978, the committee has fought to prevent the Los Angeles Department of Water and Power from destroying a lake that is not only one of western North America's scenic gems but also the critical habitat for many organisms, including California gulls (*Larus californicus*) and eared grebes (*Podiceps nigricollis*). Even though the committee has made tremendous strides toward undoing the damage caused by diversion of Mono Basin waters, there is much yet to be done. If you're interested, consider joining (Membership Desk, Mono Lake Committee, P.O. Box 29, Lee Vining, CA 93541, or enroll at the website). If you are in the area, stop in at the Mono Lake Committee Information Center and Bookstore in Lee Vining, as well as the U.S. Forest Service's Mono Basin National Forest Scenic Area Visitor Center (www.fs.fed.us/r5/inyo/recreation/rec-reports/mono.shtml). The latter has excellent exhibits on local natural history, the native peoples of the area, and Mono Lake itself.

The information on recent Mono Lake and Owens Lake history that I provide is from Cutting (2007), Prather (2008), Reis (2008), and both the Mono Lake Committee and the Owens Valley Committee (www.ovcweb.com) websites. On Owens Lake dust, see Dahlgren, Richards, and Zu (1997) and Reheis (1997).

On the Truckee and Carson River basins, see Townley (1977); on water history in the western Great Basin in general, see Harding (1965), though parts of this reference are dated. Table 5-1 contains the references I used for estimates of lake area. The information I have presented on Great Basin climates is drawn largely from Houghton (1969) and Houghton, Sakamoto, and Gifford (1975); see also Benson and Thompson (1987a, 1987b), Hidy and Klieforth (1990), and Wigand and Rhode (2002). Mock (1996) provides a series of important illustrations showing precipitation by seasons in western North America. D. K. Adams and Comrie (1997) present a superb discussion of western North American monsoons. Data on precipitation in the Ruby Mountains come from R. S. Thompson (1984); water budgets for Great Salt Lake and Walker Lakes are presented in Everett and Rush (1967), Rush (1972), and Arnow (1980, 1984). Evaporation rates for Nevada are provided by B. R. Scott (1971), while Shevenell (1996) provides an important discussion of potential evapotranspiration rates in Nevada. The potential evaporation rate (PET) measures the total amount of water that could be evaporated and transpired by plants were there no limits on the amount of water available to be lost in these ways; it represents the maximum possible amount of water that could be lost to the atmosphere (see Lu et al. [2005] if you are interested in a detailed discussion of the calculation of these rates).

The recent history of Goose Lake, including periods of overflow into the Pit River drainage, is provided by K. N. Phillips and Van Denburgh (1971). Interbasin groundwater flow in the Great Basin is mapped in a number of places, including Division of Water Resources, State of Nevada (1972), Bedinger, Harrill, and Thomas (1984), and J. R. Harril, Gates, and Thomas (1988).

The best maps of Pleistocene lakes in the Great Basin are those by Williams and Bedinger (1984) and, for the western Great Basin, Reheis (1999a); these tend to supplant the earlier effort by C. T. Snyder, Hardman, and Zdenek (1964). Other helpful maps are to be found in G. I. Smith and Street-Perrot (1983) and Orme (2008b). Mifflin and Wheat (1979) on the Pleistocene lakes of Nevada is an indispensable source. Papers in Hershler, Madsen, and Currey (2002) provide important overviews of Great Basin Pleistocene lakes, as does Benson (2004b). For an important discussion of the history of research of the Great Basin's pluvial lakes, see Orme (2008b).

On the modern Great Salt Lake, see Arnow (1980, 1984), Sturm (1980), Alder (2002), Atwood (2002), J. L. Mason and Kipp (2002), and other papers in Gwynn (1980, 2002). The recent fluctuations of Great Salt Lake are also discussed by Lindskov (1984), Mabey (1986, 1987), and Arnow and Stephens (1990). The economic damage caused by the increased precipitation received by Utah in 1983 is detailed by Kaliser and Slosson (1988), Kaliser (1989), and the papers in Gwynn (2002). For a concise and readable discussion of crustal deformation in the Bonneville Basin caused by the lake itself, see B. G. Bills, Wambeam, and Currey (2002).

There is an immense literature on Lake Bonneville. The classic reference on Lake Bonneville is G. K. Gilbert (1890); my presentation has depended heavily on O'Connor (1993); Oviatt (1997); Oviatt and Miller (1997); Sack (1999, 2002); Oviatt, Madsen, and Schmitt (2003); Godsey, Currey, and Chan (2005); and Patrickson et al. (2010). The Lake Bonneville shoreline elevations used in the text are from Currey and Oviatt (1985) and Oviatt, Currey, and Miller (1990); different elevations are to be found in Benson and Currey et al. (1990; see their figure 7).

The Homestead Cave fishes are discussed by Broughton (2000a, 2000b); Broughton, Madsen, and Quade (2000); Madsen et al. (2001); and Broughton et al. (2004); on Homestead Cave, see Madsen (2000). This site is discussed in later chapters.

O'Connor (1993) discusses the Bonneville Flood in detail. On the largest known historical floods, see Herschy (2001, 2003). On prehistoric megafloods in general, see V. R. Baker (2002). On the Missoula floods, see Waitt (1984, 1985), O'Connor and Baker (1992), Clague et al. (2003), and Booth et al. (2004). When Harlan Bretz first suggested that what he called the Channeled Scablands were created by a massive flood, scientists found the idea absurd. Bretz was right, and Baker (2008) provides a fascinating history of the lengthy debate over this issue. The Altai Mountains floods are discussed in V. R. Baker, Benito, and Rudoy (1993) and Carling et al. (2002).

My discussion of Lake Bonneville history during Provo times is based on Godsey, Currey, and Chan (2005) and on unpublished data and interpretations by Holly Godsey, Jack Oviatt, David Miller, and Margaret Chan. The 2005 paper suggests that there may have been a steep decline in Provo levels sometime before 12,500 years ago, but this no longer appears to be the case, and I have not included that decline in figure 5-10 or elsewhere. The 12,500-year date for the fall from Provo levels also comes from unpublished work by Godsey and her colleagues. Oviatt (1988) proposed the monsoonal explanation for Lake Gunnison overflow; my account of his more recent explanation for this overflow is based on discussions with him.

The place to start to learn more about the Gilbert-aged lake in the Bonneville Basin is Oviatt et al. (2005). This paper also places the Gilbert episode within the wider frame of Younger Dryas–aged events in western North America as a whole. I have augmented this paper with ideas and facts gleaned from a memorable discussion of Lake Bonneville history by Oviatt at the

Dugway Proving Grounds, Utah, in July 2008, and subsequent discussions with him. The Old River Bed sequence is discussed in detail in chapter 8.

I have used a value of 19,800 square miles for the surface area of Lake Bonneville at its maximum, following Currey, Atwood, and Mabey (1984; see also figure 16 in Currey 1990). As is often the case for estimates of the surface areas of late Pleistocene Great Basin lakes, other values exist. Orme (2008b), for instance, suggests 19,960 square miles, slightly higher than the 19,940 calculated by T. R. Williams and Bedinger (1984). Such differences are far less than the errors likely to be involved in making these calculations in the first place.

The details that I have provided on the Younger Dryas are from Alley et al. (1993), Taylor et al. (1997), Severinghaus et al. (1998), Alley (2000), Friele and Clegg (2002), McManus et al. (2004), Broecker (2006), Birks and Birks (2008), Bradley and England (2008), Steffensen et al. (2008), Broecker et al. (2010), and Murton et al. (2010). The best recent review of the end of the last glaciation, and the place of the Younger Dryas within it, is Denton et al. (2010). The dates I give for the Younger Dryas are converted from the calendrical dates given in Steffensen et al. (2008, table 1); those calendrical dates are 12,712 ± 74 to 11,711 ± 12. There are some very good websites explaining the intricacies of the thermohaline circulation; www.windows.ucar.edu/tour/link=/earth/Water/thermohaline_ocean_circulation.html is an excellent place to start.

Pre–Lake Bonneville lakes in the Bonneville Basin are discussed by Oviatt, McCoy, and Reider (1987); R. S. Thompson et al. (1995); Bouchard et al. (1998); O. K. Davis (1998, 2002); Oviatt et al. (1999); Kaufman, Forman, and Bright (2001); Oviatt and Thompson (2002); and Hart et al. (2004). On Bear River history, see Bouchard et al. (1998), O. K. Davis (1998, 2002), Hart et al. (2004), and Reheis, Laabs, and Kaufman (2009); on today's Bear Lake, see Lamarra, Liff, and Carter (1986); Colman et al. (2006); and Dean, Wurtsbaugh, and Lamarra (2009). For an animated history of Lake Bonneville during the last 30,000 years, see the Utah Geological Survey's superb website (many of the survey's publications are available electronically, for free, at this site; the animation itself is at http://geology.utah.gov/utahgeo/gsl/flash/lb_flash.htm).

I have used an area of 8,284 square miles for Lake Lahontan at its maximum, following Reheis (1999b). As with Lake Bonneville, other estimates abound. For instance, adding up the surface areas for the separate Lahontan subbasins provided in Benson, Kashgarian, and Rubin (1995, table 6) with a lake highstand of 4,380 feet provides an estimate of 8,794 square miles. Benson and Mifflin (1986) give the lake a surface area of 8,610 square miles, using a highstand of 4,298 feet.

Very early lakes in the Lahontan Basin are discussed in Lao and Benson (1988), Reheis and Morrison (1997), Reheis (1999b), Negrini (2002), and Reheis et al. (2002). The history of ancient lakes in the Great Basin in general is too rich and too complex to discuss here. Reheis et al. (2002) is the best place to start, but Reheis et al. (1993) and Reheis and Morrison (1997) provide important additional details.

Russell (1885b) will always be the classic source on Lake Lahontan history, but an enormous number of skilled scientists have contributed to our knowledge of Lake Lahontan since that time. Those I have relied on most heavily here are Benson and Mifflin (1986); Benson and Thompson (1987a); Lao and Benson (1988); Benson and Paillet (1989); Benson and Peterman (1995); Benson, Kashgarian, and Rubin (1995); Benson, White, and Rye (1996); K. D. Adams and Wesnousky (1998, 1999); Rhode, Adams, and Elston (2000); Briggs, Wesnousky, and Adams (2005); and K. D. Adams (2010). For more on the diversion of the Humboldt, Walker, and Truckee rivers, see K. D. Adams, Wesnousky, and Bills (1999). Jonathan Davis discussed possible diversions of the Humboldt in a number of places; the publications I refer to in the text are J. O. Davis (1982, 1987). The Great Basin lost a major scholar and a friend to many when he was killed by a drunk driver near Virginia City, Nevada, on December 14, 1990, at the age of forty-two.

Larry Benson's work has emphasized the analysis of the geological deposits known as tufas. Tufas are deposits of calcium carbonate that form at the mouths of springs, from mixtures of lake and spring water, and from lake water alone. Occurring in many different forms, tufa is abundant in Great Basin Pleistocene lake basins; the spiky projections along the edges of Mono Lake in eastern California are all tufas. The best place to start learning about tufas is Benson (2004a), a piece of cooperative work with the U.S. Geological Survey, the Pyramid Lake Paiute Tribe, and the Pyramid Lake Visitor Museum Center.

Winograd and Roseboom (2008) reviewed the Yucca Mountain project as it stood shortly before it was set aside. Ewing and von Hippel (2009) provide a concise review of the reasons that the project has been shelved or, more precisely, "scaled back to those costs necessary to answer inquiries from the Nuclear Regulatory Commission, while the Administration devises a new strategy toward nuclear waste disposal" (Office and Management and Budget 2009:65).

Important analyses of Walker Lake and Walker River, including their Pleistocene histories, are found in the Lake Lahontan references cited above, as well as in G. Q. King (1978, 1993, 1996), Benson (1988), Bradbury (1987), and Bradbury, Forester, and Thompson (1989); references on the Holocene history of the lake and river are provided in chapter 8. Data on stream inflow into the Lahontan Basin as a whole are taken from Benson and Paillet (1989); this reference also discusses what would have happened in the Lahontan Basin had inflow rates remained at 1983 levels. R. B. Morrison (1964) is a key reference on the geology of the southern Carson Desert and of Lake Lahontan in this area. This work also contains Morrison's definition of what he called the First Fallon Lake. Currey (1988); Elston, Katzer, and Currey (1988); and Benson, Currey, et al. (1990) argue that this lake might be contemporary with the rise of Lake Bonneville to the Gilbert shoreline; see also Dansie, Davis, and Stafford (1988).

Negrini (2002) provides an excellent introduction to the Pleistocene lakes of the northwestern Great Basin, including the Oregon lakes, on which I have relied heavily. If you are interested in those lakes, this is where I would start. I. S. Allison (1982) provides a valuable introduction to the Summer Lake Basin and to Lake Chewaucan, but many aspects of that source are now dated. J. O. Davis (1978) provides an important analysis of the tephras of this part of the world, while J. O. Davis (1985a) got the ball rolling on the analysis of the tephra layers in the deposits of Lake Chewaucan; Negrini and Davis (1992) is the work I mention in the text. Gobalet and Negrini (1992) discuss the remains of tui chub (*Gila bicolor*) recovered from the sediments of this lake, estimated to be 98,000 years old. The works I have relied on most heavily in my discussion of the Chewaucan and Fort Rock basins are Friedel (1993, 1994, 2001), Palacios-Fest et al. (1993), Negrini et al. (1994, 2000), Cohen et al. (2000), and Licciardi (2001).

On the tephras at Summer Lake, see, in addition to the papers by Negrini and his colleagues, Kuehn and Foit (2001, 2006); among many other things, these authors provide strong evidence that the Wono tephra comes from Newberry Crater. Kuehn and Foit (2006) also provide a concise and very readable introduction to the ways in which tephra layers can be identified. On the Mono Lake excursion, see S P. Lund et al. (1988); Liddicoat (1996); Benson, Liddicoat, et al. (2003); and Liddicoat and Coe (2008). Zimmerman et al. (2006) have recently suggested that the excursion represented in the Mono Lake sediments was incorrectly dated, falling not at 28,600 radiocarbon years ago (about 32,000 calendar years ago), but instead at around 40,000 calendar years ago, and that it represents an earlier magnetic event, the Laschamp excursion, known from other sequences. Among other things, this requires that previous dates for the late Pleistocene sediments of Mono Lake be incorrect and that the tephra used to correlate the Mono Lake excursion with deposits in Pyramid Lake have been incorrectly identified. It also requires that the relationship between the Mono Lake excursion and the identifications of a similar excursion elsewhere on the planet and dated to the same time be coincidental (e.g., Channell 2006, S. Lund et al. 2006). For these reasons, I have not accepted the Zimmerman et al. (2006) conclusions here; for a brief response to the Zimmerman et al. (2006) argument, see Liddicoat and Coe (2008).

Dates for the Wono and Trego Hot Springs tephra that I use here are from the work of the ubiquitous Larry Benson and his colleagues (Benson, Smoot, et al. 1997). For a general introduction to luminescence dating, see Lian and Roberts (2006); for a splendid discussion of luminescence dating in archaeology, see Feathers (2003).

The figures I provide on overflow depths for lakes in the Mono Lake–Death Valley systems are calculated from those in G. I. Smith and Bischoff (1997); see also Benson, Currey, et al. (1990). My discussion of the history of the Mono Lake Basin relies heavily on Benson, Lund, et al. (1998) and Reheis et al. (2002); the information on the modern levels of Mono Lake is drawn from the website of the Mono Lake Committee. Reheis et al. (2002) have also shown that when Lake Russell overflowed sometime prior to about 760,000 years ago, it did so not to the east, into Adobe Valley, but to the north, into the Walker Lake drainage system.

The 1,060-foot-deep core from Owens Lake is discussed in G. I. Smith and Bischoff (1997). Benson, Kashgarian, et al. (2002); Benson (2004); and Benson, Lund, et al. (2004) discuss the use of oxygen isotopes to reconstruct Great Basin lake history; I have made heavy use of those discussions here. The results of this work for Pleistocene Owens Lake are presented in Benson, Burdett, et al. (1996, 1997); Benson (2004b); and Benson, Lund, et al. (2004). Bacon et al. (2006) and Orme and Orme (2008) discuss and analyze Owens Lake shoreline data. The correlations across the independent work conducted by these groups are in general impressively strong; in cases where they differ, I have split the differences between the dates implicated by the two sets of analyses. Both Bacon et al. (2006) and Orme and Orme (2008) use calibrated ages; those I give here are either taken from their radiocarbon dates or translated back into radiocarbon ages. For a slightly different interpretation of the latest Pleistocene history of Owens Lake, see F. M. Phillips (2008).

The key references for Searles Lake are G. I. Smith (1979, 1984); G. I. Smith et al. (1983); Benson, Currey, et al. (1990); F. M. Phillips et al. (1994); and Lin et al. (1998); see also G. I. Smith and Street-Perrott (1983) and G. I. Smith, Benson, and Currey (1989). The history of Lake Panamint that I provide depends very heavily on Jayko et al. (2008) and F. M. Phillips (2008); other important discussions are provided by Fitzpatrick and Bischoff (1993), Knott (1997), Roberts and Spencer (1998), and Jayko (2005). Evidence for a lake in Owens Valley during MIS 6 is provided by Jayko and Bacon (2008).

Work by Hooke (1972) began to tease apart the latest Pleistocene history of lakes in Death Valley; my discussion of our current knowledge of this history has relied heavily on J. Li et al. (1996); J. Li, Lowenstein, and Blackburn (1997); Knott (1997); D. E. Anderson (1998); Ku et al. (1998); Lowenstein (2002); and D. E. Anderson and Wells (2003); see also Yang et al. (2005). The 13,450-year date I mention in the text is from Klinger (2001) and might represent a spring, rather than a lake, deposit. There are significant debates over some aspects of Death Valley lake history that I have not mentioned here; for those interested, easy access to the relevant literature is provided by Machette, Klinger, and Knott (2001); Hooke (2002, 2005); and Knott, Tinsley, and Wells (2004). On the Pleistocene history of the Amargosa River, see D. E. Anderson (1998) and Menges (2008). I have relied heavily on the latter reference in my discussion of Lake Tecopa (on which, see R. B. Morrison [1999] and R. B. Morrison and Mifflin [2000]). For those interested in the full sweep of geological history in the Owens River system, from Mono Lake to Death Valley during the past 12,000,000 years or so, I recommend the masterful synthesis by F. M. Phillips (2008). For a superb field guide to the Pliocene and Pleistocene lakes of the western Great Basin in general, including much information not available elsewhere, see Reheis et al. (2008).

There is an extraordinarily rich, and at times feisty, literature on virtually every aspect of the history of the Mojave River Basin. For river and lake history, the best place to start is with Enzel, Wells, and Lancaster (2003b) and Reheis and Redwine (2008); the works I have relied on most heavily here include these and S. G. Wells et al. (1989, 2003); Knott (1997); Meek (1999, 2000, 2004); Tchakerian and Lancaster (2002); K. C. Anderson and Wells (2003); B. F. Cox, Hillhouse, and Owen (2003); Enzel, Wells, and Lancaster (2003a); Jefferson (2003); and Reheis, Miller, and Redwine (2007). For those interested in exploring the Mojave River area, every issue of the *San Bernardino County Museum Association Quarterly* includes an enormous amount of information on local geology, paleontology, archaeology, and history.

In presenting their dates for Lake Manix, Reheis and Redwine (2008) use a calibration curve constructed by Fairbanks et al. (2005). As they note, this curve is experimental, and I have converted their dates back to radiocarbon years using the software found at http://radiocarbon.ldeo.columbia.edu/research/radcarbcal.htm.

Dugas (1998) discusses the late Pleistocene and Holocene history of Malheur Lake; on Fort Rock Lake's possible external drainage, see I. S. Allison (1979) and Friedel (1993, 2001). My discussion of Pleistocene Lake Alvord is based on D. T. Carter et al. (2006); see Hemphill-Haley, Lindberg, and Reheis (1999b) and Lindberg (1999) for the origins of the idea that there was a massive flood here.

The first mention of the glacier on Wheeler Peak (then called Jeff Davis Peak) is in Russell (1885a); for a more poetic account, see Heald (1956). G. Osborn and Bevis (2001) and G. Osborn (2004) provide a review of the history of our knowledge of this glacier. The recent general reviews of the montane glaciers of the western United States to which I refer in the text are Pierce

(2004) and Kaufman, Porter, and Gillespie (2004). By far the most important recent reviews of Great Basin glaciation are those provided by G. Osborn and Bevis (2001) and G. Osborn (2004); those papers also provide access to the earlier literature on this topic. Bevis (1995) provides important information not available elsewhere on the glaciation of the mountains listed in table 5-6. My discussion of glaciation in the Ruby Mountains and the East Humboldt Range relies heavily on those sources, but also on R. P. Sharp (1938) and Wayne (1984). Deglaciation of Steens Mountain is discussed by Mehringer (1986), while Bevis (1995, 1999) provides additional discussions of the glacial episodes here. R. S. Thompson (1984) discusses his attempt to core lakes in the Ruby Mountains.

A number of years ago, geologists defined and dated a series of glacial advances and retreats in a set of drainages on the east face of the White Mountains overlooking Fish Lake Valley. Significant questions have been raised about this work, ranging from the chronology of the glacial sequence to whether the earliest deposits involved are glacial in origin. On all of this, see Elliot-Fisk (1987); Swanson, Elliot-Fisk, and Southard (1993); Reheis (1994); G. Osborn and Bevis (2001); and G. Osborn (2004). The 16,000-radiocarbon-year date I provide for the Perry Aiken glaciation is converted from the age in calendar years provided by Zreda and Phillips (1995); see also F. M. Phillips et al. (1996). It is not at all clear whether the Spring Mountains of southern Nevada were glaciated, but current arguments suggest that they were not. The debate over this issue is discussed in detail by J. Osborn, Lachniet, and Saines (2008); this reference also provides a guided tour to the relevant geology.

Porter (1977) and Porter, Pierce, and Hamilton (1983) provide the information I have used on ELAs, including those in the Cascade Range, while Dohrenwend (1984) provides the ELA analysis for the Great Basin (see also Zielinski and McCoy [1987], but this paper depends heavily on Blackwelder's early data on glaciation in the Great Basin).

Putnam (1949) named Lake Russell; Putnam (1950) argued that this lake reached its highstand at the same time as local Sierran glaciers were reaching their maximum. Lajoie (1968) remains a critical reference on the Quaternary history of this area. For evidence that the highstand of Lake Russell came after the maximum expansion of local glaciers, see Lajoie and Robinson (1982); Benson, Currey, et al. (1990); and Benson, May, et al. (1998). Bursik and Gillespie (1993) and F. M. Phillips et al. (1996) discuss the chronology of these glaciers.

The age of the Bull Lake glaciation is discussed in detail in Pierce (2004); on Little Cottonwood and Bells canyons, see Richmond (1964, 1986); Madsen and Currey (1979); W. E. Scott (1988); Godsey, Currey, and Chan (2005); and Lips, Marchetti, and Gosse (2005). On the relationship of glaciers along the western Wasatch Range to those in nearby areas and the role that lake-derived precipitation played in the history of those glaciers, see Munroe and Mickelson (2002); Laabs, Plummer, and Mickelson (2006); Munroe et al. (2006); Laabs et al. (2007); and Refsnider et al. (2008). On the degree to which Lake Bonneville and Lake Lahontan created their own climates, see Hostetler et al. (1994). Licciardi et al. (2004) discuss variability in the history of western North American glaciers during the Last Glacial Maximum.

My discussion of the implications of Owens Lake sediments for glaciation in the Sierra Nevada is based on Benson, Burdett, et al. (1996); Benson, May, et al. (1998); Bischoff and Cummins (2001); and Benson (2004b). Bischoff and Cummins (2001) use calendar years; I have converted these to radiocarbon years.

Crucial references on the late Pleistocene history of Bear Lake include Laabs and Kaufman (2003); J. Bright et al. (2006); Kaufman et al. (2009); Reheis, Laabs, and Kaufman (2009); Rosenbaum and Heil (2009); Smoot (2009); and Smoot and Rosenbaum (2009). As elsewhere, I have converted the calendar dates used by these authors into radiocarbon years.

The simulations showing the conditions under which Lake Lahontan might (and might not) reappear are provided by Benson and Paillet (1989) and Hostetler and Benson (1990). References to Last Glacial Maximum temperature estimates are provided in table 5-8. The global estimates that I discuss are from Schneider von Deimling et al. (2006, 2008); the Great Basin estimate that I derived from their work is taken from figure 2A in Schneider von Deimling et al. (2006).

Matsubara and Howard (2009) present their important analysis of the relationship between climatic variables and the sizes of Pleistocene lakes in the Great Basin. For more on Ernst Antevs, see chapter 8 and www.geo.arizona.edu/Antevs/antevs.html. On the jet stream hypothesis and its possible effects on the Great Basin, see Antevs (1925, 1948); Kutzbach and Guetter (1986); Kutzbach (1987); Benson and Thompson (1987a, 1987b); COHMAP Members (1988); Benson and Klieforth (1989); Hostetler and Benson (1990); Oviatt, Currey, and Miller (1990); R. S. Thompson et al. (1992); Oviatt (1997); Benson, Kashgarian, and Rubin (1995); Benson, White, and Rye (1996); Negrini (2002); Benson (2004b); Enzel, Wells, and Lancaster (2003b); and Munroe et al. (2006). I would start with Negrini (2002), which provides a superb overview of the issues involved.

On Lake Cloverdale, see Krider (1998); on Lake Estancia, B. D. Allen and Anderson (2000); R. Y. Anderson, Allen, and Menking (2002); and Menking et al. (2004); on Lake San Agustin, F. M. Phillips et al. (1992); on Lake Cochise, Waters (1989).

Jewell (2008) presents a very intriguing alternative late Pleistocene jet-stream history, with the jet remaining far to the south until after the retreat of Lake Bonneville from the Provo shoreline (which, following Godsey et al. [2005], he places at about 12,000 years ago). This, however, would not seem to account for the fact that some far southern lakes—San Agustin and Cloverdale—were low during this interval.

Regarding table 5-8, I note that Forester et al. (1999) estimate a glacial-age mean annual temperature (MAT) of 7.2 to 9°F in the Yucca Mountain area; I used the Yucca Mountain MAT given by R. S. Thompson, Anderson, and Bartlein (1999), of 24.1°F, to calculate the temperature difference given in this table.

CHAPTER SIX

Late Pleistocene Vegetation of the Great Basin

Learning about Ancient Vegetation

William Lewis Manly and his traveling companions spent the Christmas of 1849 in Death Valley. Manly's companions did not escape until mid-February, and then only because Manly and his friend John Rogers walked out, returning with supplies, directions, and hope. Not until March 7 did the whole group reach Rancho San Francisco, near Los Angeles, and full safety. On their way out, they gave Death Valley its name, even though only one of their immediate group had actually died there: "We took off our hats, and then overlooking the scene of so much trial, suffering and death spoke the thought uppermost saying: — 'Good bye Death Valley!' then faced away and made our steps toward camp" (Manly 1894:216).

Manly and his group had been lured into Death Valley by talk of a cutoff that would save them many days in the trip from Salt Lake City to California. The shortcut they tried left the Old Spanish Trail in southwestern Utah and then cut southwest across southern Nevada, leading the travelers to Ash Meadows, then into the heart of what is now Death Valley National Park.

Seventy miles northwest of Las Vegas, Manly's group passed by Papoose Lake, now close to the boundary between Nellis Air Force Base and the Nevada Test Site. Realizing that continuing west might see them die of thirst, the small band of travelers headed south:

> We turned up a cañon leading toward the mountain and had a pretty heavy up grade and a rough bed for a road. Part way up we came to a high cliff and in its face were niches or cavities as large as a barrel or larger, and in some of them we found balls of a glistening substance looking something like pieces of varigated [sic] candy stuck together. The balls were as large as small pumpkins. It was evidently food of some sort, and we found it sweet but sickish, and those who were so hungry as to break up one of the balls and divide it among the others, making a good meal of it, were a little troubled with nausea afterwards. (Manly 1894:126).

Manly guessed that what they had found was a food cache belonging to Indians, and was concerned that what they had done might cause them serious problems. "I considered it bad policy to rob the Indians of any of their food," he went on to say, "for they must be pretty smart people to live in this desolate country and find enough to keep them alive . . . they were probably revengeful, and might seek to have revenge on us for the injury" (Manly 1894:126).

The Manly party wasn't the only one to take notice of this "glistening substance," though it may have been the only group to have mistaken it for food. In 1843, John C. Frémont found the same stuff. Exploring the canyon of a small tributary of the Bear River in far southern Idaho on August 29 of that year, he found "several curious caves" on the roofs of which he noted "bituminous exudations from the rock" (Frémont 1845:141). Sixteen years later, the U.S. Army's Corps of Topographical Engineers assigned Captain James H. Simpson the task of discovering a better wagon route across the Great Basin. Simpson gladly accepted the task—no surprise, since he had suggested it in the first place—and traveled from Salt Lake City to Genoa, Nevada, and back during the spring and summer of 1859. By the time he was done, he not only had found a better wagon route but had also blazed a path that the Pony Express was to follow closely in 1860, that the telegraph was to use in completing its transcontinental service in 1861 (putting the Pony Express out of business), and that U.S. Highway 50 follows closely today.

On July 16, 1859, Simpson was exploring Dome Canyon in the House Range, just west of the Sevier Desert and south of what is now Fish Springs National Wildlife Refuge. He found the "walls of the cañon full of small caves, and as usual showing a great deal of the resinous, pitchy substance, that seemingly oozes out of the rock" (Simpson 1983:125). Unlike the members of Manly's party, however, Simpson did not taste it; he guessed that it might have been "the dung of birds or of small animals" (Simpson 1983:125).

Simpson was very close to being right. These hard, shiny deposits so often found covering the walls of caves and rock crevices in the Great Basin are made by a group of squirrel-sized rodents that belong to the genus *Neotoma*. These active little mammals are found throughout much of North America, but they are especially common, and rich in species, in the Arid West. The first members of the genus to be described scientifically were from the woodlands of the East: they were given the scientific name *Neotoma floridana* and came to be called woodrats. That common name was so well established by the time the western species came to be described that the western forms have been called woodrats as well. Since that is their official common name, that is what I use here.

Western "woodrats," however, may or may not live in woodlands, so other common names have come into use for them. The most widely accepted of these names is "pack-rat," which comes from the fact that these rodents often pick up whatever small items they find interesting and pack them off to their dens. Because they use their mouths to do this, they often leave behind whatever it was they were carrying when they discovered the new item they found more appealing. The third common name applied to them, "trade rat," comes from this behavior.

Today, there are three species of woodrats in the Great Basin. One of these, the white-footed woodrat (*Neotoma fuscipes*), is essentially a far western species, found along the West Coast from Oregon to Baja California. It enters the Great Basin only on its far northwestern edge, in northeastern California and adjacent south-central Oregon. The desert woodrat (*Neotoma lepida*), on the other hand, is found almost throughout the Great Basin. The bushy-tailed woodrat (*Neotoma cinerea*) also occupies much of the Great Basin, but it is missing from large parts of southern Nevada and southeastern California.

Although the general ranges of these two latter woodrats overlap substantially, the desert woodrat tends to live in and adjacent to the valleys of the Great Basin, while the bushy-tailed woodrat is essentially an animal of higher altitudes. Only in the northern Great Basin—south-central Oregon, for instance—are bushy-tailed woodrats able to live at elevations low enough that their ranges routinely overlap those of desert woodrats. Even here, competition between the two is lessened by their tendency to focus attention on different habitats. Desert woodrats routinely den beneath isolated boulders, though they also build dens at the base of shrubs and trees. Bushy-tailed woodrats, on the other hand, prefer caves and crevices. They also take readily to buildings: the highest known record for bushy-tailed woodrats in the Great Basin was provided by an animal living in the summit hut on White Mountain Peak in eastern California, at an elevation of 14,276 feet.

Packrat Midden and Pollen Analysis

What Manly and his friends ate, and what Simpson guessed may have had something to do with the dung of small mammals, was part of a packrat midden (here, "packrat" is the term almost always used). Woodrats bring a wide variety of materials back to their dens. These objects routinely include twigs, leaves, seeds, stones, and bones, but they may include anything that piques a woodrat's interest and is small enough to be carried or dragged along. Much of what they retrieve is plant material, and much of this is either consumed or used in den construction. A good deal of it, however, simply accumulates, along with everything else, in the areas near the animals' dens. The accumulated detritus of a woodrat's life grows to often-substantial heaps of litter. These heaps are packrat middens.

If this were all there were to it, packrat middens would appear comparable to our basements, attics, or garages, places in which the detritus of our lives tends to accumulate. However, unlike most people I know, woodrats also urinate and defecate on their middens. If the midden on which a woodrat is urinating is in the open, the urine just washes away. But if the midden is in a sheltered spot—in a dry cave or crevice, for instance—the urine crystallizes and becomes about as hard as rock candy (recall Manly's description).

This crystallized material is called amberat, referring both to its source and to its consistency. Because the urine saturates the midden, everything in it becomes encased in this hardened material. As long as the midden does not become wet, it can last for tens of thousands of years. As Geof Spaulding and his colleagues put it, many old, hardened packrat middens "resemble blocks of asphalt with the consistency and mass of an unfired adobe brick" (Spaulding et al. 1990:60). The oldest known middens are too old to date by the radiocarbon method—that is, they are more than 40,000 years old or so.

Manly and his friends were thus eating a packrat midden—a mass of plant fragments, bones, stones, dirt, fecal pellets, and other items cemented together by crystallized packrat urine. There is no way of knowing the age of the midden they ate but it is the potential age of such middens, coupled with their contents, that has earned them significant amounts of scientific attention during the past few decades.

That ancient packrat middens could teach us a tremendous amount about the past was dramatically demonstrated by botanist Philip V. Wells and zoologist Clive D. Jorgensen. In 1961, Wells and Jorgensen had gone to the Nevada Test Site, in southwestern Nevada. Hiking down Aysees Peak, near Frenchman Flat, Jorgensen broke off a piece of packrat midden and found that it was full of juniper, even though Aysees Peak is too hot and dry for juniper today. One of the advantages of packrat middens is that woodrats do not go far from their dens to accumulate the material that ends up in the middens: a few hundred feet is their normal cruising range. As a result, the juniper in the Aysees Peak midden that Jørgensen had found must have grown on that peak, even though the plant was no longer there.

Wells and Jorgensen ended up collecting, analyzing, and dating the material from nine packrat middens in the Frenchman Flat area. All of them contained Utah juniper. They were able to show that the youngest of these middens dated to 7,800 years ago, while the oldest was more than 40,000 years old. One of the middens, from Mercury Ridge on the Spotted Range, even held a marmot skull, dated to 12,700 years ago. Like the juniper, the marmot was out of place: today, marmots are found no closer than one hundred miles away (see chapters 7 and 8).

Wells and Jorgensen published their results in *Science* in 1964. In this short paper, they documented that packrat middens can contain ancient plant and animal remains, that these remains can often be readily identified, and that the middens themselves can be dated by the radiocarbon method. They also used the data they had obtained to discuss what the late Pleistocene vegetation of the Frenchman Flat area must have been like.

This was a major contribution. Wells and Jorgensen had discovered a novel source of readily dated ancient plant material. The years after the appearance of their paper saw an explosion of interest in this source of information about plant and, through the plants, climate, history in the Desert West. Much of the earlier work was done either by Wells or by the students of Paul S. Martin of the University of Arizona—the same Paul Martin who has so strongly championed human hunting as the cause of the demise of North America's Pleistocene mammals. Martin's students, in particular, have been avid analysts of packrat middens, especially those in the southwestern United States and in the central and southern Great Basin. As a result, a tremendous amount has been learned about plant history in the Arid West.

This is not to say that we were totally ignorant of the late Pleistocene and Holocene history of Great Basin vegetation prior to Wells and Jorgensen's discovery. In fact, two prime sources offer information about plant history in the Desert West. On the one hand, there are the ancient seeds, leaves, twigs, and other larger parts of plants (called plant macrofossils) found in such abundance in packrat middens. On the other hand, there is pollen. Packrat middens also contain pollen, but the more common source of ancient pollen in the Great Basin is provided by the sediments of lakes that have continually held water since those sediments were deposited (see chapter 3). Other sources also exist: the sediments that surround springs, for instance, and those that fill caves. Even the deposits of playa lakes have some potential for providing ancient pollen.

Palynologists prefer to study the pollen from the deposits of lakes that have been continually wet for simple reasons. If a lake in the Arid West dries, the exposed deposits of that lake are inevitably eroded by wind, leaving often-massive gaps in the sequences represented by those deposits. Sometimes the gaps created in this fashion are obvious; sometimes, they are extremely difficult, if not impossible, to detect. In addition, the pollen in lake deposits that have been continually wet are usually be better preserved than that in sediments that have been alternately soaked and dried. As a result, if palynologists have a choice, it is a lake that has continually been a lake that draws their attention.

Detailed analyses of pollen from Great Basin contexts began in the 1940s, with now-classic work by Henry P. Hansen, then of Oregon State University. It has continued ever since, and we now have lengthy pollen-derived vegetational histories from areas as far-flung as Tule Springs (just north of Las Vegas) in the south to Steens Mountain (Oregon) in the north, and from the Wasatch Range in the east to the Lake Tahoe Basin in the west.

When Wells and Jorgensen introduced packrat midden analysis to the scientific community, they modestly noted that "*Neotoma* middens may have unique value as a check on the palynological approach to Pleistocene ecology" in the Desert West (Wells and Jorgensen 1964:1172). They felt such a check was important because the pollen of many wind-pollinated plants disperses widely, producing a "pollen rain" that reflects not the plants of the particular patch of ground on which it fell, but of the regional wind-pollinated vegetation as a whole. In addition, pollen that ends up in a lake comes not just from the air but also from all the tributaries that feed it, adding another regional aspect to the pollen in the lake's deposits. In the Great Basin, most valley-bottom lakes receive water from streams that originate in the surrounding mountains. As a result, the pollen in those lakes often reflects both the vegetation of higher elevations as well as the plants that lived closer to the lake itself.

We now know, with more than forty years of work behind us, that Wells and Jorgensen could have claimed even more for the approach they pioneered. The analysis of pollen from the stratified deposits of lakes, springs, caves, and other settings provides a fundamentally different kind of information than is provided by the analysis of the plant macrofossils provided in abundance by packrat middens. Each approach has strong advantages and some disadvantages; the two together provide a powerful means of unraveling plant history.

The study of well-preserved pollen from deep, stratified deposits results in a reconstructed vegetational sequence that is virtually continuous, unless erosion has removed part of the sequence. Further, since the pollen that becomes embedded in ever-deepening layered deposits often reflects the vegetation of fairly broad areas, the effects of strictly local phenomena—for instance, the local extinction of one species of woodrat with one set of dietary preferences and its replacement by another species with a different set—can be muted. In addition, in favorable depositional environments, pollen can be truly abundant: it is not uncommon for a single cubic centimeter of lake deposits to contain tens of thousands of pollen grains. In the right kind of setting, pollen analysis can provide continuous, detailed records of plant history over substantial amounts of time, records based on rich samples of the pollen produced by that vegetation.

Packrat midden analysis doesn't come close to doing that. Although generations of woodrats may occupy the same area and their middens may thus accumulate over long spans of time, the deposition of those middens is not continuous in the same way that sedimentation in a lake basin is continuous. In addition, many middens represent only brief periods of time. In some cases, middens that began to grow tens of thousands of years ago may have been inactive for thousands of subsequent years before new material began to be deposited on them. As a result, each plant assemblage, and sometimes each species of plant, has to be radiocarbon dated if its age is to be known accurately.

Unlike pollen in lake or cave sediments, plant fragments in packrat middens represent the vegetation that was growing in the area within a few hundred feet of that midden. Woodrats do not travel as far as the wind, nor are they routinely blown across the landscape or washed into the areas in which their middens accumulate, mouths full of pieces of plants.

So, in sharp distinction to most pollen records from places like lakes and springs, the plant assemblages from packrat middens represent "spot samples," snapshots of the vegetation at a particular place at a particular time. As a result, much labor is required to build long sequences of vegetational change from packrat middens. First, middens of the right age have to be found and collected. Then, the plants of those middens have to be extracted and identified. Then, they have to be dated. And many middens, representing the appropriate times, have to be analyzed if vegetational change on a regional basis is to be understood.

If pollen sequences can provide a continuous picture of vegetational change through time with far less effort, why bother with middens at all? There are, in fact, many reasons to bother. For openers, the pollen of many plants often cannot be identified beyond the genus level; indeed, pollen often cannot be identified beyond the family level and sometimes not even to that. To take a prime Great Basin example, the pollen of the various species of saltbush, genus *Atriplex*, that are so abundant in many parts of the Great Basin can be identified only as "cheno-ams." That group includes almost all members of the goosefoot family, the Chenopodiaceae (which, in the Great Basin, includes *Atriplex*, the goosefoot genus *Chenopodium*, and others), as well as the pigweed genus *Amaranthus* (family Amaranthaceae). I said "almost all" because greasewood, genus *Sarcobatus*, is a chenopod, but its pollen is distinct enough that it can be identified. As a result, when Great Basin pollen analysts refer to cheno-ams—a term frequently encountered in what follows—they generally mean all the plants I just referred to, minus greasewood.

Well-preserved plant fragments from packrat middens can frequently be identified to the species level. Thus, while pollen analysts can discuss whether the abundance of pine trees in an area decreased or increased through time, midden analysts can tell you whether the pine involved was singleleaf pinyon, Colorado pinyon, or limber pine. Having access to the species of plant involved opens the possibility of extracting species-by-species plant histories in the Arid West. Indeed, midden analysts have been phenomenally successful at doing just that.

In many ways, it is also a benefit that the plants in packrat middens almost necessarily represent local vegetation. Pines, for instance, are notoriously prolific producers of easily dispersed pollen. As a result, an ancient pollen assemblage from lake sediments composed of 20 percent pine pollen doesn't necessarily mean that pine trees were growing around that lake. But a plant assemblage from an ancient packrat midden composed of 20 percent pine leaves, twigs, and seeds does mean that pines were growing nearby—and you would know what species it came from.

In a splendid example of long-distance wind transport of pollen, paleobotanist Louis Maher of the University of Wisconsin once demonstrated that the pollen of jointfir (genus *Ephedra*) can be found in the Great Lakes region, at least seven hundred miles from its nearest possible source. If you find an *Ephedra* seed in a packrat midden, you can be sure that it was growing nearby at the time it was deposited. With the right material, you can also tell what species of *Ephedra* it was and the *Ephedra* fossils can be radiocarbon-dated directly. Although pollen grains can also be directly dated (see chapter 3), that process is extremely laborious and expensive and unlikely to come into common use. In contrast, it is easy to date seeds, leaves, and twigs.

Pollen analysis of sediments from lakes, springs, and caves, and the analysis of plant macrofossils from packrat middens, can, just as Wells and Jorgensen noted, serve as checks on one another. But even better, these approaches provide two complementary, independent sources of information about past vegetation. Each used alone is powerful. When used together, they provide an extremely sensitive means of learning about the past history of desert western vegetation. This becomes even truer when it is realized that packrat middens routinely occur on different parts of the landscape than do most of the deposits that provide pollen. Ancient packrat middens are not often found on the floors of Great Basin valleys, since those valleys rarely provide the rocky outcrops that the rodents prefer and that the middens need if they are to survive the centuries.

Our knowledge of the late Pleistocene and Holocene vegetational history of the Great Basin thus comes primarily from the analysis of pollen from stratified lake and cave deposits and from the analysis of plant macrofossils from packrat middens. The Great Basin is a vast place, the number of paleobotanists working here fairly small, and the number of truly acceptable lakes for pollen work low. Middens are abundant, but intense midden-based investigations of the plant history of this area go back no more than forty-five years or so. Much is to be learned, and huge surprises are probably in store for us. In fact, pinyon pine provided one not long ago.

But we do know enough to have a general picture of the late Pleistocene and Holocene vegetation of the Great Basin. In the pages that follow, I focus on the late Pleistocene part of that picture, reserving the Holocene for chapter 8. I begin by reviewing five areas that, as a whole, provide a good feel for exactly how different the plant communities of the Great Basin were toward the end of the Pleistocene: the northwestern Bonneville Basin, Ruby Valley on the eastern edge of the central Great Basin, the Carson Sink in the Lahontan Basin, Owens Valley, and the Mojave Desert's Death Valley.

Five Regional Pictures

The Northwestern Bonneville Basin

Our most detailed knowledge of the late Pleistocene plant history of the northwestern Bonneville Basin has been built through the analysis of packrat middens by Dave Rhode and David Madsen, and of pollen from a deep core taken from Blue Lake, on the Utah-Nevada border just south of Wendover, by Dave Rhode and his colleagues (figure 5-7). Insofar as they overlap in time, these two very different records provide superb complements to one another.

Only a single known packrat midden predates 14,000 years ago from this area, but what it tells us is remarkable. This midden, from a small rockshelter named Top of the Terrace, is from an elevation of 6,600 feet in the Goshute Mountains, to the immediate west of the Bonneville Basin itself. The oldest samples that it has provided are so old that they cannot be dated by the radiocarbon method, but they certainly predate 40,000 years ago, the practical limit of radiocarbon dating These oldest samples tell us that this part of the Goshute Mountains was, sometime before 40,000 years ago, characterized by vegetation that included, among other things, Utah juniper, sagebrush, horsebrush (*Tetradymia*), and snowberry. Today, the site is surrounded by some of these same plants, notably Utah juniper and sagebrush, but the snowberry (which has a wide ecological range) and horsebrush (which prefers lower elevations today) are not known from here today. While this might not sound like a particularly exciting discovery, the Utah juniper does tell us something very important. This is a tree that prefers relatively warm and dry environments. It is also a plant that doesn't become common again in this area until the middle Holocene, about 8,000 years ago. The earliest samples from the Goshute Mountains midden seem to catch something of the end of its late Pleistocene abundance in this region, at the same time as it suggests that the area may have been fairly warm and dry when these samples accumulated.

The same samples that provided the Utah juniper, suggesting a fairly warm and dry climate, also contained fecal pellets left behind by pikas. I have much more to say about pikas in later chapters. Here, I just mention again that these small mammals require habitats that are quite cool; even brief exposure to temperatures in excess of about 80°F can be fatal to them. Great Basin pikas are now found almost exclusively at high elevations, living in talus slopes that provide them with protection from predators and refuge from warm temperatures.

Packrat middens located in places where pikas live, or once lived, often contain pika pellets. Why this is the case is known only to the woodrats themselves. Perhaps the woodrats collect them, but it is also possible that pikas leave their pellets behind while stealing plants the woodrats had collected, a kind of scatological calling card. Or perhaps woodrats drag in pika pellets adhering to plants that the woodrats steal from pikas. No matter how the pellets get there, however, their presence in a midden mean that pikas must have been living nearby.

So pikas require cool habitats, while Utah junipers are found where it is fairly warm and dry. Both were living at the same place at the same time, and at an elevation of 6,600 feet, when the earliest Top of the Terrace midden sample was deposited. From a modern perspective, this seems quite odd. There are no places, at least to my knowledge, where this happens today. If you see a pika, there are sure to be no Utah junipers nearby. Paleoecologists are quite used to finding such odd cohabitations. They were so common during the late Pleistocene that they have earned their own name. They are referred to as nonanalog species associations, simply meaning that since they do not occur today, they have no modern analogs. I return to them shortly.

The next-oldest samples from the Top of the Terrace midden also accumulated prior to about 40,000 years ago. These samples, however, lack trees of any kind but are full of bits of shrubs that prefer cool, high elevation settings—sagebrush, currant (probably *Ribes montigenum*), and shrubby cinquefoil (*Dasiphora fruticosa*), for instance. The obvious comparison, Rhode observes, is to the montane meadows now found toward the tops of such places as the Ruby Mountains and the Jarbidge Range, suggesting that the fairly warm and dry climate indicated by the earliest samples had now given way to cooler and moister conditions. Limber pine does show up in some of these samples, where it is soon joined by shadscale—another apparently odd combination, coupling a species that today forms the upper treeline in many Great Basin ranges with a shrub now characteristic of hot and dry valley bottoms. At the same time, the shrubs that prefer cool and moist settings disappear. This combination, Rhode observes, is best explained by cold and dry climates between about 40,000 and 30,000 years ago, consistent with the fact that Lake Bonneville was at very low levels at this time, as I discussed in the previous chapter.

Lake Bonneville began to deepen around 28,000 years ago. The Top of the Terrace midden shows that limber pine is replaced by Engelmann spruce after this time, while shadscale gives way to the same kinds of montane meadow plants that were found here earlier, including currant and cinquefoil. By 17,000 years ago, when the lake was becoming ever deeper, the area around Top of the Terrace was surrounded by vegetation dominated by sagebrush and grass, with both limber pine and spruce now rare, suggesting very cold conditions. As if to confirm that lakes had grown in the area, the midden sample of this age even contains fish bones, presumably carried here by predators from the waters below.

The Top of the Terrace samples that fall between 17,000 and 14,000 years ago contain greater amounts of spruce, currant, and cinquefoil, strongly suggesting that while things had gotten warmer than they had been at 17,000 years ago, they were still both markedly cool and moist. This, of course, is the same interval that saw Lake Bonneville reach its peak and glaciers descend into the Bonneville Basin on its eastern side.

Rhode (2000b:141) notes that the correspondence between the plant record gained from Top of the Terrace and the history of Lake Bonneville is "strikingly close." He is absolutely right. When the lake was low, the midden suggests dry conditions. As the lake grew, the Top of the Terrace record suggests very cold conditions and even provides us with the remains of fish that must have come from the lake. During the interval when the lake reached its maximum, the plants from the midden show us that the landscape was both cool and wet. From a scientific perspective, all of this is quite heartening.

Also heartening is the fact that starting at about 14,000 years ago, the packrat midden record for the northwestern Bonneville Basin becomes much stronger, since Top of the Terrace is now joined by eight other middens arrayed along a 35-mile stretch along the Utah-Nevada border, beginning just north of Wendover and moving south from there. To make things even better, this information can be compared to the pollen record from Blue Lake.

The midden record shows that from 14,000 to 13,000 years ago, the vegetation of this area, up to an elevation of at least 6,600 feet, was dominated by shrubs now characteristic of cool and moist Great Basin subalpine highlands, with conifers rare at best and absent in many areas. The trees return at about 13,000 years ago, with limber pine becoming extremely common and remaining so until about 10,700 years ago. In fact, limber pine is almost the only conifer present below about 5,900 feet in elevation, the exception being the shrubby prostrate juniper. This cold-loving exception was part of the understory to widespread limber pine woodlands that developed in this area after 13,000 years ago, along with sagebrush, buffaloberry and other shrubs that do well in cold conditions. Indeed, Rhode and Madsen estimate that, between 13,000 and 11,000 years ago, summer temperatures here must have been at least 11.9°F colder than they are today, with precipitation only slightly above modern levels, to have supported this plant community. That a lake still existed somewhere in the area is shown by the fact that middens dating to these interval occasionally contain the remains of lake dwelling fishes, again presumably transported by predators to places where woodrats were waiting to incorporate fish bones into their middens.

The midden record begins to change around 11,500 years ago, as plants that like it hotter and drier begin to thrive in the region. Shadscale begins to increase in abundance after this time as limber pine begins its decline. By 11,000 years ago, the vegetation of the lower elevations of the northwestern Bonneville Basin was dominated by plants at home in the lower elevations of the Great Basin today—shadscale, sagebrush, and rabbitbrush, for instance. After 10,700 years ago, the limber pine is gone from these settings, marking a far warmer and drier landscape than this area had seen for tens of thousands of years. While inquiring minds might want to know what the plants were doing when Lake Bonneville had risen to the Gilbert shoreline between about 10,500 and 10,000 years ago, the midden record is mute on this score since no middens of this age have been found.

The Blue Lake pollen record, however, is anything but mute here. Blue Lake, and the wetlands that surround it, is supported by springs that emerge east of the Lead Mine Hills some fifteen miles south of Wendover, on the border of Utah and Nevada (figure 5-7). Dave Rhode and his colleagues extracted a twenty-nine-foot-long sediment core from these wetlands; Lisbeth Louderback and Rhode then used the pollen from this core to build a vegetational sequence that extends back some 12,700 years.

The strongest part of the Blue Lake record covers the past 10,000 years and so is discussed in chapter 8. However, it does show that at about 12,700 years ago, regional plant communities were dominated by pine and sagebrush and that this vegetation persisted to about 10,200 years ago. Soon after, the abundance of sagebrush declined rapidly but the pine persisted at relatively high abundances until sometime before 9,900 years ago. As this occurred, the pollen group that includes saltbush—the cheno-ams—increased dramatically, as did greasewood, grasses and such aquatic plants as bulrush and cattail.

The packrat midden record suggests that the late Pleistocene pine pollen in the Blue Lake sediments must have been limber pine, the only member of the genus known to have been in the area at that time. Louderback and Rhode observe that their Blue Lake pollen record is consistent with the packrat midden samples available for this general area, with limber pine and sagebrush common until around 11,000 years ago, after which the midden record quickly disappears. The Blue Lake record, they suggest, shows that limber pine and sagebrush remained fairly common in this area until 10,200 years ago, after which they declined in favor of such plants as saltbush.

How might this relate to the rise of Lake Bonneville to the Gilbert shoreline? There is only one reasonable explanation for the increase in greasewood, grasses, aquatic plants, and the group that contains saltbush in the Blue Lake core after 10,200 years ago. As Lake Bonneville declined from the Gilbert shoreline, desert vegetation (the saltbush and greasewood) replaced the sagebrush-dominated plant community there before. At the same time, shallow-water marshes developed in the Blue Lake area, increasing the abundance of such plants as bulrush and cattail. Once again, the independently derived plant and lake histories coincide strikingly well.

Note, by the way, that none of the Pleistocene packrat middens from this area have provided the remains of pinyon pine. This is not some sampling accident or the result of some mutual dislike between woodrats and pinyon. Pinyon pine was simply not in this area until some 3,000 years after the Pleistocene had ended. This remarkable fact I discuss in chapter 8.

Ruby Valley

Northeastern Nevada's Ruby Mountains rise to 11,387 feet, intercepting significant amounts of moisture brought by Pacific westerlies that travel relatively unimpeded along the Humboldt Valley until this point. The uppermost elevations of the Rubies may, as a result, receive as much as 50 inches of precipitation a year, nearly all of which falls in the winter. Some of this precipitation feeds a large series of springs that emerge from the east side of the southern Ruby Mountains. Those springs, in turn, provide substantial amounts of groundwater to Ruby Valley, just east of the mountains.

Because of all this, Ruby Valley is well watered, containing two lakes as well as substantial marshes. Although Franklin Lake often dries during the summer, Ruby Lake (also called the Ruby Marshes) has remained wet even during most historic droughts. Indeed, Ruby Lake forms the heart of the 37,600 acre Ruby Lake National Wildlife Refuge, which borders the southern Ruby Mountains (see figures 6-1, 6-2, and 6-3). Only about 25 percent of the water that supports the 15,000 acres of Ruby Lake and its marshes is provided by precipitation directly on the lake; the rest comes from the springs and, ultimately, from the Pacific.

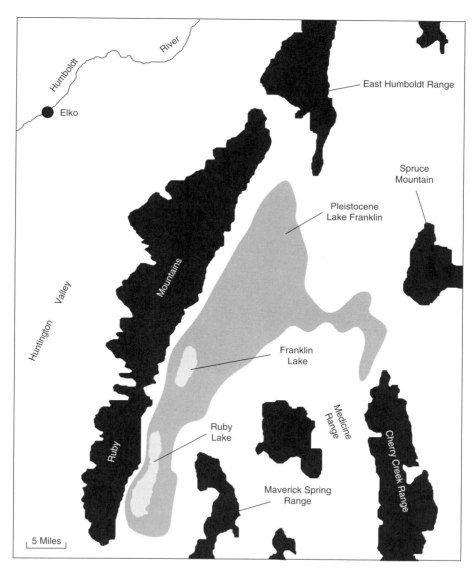

FIGURE 6-1 Ruby Valley region.

FIGURE 6-2 Ruby Valley from the eastern end of Overland Pass, part of the Pony Express route.

FIGURE 6-3 Ruby Marshes from the west.

The vegetation of Ruby Valley and the adjacent Ruby Mountains has been well described by, among others, hydrologist R. F. Miller and his colleagues, by botanists Lloyd Loope and David Charlet, and by paleobotanist Bob Thompson. Miller and his colleagues have shown that the nature of the valley-bottom vegetation here is sensitively determined by the local position of the water table. In those parts of the valley floor where the water table periodically rises to the surface and causes flooding, the dominant plants include saltgrass, shrubby cinquefoil, and Baltic rush (*Juncus balticus*). Wet meadows commonly contain beaked sedge (*Carex rostrata*) and silverweed cinquefoil (*Argentina anserina*), while the marshes themselves support both hardstem bulrush (*Schoenoplectus acutus*) and broadleaf cattail (*Typha latifolia*).

A very different set of plants is found where the water table is too low to reach the surface directly but is high enough for water to rise to the surface by capillary action. Saltgrass also occurs here, but its companions now include big greasewood, rubber rabbitbrush (*Ericameria nauseosa*), and Great Basin wildrye (*Leymus cinereus*).

Where the water table is still farther removed from the surface but capillary action is able to carry water to the roots of plants, yet another plant association is found. Rubber rabbitbrush and big greasewood occur in this setting as well, but they are now accompanied by big sagebrush or shadscale, the exact combination of plants depending on how finely textured the sediments that make up a given patch of ground happen to be.

Finally, when the water table becomes so far removed from the surface that plants must receive all their moisture from precipitation, a different set of shrubs appears. Gravel bars produced by Pleistocene Lake Franklin (see below) are marked by black sage and winterfat (*Krascheninnikovia lanata*). In lower settings between the bars, big sage, shadscale, winterfat, and yet another species of saltbush, the low Nuttall's saltbush (*Atriplex nuttallii*), are common.

To the west of Ruby Lake, along the eastern flanks of the Ruby Mountains, big sagebrush is abundant, and continues to be so into the pinyons and junipers. The pinyon-juniper here extends from about 6,100 feet to about 8,200 feet in elevation, and is often dense. Mountain mahogany becomes increasingly common within the pinyon-juniper as elevation increases, and then continues, often in virtually pure stands, to an elevation of about 9,500 feet.

The northern Ruby Mountains support an impressive number of subalpine conifers—limber pine, whitebark pine, Engelmann spruce, white fir, and a small stand of bristlecone pine. The southern Rubies, however, have only limber pine and bristlecone pine, both primarily found above an elevation of about 8,500 feet and continuing to elevations in excess of 10,500 feet, above which true alpine vegetation may occur.

In a very general sense, the vegetation of Ruby Valley and the southern Ruby Mountains is well characterized by Billings's set of six Great Basin plant zones (see chapter 2). The Shadscale and Sagebrush-Grass Zones are present in the valley bottom and along the lower flanks of the Ruby Mountains; a Pinyon-Juniper Zone occurs in the mid-elevations of the eastern slope of the southern Rubies; an Upper Sagebrush-Grass Zone occurs above it, followed by a Limber Pine-Bristlecone Pine Zone, and then, finally, by a fairly weakly developed Alpine Tundra Zone.

These plants zones, however, did not exist here during the late Pleistocene. Our most direct knowledge of the late Pleistocene vegetation of Ruby Valley and the adjacent Ruby Mountains comes from the work of Bob Thompson. His efforts to understand the vegetational history of this area were multi-pronged: he collected packrat middens from the southern Ruby Mountains, he attempted to core upland lakes, and he extracted two sediment cores from Ruby Lake itself. Of these multiple approaches, however, only one of the Ruby Lake cores reached the Pleistocene. Thompson found more than one hundred packrat middens, but those he dated were at the most 3,000 years old. Of the upland lakes he attempted to core, only Upper Dollar Lake provided a lengthy vegetation record, but it did not reach the Pleistocene, and even the Holocene sediments he retrieved here had been significantly disturbed, perhaps by landslides.

The sediment core that Thompson pulled from Ruby Lake, however, reveals quite a bit about the late Pleistocene. He extracted a core 23.5 feet long from the deposits of this lake and obtained seventeen radiocarbon dates on the organic materials contained within that core. Those dates, plus the fact that the core also contained 6,730-year-old Mazama ash at a depth of 6.2 feet, provide good temporal control over the information on lake and vegetation history that his core provided.

Ruby Valley, and the northern end of nearby Butte Valley, contained a substantial lake during the Pleistocene. This lake, called Lake Franklin, reached a maximum depth of about 130 feet, and covered some 471 square miles (figure 6-1). Thompson's Ruby Lake core provided some of the first detailed information available on the history of this substantial lake.

Thompson's data on the history of Lake Franklin come not only from the pollen contained within the core but also from the remains of green algae and ostracodes that it contained. Although changing frequencies of green algae through time in ancient lake sediments can be difficult to interpret securely, the analysis of changing abundances of ostracodes is more straightforward. Ostracodes are tiny—roughly 0.05 inch across—crustacea that live near the junction between water and sediment at the bottom of streams, lakes, and oceans. These minute animals have bivalved shells, as if they were tiny molluscs. They are not molluscs, however, but crustacea, and as such, they molt as they grow, replacing their old shells with new ones. Because their shells, composed largely of calcium carbonate, preserve extremely well in congenial settings, and because their habitat requirements are fairly well known, the study of the history of ostracodes has become an important tool in understanding Pleistocene and Holocene lake histories.

Rick Forester of the U.S. Geological Survey identified and analyzed the ostracodes from Thompson's Ruby Lake core. Between the ostracodes, on the one hand, and the algae and pollen (identified and analyzed by Thompson), on the other, Thompson was able to reconstruct a significant part of the late Pleistocene history of the Ruby Lake Basin.

The base of the Ruby Lake core, 23.5 feet down, dated to about 37,000 years ago. The algae and ostracodes—or, more exactly, the lack thereof—strongly suggest that from this time until about 28,000 years ago, the Ruby Lake Basin was at most seasonally wet and perhaps received only groundwater discharge onto a playa surface. Between 28,000 and 23,000 years ago, the water deepened somewhat, to form a shallow, saline, and perhaps ephemeral lake. Not only do the ostracodes suggest such a lake, but so does the high abundance of the pollen of ditchgrass (*Ruppia*), a plant that thrives in brackish or saline water.

Sometime, and perhaps soon, after 23,000 years ago, ditchgrass pollen virtually disappeared from the Ruby Lake core.

This, combined with the information provided by algae and ostracodes, suggests that the water level had risen to create a permanent, yet still highly saline, lake. This episode is not well dated, but falls between 23,000 and 18,500 years ago.

When, then, did Lake Franklin rise to flood nearly five hundred square miles of Ruby Valley to a depth of some 130 feet? Algae and ostracodes both show that that happened soon after 18,500 years ago. Fresh, deep water filled the basin at this time, and, with the possible exception of a brief interlude when the lake may have declined substantially, deep water conditions lasted until at least 15,400 years ago.

Geologist Karl Lillquist's efforts provide more precision here. By obtaining radiocarbon dates from the geological features left around the edges of Lake Franklin, he showed that the lake had reached its highstand sometime before 16,800 years ago and remained here until about 15,070 years ago. After that, the lake spent some 2,300 years going down, up, and down again, with a major surge upward between about 12,900 and 12,700 years ago. After this time, the lake fell so far that, Lillquist suggests, it may have dried entirely, only to return in a shallow way about 11,500 years ago and then fall again.

This record matches Thompson's fairly well, though not perfectly. The shoreline record suggests that the lake had reached its highstand pretty much when the core data suggest the same thing, though Lillquist's directly dated evidence is more precise here. In addition, the period between about 15,400 and 10,800 years ago is not well represented in the core. That lack may represent a period of time when lake deposition stopped and existing deposits were eroded by the wind. If that is correct, the erosion may have occurred during the possible desiccation events noted by Lillquist, after 12,700 and then again after 11,500 years ago. When Thompson's core record kicks back in, at 10,800 years ago, the fossil record continues to suggest the presence of a freshwater lake, perhaps moderate in size. By 10,400 years ago, this lake had declined to low levels and had become more saline. This, of course, is roughly the time that Lake Bonneville rose to the Gilbert shoreline, between about 10,500 and 10,100 years ago. Lillquist, however, found no evidence of a Gilbert-aged lake during the course of his work.

The entire Pleistocene sequence from the Ruby Lake core, from 37,000 to 10,000 years ago, is dominated by two terrestrial plants—pine and sagebrush. Sagebrush contributed about 50 percent of all the pollen present in the Pleistocene levels of the core. Pine contributed about 30 percent, sometimes more, sometimes less. Unlike the case for Blue Lake, however, we have no late Pleistocene packrat middens from the Ruby Mountains area to tell us which species of pine might have been in this area at this time. What is not abundant in Ruby Lake pollen profiles is as interesting as what is. Greasewood provided less than 5 percent of the late Pleistocene pollen identified by Thompson, while the remaining chenopods, including *Atriplex*, provided less than 10 percent. The pollen group that includes the junipers also contributed less than 5 percent of the total.

How do these numbers compare to the modern pollen rain in the Ruby Marshes? Thompson collected surface pollen from the marsh in order to answer this question. He found that these modern samples contained about 30 percent sagebrush pollen (compared to 50 percent in the late Pleistocene), and from 30 to 40 percent pine pollen (compared to about 30 percent in the late Pleistocene). Greasewood accounted for about 20 percent of the modern pollen rain in the marsh (compared to less than 5 percent during the late Pleistocene), *Atriplex* for about 15 percent (compared to less than 10 percent for all the cheno-ams, minus greasewood), and the juniper group, nearly 20 percent (compared to the late Pleistocene total of less than 5 percent).

Thompson was cautious when it came to deciphering the meaning of the late Pleistocene pine pollen represented in his core. He observed that modern marsh samples contain more pine pollen than the late Pleistocene lake samples. This is the case even though there are now no pines in Ruby Valley itself; they are confined to elevations above 6,100 feet on the eastern flanks of the Ruby Mountains. As a result, Thompson argued that while pines existed somewhere in the region, they certainly did not approach the edge of the lake and may not have been very common, or even present, in the southern Ruby Mountains as a whole. In addition, he suggested that woodland junipers, now abundant in the southern Rubies, must have been rare here during the late Pleistocene and may also have been absent.

Thompson pictures the late Pleistocene vegetation of Ruby Valley as having been a sagebrush steppe in existence from at least 37,000 years ago to, and beyond, the end of the Pleistocene. In the absence of Pleistocene-aged packrat middens, or some other source of equally ancient plant macrofossils, it is not possible to know what species of sagebrush is represented in the Ruby Lake core. It is, however, fully reasonable to suggest that big sage was heavily involved in producing the pollen so abundant in the late Pleistocene sediments of the Ruby Lake core.

Thompson's data thus show that pines were probably less abundant, and perhaps far less abundant, in the Ruby Valley region during the late Pleistocene than they are today, that junipers were rare and perhaps even absent, and that the vegetation of Ruby Valley at this time is best characterized as having been a sagebrush steppe. The saltbushes, so common here now, must have been in the area, but they, too, appear to have been rare.

Thompson suggests that the long period during which sagebrush steppe dominated this region implies a stable, cold, and relatively dry climatic regime. He also suggests that this entire period, from 37,000 years ago to beyond 10,000 years ago, was marked by precipitation that fell mainly in the winter.

Thus, Billings's six Great Basin plant zones did not exist in Ruby Valley and the adjacent Ruby Mountains during the late Pleistocene, although they exist here today. Admittedly, we do not know much about the vegetation of the mountains themselves, but there was no well-defined

Shadscale Zone in the valley bottom, and perhaps hardly any shadscale at all. There was no Pinyon-Juniper Zone in the mountains, and perhaps hardly any juniper trees at all. Indeed, as may be seen in chapter 8, there was almost certainly no pinyon either, just as there wasn't in the northwestern Bonneville Basin at this time. Of Billings's six modern plant zones, only the Sagebrush-Grass Zone flourished here during the late Pleistocene, and this it did for at least some 20,000 years.

The Lahontan Trough: Carson Sink

During the mid-nineteenth century, emigrants to California crossed Nevada by following the Humboldt River west as far as they could, to Humboldt Lake and the Humboldt Sink. Once here, they had two choices. They could continue to the southwest and cross forty miles of desert to reach the Truckee River near modern Wadsworth, Nevada (and along the route now followed by Interstate 80). Or, they could drop south to cross forty miles of desert to reach the Carson River near modern Fallon (and close to the route followed by State Highway 95; see figure 6-4).

Neither choice was pleasant. Both of these Forty Mile Deserts not only were hot but, far more important, lacked any predictable source of potable water. In addition, they became so heavily used that they were routinely devoid of food for livestock. Emigrants' diaries are full of the horrors encountered here:

> We now begin to meet with the destruction of property and stock; the road being almost lined with wagons, the dead and dying horses, mules, and cattle . . . we went to victory, stalking our way through indescribable scenes of suffering and want . . . hundreds of horses, mules and cattle dead, dying, and suffering, are laying thick around, and wagons, carts, and carriages line both sides of the road . . . the worst of it all now is to see, every few hundred yards, the grave of some kind brother, father or mother, and even some who have not been buried, but have probably been foresaken when sick or faint, and left to die and waste away in the winds and rains of heaven. (John Wood in the Carson Desert, 1850; from Curran 1982:182–183)

The southerly route passed by the west end of the West Humboldt Range, then continued through the Carson Desert to the Carson River. This area is, of course, within the basin of Pleistocene Lake Lahontan. It is also part of Reveal's Lahontan Trough, that relatively low-lying area along the western edge of the Great Basin that I discuss in chapter 2.

Vast expanses of the Carson Desert region, and in particular the well-drained sediments of Lake Lahontan and the surrounding hills above the valley floor, are covered by saltbush,

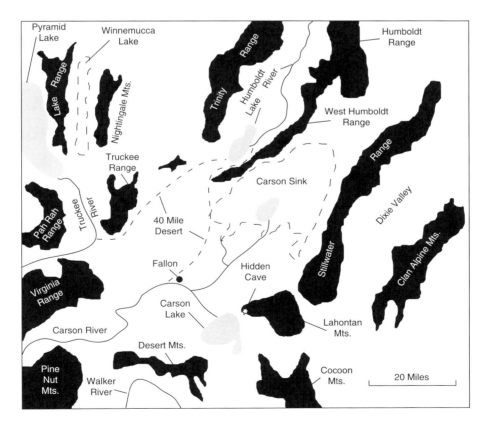

FIGURE 6-4 Carson Desert region.

FIGURE 6-5 Eetza Mountain, in the northern Carson Sink. Hidden Cave is located in the canyon to the left, just above the light-colored patch.

the small and grayish Bailey's greasewood (*Sarcobatus baileyi*), and the low, spiny bud sage (*Picrothamnus desertorum*). Dwight Billings has shown that this association of plants replaces big sagebrush and its companions in this region where annual rainfall drops beneath about 7 inches (today, Fallon receives about 5 inches a year). He has also shown that Bailey's greasewood provides about 50 percent of the shrubby cover within this complex of plants, while shadscale provides an additional 20 percent. Winterfat, an important winter food plant for cattle and deer, can be found in this association as well, as can Indian ricegrass (*Achnatherum hymenoides*), an important source of edible seeds for Native Americans throughout the Great Basin.

Stabilized sand dunes in the Carson Desert region often support significant stands of smokebush (*Psorothamnus polydenius*), a plant that had a variety of medicinal uses among the Northern Paiute of western Nevada. Fourwing saltbush is commonly found with smokebush on these dunes, as are two species of horsebrush.

But more obvious to the casual observer are the plants that surround the edges of the playas in this region. Here, big greasewood is common, often in almost pure stands. This branchy, light-green shrub is the one that most travelers notice from their cars as they drive on roads that cut through or next to playas, not only in the Carson Desert but also in much of the Great Basin. When big greasewood occurs with another shrub in the Carson Desert, that companion is often yet another species of *Atriplex*, the large-leaved Torrey saltbush (*Atriplex torreyi*), though in certain settings, big greasewood can be found alongside shadscale and bud sage as well.

Even today, after more than a century of diversion of the waters that flow into the Carson Sink, substantial marshes usually exist in the Sink and elsewhere in the area. The watercourses that feed these marshes support Fremont cottonwoods (*Populus fremontii*—named for Frémont, who found it near Pyramid Lake in 1844); the marshes themselves support significant stands of both bulrush and cattail.

The Carson Desert region ranges from about 3,900 feet in elevation (for instance, at Carson Lake) to about 4,900 feet in the bordering hills. Hills not much farther removed reach above 5,000 feet, and the West Humboldt Range, which borders the Carson Desert to the north, reaches 6,349 feet. All parts of this region that lie beneath an elevation of about 4,380 feet were beneath the waters of Lake Lahontan at its highstand. And much of it is now marked by plants that belong to just two genera—*Atriplex* and *Sarcobatus*, the genera to which shadscale and greasewood belong.

Our knowledge of the late Pleistocene vegetation of the Carson Desert is not particularly strong, though what we know we seem to know securely. Attempts to extract pollen-rich sediment cores from Carson Lake by the U.S. Geological Survey have come to naught, and very few late Pleistocene packrat middens have been analyzed from this area. Instead, much our understanding of the vegetation of this part of the Lahontan Basin toward the end of the Pleistocene comes from detailed analyses, by paleobotanists Peter Wigand and Peter Mehringer, of pollen from the deep sediments of Hidden Cave. Fortunately, the work that they did is of very high quality.

Hidden Cave sits on the northern edge of Eetza Mountain, one of a series of low (maximum elevation 4,281 feet) hills that form the enthusiastically named Lahontan Mountains in the southern Carson Desert (figures 6-4, 6-5, and 6-6). The name of the site is quite apt: the opening to the cave is

FIGURE 6-6 The hidden Hidden Cave. Not visible here, the entry to the cave is located just above the bare spot in the center of the photograph, formed by sediments removed from the cave during the 1979–1980 American Museum of Natural History excavations.

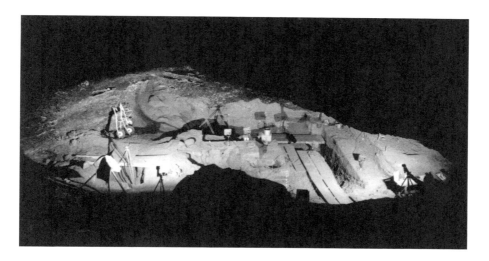

FIGURE 6-7 Interior of Hidden Cave during the 1979–1980 excavations. Photograph by Albert Alcorn, courtesy of David H. Thomas and the Division of Anthropology, American Museum of Natural History, New York.

simply a small hole in the face of Eetza Mountain. The site is so hidden that it was not until the mid-1920s that it was discovered, by a group of boys searching for stolen money that had reportedly been stashed in a cave somewhere in the area. Even today, the opening to the site can be difficult to find, though the task is made much easier by the fact that a well-marked trail now leads to it.

Although the opening is small, the cave itself is quite large—approximately 150 feet long and 95 feet wide. Initially, the distance from the floor to the ceiling was about 15 feet. Today, after several archaeological projects, that distance is much greater.

The first of these projects took place in 1940, conducted by S. M. and Georgia N. Wheeler, of the Nevada State Parks Commission. Norman Rouse and Gordon Grosscup, of the University of California (Berkeley), conducted additional archaeological work within the cave in 1951, but the most important excavations took place here in 1979 and 1980, directed by Dave Thomas of the American Museum of Natural History (figure 6-7).

These most recent excavations were interdisciplinary in nature, as is characteristic of modern archaeology in general. The team assembled by Thomas included geologists, whose job it was to guide the stratigraphic excavations of the site and to interpret the geological history of those deposits; archaeological zoologists, whose task was to identify and analyze the animal remains from the site; paleobotanists, whose goal was the identification and interpretation of the plant remains recovered by the excavations; and, of course, a series of archaeologists whose interests focused on the cultural material the site provided.

Geological work on the sediments revealed by Thomas's excavations showed that Hidden Cave was formed sometime after 21,000 years ago, presumably as a result of wave action by the waters of Lake Lahontan. Because the mouth of the cave falls at 4,104 feet, well beneath the 4,380 feet highstand reached by Lake Lahontan, much of the Pleistocene-aged depositional sequence within the cave consists of water-deposited sands and clays and water-worn gravels.

I have much more to say about Hidden Cave in later chapters. Here, I am concerned only with the plant remains provided by the late Pleistocene deposits of the site, which date to between about 18,000 and 10,000 years ago.

Those remains proved to be fairly scanty, in contrast to the richer pollen and seed record provided by the younger, Holocene-aged sediments of the cave. Indeed, many of the Pleistocene samples carefully extracted and studied by Wigand and Mehringer proved to be so devoid of pollen that they could not be used to say anything secure about the late Pleistocene vegetation of the area. Others, however, were richer, and those more productive samples suggest that the late Pleistocene vegetation of this area was very different from what it is now.

Today, the vegetation of the Eetza Mountain area is much like that which I described for the Carson Desert in general. The slopes and rolling crest of Eetza Mountain are covered by bud sage, Bailey's greasewood, and fourwing saltbush. In the playa beneath Hidden Cave, big greasewood is abundant both on the flats and on the low dunes, with Torrey saltbush sharing the dunes. The late Pleistocene Hidden Cave samples analyzed by Wigand and Mehringer, however, contained as much as 60 percent pine pollen. Another 30 percent or so of that pollen was contributed by sagebrush.

The sagebrush pollen from the late Pleistocene deposits of Hidden Cave could be identified only to the genus level, as *Artemisia*. It would be of great value to know whether this was bud sage, which grows on the slopes of Eetza Mountain today, or big sage, which is absent, or some other species entirely. As I discussed, Billings has observed that big sage now grows only in those parts of the Carson Desert region that receive more than 7 inches of precipitation a year. Today, the Carson Desert itself receives only some 5 inches of precipitation annually, with evaporation above 50 inches a year. The remains of the birds and mammals whose remains were excavated from Hidden Cave provide subtle hints that the sage so well represented in Wigand and Mehringer's late Pleistocene samples at least includes big sage. The pollen itself, however, cannot tell us that.

Because pines can produce copious quantities of pollen that can be spread far and wide by the wind, high frequencies of pine pollen in cave or lake sediments do not necessarily mean that pines grew in the immediate neighborhood. In the Holocene deposits of Hidden Cave, pine never contributed much more than 30 percent of the total amount of pollen in Wigand and Mehringer's samples, compared to more than 50 percent in some of the late Pleistocene samples. The contrast may suggest that pines grew far closer to Hidden Cave during the late Pleistocene than they do today. Because of the difficulties involved in identifying species of pine from their pollen, the kind of pine represented in the deposits of Hidden Cave is not known. Even more problematic, however, is the fact that insofar as the pine pollen in the Pleistocene levels of Hidden Cave came from sediments associated with the waters of Lake Lahontan, it may have been derived from great distances. As a result, the high levels of pine pollen in those sediments may say little about the local distribution of the trees themselves.

The high representation of pine and sagebrush pollen in the late Pleistocene deposits of Hidden Cave is of high interest, but equally interesting is what is not represented. Here, the chenopods—the plant family that includes the saltbushes, greasewood, and iodinebush (*Allenrolfea occidentalis*)—stand out.

Even though *Atriplex* pollen cannot be distinguished from most of the other chenopods, Wigand and Mehringer were able to identify the greasewood pollen from their late Pleistocene samples and thus remove it from the rest of this group. The remaining cheno-am pollen suggests that *Atriplex* could not have been very abundant during the late Pleistocene in the Hidden Cave area, perhaps in part a reflection of the fact that a significant fraction of the potential saltbush habitat was under the waters of Lake Lahontan during much of this time. All the cheno-ams combined (excluding *Sarcobatus*) provided only about 10 percent of the pollen grains in Wigand and Mehringer's late Pleistocene Hidden Cave samples. In contrast, the late Holocene samples they analyzed ran to about 40 percent cheno-ams (again without *Sarcobatus*), much of which is undoubtedly from *Atriplex*.

What does all this suggest about the vegetation of the southern Carson Desert during the time represented by Wigand and Mehringer's late Pleistocene samples, between about 18,000 and 10,000 years ago? First, it may suggest that pines extended to much lower elevations than they extend here today. Second, it suggests that sagebrush—species unknown, but I would guess that much of it was big sage—was common on the slopes surrounding the site, and perhaps extended well toward Lake Lahontan itself. Finally, it suggests that saltbushes, now such an important component of the local vegetation, were at best rare. In short, the late Pleistocene vegetation of the Lahontan Mountains and adjacent Carson Desert during the late Pleistocene appears to have been a sagebrush-dominated steppe, perhaps with pines not far off.

Peter Wigand and Cheryl Nowak have analyzed a packrat midden from the edge of the Hot Springs Mountains, just north of Hazen and at about the elevation of the highest Lake Lahontan shoreline on the edge of the Carson Sink. Today, the Hot Springs Mountains are covered by the same kind of treeless vegetation found around Hidden Cave, but this was not the case toward the end of the Pleistocene. In three separate midden samples, dated to about 20,500, 12,300, and 12,000 years ago, Wigand and Nowak discovered the remains of Utah juniper and big sage. In fact, a large series of middens from along the western edges of Lake Lahontan have shown that the late Pleistocene saw Utah juniper thriving at elevations well below elevations where it now exists.

Wigand and Nowak have shown that whitebark pine grew along the western edge of the Lahontan Basin at 23,000 and at 12,000 to 11,000 years ago. It was not here between these times, probably because glacial maximum climates were too cold for it, and it does not show up after 11,000 years ago, probably because it became too warm. The tree is also not here today; in fact, it comes no closer than thirty miles to the east and at elevations 3,700 feet higher than the midden samples that Wigand and Nowak analyzed.

All of this is, of course, quite different from what is here now. In fact, Wigand and Rhode estimate that during the times whitebark pine grew along the western edge of Lake Lahontan, mean annual temperatures must have been at least 12.6°F colder than they are now, and that during the glacial maximum, when it became too cold for this tree to survive here, it must have been 1.8°F to 3.6°F colder than this.

Owens Valley

In the previous chapter, I discuss two cores drilled by the U.S. Geological Survey that have greatly increased our understanding of the history of Owens Lake: core OL-92, which extracted some 800,000 years of lake history, and core OL-84B, which Larry Benson used to help decipher the last 50,000 years or so of that history. These cores, it turns out, contained significant amounts of pollen. U.S. Forest Service paleobotanist Wally Woolfenden used the first of them to help decipher the past 180,000 years of vegetation history in this part of Owens Valley. University of Nevada paleobotanist Scott Mensing focused his efforts on the second of these cores, providing a detailed look at the sediments that dated to between 13,500 and 7,000 years ago. Thanks to work by a series of scholars, we can compare the results of the pollen-based analyses to data independently derived from nearby packrat middens

The valley floor that surrounds Owens Lake is now covered by plants that mark Billings's Shadscale Zone. Among other things, these include shadscale and Parry's saltbush (*Atriplex parryi*) as well as bud sagebrush and big greasewood. No juniper, no pinyon, no Joshua tree may be seen.

Had you stood here 21,000 years ago, however, you quite likely would have been able to see all three of these things at once. We know this because Peter Koehler and Scott Anderson found two packrat middens on a rocky outcrop, called Haystack Mountain, that was about 330 feet north of Owens Lake at its late Pleistocene highpoint (figures 6-8 and 6-9). They extracted six separate samples from these middens and discovered that those samples had been deposited between 22,900 and 14,870 years ago—that is, they span the Last Glacial Maximum and, by doing so, include the period of time when the lake was at its own maximum. All of these samples contained Utah juniper and all of the samples between 22,900 and 17,680 years ago also contained single-needle pinyon.

This makes it clear that at least parts of the valley floor adjacent to Owens Lake supported pinyon-juniper woodland while Owens Lake was at its maximum some 23,000 to 16,400 years ago. As if to confirm that Owens Valley had cool summers during the lake's highstand, the samples dated to about 17,700 and 16,000 years ago also had Rocky Mountain juniper in them. This tree does not exist in California today; the nearest it comes to Owens Valley today is in the Spring Mountains of southern Nevada, some 140 miles to the east, and is not found in areas marked by pronounced summer droughts. Only the sample dated to 21,000 years ago provided Joshua tree, but although these trees do not occur here today, one does not have to go far to find them, since they occur to the southeast of Owens Lake.

The Alabama Hills are located ten miles to the northeast of Owens Lake, overlooking the town of Lone Pine (figure 6-10). This is surely one the most recognizable places within the Great Basin, made so by all the films that have been shot here. *Riders of the Purple Sage* (1925, 1941), *The Lives of a Bengal Lancer* (1935), *The Charge of the Light Brigade* (1936), *The Lone Ranger* (1938), *Gunga Din* (1939), *High Sierra* (1941), *The Ox-Bow Incident* (1943), *Bad Day at Block Rock* (1955), *Star Trek 5: The Final Frontier* (1989), and, to add a personal favorite, *Tremors* (1990) were all made, in whole or in part, in the Alabama Hills, as were hundreds of others (film historian Dave Holland lists 362). While film buffs may not know the name of the place, they certainly know what it looks like. And while Great Basin paleoecologists might not know the cinematic importance of this area, they all know the name of this isolated Sierran outcrop.

That is because in addition to working immediately adjacent to Owens Lake itself, Koehler and Anderson looked for, and found, two ancient packrat middens in the Alabama Hills as well. And, unlike some of those who encountered the alien life-forms in *Tremors*, they survived to tell the story.

The vegetation of the Alabama Hills is perhaps best characterized as part of Billings's Shadscale Zone, as is the vegetation that now surrounds Owens Lake. Desert saltbush (*Atriplex polycarpa*), shadscale, big sagebrush, and jointfir are found here, as are plants more characteristic of the cooler parts of

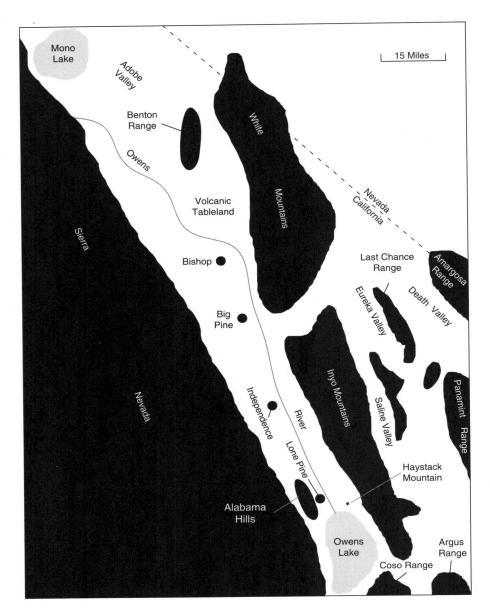

FIGURE 6-8 Owens Valley region.

the Mojave Desert—blackbrush (*Coleogyne ramosissima*) and green rabbitbrush (*Ericameria teretifolia*), for instance.

The samples from Koehler and Anderson's Alabama Hills middens span some 28,000 years, from 31,450 to 2,800 years ago. Given that they also collected material from a modern midden, they ended with a series of plant samples that intermittently span the past 31,450 years. Given what they found at Owens Lake, it comes as no surprise to learn that the late Pleistocene vegetation of this area bears few similarities to what is found here today.

Their most ancient midden, from the Two Goblin site, showed that between about 31,500 and 19,000 years ago, the Alabama Hills supported both Utah juniper and Joshua tree with bitterbrush (*Purshia tridentata*) an important part of the understory. The last secure record for Joshua tree falls at 19,000 years ago, but Utah juniper continues to 9,500 years ago, after which it disappears.

Some of the dated midden samples from the Alabama Hills coincide very closely in time with the dated samples that Koehler and Anderson retrieved from the edge of Owens Lake. As a result, we know with some certainty that at the same time as the Haystack Mountain middens that fall at 3,790 feet and date to between 23,000 and 17,700 years ago had both pinyon and juniper, the area surrounding the Two Goblin midden, at an elevation of 4,790 feet, had the juniper but not the pinyon. In fact, the only evidence

FIGURE 6-9 Haystack Mountain, near Owens Lake. Between 23,000 and 17,700 years ago, Utah juniper and single-needled pinyon grew at this spot.

FIGURE 6-10 The Alabama Hills, in the Owens Valley near Lone Pine. Single-needled pinyon pine occurred in this area about 11,500 years ago; packrat middens from this area, dated to between 31,450 and 13,350 years ago, provided the remains of Utah juniper but no pinyon.

for Pleistocene-aged pinyon in the Alabama Hills dates to 11,450 years ago, from a midden analyzed by Peter Wigand that falls within an interval not sampled by Koehler and Anderson.

This might not seem odd, but it is. As I discuss in chapter 2, as you move upslope in today's pinyon-juniper woodland in the Great Basin, you first encounter Utah junipers, then pinyon; as you gain more elevation, the juniper drops out before the pinyon does. In Owens Valley during full glacial times, however, it seems that as you moved upslope, it was the pinyons that were lost first.

Why this was the case is not obvious, but it is obvious that it was the case. We know this because of packrat midden work done by Steven Jennings and Deborah Elliot-Fisk in an area toward the northern end of Owens Valley known as the Volcanic Tablelands, some thirty miles north of the Alabama

FIGURE 6-11 The Volcanic Tablelands, Owens Valley. Low volcanic features of the sort seen in the distance provided the packrat middens discussed in the text.

Hills (figure 6-11). The one Pleistocene-aged midden sample here dates to 19,300 years ago, close in time to samples from Owens Lake and the Alabama Hills. This sample came from an elevation of 4,400 feet and from a location today covered with shadscale and big sagebrush. The midden sample, though, was full of Utah juniper, along with such understory plants as antelope bitterbrush and spineless horsebrush (*Tetradymia canescens*). None of these plants occur at the site today, and you have to climb 1,970 feet higher than the midden to find Utah juniper in this area. Just as in the Alabama Hills, there is no pinyon in this sample. The next-youngest Volcanic Tableland sample, which dates to 9,800 years ago, has Utah juniper but also lacks pinyon (there are no samples younger than this from here). Pollen samples from these two middens confirm the plant macrofossil results: those samples are loaded with juniper pollen (actually, from the larger group that includes juniper, but this is the plant that must have provided it), with pine pollen extremely rare in both.

The Owens Lake pollen work done by Wally Woolfenden falls neatly in line with the chronologically spottier records provided by the packrat middens. His work shows that between 25,700 and 17,700 years ago, the most common pollen in the lake's sediments was provided by juniper, a perfect match for the midden record. Pine pollen was also common, matching the presence of pinyon in the nearby packrat middens at this time. As usual, though, it is hard to know exactly what to make of this because pine pollen is so abundant and so widely distributed by the wind. In addition, because the waters that fed the lake brought in pollen from a wide surrounding area, including the Sierra Nevada, this mechanism may also have introduced pollen from afar, though Woolfenden suspects this source may have been negligible compared to the amount of wind-distributed pollen that fell on the lake's surface.

As an aside, I note that the same thing can't be said about the pollen from the world's largest tree, the giant sequoia (*Sequioadendron*), in Woolfenden's Owens Lake record. Giant sequoia pollen is found scattered throughout this record, from about 175,000 to about 15,000 years ago. Giant sequoias are today found west of Owens Valley, but only on the west slope of the Sierra Nevada. Scott Anderson's studies of modern giant sequoia groves has shown that the pollen of this tree does not disperse very far at all. Move 1,000 feet from the grove, and nearly all of it is gone. If giant sequoia pollen blew into Owens Lake, it must have come from very close by. It is far more likely that it washed in, but the only way that could happen is if giant sequoias grew on the east slope of the Sierra Nevada during the Pleistocene, where they do not now occur. That this must have been the case is also suggested by Owen Davis's pollen work at Mono Lake, to the north. Here, he has shown, giant sequoia pollen is extremely abundant at the end of the Pleistocene—from 5 to 20 percent of all pollen deposited here between about 11,600 (the base of the record) and 10,000 years ago, then dropping to less than 1 percent after 10,000 years ago (and disappearing by 7,800 years ago or so). Barring some massive change in either wind strength or the ability of giant sequoia pollen to be wind-distributed, Davis's work also strongly suggests that these trees were living on the east slope of the Sierra Nevada during the late Pleistocene.

Woolfenden interprets his Owens Lake pollen record as suggesting nearby pinyon-juniper woodland with a sagebrush-bitterbrush-grass understory. He suggests that that woodland began to thin sometime after 21,000 years ago, with an increase in cheno-ams after about 14,000 years ago indicating the increasing abundance of warm-desert plants, though declines in the level of Owens Lake might have helped cause this increase by increasing the amount of habitat appropriate for such plants.

Woolfenden's pollen work was aimed toward establishing the general nature of the vegetation surrounding Owens Lake during the past 180,000 years and thus did not focus in great detail on the latest Pleistocene. Scott Mensing's work on the second Owens Valley core, on the other hand, did just that for the very end of this period. Pine pollen tends to be very abundant throughout the sequence he extracted, but, for reasons I just discussed, the implications of this for the nature of the local vegetation are not evident. His work does, however, show that at around 13,500 years ago, junipers were still common on the landscape, with an understory of abundant sagebrush and, to a lesser extent, cheno-ams. His record suggests that juniper began a long but steady decline shortly after 13,000 years ago as it began its march upslope, to be replaced first by sagebrush steppe and then, as warming and drying continued, by increasing amounts of cheno-ams mixed in with the sage. It is this mixture that characterized the valley floor as the Pleistocene came to an end.

All of these sequences are gratifyingly consistent and, once again, shows how much can be learned by combining high-quality pollen and packrat midden research. There is one question that they don't answer, though, and that is about the density and distribution of Utah juniper on the Owens Valley landscape during the late Pleistocene. Two interpretations are possible here. One is that the floor of Owens Valley was actually covered by juniper woodland in the late Pleistocene. This is the interpretation offered, for instance, by Scott Mensing. The other possibility, however, is that juniper woodlands were not continuous across the valley floor but instead occupied the lower slopes of the mountains that edge the valley, the valley floor adjacent to those slopes, and rocky outcrops within the valley itself. I return to this issue later in the chapter.

Note, by the way, that this is the first time I have mentioned Pleistocene-aged pinyon pine in the Great Basin. This is no accident, as we shall see.

The Mojave Desert: Death Valley

High temperatures and low precipitation combine with soils that are often extremely saline to greatly limit the kinds and densities of plants that can now survive in Death Valley (figure 6-12). In fact, some two hundred square miles of the central part of the valley are covered by a playa that has ample water for plants but whose salinity is so high that they simply do not grow there. Detailed studies by George B. Hunt of the U.S. Geological Survey have shown that as plants descend toward this saltpan, their advance stops fairly abruptly as groundwater salinity exceeds 6 percent.

A series of gravel fans, whose permeability to water varies but which generally allow water to pass readily, slopes toward the saltpan from the adjacent mountains. Between the toes of these fans and the saltpan itself is a zone about a mile wide where, Hunt has shown, the ground is sandy, the water table is fairly high, and salinity levels do not exceed 6 percent. Most of the plants that occur here depend on the availability of groundwater to survive (such plants are called phreatophytes). Their precise distribution in this zone is dictated largely by their tolerance to the salinity of the water they utilize. The most tolerant of them is iodinebush, a greenish succulent that covers large areas just above the playa, often growing on large mounds.

Moving back from the saltpan, other plants appear, most of which also appear to be dependent on the availability of groundwater: saltgrass Mojave seablite (*Suaeda moquinii*), alkali sacaton (*Sporobolus airoides*), arrowweed (*Pluchea sericea*), and fourwing saltbush. There are even trees in this zone, most notably honey mesquite (*Prosopis glandulosa*) but also introduced tamarisks (*Tamarix*).

These plants are quickly left behind as you move above the sandy substrate that surrounds the saltpan and onto the gravel matrix of the fans. Here, most plants survive without access to the water table and instead use moisture made available by surface runoff. Toward the northern end of the valley, the lower parts of the gravel fans are covered by stands of desert holly (*Atriplex hymenelytra*).

To the south, the lower parts of the gravel fans are marked by desert saltbush. Creosote bush makes its appearance on the fans as well. As I discuss in chapter 2, creosote bush is one of the most widespread plants of the Mojave Desert and is also the species whose northern boundary marks the border between this desert and the botanical Great Basin. In Death Valley, it is found from near the bases of the fans, as low as −240 feet in elevation, up onto the flanks of the mountains themselves, to elevations as high as about 4,000 feet.

While creosote bush covers much of the fans, two other shrubs tend to be found only along their upper reaches. In the north, white bursage or burrobush (*Ambrosia dumosa*), a common Mojave Desert shrub also known as burro weed, replaces desert holly along higher elevations of the fans. To the south, brittlebush (*Encelia farinosa*) tends to replace desert saltbush in the same fashion. Creosote bush can be found with them all.

Springs that emerge on the gravel fans often support dense vegetation, including plants that otherwise grow in the valley bottom where they have access to groundwater—arrowweed, alkali sacaton, and fourwing saltbush, for example. Mesquite is found here as well, both honey mesquite, and, at higher elevations, screwbean mesquite (*Prosopis pubescens*).

To the east and west of Death Valley, the creosote bush of the gravel fans is replaced by shadscale in the Panamint, Black, and Funeral ranges. In the Panamint Range, well-developed pinyon-juniper woodland occurs between elevations of about 6,000 feet and 9,600 feet. As is common elsewhere in the pinyon-juniper woodlands of the Great Basin, Utah juniper both appears and disappears first with increasing elevation. For instance, along the road from Wildrose Campground to Mahogany Flat, on the west slope of the Panamint Range, the first Utah junipers occur at an elevation

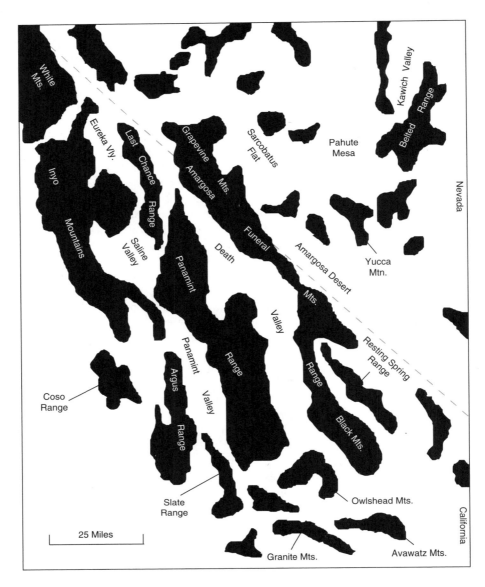

FIGURE 6-12 Death Valley region.

of about 5,850 feet, while the first singleleaf pinyons occur at 6,100 feet. By the time Mahogany Flat (so named because it is covered with mountain mahogany) has been reached, at an elevation of 8,133 feet, the junipers are gone. The pinyons, however, continue upward to an elevation of about 9,600 feet, at which point krummholz pinyons can be seen.

The juniper in this woodland includes not only the familiar Utah juniper but also, as botanist Frank Vasek has shown, the primarily Sierran western juniper. Western juniper is not common here, but it is found at the very upper reaches of the pinyon-juniper zone and in the lower reaches of the bristlecone pine–limber pine association, which comes next. The shrubs in the pinyon-juniper include plants that would be entirely at home in the botanical Great Basin to the north: yellow rabbitbrush, Mormon tea, black sagebrush (*Artemisia nova*), and Stansbury cliffrose (*Purshia stansburiana*).

Even though the Panamint Range borders Death Valley, it is sufficiently massive and tall to support subalpine conifers on its higher elevations. Limber pine is found above 9,600 feet, bristlecone above 9,800 feet, with big sage and yellow rabbitbrush as accompaniments. The uppermost 80 feet or so of the slopes of the ridge along Telescope Peak, for instance, are marked by both of these subalpine conifers, although as the top of the Panamints is approached here, limber pines become more and more scattered and bristlecones become thicker. There are even scattered pockets of white fir in the moister, more sheltered, and higher canyons of the Panamint Range.

Toward its northern end, Death Valley narrows considerably. Here, the Panamint Range trends northward, while the Grapevine Mountains trend toward the northwest. Toward the far northern tip of the Panamint Range, the two mountains come within a few miles of one another, a narrow strip of Death Valley wedged in between.

Along the western flank of the Grapevine Mountains, creosote bush, along with desert holly at lower sites and white bursage at higher ones, dominates up to elevations of about 4,400 feet. Above about 5,000 feet, big sagebrush becomes the dominant shrub, accompanied by yellow rabbitbrush and jointfir. Singleleaf pinyon makes its appearance at about 5,800 feet, accompanied by shrubs from the sagebrush zone beneath, including yellow rabbitbrush and jointfir. There is Utah juniper amongst the pinyon here, but, unlike the situation in the Panamint Range to the west, it is rare in the Grapevine Mountains. The pinyon thins considerably above 8,000 feet; by 8,200 feet, limber pine makes its appearance, and continues upward to the summits of both Grapevine (8,738 feet) and Wahguyhe (8,628 feet) peaks.

The floor of Death Valley is so inhospitable to plants that not even creosote bush can survive there. Excluding mesquite, whose roots can penetrate 150 feet or more in search of water, and the introduced tamarisk, there are no trees on the valley floor, on the gravel fans, or even on the lower flanks of the mountains. Trees don't begin to appear until elevations of about 6,500 feet have been reached, 6,800 feet above the valley bottom itself.

Our knowledge of the late Pleistocene vegetation of this area comes primarily from the work of Philip Wells and his student Deborah Woodcock, who have analyzed a series of packrat middens from the flanks of the valley. Augmenting that work are excellent analyses done by Geof Spaulding on middens from Eureka Valley to the immediate west of Death Valley and from the Amargosa Desert to the immediate east.

Nine dated late Pleistocene samples, from four packrat middens, have been analyzed from Death Valley itself, and from its adjacent mountains. Table 6-1 lists key plant species from these samples; many more species have been identified from these middens than are listed here, but these are the ones on

TABLE 6-1
Late Pleistocene Packrat Midden Plant Assemblages from Death Valley and Adjacent Mountains

Elevation (feet)	Age (years ago)	Key Plant Species
Amargosa Range		
4,200	11,600	Curlleaf Mountain Mahogany
		Utah Juniper
		Whipple Yucca
3,700'	9,680	Utah Juniper
		Joshua Tree
Panamint Range		
2,540	13,060	Shadscale
		Utah Juniper
	11,210	Beavertail Cactus
		Whipple Yucca
	9,455	Brittlebush
		Beavertail Cactus
	9,090	Beavertail Cactus
1,395	19,550	Shadscale
		Utah Juniper
		Beavertail Cactus
		Joshua Tree
		Whipple Yucca
	17,130	Shadscale
		Beavertail Cactus
		Whipple Yucca
	10,230	White Bursage
		Brittlebush

SOURCE: From P. V. Wells and Berger 1967, Van Devender 1977, P. V. Wells and Woodcock 1985, and Woodcock [1986])

which I focus. The table presents this information according to the elevation of the midden, and then, within each midden, by the age of the samples from oldest to youngest.

Look first at the Panamint Range sample that is both lowest in elevation (1,395 feet) and oldest (19,550 years). The midden that provided this sample is on a south-facing slope whose vegetation today is dominated by creosote bush and white bursage, a classic Mojave Desert combination. However, this midden sample suggests—and a wealth of other data confirm—that neither of these two widespread Mojave Desert plants were here during full-glacial times. Instead, at 19,550 years ago, this site was surrounded by shadscale, Joshua trees, Utah juniper, and Whipple yucca (*Hesperoyucca whipplei*). The next-youngest assemblage from this midden (17,130 years old) still has the shadscale and Whipple yucca, but the Joshua trees and juniper are gone.

Today, Joshua trees occur no lower than about 4,000 feet in elevation in the Death Valley area, some 2,600 feet higher than this site. Juniper occurs as low as 4,700 feet in the northern Panamint Range today, but it is generally found no lower than about 6,000 feet in this region, some 4,600 feet higher than the 19,550-year-old sample.

Because this midden is on a south-facing slope, the woodrats that built it occupied a warmer and drier setting than they would have occupied had they chosen a north-facing slope. Perhaps this is why neither Joshua trees nor juniper are present in the 17,130-year-old sample from this midden, which retains the shadscale and Whipple yucca but loses the trees. That is, it is possible that these trees were still present on north-facing slopes at this elevation 17,000 years ago, but gone from south-facing ones. However, it is also possible that the loss of juniper here reflects a general upslope movement of trees sometime after 19,000 years ago.

Whether or not the loss of juniper at this spot is a local or regional phenomenon, the 19,550-year-old sample from the Panamint Range shows that juniper extended down to 1,395 feet. The fact that juniper was not abundant in this midden suggests that this site may have been near the lower elevational reaches of this tree at that time. The 4,600 feet elevational displacement for Utah juniper, from full-glacial times to today, that this midden documents is impressive, though it is topped by the Pleistocene displacement of bristlecone pine in southern Nevada (see below).

Juniper woodland thus descended nearly to the floor of Death Valley during the full glacial, after which it began to move upward. Exactly when it reached its current elevational limits is not known, but that did not happen until the Holocene, and perhaps well into the Holocene.

Wells and Woodcock have suggested that the upslope movement of juniper was slow. While this may have been the case, we simply do not have enough middens from this area to know. As Geof Spaulding has observed, it has not been possible to fix the time of the loss of woodlands at any one site because the transition from the presence to the absence of trees is routinely lost in the large temporal gaps that separate the midden samples.

What is absent from the Death Valley middens is as interesting as what is in them. Today, white bursage, brittlebush, and creosote bush are abundant on the gravel fans that connect the mountains to the valley floor. The late Pleistocene middens, however, contain no trace of these plants until the Pleistocene is essentially over. The earliest midden date for white bursage and brittlebush here falls at 10,230 years ago. The earliest record for creosote bush is at 1,990 years ago, but that is undoubtedly because no earlier Holocene middens are known from this area; it is known from just northwest of Death Valley as early as 5,400 years ago (see chapter 8).

There is little reason to think that the Pleistocene absence of these plants reflects bias on the part of the woodrats who were collecting the plants. There is a chance, and perhaps a good one, that the species of woodrats doing the collecting shifted as the Pleistocene came to an end (see chapter 7). Since different species of woodrats are known to have different dietary preferences, such a shift could make vegetational change look more pronounced than it was. However, there is no reason to think that if white bursage and brittlebush were in the area at the time the earlier middens were accumulating, local woodrats would not have at least occasionally introduced the remains of those plants into their middens, no matter what species of woodrat was involved.

There is, perhaps, a slightly better chance that these two species of plants were in Death Valley during the late Pleistocene, but beneath the elevations from which the middens come. That middens are generally restricted to sheltered areas above valley floors is one of the difficulties with which packrat midden analysts must contend. However, as I discuss in the previous chapter, during at least some of the late Pleistocene, the lower elevations of Death Valley were covered by water.

Whenever the lake was here, its presence limited the lowest elevation that plants could reach in the valley. This in turn means that for at least some, and perhaps a substantial, period of time, plants like white bursage and brittlebush would have been limited to a narrow strip between the lowest middens that lack them (for instance, at 1,395 feet between 19,550 and 17,130 years ago) and the water itself (which reached a maximum depth of between 260 feet and 295 feet, or a maximum elevation of about 15 feet above sea level; remember that Badwater sits at –282 feet). If the plants were in that strip at that time, finding evidence for them is quite a challenge.

However, the information on plant history that is available for areas near Death Valley, amassed by Geof Spaulding, matches that assembled by Wells and Woodcock for the valley itself, and there is no reason to doubt the arguments all three of these scientists make. White bursage and brittlebush, characteristic plants of the modern Mojave Desert, are not found in these middens because they were not here during the late Pleistocene.

The same thing is true for creosote bush. A glance at table 6-1 shows that there are no records for creosote bush in Pleistocene Death Valley. The widespread abundance of this plant in this part of the world is a remarkably recent phenomenon, as I discuss in chapter 8. In fact, of all the plants listed in table 6-1, the only one that actually seems to have stayed put during the 20,000 years between the earliest dated Death Valley midden and today is *Opuntia basilaris*—the beavertail pricklypear.

Of the five Death Valley midden samples older than 11,000 years, four provided the remains of Whipple yucca. These samples range from 11,210 to 19,550 years old, and from 1,395 feet (the oldest) to 4,200 feet in elevation. This plant does not grow in the Death Valley area today. In fact, the closest it comes is the far western fringe of the Mojave Desert on the west (on the Pacific slope of California, it is called chaparral yucca), and the Grand Canyon on the east (where it is called desert Spanish bayonet). Nonetheless, it was fairly common in Death Valley during the late Pleistocene. It does not, however, appear to have survived here beyond the Pleistocene/Holocene transition, some 10,000 to 11,000 years ago.

Deborah Woodcock examined the temperature extremes of the areas in which Whipple yucca now grows, and found that it survives only where average July temperatures fall between 64.5°F and 84.2°F. In Death Valley today, July temperatures average about 102°F. The lowest midden that contains Whipple yucca falls at 1,395 feet, where July temperatures are lower than that but still substantially higher than 84°F (Woodcock estimated that they are something on the order of 93°F). The implications are obvious: to occur here, Whipple yucca probably needed summer temperatures much cooler than they are now. Woodcock, in fact, estimated that they were at least 10.8°F to 14.5°F cooler.

It also appears that this plant may have needed warmer winter temperatures than are now found in Death Valley. Woodcock found that the coolest average January temperatures in areas in which Whipple yucca now lives fall at about 39.2°F. This is above the 37°F average for the floor of Death Valley, but Whipple yucca was growing at an elevation of 4,200 feet at 11,600 years ago, very close to the modern line of winter snow accumulation.

All of this suggests that the late Pleistocene summers of Death Valley were cooler, and perhaps substantially cooler, than they are now, and the winters at least somewhat warmer. Compared to today, in other words, Death Valley's climate was not only more tolerable from the perspective of chaparral yucca but also more equable.

Woodcock made similar calculations using the known climatic values associated with the modern distribution of Utah juniper. This allowed her access not only to late Pleistocene temperatures but also to late Pleistocene precipitation. In the end, she estimated that the full glacial summer temperature decline was on the order of 14.4°F to 25.2°F, while precipitation was on the order of three to four times higher than that of today.

When did these conditions end? From the fact that white bursage and brittlebush appear at 1,395 feet by 10,230 years ago, Woodcock infers that a shift to far more modern climatic conditions in Death Valley occurred sometime between 11,000 and 10,000 years ago. Given that the youngest sample that contains Whipple yucca dates to 11,210 years ago, that argument appears fully reasonable.

Although four middens, and nine separate samples from those middens, represent little on which to base a regional vegetational reconstruction spanning some 10,000 years, Geof Spaulding's work to the east and west of Death Valley provide results fully in line with those obtained by Wells and Woodcock.

The far northern end of Death Valley is bounded on the west by the Last Chance Range. Just across this range is Eureka Valley, famed for its massive sand dunes, which, at a height of about 682 feet, are the tallest in the Great Basin (figures 6-12 and 6-13). Eureka Valley is the Mojave Desert's

FIGURE 6-13 Eureka Dunes, southern Eureka Valley, Death Valley National Park.

northernmost intermountain basin; creosote bush, and with it the Mojave Desert itself, dwindles and disappears in the hills to the immediate north.

Spaulding worked at the northern end of Eureka Valley, in the Horse Thief Hills. He extracted a single late Pleistocene assemblage (and several Holocene ones) from the far southern end of these hills at a locality he called Eureka View (figure 6-14). The vegetation of this area is now dominated by creosote bush and shadscale, but that was not the case when the late Pleistocene Eureka View midden accumulated. That midden, at an elevation of 4,690 feet and dated to 14,720 years ago, is marked by an abundance of Utah juniper twigs and seeds. Today, the closest this tree gets to Eureka View is in the Last Chance Range to the east, at elevations some 1,600 feet higher than the midden. Of the two dominant plants in the area today—creosote bush and shadscale—Spaulding found only shadscale in the midden. Just as in Death Valley, creosote bush does not seem to have been here during the Pleistocene; in fact, it does not appear in Spaulding's middens until 5,400 years ago. White bursage, which also grows in the midden locality today, was likewise missing from the late Pleistocene assemblage; it does not show up until 8,800 years ago.

Alongside the shadscale and juniper, the late Pleistocene Eureka View midden also contained a number of plants that today are characteristic not of the Mojave Desert but of the floristic Great Basin to the north. Perhaps most surprising, however, is the fact that the sample also provided a few needles of what appears to be limber pine, suggesting that this subalpine conifer might have grown nearby as well.

Spaulding examined a second late Pleistocene midden sample, dated to 10,690 years ago, from a few miles north of the Eureka View locality, at an elevation of 5,300 feet. The site that provided this midden is just north of the local edge of the Mojave Desert, the modern vegetation dominated by shadscale and yellow rabbitbrush and lacking creosote bush. The most common plant fossils in the midden sample, however, are from Utah juniper; shadscale was not far behind, but there was no rabbitbrush at all.

Spaulding has also done an impressive amount of work along the eastern edge of the Amargosa Desert, east of Death Valley. Among other things, he has analyzed Pleistocene middens from the Skeleton Hills and from the Specter Range. The rich picture of late Pleistocene vegetation that Spaulding has developed here is based on more than a dozen midden samples from locales that range in elevation from 2,600 feet to 3,900 feet, and that range in time, according to several dozen radiocarbon dates, from 36,600 years ago to the end of the Pleistocene.

Spaulding's work here shows once again that sites now dominated by Mojave Desert plants associations were, from full-glacial times to the end of the Pleistocene, marked by Utah juniper, rabbitbrush, and shadscale. Although lower-elevation samples contain higher proportions of desert shrubs and other xeric plants than do higher elevation middens, Utah juniper grew at elevations as low as 2,600 feet and perhaps lower (this is where the middens run out), and did so at some sites until beyond the end of the Pleistocene. Singleleaf pinyon even shows up in full-glacial age samples from both the Specter Range and the Skeleton Hills: a record from the Skeleton Hills, at 3,035 feet in elevation, provides the northernmost, lowest full-glacial record for this species yet known.

As in areas to the west, the junipers along the eastern edge of the Amargosa Desert were accompanied by plants

FIGURE 6-14 Northern end of Eureka Valley. The Last Chance Range is to the right, the Horse Thief Hills (the location of Spaulding's Eureka View midden) in the distance.

not here today—snowberry, for instance, which now grows only at far higher elevations or in more northerly places. And, as elsewhere, some of the most characteristic plants of the modern Mojave Desert are absent from Spaulding's middens: white bursage does not appear in his Skeleton Hills samples until 8,800 years ago; creosote bush does not make its appearance until later yet.

All of this information—from Death Valley, from Eureka Valley, and from the eastern edge of the Amargosa Desert—provides a fairly firm picture of the late Pleistocene vegetation of this region.

What we would see in Death Valley during full-glacial times, say 19,000 years ago, would be dramatically different from what we would see today. The ground on which the Furnace Creek Visitors Center now sits may have been underwater then. The flanks of the mountains to both east and west would have been covered not with such xerophytic shrubs as creosote bush but instead with Utah juniper, shadscale, and scattered Joshua trees and Whipple yucca. It is hard to know what plants would have been in the valley itself, but the typical plants that now extend toward the toes of the gravel fans—white bursage, brittlebush, and creosote bush—were absent. Shadscale was probably prominent in the lower elevations at that time, most likely accompanied by a series of plants that now grow only to the north or at higher elevations.

Even toward the end of the Pleistocene—at, say, 11,000 years ago—we would not see a botanical landscape anything like that of today. The white bursage and brittlebush would still be missing—they do not appear for another eight hundred years or so—as would the creosote bush, for which we have to wait longer yet. The trees would have moved well upslope, but Utah junipers would still be living several thousand feet lower than they do now and would be easily visible from the valley floor. Scattered chaparral yucca would probably be here as well, though not for long.

If we could actually visit Death Valley 19,000 years ago, or even much closer to the end of the Pleistocene than that, we would probably not have to worry nearly as much as we do know about the proper time of year to go. Winters, mild today, appear to have been even milder yet during the late Pleistocene. Summers, scorching today, may have been marked by temperatures similar to those of Seattle on a hot July day—80°F or 85°F, perhaps, but perhaps also associated with frequent cloud cover. Unless we happened to recognize the physical features of the landscape—Telescope Peak, for instance, or Grapevine Peak—we might not even realize that we were in Death Valley. Death Valley, hell in '49, was a far different place during the late Pleistocene.

A General Look at Late Pleistocene Great Basin Vegetation

During the late Pleistocene, Utah juniper and Joshua trees were not far from the floor of Death Valley, but there was no desert holly, no white bursage, and no creosote bush. In Owens Valley, Utah juniper and pinyon pine grew right next to the shores of the Pleistocene lake and Utah juniper was widespread throughout the valley. In the southern Carson Desert, there was abundant sage, presumably big sage, with pines nearby, but little in the way of *Atriplex*. At the same time, Utah juniper was thriving at elevations along the western edges of Lake Lahontan, well beneath the elevations at which it now occurs. In Ruby Valley, an expansive sagebrush-grass steppe lay above the lake itself, but not much in the way of saltbushes, no pinyon, and perhaps no juniper. In the northwestern Bonneville Basin, limber pine descended close to valley floors between 13,000 and 11,000 years ago, with an understory that included prostrate juniper, sagebrush, and buffaloberry.

Nearly all the plants that were in these areas are still fully at home in the hydrographic Great Basin. Today, for instance, Joshua trees and Utah junipers grow side-by-side in the lower elevations of the northern Pahranagat Mountains. Sagebrush-grass steppe characterizes much of the northern Great Basin of south-central Oregon. But while the plants are now common within the hydrographic Great Basin, during the late Pleistocene they were often in what seem to be the wrong places, and often in combinations not seen today. Had I chosen examples other than the five I have given here, the results would have been much the same.

The Mojave Desert

Our knowledge of the late Pleistocene vegetation of the Mojave Desert as a whole is remarkably detailed. In part, this knowledge was spurred by the pioneering Mojave Desert packrat midden work of Philip Wells, and by insightful early packrat and pollen work in this area by Peter Mehringer. This work not only revealed the unsuspected nature of the botanical history of the Mojave Desert but also showed the potential of this area to yield its secrets to paleobotanical efforts.

In more recent years, a series of scientists have greatly expanded on this work. We have encountered them all before—Scott Anderson, Peter Koehler, Geof Spaulding and Peter Wigand. Much, but not all, of this effort has been part of the Yucca Mountain nuclear waste storage project discussed in chapter 5.

The results of this work are very consistent. Koehler, Anderson, and Spaulding have shown that in the central Mojave Desert of California, from the Mojave River Basin to the southern end of Death Valley, packrat middens that date to between 24,400 and 11,500 years ago consistently contain the remains of both Utah juniper and singleleaf pinyon, along with jointfir and Stansbury cliffrose; two of them, 11,900 and 11,500 years old, also contain significant amount of sagebrush. The middens that contain this material all lie between about 4,000 feet and 4,330 feet; the single late Pleistocene midden that lies below this elevation here, at 3,480 feet, has juniper but no pinyon at 17,900 years ago. Today, these middens sit in landscapes dominated by creosote bush, along with such plants as bursage and Mormon tea. There is no pinyon-juniper here today. Indeed, while Utah juniper may be found at higher elevations in the general region, it is not to be found at these midden sites. All this shows that these areas, today dominated by creosote bush, supported pinyon-juniper woodland during the late Pleistocene. It is not yet clear from the macrofossil record when the pinyon disappeared, but it was still present 11,500 years ago and gone at 9,900 years ago, the next-youngest sample in the series.

Bob Thompson, among others, has shown that certain plants are routinely better-represented in packrat middens by pollen than they are by macrofossils. Of the nine groups of pollen that he analyzed, the greatest differences between pollen and macrofossil representation in middens was provided by sagebrush and cheno-ams. As Thompson notes, these are wind-pollinated plants that most commonly grow on fine substrates; the plants might be common nearby, but not on the rocky outcrops that contain the middens. As a result, the middens may often contain pollen from these plants, but not the macrofossil remains of the plants themselves. That means that analyzing both the macrofossils and the pollen can provide a picture of both local (the macrofossils) and regional (the pollen) vegetation.

This does not mean, though, that the pollen frequencies from middens are akin to those from the sediments found at the bottom of deep lakes, or lakes that, like Owens Lake, were deep at some time in the past. Sediments from the deepest part of those lakes tend to be dominated by pollen dispersed by the wind. The pollen in packrat middens, however, can have multiple origins. In addition to the pollen carried in by the wind, the woodrats themselves can and do introduce pollen into their middens. Thompson noted that the pollen of certain plants might be heavily represented in middens because it was attached to the flowers or cones that the woodrats brought home. In addition, as Owen Davis and Scott Anderson have discussed, woodrats accumulate pollen both on their fur and in their fecal pellets. To this can be added the fact that they also retrieve the fecal pellets of other animals. Of these sources, however, the two most important sources are likely to be the wind and the plant parts the furry creatures themselves retrieve. The former provides a regional pollen signal, while the latter emphasizes the local one.

With all of this in mind, Koehler and his colleagues took the important step of analyzing the pollen from their central Mojave Desert middens. They showed that, with one exception, their late Pleistocene samples include high frequencies of pollen from both juniper and pinyon (or, more accurately, from the subgroup of pines to which pinyon belongs, but pinyon is the obvious choice here). Indeed, the one exception comes from that 17,900-year-old sample midden that lacked pinyon macrofossils and which contained very low frequencies of pinyon pollen. This is a perfect match for this part of the macrofossil record.

On the other hand, these samples also contained high proportions of bursage and sagebrush pollen, neither of which are represented in the late Pleistocene middens but which must have been present, presumably at much lower elevations.

So we know that during the late Pleistocene, areas within the central Mojave Desert now covered with creosote bush and other Mojave Desert shrubs were then covered with pinyon-juniper woodland, and that they were so covered between about 24,400 and 11,500 years ago—with the exception, it seems, for the period of time focused on about 18,000 years ago. We have already seen that the late Pleistocene packrat middens of the northwestern Bonneville Basin contained no pinyon and that there was none in those from the Carson Sink and only a single known instance from Owens Valley north and west of the lake itself.

In fact, we seem to know exactly how far north pinyon, and pinyon-juniper woodland, got during the late Pleistocene in the Great Basin. The northernmost securely dated occurrences are provided by three sets of records.

First, there is the pinyon in Koehler and Anderson's Owens Valley Haystack Mountain samples, where it occurred, along with Utah juniper, between 22,900 and 17,680 years ago. These samples fall at a latitude of 36°36′ N. Then there is the pinyon in Wigand's Alabama Hills midden at 11,450 years ago; pinyon was not in the 19,300-year-old Volcanic Tablelands sample some thirty miles to the north.

Finally, there are Geoff Spaulding's Specter Range midden samples. The full run of Spaulding's late Pleistocene samples from here date from 32,100 to 18,700 years ago. All contain abundant Utah juniper macrofossils; those that date to between 19,300 and 18,700 years ago also contain large amounts of pinyon. These samples fall at a latitude of 36°40′ N.

Even though pinyon-juniper now covers some 18 million acres of the Great Basin, not a single securely dated Pleistocene-aged pinyon pine macrofossil is known from north of these places. In addition, although pinyon was present as late as 11,500 years ago in the central Mojave Desert, the northernmost records that we have for it are full-glacial in age, dating to about 23,000 to 18,700 years ago. As Koehler and his colleagues observe, there appears to have been a late Pleistocene plant boundary that fell at about 36° N latitude during the late Pleistocene. The relatively warm and moist areas to the south supported pinyon-juniper woodland as well as such mild-winter plants as Joshua tree and Whipple yucca. In the Mojave to the north of here, the pinyon, Joshua trees, and Whipple yucca disappear.

I said "securely dated" in the previous paragraph because late Pleistocene packrat midden samples to the north of the Specter Range do contain pinyon remains. These were provided by Geoff Spaulding's work in packrat middens from southeastern Nevada's Fortymile Canyon, at a latitude of about 36°57′ N. As a group, his samples dated from before 50,000 years ago well into the Holocene. These samples contained abundant pinyon macrofossils that dated to 47,000 years ago, well before the Last Glacial Maximum, and dribs and drabs of them that dated from 21,800 to the end of the Pleistocene.

It has always been hard to know what to make of either dribs or drabs in the fossil record but these particular ones might well be contaminants. There are a number of ways in which such contamination can occur, of course; perhaps the most intriguing of them in this context is the possibility that the woodrats themselves incorporated material that had been snagged by their ancient ancestors. Only directly dating each of these dribs could tell us how old they are, and Spaulding himself has not accepted them as providing evidence for pinyon north of the Specter Range. To the north of Fortymile Canyon, there is not even untrustworthy evidence of late Pleistocene pinyon anywhere in the Great Basin, not a single twig, needle, nut, or piece of wood.

More generally, from before 24,000 years ago to about 11,000 years ago, woodlands composed of Utah juniper, or of juniper and singleleaf pinyon, were widespread throughout the Mojave Desert to just north of 36° N latitude and from elevations as low as 2,000 feet to as high as 6,000 feet. The understory of this xeric woodland included plants now common at higher elevations, or in the Great Basin to the north—jointfir, rabbitbrush, and shadscale.

Juniper woodland also extended beneath these elevations during this period, but on the driest sites, treeless desert vegetation thrived, with such plants as shadscale, jointfir, and rabbitbrush again important. Death Valley and, to the east, the Amargosa Desert, have both provided examples of this low elevation, glacial-age desert vegetation. Plants that today thrive in this zone, and that are fully characteristic of the modern Mojave Desert—white bursage and creosote bush, for example—were then absent or extremely rare. Given the tremendous abundance of these plants in the Mojave Desert today, this is remarkable.

There is also ample evidence that subalpine conifers descended to very low elevations, at times overlapping Utah juniper woodlands in their lower reaches. Peter Wigand, for instance, has shown that limber pines were present at an elevation of 5,500 feet in the Pahranagat Mountains between 18,000 and 19,000 years ago. This is some 3,000 feet beneath the lower limit of limber pine here today, and it is at the lower limit of the current distribution of juniper here. Wigand has also shown that limber pine grew at 5,000 feet on the southern edge of Pahute Mesa, east of the Amargosa Range some 18,000 years ago. Geof Spaulding has shown that at 16,500 years ago, limber pine co-occurred with Joshua trees, Utah juniper, and shadscale at 6,100 feet on the western flank of the Sheep Range. If this assemblage of plants can be duplicated anywhere in the world today, I am not aware of it—another example of the nonanalog biotic assemblages of the sort I discuss earlier in this chapter.

Move back slightly earlier in time, and there are equally impressive records for the downslope movement of a number of other subalpine conifers. In the late 1960s, Peter Mehringer and tree-ring expert Wes Ferguson of the University of Arizona showed that at 23,600 years ago, bristlecone and limber pines, along with white fir and Utah juniper, grew on the south side of Clark Mountain, in southeastern California some fifty miles southwest of Las Vegas. The midden that provided this record sat at an elevation of 6,270 feet. Bristlecone pine doesn't occur on Clark Mountain today, though it is found on the massive Spring Range to the north. In the Spring Range, however, it is found at an elevation 5,600 feet higher than it was growing on Clark Mountain during the late Pleistocene. White fir does occur on Clark Mountain today, but on the north, not the south, side. It would have to descend some 1,650 feet, and survive a south-facing exposure, to reoccupy the position it held here 23,600 years ago.

The bristlecone pine was gone from this spot by 12,500 years ago. By this time, the midden site studied by Mehringer and Ferguson was within Utah juniper–singleleaf pinyon woodland, though white fir and limber pine still existed near enough by that they, too, were represented in the younger midden sample.

These are not isolated records. Wigand has 20,300-year-old white fir, limber pine, and Utah juniper at 5,500 feet in the Pahranagat Range, 3,000 feet beneath the closest white fir now to be found in this range. Spaulding has shown that at 21,700 years ago, bristlecone pine, limber pine, white fir, and singleleaf pinyon existed at an elevation of 6,500 feet in the southeastern Sheep Range. Today, limber pine and white fir are found 1,400 feet higher, and bristlecone 1,700 feet higher, in the same range.

Indeed, in many areas within southern Nevada, both limber pine and white fir extended down to about 5,300 feet during the late Pleistocene, well below their modern lower limit. The record for these lowered elevations extends from almost 35,000 to nearly 10,000 years ago. This is not to say that they were this low for this entire period, or that the two trees had identical histories for this lengthy period of time. Instead, it is to say that, in general, such now-subalpine conifers as bristlecone pine, limber pine, and white fir extended to remarkably low elevations, in some cases more than a mile beneath their current lower limits in nearby settings. They also existed in ranges where they are now extinct: bristlecone in the Clark Range, limber pine in the Eleana Range, and white fir in the Pahranagat Range, to give a few examples. Beneath it, and sometimes with it, grew a woodland composed of Utah juniper or of both Utah juniper and singleleaf pinyon, with an understory of shrubs whose affiliation is not to the Mojave Desert, but to more northern, or higher-elevation, settings.

Only in the driest of settings did true desert shrub vegetation exist. Indeed, the juniper and pinyon-juniper woodlands of the late Pleistocene Mojave Desert were so widespread that it was thought not long ago that true desert vegetation might not have existed at all here during the late Pleistocene.

As I have mentioned several times, one of the frustrations midden analysts must learn to accept is that their plant records click in and out through time. As a result, and as I have discussed, major transitions between kinds of vegetation can be lost unless the sample of available middens is extremely fine-grained temporally. A sample of that sort may eventually become available for the Mojave Desert, but it isn't available yet. On the other hand, one of the advantages of the pollen record, as I have also discussed, is that it is often chronologically detailed, even though the species or even genera of plants so carefully tracked through time may be unknown. When both kinds of information are available, these frustrations can disappear.

It is fortunate, then, that work done by Peter Mehringer at Tule Springs has provided a detailed pollen record that shows the replacement of a juniper woodland with a sagebrush understory by desert shrub vegetation dominated by sagebrush and *Atriplex*. That transition occurred about 12,000 years ago. What species of sage and saltbush were involved is unknown, but there can be no doubt that this was not a shift to modern vegetation: today, this area is dominated by white bursage, creosote bush, and *Atriplex*. As we will see later, the establishment of something like the modern vegetation of this area did not occur until well into the Holocene.

At Tule Springs, the low-elevation juniper woodland seems to have met its demise around 12,000 years ago. A wide range of packrat midden samples confirms that significant vegetational change was taking place in many parts of the northern Mojave Desert at about this time. Spaulding, for instance, has analyzed a series of middens from the Eleana Range, on the southern flank of Pahute Mesa (actually just north of the northern edge of the Mojave Desert). Between 17,100 and 13,200 years ago, one of his middens, located at an elevation of about 5,900 feet, contained abundant limber pine, along with such shrubs as rabbitbrush and sagebrush. At 11,700 years ago, the shrubs were still here, but the limber pine was essentially gone, replaced by a Utah juniper–singleleaf pinyon woodland, which still exists nearby. On the eastern edge of the Amargosa Desert, Owl Canyon supported a juniper woodland, with Utah agave (*Agave utahensis*), at an elevation of about 2,600 feet, at 12,300 years ago. At 10,300 years ago, the agave was still there, but the junipers were gone. Today, neither juniper nor Utah agave grows here. Instead, the vegetation of this site is dominated by such Mojave Desert shrubs as creosote bush and white bursage, which are not in the middens.

Thus, the low elevation woodlands of the Mojave Desert had begun to disappear by 12,000 years ago. Two thousand years later, desert vegetation was widespread in low elevations settings (that is, beneath about 3,500 feet) in the Mojave Desert, though this vegetation was still significantly different from that which is found in the area today.

The woodlands gave way at different times in different places. The earliest evidence now available for true desert vegetation in the northern Mojave Desert comes from the eastern edge of the Amargosa Desert. Here, Spaulding has shown that a desert community was in place by 14,800 years ago at an elevation of 3,000 feet. Today, the vegetation of the site that provided this late glacial midden is dominated by creosote bush and white bursage. At 14,800 years ago, however, the desert vegetation that grew here was dominated by shadscale and desert snowberry (*Symphoricarpos longiflorus*). Snowberry does not grow in this area now. In fact, it does not grow on low elevation sites in the Mojave Desert at all, though it is in the mountains of the floristic Great Basin to the north.

In contrast to the early "desertification" of this part of the eastern Amargosa Desert is the persistence of Utah juniper woodland in the central Sheep Range to the east. Here, Spaulding has shown that juniper woodland was present at 9,400 years ago in Basin Canyon, at an elevation of 5,340 feet, on a site now marked by Joshua trees and jointfir. And, on the Spotted Range between the Amargosa Desert and the Sheep Range, juniper woodland seems to have lasted as late as 7,800 years ago on a site that, today, is also marked by Mojave Desert vegetation.

In short, toward the end of the Pleistocene, the low elevation woodlands of the Mojave Desert gave way to desert vegetation, but when this happened depended on both time and place. Not surprisingly, low elevation sites tended to lose their woodland earlier than higher ones; drier sites lost woodland earlier than wetter ones. Exposure, elevation, substrate, and other environmental variables all played a role in determining when the woodland disappeared from any particular spot. Most, but not all, of the woodlands were gone by the time the Pleistocene came to an end. In their place was desert vegetation, but even that was distinctly different from the vegetation that occupies most of this area today.

CLIMATIC ESTIMATES FROM MOJAVE DESERT LATE PLEISTOCENE VEGETATION

A number of scientists have combined our knowledge of the modern climatic requirements of plants with the distribution of those plants during the late Pleistocene to estimate the temperature and precipitation of the Mojave Desert at that time.

Spaulding was one of the pioneers in this area, focusing his work on the Yucca Mountain area. In 1985, he argued that altered distributions of such plants as limber pine and Utah juniper, along with the absence of such frost-sensitive plants as creosote bush, suggested that average annual temperatures in the northern Mojave Desert (and the Yucca Mountain area in specific) during full-glacial times, around 18,000 years ago, were between 10.8°F and 12.6°F cooler than those of today. Average winter temperatures here were, he suggested, at least 10.8°F lower than their modern values, summer temperatures some 12.6°F to 14.4°F cooler. He also estimated that full-glacial precipitation here was some 30 to 40 percent higher than it is now, with most of that precipitation falling during the winter months.

Bob Thompson and his colleagues built on this pioneering effort, adding to it additional midden data as well as the much deeper knowledge that has been gained since 1985 on the climatic requirements of the plant species involved and on the modern climates of the Yucca Mountain area. Their estimates suggest an average annual temperature decrease of 14.4°F at 18,000 years ago, with precipitation at 240 percent of modern levels.

There are other estimates for the full-glacial climates of the northern Mojave Desert. Spaulding himself, for instance, has more recently suggested that the average annual precipitation increase for the northern Mojave Desert might have ranged between 30 and 90 percent. As I noted earlier in this chapter, Deborah Woodcock estimated full glacial summer temperatures in Death Valley to have been 14.4°F to 25.2°F cooler than today, winters perhaps somewhat warmer, and average annual precipitation three to four times higher. It is notable that while temperature estimates tend to fall in a very narrow range—between 10.8°F and 16.2°F below modern—precipitation estimates tend to vary widely, even when made for the same place and time.

The fact that different approaches give similar answers doesn't necessarily mean the answer is correct, but the fact that similar approaches give different answers is not heartening. If we cannot yet turn empirical data from the past into secure estimates of the precipitation patterns that produced them, one has to wonder about predictions about future precipitation, for which no such empirical data exist. With temperature, though, we seem to be on more secure ground.

ANOMALOUS LATE PLEISTOCENE PLANT ASSOCIATIONS IN THE NORTHERN MOJAVE DESERT

During the first half of the twentieth century, two formidable plant ecologists, Frederick E. Clements and Henry A. Gleason, engaged in a heated debate over the nature of plant communities. Clements viewed a climax plant community as "a complex organism inseparably connected with its climate," the inherent unity of which "rests upon the fact that it is not merely the response to a particular climate, but is at the same time the expression and indicator of it" (Clements 1936:254). To Clements, plant communities look the way they look because they have to look that way. In this view, those communities are made up of species that would respond in virtually identical ways to climatic change. The oak-hickory forests of the northeastern United States, for example, or the spruce-fir forests of Canada, were seen as being organic units that had lives of their own. After all, when patches of such communities burned down or were cut, the same association of plants grew back as long as enough time were allowed to elapse. As a result, ecologists talked about the movement of entire plant communities during periods of significant climatic change, believing that those communities moved north or south or upslope or downslope during such periods.

Gleason espoused a radically different view. In his view, "every species of plant is a law unto itself, the distribution of which in space depends on its individual peculiarities of migration and environmental requirements" (Gleason 1926:26). Gleason (1926:16) famously referred to this view as "the individualistic concept of the plant association."

Clements's view has now succumbed to the sheer weight of ecological and paleobotanical data demonstrating that it was wrong. We now know that plant (and animal) communities are not organic entities with lives of their own, but are instead historically constructed assemblages of species, each of which has its own adaptive requirements and its own particular history. The history of the oak-hickory forest in eastern North America provides a good example. During full glacial times, oaks were fairly widespread in the east but hickory was confined to the southeast. As the ice retreated, hickory began its slow move northward. As a result, while there was oak in the eastern Great Lakes area by 12,000 years ago, hickory didn't become common in this region until well into the Holocene (that is the problem with the Meadowcroft Rockshelter plant remains that I discuss in chapter 3). Today's northeastern oak-hickory forests were built from the intersecting but separate histories of different species of plants with different adaptive requirements and different migrational histories.

It is also true that the plant communities we see around us today didn't have to become what they in fact became. Historical accident may play a large role in determining the nature of biotic communities, a fact well known more than a century ago. To take but one example, cattle egrets (*Bubulcus ibis*) were first reported in South America during the 1870s, apparently carried there from Africa by a tropical storm. They had reached Florida by the 1940s, and were breeding in Canada by the early 1960s; they were in the eastern Great Basin by 1963 and the western Great Basin by 1974. Such dispersal accidents may be quite important: if a plant or animal happens to get well established somewhere, later arrivals may not be able to replace it.

In short, the plant and animal communities that now surround us reflect the results of individualistic responses of species to climate change, of differential migrational abilities, of competitive relationships among species, and of sheer chance. These species-specific responses are now often referred to as "Gleasonian individualism."

This view has a number of important implications. It implies, for instance, that we often cannot predict in detail what will happen to the members of a given biotic community in the face of future climatic change, although we can predict that each species will react in its own way. This is true because we often lack sufficient knowledge of the habitat requirements of those species, because we often lack sufficient knowledge of how a given species will react in new competitive settings, and because there is no way to predict the occurrence of chance events.

That species react in individualistic fashion to climatic change helps account for the fact that the late Pleistocene vegetation of the Mojave Desert includes assemblages of plants not to be found anywhere today. The combination of limber pine, Joshua tree, Utah juniper, and shadscale some 16,500 years ago in the Sheep Range is but one example.

Others abound. Spaulding, for instance, has found that desert almond (*Prunus fasciculata*), white fir, Utah agave, limber pine, and shadscale grew together in the southeastern Sheep Range 24,400 years ago. Three thousand years later, shadscale, white fir, and ocean spray (*Holodiscus*) occupied this same spot. Neither of these combinations is known to occur today, here or anywhere else. These and other "anomalous" plant associations ultimately resulted from the individual responses of separate species to the climatic challenges presented to them during the late Pleistocene. Yet another example is provided by the occurrence of singleleaf pinyon–Utah juniper woodland beneath Utah juniper alone, when exactly the opposite is what happens in the Great Basin today.

These facts also provide a very simple answer to the question "where was the Mojave Desert during the Pleistocene?" The answer is that, in terms of the rich plant associations that now characterize it, it simply did not exist.

The Floristic Great Basin

The woodrats of the Mojave Desert at one time led the way in accumulating middens that scientists then analyzed for their contents. The floristic Great Basin, however, is now catching up, adding to pioneering work done by Philip Wells and Bob Thompson (figure 6-15). In addition, and as I have in part already discussed, we have pollen sequences that help flesh, or maybe leaf, out the information provided by the packrat middens.

I begin with the Bonneville Basin and the area immediately adjacent to it. Working independently, Wells and Thompson have analyzed a large series of middens from in and near the Deep Creek, Snake, Confusion, and Wah Wah mountains (figure 5-7; the Wah Wah Mountains are just south of the Confusion Range), and have also provided midden-based glimpses of late Pleistocene vegetation to the north and south. With that work, and with research by Dave Rhode in the northwestern Bonneville Basin, we now have an excellent midden-based picture of the nature of late Pleistocene vegetation along the western edge of the Bonneville Basin.

The Snake, Confusion, and Wah Wah range middens document that subalpine conifers, in particular bristlecone pine and the shrubby common juniper, occupied very low elevations during the full glacial, and, with limber pine, remained low through the end of the Pleistocene. For instance, Wells's work in the Confusion Range has shown that all three of

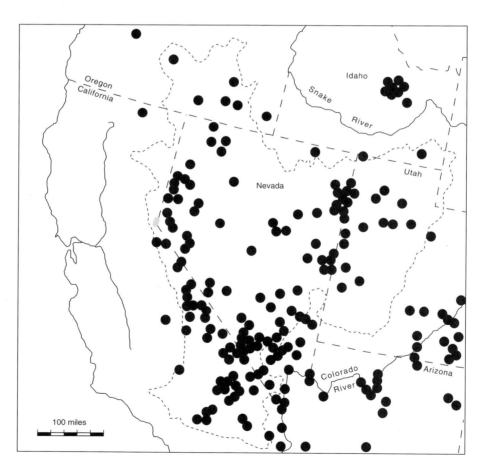

FIGURE 6-15 Distribution of packrat middens (redrawn from map provided by North American Packrat Midden Database).

these species survived until at least 11,900 years ago at elevations as low as 5,250 feet. For bristlecone pine, this represents an elevational decrease of about 2,200 feet; for common juniper, the decrease is about 2,800 feet.

In protected settings, Engelmann spruce descends to fairly low elevations today as can bristlecone pine. However, this happens only in special settings—in moist canyon bottoms that receive substantial cold air drainage. During full- and late-glacial times, Engelmann spruce descended to elevations as low as 6,100 feet in the Snake Range, and remained there until as recently as 10,200 years ago. Although present in only small amounts during latest Wisconsin times, this spruce has been detected in a number of full- and late-glacial low elevation middens in the Snake Range, leading Wells to suggest that it was then fairly abundant in the lower elevations of the Snake Range. Bristlecone pine is so abundant in the low elevation late Pleistocene middens that no question can arise as to its great abundance on the landscape at the time the middens accumulated.

In this part of the Great Basin, then, the late Pleistocene saw subalpine conifers, including bristlecone pine, common juniper, and, on the Snake Range, Engelmann spruce, growing at very low elevations on the mountain flanks and doing so until the Pleistocene was nearly over. This is quite different from the picture extracted by Dave Rhode not that far to the north. His middens contain nary a trace of bristlecone pine but are loaded with limber pine between about 13,000 and 11,000 years ago.

All this makes it very possible that the late Pleistocene distributional edge of bristlecone territory in this area fell somewhere between the Goshute Mountains and the northern edge of the Snake Range. That is because the highest elevation midden in the northwestern Bonneville Basin, Top of the Terrace (6,600 feet), lacks bristlecone pine, while the middens analyzed by both Thompson and Wells in the northern Snake Range, some eighty miles to the south, have the remains of this tree in abundance, and at elevations as low as 6,100 feet.

Dave Rhode and David Madsen have pointed out that it is possible that bristlecone pine was in the mountains adjacent to the northwestern Bonneville Basin during the late Pleistocene and that the middens they analyzed from that area are simply in the wrong place to have recorded the tree. After all, there is bristlecone pine on the Goshute Mountains today, at an elevation of 8,800 feet, and this is the same range that provided the Top of the Terrace midden. Perhaps, they suggest, the lack of bristlecone in this midden, and in the others they analyzed, reflects the placement of the middens, not the absence of the tree.

However, a few late Pleistocene-aged middens from between the Goshute Mountains and the northern Snake Range of late Pleistocene age have been analyzed and they also lack bristlecone. Those middens include Twin Peaks, at 5,200 feet and full of 12,000-year-old limber pine, and Granite Canyon, on the east face of the Deep Creek Range at 6,800 feet, full of 13,600-year-old limber pine. Neither place has limber pines today. Although it is certainly possible that the lack of bristlecone in the middens of the northwestern Great Basin is a function of some sort of sampling error, the entire pattern seems more consistent with the view that bristlecone pine reached its northern late Pleistocene limit in this area somewhere just south of the Goshute Mountains.

Thompson has calculated that the low elevations reached by bristlecone pine during the full-glacial in this area imply a summer temperature decrease between 16.2°F and 18°F (see table 5-8). This estimate is within the range of others that have been made for this period of time, although it slightly exceeds Spaulding's estimate for the Mojave Desert full-glacial summer temperature decline. Unlike Spaulding and Woodcock, Thompson has not attempted to calculate winter temperatures from his data, since the central Great Basin does not contain the variety of frost-sensitive plants found in the Mojave Desert. He does cautiously speculate that mean annual precipitation in the mountain flanks may have been some 150 to 200 percent greater than that of today, with all or nearly all of that increase falling in the winter months. To the north, as I mentioned, Rhode and Madsen have suggested that summer temperatures between 13,000 and 11,000 years ago must have been at least 11.9°F colder than they are today, with precipitation only slightly above its modern levels, to have supported the limber-pine dominated plant community that then characterized that area. Although all these estimates may be incorrect, it does raise the possibility that different patterns of precipitation between north and south may provide part of the reason that bristlecone pine peters out between the Snake Range and Goshute Mountains.

If it is easy to overlook the forest for the trees today, it is probably even easier to overlook the shrubs and grasses for the trees in the deeper past. This is especially the case in areas in which the trees had such spectacular histories and so much of our knowledge comes from packrat middens. As I have noted before, those middens accumulate in rocky areas, so even if the middens themselves are from valley settings, they tend to reflect the vegetation of the rocky outcrops on which they sit.

As we have seen, subalpine conifers also reached low elevations in the Mojave Desert. There, however, the valley bottoms are much lower than they are in the floristic Great Basin, and juniper or pinyon-juniper woodlands, and in some cases true desert vegetation, occupied still lower altitudes.

In the floristic Great Basin, valley bottoms are several thousand feet higher than those in the Mojave Desert (see figure 2-6). Basal elevations in Snake Valley, which lies between the Snake and Confusion ranges, fall at about 4,800 feet. Basal elevations in Tule Valley, east of the Confusion Range, fall at about 4,500 feet. Moreover, both Snake and Tule valleys held arms of Lake Bonneville when that lake was at the Bonneville and Provo shorelines, between about 15,500 and 12,500 years ago (see chapter 5). Indeed, geologist Dorothy Sack has shown that Tule Valley may have been integrated into the Lake Bonneville system as early as 19,500 years ago.

An intriguing question is raised by the fact that subalpine conifers descended to very low elevations in the east-central Great Basin, an area of high-elevation valleys that at times held pluvial lakes. Could it be that bristlecone or limber pine woodland, or both, was virtually continuous across the mountain flanks from the Wasatch Range to at least the central Great Basin? Is it even possible, as Bob Thompson and Jim Mead (1982:42) once asked, that subalpine conifers "formed a continuous belt from the Wasatch Front to the Sierra Nevada"?

It is here that our knowledge of the history of sagebrush and shadscale vegetation within the Great Basin becomes all-important. As Thompson and Mead also pointed out, in order for subalpine woodlands or forests to have been continuously distributed from the Rockies to the Sierra Nevada, or even from the Rockies to the central Great Basin, those trees would have had to occupy not only the mountain flanks but also the valley bottoms. Understanding the nature of the vegetation of those valley bottoms is thus critical in evaluating whether or not subalpine conifers were continuously distributed across the Great Basin during the late Pleistocene.

Fortunately, the nature of the valley-bottom vegetation in the floristic Great Basin is reasonably well understood, as the regional pictures I described show. Wigand and Mehringer's work at Hidden Cave documents that the southern Carson Desert supported sagebrush steppe during the late Pleistocene, perhaps with pines nearby. Thompson's work in Ruby Valley demonstrates that this major valley bottom also supported sagebrush steppe, and supported it at an elevation—roughly 6,000 feet—above that occupied by bristlecone pine on the slopes of the Snake, Confusion, and Wah Wah ranges to the south. Likewise, the Blue Lake record shows abundant sagebrush on the local landscape between the start of the record, around 12,700 years ago, to shortly after 10,200 years ago.

Other records follow suit, all suggesting both that conifers were far more widely distributed during the late Pleistocene than they are now and that significant amounts of sagebrush steppe existed on the valley bottoms. Swan Lake provides an example. This lake is located at the north end of Cache Valley, Idaho, at an elevation of 4,765 feet and about 4.5 miles south of Red Rock Pass, the Lake Bonneville spillway into the Snake River drainage. Today, the vegetation surrounding the lake is dominated by big sagebrush, rabbitbrush, and grasses, with big greasewood common as well. Robert Bright's analysis of the pollen from a core taken from this lake showed that the deepest sediments within the lake were laid down between about 12,100 and 10,200 years ago. During that interval, pine—perhaps limber and lodgepole—and spruce together account for up to some 90 percent of the pollen in the samples. Exactly where these trees were is not known, but the fact that no conifer leaves were found in the lake sediments suggests that they were not growing adjacent to the lake itself. Instead, they probably occupied the surrounding hills, much closer than these trees are found today.

Sagebrush pollen occurs in low percentages in the late Pleistocene deposits of Swan Lake, perhaps because it was swamped by the copious pollen produced by the nearby pines. It is present, however, and Bright suggested that sagebrush steppe must been nearby, although far more limited in extent than it is here today. At Grays Lake, some sixty miles to the northwest and outside of the Great Basin itself, Jane Beiswenger has shown that sagebrush remained abundant during the late Pleistocene, even as conifers expanded in the surrounding uplands. At Swan Lake, the abundance of sagebrush pollen increased, and that of pine and spruce decreased, dramatically as the Pleistocene came to an end, leaving no doubt that sagebrush steppe expanded markedly here at this time.

Joining all of this is a series of sequences from lower elevations within the Bonneville Basin, including several deep cores taken either from Great Salt Lake or the Lake Bonneville playa. There are good things and bad things about the information these analyses have provided. The first bad thing is that, for many reasons, the chronological control available for them is often very tenuous. The second bad thing is that the pollen they contain represents the contribution of a huge region, brought in both by the wind and by all the streams that drained into Lake Bonneville. David Madsen, for instance, estimated that the pollen in Great Salt Lake core C, taken from the southern end of the lake, may reflect the vegetation of some 50,000 square miles.

The good things about them include the fact that they can span immense amounts of time. For instance, cores drilled in the Great Salt Lake by the Amoco Production Company penetrated nearly two miles of sediment, providing a depositional record, including pollen, that extends back beyond 10,000,000 years ago. The Black Rock core, drilled by the U.S. Geological Survey in the southern Bonneville Basin, extracted nearly 1,000 feet of sediment covering the past 3 million years or so. In addition, while it may be frustrating that sediments of this sort provide pollen from such a huge area, that can also be a benefit, because it tells us what was going on in the region as a whole.

As Owen Davis has observed, analyzing the pollen from these cores has provided very consistent results. Taken as a whole, they show that between 25,000 and 14,000 years ago, the vegetation surrounding the northern Bonneville Basin consisted largely of pine woodland and sagebrush steppe, not closed canopy forest. The cores also suggest that the pine began to retreat, and the sagebrush steppe to expand, at about 10,800 years ago. The match this provides with Dave Rhode's midden record to the west is not perfect, since pine in those middens becomes abundant about 13,000 years ago, but much of the tree pollen in these cores may been introduced by streams coming from mountains to the east and south rather than having been derived from the west. But, as elsewhere, these low elevation cores suggest the presence of sagebrush steppe in valley bottoms.

Much the same is true elsewhere in the floristic Great Basin. Peter Wigand and Cheryl Nowak, for instance, have analyzed both the plant macrofossils and pollen from packrat middens samples taken from the southern end of the Virginia Range, just west of Pyramid Lake in the Lahontan Basin at elevations between 4,530 feet and 5,020 feet. All samples falling between about 25,000 and 11,000 years ago contained Utah juniper and nearly all contained both sagebrush and *Atriplex*, plants that remain in the area today. The only subalpine conifer that shows up in these middens is whitebark pine, at about 23,500 years ago and between 11,000 and 12,000 years ago. Their analysis of the pollen from these samples provided very small amounts of pine pollen during the Pleistocene, alongside larger amounts of sagebrush, and more of juniper, pollen. Subalpine conifers were surely not common here at these times. To the west, Bob Thompson has shown that sagebrush, presumably big sagebrush, grew alongside western juniper and desert peach (*Prunus andersonii*) in the Winnemucca Lake Basin between 12,000 and 11,000 years ago. Neither big sage nor desert peach grow this low here today, and western juniper no longer occurs in the basin. Instead, the dominant plants surrounding Thompson's midden sites include shadscale and bud sage.

These studies, and others like them, show that many Great Basin valley bottoms supported extensive sagebrush steppe vegetation during the late Pleistocene. Shadscale was certainly present as well, especially on the saline substrates left behind by shrinking Pleistocene lakes. Wigand and Nowak, for instance, analyzed very late Pleistocene packrat middens from the very arid Smoke Creek desert. These middens provided Utah juniper and shadscale, but no sagebrush, between 12,000 and 10,000 years ago. Given that wherever pollen evidence is available from such contexts, it shows the presence of vegetation of this sort, it appears that most valley bottoms in the floristic Great Basin were characterized by such shrubs as sagebrush and, in places, shadscale, with conifers apparently present only on the rocky outcrops that dot the valley floors.

Because these rocky outcrops happen to be the place where woodrats make their middens, those middens themselves provide a biased sample of the plants in the valley bottom as a whole. Thompson, for instance, analyzed a 12,200-year-old midden from a limestone outcrop, at 5,400 feet in elevation, in the Snake Valley, just east of the Snake Range. That midden contained limber pine, but the other plants in this midden included sagebrush, shadscale, and winterfat. He suggests, and everything we know confirms, that the limber pine here occupied only the limestone outcrop, not the finer-textured valley-bottom sediments that surrounded it.

As is probably clear, the Mojave Desert was not a "refuge" for such cold desert shrubs as sagebrush and saltbushes during the late Pleistocene. All indications are that at least sagebrush, and perhaps a diverse variety of other cold desert shrubs, had a far wider distribution during late glacial times than they do today. The abundance of sagebrush in the late Pleistocene Mojave Desert simply represents a massive increase in their range, not a place to which they were forced to retreat during glacial times. Indeed, the deep Bonneville Basin cores that I mentioned above document that sagebrush has a history at least in this area that goes back millions of years, as do cheno-ams.

In the far southeastern Great Basin, plant remains from a series of packrat middens in Meadow Valley Wash, near Caliente, Nevada, have been analyzed by David Madsen, Geof Spaulding, Tom Van Devender, and Philip Wells. These middens, ranging from 4,400 feet to 5,000 feet in elevation, contained Douglas-fir, Rocky Mountain juniper, and limber pine between 20,000 and 12,600 years ago. It is not known when those plants retreated from this area, since the next midden in the series, which does not contain them, is only 6,600 years old. Today, the only nearby conifers are singleleaf pinyon and Utah juniper, which occur at low elevations—down to about 4,400 feet—in this relatively cool and moist valley. Douglas-fir is found no closer than about sixty miles to the east.

Although our knowledge of the late Pleistocene vegetation of the Great Basin is spotty—sometimes, as Wigand and Rhode (2002:345) put it in a slightly different context, "more gap than record"—the results of all this work make it extremely unlikely that any subalpine conifers were continuously distributed across significant parts of the Great Basin during the late Pleistocene. The distributions of such subalpine conifers as bristlecone and limber pines extended to extremely low elevations in the floristic Great Basin during this time, down to the bases of mountain ranges and occupying appropriate bedrock outcrops in the valley bottoms. In many areas, subalpine and montane conifers have become locally extinct since the end of the Pleistocene, as the glacial-age Douglas-fir in Meadow Valley Wash illustrates. Nonetheless, the finer-grained sediments of those valley bottoms supported sagebrush steppe, along with plants that today occur in the Shadscale Zone, including shadscale itself. The late Pleistocene populations of bristlecone and limber pines in the mountains of the floristic Great Basin may have been far closer to one another than they are today, but in many areas, substantial amounts of desert shrub vegetation intervened, occupying the lowlands between the mountains themselves. This vegetation was often, probably usually, sagebrush steppe and the plants that comprised this sagebrush steppe extended well upward into the mountains, where they formed the understory in the subalpine conifer woodlands and forests.

During the late Pleistocene, valley bottoms today dominated by the vegetation that led Billings to define a Shadscale Zone were often dominated instead by sagebrush steppe. In fact, one has to look far and wide to see anything that looks like Billings's Shadscale Zone during the late Pleistocene although, as with big sagebrush, no late Pleistocene "refuge" is needed to explain the presence of shadscale in the Great Basin during the Holocene: it was here all along. This is not the case for pinyon-juniper in the floristic Great Basin, however. Although Utah juniper has a lengthy history in parts of the Great Basin, there was no pinyon here during the Pleistocene and so no Pinyon-Juniper Zone either.

Great Basin Conifers in Deeper Historical Perspective

As I discuss in chapter 2, the montane conifers of the Great Basin have closer affinities to those of the Rocky Mountains than to those of the Sierra Nevada. Since this is the case, some scientists suggest that the Great Basin must have received most of its montane conifers from the Rockies or from mountains to south, and not from the Sierra Nevada. In this view, the wooded mountain masses of the Great Basin, surrounded by vast expanses of desert shrub vegetation, are habitat islands that had to receive their high-elevation conifers from elsewhere. If true, the mountains of the Great Basin become targets for dispersing montane conifers, and the trick becomes understanding how and when those plants got there.

There can be no question that some conifers did disperse into the Great Basin during or after full-glacial times. Singleleaf pinyon provides an excellent example: this pine was not in the floristic Great Basin during the Pleistocene but it is decidedly here now. The same appears to be true for white fir. Although found fairly routinely in relatively low elevation, Pleistocene-aged middens in the Mojave Desert, it has yet to be found in middens of that age to the north. It is, for instance, on the Snake Range today but not in any Snake Range midden sample that reaches back to the Pleistocene. It is now in the Goshute, Pilot, and Deep Creek ranges, but not in any Pleistocene midden from in or near those places. The same is true for midden samples of this age from the western edge of the Lahontan Basin: the tree is there but Pleistocene-aged middens from the area do not contain it. Douglas-fir provides a third example, since it is not known from the floristic Great Basin until very late in the Pleistocene.

But even though some conifers now found in the floristic Great Basin are fairly recent—latest Pleistocene or Holocene—immigrants, we do not necessarily have to look to the Rockies or the Southwest as a source for most high-elevation Great Basin conifers. Instead, it is likely that many of these conifers, or their direct ancestors, were in the Great Basin throughout the Quaternary, having evolved in the general area where they are now found. If this is correct, significant aspects of the distribution of montane conifers in the Great Basin are explained not by Pleistocene immigration into the area but by long-term geologic and evolutionary processes, including the fragmentation of the distributional ranges of tree species and the extinction of species that could not adapt to new conditions.

Indeed, as Constance Millar has argued, some conifers not found in the Great Basin today may well have originated in this area and moved outward from there. One example is provided by foxtail pine (*Pinus balfouriana*), a tree so closely related to Great Basin bristlecone pine that the two can interbreed, though they do not do so naturally because they are not found in the same places. Millar observes that foxtail pines were found in the Miocene-aged floras of Nevada, as early as 23 million years ago. This was a time when the Great Basin did not yet exist and what was to become this region was marked by far wetter and cooler summers. As summers warmed and precipitation decreased, foxtail pine seems to have been forced west and into higher elevations. Today, this species is found only in the southern Sierra Nevada and in the mountains of Northern California.

Some intriguing genetic evidence suggests that bristlecone pine may have been in the Great Basin throughout the Pleistocene. Ronald Hiebert and J. L. Hamrick analyzed genetic variation in five isolated populations of bristlecones in the eastern Great Basin (from the Egan, Snake, and White Pine ranges) and from the adjacent Colorado Plateau. They discovered that levels of genetic variability within these populations are greater than they are between them. They argue that had Great Basin mountains become stocked with bristlecones as a result of the long distance dispersal of seeds by birds, the genetic pattern should have been exactly the opposite: relatively low variation within populations (a function of the small sizes of the initial populations), coupled with high genetic variation between them (a function of isolation). On the other hand, if bristlecones have a long residence time within the Great Basin, and were able to share genetic material during glacial episodes when the trees were widespread, then high genetic variation within populations, and low variation among populations, is to be expected. Bristlecones, they conclude, may have been in the Great Basin throughout the Pleistocene. Similarly, analyses of genetic variation within populations of limber pine also suggest lengthy residence times for this species within the Great Basin even if, at times, some of those populations dwindled dramatically, sometimes to the point of local extinction. While genetics cannot tell us how long any tree species has been in the Great Basin, these conclusions are consistent with the suggestion that many montane conifers have their evolutionary history in the Great Basin itself.

Biogeographers, those who describe and explain the distribution of life on earth, have long been fascinated by what are known as disjunct distributions, in which populations of a species are not in contact with other populations of the same species. The northern and southern populations of foxtail pines provide a classic example, as do pikas on isolated Great Basin mountain ranges. Every set of disjunct populations is telling us something about that organism's history; the trick is to figure out what that is.

I have more to say about disjunct distributions in the following chapter, but in general, biogeographers distinguish between what are known as vicariance explanations, in which those distributions are explained by the appearance of an uncrossable barrier within the distribution of a once-continuous set of populations, and dispersal explanations, in which they are explained by the movement of members of one population to a far-flung new location. It now appears that both of these processes were involved in producing the modern distribution of montane conifers in the Great Basin. Some, certainly including singleleaf pinyon and probably both white fir and Douglas-fir, arrived by dispersal and only relatively recently. Others, details to be worked out, have been here all along, with their disjunctions caused by the hazards of being alive in such a geologically active area.

Chapter Notes

Manly (1894) provides his own story in a Desert West classic. L. Johnson and Johnson (1987) provide a detailed discussion of the route taken by Manly and his companions in the California deserts; see also Southworth (1987). Lingenfelter (1986) provides a fascinating discussion of the history of Death Valley as a whole.

Archaeologist Don Fowler has suggested to me that the substance eaten by the Manly party may have been just what Manly thought it was—a food cache belonging to Indians. In particular, he thinks that that the "balls of a glistening substance" might have been the sweet material exuded from the leaves and stems of *Phragmites australis*, the common reed or cane (other plants produce the substance as well). Created when aphids attack the plant, this secretion, known as honeydew or bug sugar, was widely collected and eaten in the Great Basin. It tastes, according to Mark Harrington (1945:95), "something like malted milk," although this has to be tempered by the fact that it may also be full of the bodies of the insects that were associated with it. Ethnographic descriptions suggest that honeydew was stored in textile bags (C. S. Fowler 1989, 1992) or in baskets (Steward 1933, Harrington 1945; see C. S. Fowler 1990b:91, for a picture of one of these baskets). There appear to be no reports of this material being stored in rock crevices, where it would likely have been quickly consumed by rodents. As a result, it does seem more likely that the Manlys did in fact dine on packrat midden. Nonetheless, Don Fowler's point is well taken: it is not impossible that it was honeydew. Given that they reported the substance to be "sweet but sickish," the obvious test would be to take a taste of packrat midden, but I have not been able to bring myself to do this. Excellent discussions of the use of honeydew in the Great Basin are found in Heizer (1945), V. H. Jones (1945), Sutton (1988), C. S. Fowler (1992), and Rhode (2002).

See E. R. Hall (1981) for the distribution of North American woodrats; standardized common names for all of the world's mammals are in Wilson and Cole (2000). Rhode (2001) provides an excellent introduction to packrat middens and their analysis. Betancourt, Van Devender, and Martin (1990b) is required reading for anyone interested in this topic; in that volume, Betancourt, Van Devender, and Martin (1990a) discuss the history of packrat midden studies; T. A. Vaughan (1990) and Finley (1990) discuss woodrat ecology; and Spaulding et al. (1990) discuss the composition and analysis of packrat middens. The seminal paper by Wells and Jorgensen (1964) remains fascinating reading; the synthesis by P. V. Wells (1976) is dated but helps document the development of packrat midden studies. On woodrats in the White Mountains, see Grayson and Livingston (1989) and Carey and Wehausen (1991).

Hansen's earlier work in the Great Basin is readily accessed through H. P. Hansen (1947). For an introduction to pollen analysis, see P. D. Moore, Webb, and Collinson (1991); O. K. Davis (2001) provides a brief and very readable introduction as well. Maher (1969) discusses *Ephedra* pollen in the Great Lakes region.

My discussion of the vegetation history of the northwestern Bonneville Basin is drawn from Rhode and Madsen (1995), Rhode (2000b), and Louderback and Rhode (2009). My discussion of the vegetation of Ruby Valley and the Ruby Mountains is based on Critchfield and Allenbaugh (1969), Loope (1969), R. F. Miller et al. (1982), Thompson (1984, 1992), Charlet (1991, 1996), and my own fieldwork. The Ruby Lakes cores are discussed in detail in R. S. Thompson (1984, 1992) and R. S. Thompson et al. (1990). The depth of Lake Franklin provided in the text is from Mifflin and Wheat (1979); this value was used by Lillquist (1994) as well. See Delorme and Zoltai (1984) and Forester (1987) for discussions of ostracodes and their use in paleoenvironmental analysis.

G. Q. King (2003) provides an excellent introduction to the Forty Mile Desert; the emigrant trails through this area are plotted in Franzwa (1999). The plants of the Carson Desert region are discussed by Billings (1945). The history of water projects in this part of Nevada as a whole is presented by Townley (1977); see also Reisner (1993). A succinct discussion of the vegetation of the Carson Desert area, and of historic modifications to that vegetation, is provided by R. L. Kelly and Hattori (1985). As elsewhere, my discussion of the trees and shrubs of the Great Basin has benefitted from Lanner (1983b) and Mozingo (1987).

The history and archaeology of Hidden Cave is discussed in detail by Thomas (1985). You don't need a library to find this publication, or any other American Museum of Natural History publication that I cite in this book, since they are available free at http://digitallibrary.amnh.org/dspace/. Subsequent discussions of Hidden Cave are provided by Rhode, Adams, and Elston (2000) and Rhode (2003). D. D. Fowler (1973) provides a brief biographical sketch of S. M. Wheeler. The geology of the deposits of Hidden Cave was analyzed by J. O. Davis (1985b). Wigand and Mehringer (1985) provide a detailed analysis of the pollen and seeds from Hidden Cave; my discussion of the late Pleistocene vegetation of the Carson Desert depends fully on that paper. My description of the vegetation in the Hidden Cave area is based on my own fieldwork; the precipitation and evaporation figures are from Roger Morrison's classic monograph on the southern Carson Desert (R. B. Morrison 1964); see also R. L. Kelly and Hattori (1985) and, for an important discussion of historic fluctuations in water availability in the Carson Sink, R. L. Kelly (2001). Peter Wigand and his colleagues have discussed the western Lahontan Basin packrat midden record in a number of places; I have relied most heavily on Wigand and Nowak (1992), Nowak et al. (1994a, 1994b), and Wigand and Rhode (2002).

Hidden Cave is on land under the stewardship of the Bureau of Land Management. The site is easy to visit. Follow State

Highway 50 south from Fallon; after about ten miles of skirting the northern edge of the Lahontan Mountains, you will see an excellent dirt road, marked by a Hidden Cave sign, on the left (east) side of Highway 50. Turning left here, you will pass a broad opening in Eetza Mountain to the south (note the Lake Lahontan terraces here) and will then come to a second, far smaller opening. Hidden Cave is on the east side of this opening; the trail will take you there.

The Bureau of Land Management has closed the cave by means of a substantial gate across its mouth. For those who wish to do more than simply see where the site is, the BLM and the Churchill County Museum provide free tours of the site. These tours leave from the museum; participants are required to provide their own transportation. Tours by twelve or more people may also be arranged through the museum (www.ccmuseum.org). The museum is located at 1050 South Maine Street in Fallon. Whether or not you opt for the Hidden Cave tour, this facility is well worth the visit, with excellent displays on regional prehistory (Hidden Cave), Native Americans, and local history.

For a detailed discussion of the factors influencing the vegetation immediately surrounding Owens Lake, see Dahlgren, Richards, and Yu (1997). My discussion of Owens Valley plant history is drawn from S. A. Jennings and Elliot-Fisk (1993), Koehler and Anderson (1994, 1995), S. A. Jennings (1996), Litwin et al. (1997), Woolfenden (1996a, 2003), and Mensing (2001). The 11,450-year date for pinyon pine in the Alabama Hills appears in Wigand (2003), but it was Wigand himself who told me that it is associated with pinyon pine. Woolfenden (1996b) contains a superb concise summary of Pleistocene and Holocene vegetation history in the Sierra Nevada region, including Owens Valley. The records for giant sequoia that I discuss are from O. K. Davis (1999b) and Woolfenden (2003). O. K. Davis (1999a) discusses the fact that giant sequoias were, in general, more widely distributed during the Pleistocene. R. S. Anderson (1990b) presents his analysis of the dispersal abilities (or lack thereof) of giant sequoia pollen.

If you are in Owens Valley and have an interest in film, don't miss the Beverly and Jim Rogers Museum of Film History (www.lonepinefilmhistorymuseum.org/museum.htm). The museum celebrates the very rich film history of the Lone Pine area, including the Alabama Hills, and is very much worth the visit. The museum website provides a detailed account of the exhibits, but cannot match the friendliness and interest of the well-designed museum itself (though I was surprised that the staff classify *Tremors* as science fiction). For a list of the films made in this area, see Holland (2005) or search the Internet Movie Database (http://www.imdb.com/). The former provides a far more complete listing of the films made here, but the latter provides more information on locally filmed television shows. Clint Eastwood's classic *High Plains Drifter*, by the way, is also an Owens Valley film, but the town of Lago, on which the action centers, was on the south side of Mono Lake.

My presentation of the vegetation of the lower elevations of Death Valley depends heavily on C. B. Hunt (1966); see also his more popular book on the geology, ecology, and prehistory of Death Valley (C. B. Hunt, 1975). Information on precipitation and evaporation in Death Valley is from C. B. Hunt et al. (1966); see also C. B. Hunt and Mabey (1966). My discussion of the plants of the Panamint, Black, and Funeral ranges is based on Vasek (1966), D. R. Schramm (1982), Barbour (1988), and Vasek and Thorne (1988); see also R. H. Webb, Steiger, and Turner (1987). The information on the Grapevine Mountains is drawn from Kurzius (1981).

The late Pleistocene plants of Death Valley are discussed by P. V. Wells and Woodcock (1985) and Woodcock (1986); see also P. V. Wells and Berger (1967) and Van Devender (1977). Bader (2000) discusses Death Valley vegetation between 166,000 and 111,000 years ago based on pollen from a deep core drilled by the U.S. Geological Survey; unfortunately, the late Pleistocene pollen record from this core has not been reported. The record for Joshua trees at 4,000 feet in the Death Valley area is from Grapevine Canyon (Kurzius 1981); the record for Utah juniper at 4,700 feet in the northern Panamint Range is provided by Peterson (1984). Spaulding (1990b) and Koehler, Anderson, and Spaulding (2005) place aspects of the Death Valley data in the broader context of the late Pleistocene vegetational history of the Mojave Desert as a whole. That different species of woodrats have different diets, and that these differences may introduce bias into our midden-derived understanding of ancient vegetation, is discussed by Finley (1990) and Dial and Czaplewski (1990).

On the Eureka Valley sand dunes, see Bagley (1988), Norris (1988), and G. Smith (1988); on visiting Eureka Valley, see Digonnet (2007). Spaulding's work on the vegetational history of Eureka Valley and the eastern Amargosa desert can be found in Spaulding (1980, 1983, 1985, 1990b, 1990c) and in Spaulding, Leopold, and Van Devender (1983).

Koehler, Anderson, and Spaulding (2005) discuss the packrat midden–based vegetational sequence of the central Mojave Desert. R. S. Thompson (1985), O. K. Davis and Anderson (1987), Van Devender (1988), O'Rourke (1991), and Spaulding (1990a) discuss the relationships between the macrofossils and pollen found in packrat middens.

My general discussion of the late Pleistocene vegetation of the Mojave Desert has drawn heavily on Mehringer (1965, 1967); Mehringer and Ferguson (1969); Mehringer and Warren (1976); Spaulding (1977, 1981, 1985, 1990b, 1990c, 1994, 1995); Van Devender, Thompson, and Betancourt (1987); Wigand (1990b); Wigand et al. (1995); Forester et al. (1999); Wigand and Rhode (2002); and Koehler, Anderson, and Spaulding (2005). Scott Anderson kindly provided me with data on sagebrush macrofossils in the central Mojave Desert middens. Spaulding (1994) discusses the Fortymile Canyon middens that fall at about 36°57' N. Although not likely to be in your library, this reference is currently available through the DOE Office of Scientific and Technical Information Website (www.osti.gov/bridge/product.biblio.jsp?osti_id=201749), or just type the digital object identifier (DOI) 10.2172/201749 into Google. The Owl Canyon middens were originally analyzed by Mehringer and Warren (1976); in my discussion, I have used the dates for their samples provided by Wigand (1990b). Van Devender (1977) and Van Devender and Spaulding (1979) argued that the Mojave and adjacent Sonoran deserts did not become deserts until the Holocene; they later helped demonstrate that this view was incorrect. That is how good science works.

The climate estimates I discuss here are provided by Spaulding (1985), Woodcock (1986), Forester et al. (1999), and Thompson, Anderson, and Bartlein (1999); see also the references in table 5-8.

Nicolson (1990) provides an important discussion of Henry Gleason's career and of the Clements-Gleason debate. While Clements may have lost this particular debate, he made enormous contributions to our understanding of plant adaptations. My discussion on the differential history of oak and hickory in the East is derived from T. Webb, Hemmings, and Muniz (1998). On cattle egrets, see Palmer (1962), Godfrey (1966), Hayward et al. (1976), Blake (1977), and Ryser (1985).

R. S. Thompson (1978, 1984, 1990) and P. V. Wells (1983) discuss the late Pleistocene midden data from the Deep Creek, Snake, Confusion, and Wah Wah ranges; references to the northwestern Bonneville Basin were provided above. Sack (1990) outlines the Quaternary history of Tule Valley. The Twin Peaks midden is discussed by Rhode and Madsen (1995) and Rhode (2000b); the Granite Canyon midden, by R. S. Thompson (1984). The general late Pleistocene distribution of conifers and desert shrub vegetation is discussed in R. S. Thompson and Mead (1982), R. S. Thompson (1984, 1990), and P. V. Wells (1983). All of this work was of such high quality that it is as valuable now as when first published.

R. C. Bright (1966) presents the analysis of the Swan Lake sediments; the Grays Lake analysis is in Beiswenger (1991). Analyses of Pleistocene, and earlier, pollen from Lake Bonneville sediments are presented by Madsen (in Spencer et al. (1984), R. S. Thompson et al. (1995), and O. K. Davis (2002). The last of these references includes an excellent synthesis of most of this work and is the best place to start. R. S. Thompson, Benson, and Hattori (1986) discuss the Winnemucca Lake middens. The Virginia Range middens are presented in Wigand and Nowak (1992) and Nowak et al. (1994a, 1994b). Wigand and Nowak (1992) and Nowak et al. (1994b) discuss the Smoke Creek Desert work. The Meadow Valley Wash middens are presented in Madsen (1972, 1976), Spaulding and Van Devender (1980), and P. V. Wells (1983).

One of the best places to learn about the Tertiary history of western Great Basin and Sierra Nevada floras is the synthesis provided by Millar (1996); equally valuable is her discussion of the global history of pines (Millar 1998). Minnich (2007) provides a broad overview of the deeper history of the vegetation of California. For a map showing the distribution of foxtail pines, take a look at Ron Lanner's excellent *Conifers of California* (1999) or go to the Gymnosperm Database (http://www.conifers.org/) and enter *Pinus balfouriana*. Hiebert and Hamrick (1983) analyze genetic variability within five populations of eastern Great Basin and western Colorado Plateau bristlecones; Mitton, Kreiser, and Latta (2000) and Jørgensen, Hamrick, and Wells (2002) discuss the genetics of limber pine. Vicariance and dispersal explanations of disjunct distributions are discussed in lucid detail in Lomolino, Riddle, and Brown (2006).

CHAPTER SEVEN

Late Pleistocene Vertebrates of the Great Basin

As discussed in chapter 4, the extinction of an amazingly diverse variety of North American mammals toward the end of the Pleistocene has provided paleontology, and archaeology, with one of its most hotly debated mysteries. The Great Basin, however, has been largely peripheral to this debate. This is not because the Great Basin is lacking in late Pleistocene vertebrate faunas. Many fossil-bearing sites of this age have been reported from the region, and hundreds of other Great Basin localities have provided isolated bones and teeth of extinct Pleistocene mammals.

To some extent, the Great Basin's low profile in the extinction debate follows from the fact that so few paleontologists have focused their research on the late Pleistocene mammals of this area. That doesn't account for all of it, however, since some superb work has been done in the Great Basin and at least one late Pleistocene site from this region—Fossil Lake, in south-central Oregon—is internationally known.

Some very important Great Basin sites were excavated early in the twentieth century, at a time when the need for careful, fine-grained stratigraphic excavations had yet to be recognized, including Gypsum Cave in southern Nevada's Frenchman Mountains and Smith Creek Cave in eastern Nevada's Snake Range (see figure 7-1 for locations). Others—Astor Pass, northwest of Pyramid Lake, for instance—were described on the basis of fossil material for which no precise locational information at all was obtained. All of this work is inadequate by today's standards. However, problems of this sort exist wherever paleontology has a long history. Indeed, there are probably fewer problems of this sort in the Great Basin than elsewhere, simply because fewer people have been interested in the late Pleistocene paleontology of this region.

More to the point, even paleontological sites that have been collected recently, and with exquisite care, have often been impossible to date with any real accuracy. This is because most Great Basin sites known to date to the late Pleistocene come either from the surface or from caves whose deposits cannot be deciphered stratigraphically.

Fossil Lake is an excellent example of the former, largely surficial variety. This site, located in the eastern part of the basin of Pleistocene Fort Rock Lake in south-central Oregon, was brought to scientific attention in 1876. Since that time, dozens of publications on the Pleistocene fishes, birds, and mammals collected from the site have appeared; geologist Ira Allison summarized many of them in an important monograph that appeared in 1966. No one has counted the numbers of fossils that have been scientifically collected and described from this site, but there are thousands. They include the remains of such extinct mammals as Harlan's ground sloth, giant short-faced bear, horse, flat-headed peccary, large-headed llama, and mammoth. (Chapter 4 describes extinct North American Pleistocene mammals; see table 4-1 for the scientific names of those mammals.) The site also yielded huge samples of well-described bird specimens. Indeed, Fossil Lake was the first paleontological site in North America to have provided significant numbers of Pleistocene bird fossils.

Even though Fossil Lake is a very important site, and even though it has been studied for more than a century, we are still struggling to learn how old the material from this site is. Most of the specimens are from deflation basins, the fossils having been revealed as a result of wind erosion of the Pleistocene sediments in which they were once embedded. Until very recently, paleontologists assumed that the fossils from here were very late Pleistocene in age, an assumption that seemed to be supported by a radiocarbon date of 29,000 years obtained by Allison from snail shells associated with bird bones that he excavated.

Now, however, it is turning out that things are far more complicated than this. Work by paleontologist Jim Martin, with the support of the Bureau of Land Management, has begun to clarify the chronological situation here. Using

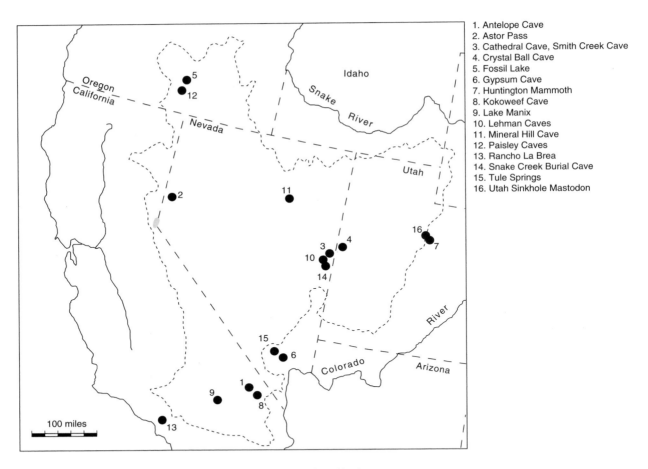

FIGURE 7-1 Locations of sites with extinct and extirpated mammals and birds.

a variety of techniques, including the identification of tephras in the Fossil Lake deposits (see chapter 5), he has shown that the oldest deposits at this site were laid down more than 650,000 years ago, that the most recent ones are Holocene in age, and that the fossils are spread across this entire time span. He has also shown that the snail dated by Allison came from a unit about 600,000 years old, a difference that probably reflects the movement of more recent snail shells downward by burrowing rodents. Martin has begun the process of tying together the remains of particular animals with the sediments from which they came, but this is a huge job, and it will likely be many years before we have anything approaching a detailed understanding of the ages of the Fossil Lake fossils.

The cave sites are a different matter. Many caves were excavated so early that detailed stratigraphic information is not available, while others were excavated much more recently but so poorly that the same problems exist. However, some Great Basin caves with late Pleistocene deposits simply don't provide the depositional context needed to provide detailed answers to chronological questions.

Crystal Ball Cave in western Utah's Snake Valley is an excellent example. Located in Gandy Mountain about three miles west of the small town of Gandy, Utah, Crystal Ball Cave is some 500 feet long, with an area of about 20,000 square feet. When the cave was first discovered in 1956, bones of extinct horses and camels were lying on the surface. These were collected by Herbert Gerisch and Robert Patterson and donated to the Natural History Museum of Los Angeles County, thus making the paleontological potential of the site known to science. Excavations were conducted at the site by Brigham Young University paleontologists Wade Miller and Tim Heaton in the late 1970s and early 1980s. An excellent discussion of the results of this work was published by Heaton, who reported that the deposits of the site contained the remains not only of extinct horse and camel but also of large-headed llama, sabertooth, and a new species of short-faced skunk. Literally thousands of bones and teeth came from this site, due to the careful work done here by Miller and Heaton, initially made possible by the insightful generosity of Gerisch and Patterson.

Exactly how old is all this material? Unfortunately, in spite of exacting excavations, we don't really know. As rich as they were in bone (some 10 percent by volume), the sediments of Crystal Ball Cave were only about eighteen inches deep and were completely unstratified. As Heaton

(1985:346) observed, "The cave seems to have been accumulating fossils continuously from some date in the past, when an entrance was formed, to the present." Four radiocarbon dates were obtained, and these range from more than 23,000 years ago to 12,980 years ago. These dates, however, were on the mineral (apatite) fraction of the bone, which often provides invalid ages. The extinct mammals themselves would seem to require a late Pleistocene age for the site, which is consistent with the radiocarbon dates. Beyond this, however, there is no way of knowing precisely how old these specimens are. Given current technology, the only way their ages could be discovered would be to radiocarbon date each specimen.

An equally frustrating example comes from Mineral Hill Cave. This remarkable site is located in central Nevada's Sulphur Spring Range, not far from the town of Carlin. Known to scientists for well more than a century, the cave became known to archaeologists and paleontologists as a result of excavations conducted there by archaeologist Kelly McGuire. Most important from our perspective, McGuire reported that he had discovered the remains of a series of Pleistocene mammals, including, horse, camel, llama (*Hemiauchenia*), and shrub-ox. This report led paleozoologist Bryan Hockett of the Bureau of Land Management to launch more expansive excavations here. The results, discussed in detail in one of the best monographs on the Pleistocene paleontology of the Great Basin ever to have appeared, were remarkable. Hockett, and the team he assembled to assist in the analysis of the bones and teeth that his painstaking excavations had recovered, were able to identify sixty-four genera, and seventy-four species, of mammals, reptiles, birds, and fish. Of those, seven species—six mammal and one bird—are extinct forms known only from the Pleistocene (table 7-1).

Exactly how old is all of this material? That's where things once again get frustrating, as they so often have in Great Basin Pleistocene paleontology. With extremely rare exceptions, the deposits of Mineral Hill Cave were not stratified. In fact, materials dating to very recent times were often found next to the bones of extinct Pleistocene mammals, with charcoal dating to the past 2,000 years mixed with bones that dated to more than 40,000 years ago. Just as bad, some of the radiocarbon dates that Hockett received were in reverse stratigraphic order, meaning that older material was lying higher in the deposits than younger material. Nice caves simply do not behave this way. As Hockett noted, this cave seems to have gone bad because both water and rodents had disturbed the deposits.

TABLE 7-1
The Extinct Pleistocene Vertebrates of Mineral Hill Cave

Species	*Common Name*	*Age (years)*
Camelops hesternus	Yesterday's camel	$48,900 \pm 3,100$
		$46,550 \pm 1,100$
		$44,600 \pm 3,000$
Hemiauchenia macrocephala	Large-headed llama	$50,190 \pm 1,940$
		$39,230 \pm 1,330$
		$36,320 \pm 320$
Navahoceros fricki	Mountain deer	$49,800 \pm 1,700$
		$37,750 \pm 440$
Equus cf. *occidentalis*	Western horse	$45,700 \pm 1,000$
		$42,420 \pm 820$
Equus cf. *conversidens*	Mexican ass	>46,400
		$35,080 \pm 280$
cf. *Miracinonyx trumani*	American cheetah	>52,200
Brachyprotoma obtusata	Short-faced skunk	—
cf. *Buteogallus fragilis*	Fragile eagle	—

SOURCE: B. S. Hockett and Dillingham 2004.
NOTE: The abbreviation "cf." stands for the word "confer," derived from the Latin verb *conferre*, "to bring together" (as in "conference"), and meaning, in this context, "compare" or "compares with". "*Equus* cf. *occidentalis*" means that the paleontological material assigned to this category certainly comes from the genus *Equus* and compares most convincingly with the species *Equus occidentalis*, but may, in fact, come from a different species. "cf. *Miracinonyx trumani*" means that the material involved compares most closely with American cheetah but might represent a different genus—in this case, the cougar, *Felis concolor* (but B. S. Hockett and Dillingham [2004] make the case that the single foot bone is extremely likely to represent American cheetah).

As a result of all this, the only way that Hockett and his colleagues could derive anything resembling a chronology for the site was by dating one bone at a time. They ended up getting fifty-five dates in this way (table 7-1). That is a lot of dates, all made possible by the very generous support of the Bureau of Land Management. These dates showed that the bones in Mineral Hill Cave had accumulated across at least the past 50,000 years or so, but, because the site was not stratified, the results still could not be used to build a picture of faunal change through time.

Other cases of this sort abound. I mention one of them in chapter 3, the Paisley Caves. There, I discuss the fact that these sites contain human coprolites securely dated to as old as 12,400 years ago, but I did not mention that the deepest layers of these sites contain materials of very different ages. A coprolite dating to 4,100 years ago, for instance, was found just above the bones of extinct camel and horse. Rodents appear to be the prime culprit, but whatever the cause, the only way to know the ages of the remains of the extinct mammals here is to date each of them separately. At the moment, the bone of an extinct horse from this site has been dated to 11,130 years ago, while that from an extinct camel or llama clocked in at 12,300 years ago. When these dates are combined with those for the human coprolites, they show that people and some of the extinct mammals were here at roughly the same time. But, just as at Mineral Hill Cave, the stratigraphy of the site itself cannot be counted on to provide us with a trustworthy faunal sequence. Site after site in the Great Basin has confronted us with the same frustrating situation.

The Extinct Late Pleistocene Mammals of the Great Basin

In short, while the Great Basin has provided a good number of late Pleistocene paleontological sites, most of these lack sound chronological control. All we can really say is that they are late Pleistocene. However, even though the chronology of these sites is not well known, the sites that have been studied document that the Great Basin supported a diverse variety of now-extinct mammals during this period. In fact, of the thirty-six genera of mammals known to have become extinct in North America toward the end of the Pleistocene, twenty have been reported from the Great Basin. The list is substantial: it includes three of the four extinct ground sloths, five of the seven extinct carnivore genera, the sole lagomorph, one of the two extinct perissodactyls, eight of the thirteen extinct artiodactyls, and two of the three extinct proboscideans.

These are listed in table 7-2, along with three extinct species that belong to genera that still exist in North America. Given that twenty of thirty-six extinct North American Pleistocene genera are known from the Great Basin, it is hard not to wonder which of the remaining sixteen may have lived here during the late Pleistocene but have simply not yet been found. My guess, though, is that the list of extinct late Pleistocene Great Basin mammals is nearly complete. Of the sixteen missing taxa, only one—the dhole—might reasonably be expected to be found here in the future. As I mention in chapter 4, dholes are now widely distributed in southern and eastern Asia. During the later Pleistocene, their distribution extended deep into southwestern Europe and into Alaska and the Yukon Territory. They are also known from San Josecito Cave in northern Mexico. Since these canids were in both eastern Beringia and northeastern Mexico, they must also have been found between these two places, and the Great Basin may well have been within their range.

Nearly all other North American extinct Pleistocene mammals that are unknown from the Great Basin had distributions far to the north (for instance, the saiga) or south and east of this region (for instance, the capybaras, pampathere, glyptodont, and flat-headed peccary). Late Pleistocene tapirs come close—Arizona and southeastern California—but there is no suggestion yet that they came close enough to have entered the Great Basin.

As mentioned, table 7-2 also lists three extinct species of mammals belonging to genera that still exist in North America. These are the dire wolf, which belongs to the same genus as both the gray wolf (*Canis lupus*) and coyote (*Canis latrans*); the American lion, belonging to the same genus as the jaguar, *Panthera onca*, which ranges into the far southern United States (and is known from the late Pleistocene deposits of eastern Nevada's Smith Creek Cave); and Harrington's mountain goat, a close relative of the living mountain goat, *Oreamnos americanus*, which is thriving in the Pacific Northwest and northern Rocky Mountains (see chapter 4).

Both the dire wolf and the American lion seem to have been widespread, though apparently not common, in the Great Basin. The first of these is known from the Mojave Desert in the south to Fossil Lake, Oregon in the north; the American lion, from sites scattered across much of Nevada. The very first recognition of Harrington's mountain goat came from a Great Basin site, Smith Creek Cave, on the eastern edge of Nevada's Snake Range. Thanks in a significant degree to the efforts of paleontologist Jim Mead, we know that it occurred from the central Great Basin into northern Mexico and that it survived, in at least some areas, until shortly after 11,000 years ago.

As I have discussed (see chapter 4), sixteen of the thirty-six extinct late Pleistocene North American mammal genera have secure radiocarbon dates that fall between 12,000 and 10,000 years ago. This, of course, is the period of time that falls immediately after the earliest secure evidence for people in the Americas (Monte Verde, at 12,500 years ago, and the Paisley Caves, at about 12,350 years ago), and that includes the well-dated archaeological phenomenon known as Clovis (11,200–10,800 years ago).

Where does our knowledge of the chronology of the latest extinctions stand in the Great Basin? Of the twenty extinct genera known from this area, seven have radiocarbon dates from Great Basin sites that place them within the same latest

TABLE 7-2
The Extinct Late Pleistocene Mammals of North America and the Great Basin, Including Three Extinct Species of Existing North American Genera (*Canis dirus*, *Panthera leo*, and *Oreamnos harringtoni*)
Forms known from the Great Basin are in bold. See table 4-1 for the orders and common names of the extinct genera.

Family/Genus	Illustrative Great Basin Site	References
Pampatheriidae		
Pampatherium		
Holmesina		
Glyptodontidae		
Glyptotherium		
Megalonychidae		
Megalonyx	Orem, UT	Gillette, McDonald, and Hayden 1999; K. A. McDonald, Miller, and Morris 2001
Megatheriidae		
Eremotherium		
Nothrotheriops	Gypsum Cave, NV	Stock 1931; Harrington 1933
Mylodontidae		
Paramylodon	Carson City, NV	Stock 1920, 1925, 1936b
Mustelidae		
Brachyprotoma	Crystal Ball Cave, UT	Heaton 1985
Canidae		
Cuon		
Canis dirus	Silver Creek, UT	W. E. Miller 1976
Ursidae		
Tremarctos		
Arctodus	Huntington Dam, UT	Gillette and Madsen 1992; Madsen 2000a, Schubert 2010
Felidae		
Smilodon	Black Rock Desert, NV	Dansie, Davis, and Stafford 1988; Livingston 1992b
Homotherium	Fossil Lake, OR	Jefferson and Tejada-Flores 1993
Miracinonyx	Crypt Cave, NV	Orr 1969; D. B. Adams 1979
Panthera leo	Tule Springs, NV	Mawby 1967
Castoridae		
Castoroides		
Caviidae		
Hydrochoerus		
Neochoerus		
Leporidae		
Aztlanolagus	Cathedral Cave, NV	Jass 2007
Equidae		
Equus	Long Valley, NV	Huckleberry et al. 2001
Tapiridae		
Tapirus		
Tayassuidae		
Mylohyus		
Platygonus	Franklin, ID	H. G. McDonald 2002
Camelidae		
Camelops	Hidden Cave, NV	Grayson 1985
Hemiauchenia	Crystal Ball Cave, UT	Heaton 1985
Palaeolama		
Cervidae		
Navahoceros	Mineral Hill Cave, NV	B. S. Hockett and Dillingham 2004
Cervalces		

TABLE 7-2 (CONTINUED)

Family/Genus	Illustrative Great Basin Site	References
Antilocapridae		
Capromeryx	Schuiling Cave, CA	Downs et al. 1959
Tetrameryx	Lark Fauna, UT	Gillette and Miller 1999
Stockoceros		
Bovidae		
Saiga		
Euceratherium	Falcon Hill Caves, NV	Hattori 1982; Dansie and Jerrems 2005
Bootherium	Logan City Cemetery, UT	Nelson and Madsen 1978, 1980; W. E. Miller 2002
Oreamnos harringtoni	Smith Creek Cave, NV	Stock 1936a; S. J. Miller 1979; J. I. Mead, Thompson, and Van Devender 1982
Gomphotheriidae		
Cuvieronius		
Mammutidae		
Mammut	Mastodon Sinkhole, UT	W. E. Miller 1987
Elephantidae		
Mammuthus	Black Rock Desert, NV	Dansie, Davis, and Stafford 1988; Livingston 1992b

Pleistocene period (table 7-3). This is statistically identical to the proportion of genera that has been dated to this period for North America as a whole. This similarity does not result from the fact that the Great Basin has provided the only terminal Pleistocene dates for these seven genera, since all but the shrub-ox *Euceratherium* also have dates from elsewhere in North America that fall between 12,000 and 10,000 years ago.

It is hard, though, to know what this similarity means. On the one hand, it might reflect the chronological structure of late Pleistocene extinctions in North America—that roughly 40 percent of the extinct late Pleistocene North American genera were lost between about 12,000 and 10,000 years ago, no matter where we look. On the other hand, it might also reflect that the most abundant of these mammals have, in general, been easiest to date to the end of the Pleistocene (see chapter 4).

It is impossible not to wonder how common these animals were on the Great Basin late Pleistocene landscape. If we sat on the flanks of the Toquima Range or the Ruby Mountains or the Stillwater Range 15,000 years ago and looked down to the valleys below, would we see herds of horses, camels, mammoths, and four-pronged antelope? Or would we have to wait days to see just a few of these animals?

Unfortunately, we have no way of assessing such things directly. There are, though, some ways of guessing at those abundances. In an insightful paper, biologists Richard Mack and John Thompson once observed that the structure of the arid steppe plant communities of intermountain western North America as a whole is not consistent with the presence of substantial herds of large mammals of any sort. This is because those communities are characterized by, among other things, grasses that grow in clumps in the midst of a

TABLE 7-3
Extinct Mammal Genera from the Great Basin with Secure Radiocarbon Dates of Fewer Than 12,000 Years Ago

Genus	Illustrative Site	Age	Reference
Nothrotheriops	Gypsum Cave, NV	11,360 ± 260	Long and Martin 1974
Arctodus	Huntington Mammoth Site, UT	10,870 ± 75	Madsen 2000a
Equus	Fishbone Cave, NV	11,210 ± 50	Dansie and Jerrems 2005
Platygonus	Franklin Peccary, ID	11,340 ± 50	McDonald 2002
Camelops	Sunshine, Long Valley, NV	11,390 ± 60	Huckleberry et al. 2001
Euceratherium	Falcon Hill, NV	11,950 ± 50	Dansie and Jerrems 2005
Mammuthus	Huntington Mammoth Site, UT	11,220 ± 110	Madsen 2000a

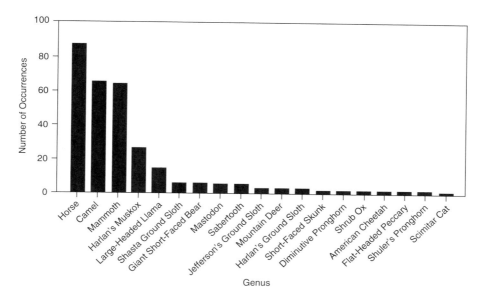

FIGURE 7-2 Relative abundance of extinct Pleistocene mammals in the Great Basin. An "occurrence" refers to a stratigraphically or geographically distinct record for the genus involved (see Grayson 2006d).

widespread biological soil crust. The grasses do not tolerate heavy levels of ungulate grazing (and were rapidly replaced by alien species after the introduction of livestock), and the biotic crusts are rapidly destroyed by trampling (see chapter 2). This suggests that although large Pleistocene mammals were widespread in the Great Basin, they may not have been particularly dense on the landscape.

On the other hand, it is certainly possible that the crusts themselves developed in response to the loss of those mammals. In addition, our knowledge of the nature and distribution of grasses in the Pleistocene Great Basin is not strong. Remember how difficult it is to identify grasses from pollen, and that the woodrats that amass plant parts that can be identified to the genus and species level do not tend to live in places that would have been grazed by significant numbers of large herbivores. As a result, it is perhaps best to treat Mack and Thompson's argument as an important hypothesis to be tested rather than as necessarily telling us about the numbers of large mammals on the Great Basin Pleistocene landscape. It is hard not to pose this as a challenge for young paleobotanists: devise ways to detect the presence of cryptogamic crusts in the deeper past, and then determine their prehistoric distribution and abundance before and after the demise of the Great Basin's Pleistocene fauna. If Mack and Thompson are right, there should not be much difference through this period of time, but if they are wrong, the Great Basin's widespread soil crusts should postdate the losses.

The Pleistocene birds of the Great Basin provide another possible way of assessing the abundances of large mammals on the landscape here. These I discuss in the next section; here I simply mention that the remarkable diversity of birds that made their living by scavenging terrestrial herbivores suggests that large mammals could not have been all that rare on that landscape. Since we do not know how abundant the birds themselves were, however, this is not that much help.

Although the absolute abundances of late Pleistocene mammals cannot be measured, the relative abundances of these animals can be assessed by counting the number of stratigraphically and geographically separate locations from which each of the extinct genera has been reported in the Great Basin. The results of doing this are shown in figure 7-2. This figure strongly suggests that horses, camels, mammoths, and helmeted muskoxen were the most abundant now-extinct mammals in the Great Basin during the late Pleistocene. These relative abundances are fairly similar to those for North America as a whole, which suggests the most common now-extinct Pleistocene mammals to have been horse, mammoth, mastodon, and camel, in that order.

Taken as a whole, these numbers may be a little deceiving, since not all of these animals were distributed throughout the Great Basin during the late Pleistocene. The helmeted muskox, for instance, has been reported only from the edges of Pleistocene Lake Bonneville, while the long-nosed peccary is known only from the far northern parts of the Great Basin—from Fossil Lake, Oregon, and southern Idaho. Similarly, ground sloths appear to have entered the Great Basin only along its edges. The Shasta ground sloth is known from southeastern California and southern Nevada (a complete individual has been reported from the Spring Mountains); Jefferson's ground sloth has been found in late Pleistocene deposits in southern Nevada and along the eastern edge of Pleistocene Lake Bonneville. Horses, camels, and mammoth, on the other hand, seem to have been widespread throughout much of the region. Indeed, mammoths are known from elevations as high as 9,000 feet, and mastodon as high as 9,800 feet, on the Wasatch Plateau, Utah.

Our understanding of the causes of late Pleistocene extinctions in the Great Basin is no more secure than it is for any other part of North America, and for the same reasons. While we do know that at least 35 percent of the now-extinct genera survived beyond 12,000 years ago in the Great Basin (table 7-3), we don't know how long these populations might have dwindled until they reached the point of extinction, and have no idea when the other genera were lost. Since we lack that information, it is unreasonable to think that we might be able to explain precisely how and why these extinctions occurred—just as is the case for the rest of North America.

In chapter 4, I mention that only mammoth and mastodon have been found associated with artifacts in North America in such a way as to show that people hunted now-extinct Pleistocene mammals. There are no such associations in the Great Basin, but this is not surprising and may not mean very much. As I explore in chapter 9, the archaeological record for the Great Basin becomes extremely rich by about 11,000 years ago, but much of that record is provided by sites exposed on the surface of the ground. This is not the kind of context likely to provide secure evidence that people hunted large Pleistocene mammals in the Great Basin. It will take the discovery of buried sites in some number to tell us whether this was the case.

What is quite certain, however, is that once the extinctions were over, the Great Basin was left with only five genera of large herbivores: bison, deer, elk (*Cervus elaphus*), mountain sheep, and pronghorn. And, as we shall see in the next chapter, not all of these were particularly common during the past 10,000 years or so.

The Huntington Mammoth and the Sinkhole Mastodon

In 1988, construction workers turned up the nearly complete remains of a Columbian mammoth on central Utah's Wasatch Plateau, about halfway between the towns of Nephi and Price (figure 7-1). Archaeologist David Madsen and paleontologist Dave Gillette, then of the Utah Geological Survey, were called in to excavate the site. In addition to the mammoth, the site provided a fragmentary skull and rib of the giant short-faced bear and a stone projectile point. These, amazingly enough, were taken from the open excavations by a guard whose job it was to protect the site from vandals; the specimens were ultimately retrieved by the U.S. Forest Service. As a result of the inappropriate actions of this "guard," the relationship between the remains of the bear, the artifact, and the mammoth cannot be known with certainty.

This location, the Huntington Mammoth site, is at the base of the Huntington Reservoir Dam, just east of the Great Basin drainage divide. Huntington Creek drains into the San Rafael River, which drains into the Green River, which drains into the Colorado. I mention it here because of the elevation at which this mammoth lived or, more accurately, died: 8,990 feet. Today, the vegetation of the area that provided the mammoth is dominated by Engelmann spruce and subalpine fir on nearby north- and east-facing slopes, while south- and west-facing slopes are characterized by big sagebrush and quaking aspen (*Populus tremuloides*).

The plant macrofossils of the lake sediments from which the mammoth was recovered were dominated by Engelmann spruce, suggesting that this animal was living in a subalpine environment at the time of its death. Madsen and Gillette also collected what appeared to be chewed organic material from between the ribs and pelvis of the animal. The condition and location of this material strongly suggests that it represents the intestinal contents of the mammoth, the mammoth's last meal or close to it. Madsen's identification of the plant macrofossils of this material showed it to include the needles of subalpine fir and, probably, sedge. Since the lake sediments in which the mammoth was found do not contain any indication that fir was then growing nearby, Madsen reasonably suggests that this material had been ingested elsewhere, then carried to the point where this elderly (fifty-five to sixty years old) and sick (the skeleton showed many pathologies) mammoth breathed its last. Geochronologist Tom Stafford extracted amino acids from both the mammoth and giant short-faced bear specimens, and had accelerator dates run on them. The mammoth dated to about 11,220 years ago; the bear, to 10,900 years ago.

Gouges on the skeleton suggest that after the mammoth had died, it had been scavenged by a carnivore. Intriguingly, the depth and spacing of the gouges match the size and spacing of the canine teeth of the short-faced bear from the site. While, as Madsen points out, the dates of the mammoth and the bear suggest that the latter was not likely to have scavenged the former, it is certainly possible that the gouges were caused by a scavenging *Arctodus*. After all, and as discussed in chapter 4, it does appear that short-faced bears made at least a substantial part of their living by scavenging.

The Huntington mammoth is not the only high-elevation proboscidean known from Utah. A scant two miles northwest of the mammoth site, paleontologist Wade Miller discovered the partial remains of two mastodons in a sinkhole that had been formed by the collapse of the underlying fine-grained sandstone. The collapse created an opening, or sinkhole, that today is some 120 feet by 150 feet in dimension. A pond developed in this opening; the mastodon remains were excavated from the sediments that formed within the pond. What makes the sinkhole mastodons so interesting is not only that they were the first to be securely documented from the Great Basin but also that they come from the highest elevation recorded for any mastodon: 9,780 feet.

The vegetation of the area that now surrounds the sinkhole is not unlike the vegetation of the area in which the Huntington mammoth was found: sagebrush in the immediate vicinity, with spruce/fir forest nearby. Paleobotanist Deborah Newman identified the pollen from the sediments in which the mastodons were embedded. She found that the vegetation of the area at the time the animal died was very similar to that found here today: sagebrush, spruce, and fir contributed 41 percent of the identified pollen, with pine contributing an additional 21 percent. Unfortunately, it is not known when all of this material accumulated. Radiocarbon dates on the mineral fraction of the bone provided dates of between 7,080 and 7,650 years, but, as I have mentioned, this fraction often provides inaccurate results.

Because of the careful work of Gillette, Madsen, and Miller, we now know that mammoth and mastodon occupied very high elevations in the Great Basin and immediately adjacent areas, and that the vegetation of these mountainous areas then included subalpine conifers, much as it does today. It appears that there is tremendous potential for high-elevation vertebrate paleontology in the Great Basin, something not recognized before these discoveries were made.

Extinct Late Pleistocene Birds

Extinction at the genus level was not confined to mammals at the end of the Pleistocene in North America. About twenty genera of birds also became extinct at this time, of which six continue to live on elsewhere (see table 7-4). The chronology of the avian extinctions is even more poorly understood than the chronology of the mammalian ones. In most instances, about all we can say is that they occurred toward the end of the Pleistocene, but there are some important exceptions that I will discuss shortly.

Things are even worse for the Great Basin. There are in general very few specialists in avian paleontology, and few of these work in the Great Basin. Indeed, much of the most interesting and important work that has been done was done years ago, a good deal of it as a result of the remarkable efforts of Hildegarde Howard of the Natural History Museum of Los Angeles County.

This work has documented the late Pleistocene presence of eight now-extinct genera of birds in the Great Basin (tables 7-4 and 7-5). Of the thirty-six genera of mammals that became extinct in North America toward the end of the Pleistocene, twenty, or 56 percent, are known from the Great Basin, and this list is probably fairly complete. For birds, the comparable figure is 40 percent (eight out of twenty), far lower than it is for mammals. Even if, as Kenneth Campbell and his colleagues have suggested, the few specimens that have been assigned to the teratorn genus *Cathartornis* actually came from birds that belong to *Teratornis*, this figure would rise to only 42 percent. Since our knowledge of the extinct late Pleistocene birds of the Great Basin is derived from a very small number of sites, there is no reason to think that our list of extinct birds is anywhere near complete. The Old World vulture *Neophrontops*, for instance, is known from both southern California and Wyoming and there is no reason to think that it was not found in the Great Basin as well.

Even as it stands, however, the list shows that the Great Basin late Pleistocene avifauna was in every way as remarkable as its mammalian counterpart. In addition to the eight genera of extinct birds known from here, four species of birds belong to genera that still exist that are very much worth mentioning. These are the Western Black Vulture (*Coragyps occidentalis*) and the Fragile Eagle (*Buteogallus fragilis*), both of which are extinct, and the California Condor (*Gymnogyps californianus*) and Crested Caracara (*Polyborus plancus* but also assigned to *Caracara cheriway*), both of which are still with us. I discuss them all here.

I begin with the storks. Most storks of the genus *Ciconia* are Old World in distribution, found from southern Africa north through Europe and Asia. The only exception is the Maguari Stork (*Ciconia maguari*), which is South American. However, the extinct Asphalt Stork (*Ciconia maltha*), first described from Rancho La Brea, is known not only from the late Pleistocene asphalt deposits or tar pits of southern California but also from Florida and from the Lake Manix Basin, in deposits approximately 20,000 years old. Only the Wood Stork (*Mycteria americana*) is now to be found in the Great Basin, and it is seen only occasionally during the summer months.

Today, the American Flamingo (*Phoenicopterus ruber*) does not breed north of the Caribbean. Post-breeding flamingos sometimes wander into the far eastern United States, and there are a series of flamingo sightings for Utah and Nevada. Ornithologists Ray Alcorn and Fred Ryser both suspect that the Great Basin sightings are of escaped birds; somehow, those seen at the artificial lake at the MGM Grand in Reno seem especially suspicious.

No such uncertainty surrounds the two extinct species of late Pleistocene Great Basin flamingos, both of which have been reported from Fossil Lake and the Lake Manix Basin. *Phoenicopterus copei*, Cope's Flamingo, appears to have been slightly larger than the modern Greater Flamingo, and had stouter legs; *P. minutus* was, as the name suggests, a smaller version, some 75 percent of the size of the Greater Flamingo.

When Hildegarde Howard identified these species from Lake Manix, no absolute dates for the deposits from which this material came were available. She suggested that the Manix Lake flamingos, and the many other birds she identified from this site, were late Pleistocene in age. However, paleontologist George Jefferson has now shown that the sediments from which the flamingo remains came are some 300,000 years old, and thus not late Pleistocene at all. Interestingly, these deposits also provided the remains of an immature Cope's Flamingo, suggesting that this large flamingo may have bred in the Great Basin at the time these deposits accumulated.

In contrast to *Phoenicopterus* and *Ciconia*, genera which still exist, the "pygmy goose" genus *Anabernicula* is extinct. Known from late Pleistocene deposits ranging from Texas to California (and, in earlier deposits, from Florida), as well as from Fossil Lake and Smith Creek Cave, *Anabernicula* was about the size of a mallard duck. However, it was related not to our common dabbling ducks (genus *Anas*), but instead to

TABLE 7-4
Extinct Late Pleistocene Genera of North American Birds
Genera in bold are known from the Great Basin.

Family/Genus	Modern Relatives	References
Ciconiidae		
Ciconia[a]	Storks	Emslie 1998
Phoenicopteridae		
Phoenicopterus[a]	Flamingos	Howard 1946
Anatidae		
Anabernicula	Shelducks	Howard 1964a, 1964b
Cathartidae		
Breagyps	Condors, Vultures	Emslie 1988; Hertel 1995; Van Valkenburgh and Hertel 1998
Teratornithidae		
Aiolornis	(Extinct Teratorns)	Campbell, Scott, and Springer 1999
Teratornis		Campbell, Scott, and Springer 1999; Fox-Dobbs et al. 2006; Hertel 1995; Van Valkenburgh and Hertel 1998
Cathartornis		Van Valkenburgh and Hertel 1998
Accipitridae		
Spizaëtus[a]	Hawks, Eagles	Hertel 1995; Van Valkenburgh and Hertel 1998
Amplibuteo		Emslie and Czaplewski 1999; Hertel 1995; Van Valkenburgh and Hertel 1998
Neophrontops	Old World Vultures	Hertel 1995; Van Valkenburgh and Hertel 1998
Neogyps		Hertel 1995; Van Valkenburgh and Hertel 1998
Falconidae		
Milvago[a]	Caracaras	Emslie 1998
Phasianidae		
Neortyx	Quails	Emslie 1998; Holman 1961
Burhinidae		
Burhinus[a]	Thick-knees	Howard 1971
Charadriidae		
Belonopterus[a]	Lapwings	Campbell 2004; Emslie 1998
Icteridae		
Cremaster	Cowbirds, Blackbirds	Emslie 1998; Hurlbert and Becker 2001; Morgan 2002
Pandanaris		Emslie 1998; Hurlbert and Becker 2001; Miller 1947; Morgan 2002
Pyeloramphus		Howard and Miller 1933; Miller 1932
Corvidae		
Protocitta	Jays	Hurlbert and Becker 2001; Morgan 2002
Henocitta		Emslie 1998; Hurlbert and Becker 2001; Morgan 2002

[a] Genus survives outside of North America.

the shelducks (genus *Tadorna*). Today, shelducks primarily occupy temperate regions within Europe, Asia, Africa, and the Pacific, though the genus includes one tropical species.

Clark's Condor (*Breagyps clarki*) belonged to the same family of birds, the Cathartidae (or, some would say, the Vulturidae), as the Turkey Vulture (*Cathartes aura*) and the California Condor (*Gymnogyps californianus*). Slightly larger than the California Condor, Clark's Condor sported a long but narrow beak, and is known only from the asphalt deposits of southern California and from Smith Creek Cave. The deposits of Smith Creek Cave, however, contained the remains of at least six of these large scavengers, one of which was immature, suggesting that this bird bred in the eastern Snake Range.

Clark's Condor was not the only condor in the Great Basin during the late Pleistocene. The famous California

TABLE 7-5
Great Basin Sites with Extinct Late Pleistocene Bird Genera

Site (species)	Reference
Fossil Lake (*Phoenicopterus copei, P. minutus*)	Howard 1946
Lake Manix (*Phoenicopterus copei, P. minutus*)	Howard 1955; Jefferson 1985, 1987, 2003
Lake Manix (*Ciconia maltha*)	Howard 1955; Jefferson 1985, 1987, 2003
Fossil Lake (*Anabernicula oregonensis*)	Howard 1946, 1964a
Smith Creek Cave (*Anabernicula gracilenta*)	Howard 1952, 1964b
Smith Creek Cave (*Breagyps clarki*)	Howard 1935, 1952
Smith Creek Cave (*Aiolornis incredibilis*)	Howard 1952; Campbell, Scott, and Springer 1999
Crystal Ball Cave (*Teratornis merriami*)	Emslie and Heaton 1987
Tule Springs (*Teratornis merriami*)	Mawby 1967
Fossil Lake (*Spizaëtus pliogryps*)	Howard 1946
Smith Creek Cave (*Spizaëtus willetti*)	Howard 1935, 1952
Smith Creek Cave (*Neogyps errans*)	Howard 1952

Condor was found here as well, reported from Antelope Cave in southeastern California's Mescal Range, from Gypsum Cave in southern Nevada, and from Smith Creek Cave. In fact, paleontologist Steve Emslie has obtained radiocarbon dates for condor remains from both Antelope and Gypsum caves: they are 11,080 and 14,720 years old, respectively.

Today, these huge scavengers—they can have wing-spans of more than nine feet—are nearly extinct in the wild. In fact, by 1981, just twenty-one wild birds were left. In 1987, the six that remained were captured to become part of a then-controversial program to breed and reintroduce them into the wild. Progress has not been smooth, with many mortalities due to the ingestion of such things as lead shot from scavenged prey and indigestible metal and other "junk" (as it is properly called by condor experts) by nestlings. Nonetheless, this is a gamble that has paid off, with some 130 birds now living in the wild, centering on release sites in California, Arizona, and Baja California.

During early historic times, California Condors were largely restricted to the Pacific Coast. This was not the case, however, during the late Pleistocene. Then, condors were found from California to Florida and as far north as New York state; there is a good chance they were found throughout much of North America south of glacial ice. As a result, it is not surprising that they also lived in the Great Basin at that time, and it will not be surprising when more late Pleistocene examples of them are reported from this area. It will be surprising, however, if many of those new discoveries post-date 10,000 years ago by any significant degree. This is because exacting work by Steve Emslie suggests that their numbers had dwindled dramatically by this time, with the latest radiocarbon dates from the Grand Canyon area falling at about 9,600 years ago. It was, Emslie very reasonably suggests, at about this time that their distribution became more tightly confined to the Pacific Coast region.

Smith Creek Cave also provided a number of specimens of the Western Black Vulture, *Coragyps occidentalis*. This species, well known from Rancho La Brea, is extinct, but the closely related Black Vulture, *Coragyps atratus*, comes no closer to the Great Basin than southern Arizona. Compared to its modern relative, the late Pleistocene Western Black Vulture was a bulkier bird, with shorter and stouter legs, but was probably much like it in adaptations. Superb work done by Kena Fox-Dobbs and her colleagues on Western Black Vulture bone chemistry has shown that these birds were terrestrial scavengers, a conclusion that matches one reached earlier by biologists Blaire Van Valkenburgh and Fritz Hertel. I return to this work below.

The teratorns are closely related to the vultures and condors, but are placed in their own family, the Teratornithidae. The first of these huge, condor-like birds was discovered at Rancho La Brea; the same species, *Teratornis merriami* or Merriam's Teratorn, is now known from other asphalt deposits in southern California, from Florida, and from northern Mexico, among other places. These other places include the Great Basin. Here, it has been found in the late Pleistocene deposits of Tule Springs, where it is associated with radiocarbon dates that fall between 11,500 and 13,100 years ago, and from Crystal Ball Cave, Utah, where it is undated. In life, Merriam's teratorn would have looked something like a California Condor, but it weighed about a third more (probably around thirty pounds), and had a wingspan of thirteen feet or more (compared to a maximum of slightly more than nine feet for the California Condor) as well as a hooked, somewhat eagle-like beak.

As impressive as this bird was, however, it was almost dwarfed by its relative, the Incredible Teratorn, *Aiolornis incredibilis*. First described by Hildegarde Howard on the basis of a single wrist bone from Smith Creek Cave, this bird is now also known from a handful of specimens from California, Mexico, and Florida that range in age from more than 3 million years ago to the late Pleistocene. Aptly named, the Incredible Teratorn had a wingspan estimated at sixteen to eighteen feet, nearly twice that of the California Condor. Although this wingspan places the Incredible Teratorn among the largest flying birds known, place of honor in this category belongs to a much earlier member of the teratorn group, *Argentavis magnificens*. This teratorn, from late Miocene deposits in Argentina, had a wingspan of some 22 feet. Weighing an estimated 150 pounds, this is the largest flying bird known to have existed.

Detailed analyses of the bones of Merriam's teratorn by teratorn experts Kenneth Campbell and Eduardo Tonni suggest that these birds were agile on the ground, well adapted for walking but far less so for running; they compare their walking gaits to those of storks and turkeys. They were certainly capable flyers, though how they flew has been the subject of some debate. Campbell and Tonni make a compelling argument that they flew much like condors, soaring on rising air currents and perhaps being sufficiently decent runners that they were able to take flight by getting a running start.

What teratorns did for a living has been the subject of a healthy discussion. In contrast to many who assumed that Merriam's Teratorn must have been a scavenger, Campbell and Tonni suggested that they were active predators, stalking and killing such small vertebrates as rodents, fledgling birds, and lizards. How such a large bird could obtain enough food in this way is not at all obvious, however, and others have disagreed with this conclusion. Fritz Hertel suggested that they were fish-eaters, but the work conducted by Fox-Dobbs and her colleagues showed that only one of the ten teratorns from Rancho La Brea that they analyzed had a diet that seems to have included any marine component at all. The other nine individuals fed on terrestrial browsing and grazing mammals. How they obtained those mammals remains unknown but Fox-Dobbs and her co-workers conclude that it was most likely a scavenger, just as others have concluded.

The Errant Eagle, *Neogyps errans*, is known from the tar pits of southern California, from northern Mexico, and from Smith Creek Cave. This bird belongs to the widespread family of hawks and eagles (the Accipitridae), but proper placement within that family is not clear. Most often, they have been placed with the subfamily of Old World Vultures—with such birds as the Egyptian Vulture (*Neophron percnopterus*), the Griffon Vulture (*Gyps fulvus*), and the Bearded Vulture or Lammergeier (*Gypaetus barbatus*). Even if this placement is correct, however, aspects of the skeleton of the Errant Eagle—for instance, its eagle-like skull and feet—suggest that it was far more capable of taking live prey than are modern Old World Vultures. An analysis of its skull led Hertel to conclude that it was primarily a scavenger, though he and Van Valkenburgh suggest that it was likely an active hunter as well.

Hawk Eagles, genus *Spizaëtus*, are also accipiters, but they belong to the subfamily that includes the true eagles. Two extinct species of this genus have been reported from the Great Basin, from Fossil Lake and Smith Creek Cave. Today, Hawk Eagles can be found in both the Old World and the New; in the Americas, they come no farther north than tropical Mexico, where they exist on a diverse variety of small vertebrates. They have never been reported on the wing in North America. Like their modern relatives, the extinct forms are thought to have been active predators.

The extinct Fragile Eagle has been reported from both Fossil Lake and Mineral Hill Cave. Its close relative, the Common Black Hawk (*Buteogallus anthracinus*) now breeds as far north as the Utah-Arizona border. Its extinct relative, however, is known not only from the Great Basin but also from a number of other places in the West as well, including the asphalt deposits of Southern California; New Mexico; and Hawver Cave, northern California. Much earlier examples have also been reported from Florida. Modern Black Hawks rely heavily on small animals for their diet, ranging from crabs, lizards, and frogs to small birds.

Finally, there is the Crested Caracara, *Polyborus plancus*. Today, this bird, which makes its living both as a scavenger and as a predator on small animals, makes it no farther north than the very southern edge of Arizona and Texas, with an isolated population in central Florida. It was more widespread during the late Pleistocene, found as far north as Nebraska and Wyoming and as far west as southern California. It also occurred in the Great Basin, having been identified by Steve Emslie and Tim Heaton at Crystal Ball Cave.

Caracaras and teratorns in the Great Basin, perhaps dining on camels and sloths next to a lake dotted with flamingoes and storks. Hard to believe, perhaps, but it is true.

An Abundance of Scavengers

North America now supports only three species of scavenging vultures and condors (birds of the family Cathartidae). These three are the Turkey Vulture, the Black Vulture, and the California Condor. The Great Basin supports only the Turkey Vulture. During the late Pleistocene, however, the Great Basin supported at least five species of vultures, condors, and teratorns, scavengers all: Clark's Condor, the Western Black Vulture, the California Condor, the Incredible Teratorn, and Merriam's Teratorn. It is extremely likely that the Errant Eagle scavenged as well (table 7-6).

The Turkey Vulture has yet to be reported from the late Pleistocene Great Basin, even though it is common here today. This is probably no accident. Many years ago, Hildegarde Howard and Alden H. Miller observed that the remains of the now-extinct Western Black Vulture far outnumber those of the Turkey Vulture in the late Pleistocene deposits of Rancho La Brea. Turkey Vultures became abundant only as the Western Black Vulture decreased in number. Although Great Basin data are scanty, the same may well have been true here. Since the Turkey Vulture is present in the late Pleistocene asphalt deposits of southern California, it was probably present, albeit relatively uncommon, in the late Pleistocene Great Basin as well. That it has yet to

TABLE 7-6
Late Pleistocene Vultures, Condors, and Teratorns Known from the Great Basin

Family	Species	Common Name
Cathartidae	*Breagyps clarki*	Clark's Condor
	Coragyps occidentalis	Western Black Vulture
	Gymnogyps californianus	California Condor
Teratornithidae	*Aiolornis incredibilis*	Incredible Teratorn
	Teratornis merriami	Merriam's Teratorn
Accipitridae	*Neogyps errans*	Errant Eagle

be found here follows from its relative rarity and from the small sample of late Pleistocene birds that is available from this region: rare things require relatively large samples to be detected.

When the Turkey Vulture is found in the late Pleistocene Great Basin, as it almost certainly will be, that will bring the total number of vultures, condors, and teratorns known from this region to seven. Exactly how rich this aspect of the Great Basin avifauna was can be judged from the fact that only seven species of vultures and condors are found in the Americas today.

The large number of species of vultures, condors, and teratorns in the late Pleistocene Great Basin raises a series of interesting ecological questions. The presence of so many species of these birds here suggests that the mammal fauna of the time not only was rich in species but had to be large enough in numbers of individual animals to support them. The large size of Clark's Condor, the California Condor, and the teratorns suggest not only that they must have been birds of open terrain but also that there must have been thermal updrafts substantial enough to allow the soaring flight of these birds, unless their soaring depended totally on upslope winds. Since many have suggested that the growth of the pluvial lakes was accompanied by cooler summers and greater cloud cover (see chapter 5), it would be of interest to know whether the number of vultures and condors decreased as the pluvial lakes grew, and increased as they shrank. It would be of value to know whether there might have been a north-south decline in both the number of species and number of individuals of these birds, just as there may have been a north-south decline in intensity of cloud cover (if that may be read from the evidence of the pluvial lakes themselves). On the other hand, since intense winds can play the same role as thermals in supporting the flight of these birds, this intriguing question may not have a meaningful answer.

It is hard not to wonder about the competitive relationships among these birds. Today, as many as seven different species of Old World vultures may be found feeding on a single carcass in Serengeti National Park, with different sets of them feeding in different ways and arriving at different times. Fritz Hertel has aptly summarized these as rippers (which feed mainly on the tougher outer portions of a carcass), gulpers (which feed on the softer innards), and scrappers (which feed on small scraps on and near the carcass). Much the same was likely true of the Great Basin's large avian scavengers. Indeed, Hertel has used the skull morphology of these birds to argue that *Neogyps errans* was a ripper, *Breagyps clarki* and *Gymnogyps californianus* gulpers, and *Neophrontops americanus* and *Coragyps occidentalis* scrappers.

The vultures and condors of the late Pleistocene differed dramatically in size, with the smaller Black Vulture and Turkey Vulture (assuming it was there) on the smaller end, the huge Clark's Condor and California Condor in the middle, and teratorns at the larger end of the scale. In Africa today, larger vultures take wing later in the day than smaller ones, as they wait for the development of the more substantial thermal updrafts they require. In communal roosts of Turkey and Black vultures in eastern North America, both species wait for the development of thermals before taking flight, but Turkey Vultures leave well before Black Vultures. In addition, dominance among species of vultures tends to be determined by body size. If we waited by the carcass of a camel in the Snake Valley during the late Pleistocene, would the smaller vultures arrive first, followed later by the larger, and probably dominant, condors, and then last of all by the teratorns?

It is routinely suggested, with very good reason, that the decline of the vultures, condors, and teratorns in western North America somewhere near the end of the Pleistocene followed almost automatically from the extinction of the mammals on which they depended for food. While this most certainly provides part of the answer, it is hard to imagine that it provides all of it. There were bison in some parts of the Great Basin during the Holocene (chapter 8); there were mountain sheep, pronghorn, and black-tailed deer, and although elk have never been abundant here, they did exist on the area's northern margins. Perhaps these sizable mammals could not support six species of large scavenging birds, but why not two or three? Where Black and Turkey Vultures occupy the same region today, they routinely roost and feed together. There must be more to the depauperate nature of the modern Great Basin's scavenging bird fauna than simply the loss of the large mammals. The climates to which these birds were adapted disappeared as well, and a full account of the loss of the vultures, condors, and teratorns will most likely have to take climatic change into account.

Why Did California Condors Survive?

Another obvious thing to wonder about is why the California Condor survived at all, even if its distribution was far more restricted after 10,000 years ago than before. Insightful work by two different research groups, one led by C. Page Chamberlain and the other by Kena Fox-Dobbs, has shown that the answer appears to be surprisingly simple.

Because different kinds of diet leave different chemical signatures in bone and other tissues, these teams were able to decipher the kinds of things that Western Black Vultures, Merriam's Teratorns, and California Condors were eating toward the end of the Pleistocene. I have already mentioned some of the conclusions reached by Fox-Dobbs's team, but there is more. These teams together demonstrated that whereas Merriam's Teratorn and Black Vultures focused their diets on terrestrial herbivores, California condors had more options. In areas where they became extinct, including the Southwest, Texas, and Florida, their diets were entirely terrestrial. In California, however, some had terrestrial diets but others were consuming large amounts of marine food, presumably such things as whales and seals found dead along the shore. This conclusion matches observations of condor feeding behavior made during early historic times, before commercial hunting had reduced the numbers of marine mammals washing ashore. When the Pleistocene ended and the terrestrial meat market collapsed, only those condors living along the Pacific Coast that were adapted to using marine food sources were able to survive.

Altered Late Pleistocene Distributions of Existing Great Basin Mammals

I have mentioned that our understanding of the details of the history of the now-extinct late Pleistocene mammals and birds of the Great Basin is harmed by the lack of chronologically well-controlled sequences that might tell us about the details of that history. Fortunately, the same is not true for the latest Pleistocene small mammals of the Great Basin. In fact, this area may have the most detailed latest Pleistocene and Holocene small mammal record available for any place in the world.

There are multiple reasons for this. First, the general aridity of the Great Basin provides an excellent setting for the preservation of fossil accumulations in sheltered settings. Even such poorly stratified sites as Crystal Ball and Mineral Hill caves, discussed earlier, tend to contain extremely well-preserved bones and teeth. Second, for all that I said before about the lack of stratified deposits containing the remains of extinct large mammals, the Great Basin does have an abundance of caves with stratified deposits rich in remains of small vertebrates. The careful excavation and analysis of these has provided exquisite data on small mammal histories. Open sites have provided such information as well, but these sites have provided far shorter chronological sequences and reduced chances for organic preservation. Finally, because the Great Basin is dotted with packrat middens that can endure on the landscape for tens of thousands of years, and because they at times contain the bones, teeth, and fecal pellets of small mammals, analyses of these middens have contributed remarkably precise data about the history of Great Basin small mammals.

It might seem strange that our knowledge of the late Pleistocene history of extinct mammals and birds is so weak while exactly the opposite is true for small mammals. How can these two facts—and they are facts—possibly be reconciled? How can the record for large, extinct mammal history be so shoddy at the same time as the records for small mammals (none of which are extinct) be so strong? In fact, the reasons for this huge difference are fairly straightforward.

I carefully used the words "latest Pleistocene" when I stressed how good our record for this period of time is. I did that because the detailed small mammal record begins after about 11,500 years ago, very late in the Pleistocene, and perhaps well after the now-extinct large mammals and birds that still existed had begun to dwindle in abundance, thus reducing their chances for becoming incorporated in the cave deposits involved.

Thank Heavens for Little Birds

In addition, the stratified small mammal sequences that form the basis of our knowledge of these animals accumulated largely as a result of the activities of owls. These birds have the happy habit of feeding on a diverse variety of small mammals, digesting all that is digestible then regurgitating the leftovers in the form of what are known as owl pellets. These roughly thumb-sized (though how big the thumb depends on how big the owl) conglomerations are composed mostly of the fur, bones, and teeth of leftovers from the birds' meals. If the owl happens to be roosting in a cave, the deposits in that cave can become rich in pellets and the bones and teeth within them. If the cave is in an arid setting, those bones and teeth can endure for tens of thousands of years. The activities of owls go a long way to explain why we have such a detailed set of small mammal histories for the Great Basin. It might be fun to think about the possibility of teratorn-sized owls scooping up ground sloths, horses, and camels and regurgitating their remains beneath their cave roosts, but such things did not exist. Great Basin owls were of a normal size and fed on such normal owl things as kangaroo rats and voles.

In other parts of the world—France, for instance, and Spain—late Pleistocene sites tend to be chock full of the bones of large mammals. This is because these areas had carnivores that brought the remains of animals back to those caves; hyenas provide a prime example. Another example, however is provided by the human hunters who lived in this part of the world. Indeed, our very detailed knowledge of the later Pleistocene large mammal history of Europe is to a significant extinct built on the backs of those hunters, who filled cave after cave with the bony results of their efforts.

Such deposits also accumulated in the Great Basin. For instance, the famous archaeological site of Gatecliff Shelter, which I discuss in the following chapter, contained the remains of some two dozen butchered mountain sheep, the remains of human hunting some seven hundred years ago. However, the late Pleistocene caves and rockshelters of the Great Basin contain no such material. This may reflect the fact that large, now-extinct mammals were rare on the landscape when people arrived, but it may also result from the fact that any truly large mammals they killed were processed in the open, where they had been killed. It is also possible, of course, that there simply weren't very many human hunters in the Great Basin during the late Pleistocene. This possibility I explore in chapter 9.

For all these reasons, and perhaps more, the caves and rockshelters of the Great Basin have provided a frustratingly weak late Pleistocene large mammal and bird record but a very strong one for latest Pleistocene small mammals.

Rather than discussing all that we know about latest Pleistocene small mammal history in the Great Basin, I am going to provide a series of vignettes, individual species histories, that, when taken together, provide a strong feel for what was happening in the Great Basin in general as the Pleistocene came to an end. I continue these histories to the end of the early Holocene in this chapter, and then complete them in the next, where I discuss what came next.

Homestead Cave

I begin with a reminder about Homestead Cave, discussed in chapter 5 in association with Jack Broughton's work on the fishes found in its late Pleistocene strata (figures 5-7 and 5-11). As I said in that chapter, this site is located just west and south of Great Salt Lake, overlooking the Bonneville Salt Flats from an elevation of 4,612 feet, about 330 feet above the valley floor. The site was exquisitely stratified, with well-defined strata at times so thin that they could not be excavated separately. In the end, the deposits were removed according to eighteen separate layers. Chronological control was originally provided by twenty-one radiocarbon dates, then augmented by a dozen more. Very recent work, by biologist Rebecca Terry, has added additional dates for three of the strata. As a result, Homestead Cave has one of the best-controlled chronologies of any cave site in the Great Basin, with a record that begins about 11,300 years ago and that continues into historic times.

Not only was the site exquisitely stratified, but it was also full of the bones of small mammals, a result of the fact that the assemblages accumulated mainly as a result of the foraging of owls and secondarily as a result of the collecting activities of woodrats. A total of about 184,000 mammal bones and teeth were identified to at least the genus level from fifteen of the eighteen Homestead strata (four of the strata were not completely studied). It is this set of material that provides the underpinning for many of the histories that follow.

Pygmy Rabbits (*Brachylagus idahoensis*)

Pygmy rabbits are the smallest North American member of the family of mammals known as the Leporidae, or rabbits and hares. As I discuss in chapter 2, they are now discontinuously distributed throughout the botanical Great Basin and parts of immediately adjacent areas; there is also a disjunct population in southeastern Washington. In all these areas, pygmy rabbits require tall, dense stands of big sagebrush to thrive. As the amount tall sagebrush has decreased in the Great Basin during recent decades, so has the abundance of these rabbits: all Great Basin states list them as species of special concern.

The prehistoric record for pygmy rabbits in the Great Basin shows that their history in this region since the end of the Pleistocene has not necessarily been a happy one for them. This history is documented in detail at Homestead Cave and, with less precision, at a number of other sites in arid western North America (see figure 7-3).

At Homestead Cave, pygmy rabbits were common at the end of the Pleistocene, between about 11,300 and 10,200 years ago. As the Pleistocene came to an end, though, their numbers declined sharply (figure 7-4), due to a decrease in the abundance of the sagebrush habitat on which these animals depend.

We don't have direct information on the history of sagebrush in the Homestead area at this time, but we do have compelling indirect evidence that shows that sage must have declined at this time. In particular, as pygmy rabbits became less common, so did Ord's kangaroo rat. At the same time chisel-toothed kangaroo rats became far more abundant (figures 7-5 and 7-6).

As discussed in chapter 2, the specialized teeth of the chisel-toothed kangaroo rat allow it to thrive in vegetation dominated by saltbush (figure 2-12). When Ord's and chisel-toothed kangaroo rats occur in the same area, the former outcompetes the latter in sagebrush, but the opposite happens in saltbush habitats. The fact that the chisel-toothed version increased in number at the same time as pygmy rabbits were declining at Homestead strongly suggests that the end of the Pleistocene here saw the replacement of sagebrush-dominated habitat by vegetation dominated by shadscale.

The decline in pygmy rabbit abundance at the end of the Pleistocene at Homestead Cave is matched by data provided by Danger Cave, located across the Bonneville Salt Flats about sixty-five miles to the west. The Danger Cave faunal sequence, obtained from excavations conducted during the 1950s, is not nearly as rich as that available from Homestead Cave, but it does contain thirty-seven specimens of pygmy rabbit. All but two of these were recovered from the oldest stratum, deposited between about 10,600 and 10,100 years ago.

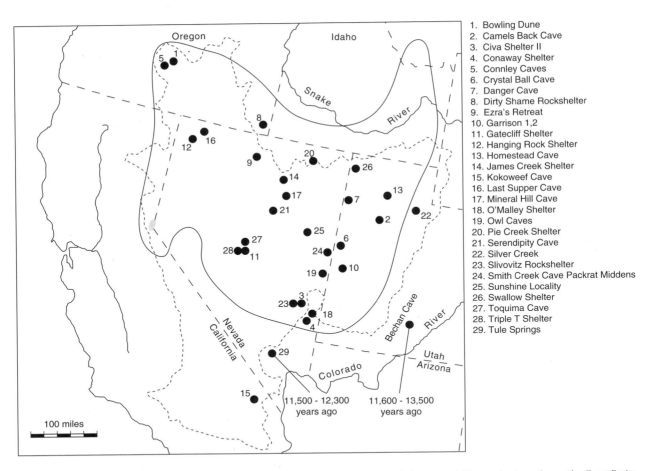

FIGURE 7-3 Distribution of dated archaeological and paleontological sites with identified pygmy rabbit remains in and near the Great Basin. The solid line represents the boundary of pygmy rabbit distribution in the Great Basin.

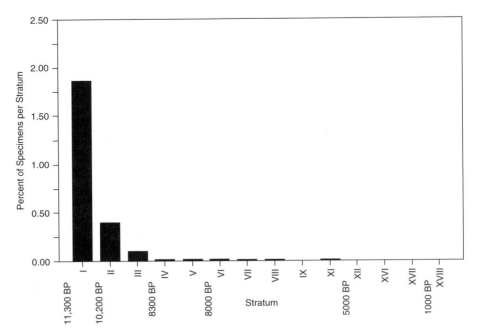

FIGURE 7-4 Changing relative abundance of pygmy rabbits through time at Homestead Cave. The higher the bar, the more abundant were pygmy rabbits in each stratum.

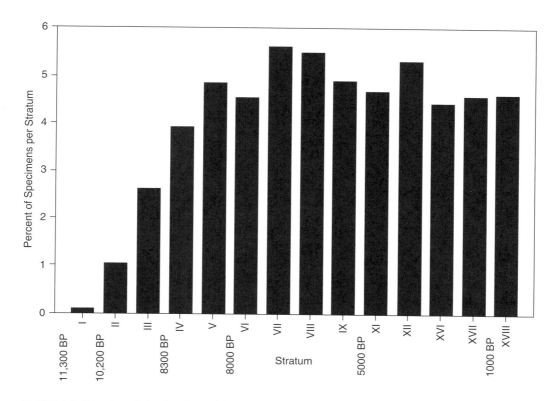

FIGURE 7-5 Changing relative abundance of chisel-toothed kangaroo rats through time at Homestead Cave. The higher the bar, the more abundant were chisel-toothed kangaroo rats in each stratum.

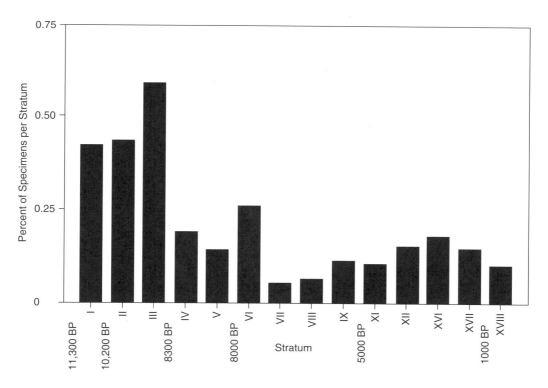

FIGURE 7-6 Changing relative abundance of Ord's kangaroo rats through time at Homestead Cave. The higher the bar, the more abundant were Ord's kangaroo rats in each stratum.

The pygmy rabbit histories provided by Homestead and Danger caves match what we know of the history of this animal from the Arid West in general. Figure 7-3 shows the location of all archaeological and paleontological records for pygmy rabbits from within and near the Great Basin that can be placed in at least a general chronological framework. Figure 7-7 eliminates all of the sites that contain only late Pleistocene–aged pygmy rabbit specimens and shows that once these sites are removed, all pygmy rabbit records in the intermountain area south of Washington State are from within the current distributional range of the species.

Pygmy rabbits have also been identified from Isleta Cave 2, in west-central New Mexico and from Sheep Camp Shelter, northwestern New Mexico (see table 7-7), well outside the Great Basin. Although the age of these specimens is not known (both sites contain a combination of Pleistocene and Holocene material), they do document that the range of the species once extended far to the south and east of its current distribution. Given the fate of other pygmy rabbit populations south of where they now live, it seems most likely that the pygmy rabbit specimens from these sites are Pleistocene in age, but only by directly dating the specimens themselves can we learn their ages.

Although pygmy rabbit abundance in the Homestead Cave area declined dramatically at the end of the Pleistocene, they were still present during the early Holocene (figure 7-4). In fact, they lingered on here until about 8,000 years ago, after which time they disappear. Why this was the case we will see in the next chapter.

Yellow-Bellied Marmots (*Marmota flaviventris*)

Yellow-bellied marmots are essentially big, fat squirrels, closely related to the woodchucks, or groundhogs, of eastern and northern North America (think Punxsatawney Phil of weather-forecasting fame). They are widely, but again discontinuously, distributed across intermountain western North America, reaching as far north as central British Columbia. Within the Great Basin, they are primarily confined to the higher elevations of mountains, though they descend into relatively low altitudes in the northernmost reaches of this area and along the eastern edge of the Sierra Nevada. They are absent from the southern Great Basin, most likely a result of the intense summer temperatures and dryness of this area: marmots tend to like it fairly cool and moist and require green grasses and similar vegetation to

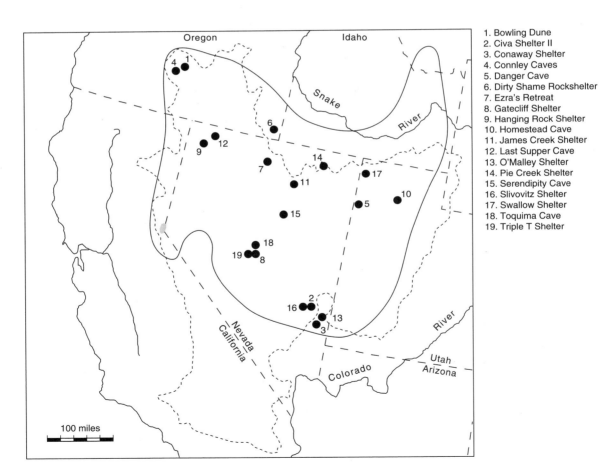

FIGURE 7-7 Distribution of dated Holocene-aged archaeological and paleontological sites with identified pygmy rabbit remains in and near the Great Basin.

TABLE 7-7
Southwestern Extralimital Records for *Brachylagus idahoensis*, *Marmota flaviventris*, and *Neotoma cinerea*
LP = late Pleistocene (ca. 40,000–10,000 years ago); ND = no secure age known; LP/H? = late Pleistocene or possibly Holocene; LH = late Holocene

Site	Age	*Brachylagus idahoensis*	*Marmota flaviventris*	*Neotoma cinerea*	Reference
Algerita Blossom Cave, NM	LP			X	Harris 1993a
Anthony Cave, NM	LP			X	Harris 1977, 1984
Baldy Peak Cave, NM	LP/H?		X	X	Harris 1984, 1990
Big Manhole Cave, NM	LP		X	X	Harris 1984, 1993a
Burnet Cave, NM	ND		X	X	Schultz and Howard 1936; Murray 1957
Conkling Cavern, NM	ND			X	Harris 1977, 1984, 1993a
Cylinder Cave, AZ	ND		X		Lange 1956
Dark Canyon Cave, NM	LP		X	X	Harris 1977, 1984, 1990; Lundelius 1979
Dry Cave, NM	LP		X	X	Harris 1970, 1977, 1984, 1985, 1987, 1990, 1993a, 1993b
Dust Cave, TX	LP		X		Harris 1990
Fowlkes Cave, TX	LP		X		Dalquest and Stangl 1984
Government Cave, AZ	ND		X		Lange 1956
Hermit's Cave, NM	LP		X		Schultz, Martin, and Tanner 1970
Howell's Ridge, NM	ND			X	Harris 1977, 1984, 1985
Isleta Cave 1, NM	ND			X	Harris and Findley 1964; Harris 1984, 1985
Isleta Cave 2, NM	ND	X	X	X	Harris and Findley 1964; Harris 1984, 1985, 1993a
Jimenez Cave, Chihuahua, Mexico	ND			X	Harris 1984, Messing 1986
Keet Seel, AZ	LH		X		Lange 1956
Lower Sloth Cave, TX	ND		X	X	Logan 1983
Manzano Mountains Cave, NM	ND		X		Howell 1915; Harris 1993a
Muskox Cave, NM	ND		X	X	Logan 1981
New La Bajada Hill, NM	ND		X		Stearns 1942
Papago Springs, AZ	LP		X		Czaplewski et al. 1999
Pendejo Cave, NM	LP		X	X	Harris 2003
Pit Stop Quarry, AZ	LP	X			Murray et al. 2005
Pratt Cave, TX	ND		X	X	Lundelius 1979, pers. comm.
Rampart Cave, AZ	LP		X		Van Devender, Phillips, and Mead 1977; Mead and Phillips 1981
San Josecito Cave, Mexico	LP		X		Arroyo-Cabrales and Polaco 2003
Sheep Camp Shelter, NM	ND	X	X	X	Gillespie 1984; Harris 1990, 1993a
Shelter Cave, NM	ND			X	Harris 1977, 1984, 1993a
Tse'An Kaetan Cave, AZ	ND		X		Lange 1956
Tse'An Olje Cave, AZ	ND		X		Lange 1956
Tularosa River Cave, NM	ND		X		Harris 1993a
U-Bar Cave, NM	LP		X	X	Harris 1987
Upper Sloth Cave, TX	ND		X	X	Logan and Black 1979
Vulture Cave, AZ	LP/H?		X		Mead 1981; Mead and Phillips 1981
Woodchuck Cave, AZ	LH		X		Lockett and Hargrave 1953; Lange 1956

fuel themselves. They also have an appealing lifestyle. They emerge from hibernation in the spring or early summer, eat and make baby marmots, then quickly retreat underground to wait until the following spring to start all over again. Were there a television show on the fabulous lifestyles of the fat and furry, they would surely be featured.

The history of marmots at Homestead Cave is similar to that of pygmy rabbits (figure 7-8). They were fairly common in the latest Pleistocene deposits of this site, after which their numbers plummeted. They managed to hang on here about as long as pygmy rabbits managed to do so, becoming locally extinct sometime around 8,000 years ago or so. As with pygmy rabbits, we know exactly why this extinction event occurred (see chapter 8).

Aspects of this history are replicated by other sites within the Bonneville Basin. At Danger Cave, marmots were present only in the earliest stratum, between about 10,600 and 10,100 years BP. There is a third Bonneville Basin record for them at Camels Back Cave, provided by the work of archaeologists Dave Schmitt and Karen Lupo. Here, marmots are represented in sediments laid down around 8,800 years ago but not after that time.

During the late Pleistocene, marmots were found well south and east of the Great Basin, into northern Mexico, far south of their modern limits (table 7-7). The most recent of these extralimital records is from Dry Cave, New Mexico, dated to about 14,500 years ago. As figures 7-9 and 7-10 show, the decline in marmot abundance at the end of the Pleistocene at Homestead Cave is associated with the loss of these southern populations, just as was the case with pygmy rabbits.

In short, the abundance of yellow-bellied marmots in the Great Basin declined dramatically at the end of the Pleistocene. That decline seems to have occurred at about the same time as the northward retreat of marmots to within their current distributional boundary. In some areas of the Great Basin where marmots no longer occur, those animals survived through the early parts of the Holocene, only to become extinct at about 8,000 years ago. In fact, it seems very likely that the final disappearance of marmots from most of the area once covered by the waters of Lake Bonneville occurred at about this time.

Bushy-Tailed Woodrats (*Neotoma cinerea*)

We have encountered bushy-tailed woodrats many times before, primarily in their role as the manufacturers of the middens that have been so important in providing information about the nature of Great Basin vegetation during the past 40,000 years or so. What I haven't mentioned is that these are animals of relatively cool and moist environments, and that they are widely distributed in western North America, from northern New Mexico northwest to the Pacific Coast and northward to the Northwest Territories. They are also widespread within much, though not all, of the Great Basin. As with the other mammals discussed here, bushy-tailed

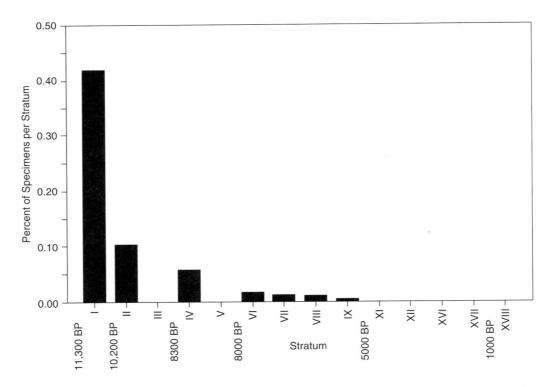

FIGURE 7-8 Changing relative abundance of yellow-bellied marmots through time at Homestead Cave. The higher the bar, the more abundant were yellow-bellied marmots in each stratum.

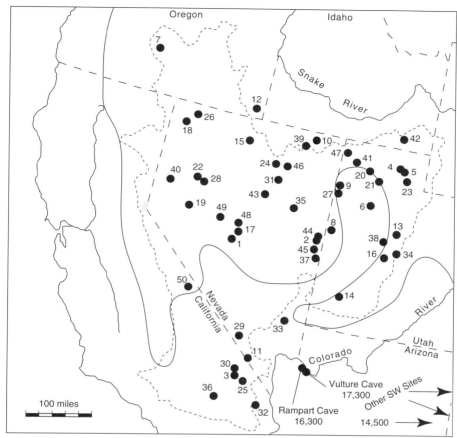

Sites: 1. Alta Toquima Village; 2. Amy's Shelter; 3. Antelope Cave; 4. Bear River 1; 5. Bear River 2; 6. Camels Back Cave; 7. Connley Caves; 8. Crystal Ball Cave; 9. Danger Cave; 10. Deer Creek Cave; 11. Devils Peak; 12. Dirty Shame Rockshelter; 13. Ephraim; 14. Evans Mound; 15. Ezra's Retreat; 16. Five Finger Ridge; 17. Gatecliff Shelter; 18. Hanging Rock Shelter; 19. Hidden Cave; 20. Hogup Cave; 21. Homestead Cave; 22. Humboldt Lakebed Site; 23. Injun Creek; 24. James Creek Shelter; 25. Kokoweef Cave; 26. Last Supper Cave; 27. Lead Mine Hills Cave; 28. Lovelock Cave; 29. Mercury Ridge; 30. Mescal Cave; 31. Mineral Hill Cave; 32. Mitchell Caverns; 33. Mormon Mountain Cave; 34. Nawthis Village; 35. Newark Cave; 36. Newberry Cave; 37. Owl Caves; 38. Pharo Village; 39. Pie Creek Shelter; 40. Pyramid Lake Fishway, 26Wa1010 and 1020; 41. Remnant Cave; 42. Rock Springs Bison Kill; 43. Serendipity Cave; 44. Smith Creek Cave; 45. Snake Creek Burial Cave; 46. South Fork Shelter; 47. Swallow Shelter; 48. Toquima Cave; 49. Wagon Jack Shelter; 50. White Mountains Villages

FIGURE 7-9 Distribution of dated archaeological and paleontological sites with identified yellow-bellied marmot remains in and near the Great Basin. The solid line represents the distributional boundary of marmots in the Great Basin. The numbers on the bottom right corner provide the radiocarbon dates for the indicated sites.

woodrats are found discontinuously within those parts of the Great Basin that fall within their general range. Toward the northern parts of the Great Basin, they can be found at nearly all elevations but they become increasingly restricted to higher elevations toward the south. For many years it had been assumed that bushy-tailed woodrat populations in the mountains of the hotter and drier portions of the Great Basin were completely isolated by the valleys that separate them but we now know that this is not always the case.

The Homestead Cave record for these woodrats differs in some significant ways from that for pygmy rabbits and marmots at this site. Given their preferences for fairly cool and moist settings, it is not surprising that they are common in late Pleistocene deposits here (figure 7-11). In fact, they are the most abundant mammal in the deposits that accumulated between about 11,300 and 10,200 years ago. Their numbers decline across the transition from the Pleistocene to the Holocene, but the decline is much less dramatic than those seen for pygmy rabbits and marmots at this time (figures 7-4 and 7-8). Instead, their steepest decline occurs at 8,300 years BP—about the same time as pygmy rabbits and marmots call it quits here. Bushy-tailed woodrats do occur in low numbers in the Homestead Cave area today, but that is a story for the following chapter.

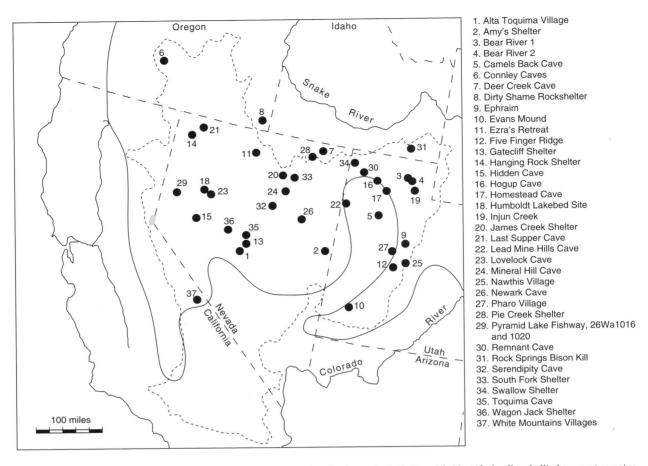

FIGURE 7-10 Distribution of dated Holocene-aged archaeological and paleontological sites with identified yellow-bellied marmot remains in and near the Great Basin.

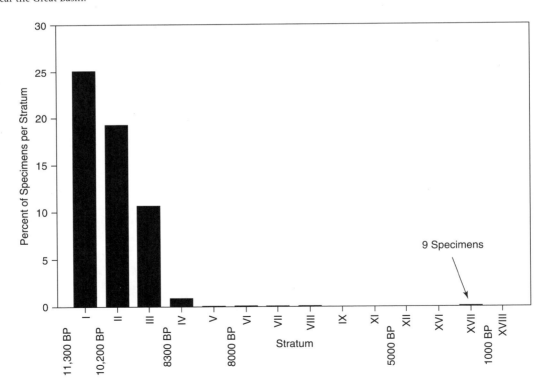

FIGURE 7-11 Changing relative abundance of bushy-tailed woodrats through time at Homestead Cave. The higher the bar, the more abundant were bushy-tailed woodrats in each stratum.

Other histories for bushy-tailed woodrats within the Bonneville Basin parallel the Homestead Cave record reasonably well. At Danger Cave, these animals were common in the earliest deposits (10,600–10,100 years ago) but, unlike the situation at Homestead Cave, their abundance at Danger Cave dropped sharply at the end of the Pleistocene. Just as they did at Homestead, however, they seem to have disappeared here around 8,000 years ago. These animals are not known from the Silver Island Mountains, in which Danger Cave is located, though they are present in the nearby Pilot Range.

Dave Schmitt and Karen Lupo's work at Camels Back Cave has provided a bushy-tailed woodrat record quite similar to that provided by Homestead Cave. They are common in the deepest Camels Back deposits, which appear to span the Pleistocene/Holocene transition and date to as late as 9,560 years BP, and then decline sharply in abundance at about 8,800 years BP, shortly before this happens at Homestead Cave. Finally, Bryan Hockett has shown that at Pintwater Cave, on the eastern edge of the Great Basin in the northern Mojave Desert, small numbers of bushy-tailed woodrats were deposited between 32,000 and 8,300 years BP, but none after that time. Neither the Camels Back nor the Pintwater Cave areas support these animals now.

All of these records are gratifyingly consistent. Bushy-tailed woodrats were common in the late Pleistocene and, to a lesser degree, in the early Holocene, in areas where they either do not now occur or, if they do occur, do so only rarely. Something then happened at about 8,300 years ago to change all this.

Bushy-tailed woodrats have left few records directly south of their southern Great Basin distributional boundary: only Mescal and Kokoweef caves in the Mojave Desert of southeastern California have provided them (figure 7-12). This is somewhat surprising, since bushy-tailed woodrats were found throughout much of New Mexico and into at least western Texas during the Pleistocene (table 7-7). We don't know much about the ages of the Mojave Desert specimens, but none has been reliably dated to the Holocene. What we do know is that, with the marginal exception of Pintwater Cave, which is today on the very edge of bushy-tailed woodrat range, these animals do not seem to have occurred south of their modern distributional boundary since the end of the Pleistocene, with the youngest such record falling at about 10,700 years ago. All dated records

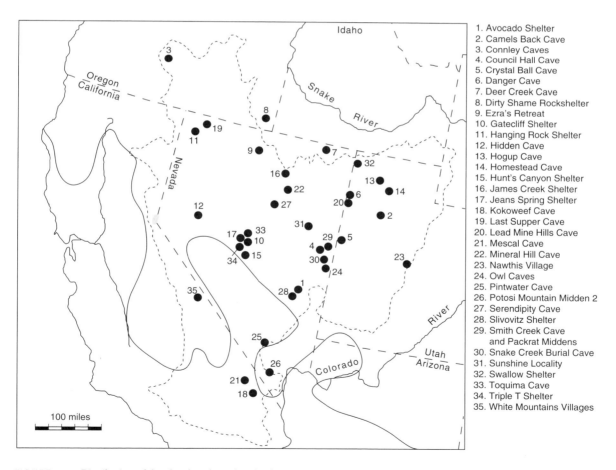

FIGURE 7-12 Distribution of dated archaeological and paleontological sites with identified bushy-tailed woodrat remains in and near the Great Basin.

that younger than this are within, or just on the edge of, that boundary. This, of course, is the same situation we saw for pygmy rabbits and marmots.

Northern Pocket Gophers (*Thomomys talpoides*)

On January 5, 2008, a levee that helps contain the Truckee Canal, which brings Truckee River water to farms near Fallon, Nevada, gave way in the midst of a major winter storm. Massive flooding resulted, forcing about 1,500 people from their homes and causing formidable physical damage. The flooding destroyed any evidence that might have existed for the cause of the breach, but it was quickly suggested that pocket gophers might have been to blame.

I have no idea if pocket gophers were involved in causing this flood but they may have been, since they are superb excavators. Four species of pocket gophers are currently found within the Great Basin, all of which belong to the genus *Thomomys*. Of these, the northern pocket gopher is confined either to the northern, cooler parts of this region, or, in the more southerly parts of its range, to higher, more mountainous settings. When restricted to mountains within its general range, these animals are routinely replaced in adjoining valleys by the larger Botta's pocket gopher, *Thomomys bottae*.

Aspects of the distribution of these rodents have long been recognized as curious. In 1946, E. R. Hall observed that the pocket gopher of the tall and massive Snake Range of eastern Nevada is not the northern pocket gopher, which one would expect, but instead Botta's pocket gopher, the valley-bottom species, which one would not expect. "I suppose," Hall (1946:437) wrote, Botta's pocket gopher "was the only kind of pocket gopher around the bases of Wheeler Peak and Mount Moriah [on the Snake Range] when the higher elevations of these peaks became available for gopher-occupancy." In the absence of northern pocket gophers, he continued, it is likely that Botta's pocket gopher "worked slowly upward, developing populations adapted to living at these higher elevations as it went."

Much the same can be said for the Toquima Range of central Nevada. Here, Hall observed, Botta's pocket gopher is also present, even though the northern pocket gopher is the only member of the genus in the Toiyabe Range to the west and the Monitor Range to the east. In both cases, Hall suggested, the early absence of *Thomomys talpoides* must have allowed *Thomomys bottae* to occupy elevations and habitats it otherwise could not have occupied.

Recent work has documented that Hall was right in suggesting that Botta's pocket gopher can occupy higher elevations when northern pocket gophers are absent. Biologist Eric Rickart has shown that in central Utah's Stansbury Mountains, where there are no northern pocket gophers, Botta's pocket gopher extends 1,300 feet higher than it does in the Uinta Mountains, where northern pocket gophers are present. In the Deep Creek Range just east of the Utah-Nevada border, that figure is 3,800 feet; in the Snake Range, it is 4,200 feet. That is, when the northern pocket gopher is not around to provide competition, Botta's pocket gopher is able to survive quite well in mountainous settings.

When Hall (1946) made his insightful suggestions concerning the possible history of these pocket gophers in the Great Basin, nothing was known of that history. He could conclude that the "antiquity of a species, a presence of one species and absence of another, seems to have been one factor responsible" for the current pattern of species distributions (1946:437) but he could not turn to the fossil record to support or refute this supposition. Today, this is no longer the case, even if we do not have answers to all the distributional questions that Hall raised.

Northern pocket gophers are now found as far east as the North American Plains and as far south as northern New Mexico and Arizona. During the late Pleistocene, however, their distribution was far broader: they were found as far east as Wisconsin and Missouri and as far south as Texas and southern New Mexico.

The late Pleistocene record for northern pocket gophers directly south of their modern Great Basin distributional limit is sparse but informative (figure 7-13), given that all of the southern specimens that are well dated and securely identified are at least 9,000 years old. At Homestead Cave, northern pocket gophers are the only members of the genus present between about 11,300 to 10,200 years BP. All later gopher specimens from this site represent Botta's pocket gopher, the same one that occurs in the region today. Homestead Cave, in fact, provided the first sequence from the Great Basin to suggest that Botta's pocket gopher replaced northern pocket gophers in a low elevation setting, presumably in response to climate change.

The Homestead Cave sequence also suggests that the montane distributional oddities noted by Hall, involving the presence of Botta's pocket gophers in mountains where the northern pocket gopher would otherwise be expected, may not be due to the fact that northern pocket gophers never colonized these areas. After all, northern pocket gophers were in the Homestead Cave area, only to be replaced by Botta's pocket gopher after the Pleistocene came to an end.

Pintwater Cave, southern Nevada, has provided a generally similar sequence. Here, northern pocket gophers are found in deposits that date to between 32,000 and 9,000 year ago; after that time, only Botta's pocket gophers were present. Unlike Homestead Cave, however, *T. bottae* is also present in strata that contained *T. talpoides*, documenting that both mammals were living close enough to this low-elevation (4,160 feet) site for their remains to have been incorporated into it. This is not all that surprising, since the distributions of the two animals can overlap today.

As I mentioned, northern pocket gophers are unexpectedly absent from both the Toquima and Snake ranges. While there are no Pleistocene faunal sequences from in or

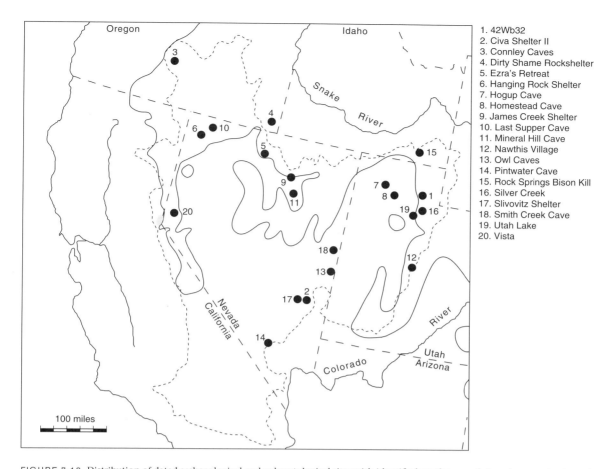

FIGURE 7-13 Distribution of dated archaeological and paleontological sites with identified northern pocket gopher remains in and near the Great Basin.

near the Toquima Range, such data do exist for the Snake Range. Gopher specimens from Owl Caves 1 and 2, in the Snake Valley just east of the Snake Range, were identified as most likely representing northern pocket gophers, all of which are of late Pleistocene age. Owl Cave 2 does not contain Botta's pocket gopher, but Owl Cave 1 does, and all the remains of this animal from this site are from levels higher, and so presumably younger, than those that provided northern pocket gophers.

Not far to the south, Smith Creek Cave—the same cave on the eastern edge of the Snake Range that provided the remains of some of the extinct birds I discuss earlier in this chapter—also provided *T. talpoides*, but not *T. bottae*, of late Pleistocene age (there are no stratigraphically secure records for either species from later deposits at the site). Excavations at the Lehman Caves, in the southern Snake Range, have also provided the remains of northern pocket gophers but the age of these specimens is unknown.

This set of records from the general vicinity of the Snake Range shows that Hall's intriguing speculations on the history of northern and Botta's pocket gophers here cannot be correct. Northern pocket gophers occupied at least the eastern edge of the Snake Range during the late Pleistocene, only to become extinct here, much as they were extirpated from the Mojave Desert in the vicinity of Pintwater Cave and from that part of the Bonneville Basin in which Homestead Cave sits. While it makes sense that northern pocket gophers do not occupy the hot and dry Mojave Desert or the lower elevations of the Bonneville Basin, it is not at all obvious why they do not inhabit the Snake Range.

The Toquima Range presents a similar biogeographic mystery, since it is also marked by the presence of *T. bottae*, not *T. talpoides*. Unfortunately, the only deep fossil record we have from this area, from Gatecliff Shelter, does not go back far enough in time to help much. Gatecliff shows us that Botta's pocket gopher was the only member of the genus in the vicinity of the site during the past 7,300 years or so. This, however, does not take us back far enough in time to know whether northern pocket gophers occurred in this area during the late Pleistocene or early Holocene, only to be replaced by Botta's version later in time. Given the sequences in the southern and eastern Great Basin, however, this now seems more likely than Hall's initial idea.

LATE PLEISTOCENE VERTEBRATES 197

Dark Kangaroo Mice (*Microdipodops megacephalus*)

People living in or near the Great Basin tend to know something about the small mammals I have talked about so far. They have lost socks to woodrats, read that gophers have been blamed for floods, heard that there is concern over the future of pygmy rabbits, and if they hike in the right places, seen marmots in the wild. If they have walked around the desert at night, they have probably seen kangaroo rats—one of the Great Basin's most beautiful mammals—hopping about. Few, however, know about the dark kangaroo mouse.

This is unfortunate, since the genus to which these animals belong, *Microdipodops*, is the only genus of mammal virtually confined to the Great Basin. Their closest relatives include the kangaroo rats (they are placed in the same rodent family, the Heteromyidae), and, like kangaroo rats, have hind legs much longer than their front ones, and tails as long as, or longer than, their bodies. And, like kangaroo rats, they are essentially bipedal, getting around by hopping on their long hind limbs. It is these characteristics—hoppers with long hind legs and long tails—that have led them to be called kangaroo rats and mice.

Of the two species in the genus, only the dark kangaroo mouse (*Microdipodops megacephalus*) has a prehistoric record and even it is quite sparse. It is, though, worth talking about, both because this is almost a strictly Great Basin mammal (it does sneak into central Oregon just north of the Great Basin), and because the fossil record that we do have is so interesting.

Today, dark kangaroo mice are found throughout much of the lower elevations of the central Great Basin but they also occupy two areas geographically apart from the main range of the species. As I mentioned at the end of chapter 6, populations of this sort are in general termed "disjunct" by biogeographers, simply meaning that they occur apart from the main range of the species. The disjunct populations of dark kangaroo mice occur along the California-Nevada border and in the Bonneville Basin of western Utah (see figure 7-14). No matter where they live, these animals prefer habitats that range from the sandy edges of dunes to gravelly soils, and in vegetation ranging from sagebrush to greasewood. In those places, they depend heavily on seeds and insects for their diets.

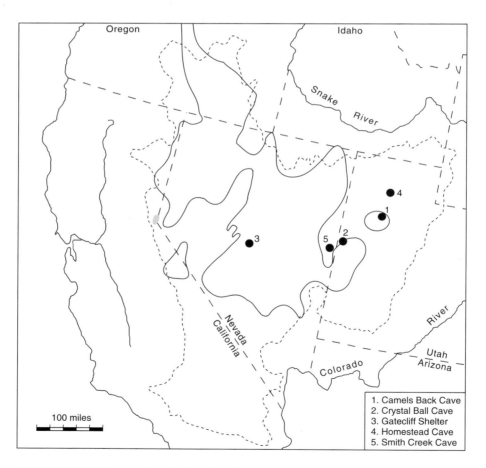

FIGURE 7-14 Distribution of dated archaeological and paleontological sites with identified Great Basin kangaroo mouse remains in and near the Great Basin.

As I mention in chapter 6, disjuncts can occur in only two ways. First, a disjunct population can result from animals that manage to establish themselves in an area they did not occupy before. An obvious example is provided by the Norway rat (*Rattus norvegicus*), whose original homeland is in Asia but which colonized North America with ease. A second is provided by the English (or house) sparrow (*Passer domesticus*), which was purposefully introduced in 1852 in Brooklyn, New York, and which managed to reach Death Valley by about 1914, perhaps by following railroad construction camps. Cattle Egrets, discussed in chapter 6, provide a third example. Histories of this sort provide the dispersal explanations I mention in the previous chapter.

Disjuncts can also occur when once-continuous distributions are somehow disrupted, the basis of the vicariance explanations that I mentioned earlier. Tasmania, for instance, is now an island off the coast of southeastern Australia. During times of lowered sea level during the Pleistocene, the two land masses were connected. This connection allowed the famous Tasmanian devil (*Sarcophilus harrisii*) to roam continuously from what is now Tasmania through much of Australia. When sea levels rose again, the terrestrial connection was severed and the Tasmanian populations became disjunct from those on the mainland. The mainland populations then became extinct during very late prehistoric times, leaving the Tasmanian populations behind.

How, then, did the disjunct populations of dark kangaroo rats in the Great Basin come about? As scant as it is, the fossil record for these animals provides some strong suggestions. Of the five sites that have provided this record, four are within, or on the edge of, its current range. Dave Schmitt and Karen Lupo have shown that at Camels Back Cave, which is within the modern range of the animal, the remains of *M. megacephalus* are scattered throughout the sequence, dating to between the latest Pleistocene and the past few hundred years. The temporal depth of these records suggest that the Utah disjunction is not recent.

Homestead Cave may reveal when this disjunction occurred. Dark kangaroo mice have never been reported from this part of Utah but Homestead Cave contains their remains in deposits that date to between 11,300 and 8,500 years ago but not after that time. These animals seem to have become locally extinct here toward the end of the early Holocene, suggesting that this may also have been the period of time that saw the disjunction of the now-isolated Bonneville Basin populations of this animal.

Thus, the Camels Back Cave record suggests the long-term existence of the disjunct Utah kangaroo mice, while the Homestead Cave sequence suggests that this disjunction occurred toward the end of the early Holocene. More sites with more specimens will someday tell us whether or not this reconstruction is correct, and, if it is, whether or not the disjunct population that occurs along the California-Nevada border dates to the same period.

A Note on American Pikas (*Ochotona princeps*)

Pikas have recently become the Great Basin's most-discussed animal. This is not because they are so attractive that they can cause cute-shock in the unwary, but because there are very good reasons to think that global warming is greatly diminishing their chances for a continued existence here. Both the fossil and historical records for these animals have prompted these concerns, but this is a matter I discuss in detail in the following chapter. Here, I simply want to note how widespread these animals were in the Great Basin during the Pleistocene.

American pikas are small, diurnal herbivores related to rabbits and hares that are now found across the mountains of western North America, from the southern Sierra Nevada and Rocky Mountains to central British Columbia. Within this general range, they are found in talus slopes near the vegetation that provides their diet. These talus slopes not only offer protection from predators but can also provide refuge from warm temperatures: even brief exposure to temperatures in excess of about 80°F can be fatal to them.

The paleontological record, however, shows that these geographic and habitat restrictions are fairly recent. Pikas were found as far east as Ontario, Canada as recently as 9,000 years ago, though the species involved was larger than the American pika. *Ochotona princeps* itself lived in the eastern United States until some 30,000 years ago and perhaps later. In addition, American pika remains from western North American packrat middens show that these animals were at one time living at relatively low elevations in areas that lacked any trace of talus. As a result of all this, the geographic and habitat restrictions of today's American pikas cannot be used to infer where they may have lived in the past.

Because summer temperatures in Great Basin valleys routinely exceed those which pikas can tolerate, it is no surprise that they are now confined to a small subset of Great Basin mountain ranges. Because pikas have weak dispersal abilities, it is very likely that the ancestors of today's pikas colonized these and other Great Basin ranges during the Pleistocene. If this is the case, their current distribution very likely reflects the results of differential extinction across the full set of ranges they once occupied, just as biogeographer Jim Brown has argued. To the extent that this is true, it follows that pikas must have been far more widespread in the Pleistocene than they are now, and that their history since that time must primarily be one of extinction.

The late Pleistocene fossil record shows that this was the case, with sites scattered throughout much of the Great Basin where pikas no longer occur, including many places far south of their current distribution and in areas where talus is not present (table 7-8 and figure 7-15). Not only are

TABLE 7-8
Prehistoric Great Basin Sites Containing Pika Remains, from North to South

Sites yielding pika dung pellets are indicated in italics; all others contain skeletal materials. LP = late Pleistocene; EH = early Holocene; MH = middle Holocene; LH = late Holocene; ND = no secure age known.

Site and Excavation Level (if appropriate)	Elevation (feet)	Distance (miles) to Nearest Current Population	Age	References
Connley Cave 4, Fort Rock Basin, OR: Stratum 3	4,450	35: Paulina Lake, OR	ND	Grayson 1979; Jenkins, Aikens, and Cannon 2002
Connley Cave 4, Fort Rock Basin, OR: Stratum 4	4,450	35: Paulina Lake, OR	ND	Grayson 1979; Jenkins, Aikens, and Cannon 2002
Connley Cave 5, Fort Rock Basin, OR: Stratum 3	4,450	35: Paulina Lake, OR	ND	Grayson 1979; Jenkins, Aikens, and Cannon 2002
Deer Creek Cave, Jarbidge Mountains, NV: 84–90 in.	5,805	70: Steels Creek, NV	EH	Shutler and Shutler 1963
Deer Creek Cave, Jarbidge Mountains, NV: 90–96 in.	5,805	70: Steels Creek, NV	EH	Shutler and Shutler 1963
Deer Creek Cave, Jarbidge Mountains, NV: 108–114 in.	5,805	70: Steels Creek, NV	LW	Shutler and Shutler 1963
Deer Creek Cave, Jarbidge Mountains, NV: 120–126 in.	5,805	70: Steels Creek, NV	LW	Shutler and Shutler 1963
Deer Creek Cave, Jarbidge Mountains, NV: 126–130 in.	5,805	70: Steels Creek, NV	LW	Shutler and Shutler 1963
Hanging Rock Shelter, Hanging Rock Canyon, NV: Stratum 2/4	5,660	40: Hays Canyon, NV	ND	Grayson and Parmalee 1988
Hanging Rock Shelter, Hanging Rock Canyon, NV: Stratum 4	5,660	40: Hays Canyon, NV	ND	Grayson and Parmalee 1988
Hanging Rock Shelter, Hanging Rock Canyon, NV: Stratum 5	5,660	40: Hays Canyon, NV	EH	Grayson and Parmalee 1988
Raven Cave 1a, Raven Cave, NV	4,955	50: Steels Creek, NV	LW	Rhode and Madsen 1995
Mad Chipmunk Cave, Toano Range, NV	6,100	45: Steels Creek, NV	ND	Dively-White 1989, 1990; White 1991
Top of the Terrace 4A, Goshute Range, NV	6,600	45: Steels Creek, NV	LW	Rhode and Madsen 1995
Bronco Charlie Cave, Ruby Mountains, NV	7,000	25: Long Creek, NV	ND	Spiess 1974
Mineral Hill Cave, Sulphur Spring Range, NV	6,760	45: Long Creek, NV	LW	Hockett and Dillingham 2004
Serendipity Cave, Roberts Mountains, NV: Pre-Mazama	7,000	80: Long Creek, NV	EH	Livingston 1992a
Granite Canyon 1, Deep Creek Range, UT	6,790	100: Currant Mountain, NV	LW	Thompson and Mead 1982; R. S. Thompson, pers. comm.
Crystal Ball Cave, Snake Valley, UT	5,775	85: Currant Mountain, NV	LW	Heaton 1985
Smith Creek Cave, Snake Range, NV: Grey	6,400	75: Currant Mountain, NV	ND	Miller 1979; Mead, Thompson, and Van Devender 1982
Smith Creek Cave, Snake Range, NV: Reddish-Brown	6,400	75: Currant Mountain, NV	LW	Miller 1979; Mead, Thompson, and Van Devender 1982
Smith Creek Cave 4, Snake Range, NV	6,400	75: Currant Mountain, NV	LW	Thompson and Mead 1982; R. S. Thompson, pers. comm.
Smith Creek Cave 5, Snake Range, NV	6,400	75: Currant Mountain, NV	LW	Thompson and Mead 1982; R. S. Thompson, pers. comm.
Council Hall Cave, Snake Range, NV	6,695	75: Currant Mountain, NV	LW	Miller 1979; Mead, Thompson, and Van Devender 1982

TABLE 7-8 (CONTINUED)

Site and Excavation Level (if appropriate)	Elevation (feet)	Distance (miles) to Nearest Current Population	Age	References
Streamview Rockshelter 1, Snake Range, NV	6,100	75: Currant Mountain, NV	LW	Thompson and Mead 1982; R. S. Thompson, pers. comm.
Streamview Rockshelter 2, Snake Range, NV	6,100	75: Currant Mountain, NV	LW	Thompson and Mead 1982; R. S. Thompson, pers. comm.
Streamview Rockshelter 3, Snake Range, NV	6,100	75: Currant Mountain, NV	MH	Thompson and Mead 1982; R. S. Thompson, pers. comm.
Arch Cave 2A, Snake Range, NV	6,495	70: Currant Mountain, NV	LW	Mead and Spaulding 1995; http://esp.cr.usgs.gov/data/midden
Gatecliff Shelter, Toquima Range, NV: Stratum 1, H3	7,610	20: Mt. Jefferson, NV	LH	Grayson 1983
Gatecliff Shelter, Toquima Range, NV: Stratum 9	7,610	20: Mt. Jefferson, NV	LH	Grayson 1983
Gatecliff Shelter, Toquima Range, NV: Stratum 20	7,610	20: Mt. Jefferson, NV	MH	Grayson 1983
Gatecliff Shelter, Toquima Range, NV: Stratum 22	7,610	20: Mt. Jefferson, NV	MH	Grayson 1983
Gatecliff Shelter, Toquima Range, NV: Strata 24–25	7,610	20: Mt. Jefferson, NV	MH	Grayson 1983
Gatecliff Shelter, Toquima Range, NV: Strata 31–32	7,610	20: Mt. Jefferson, NV	MH	Grayson 1983
Gatecliff Shelter, Toquima Range, NV: Stratum 33	7,610	20: Mt. Jefferson, NV	MH	Grayson 1983
Gatecliff Shelter, Toquima Range, NV: Stratum 37	7,610	20: Mt. Jefferson, NV	MH	Grayson 1983
Gatecliff Shelter, Toquima Range, NV: Stratum 54	7,610	20: Mt. Jefferson, NV	MH	Grayson 1983
Gatecliff Shelter, Toquima Range, NV: Stratum 56	7,610	20: Mt. Jefferson, NV	MH	Grayson 1983
Garrison 1, Snake Valley, UT	5,380	70: Currant Mountain, NV	LW	Thompson and Mead 1982; R. S. Thompson, pers. comm.
Garrison 2, Snake Valley, UT	5,380	70: Currant Mountain, NV	LW	Thompson and Mead 1982; R. S. Thompson, pers. comm.
Snake Creek Burial Cave, Snake Valley, NV	5,680	70: Currant Mountain, NV	LW	J. I. Mead, pers. comm.
Owl Cave 2, Snake Valley, NV: Level 2	5,575	70: Currant Mountain, NV	LW	Turnmire 1987
Owl Cave 2, Snake Valley, NV: Level 4	5,575	70: Currant Mountain, NV	LW	Turnmire 1987
Owl Cave 2, Snake Valley, NV: Level 5	5,575	70: Currant Mountain, NV	LW	Turnmire 1987
Owl Cave 2, Snake Valley, NV: Level 9	5,575	70: Currant Mountain, NV	LW	Turnmire 1987
Rock Shelter, Mineral Mountains, UT	6,190	30: Britts Meadows, UT	ND	Schmitt and Lupo 1995; D. N. Schmitt, pers. comm.
Corral North, White Mountains, CA	10,990	0: Site Vicinity	LH	Grayson 1991
Corral South, White Mountains, CA	10,990	0: Site Vicinity	LH	Grayson 1991
Enfield, White Mountains, CA	10,400	0: Site Vicinity	LH	Grayson 1991
Midway, White Mountains, CA	11,285	0: Site Vicinity	LH	Grayson 1991
Rancho Deluxe, White Mountains, CA	11,680	0: Site Vicinity	LH	Grayson 1991
Crooked Forks, White Mountains, CA	10,335	0: Site Vicinity	LH	Grayson 1991
Eleana Range 2-7, Eleana Range, NV	5,940	115: Greenmonster Canyon, NV	LW	Mead and Spaulding 1995 http://esp.cr.usgs.gov/data/midden

(continued)

TABLE 7-8 (CONTINUED)

Site and Excavation Level (if appropriate)	Elevation (feet)	Distance (miles) to Nearest Current Population	Age	References
Eleana Range 2-10, Eleana Range, NV	5,940	115: Greenmonster Canyon, NV	LW	Mead and Spaulding 1995; http://esp.cr.usgs.gov/data/midden
Eleana Range 2-11r, Eleana Range, NV	5,940	115: Greenmonster Canyon, NV	LW	Mead and Spaulding 1995; http://esp.cr.usgs.gov/data/midden
Eleana Range 3-10, Eleana Range, NV	5,905	115: Greenmonster Canyon, NV	LW	Mead and Spaulding 1995; http://esp.cr.usgs.gov/data/midden
Mormon Mountain Cave, Mormon Mountain, NV	4,500	110: Brian Head, UT	LW	Jefferson 1982
Fortymile Canyon 11, Yucca Mountain, NV	4,300	110: Cottonwood Lakes, CA	LW	Mead and Spaulding 1995; http://esp.cr.usgs.gov/data/midden
Pintwater Cave, Pintwater Range, NV: Unit 3, Level 2	4,160	160: Brian Head, UT	MH	Hockett 2000; http://esp.cr.usgs.gov/data/midden
Pintwater Cave, Pintwater Range, NV: Unit 3, Level 4	4,160	160: Brian Head, UT	EH	Hockett 2000; http://esp.cr.usgs.gov/data/midden
Pintwater Cave, Pintwater Range, NV: Unit 3, Level 5	4,160	160: Brian Head, UT	LW	Hockett 2000; http://esp.cr.usgs.gov/data/midden
Spires 2, Sheep Range, NV	6,695	155: Brian Head, UT	LW	Mead and Spaulding 1995; http://esp.cr.usgs.gov/data/midden
Flaherty Mesa FM1, Sheep Range, NV	5,805	155: Brian Head, UT	LW	Mead and Spaulding 1995; http://esp.cr.usgs.gov/data/midden
Willow Wash 4E, Sheep Range, NV	5,200	155: Brian Head, UT	LW	Mead and Spaulding 1995; http://esp.cr.usgs.gov/data/midden
Penthouse 1, Sheep Range, NV	5,250	155: Brian Head, UT	LW	Mead and Spaulding 1995; http://esp.cr.usgs.gov/data/midden
Corn Creek PR3, Las Vegas Valley, NV	3,480	160: Cottonwood Lakes, CA	LW	Mead and Spaulding 1995; http://esp.cr.usgs.gov/data/midden
Potosi Mountain 2A1, Spring Range, NV	6,170	160: Cottonwood Lakes, CA	LW	Mead and Murray 1991
Potosi Mountain 2A2, Spring Range, NV	6,170	160: Cottonwood Lakes, CA	LW	Mead and Murray 1991
Potosi Mountain 2C2, Spring Range, NV	6,170	160: Cottonwood Lakes, CA	LW	Mead and Murray 1991
Mescal Cave, Mescal Range, CA	5,085	165: Cottonwood Lakes, CA	LW	Jefferson 1991
Antelope Cave, Ivanpah Mountains, CA	5,800	165: Cottonwood Lakes, CA	LW	Reynolds et al. 1991b; FAUNMAP Working Group 1994
Kokoweef Cave, Ivanpah Mountains, CA	5,805	170: Cottonwood Lakes, CA	LW	Goodwin and Reynolds 1989; Reynolds et al. 1991a; FAUNMAP Working Group 1994

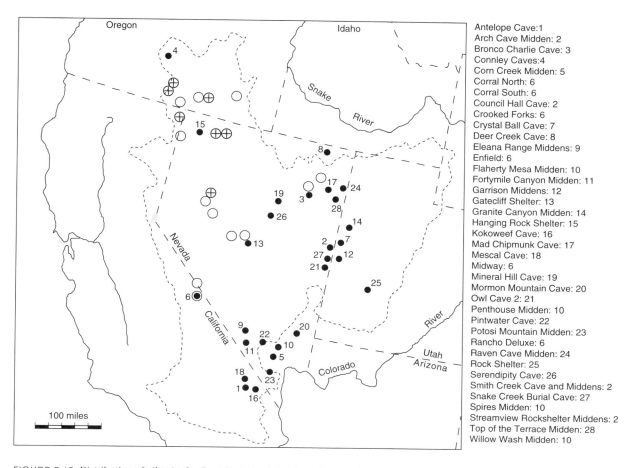

FIGURE 7-15 Distribution of pikas in the Great Basin (updated from Grayson 2005). Closed circles show prehistoric archaeological and paleontological records; large open circles mark living populations; large open circles with crosses mark populations lost during historic times. (Each circle may represent more than one population; populations in the Sierra Nevada and Rocky Mountains are not shown.)

these sites found in parts of the Great Basin where pikas do not exist today, but they tend to be at far lower altitudes, averaging about 5,750 feet in elevation, some 2,500 feet beneath the average elevation at which Great Basin pikas now live. This is a huge difference. The fact that pikas are now largely confined to talus slopes on a small subset of Great Basin mountains is very much a product of the past 10,000 years, as I discuss in chapter 8.

Some Other Late Pleistocene Mammalian Surprises

Heat-intolerant pikas living a half-mile beneath their current average elevation in the Great Basin, pygmy rabbits in what is now shadscale vegetation, marmots in the Mojave Desert, northern pocket gophers and bushy-tailed woodrats living in areas that today cannot possibly support them—all this shows a late Pleistocene Great Basin dramatically different from what is encountered here now. Had I chosen to start my discussion of the Pleistocene with mammals, some of this might have come as a big surprise. However, given what I have said about the history of Great Basin lakes, glaciers, and plants, with the implications of all that for a colder and wetter late Pleistocene, the altered distributions of Great Basin small mammals should not be not surprising at all.

What would be surprising is if no other mammals showed similar altered late Pleistocene distributions. In fact, many of them do. For instance, Mescal and Kokoweef caves in southeastern California both contain specimens of golden-mantled ground squirrels (*Spermophilus lateralis*), apparently of late Pleistocene age, though the dating is not secure. Geographically, these caves fall between two modern, isolated subspecies of golden-mantled ground squirrels, one of which lives in the Spring Range of southern Nevada, the other occupying the San Bernardino Mountains of southern California. The Mescal and Kokoweef cave specimens may well reflect the time when, prior to the isolation of these two recent subspecies, golden-mantled ground squirrels were continuously distributed across this region, making the populations in the Spring Range and San Bernardino Mountains relics of that prior time.

Golden-mantled ground squirrels are still found in the Great Basin, and thus differ from a series of now-northern mammals not found here today (table 7-9). Not long ago,

TABLE 7-9
A Sampler of Pleistocene Records for Northern Mammals That No Longer Occur in the Great Basin
Sites shown in figure 7-1.

Location	Elevation (feet)	Species	Age (years ago)	References
Crystal Ball Cave	5,775	*Martes americana*	Late Wisconsin?	Heaton 1985
Snake Creek Burial Cave	5,680	*Martes americana*	Late Wisconsin	E. M. Mead and J. I. Mead 1989
		Mustela nivalis	Late Wisconsin	
Smith Creek Cave	6,400	*Phenacomys intermedius*	>12,600	Mead, Thompson, and Van Devender 1982; Jass 2007
		Synaptomys borealis?	>12,600	

it was suggested that northern bog lemmings (*Synaptomys borealis*) had occurred in the Great Basin during the late Pleistocene. These small rodents are primarily northern in distribution, barely entering the far northern United States along the Canadian border. They generally inhabit bogs in spruce forests but can also be found in such settings as subalpine meadows and alpine tundra. It has been known for quite some time that they were found well south of their current distribution toward the end of the Pleistocene, having been discovered in cave deposits in, for instance, Tennessee. They have also reported from Cathedral Cave and Smith Creek Cave, nearby sites in eastern Nevada's Snake Range. However, more recent analysis of the Cathedral Cave specimens by Chris Bell and Chris Jass suggests that these come instead from an extinct species of bog lemming. Newly derived dates obtained by Jass from this site show that the fossils here may be as much as 150,000 years old. The Smith Creek Cave material has not been reanalyzed but this needs to be done if we are to be sure which species of lemming is involved.

Jim Mead has identified the heather vole (*Phenacomys intermedius*) in late Pleistocene deposits at Smith Creek Cave. Specimens were also reported from Cathedral Cave but, as with the lemmings, these appear to belong to an extinct species. Today, heather voles are found only on the easternmost and westernmost fringes of the Great Basin: in the Carson Range in western Nevada and in the Uinta and Wasatch ranges of central Utah. Although these mouse-sized rodents occupy a variety of habitats, they are primarily animals of coniferous forests, regularly living at and above treeline. While this site provides the only late Pleistocene record for the heather vole in the Great Basin, it survived in the Great Basin well into the Holocene. In fact, the first record from this area was not from the Pleistocene, but from deposits that date to about 5,300 years ago from Gatecliff Shelter in the Toquima Range. I presume, but do not know, that it was fairly widespread in the Great Basin during the later Pleistocene, and that it began to be lost from this area as the Pleistocene ended, with the Toquima Range perhaps harboring some of its last representatives. On the other hand, I have long wondered whether it might not still be living at the top of one or more isolated Great Basin mountains, though the chances of this seem very slim.

Many other small mammals could be discussed here. For instance, Thomas Goodwin and Robert Reynolds have shown that Townsend's ground squirrels (*Spermophilus townsendii*) and least chipmunks (*Tamias minimus*) occupied low elevations in the Mojave Desert of southeastern California toward the end of the Pleistocene. These squirrels would presumably have lived in the coniferous woodlands and sage-covered valleys found in the Mojave at this time (see chapter 6). Today, both species are widespread in the sagebrush communities of the Great Basin far to the north, but neither occurs in the Mojave Desert.

Small, cold-country carnivores also occurred in the Great Basin during the late Pleistocene. The first of these is the pine marten (*Martes americana*), a mink-sized predator of boreal forests. Today, it is found in both the Sierra Nevada and Rocky Mountains, but it has never been seen in the flesh in the Great Basin. The Great Basin late Pleistocene records come from two Snake Valley sites: Crystal Ball Cave, and Snake Creek Burial Cave. Then there is the least weasel (*Mustela nivalis*). This species can be found in the farmlands of the northern Midwest but in western North America, it is found only in the north. Indeed, west of 104° longitude (approximately the eastern border of Wyoming and Montana), it comes no farther south than northern Montana. It is known from one late Pleistocene site in the Great Basin, Snake Creek Burial Cave, in Snake Valley, and will surely be found elsewhere as our knowledge of the fossil record increases.

Small Mammal Histories and Pleistocene Extinctions in the Great Basin

It is very tempting to connect our knowledge of late Pleistocene and early Holocene small mammal histories in the Great Basin with the extinctions of the mammals and birds that I discussed earlier in this chapter. After all, at least seven of the large mammals were on their way out at about the same time as many small mammals were undergoing, or about to undergo, very significant shifts in their distribution and abundance (table 7-3). Indeed, there are dramatic changes in the Great Basin's latest Pleistocene small mammal fauna in addition to those I have discussed here. For instance, a distinct form of the sage vole (*Lemmiscus curtatus*) appears to have become extinct in the Great Basin (and perhaps everywhere that it occurred) at the very end of the Pleistocene. However, as tempting as it might be to link the extinctions with these dramatic late Pleistocene changes in the small mammal fauna, and as much as I suspect the two are related in more than a coincidental way, this is a temptation perhaps best resisted for now.

Our understanding of Great Basin small mammal history as the Pleistocene came to an end shows that these changes were part of a lengthy process, as opposed to an event. Marmots did not disappear from the Homestead Cave area at the end of the Pleistocene; they simply became less abundant, and the same is true for pygmy rabbits and bushy-tailed woodrats. Of course, the climate changes that caused this process might also have at least helped cause the extinctions; in fact, I would be astonished if they didn't. However, we certainly do not know this. In addition, we will have no real opportunity to know it until we have a far better understanding of the changing abundances, and of the chronology of those changing abundances, of each of the extinct birds and mammals across this period of time, as well as a better understanding of the details of climate change within the Great Basin itself. As I said at the outset of this discussion, the depth of our knowledge of the latest Pleistocene and Holocene history of Great Basin small mammals is among the most detailed available anywhere in the world. Our knowledge of the nature of the bird and mammal extinctions, on the other hand, is frustratingly weak. We will not be able to put the two records together in any compelling way until this situation changes.

Mammals on Mountaintops: J. H. Brown's Model of Great Basin Montane Mammal Biogeography

Pikas were not present at Homestead or Camels Back caves although they are currently found in the Rocky Mountains to the east and are known from a series of Pleistocene localities along the Utah-Nevada border just to the west of the Bonneville Basin. Indeed, there are no prehistoric records at all for pikas in the Bonneville Basin except for those along its far western edge, nor do any mountains within the Bonneville Basin support them today. It now seems possible that pikas either did not colonize the Bonneville Basin during the Pleistocene, or, if they did, did not survive into the latest Pleistocene and early Holocene, when fossil records become thicker on the landscape.

If this were the case, it may be related to the fact that of the forty-six species of mammals found in the Uinta Mountains of northeastern Utah and surveyed by biologist Erik Rickart, pikas are one of only four species that today do not descend to elevations of less than about 7,550 feet. The other three—the water vole, *Microtus richardsoni*; the snowshoe hare, *Lepus americanus*; and the western heather vole, *Phenacomys intermedius*—are also unknown from within the Bonneville Basin. All four, Rickart pointed out, seem to be completely confined to mountains in this part of the world.

Pikas were one of a series of mammals used by Jim Brown in 1971 to produce what soon became the most important analysis of Great Basin mammal biogeography ever to have appeared. Brown examined the distribution of fifteen species of "boreal" or "montane" mammals—mammals that "occur only at high elevations in the Rockies and Sierra Nevada and are unlikely to be found below 7,500 feet at the latitudes of the Great Basin" (1971:469)—across seventeen Great Basin mountain ranges. His quantitative analyses of this distribution led him to conclude that these mammals had reached the mountains during the Pleistocene, that opportunities for colonization ended when the Pleistocene ended, and that since then, montane mammals have become differentially extinct across the ranges. As I mentioned above, this view seems to explain pika history on the Great Basin fairly well; we will see exactly how well in the following chapter.

Biogeographer Mark Lomolino has pointed out that Brown's Great Basin analysis helped spur the study of the biogeography of mountains in many parts of the world. As part of this, it either directly stimulated, or strongly influenced, an amazingly wide variety of detailed analyses of the past and present distributions of mammals in arid western North America.

The end result of this work in the Great Basin was to show that Brown's model was incorrect. When he built that model, he could not have known that the distribution of montane mammals across Great Basin mountains was inadequately understood. As mammalogist Tim Lawlor has elegantly shown, as those distributions became better understood, support for Brown's model dissolved. At the same time, analyses of the fossil record made it clear that the late Pleistocene and Holocene history of Great Basin mountain faunas has been a dynamic one, shaped by both colonization and extinction events, some of which may have been quite recent. In line with this, genetic analyses of living populations of yellow-bellied marmots—one of the mammals that Brown treated as montane—have shown that these animals have moved between mountain ranges in the Great Basin in relatively recent times, probably by taking advantage of the hilly areas that connect them.

Second, rather than acting as a homogeneous unit, each of the species within Brown's set of montane mammals can be expected to have its own independent history. To use an example from Lawlor's work, Nuttall's cottontails (*Sylvilagus nuttallii*) are routinely found below elevations of about 7,500 feet in the Great Basin, but pikas are not; both were treated as montane in Brown's approach. Different environmental tolerances of this sort can be expected to lead to very different colonization and extinction histories.

Homestead Cave provides an example of these processes in action, since it provides histories for a number of the taxa included in Brown's initial analysis, including northern pocket gophers, yellow-bellied marmots, and bushy-tailed woodrats. Each has a distinctly different history here. Northern pocket gophers appear to have been become locally extinct at the end of the Pleistocene but marmots and bushy-tailed woodrats lasted until the end of the early

Holocene. Marmots have not returned, but, in a story I tell in the following chapter, bushy-tailed woodrats have. These are precisely the kind of individualistic species responses that I discuss in the previous chapter where I contrasted the views of Frederick Clements and Henry Gleason. Responses of this sort are very well known paleontologically; the Great Basin is not some kind of exception in this regard. Because they are the rule rather than the exception, I suspect that every paleoecologist would agree with Brown's conclusion that "with the notable exception of rare obligate interdependencies, each kind of organism tends to shift its abundance and distribution individualistically as environmental changes affect its unique tolerances and requirements" (2004:364).

Why Are There Fishes in Devils Hole but None in the Great Salt Lake?

The Fishes of Devils Hole

Unlike some of the montane mammals of the Great Basin, the fish that occupy separate and often minute drainages within the Great Basin are obviously fully isolated from their nearest neighbors in other such drainages. Examples can be drawn from any of the fifty-two species of fishes native to the Great Basin. Indeed, ichthyologists Carl L. Hubbs and Robert Rush Miller wrote a series of classic monographs in which they detailed the distribution of fishes in the modern Great Basin and used those distributions to reconstruct past relationships among now-isolated Great Basin drainage systems. But the most remarkable instance of isolation among Great Basin fishes is provided by the pupfishes and springfishes that survive in the various drainages that once flowed into Death Valley.

These fish belong to two different families, the Cyprinodontidae and the Empetrichthyidae. Fishes belonging to the latter, springfish, family are found only in Nevada, but the cyprinodonts are very widespread, found in Africa, Asia, Europe, and both North and South America. The Great Basin supports, or supported until very recently, nine species of these two families of fish.

The four empetrichthyids include two that still exist in their native habitat: the White River springfish (*Crenichthys baileyi*) of the White River drainage system in eastern Nevada, and the Railroad Valley springfish (*Crenichthys nevadae*) in south-central Nevada's Railroad and Duckwater valleys. The Pahrump poolfish (*Empetrichthys latos*) was once found in springs in the Pahrump Valley but now exists only in refuges elsewhere. The Ash Meadows poolfish (*Empetrichthys merriami*) is simply gone.

My emphasis here, however, is going to be on the pupfishes of the genus *Cyprinodon*. There are about thirty species of this genus in arid southwestern North America, extending from the southern Great Basin into Mexico and western Texas. Genetic analysis has shown that that nine of these species form a closely related group that shares a common ancestor—ten if we include the cottonball marsh pupfish as a separate species. Five of these are found in the Great Basin, all of them in drainages that at one time contributed their waters to Death Valley's Lake Manly (for a discussion of this drainage system, see chapter 5).

These five species, and their locations, are listed in table 7-10. All are tiny (1 to 2.5 inches long), the males often, though not

TABLE 7-10
The Ten Closely Related Pupfishes of the Great Basin and Other Parts of the Arid West

Species	Common Name	Distribution
Great Basin Species		
Cyprinodon radiosus	Owens Pupfish	Owens River Valley
Cyprinodon nevadensis	Amargosa Pupfish	Amargosa River Drainage
Cyprinodon salinus	Salt Creek Pupfish	Salt Creek, Death Valley
Cyprinodon milleri[a]	Cottonball Marsh Pupfish	Death Valley
Cyprinodon diabolis	Devils Hole Pupfish	Devils Hole (Ash Meadows)
Closely Related Arid Western Species		
Cyprinodon fontinalis	Carbonera Pupfish	Guzmán Basin, Chihuahua
Cyprinodon pisteri	Palomas Pupfish	Guzmán Basin, Chihuahua
Cyprinodon albivelis[b]	Whitefin Pupfish	Papigóchic Basin, Chihuahua
Cyprinodon eremus	Sonoyta Pupfish	Sonoyta Basin, Sonora
Cyprinodon macularius	Desert Pupfish	Lower Colorado River Basin

SOURCE: After Smith et al. 2002 and Echelle 2008; common names are from Froese and Pauly 2008.
[a]Often treated as a subspecies of *C. salinus*.
[b]Also found in the Guzmán Basin, where it may have been introduced.

always, brightly colored and aggressively territorial. They are denizens of springs and slow-flowing streams with the ability to live in remarkably warm and highly saline waters.

As fascinating as the adaptations of these fishes are, the most evident mystery they present is the fact that they are there at all. The most extreme example is provided by the smallest pupfish known, the one-inch-long Devils Hole pupfish, *Cyprinodon diabolis*. As Edwin Pister has pointed out, this fish occupies the smallest habitat of any viable population of vertebrates in the world.

Devils Hole is a narrow crevice in the Devils Hole Hills, a low range in southwestern Nevada's Ash Meadows, about thirty miles east of Death Valley, and some 100 feet to 150 feet above the adjacent floor of the Amargosa Desert. Today, it is within a detached part of Death Valley National Park, surrounded by Ash Meadows National Wildlife Refuge. Approximately 55 feet beneath the rim of this crevice lies the surface of a pool that measures some 23 feet by 10 feet; the uppermost waters of this pool maintain a nearly constant temperature of about 93°F, making it easy to understand why the old name for Devils Hole was the Miner's Bathtub. The pool itself is fed by a vast underground aquifer that draws much, though not all, of its water from the Spring Mountains to the east. It would be nice to know how deep it is, but that is not a question that can be answered, since the bottom has never been reached. Divers have gotten as far down as 436 feet and, from there, could see down to about 500 feet, but it keeps going beyond that.

The Devils Hole pupfish, however, care little about the depth of the pool. They descend no farther than about 85 feet, the depth to which sunlight penetrates, and the bulk of their activities is concentrated on a limestone shelf, some 6.5 feet by 13 feet across, which is covered by a few inches to a few feet of water. It is here that the fish feed and breed, and to which they quickly return after disturbance.

Counts provided by Alan Riggs and James Deacon show that between 1972 and 2003, an average of about 330 pupfish lived here. The highest number, 577, was reached in 1990, but the population began to decline in 1997 and, for reasons unknown, continued to decline after that. By April 2006, only 38 were left, the lowest number ever counted. In response, the U.S. Fish and Wildlife Service began an artificial feeding program. As I write this, there are 126 pupfish in Devils Hole.

The decline is depressing on many levels. Late in the 1960s, the pumping of groundwater in the region began to cause a potentially disastrous decline in the Devils Hole pool. Fortunately, a 1976 U.S. Supreme Court decision required that the pool be kept sufficiently high to insure the continued existence of this remarkable fish. Eight more years were to pass before the necessary safeguards were actually in place, but they did come to be, including the establishment, in 1984, of Ash Meadows National Wildlife Refuge. This refuge has helped to protect the Devils Hole pupfish, but also supports more than two dozen kinds of plants and animals found nowhere else. All of these efforts led to an increase in the water level in Devils Hole as well as a rebound in the number of fishes. But whatever has been causing the recent decline in pupfish may negate many of these important successes.

How did the pupfish get to Devils Hole in the first place? How did Death Valley's Salt Creek Cottonball Marsh pupfishes get to where they live? How, in fact, did any of the pupfishes and poolfishes of the Death Valley drainage system get to where they are now?

Certain biogeographic conundrums permit fairly simple solutions. The isolated populations of dark kangaroo mice that I discuss earlier in this chapter seem to provide an example. But there are other such conundrums whose answers are completely elusive. The pupfishes of the Mojave Desert provide one of the best examples I know.

I start with Devils Hole itself, beautifully described by Alan Riggs and James Deacon as a desert "skylight to the water table" (2004:1). No one thinks that pupfishes could have arrived here by transporting themselves through the underground aquifer that supplies the water in which the fishes now live. The fish today confine themselves to areas that receive sunlight, they lack any adaptations to an underground existence, and the waters they would have had to use to get here would not have provided the food they needed to survive. If all this is correct, the fish must somehow have gotten here by surface waters that once connected Devils Hole to other parts of the Amargosa Basin.

Several seemingly well-established facts come into play here. Work by geologist Isaac Winograd and his colleagues has shown that Devils Hole has had water in it for at least the past 560,000 years. However, it most certainly has not been open to the surface for all that time. In fact, as Alan Riggs has observed, the geology of the place strongly suggests that it became a skylight to the water table around 60,000 years ago, apparently as a result of roof collapse. If this is correct, and it certainly seems to be, pupfish could not have arrived here until after that time.

So far so good. Devils Hole is ancient and the fish must have gotten here within the past 60,000 years. Obviously, the water must have risen high enough during that time to overflow, providing a connection with local surface waters and thus a transportation corridor for the fish. The water level then declined, leave the poor fish stranded. This, in fact, is exactly what ichthyologist Robert Miller suggested in 1981.

This would be the perfect explanation were it not for the fact that it doesn't seem to have happened. Barney Szabo and his colleagues have shown that during the past 116,000 years, the water in Devils Hole has risen no higher than about thirty feet from where it sits today. If this is correct, the water level would still have been some twenty-six feet lower than its overflow level. Of course, this may not be correct; the reconstruction by the Szabo team does include a few intervals during the past 60,000 years during which the water level is not well constrained. However, if there had been an overflow here that lasted long enough to provide a substantial connection with other local surface waters, one that allowed pupfish to colonize Devils Hole itself, then one would expect there to be geological evidence around Devils Hole—rocks eroded by water along the edges of the feature, for instance. Since no such evidence exists, we lack an obvious way for the fish to have gotten here. One is led to think seriously about the possibility of pupfish eggs attached to the feet or feathers of birds or some other highly unlikely means of dispersal, the kind of low-odds mechanism that paleontologist George Gaylord Simpson once called "sweepstakes dispersal." More likely, though, we are missing something obvious, the most likely being evidence for overflow significant enough to provide a connection to other local surface waters but brief enough to leave little geological evidence of its prior existence. But the bottom line is that we don't have a clue as to how the fish got here.

As perplexing as this is, though, it is not much more perplexing then the puzzle posed by what we know of the genetics of North American pupfishes as a whole. As I mention above, nine arid western North American pupfishes are so closely related that they must share a common ancestor. Some would say that there are ten, but the Salt Creek Marsh (*Cyprinodon salinus*) and Cottonball Marsh (*C. milleri*) forms are so closely related that many treat the latter as a subspecies of the former. Since the two are so genetically similar, how these species are treated taxonomically does not matter at all for what follows.

The undisputed fact that these species all share a common ancestor means that there must have been a way for pupfishes to distribute themselves from Owens and Death Valleys on the north to northern Mexico on the south. That, in turn, would seem to require that the Owens-Death Valley system was at one time connected to the Colorado River system. When that "some time" was can to some shaky degree be estimated from the genetic data, under the basic assumption that the longer groups have been separated, the more divergent the genetic composition of those groups (many other things equal). Anthony Echelle and his colleagues have done this for the nine western pupfish, deriving a date of between 2.7 and 3.8 million years ago for their last common ancestor, and of between 5.4 and 6.3 million years ago for the origin of that ancestor. As these researchers emphasize, these dates have to be used with great care, but they at least give us a ball park in which to play.

As a result of all this, biologist Gerald Smith, whose work on the history of Great Basin fishes has been of tremendous importance, has, with his colleagues, concluded that "connections from the lower Colorado River to the Owens, Mojave, and Amargosa Rivers occurred about 3 ma [million years ago]" (2002:195). This is a completely reasonable conclusion, one made by Robert Miller before him and one that Echelle has very tentatively made as well, though he also suggests that the connection may have been as early as 6 million years ago.

Interestingly enough, a fossil pupfish, *Cyprinodon breviradius*, is known from Death Valley, and it seems to date to somewhere within this time period, establishing that pupfishes were actually here by then. Gerald Smith's analysis of it suggests to him that it is likely to be ancestral to a group that includes the Amargosa and Death Valley pupfishes (*C. salinus*, *C. diabolis*, and *C. nevadensis*) as well as one of the Chihuahuan species, *C. fontinalis*. Despite the seeming oddity of the connection Smith builds between the Death Valley pupfishes and one of the northern Mexican forms, the genetic data suggest exactly the same thing: that these species are more closely related to one another than they are to the other species in the western group. In fact, a similar relationship is shown by the water bugs (Heteroptera) of this part of the world, with forms from the Amargosa Desert also found in northern Mexico but not in-between.

Also helping out is a group of tiny mollusks, called springsnails, that belong to the genus *Pyrgulopsis*. Scientists have described 131 species of this genus, living from northern Oregon and Idaho south into Mexico. Of these, 80 occur in the Great Basin. As their name suggests, these animals primarily live in springs, though they can be found in other watery settings. A great deal of effort has been directed toward understanding springsnail biogeography, with Robert Hershler and his colleagues having done much of the recent work. Among many other things, they have shown that some of the Amargosa River Basin species have their closest relative in the lower Colorado River Basin, a situation very similar to the one we have already seen for the pupfish.

Some of these disjunct distributions might be explained by sweepstakes dispersal. Birds with snails stuck to their feet could do the job, as is thought to have been the case for a tiny snail, belonging to the genus *Assiminea*, that has very closely related forms living in both Death Valley and the lower Colorado River Valley. However, the full set of disjunct populations makes it far more likely that many aspects of the modern distributions of Great Basin pupfish are best explained by an ancient connection between the Owens Valley–Death Valley system and the lower Colorado River Basin.

There is just one problem with this explanation: no geological evidence supports it. In fact, many geologists are insistent that there was no such connection during the past 3 million years, and suggest that the last time this might have happened was on the order of 6 million years ago. Six million years ago would be fine, except for that fact that there is no evidence for this either.

So the pupfishes of the Great Basin pose unsolved problems every step of the way, from understanding their general distribution to understanding how they managed to get into Devils Hole. The biological data strongly suggest that there must have been a connection between the Death Valley and the Colorado drainages, to get pupfish here in the first place. Those data also suggest that there must have been a surface-water connection between Devils Hole and some other part of the Amargosa Basin that contained the fish. All we need are geological data to either give these biological hypotheses more credence than they have now or to show us that they are wrong

Perhaps the most important problem regarding pupfish, though, is keeping them alive in the face of all the problems we are creating for them.

The Fishes of Great Salt Lake and Lake Bonneville

It is far easier to explain why there are no fishes in the Great Salt Lake. Although twenty species of native fish live in the Bonneville drainage as a whole (table 7-11), and although fish can be found near the mouths of tributaries that drain into Great Salt Lake, the lake itself has none because it is too saline to support them.

This was not the case during the late Pleistocene. We know this because eleven species of fishes are known from the deposits of Lake Bonneville: the ten indicated in table 7-11, plus one I will mention soon. The exact age of the deposits that have provided the identified fish remains is not always well controlled, but none appear to be significantly older than the Stansbury terraces, which date to between 22,000 and 20,700 years ago (see table 5-3).

Table 7-12 lists the native fishes of Bear Lake, to the northeast of the Lake Bonneville Basin, and Utah Lake, in the eastern part of that basin (figure 2-2). It also shows which of these fishes have been reported from late Pleistocene contexts in the Bonneville Basin. During much of its late Pleistocene history, Bear Lake was connected to Lake Bonneville, while Utah Lake was part of Lake Bonneville. As a result, it is perhaps no surprise that of the ten species of fishes that now exist in Bear Lake, eight are also known from deposits laid down when Lake Bonneville still existed. On the other hand, of the eleven species that existed historically in Utah Lake (many are now gone), only four are known from deposits of this age. All four of the species that exist only in Bear Lake are known from late Pleistocene Lake Bonneville, but neither

TABLE 7-11
Native Fishes of the Modern Bonneville Drainage System
Species in bold are known from Pleistocene-aged Lake Bonneville deposits; the last column identifies the late Pleistocene site or sites from which the species has been reported.

Family and Species	Common Name	Sites (see table notes)
Salmonidae (Trouts)		
Prosopium abyssicola	Bear Lake Whitefish	HC
Prosopium gemmifer	Bonneville Cisco	BRC, HC, HS
Prosopium spilonotus	Bonneville Whitefish	BRC, HC, HS
Prosopium williamsoni	Mountain Whitefish	
Oncorhynchus clarki	Cutthroat Trout	BRC, HC, ORB, SCC
Cyprinidae (Minnows)		
Gila atraria	Utah Chub	BRC, HC, HS, PSG, SCC, ORB
Snyderichthys copei	Leatherside Chub	
Iotichthys phlegethontis	Least Chub	
Rhinichthys cataractae	Longnose Dace	
Rhinichthys osculus	Speckled Dace	
Rhinichthys sp.	Deep Creek Dace (undescribed)	
Richardsonius balteatus	Redside Shiner	HC
Catostomidae (Suckers)		
Catostomus ardens	Utah Sucker	HC, HS
Catostomus discobolus	Bluehead Sucker	HC
Catostomus platyrhynchus	Mountain Sucker	
Chasmistes liorus	June Sucker	
Cottidae (Sculpins)		
Cottus bairdi	Mottled Sculpin	BRC, HC, HS
Cottus beldingi	Paiute Sculpin	
Cottus echinatus[a]	Utah Lake Sculpin	
Cottus extensus	Bear Lake Sculpin	BRC, HC, HS

SOURCES: From Sigler and Sigler 1987, 1996; Smith et al. 2002.

SITES AND SOURCES: BRC (Black Rock Canyon), Smith, Stokes, and Horn (1968); HC (Homestead Cave), Broughton (2000a, 2000b), Broughton, Madsen, and Quade (2000); HS (Hot Springs), Smith, Stokes, and Horn (1968); ORB (Old River Bed), Smith et al. (2002), Oviatt, Madsen, and Schmitt (2003); PSG (Public Shooting Grounds), Murchison (1989a, 1989b); SCC (Smith Creek Cave), Mead, Thompson, and Van Devender (1982).

[a]Became extinct during the 1930s.

TABLE 7-12

Native Fishes of Modern Bear Lake and Utah Lake and from Late Pleistocene Lake Bonneville

The exclamation point (!) indicates a species found only in Bear Lake or Utah Lake.

Species	Bear Lake	Utah Lake	Lake Bonneville
Bear Lake Whitefish (*Prosopium abyssicola*)	!		x
Bonneville Cisco (*Prosopium gemmifer*)	!		x
Bonneville Whitefish (*Prosopium spilonotus*)	!		x
Mountain Whitefish (*Prosopium williamsoni*)	x	x	
Cutthroat Trout (*Oncorhynchus clarki*)	x	x	x
Utah Chub (*Gila atraria*)	x	x	x
Leatherside Chub (*Snyderichthys copei*)		x	
Least Chub (*Iotichthys phlegethontis*)		x	
Longnose Dace (*Rhinichthys cataractae*)		x	
Speckled Dace (*Rhinichthys osculus*)	x		
Redside Shiner (*Richardsonius balteatus*)	x		x
Utah Sucker (*Catostomus ardens*)	x	x	x
Mountain Sucker (*Catostomus platyrhynchus*)		x	
June Sucker (*Chasmistes liorus*)		!	
Mottled Sculpin (*Cottus bairdi*)		x	x
Utah Lake Sculpin (*Cottus echinatus*)		!	
Bear Lake Sculpin (*Cottus extensus*)	!		x

SOURCES: Bear Lake and Utah Lake, after Heckmann, Thompson, and White 1981; Sigler and Sigler 1996; and Smith et al. 2002. Late Pleistocene Lake Bonneville, see references for paleontological sites in table 7-11.

of the two Utah Lake specialists are. This difference probably reflects the fact that our best Pleistocene fish records come from the Great Salt Lake area, the very lake into which the Bear River now flows. If so, it also reflects some degree of localization of the Lake Bonneville fish fauna.

Of the four species of fish that find their only home in Bear Lake, three belong to the whitefish genus *Prosopium* (table 7-12). Biologist Becky Miller has shown that the genetics of these three species are consistent with their evolution from an ancestral form within Bear Lake. Subsequently, they reached Lake Bonneville via the Bear River, leaving their remains to be identified by paleontologists.

Their last gasp in Lake Bonneville is probably recorded at Homestead Cave, where, Jack Broughton has shown, Stratum I, deposited between 11,300 and 10,200 years ago, contains a rich fish fauna that includes all of the endemic Bear Lake fishes. This, as discussed in chapter 5, was the general period that saw the multiple fish die-offs that provided the owl-scavenged carcasses that ended up in Homestead Cave. In fact, these multiple death events confirm another set of suggestions made well before the Homestead Cave work began: that a fairly rich fish fauna was probably lost from Lake Bonneville during the early part of this interval, that some fishes likely recolonized the lake when it rose to the Gilbert shoreline, and that another round of extinctions occurred as Lake Bonneville retreated for a final time. In fact, our current knowledge of the history of the Bear River suggests that the last Pleistocene connections between Bear Lake and the Bonneville Basin occurred at about 12,700 and 10,000 years ago. This latter date coincides fairly well with the Gilbert interval of Lake Bonneville history.

I mentioned at the outset of this section that late Pleistocene deposits in the Bonneville Basin have provided the remains of eleven species of fishes, of which only ten are indicated in table 7-11. The missing fish is the bull trout, *Salvelinus confluentus*, a fish that has never been securely reported from the Bonneville Basin. Broughton reported this fish from Homestead Cave stratum I. Although he was very cautious in making the identification, other fish experts who have examined the specimen have confirmed it. Broughton also pointed out that there is an 1834 account of bull trout from the Bonneville Basin, one that recent scientists have doubted. The Homestead Cave data suggest that bull trout really did exist in the Bonneville Basin in very recent times. Today, the closest it is found is in the Snake River drainage—the very same drainage into which Lake Bonneville, and the Bear River, once flowed (see chapter 5).

Other Vertebrates

There were, of course, many vertebrates in the Great Basin during the late Pleistocene in addition to the ones I have discussed here. We are, for instance, beginning to have some understanding of the late Pleistocene distribution of amphibians and reptiles in this region. The most obvious animal that I have not discussed, however, is ourselves, since people were widespread in the Great Basin as the Pleistocene drew to a close. That history I leave for chapter 9.

Chapter Notes

References to the extinct late Pleistocene mammals of North America and to explanations for those extinctions are provided in chapter 4. On early explorations at Fossil Lake, see Cope (1889) and Sternberg (1909:144–170). During his 1877 expedition here, Sternberg not only gave the site its name but also collected the first specimens of the extinct flamingo *Phoenicopterus copei*. My discussion of recent work at Fossil Lake has depended heavily on J. E. Martin et al. (2005) and J. E. Martin (2006). Crystal Ball Cave is discussed by Heaton (1985) and Emslie and Heaton (1987). My comments on the distribution of the dhole follow Crégut-Bonnoure (1996), Youngman (1993), and Arroyo-Cabrales and Polaco (2003). On North American tapirs, see B. Graham (2003). Hockett and Dillingham (2004) present the results of their work at Mineral Hill Cave. References for the Paisley Caves are provided in chapter 3.

Jefferson (2003) provides a Mojave Desert record for the dire wolf (as *Canis* cf. *dirus*); the Fossil Lake record is from Elftman (1931); North American dire wolf records are reviewed by R. N. Nowak (1979) and Dundas (1999). Jefferson, McDonald, and Livingston (2004) provide references for American lions in Nevada. The current status of jaguars in southern Arizona is discussed in McCain and Childs (2008); S. J. Miller (1979) records the presence of jaguar at Smith Creek Cave. Stock (1936a) provided the first record for Harrington's mountain goat from the Great Basin but it is work by Jim Mead and his colleagues that has provided our most detailed knowledge of this animal (Mead 1983; Mead et al. 1986, 1987; Mead and Lawler 1994); Jass (2007) provides a new record from Cathedral Cave in the Snake Range, eastern Nevada.

For more details on the chronology of late Pleistocene extinctions in North America, see Grayson (2001, 2007) and Grayson and Meltzer (2002, 2003). R. N. Mack and Thompson (1982) offer their thoughts on the abundances of Pleistocene mammals in the Great Basin; see also Knapp (1996).

Table 7-2 lists Huntington Dam as an illustrative site for *Arctodus* in the Great Basin; a second is provided by Labor-of-Love Cave, on the east flank of the Schell Creek Range in eastern Nevada (Emslie and Czaplewski 1985). This site provided partial remains of two short-faced bears, one an adult, the other immature. The specimens are in the Natural History Museum of Los Angeles County, but the White Pine County Public Museum in Ely, Nevada, has an excellent cast of a complete *Arctodus* individual, along with excellent, though now somewhat dated, interpretive material (www.wpmuseum.org/).

Casts of mammoth remains from the Black Rock Desert may be seen at the Nevada State Museum (see notes for chapter 9), and at the Humboldt Museum in Winnemucca, Nevada. This latter museum has a reconstruction of the Black Rock Desert's Delong mammoth, excavated by Stephanie D. Livingston (see figure 4-1 and Livingston 1992b), shown as it appeared on the ground. Although the distribution of the bones in the exhibit differs somewhat from the way they were actually found (I participated in the excavations), the differences are minor, and the display gives a good feel for the way a late Pleistocene paleontological site looks during excavation. The museum also has an amazing display of motorized vehicles dating to between 1903 and 1918.

The abundance counts in figure 7-2 were compiled primarily from FAUNMAP Working Group (1994), Jefferson et al. (1994), Gillette and Miller (1999), Jefferson et al. (2002), W. E. Miller (2002), and Jefferson, McDonald, and Livingston (2004). For Great Basin long-nosed peccaries, see Elftman (1931) and H. G. McDonald (2002); for sloths in the Great Basin, see H. G. McDonald (1996, 2003); Gillette, McDonald, and Hayden (1999); and H. G. McDonald, Miller, and Morris (2001). The Spring Mountains sloth is reported by Reynolds, Reynolds, and Bell (1991). For high-elevation mammoths and mastodons, see W. E. Miller (1987), Gillette and Madsen (1993), and Madsen (2000a).

On the early archaeology of the Great Basin, see chapter 9. On the lack of association between people and extinct Pleistocene mammals in the Great Basin, see Huckleberry et al. (2001) and Grayson and Meltzer (2002).

The Huntington mammoth is discussed in Gillette and Madsen (1992, 1993) and Madsen (2000a). In addition to the *Arctodus* radiocarbon date for this site listed in table 7-3, there is also a date of 10,976 ± 40 years BP (Schubert 2010). For the sinkhole mastodons, see W. E. Miller (1987). Although it is not relevant to the period of time we deal with here, it is worth mentioning that Bryan Hockett and his colleagues excavated a 2 million–year-old mastodon from near Elko, Nevada, and wrote up the results in a very informative report directed toward the general public (B. S. Hockett, Brothers, and Seymour 1997). The specimens are on display in an extremely well prepared exhibit in the Northeastern Nevada Museum in Elko; the museum as a whole is well worth the visit (1515 Idaho Street in downtown Elko; www.museum-elko.us). It is, though, surprising that the museum's exhibit "The First Ones" begins with Peter Skene Ogden and not the Native Americans who preceded him in this area by 12,000 years or more.

Purists always capitalize the common names of birds, and I have followed that path here. My comments on the modern distribution of birds in the Great Basin has depended on Ryser (1985), Hayward et al. (1976), Alcorn (1988), Csuti et al. (1997), and T. Floyd et al. (2007). I have followed Gill, Wright, and Donsker (2008) for the classification and common names of modern birds. K. E. Campbell, Scott, and Springer (1999) suggest that *Cathartornis* might actually be based on specimens that belong to *Teratornis*. The Wyoming *Neophrontops* is reported by Emslie (1985); this bird has long been known from the late Pleistocene paleontological sites of Southern California (e.g., L. Miller and deMay 1942).

The chronology of the Manix Lake birds is discussed by Jefferson (2003). On the feeding habits and competitive relationships of modern vultures, see Kruuk (1967), J. H. Brown (1971), Houston (1975), J. H. Wallace and Temple (1987), Hertel (1994), and J. S. Hunter, Durant, and Caro (2006). The reconstruction of the diets of Pleistocene Western Black Vultures, California Condors, and Merriam's Teratorns that I review here is based on Chamberlain et al. (2005) and Fox-Dobbs et al. (2006). On the diets of the extinct raptors in general, see Hertel (1995) and Van Valkenburgh and Hertel (1998). On California Condors, see Emslie (1987, 1988, 1990), Steadman and Miller (1987), N. F. Snyder and Schmitt (2002), Fox-Dobbs et al. (2006), Mee et al. (2007), and BirdLife International (2008). L. Miller (1931) discusses the California Condor at Gypsum Cave; Jehl (1967) provides additional information on the birds of Fossil Lake. Emslie (1990) presents the radiocarbon dates for the Antelope and Gypsum cave California Condors. On Teratorns, see F. I. Fisher (1945);

H. Howard (1972); K. E. Campbell and Tonni (1981, 1983); K. E. Campbell (1995); Hertel (1995); Hertel and Van Valkenburgh (1998); K. E. Campbell, Scott, and Springer (1999); Fox-Dobbs et al. (2006); and Chatterjee, Templin, and Campbell (2007). The Mineral Hill Cave *Buteogallus fragilis* was reported by James (2004). For caracaras, see Emslie (1985), Emslie and Heaton (1987), Chandler and Martin (1991), and J. L. Morrison (1996). H. Howard and Miller (1939) discuss the relative abundances of Western Black Vultures and Turkey Vultures at Rancho La Brea; see also H. Howard (1962). The other late Pleistocene Southern California asphalt deposits from which Turkey Vultures have been reported are Carpinteria (DeMay 1941a) and McKittrick (DeMay 1941b). Other references for the extinct birds of the Great Basin are provided in tables 7-4 and 7-5.

Those who read the literature on Fossil Lake dating to the 1940s and 1950s will find two genera that I have not mentioned here: *Hypomorphnus* and *Palaeotetrix*. *Hypomorphnus* is now treated as a synonym of *Buteogallus* by ornithologists, while Jehl (1967) showed that the material assigned to *Palaeotetrix* is best treated as belonging to the genus to which the Spruce and Blue Grouse belong, *Dendragapus*. And, on the same topic, the genus *Wetmoregyps* appeared on the list of extinct North American late Pleistocene birds until very recently. Olson (2007), however, has shown that this bird is actually a member of the genus *Buteogallus*, which still occurs in North America (see also Suárez 2004). It is not known from the Great Basin.

Throughout, I follow Wilson and Reeder (2005) for the classification of modern mammals, and Wilson and Cole (2000) for the common names of those mammals. E. R. Hall (1946) remains the standard reference on the mammals of Nevada, much as Durrant (1952) does for Utah; see also E. R. Hall (1981). The standard reference for Oregon is Verts and Carraway (1998); I have also drawn on Hoffmeister (1986). Updated versions for Nevada and Utah are badly needed. Zeveloff and Collett (1988) provide a general and readable introduction to Great Basin mammals as a whole.

Grayson (2006a) provides the data on which my general discussion of the late Pleistocene and Holocene histories of pygmy rabbits, yellow-bellied marmots, bushy-tailed woodrats, and northern pocket gophers is based. References to the ecology of modern pygmy rabbits are provided in chapter 2. The key reference for Homestead Cave is Madsen (2000b); on the mammals from this site, see Grayson (2000a, 2000b). The Danger Cave mammals are discussed in Grayson (1988). Schmitt and Lupo (2005) present the results of their Camels Back Cave analysis; B. S. Hockett (2000) discusses Pintwater Cave.

Saysette (1996) observes that marmots are found both above and below a date of 9,830 years ago at Kokoweef Cave, southeastern California, but the stratigraphy of this site has never been described in any detail, and the date itself is problematic in a number of ways (Bell and Jass 2004). On the Dry Cave date for marmots, see Harris (1984).

There are a few late Holocene records for marmots south of their modern distribution, all mandibles from archaeological sites in Tsegi Canyon in northeastern Arizona. Because small mammal mandibles and teeth were used for various purposes by western Native Americans, from tools for stone-tool making to gambling devices (Echlin, Wilke, and Dawson 1981, Grayson 1988, and DeBoer 2001), these specimens do not provide secure evidence of very late Holocene marmots living outside the areas in which they occur today. This is especially true since at least some of the specimens were associated with objects that we know had to be imported long distances, including shells from the Pacific Coast.

The January 5, 2008, Fernley flood was widely reported at the time; the details I present here are from the Associated Press report (cbs5.com/local/Nevada.levee.break.2.623832.html). The Gatecliff faunal record, including that for northern pocket gophers, is discussed in Grayson (1983).

For Tasmanian Devils, see Strahan (1983). Grinnell (1919) provided the first published discussion of the arrival of the English Sparrow in Death Valley. On the history of introduced bird (and other) species in North America, see the superb book by Coates (2006); its natural companion volume is the earlier, classic book by Crosby (1986). On introduced mammals in general, see J. L. Long (2003). Basic information on kangaroo mice is provided by E. R. Hall (1946, 1981), O'Farrell and Blaustein (1974), and Verts and Carraway (1998); Genoways and Brown (1993) can tell you almost everything you want to know on heteromyids in general.

Schmitt and Lupo (2005) suggest that dark kangaroo mice may actually have become extinct in the Camels Back Cave area at the end of the Pleistocene, then recolonized during the past 1,500 years. To me, the small numbers of *M. megacephalus* specimens present in the site—a total of only nine from this long sequence—make it equally likely that they were here all along but just not incorporated into the deposits of the site. I am aware of no prehistoric records for kangaroo mice outside the Bonneville Basin, but J. C. Hafner, Reddington, and Craig (2006) have shown that the isolated populations of *Microdipodops megacephalus* in the Mono Basin are genetically most closely related to populations in central Nevada and suggest that they colonized the Mono area during the late Pleistocene or perhaps even during the Holocene.

There is a huge literature on pikas in general, and on the fossil record for pikas in western North America. For the latter, see J. I. Mead (1987), J. I. Mead and Spaulding (1995), Rhode and Madsen (1995), J. I. Mead and Grady (1996), and Grayson (2005). My comments on pika adaptations are drawn primarily from A. T. Smith (1974a) and A. T. Smith and Weston (1990).

On the heather vole in general, see McAllister and Hoffman (1988); on the heather vole in the Great Basin, see Grayson (1981) and J. I. Mead, Thompson, and Van Devender (1982). J. I. Mead, Bell, and Murray (1992) and Bell (1993) reported heather voles from Cathedral Cave; Bell (1995) and Jass (2007), on the other hand, find evidence only for an extinct species of *Phenacomys* from this site. Heaton (1985) and E. M. Mead and Mead (1989) provide the records for pine martens in the Great Basin, while the latter reference also discusses the least weasel at Snake Creek Burial Cave. Heaton (1987) and Bell and Mead (1998) provide additional Snake Creek Burial Cave identifications. Ziegler (1963) identified pine marten from Deer Creek Cave, but this was before recognition of the existence of the very similar, but also very extinct, noble marten (*Martes nobilis*). The specimen should be revisited to determine which species it is; I discuss the noble marten in the following chapter.

On northern bog lemmings in Tennessee and elsewhere in the east, see Guilday et al. (1978) and Lundelius et al. (1983). J. I. Mead, Thompson, and Van Devender (1982) and J. I. Mead, Bell, and Murray (1992) discuss bog lemmings at Smith Creek Cave; J. I. Mead, Bell, and Murray (1992) discuss the Cathedral Cave examples. Bell (1995) reassigned the Cathedral Cave specimens to an extinct species; these are discussed in detail in Jass (2007). Jass (2007) is very cautious about the Smith Creek Cave material, assigning these specimens to *Mictomys* sp. and noting that the chronology of the late Pleistocene deposits at this site is very much in need of clarification. *Synaptomys* has also been identified from very late Holocene deposits at O'Malley Shelter in southeastern Nevada's Clover Valley, but this record is very much in need of being reexamined (D. D. Fowler, Madsen, and Hattori 1973). Paleontologists tend to assign bog lemmings to *Mictomys borealis*, but *Synaptomys borealis* remains the official name (Wilson and Reeder 2005).

The distribution of late Pleistocene ground squirrels and chipmunks in the Mojave Desert is discussed by Goodwin and Reynolds (1989b). The late Pleistocene or early Holocene extinction of a morphologically well-defined form of sage vole is discussed by Bell and Mead (1998), Bell and Jass (2004), and Jass (2007).

For Brown's ideas on the history of montane mammals in the Great Basin, see J. H. Brown (1971, 1978) and Lomolino, Riddle, and Brown (2006). The western North American studies generated by Brown's work include Grayson (1977a, 1988, 2000a, 2006a); J. I. Mead, Thompson, and Van Devender (1982); R. S. Thompson and Mead (1982); R. Davis and Dunford (1987); R. Davis and Callahan (1992); Grayson and Livingston (1993); Lomolino and Davis (1997); Lawlor (1998); Grayson and Madsen (2000); and Rickart (2001). The studies that began to provide far more detailed information on the distribution of montane mammals in the Great Basin include Grayson and Livingston (1993) and Lawlor (1998). Those showing that each of the species in Brown's set of montane mammals had their own independent and distinct histories include Grayson et al. (1996), Lomolino and Davis (1997), Lawlor (1998), Rickart (2001), and Grayson (2006a). For examples of the individualist responses of mammals to climate change in settings other than the Great Basin, see R. W. Graham (1985, 1988, 1992). The genetic analysis of yellow-bellied marmots that I discuss was done by C. H. Floyd, Van Vuren, and May (2005).

The essential references on the modern fishes of the Great Basin are W. F. Sigler and Sigler (1987, 1996), though my count of the native species of Great Basin fishes comes from G. R. Smith et al. (2002). J. W. Sigler and Sigler (1994) provide an excellent brief introduction to those fish; for areas to the south, see W. A. Minckley and Marsh (2009). G. R. Smith et al. (2002) provide a superb discussion of the deeper history of Great Basin fish. The four species of Empetrichthyidae I discuss here are included by some in the family Goodeidae (see R. F. Miller et al. 2005 and W. A. Minckley and Marsh 2009). The classic studies on the distribution of Great Basin fishes to which I refer are Hubbs and Miller (1948) and Hubbs, Miller, and Hubbs (1974). Recent extinctions of Great Basin pupfishes and springfishes are discussed in R. R. Miller, Williams, and Williams (1989). Information on the distribution and adaptations of cyprinodont fishes has been taken from Soltz and Naiman (1978), R. R. Miller (1981), Naiman (1981), W. F. Sigler and Sigler (1987), and Riggs and Deacon (2004). Information on the Devils Hole pupfish is drawn from R. R. Miller (1948, 1981), La Rivers (1962), Deacon and Deacon (1979), Soltz and Naiman (1978), Pister (1981), Baugh and Deacon (1983), Deacon and Williams (1991), and Riggs and Deacon (2004); this last reference provides the perfect introduction to Devils Hole and its fishes. The population numbers I provide for Devils Hole pupfish are from M. E. Anderson and Deacon (2001), Nevada Fish and Wildlife Office (2008), and Boxall (2008). The geological history of Devils Hole is discussed by Riggs (1992), Winograd et al. (1992), Szabo et al. (1994), and Riggs and Deacon (2004). For detailed information on pupfish genetics, see Echelle and Echelle (1993), Echelle et al. (2005) and Echelle (2008). On the Great Basin pupfish paleontological record, see R. R. Miller (1945, 1948, 1981) and G. R. Smith et al. (2002). Polhemus and Polhemus (2002) synthesize our knowledge of the Great Basin water bug biogeography and its possible implications; on springsnails, see the reviews by Hershler and Sada (2002) and Hershler and Liu (2008a). Hershler and Liu (2008b) suggest bird transport to account for the distribution of assimineid snails in Death Valley and the Lower Colorado region.

For discussions of possible connections between the Owens Valley–Death Valley drainages and the Colorado River, see Knott et al. (2008) and F. M. Phillips (2008). On threats to the continued existence of native fishes in the Arid West in general, see W. L. Minckley and Deacon (1991) and W. L. Minckley and Marsh (2009); on springs in particular, see Unmack and Minckley (2008).

In addition to the references provided in tables 7-11 and 7-12, my discussion of the history of Bonneville Basin fishes depends heavily on Broughton (2000a, 2000b), G. R. Smith et al. (2002), Miller (2006), and Smoot (2009). Grayson (1993) suggested that the fishes of Lake Bonneville had likely undergone two episodes of late Pleistocene extinction. J. I. Mead and Bell (1994) review the history of the amphibians and reptiles of the Great Basin.

PART FOUR

THE LAST 10,000 YEARS

CHAPTER EIGHT

The Great Basin during the Holocene

In 1975, geologist Dave Hopkins announced the decision by the Holocene Commission of the International Quaternary Association to place the Pleistocene-Holocene Boundary at 10,000 years ago. The 10,000-year figure, he said, was chosen "simply because that's a nice round number." And, he added, "no one agreed with me, nor with anyone else" (Hopkins 1975:10). The date was accepted, and is, by definition, the temporal boundary that separates the Pleistocene from the Holocene.

Even though the date chosen for the boundary is arbitrary, all that we have learned since shows that it was appropriate. Most generally, the interval between 11,000 and 10,000 years ago saw the Younger Dryas cold snap, the loss of the last North American Pleistocene mammals, the retreat of Lake Bonneville from the Gilbert shoreline, and a wide variety of other phenomena that seem better associated with the Pleistocene than with the Holocene. Most notably, and certainly most importantly from the standpoint of defining the end of the Pleistocene, 10,000 years ago almost exactly marks the end of the Younger Dryas. As I discuss in chapter 5, this extreme cold snap began around 10,600 years ago and ended about 10,100 years ago, with the transition to what followed—the Holocene—occurring extremely rapidly.

While all this is the case, crossing the Pleistocene/Holocene boundary does not cross a divide between "Ice Age" and modern environments. In the Great Basin, it is not until thousands of years into the Holocene that environments markedly like those of modern times come into being. It is the Holocene development of these modern environments that I examine in this chapter.

In doing that, I will divide this geological epoch into three parts: the early (10,000 to 7,500 years ago), middle (7,500 to 4,500 years ago), and late (the last 4,500 years) Holocene. These divisions are both unofficial and arbitrary;

other divisions of the Holocene are in common use. Nonetheless, Great Basin environments during these times were in many ways distinct. As long as the bounding dates I have chosen are not taken as real edges, they provide a convenient way to examine the history of Great Basin environments during the last 10,000 years.

The Early Holocene (10,000 to 7,500 Years Ago)

Shallow Lakes and Marshes: The Mojave Desert

While the pluvial lakes of the Great Basin were gone by 10,000 years ago, many Great Basin valleys that are today dry, or nearly so, supported shallow lakes and marshes during the early Holocene. In addition, some valleys that today contain lakes, then contained lakes much higher than they are now. The Great Basin during the early Holocene was warmer and drier than what had just preceded it, but it was nonetheless very different from today.

The late Pleistocene and early Holocene history of Las Vegas Valley provides an excellent example. This broad valley trends northwest and southeast, between the Sheep Range to the north and the Spring Range to the south; the city of Las Vegas sits toward the southern end of the valley. Today, Las Vegas Valley is drained by Las Vegas Wash, which ultimately connects with the Virgin River, in turn a tributary of the Colorado. The area itself, however, is extremely arid, the wash routinely dry. Work by geologists Vance Haynes and Jay Quade and his colleagues has documented that this was decidedly not the case during the late Pleistocene and early Holocene.

Haynes's work focused primarily on the Tule Springs area in the Las Vegas Valley, some ten miles northwest of Las Vegas. Quade's efforts were directed some ten miles to the northwest of this, near Corn Creek Springs. Although these studies were directed toward different parts of the valley and were done many years apart—Haynes's work was published in 1967, and Quade's was done in the 1980s and 1990s—the results they obtained are entirely consistent.

The picture painted by their research extends well beyond the range of radiocarbon dating, and perhaps to as early as 60,000 years ago. I begin my discussion, however, with Hayne's geological Unit D, a sixteen-foot-thick set of deposits that is widespread in the valley and that dates to between about 30,000 and 15,000 years ago.

Near the margin of the valley, Unit D consists of fine-grained sediments that grade into "mudstones" (hardened deposits of clays and silts laid down in a moist environment) toward the valley floor. In many places, Unit D is riddled with the casts, or "pseudomorphs," of cicada burrows formed as a result of the deposition of carbonates in the original burrows themselves. These burrow casts are so common that in some places where the deposits of Unit D are exposed, they cover the surface of the ground. The significance of these burrows seems clear: today, cicadas and their burrows are common in the far cooler and moister sagebrush environments of the floristic Great Basin to the north.

Equally clear is the more general environmental picture that can be extracted from Unit D and its contents. Peter Mehringer showed that the mudstones of this unit at Tule Springs contained both cattail and sagebrush pollen. This, combined with the nature of the sediments themselves, suggests that between about 30,000 and 15,000 years ago, Las Vegas Valley supported small, shallow bodies of water often fringed by cattail. Above the marshes and ponds, the alluvial flats supported sagebrush and cicadas. Neither are to be found here today.

In many areas, the top of Unit D is eroded, suggesting that at least some of the marshes and ponds dried before the next depositional unit, E, began to be laid down some 14,000 years ago. Unit E is divided into two parts, with an erosional break separating the two: E_1 dates to between about 14,000 and 11,500 years ago, while E_2 began to form by about 11,200 years ago and stopped forming between 8,500 and 7,200 years, with different dates provided by different places. The sediments that make up the two, however, are quite similar: greenish clays, black organic mats, and water-deposited sandy silts that fill depressions created by springs and stream channels.

The black mats in these units are much like the one that overlies the Clovis level at the Lehner site in southeastern Arizona (chapter 4), and which dates to about 10,800 years ago. Like the Lehner black mat, those in Las Vegas Valley represent the decay of organic material in a moist environment and indicate the elevation of the water table at the time the mats formed. Unlike the Lehner black mat, however, the late Pleistocene and early Holocene versions in the Las Vegas Valley date to between 11,800 and 6,300 years ago. There were certainly black mats here before then—there is one dated to 42,100 years ago from Pahrump Valley on the other side of the Spring Mountains from Las Vegas Valley—but most of these earlier mats appear to have been destroyed by various geological processes. After 6,300 years ago, however, the picture changes, as Quade and his colleagues have shown that black mats simply did not form in southern Nevada between about 6,300 and 2,300 years ago.

Although cicada burrows are not as common in Unit E as they are in Unit D, they are nonetheless present. In addition, E_1 contains the bones of extinct Pleistocene mammals, including mammoth, horse, and camel. Unit E_2, on the other hand, does not, and the erosional break between these two depositional units may coincide with the time of extinction of those mammals here.

All of this information suggests that the water table during E_1 and E_2 times (roughly 14,000 to between 8,500 and 7,200 years ago, excluding the erosional break that separates the two) was lower than it was when Unit D was deposited: the earlier complex of marshes and ponds had diminished considerably. But the green clays in Unit E show that there was at least some standing water in the valley, the cicadas imply that cool and moist conditions continued, mollusks from these deposits show that marshes continued to exist in the area, and the black mats suggest that these marshes were fairly widespread. Indeed, although Mehringer's work revealed only traces of cattail pollen in the sediments of Unit E, it was there, and his work also showed that other moisture-loving plants grew in the area at this time. Sagebrush, now absent on the valley floor, was also present throughout the deposits that make up Unit E.

Much of E_2 is Holocene in age. The complex of marshes and active springs that characterized Las Vegas Valley while E_1 was accumulating became increasingly smaller during E_2 times, but perennial streams with marshy edges persisted here until sometime between 8,500 and 7,200 years ago. Only two dates for black mats in the Las Vegas Valley are younger than 8,640 years ago. Both fall at about 6,300 years ago, but Jay Quade, who extracted them, suggested that one of the dates may have been obtained from material that had been contaminated by modern rootlets. The water table, which was near the surface some 8,500 years ago, had dropped dramatically. The cicadas became locally extinct, the sagebrush retreated, black mats stopped forming, and the deposits of Unit E_1 began to be heavily eroded. With the end of E_2, something closer to modern conditions had arrived.

A similar sequence appears to have characterized the Mojave River drainage. In chapter 5, I discuss the fact that Pleistocene Lake Mojave reached high levels between 18,400 and 16,600 years ago (the Lake Mojave I phase), and between 13,700 and 11,400 years ago (Lake Mojave II). The end of the Lake Mojave II phase, however, did not see an end to Lake Mojave as a whole.

Today, water rarely travels the entire course of the Mojave River to reach the Lake Mojave Basin. It routinely did so, however, between 11,400 and 8,700, and perhaps as recently as 8,400, years ago. During that period, a series of shallow lakes formed, leaving behind as evidence clays blue to green in color (the lake sediments) that are interrupted by cracks caused by desiccation (the drying events that separated the intermittent lakes). This final stage in the history of Lake Mojave is referred to as Intermittent Lake III, previous intermittent lakes having formed just before Lake Mojave I and between Lake Mojave phases I and II. Although shallow lakes have formed in this basin since 8,700 years ago, none were of the magnitude and duration of those that formed during Intermittent Lake III. When this lake ended, pluvial Lake Mojave ended as well. And, as Jay Quade and his colleagues have pointed out, the life span of Intermittent Lake III correlates very well with the increased spring discharge represented by Unit E_2 in Las Vegas Valley to the east.

OWENS LAKE

When we left Owens Lake, in chapter 5, we saw that it had reached levels so high around 10,750 years ago that it might have overflowed. It fell during the Younger Dryas, distinguishing it from some of the other lakes we have discussed, then rose again to reach a depth comparable to its historic high. This, though, was not the end of a pluvial lake in this basin. Steven Bacon and his colleagues suggest that the lake rose and fell multiple times between 10,000 and about 7,000 years ago. They also suggest that even at its lowest early Holocene levels, it was still very high compared to historic standards, not falling much beneath 3,675 feet, at which point it would have been slightly over 100 feet deep.

They also argue that the lake had a major gasp between 7,000 and 6,800 years ago, reaching a surface elevation of about 3,735 feet. Antony and Amalie Orme, however, disagree strongly that the lake reached a level this high, this late, since they were unable to find any secure evidence for such a stand. Instead, they argue, the lake's last gasp may have occurred around 8,300 years ago, after which the lake fell dramatically. If so, this accords strikingly well with the sequences known from the Lake Mojave Basin and Las Vegas Valley. In all these places, the early Holocene was far wetter than what followed.

Shallow Lakes and Marshes: The Floristic Great Basin

THE BONNEVILLE BASIN

One of the most intriguing paleoenvironmental and archaeological projects that the Great Basin has seen in recent years has taken place in the heart of the Bonneville Basin, conducted by scientists whose names we have encountered before—David Madsen, Jack Oviatt, Dave Rhode, and Dave Schmitt. This work has focused on the Dugway Proving Ground, a remote military base of very restricted access in western Utah (figure 5-7). The combination of its remote location and the restricted access provided by the military means that the archaeological record of this area remains essentially pristine, with some of the earliest archaeological materials known from the Great Basin present in almost dramatic abundance.

Today, this area is exceptionally arid—get stranded in the middle of the Dugway Proving Ground without water in the middle of the summer and the remainder of your life may be short and unhappy. But this very same arid area contains rich archaeological sites that date to somewhere between 11,500 and 8,000 years ago. Obviously, access to water was not an issue at this time. One of the many accomplishments of Madsen's team was to show why this was the case.

In chapter 5, I note that toward the end of the Pleistocene, Lake Bonneville had declined to the point that two separate, large lakes existed in the basin that had previously been covered with the waters of one huge lake: the shrunken remnant of Lake Bonneville in the Great Salt Lake Desert to the north, and its counterpart, Lake Gunnison, in the Sevier Lake Basin to the south. Lake Gunnison then overflowed northward, creating the now-dry valley called the Old River Bed. The Dugway Proving Ground lies at the northern end of the Old River Bed, and it was here that Madsen's team concentrated its work.

Even holding aside the astonishing archaeology of this area (see chapter 9), what they discovered was remarkable. They found that this region was marked by a complex series of stream channels, some filled largely with gravel and coarse sands (deposited by high-energy streams), others filled primarily by sands that range from coarse to fine in texture (deposited by lower-energy streams), and many others that are intermediate between the two. Using a combination of radiocarbon dating and detailed geological analyses, they showed that nearly all of these channels were created between 10,500 and 8,800 years ago. The channels created by the high-energy streams came first, created when Lake Bonneville stood at the Gilbert shoreline and perhaps for a short while afterward. The lower-energy ones followed these in time.

The earlier, higher-energy channels were created by overflow from the Sevier Basin to the south. The lower-energy versions originated more locally, fed by a far higher water table than exists in this area today. The full set of channels documents that water was flowing in this now-dry area continuously or almost continuously from about 10,500 years ago to about 8,800 years ago. That water, Madsen and his colleagues have shown, also created extensive wetlands, and, in a part of the story I will pursue in the following chapter, it was these wetlands that allowed people to create the archaeological record that his team discovered. Earlier streams may have existed as well. In fact, given that Lake Gunnison seems to have begun overflowing to the north as early as 12,500 years ago, and certainly by 11,400 years ago (chapter 5), the streams that mark this overflow must have existed somewhere. In those areas in which Madsen's team worked, there is some evidence for at least one that existed sometime before 11,000 years ago, but such evidence is very sparse. This may be because the older channels are to found elsewhere, remain buried, or were destroyed by those that came later.

THE GREAT BASIN DURING THE HOLOCENE   219

In the previous section, I note that Intermittent Lake III existed in the Lake Mojave Basin until about 8,700 years ago, that this coincided with the early Holocene springs and marshes indicated by Unit E_2 in Las Vegas Valley and with the black mats that go along with them, and that Owens Lake seems to have remained relatively high until about 8,300 years ago or so. Thanks to the work that has been done in the Old River Bed, we now know that substantial streams continued to flow in what is now the heart of the hyperarid Bonneville Basin until almost exactly the same time.

THE LAHONTAN BASIN

Although the precision of the data that we now have available for the early Holocene water history of the Lahontan Basin cannot yet match those available for the Old River Bed, what we do have shows that this area was also far wetter for much of the early Holocene than it is now.

This history, though, is quite different from those that we have seen to this point. As I discussed in chapter 5, the latest Pleistocene highstand in the Lahontan Basin was reached during Younger Dryas times, sometime between 10,800 and 10,000 years ago, when Lake Russell, the western counterpart to the Bonneville Basin's Lake Gilbert, formed. This lake merged the Smoke Creek, Black Rock, Winnemucca, and Pyramid basins (see figures 5-12 and 5-14) under a single body of water, only to fall dramatically shortly before 10,000 years ago. Work by Ken Adams and his colleagues suggests that by 9,300 years ago, elevations as low as 3,780 feet—about the same as the playa of Winnemucca Lake and slightly beneath the lowest historic elevation of Pyramid Lake—were dry. This, though, did not last long. By 9,000 years ago, the separate basins that comprise the Lahontan Basin as a whole were once again beginning to fill with lakes, apparently reaching highpoints, at about 3,950 feet, between about 7,000 and 7,500 years ago. Then they fell dramatically, with severe consequences for the people who were living here at this time (chapter 9).

Why the early Holocene history of lakes in the Lahontan Basin is so different from those that we have encountered to this point is not known. It is fair to say that the chronological control that we have for the Lahontan Basin during this period is not nearly as good as it is for the other areas I have surveyed to this point, so perhaps surprises are in store for the future. Nonetheless, just as in those other areas, the early Holocene of the Lahontan Basin was, in general, far wetter than anything that has been seen since.

A SAMPLING OF OTHER PLACES

In chapter 6, I discuss Bob Thompson's analysis of the deep and stratified sediments that he was able to extract from northeastern Nevada's Ruby Marshes, and the implications of his work for our understanding of the late Pleistocene history of both lakes and plants in Ruby Valley.

The cores he extracted also provided important information on the lakes and marshes that have existed in this area during the last 10,000 years. Basing his arguments primarily on the ostracodes and green algae preserved in the sediments in those cores, Thompson showed that while the lake that occupied this basin had reached fairly low levels by 10,400 years ago, the lake was deeper than it is today and lasted well into the Holocene. In fact, between about 8,800 and 8,700 years ago, the lake rose briefly but then began to decline, becoming a sedge-rich marsh by 8,700 years ago. By 6,730 years ago, when Mazama ash fell in Ruby Valley, the basin was nearly dry.

To the south, eastern Nevada's now-dry Long Valley held a substantial lake, Lake Hubbs, during the late Pleistocene. Just as in Ruby Valley, desiccation of this lake did not result in desiccation of the valley. Instead, geologist Gary Huckleberry and archaeologists Charlotte Beck and Tom Jones have shown, at least the southern part of the valley remained moist until as late as 8,600 years ago. Evidence for this is provided by an organically rich deposit that represents either a pond or wetlands; this deposit has dates that range from about 10,100 to 8,600 years ago. The next date up in the sequence falls at 8,100 years ago and comes from a layer of windblown sands suggestive of a far drier valley.

Many other places within the more northerly parts of the Great Basin fall in line. The Malheur National Wildlife Refuge, which recently celebrated its one-hundredth birthday, contains one of the largest freshwater marsh systems in the United States. It also provides one of the best places in the Great Basin to watch birds, although it is also one of the best places that I have ever encountered, Great Basin or not, for donating blood to mosquitoes.

The Refuge, and the Harney Basin which contains it, holds a series of lakes—Malheur Lake, Harney Lake, and, between the two, the ephemeral Mud Lake. Although all are officially within the Great Basin, Malheur Lake has an outlet in its southeastern corner, known as Malheur Gap. When water in the lake reaches an elevation of 4,114 feet, it flows through this gap into the Malheur River, which then flows into the Snake River and thence to the Columbia. As a result, when the lake reaches this level, it is no longer a part of the hydrographic Great Basin. Since the lowest point in the Malheur Lake Basin falls at 4,087 feet, the lake has to reach a maximum depth of only 27 feet to accomplish this. As mentioned in chapter 5, the last time this happened seems to have been about 1,000 years ago.

During the Pleistocene, a substantial lake grew in the Harney Basin, with its highest shoreline some eleven feet higher than the current level of Malheur Gap. How this could be is not at all clear, given that the outlet must control the elevation of the lake. An obvious guess is that the outlet was at one time higher and has subsequently been eroded to its current level, though when this might have occurred is unknown since the high shoreline is undated. In fact, surprisingly little is known about the history of Pleistocene Lake Malheur.

Thanks to work by archaeologist Keith Gehr, geologist Dan Dugas and others, however, we do know quite a bit about the Holocene history of the lakes that formed in the Malheur Basin. In particular, we know that a lake or series of lakes existed in this basin between about 9,600 and 7,400 years ago, with the lake apparently high enough to overflow the Malheur Gap at about 8,700 years ago. Since the basin contains a lake today, this wouldn't mean a lot except for the fact that for much of this time the lake was near or above its highest historic stand. That highstand reached a maximum of 4,103 feet between 1982 and 1986, the same time as the Great Salt Lake reached its historic high (see chapter 5). What was an extraordinary event in the Malheur Basin a few decades ago seems to have been at least the norm for much of the early Holocene. After 7,400 years ago, however, the lake declined substantially—how substantially we shall see shortly.

To the west, in the Fort Rock Basin, there are also suggestions that in at least some places, the early Holocene had surface water resources more impressive than those that followed later in time. The Connley Caves provides one of those suggestions.

These caves sit on the southwestern slope of the Connley Hills in the western Fort Rock Basin, about ten miles southeast of Fort Rock Cave (figure 7-3). A mile to the southwest lies the fluctuating shoreline of Paulina Marsh, now the largest natural body of standing water in the basin as a whole, fed by the only perennial streams that exist in the Fort Rock Basin.

The Connley Caves were first excavated by archaeologist Steve Bedwell in 1967, using techniques that would not be used today. Fortunately, Dennis Jenkins, whose work on the Paisley Caves I discuss in chapter 4, returned to these sites, conducting additional excavations, obtaining new radiocarbon dates, examining Bedwell's original notes, and even interviewing some of the excavators who labored under Bedwell's command. His work has gone far to clarify Bedwell's initial results.

These sites provided an archaeological sequence that spans the period from 10,600 to perhaps 1,000 years ago or so, though the interval between 7,200 and 4,700 years ago is not represented. While Jenkins's work showed that it would be a mistake to place too much faith in the stratigraphic position of particular specimens, the fact that the radiocarbon dates he obtained fall in line with those obtained by Bedwell suggests that there is no reason to distrust the general nature of the sequence that Bedwell obtained.

Bedwell was primarily interested in the archaeological content of these caves, but in addition to providing an important series of artifacts, the rockshelters also proved to be fairly rich in bone. Importantly, the nature of this fauna changed dramatically through time. Nearly all the remains of birds found here were deposited between 10,600 and 7,200 years ago (95 percent of them, to be exact). In addition, most of those birds are tightly associated with water—grebes, ducks, shorebirds, and to a lesser extent, Sage Grouse (within their general habitat, these birds are most common in areas with abundant surface water). The steep decline in such birds after 7,200 years ago strongly suggests that the levels of Paulina Marsh dropped dramatically after that time. Precisely when this reduction occurred is not known because of the gap in the record between 7,200 and 4,400 years ago, but when the record kicks back in, the water-dependent birds are gone.

It would be nice to compare the apparent early Holocene history of Paulina Marsh with other lake and marsh sequences from nearby areas, but these are not available. A single radiocarbon date from the Lake Abert Basin suggests a high lake here about 9,400 years ago, but a single date is not much to go on. In addition, we know almost nothing of the early Holocene history of lakes and marshes in the Chewaucan Basin, in stark contrast to what we know about the history of Pleistocene lakes here (see chapter 5).

AN EXCEPTION...

Not all lower elevation settings in the Great Basin were wetter during much of the early Holocene than they are now. An obvious example is provided by Walker Lake, which appears to have been dry between about 14,000 and 4,700 years ago (see chapter 5). The desiccation of Walker Lake, however, seems to have been due to the behavior of the Walker River, which, during this 9,000-year interval, apparently flowed north, into Carson Sink, rather than south, into Walker Lake.

More to the point, then, is the history of the Dietz, or northern Alkali Lake, Basin, located in the easternmost corner of the basin that once held Pleistocene Fort Rock Lake. This dry basin contains latest Pleistocene and early Holocene archaeological materials and, as a result, has attracted a good deal of scientific attention. Work by geoarchaeologist Ariane Pinson shows that a series of late Pleistocene lakes formed here, but no radiocarbon dates exist to tell us when the last of them occurred. Once the lakes disappeared, dunes built from sediments that were alternatively wet and dry began to form, suggesting that surface water still existed in the basin. After this initial episode of dune building, the water table rose, creating a wet meadow sometime before about 9,600 years ago; to the south, a shallow lake may have formed in the Alkali Lake Basin. A second episode of dune construction followed, again suggesting sediments that were alternatively wet and dry; this seems to have happened at around 9,500 years ago. When Mazama ash fell in this basin, around 6,730 years ago, it fell onto a basin floor that was dry, and the subsequent history of the basin suggests that it has remained dry since that time.

The history of surface water in the Dietz Basin thus suggests that while the earliest Holocene here was moister than what followed, there appears to have been nothing comparable to what we have seen in such places as the Lahontan Basin, the Ruby Marshes, the Old River Bed, or, much closer by, Paulina Marsh. There are hints that other parts of the

Fort Rock Basin may also have lacked substantial amounts of surface water during much of the early Holocene but we are very much in need of compelling early Holocene surface-water histories for the Fort Rock Basin as a whole, as well as for the Chewaucan Basin to the south.

... TO THE RULE

The general rule, however, seems to be that the early Holocene in the Great Basin was a much better-watered place than it has been since then. Lakes and marshes were found in areas that do not support them today and some areas that today do support them—for instance, the Harney Basin—then had more substantial versions of them. When this wetter version of the Great Basin disappeared depends on where we look. In the Las Vegas Valley, substantial marshes and springs existed until at least 8,500, and perhaps as late as 7,200, years ago (Vance Haynes's Unit E_2), and black mats formed in abundance until at least 8,600 years ago. In the Lake Mojave Basin, intermittent lakes formed until about 8,700 years ago. Owens Lake reached relatively high levels until about 8,300 or 7,000 years ago, depending on which chronology you accept. Water continued flowing in the Old River Bed until around 8,800 years ago, an area that today is extraordinarily dry. The Lahontan Basin saw the formation of lakes in areas now dry until 7,500 to 7,000 years ago and perhaps until Mazama ash fell here, 6,730 years ago. There was a substantial lake in Ruby Valley until around 8,700 years ago, when it became a sedge-rich marsh. Substantial lakes formed in the Harney Basin between 9,600 and 7,400 years ago and Paulina Marsh seems to have been more substantial between 10,600 and 7,200 years ago than it has been since that time. Even the apparent exception, the Dietz Basin, seems to have been wetter until at least 9,500 years ago, and perhaps later, than it has been since that time.

Conifers, Sagebrush, Shadscale, Hackberry, and Creosote Bush

In many areas, our chronologies of early Holocene lakes and marshes are in need of improvement, but what we know suggests that different areas dried out at different times. No place, however, is there convincing evidence that valleys that today lack substantial bodies of water maintained lakes and marshes much beyond 7,000 years ago. When Mazama ash fell in such valleys, it routinely fell on ground that was dry or nearly so.

Obviously, if the Great Basin during the early Holocene was wetter than what followed, the plants should have followed suit. In fact, they did, differing from what is now found across this area in very substantial ways.

THE BONNEVILLE BASIN AND NEARBY AREAS

Netleaf hackberry (*Celtis reticulata*) is a widespread plant in western North America, found as a large shrub or small tree from Washington state to southern Mexico and from California to the Great Plains. It is not abundant in the Great Basin as a whole but it does occur, and where it occurs at low elevations it is often found on rocky outcrops. The species' fondness for rocky substrates seems to reflect the fact that rock fissures can hold water, providing the plants with the moisture they need.

This habitat preference means that in the right low-elevation settings, the history of hackberry can tell us something about the history of moisture in those contexts. Hackberry has another attribute that is helpful in this regard. Its seed, or endosperm, is embedded in a hard casing, called an endocarp, which is in turn covered with a sweetish, edible enticement (the mesocarp) for the birds and mammals that help to disperse it across the landscape. The endocarp is the important thing here: when deposited in the right context, it can last for thousands of years, as can the seed inside it.

Since the hackberry endocarp-endosperm combination is essentially a small nut, I will refer to the combination as a nutlet. The lower Holocene strata of Homestead Cave (chapter 5) are full of them. Strata II, III, and IV, which date to between about 8,800 and 8,100 years ago, held more than 80,000 hackberry nutlets each, with stratum IV having an estimated 290,000 of them. Their numbers then begin to dwindle. Stratum VI, which dates to 7,100 years ago, has 2,450 hackberry nutlets, but the rest of the sequence coughed up only 134 of them.

How the nutlets got in the site is not known. Rodents, including woodrats, are a good guess given the edible fruit, but there are other possibilities. Those possibilities include small birds that had feasted on the fruits only to be feasted on by raptors, which then deposited their remains, nutlets included, in the site. People cannot account for their abundance in the lower reaches of Homestead Cave since the cave never saw a significant prehistoric human presence. No matter how they got here, though, there are enormous numbers of them—until sometime right after 7,100 years ago, when they almost, though not quite, disappear.

The meaning seems evident. We don't know what was happening here during the very earliest Holocene, since that period doesn't seem well represented in the part of the site that was excavated. However, hackberry was growing in abundance on the rocky slopes above Homestead Cave between 8,800 and about 8,100 years ago, producing fruits that were harvested by happy rodents, or perhaps unhappy birds, which then introduced them into the site. Then the trees, or shrubs, began to dwindle in number and disappeared soon after 7,100 years ago. While they are found in nearby mountains today, they no longer exist in the Homestead area. This part of the Bonneville Basin was clearly far moister during the early Holocene than it has been since then.

In chapter 5, I mention that the work done by David Madsen and Don Currey at Snowbird Bog shows that elevations as high as 8,100 feet in Little Cottonwood Canyon, in the Wasatch Range just southeast of Salt Lake City, were free of glacial ice by 12,300 years ago. They did far more, though, than just core the bog and get a date from its base. They also presented a detailed analysis of the pollen contained by that bog, generating a vegetation history that covers the past 12,300 years. As part of this work, they calculated two ratios, one showing the changing abundance of spruce pollen relative to pine pollen through time, the other the changing abundance of all conifer pollen to all other pollen through time. They did this because they felt that the spruce/pine ratio most likely reflects changing moisture history, with spruce preferring wetter habitats than pine. The ratio of conifer to all other pollen, on the other hand, they suggested most likely reflected temperature history.

While the chronological control over these curves is not as strong as one might want (the work was, after all, done in 1979), the results are consistent and the implications obvious (figure 8-1). The early Holocene, until sometime before about 8,000 years ago, appears to have been both cooler and moister than what has followed.

Blue Lake lies on the Utah side of the Utah-Nevada border almost directly across the Bonneville Basin from Snowbird Bog (figure 5-7). As discussed in chapter 6, the late Pleistocene vegetational history constructed from the deposits of this lake by Lisbeth Louderback and Dave Rhode showed that regional plant communities were dominated by pine (almost certainly limber pine) and sagebrush between 12,700 and 10,200 years ago (figure 8-2).

What is now Blue Lake was covered by some five feet of water when Lake Bonneville was at the Gilbert shoreline. After the lake retreated from this elevation, sometime after about 10,200 years ago, marshes developed here; the work conducted by Louderback and Rhode shows that those marshes were in place by at least 9,900 years ago. The retreat of the Gilbert-level lake changed the nature of the vegetation recorded at this spot, from one that represented the broader catchment that contributed to the Gilbert-aged lake, to the more local catchment that resulted when the lake retreated. In addition, once the lake was gone, broad areas became available for colonization by plants adapted to the saline surfaces produced by its desiccation. These complexities have to be kept in mind when trying to understand what the Blue Lake pollen record is telling us. It can, for instance, be dangerous to compare the pollen record that accumulated while the Gilbert-aged lake covered the area with the record that accumulated when that lake was gone.

That said, what it tells us is clear enough. Sagebrush pollen declines, and cheno-am pollen increases, in the Blue Lake sediments as we enter the Holocene. In fact, the former declines and the latter increases almost continuously throughout the early Holocene. By the time we reach about 8,500 years ago, cheno-ams have become the dominant plant type of the two, strongly suggesting that the early Holocene in this area was a period of both drying and warming.

Unlike most of the other places I have discussed to this point, Blue Lake provides no compelling suggestion of an

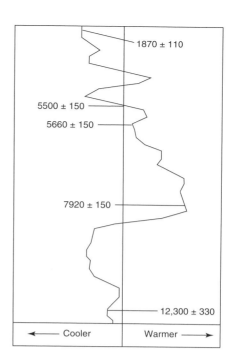

FIGURE 8-1 Ratio of conifer to all pollen through time at Snowbird Bog, Little Cottonwood Canyon, Wasatch Range, Utah (after Madsen and Currey 1979 and Madsen 2000b).

early Holocene landscape significantly wetter than anything that has been seen here since. This, at least, is what the history of sagebrush and cheno-ams extracted from the lake's sediments by Louderback and Rhode suggests. Sagebrush, in general, prefers habitats both cooler and moister than those in which such cheno-ams as shadscale thrive. As a result, paleobotanists often use the ratio of sagebrush to cheno-am pollen from ancient arid-western sediments as a measure of relative temperature and moisture. In this measure, the greater the amount of sagebrush, the cooler and/or wetter the landscape must have been. And, in the last 10,000 years, the only time that sagebrush pollen was substantially more abundant than cheno-am pollen in the Blue Lake core was before about 9,500 years ago. The early Holocene sagebrush/cheno-am ratios that follow are not that much different from those that have marked the place during the past few thousand years. This sequence is far more similar to that found by Ariane Pinson in the Dietz Basin than to the other sequences we have seen to this point.

In the previous chapter, we saw that subalpine conifers extended to the bases of mountain flanks in the northwestern Bonneville Basin until sometime after 11,000 years ago. When they retreated upward is not known, but by 10,000 years ago the low elevations that had supported them now tended to be covered with shrubs characteristic of more arid settings—sagebrush and shadscale, for instance. At Blue Lake, pine pollen, almost certainly from limber pine, drops rapidly between about 10,100 and 9,900 years ago, probably reflecting the movement of these trees upslope.

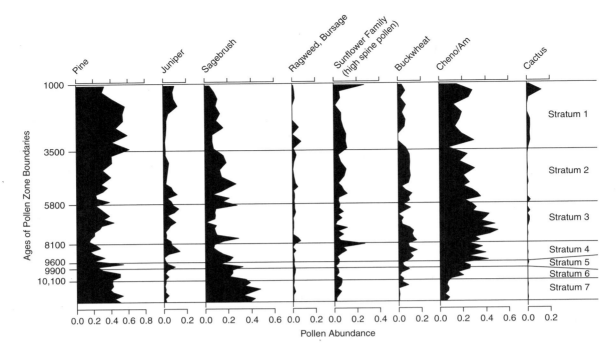

FIGURE 8-2 Histories of selected plants at Blue Lake, Utah (after Louderback and Rhode 2009).

In fact, only one site in this area, Danger Cave, might seem to suggest the existence of populations of limber pine at fairly low elevations during the early Holocene. This site is located at the base of the east-facing slope of the Silver Island Range at an elevation of about 4,315 feet, some twenty miles north of, and 50 feet higher in elevation than, Blue Lake (figure 5-7). The vegetation that surrounds the site today is marked by such things as shadscale and greasewood. While there are trees in the Silver Island Range, they are Utah junipers and pinyon pines, not limber pine, and not even these are particularly close to the site. In fact, the closest significant stands of limber pine are found in the Toano, Goshute, and Pilot Ranges, some fifteen miles away. Work by Dave Rhode and David Madsen, however, have identified the hulls of limber pine nuts from the deposits of Danger Cave that date to 7,400 years ago.

It is difficult to know what to make of these, but there is no reason to suggest that limber pine was then growing near this low-elevation, arid setting. Instead, there is every reason to believe that they were transported by people to this archaeological site. At the most, they suggest that at 7,400 years ago, limber pine may have grown closer to the site, and at lower elevations, than today. However, given that people might easily have carried limber pine nuts to this site from great distances, even that is not clear.

On the other hand, the packrat midden work done by Bob Thompson and Phillip Wells in the Snake Range and nearby mountains on the edge of the southern Bonneville Basin shows that some subalpine conifers did remain significantly beneath their modern elevational limits well into the early Holocene. In the Snake Range, for instance, both Thompson and Wells have shown that limber and bristlecone pines were widespread, even on south-facing slopes, at elevations as low as 6,500 feet as recently as 9,500 years ago. In the massive Snake and Schell Creek ranges, these trees remained at fairly low elevations throughout the early Holocene. Thompson, for instance, has limber and bristlecone pine at 6,500 feet on the east slope of the northern Snake Range at 7,350 years ago and bristlecone pine alone at 6,100 feet and 6,500 years ago in Smith Creek Canyon, again on the east slope of the Snake Range. In the Schell Creek Range, he has documented limber pine as late as about 6,500 years ago (and perhaps even somewhat later) at an elevation of 7,700 feet. All of these sites are in areas now dominated by pinyon-juniper woodland. In the smaller, lower Confusion Range to the east, however, limber and bristlecone pines, which had occupied sites as low as 5,450 feet at 11,880 years ago, were gone from these locations by 8,600 years ago; Thompson has documented a similar sequence, from Carlin's Cave (6,000 feet), at the southern end of the Egan Range.

In general, the packrat midden record from these ranges suggests that bristlecone pine retreated upslope as the early Holocene progressed, with limber pine following somewhat later. Xeric (warm and dry) sites lost their subalpine conifers first, as did the lower elevations on smaller mountain masses. Lower elevation mesic (cool and moist) habitats on more massive ranges retained their subalpine conifers far longer. Indeed, they retained them throughout the early Holocene.

As the subalpine conifers of this area moved upward between 10,000 and 7,000 years ago, a variety of plants increased in abundance. Thompson's analyses have shown that plants of the Upper Sagebrush-Grass Zone became common on lower mountain slopes as the subalpine conifers retreated. These shrubs included not only sagebrush but also rabbitbrush and ocean spray. Mountain mahogany became common in these habitats at the same time. A variety of junipers were also present on these lower slopes: most, though not all, early Holocene packrat middens analyzed from these slopes have provided the remains of Utah, Rocky Mountain, or common junipers.

In short, the record for the northwestern Bonneville Basin suggests that subalpine conifers had moved upslope by about 10,000 years ago, to be replaced by shrubs characteristic of more arid settings in the Great Basin. In the more southern Bonneville Basin, these trees, and in particular limber and bristlecone pines, remained at low elevations in cool and moist settings throughout the early Holocene. In warmer and dryer contexts, they moved upslope during this interval, to be replaced by a variety of shrubs now common in the Upper Sagebrush-Grass Zone, and by mountain mahogany and junipers. These differences are not nearly as substantial as they might seem, since the early Holocene records from the northwestern Bonneville Basin tend to be from lower elevations and less massive mountain ranges than the records provided by Thompson and Wells from the southern part of this area.

In chapter 6, I mention that Bob Thompson's analysis of sediment cores from Ruby Lake, west of the Bonneville Basin, suggested that Ruby Valley supported a sagebrush steppe during the late Pleistocene. In fact, the sagebrush steppe that existed here at the end of the Pleistocene continued to exist through the early Holocene, until about 7,700 years ago. Then, sagebrush pollen declined sharply, ending at least 30,000 years of sagebrush dominance in the vegetation of this valley. Accompanying this decline was a sharp increase in cheno-am pollen. That increase, which probably accompanied the shrinkage of the lake in this basin, must represent an expansion of saltbush vegetation in the valley.

THE LAHONTAN BASIN

To the west, in the Lahontan Basin, we have already seen that Lake Lahontan had fallen dramatically by about 10,000 years ago and that elevations equivalent to the Winnemucca Lake playa (3,780 feet) were dry by 9,300 years ago. We also saw that the separate basins that make up the Lahontan Basin began to fill with water by 9,000 years ago, reaching their early Holocene highs between 7,500 and 7,000 years ago, only to fall again soon thereafter.

This sequence is matched almost perfectly by the plant history extracted by Peter Wigand and Peter Mehringer from Hidden Cave on the southern edge of the Carson Desert. The late Pleistocene sediments of this site are dominated by pine and sagebrush (chapter 6), but this dominance ends fairly abruptly at around 10,000 years ago, as pine pollen declined to levels no greater than those that characterized the uppermost, latest Holocene, deposits in the site. Pines may have become less abundant on the local landscape at about this time, though this conclusion is clouded by the fact that the pine pollen in the Hidden Cave sediments may have come from the waters of Lake Lahontan. Sagebrush pollen also declined in the deposits of Hidden Cave at about 10,000 years ago, but the decline it underwent was not nearly as dramatic as that suffered by pine. While sagebrush pollen becomes less abundant at this time than it had been during the late Pleistocene, it was still far more abundant between 10,000 and 6,800 years ago or so than it was to be after this interval.

As the amount of pine pollen in the sediments of Hidden Cave plummeted and sagebrush pollen declined, cheno-am pollen increased. As I have discussed, much of the cheno-am pollen in Hidden Cave is probably from saltbush, and pollen of this sort is not at all abundant in the late Pleistocene deposits of the site. Soon after 10,000 years ago, however, it became so, rising to form some 40 percent of all Holocene pollen samples.

Today, the vegetation surrounding Hidden Cave is dominated by species of *Atriplex* and greasewood. The rise in cheno-am pollen at 10,000 years ago, however, does not imply that something like the modern vegetation of this area came into being at this time: sagebrush pollen is simply too abundant in most of the early Holocene deposits of Hidden Cave to suggest that. As Wigand and Mehringer note, a more modern vegetational regime does not appear to have become established until shortly after 6,800 years ago, when the abundance of sagebrush pollen drops to levels comparable to those in the latest Holocene deposits of the site. Between 10,000 and about 6,800 years ago, the pollen data from Hidden Cave seem to suggest a generally wetter landscape than is to be found here today, with a higher sagebrush component in the vegetation than has existed since then.

We can be more precise than this, however, because Peter Wigand and Dave Rhode have recently taken a closer look at the changing relative abundances at Hidden Cave of sagebrush and grasses versus cheno-ams (figure 8-3). Their analysis show that the ratios of these two groups of plants fluctuated dramatically during the early Holocene, with cheno-ams relatively more abundant between about 10,000 and 9,200 years ago and sage and grasses far more abundant from about 9,200 to 6,800 years ago, after which the relative abundance of sagebrush and grass drops dramatically. This history matches the independently derived early Holocene lake levels in this area extremely well, with cheno-ams relatively more abundant during the earliest Holocene when the lakes were low (10,000 to just before 9,300 years ago), and sagebrush and grasses relatively more abundant when the lakes were high (9,300 to just beyond 7,000 years ago). While the early Holocene in the Carson Desert seems to have been wetter than most of what came later, both lakes and pollen strongly suggest that the earliest early Holocene seems to have been the driest part of this interval here.

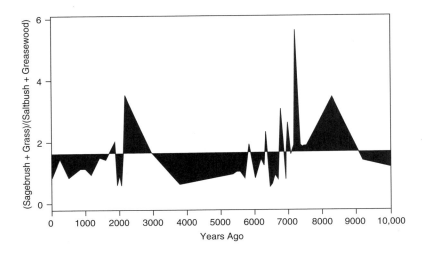

FIGURE 8-3 Plant-based aridity index for the last 10,000 years at Hidden Cave, Nevada. The lowest values indicate the highest relative abundances of saltbush and greasewood compared to sagebrush and grasses as well as the most arid intervals in this area (after Wigand and Rhode 2002).

A fairly cool and moist early Holocene in the Lahontan Basin is also indicated by packrat midden data gathered by Cheryl Nowak and her colleagues from the Painted Hills, west of Pyramid Lake. Those data show that such plants as Woods' rose (*Rosa woodsii*) and currant (*Ribes*), characteristic of cooler and wetter habitats, were able to exist at relatively low elevations here until about 8,000 years ago but not after that time.

THE NORTHWESTERN FRINGE

Table 8-1 summarizes the evidence that I have reviewed to this point for a relatively moist early Holocene in the Great Basin, as well as additional evidence that I will present shortly. From the Mojave Desert on the south through the Lahontan and Malheur basins on the north and to the Bonneville Basin on the east, a broad variety of climatic indicators shows that the early Holocene, roughly 10,000 to about 8,000 years ago was, in general, wetter than anything that has come since. How long this episode lasted depends on which indicators are examined and the particular area involved, but there is remarkable consistency here.

There are exceptions. Neither the Dietz Basin nor Blue Lake suggest a climate any wetter than today's after about 9,500 years ago. The biggest exception, though, comes from the far northwestern Great Basin, provided by work done by paleobotanists Tom Minckley, Cathy Whitlock, and Patrick Bartlein.

These scientists developed pollen sequences from Patterson (9,000 feet elevation) and Lily (6,700 feet) lakes in the Warner Mountains of northeastern California, and from Dead Horse Lake (7,375 feet), located in the mountains southwest of the Paisley Caves on the very western edge of the Chewaucan River drainage (chapter 5). Those sequences allowed them to reconstruct past climates in these areas during the past 10,500 years, though the Lily Lake record doesn't click in until around 8,900 years ago. To build those reconstructions, they compared their pollen data with a network of modern pollen samples from across western North America, a quantitative approach based on the assumption that the climate data that characterize the closest modern pollen analogs also characterized the environment at the time the ancient samples accumulated.

The results of this effort are quite different from anything we have seen before. From about 10,250 years ago to well into the middle Holocene, their records suggest that this area received precipitation up to eight inches less than modern averages in areas that today receive between sixteen and twenty-five inches per year. This is a substantial difference.

Of course, if it were also far cooler than today, reduced evaporation could have meant that it was still wetter, even with significantly less precipitation. However, their reconstructions do not suggest this at all. Instead, they suggest that, between 10,500 and at least 7,700 years ago, the temperatures of the warmest month in the Patterson and Dead Horse lakes areas were from 1.8°F to 5.4°F warmer than today. Their reconstructions of temperatures during the coldest month do not help much here since they conflict: they calculated temperatures of up to 5.4°F warmer than today for Patterson Lake between 10,500 and 6,100 years ago, but temperatures up to 5.4°F colder at both Dead Horse and Lily Lakes.

The chronological control they were able to provide for the early Holocene part of their sequences is not as good as one might want, since the three lakes have only one date that falls between 10,080 and 8,260 years ago. Nonetheless, the sediments that accumulated between these times are early Holocene in age even if we might not know with any precision when during the early Holocene they accumulated.

TABLE 8-1
Summary of Indicators of Greater-Than-Modern Moisture during the Early Holocene in the Great Basin
Numbers in parentheses indicate alternative end dates.

Location	Indicator	Age (years ago)
Mojave Desert		
Las Vegas Valley Black Mats	Marshes, Springs	11,800–8,640 (6,300)
Las Vegas Valley, Tule Springs Unit E_2	Marshes, Springs	11,800–8,500/7,200
Mojave Basin Intermittent Lake III	Lake Levels	11,400–8,700 (8,400)
Owens Lake	Lake Levels	10,000–8,300 (7,000)
Floristic Great Basin		
Bonneville Basin: Homestead Cave	Hackberry Abundance	8,800–8,100 (7,100)
Bonneville Basin: Old River Bed	Marshes, Streams	10,500–8,800
Bonneville Basin: Little Cottonwood Canyon	Conifer Ratios	10,000–ca. 8,000
Bonneville Basin: Blue Lake	Sagebrush/Cheno-am Ratios	10,000–9,500
Bonneville Basin: Southern Mountains	Low Elevation Subalpine Conifers	To ca. 7,000
Ruby Valley	Lakes, Marshes	10,400–<8,700
Ruby Valley	Sagebrush Abundance	To 7,700
Long Valley	Wetlands	10,100–8,600
Carson Sink: Hidden Cave	Cattail (Marshes)	To 9,600
Carson Sink: Hidden Cave	Sagebrush Abundance	To ca. 6,800
Carson Sink: Hidden Cave	Sagebrush/Cheno-am Ratios	9,200–6,800
Lahontan Basin: Painted Hills	Shrub histories	To ca. 8,000
Lahontan Basin	Marshes	10,000–9,600
Lahontan Basin	Lake Levels	9,000–7,500 (7000)
Harney Basin	Lake Levels	9,600–7,400
Dietz Basin	Wetlands	ca. 9,600–9,500
Fort Rock Basin: Connley Caves	Marshes	10,600–7,200

In addition, these are not the only studies from either at or beyond the far western edge of the Great Basin that conflict with studies from within the Great Basin itself. I will return to all of this at the end of my discussion of the early Holocene.

THE MOJAVE DESERT: WHITE BURSAGE AND CREOSOTE BUSH

The general sequence of vegetational change in the Mojave Desert is quite similar to that in the floristic Great Basin to the immediate north, even though many of the plants involved differ. Instead of subalpine conifers retreating upslope, for instance, it is pinyon and junipers that moved upward or that became locally extinct. Instead of sagebrush expanding into areas once occupied by conifers, it is such classic Mojave Desert plants as white bursage and creosote bush. The timing of it all, however, is much the same in both regions.

The woodlands that marked so many low-elevation areas in the Mojave Desert during the late Wisconsin tended to disappear first from lower elevation, xeric habitats, and later from higher elevation, relatively mesic ones. In many areas, pinyon and juniper were gone from less congenial settings by 10,000 years ago. In Death Valley, for instance, only a single midden sample documents the continued existence of Utah juniper beneath its modern elevational range after 10,000 years ago, and that single sample, from an elevation of 3,700 feet, dates to 9,680 years ago. In more favorable settings, these trees lingered longer, though apparently not nearly as long as they lingered in the vicinity of the Snake Range. The latest records for low-elevation conifers in the Mojave Desert are for Utah juniper, and both date to 7,800 years ago, from Mercury Ridge in the north-central Mojave Desert, and from Lucerne Valley in the far southwestern Mojave Desert. Today, both sites support creosote bush and other shrubs, but not junipers.

As these "pygmy conifers" moved upslope, plants fully characteristic of the modern Mojave Desert began to arrive, though at different times and at different places. White bursage and creosote bush provide excellent examples.

Both of these shrubs are now widespread within the Mojave Desert, but neither of them appears to have been here before, or much before, the end of the Pleistocene. The earliest record for white bursage in the Mojave Desert is from Death Valley, where a midden analyzed by Wells and Woodcock contained macrofossils of this plant dated to 10,230 years ago at an elevation of 1,400 feet. Geof Spaulding has white bursage from the Skeleton Hills at 9,200 years ago but not in a series of earlier samples from here that go back to nearly 18,000 years ago. In his Eureka View samples (chapter 6), white bursage appears at 6,795 years ago, even though he has samples from here that extend as far back as 14,700 years. In the central Mojave Desert as a whole, this plant does not appear until 8,430 years ago, even though the midden record for this region extends back some 24,000 years. In Lucerne Valley, Tom King's midden samples first show the presence of white bursage at 7,900 years ago; his oldest samples are 12,100 years old. Nowhere has this plant been found in the Mojave Desert in a context older than 10,230 years. In other places it appears later, and sometimes much later. Today, it is one of the Mojave Desert's most abundant shrubs at elevations beneath 2,500 feet.

Creosote bush is also one of the Mojave Desert's most common shrubs; as I discuss in chapter 2, the northern limit of this species sets the boundary between the Mojave Desert and the floristic Great Basin. The high abundance of creosote bush is not confined to the Mojave Desert, of course: this plant is extremely common throughout the Chihuahuan and Sonoran deserts as well (see figure 2-10). "Oceans of creosote bush" is the appropriate phrase used by Julio Betancourt and his colleagues (1990:6) to describe the abundance of this plant in the warm deserts of North America.

The genus to which creosote bush belongs has four other species, all of which occur in the deserts of South America. One of these, *Larrea divaricata*, is so similar to creosote bush that the two can interbreed and produce partially fertile offspring. This is the case even though the northernmost populations of *Larrea divaricata* are in southern Peru and the southernmost populations of creosote bush are in central Mexico. As a result, some place the two in the same species and many assume the South American form to be ancestral to the North American one. How and when the North American form came to be where it is today remains a mystery, given the huge geographic gap between the two. Recent estimates based on the amount of genetic divergence shown by the two species suggest that this might have happened between about 4 million and 8 million years ago, though others have suggested a far more recent date. No matter when it happened, the best guess is that it occurred during a time when tropical habitats shrank, decreasing the distance between arid habitats to the north and south, and that birds were the dispersal mechanism.

While we do not know when North America got its *Larrea*, the bulk of earliest dates for creosote bush in the Chihuahuan, Sonoran, and Mojave are quite recent and, for the Mojave, remarkably recent.

Two different kinds of dates need to be dealt with here. First, dates from packrat middens have provided the remains of creosote bush but come from plants other than creosote bush itself. There is, for instance, a date of 30,470 years ago from the western edge of the Sheep Range in southwestern Nevada's Mojave Desert, a date of 26,400 years ago from the Rio Grande Village midden in the Chihuahuan Desert of far southern Texas, and dates of 21,000 and 18,300 years ago from the Artillery Mountains on the northern edge of the Sonoran Desert in western Arizona. All these dates come from material associated with creosote bush, but not from creosote bush itself. As a result, they may all reflect younger contaminants.

Second, there are dates that either are directly on creosote bush plant parts or are so closely associated with those parts that there is no reason to question them.

The earliest of these dates comes from the dung of Shasta ground sloths from Rampart Cave, in Arizona's Grand Canyon. In 1961, Paul Martin and his colleagues analyzed the content of that dung and discovered that some samples contained creosote bush pollen, including one whose pollen content was dominated by this species and that dated to more than 35,500 years ago. More recently, Richard Hansen's analysis of sloth dung samples from this cave found the remains of this shrub in specimens that dated to 32,560 years ago and to between 12,000 and 13,100 years ago. These are all biogeographic oddities since creosote bush does not come close to the Grand Canyon today and no other pollen data or packrat middens have revealed the presence of creosote bush in this area. Given the close match between the results obtained by Martin's team and those obtained by Hansen, there seems no compelling reason to think that they are problematic.

All other trustworthy early dates for creosote bush from North America's warm deserts come from packrat middens, the very earliest of which falls at 18,700 years ago from the Tinajas Altas Mountains of southwestern Arizona, a date obtained by Tom Van Devender from a creosote bush twig. Other dates, provided by paleobotanist Ken Cole from Picacho Peak in far southeastern California, go back to 12,730 years ago, and still others, from the Butler Mountains, near the Tinajas Altas Mountains, are as old as 12,390 years ago. There is also a strong suggestion that creosote bush was in the Chihuahuan desert of southern Texas by about 21,000 years ago.

Holding aside the currently inexplicable records from the Grand Canyon, these dates suggest that creosote bush was present during and after full-glacial times in the Chihuahuan Desert along the border between the United States and Mexico, as well as in the southern Sonoran Desert.

Although this is hardly a controversial conclusion, it is good to know that evidence not only supports it but also suggests that creosote bush was actually in these areas before, perhaps well before, the plant macrofossil records are telling us it was here. That evidence comes from the pollen content of packrat middens.

As I have discussed before, woodrats gather plants in the immediate vicinity of the areas in which they live. That means that the first appearances of plants in those middens reflects what is happening at that particular spot. Pollen that ends up in those middens, however, may sample a much broader area. As a result, Julio Betancourt and his colleagues have pointed that the first appearances of plants in packrat midden records as judged from pollen may be much earlier that as judged from the dates on the earliest plant macrofossils.

A prime example is provided by work done by Tom Van Devender, Owen Davis, and Scott Anderson on packrat middens from Organ Pipe National Monument in far southern Arizona. Van Devender's work on the macrofossils from these middens provided the first hint of creosote bush here at about 8,000 years ago and the first truly secure evidence at 3,400 years ago. The pollen record from these middens that was developed by Davis and Anderson, however, shows significant amounts of creosote bush pollen at about 14,000 years ago. This pollen is so abundant that it cannot be explained by long-distance transport of the pollen of this insect-pollinated plant (bees love it) and seems unlikely to be accounted for by the movement of pollen downward through cracks in the midden. The pollen from these sites thus suggests an appearance of creosote bush here far earlier than is suggested by the macrofossils from the same middens.

It seems certain, then, that creosote bush was in this region during the late Pleistocene, but the situation is more complicated for the Mojave Desert to the north. Table 8-2 presents the secure radiocarbon dates for the earliest known records of creosote bush in the Mojave Desert. Several things about these dates are remarkable. First, even though the northern edge of the distribution of creosote bush defines the boundary between the Mojave Desert and the floristic Great Basin, nothing whatsoever suggests that creosote bush was in today's Mojave Desert during the Pleistocene. The earliest date for this plant here comes from one of Geoff Spaulding's Point of Rocks middens, in the Amargosa Desert, at 9,650 years ago. The second earliest date, at 9,500 years ago, is from Peter Wigand's Owl Canyon samples, in the Devils Hole area of the Amargosa Desert. The third earliest, from Little Skull Mountain just east of the Amargosa Desert, falls at 8,480 years ago. The Penthouse midden, on the southeastern edge of the Sheep Range, provided small amounts of creosote bush at 8,100 years ago. The Marble Mountains midden, just east of southeastern California's Lucerne Valley, has creosote bush at 7,930 years ago. All other dates are much younger than this. In fact, they are middle Holocene in age.

Second, most of the places with the younger dates—Ibex Pass, Eureka Valley, Granite Mountains, Lucerne Valley, and Nelson Basin—are in the more westerly parts of the Mojave Desert or, in the case of Pintwater Cave, in the far eastern

TABLE 8-2
Earliest Dates for Creosote Bush in Selected Parts of the Mojave Desert

Location	Midden Sample	Age (years ago)	Abundance	Reference
Point of Rocks, Amargosa Desert, CA	PR 2(2)	9,650	NISP = 30	Spaulding 1985
Owl Canyon, Devil's Hole Hills, NV	DHOC020469PJM1(1)	9,500	Common	Wigand in Forester et al. 1999; NAPMD
Little Skull Mountain, Jackass Flats, NV	LSM151182LP1(4)	8,480	Common	Wigand 1990b; NAPMD
Penthouse, Willow Wash Canyon, NV	Ph 1(1)	8,100	Rare	Spaulding 1981
Marble Mountains, Cadiz Valley, CA	MM 6	7,930	Common	Spaulding 1980
North Silver Lake, Silurian Valley, CA	NSL 1	6,910	Common	Koehler, Anderson, and Spaulding 2005
McCullough Range, NV	MR 3(4)	6,800	Abundant	Spaulding 1991
Pintwater Cave, Pintwater Range, NV	Not Provided	6,480	Common	Wigand 1997b
Granite Mountains, Fort Irwin, CA	NNE 4C	5,960	Common	Koehler, Anderson, and Spaulding 2005
Sunset Cove, Lucerne Valley, CA	SC 1	5,880	Common	King 1976
Ibex Pass, southern Death Valley, CA	Ibex	4,760	Common	Koehler, Anderson, and Spaulding 2005
Nelson Basin, CA	NBC2	4,580	Common	Koehler, Anderson, and Spaulding 2005

NOTE: NISP = number of identified specimens; NAPMD = North American Packrat Midden Database (discussed in chapter 8 notes).

Mojave. This may in part be an accident, reflecting where the work has been done, but even if this is the case, it is unlikely to be accidental that these western, and perhaps eastern, regions received their creosote bush later than the Amargosa Desert area.

All this seems to provide a consistent picture. Creosote bush was in the Chihuahuan Desert, along the border between the United States and Mexico and in the northern and central Sonoran Desert before, and often well before, the end of the Pleistocene. As the Pleistocene ended, it dispersed northward. Some parts of the Mojave Desert, including the Amargosa Desert, were occupied by this plant during the very early Holocene, but the more remote reaches of the Mojave Desert—for instance, Lucerne Valley and northern Eureka Valley—were not colonized until the middle Holocene, and sometimes not until the end of the middle Holocene.

There is, though, a significant complication here. Table 8-2 doesn't present all the dates for creosote bush in the Mojave Desert, only the secure ones. Others are much earlier. Geoff Spaulding has creosote bush from a 30,740-year-old sample from his Eyrie midden, on the western edge of the Sheep Range in southern Nevada. He also found this species in his Blue Diamond Road 3 midden, near the southeastern Spring Range, in a sample dated to 15,040 years ago, and in Owl Canyon Midden 1, from the eastern edge of the Amargosa Desert, at 13,150 years ago. His Penthouse samples actually first contain this shrub at 11,550 years ago.

In all of these cases, the creosote bush fragments are rare and might well be contaminants. That, in fact, is the reason why recent discussions of creosote bush history generally do not mention them (it is not so clear why those discussions do not mention the Rampart Cave data). But what if they are not contaminants and really do represent creosote bush growing in the heart of the Mojave Desert during the late Wisconsin? If so, the implication is that this species was present in the Mojave Desert during this period but was so uncommon that it is difficult to find packrat middens that contain it. If that is correct, the early Holocene dispersal of this plant across the Mojave Desert may have been in large part an internal affair, with the species dispersing widely from locally established populations.

By far the best way to tell which of these options—an early Holocene arrival from the south (the accepted view) or dispersal from local populations—is correct would be by dating the possible late Wisconsin-aged creosote bush fragments from these middens directly. Since this hasn't been done, we can't be sure which possibility is correct. However, there are some hints from pollen taken from within the middens themselves. As I discuss in chapter 6, Peter Koehler and his colleagues analyzed the pollen from their middens in the central Mojave Desert and the results of that work coincide perfectly with the results of the macrofossil analysis: creosote bush pollen doesn't show up in their middens until the macrofossils show up as well. That makes it fairly unlikely that the plant was in the area long before the macrofossils say it was there.

Similarly, Geoff Spaulding has analyzed the pollen from some of his middens in the Yucca Mountain area. One of those middens, from Sandy Valley, has a piece of creosote bush at 9,250 years ago, a piece that he suggests was a contaminant. In support of this, his analysis of the pollen from this sample showed no creosote bush at all. In fact, none of ancient Sandy Valley midden samples, which date from 8,490 to 9,430 years ago, have any *Larrea* pollen in them at all, even though modern samples have up to about 5 percent of it. Clearly, he was right in suggesting that the creosote bush fragment in the 9,250-year-old Sandy Valley midden sample was a more recent contaminant.

What the pollen data seem to tell us with some certainly is that creosote bush was not in the central Mojave Desert before the middle Holocene, while at the same time suggesting that the rare bits of creosote bush in Pleistocene-aged samples from here are contaminants. The compulsive scientist, though, would be happiest if those rare bits were dated directly.

Not only are odds strongly in favor of creosote bush being a Holocene immigrant into the Mojave Desert from the south, but there is no doubt that this species began to spread across the Mojave Desert during the early Holocene, and that much of its current range has been occupied much more recently than that.

Sagebrush in the Mojave Desert

As this was happening, it also appears that sagebrush was becoming less abundant here, although the early Holocene history of sagebrush in the Mojave Desert cannot be said to be well understood.

That we do not have a good understanding of the history of this plant here may simply reflect the fact that sagebrush had become relatively uncommon in this area by 10,000 years ago. Spaulding's Point of Rocks midden samples from the Amargosa Desert, for instance, contain sagebrush macrofossils as late as 9,840 years ago, but not in two later samples 9,560 and 9,260 years old. His Last Chance Range samples not far to the south contain sage at 11,760 years ago, but not at 9,280 years ago. Other early Holocene middens do contain sagebrush macrofossils, but only in very small amounts—in the Basin Canyon series (central Sheep Range) at 9,365 years ago, for instance, and in the Penthouse (southeastern Sheep Range) series at 8,070 years ago. In the extensive series of central Mojave Desert middens analyzed by Peter Koehler and his colleagues, sagebrush appears in only four. Two of these, mentioned in chapter 6, are late Pleistocene in age; the other two date to between 4,900 and 4,400 years ago. None of their early Holocene samples contain this plant. The macrofossil data might be taken to suggest that this shrub was rapidly disappearing from the landscape as the early Holocene began.

On the other hand, superb evidence shows that in at least parts of the Mojave Desert, sagebrush was present in some abundance through much of the early Holocene. Tom King analyzed the pollen from the middens he collected from the Lucerne Valley area and found high levels of *Artemisia* pollen in samples dating to between 12,100 and 8,300 years ago, but not after that time. The sequence agrees quite well with independent data suggesting that Lake Mojave gasped its last at around 8,700 years ago.

A similar phenomenon is known from the Tule Springs region in Las Vegas Valley. As I discuss in chapter 6, Peter Mehringer's work here suggested that, at about 12,000 years ago, juniper woodland with a sagebrush understory was replaced by desert shrub vegetation dominated by sagebrush and saltbush. In fact, sagebrush pollen is abundant in all of the samples he analyzed from Unit E_1, dated to between 14,000 and 11,500 years ago. Sagebrush pollen is also common in his samples from Unit E_2, deposited between 11,200 and 7,200 years ago, with a fairly pronounced decline in abundance at about 7,500 years ago. Samples from above E_2 contain virtually no sagebrush pollen at all. The correlation between this vegetational sequence and the sequence derived by Vance Haynes and Jay Quade for the history of shallow lakes and marshes in Las Vegas Valley is striking.

Pollen evidence from two parts of the Mojave Desert—the Lucerne Valley region and Las Vegas Valley—thus suggests that in at least some areas, sagebrush remained fairly common until about 8,000 years ago or so, after which it declined. The packrat middens, however, tend to contain significant numbers of sagebrush macrofossils only during the late Pleistocene.

Why might this be the case? One answer is suggested by King's samples from the Lucerne Valley region. While these samples contain sagebrush pollen, documenting that sagebrush was present in the area, they provided no identifiable sagebrush macrofossils. The pollen and the macrofossils in these middens seem to be telling us different things.

As I discuss in Chapter 6, Bob Thompson has shown that sagebrush macrofossils are often absent or rare in packrat middens that contain substantial amounts of sagebrush pollen. As Thompson notes, sagebrush is a wind-pollinated plant that most commonly grows on fine substrates; the plant might be common nearby, but not on the rocky outcrops that contain the middens. As a result, the middens may often contain sagebrush pollen, but not sagebrush macrofossils.

An excellent example is provided by the central Mojave Desert middens analyzed by Koehler and his colleagues. Although none of their early Holocene middens contained sagebrush plant fragments, they found that sagebrush pollen is fairly abundant in their middens from the start of the sequence, about 24,000 years ago, to almost 8,000 years ago. It then almost disappears, returns to hit a small peak at about 7,000 years ago, and then slowly fades to almost nothing by the late Holocene. In fact, the two highest peaks in sagebrush abundance, at about 24,000 and 10,000 years ago, come from samples that lacked sagebrush macrofossils. Geoff Spaulding provides similar evidence from the Fortymile Canyon area in southeastern Nevada, where, he notes, generally higher percentages of sagebrush pollen are found in the late Pleistocene and early Holocene packrat middens of this area compared to both the single late Holocene and modern samples.

In short, the pollen records suggest that sagebrush was fairly common in at least some parts of the Mojave Desert during the early Holocene, while the packrat midden records suggest that it was rare to absent. The reason for this difference is likely to lie largely in the very different ways in which macrofossils and pollen become embedded in the deposits that paleobotanists analyze.

This cannot be the whole answer, however. Spaulding's Point of Rocks and Last Chance Range middens, for instance, may lack early Holocene sagebrush macrofossils, but they did provide late Pleistocene examples. In addition, some of his late Pleistocene middens—for instance, from the Eleana Range—contain them in impressive abundance. Something more is going on here than can be explained by the different ways in which pollen and macrofossils accumulate in packrat middens.

The rest of the answer is probably provided by the late Wisconsin and early Holocene history of sagebrush itself. During the latest Wisconsin, this shrub was probably so abundant on many parts of the Mojave Desert landscape that it ended up, in the form of macrofossils, in many packrat middens. By the early Holocene, sage was sufficiently reduced in abundance that the location of the middens, on rocky outcrops, had become critical. It was still there, and common in some places, but no longer growing immediately adjacent to the middens. As a result, while the Las Vegas Valley pollen profiles show sagebrush on the floor of this valley through much of the early Holocene, middens from higher settings contain only small amounts of it. If this is correct, there is an excellent chance that the pollen from early Holocene midden samples that lack, or contain only very small amounts of, sagebrush would show that sagebrush was still common in many lower-elevation parts of the Mojave Desert through much of the early Holocene, just as records from the Lucerne Valley, central Mojave Desert, and Las Vegas Valley imply.

In suggesting a relatively moist early Holocene, the presence of sagebrush in these areas during the early Holocene matches the evidence provided by deeper lakes, more expansive marshes, and more active springs in the Mojave Desert.

It also matches the evidence that moisture-loving netleaf hackberry was reasonably common in this area at that time. There are records for this plant from a variety of packrat middens in the more westerly parts of the Mojave Desert. Geoff Spaulding identified a single 9,400-year-old specimen from Sandy Valley in southwestern Nevada, though he notes that this specimen may have been transported here from a significant distance. However, he also has hackberry in the Sheep Range north of Las Vegas at about 9,600 years ago and Peter Wigand found it to be common in an 8,500-year-old midden sample from Little Skull Mountain, west of Pintwater Cave. These specimens indicate wetter landscapes than currently exist in these areas.

Netleaf hackberry seeds are also fairly common in the deposits of Pintwater Cave, located on the west side of the Pintwater Range near Indian Springs in the eastern Mojave Desert. These seeds, Wigand has shown, occur in Holocene sediments laid down between 10,000 and 8,600 years ago. It is not likely to be coincidental that these sediments also contained significant amounts of sagebrush pollen. To him, the combination of pollen and packrat midden data from the Pintwater area suggest that early Holocene springs were active here until about 8,500 years ago. Today, the closest spring is some two miles to the north and the immediate vicinity of the site is marked by such plants as white bursage, creosote bush, shadscale, Nevada jointfir, and teddybear cholla (*Cylindropuntia bigelovii*).

Small Mammals

During the early Holocene, shallow lakes, marshes, and springs existed at low elevations in many parts of the Great Basin where they do not exist, or do not exist to the same extent, today. In the floristic Great Basin, sagebrush steppe covered areas now dominated by saltbush. Subalpine conifers, while generally moving upslope, persisted in favorable habitats at low elevations. In the Mojave Desert, junipers lasted at favorable low-elevation sites until at least 7,800 years ago. The pollen evidence shows that in at least some areas, sagebrush grew at low elevations in the Mojave until about this time as well, even as such plants as white bursage and creosote bush spread across the landscape. Although, as I explain below, there is some argument about temperatures during this period, there is no denying that conditions were wetter than those of today.

The early Holocene histories of small mammals in the Great Basin match these other environmental indicators. Because I reviewed those histories in the previous chapter, here I simply mention that during the early Holocene, a broad series of small mammals that today are found in the cooler and moister settings provided by either higher elevation or more northerly contexts in the Great Basin were then found in low-elevation and often more southerly settings.

Pikas provide a good example. As we have seen, pikas were found at very low elevations during the late Pleistocene in the Great Basin, some 2,500 feet lower than their current average elevation. There are far fewer early Holocene records for pikas in this area. This may in part be a function of the fact that the paleontological record is actually stronger for the late Pleistocene than it is for the early Holocene, but it is also very likely to be a result of the fact that the warming and drying that occurred during the early Holocene decreased the amount of pika habitat available in the Great Basin, especially at the lower elevations that have provided so many of the paleontological sites. However, even though there are far fewer early Holocene than late Pleistocene sites with pika remains in them, the average elevations of those sites is almost identical to the average elevation of the late

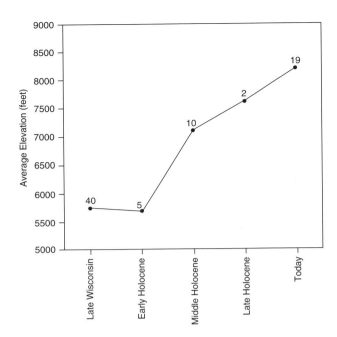

FIGURE 8-4 Changing elevations of pikas through time in the Great Basin (after Grayson 2005).

Pleistocene ones. Not until the middle Holocene does that average elevation spike upward (figure 8-4).

Other species fall in line, though we are hampered here by the fact that our most detailed and best-dated records are largely restricted to the more eastern parts of the Great Basin—Homestead Cave in the northern Bonneville Basin, Camels Back Cave in the central Bonneville Basin, and Pintwater Cave in the eastern Mojave Desert.

Although pygmy rabbits were far more abundant in the Homestead Cave area during the late Pleistocene than during the early Holocene, they continued to exist here until shortly after 8,300 years ago, suggesting the continued existence of sagebrush habitat until this time as well (figure 7-4). After this time, pygmy rabbits are simply gone from the area. At the same time as pygmy rabbits decline to disappearance at Homestead Cave, the shadscale-loving chisel-toothed kangaroo rat increased dramatically in abundance, at the expense of Ord's kangaroo rat, which prefers habitats marked by sagebrush (figures 7-5 and 7-6). To the south, at Camels Back Cave, the faunal sample for pygmy rabbits is not nearly as substantial as it is at Homestead, but their remains are found in deposits as young as 8,800 years old.

Much the same is true of yellow-bellied marmots and bushy-tailed woodrats. Marmots were in the area surrounding Homestead Cave until around 8,000 years ago (figure 7-8) and until around 8,800 years ago at Camels Back Cave, though this latter record is less convincing because of the very small number of specimens involved. Neither area supports these animals today. Bushy-tailed woodrats were common around Homestead Cave until around 8,300 years ago; they then disappeared, only to return in far reduced numbers during the very late Holocene (figure 7-11). These same animals were apparently fairly common in the Camels Back Cave area until at least 8,800 years ago, disappearing from here sometime after this date but certainly before 7,000 years ago. Bushy-tailed woodrats are never very common in the deposits of Pintwater Cave but they are there, with the most recent specimen deposited around 8,300 years ago.

Similar patterns are shown by other mammals at Homestead. For instance, the western harvest mouse (*Reithrodontomys megalotis*), which prefers moist, grassy habitats, was common at Homestead until shortly after 8,000 years ago, after which its numbers plummeted (figure 8-5). The same is true of Great Basin pocket mice (*Perognathus parvus*), an animal whose abundance in the Great Basin is positively correlated with both the amount of ground cover and winter precipitation. It was common in the Homestead Cave deposits until about 8,300 years ago when it, too, underwent a very steep decline in abundance (figure 8-6). And ditto the sage vole (*Lemmiscus curtatus*), which is often associated with habitats marked by sagebrush and grasses and which had been common in the late Pleistocene and earliest Holocene here. It, too, seems to have disappeared here by 8,300 years ago, not to return, and then in much reduced abundance, until the late Holocene (figure 8-7).

There are other low-elevation, early Holocene records for small mammals that today prefer cool and moist habitats, but these are not so easily interpreted as those from recently excavated and better-dated sites. The yellow-bellied marmot and bushy-tailed woodrat specimens from Lovelock Cave are a case in point. This site is located on the southern edge of the Humboldt Sink in the Lahontan Basin, not far from Lovelock, Nevada. The earliest excavations here were done at the beginning of this century and continued sporadically through the 1960s. Unfortunately, those excavations did not pay much attention to stratigraphic detail. As a result, and in spite of intense efforts, archaeologist Stephanie Livingston, in her detailed analysis of the bird and mammal remains from Lovelock Cave, was unable to place the marmot specimens from this site (only two bones, but also a complete, desiccated individual) in their original stratigraphic context. All we know is that these remains are Holocene in age; only by dating the specimens themselves can we now learn more from them.

Much the same is true for both the marmots and bushy-tailed woodrats from Hogup Cave, though for very different reasons. This site was excavated with great care by archaeologist C. Melvin Aikens during the 1960s, but a variety of evidence, including radiocarbon dates out of proper stratigraphic order, shows that some of the deposits within Hogup had become mixed. As a result, we do not know exactly how old the specimens of these species from this site are.

Danger Cave is almost directly across the Bonneville Basin from Homestead Cave. The faunal record available from this site would seem to document the persistence

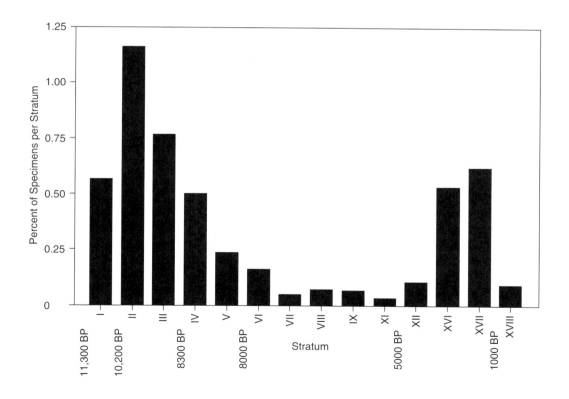

FIGURE 8-5 Changing abundance of harvest mice through time at Homestead Cave, Bonneville Basin, Utah (after Grayson 2000a).

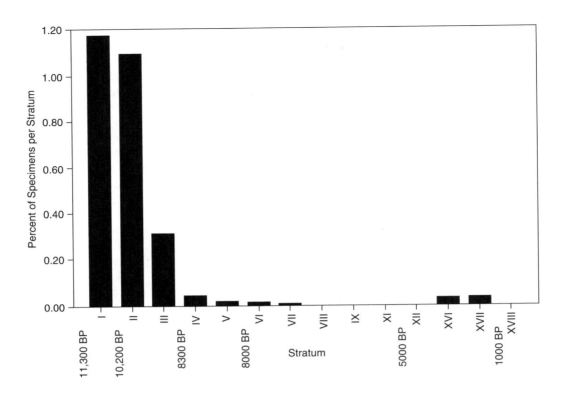

FIGURE 8-6 Changing abundance of Great Basin pocket mice through time at Homestead Cave, Bonneville Basin, Utah (after Grayson 2000a).

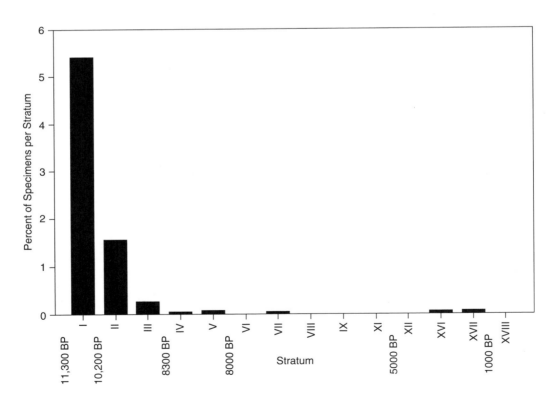

FIGURE 8-7 Changing abundance of sage voles through time at Homestead Cave, Bonneville Basin, Utah (after Grayson 2000a).

of bushy-tailed woodrats at low elevations from the latest Pleistocene down to about 8,000 years ago. While this is almost a perfect match for the Homestead Cave sequence, the lowest levels of this site are known to have suffered significant stratigraphic disturbance. As a result, it is hard to know how much faith to have in this information.

Our knowledge of early Holocene small mammals in the Great Basin is hindered not only by possible or certain stratigraphic problems but also by the fact that relatively few deposits of this age have been carefully excavated. The packrat midden record for many parts of the Great Basin is impressively strong, thanks to a small group of productive and dedicated scientists, but these middens can be counted on to provide us with information only on pikas, primarily because these rodents so often incorporated pika fecal pellets into their middens. The richest source of small mammal remains has been provided by the sediments of caves and rockshelters but very few such sites with early Holocene deposits have been excavated. Those that have been excavated—Homestead, Camels Back, and Pintwater caves, for instance—have dramatically improved our understanding of small mammal history during this period. Unfortunately, other sites that have been carefully excavated do not have early Holocene deposits. Gatecliff Shelter in the Toquima Range of central Nevada is an excellent example. This site provided some 13,000 bones and teeth of small mammals, all from well-controlled and well-dated stratigraphic contexts (figure 8-8). However, the oldest sediments in this site were deposited only shortly before 7,000 years ago. Even as it stands, though, it is clear that, during the early Holocene, low-elevation habitats in the Great Basin that are now hot and dry then supported a broad range of mammals now restricted to either higher elevations or more northerly settings in the Great Basin.

Large Mammals

The extinction of the Great Basin's large Pleistocene animals left behind five genera of sizeable mammals in this region: deer, mountain sheep, pronghorn, elk, and bison. All of these animals are what biologists refer to as artiodactyls, or even-toed ungulates (the odd-toed ungulates include such animals as horses and rhinos). In addition to these, there were bears and cougars here prehistorically, but they were quite rare (see chapter 10 for cougars).

Mule deer (*Odocoileus hemionus*), mountain sheep, and pronghorn were widespread in the Great Basin during early historic times and, to a large extent, remain so today. There are also questionable hints that white-tailed deer may have existed in far northeastern California and immediately adjacent Nevada and Oregon. These three were the common genera of larger mammals in the Great Basin.

The other two, elk and bison, were not nearly as widespread and much of the secure information we have about their distributions and abundances comes not from historic observations but from the archaeological record. Because our most secure knowledge of the Great Basin histories of these animals is for the late Holocene, I discuss them toward the end of this chapter.

How about the early Holocene record of Great Basin artiodactyls as a whole? Unfortunately, it would be generous to describe that record as pitiful. While Homestead Cave, for instance, provided an extraordinary small mammal record—some 184,000 specimens that could be identified

FIGURE 8-8 Gatecliff Shelter, Toquima Range, Nevada, during the excavations.

to at least the genus level—it provided a grand total of only twenty-two artiodactyl specimens for the entire sequence. This is not because large mammal bones were so badly broken that they couldn't be identified: they simply weren't there in any number.

It might be easy to explain why Homestead Cave had so few artiodactyls. Although large mammals can take shelter in caves during periods of bad weather and die there, that is not a common occurrence. To get their remains in such settings requires that something bring them there—people are a prime example. But Homestead Cave was not a place where people lived. Instead, its fauna accumulated largely as a result of predation by owls, and the large mammals are not there.

As I have noted, not many Great Basin cave sites with early Holocene deposits have been carefully excavated, which certainly helps explain why we have so little bone-derived information on the history of artiodactyls here. But some sites with early Holocene deposits in them did serve as dwelling places for people—Danger Cave is an obvious example—and these, too, tend to have small numbers of artiodactyl specimens.

Given that we know that people were in the Great Basin during the late Pleistocene, their absence from the landscape cannot account for this. Some other things can, however.

First, it is possible that when early Holocene hunters killed large animals, they simply butchered them on the spot and consumed them nearby. A dead bison, for instance, provides a lot of meat and the costs of transporting that meat, and any associated bones, to the nearest cave might well be prohibitive. In the Great Plains, for instance, the rich archaeological record for bison hunting comes almost entirely from sites in the open air, and it is no surprise that all the evidence we have for late Pleistocene mammoth-hunting also comes from such sites.

Unfortunately, early Holocene (and late Pleistocene) open-air but now-buried archaeological sites in the Great Basin are vanishingly rare, as we will see in the following chapter. In the rare instances we do have, artiodactyls tend to be uncommon in them as well but this may simply reflect the fact that the kinds of open sites of this age we have found so far are not large-mammal kill and processing sites.

So that is one possibility: our sample of the early Holocene archaeological record is so weak that we just do not have the kinds of sites that can tell us anything about the abundance of large mammals on the early Holocene landscape. Certainly, the evidence that we have seen for a fairly wet early Holocene might seem to suggest that this should have been a time when artiodactyls were far more abundant than they have been since then.

This, in fact, is exactly what has been assumed by many Great Basin scientists. There is, though, a second possibility. Perhaps the very sparse evidence we have for artiodactyls on the early Holocene landscape is to be read in a far more straightforward manner. Perhaps we have so little evidence for them at this time because they were, in fact, rare.

It is just this possibility that Jack Broughton and his colleagues have suggested. They observe that if we wish to understand the interrelationship of artiodactyl abundance and climate for any given area, it is not so much yearly averages that count, but seasonal extremes. Hot and dry summers decrease the amount of high-quality forage on the landscape. Severe winters with deep snow packs can make it difficult or impossible for large herbivores to gain access to food and harder for them to avoid predators, at the same time as greater energy expenditure is required for locomotion and maintaining body warmth. Animals that manage to survive these extremes may have decreased reproductive success.

In addition, large mammals have longer gestation periods than smaller ones: about two hundred days, for instance, for mule deer, compared to about thirty days for kangaroo rats. This difference has vital consequences for potential reproductive success. Animals with short gestation times may have multiple litters a year, but animals with long gestation times cannot. The latter species use long-distance cues to time the onset of reproduction, most notably the number of hours of daylight. Many of the former, quick gestation, species can use more direct ways to decide when breed—for instance, the availability of appropriate food on the landscape. Biologists Jim Kenagy and George Bartholomew have contrasted these approaches to breeding as "pulse averagers" and "pulse matchers". Pulse averagers, with long gestation times, place reproductive bets by depending on predictable seasonal change. As Kenagy and Bartholomew (1985:394) put it, they gamble that "environmental conditions taken as a whole over winter, spring, and early summer will be suitable for a single, protracted reproductive attempt." The pulse matchers, on the other hand, keep an eye out for acceptable environmental conditions and then have at it.

These differences are important because pulse averagers can be at great risk in times of adverse environmental change. If you bet that next spring is going to be a good time to raise your young, but things go wrong, those young will be lost. If this continues for a significant number of years, the long-term consequences can be grave for the species as a whole. And, because the gamble you are taking is genetically determined—it was your evolutionary background that determined the nature of the bet—you cannot easily change it. Pulse matchers don't have this problem, since their decision to breed is based on more proximate indicators of habitat quality. This doesn't mean that these species have nothing to worry about, since bad habitat is bad habitat; if the shadscale is no longer there, the kangaroo rats that depend on it are not going to breed, and their numbers will decline. But it does mean that pulse matchers can track environmental quality much more closely than pulse averagers.

For quite some time, these issues have played a major role in attempts to understand the causes of Pleistocene extinctions in North America, but Broughton and his colleagues argue that these biological facts may well account for a number of aspects of the early Holocene in the Great Basin. In particular, they argue that if seasonal extremes were greater during the early Holocene, with hotter, drier summers and colder, wetter winters, artiodactyls would have been at a great reproductive disadvantage in this region. The question then becomes whether or not the early Holocene in the Great Basin was a time of such seasonal extremes.

They call on a number of empirical studies to argue that this was the case. For example, they note the evidence, discussed above, that Tom Minckley and his colleagues provide for a hot and dry early Holocene along the edges of the far northwestern Great Basin. In addition, they point out that the early Holocene spring and stream activity in the Old River Bed area that I discussed earlier in this chapter might well have resulted from heavy winter snowfalls, and subsequent warm-season runoff, in the adjacent mountains. They suggest that these, and similar, studies are consistent with a Great Basin with far greater seasonal extremes than exist today—colder and wetter winters with heavy snowpack in the mountains and hotter and drier summers.

Because we have very little bone-based evidence for artiodactyl abundance in the Great Basin during the early Holocene, Broughton and his team turned to a very different source of information on those abundances: the artiodactyl fecal pellets from Homestead Cave.

Homestead Cave is full of those pellets. The Homestead research team counted a huge sample of them and, while they did not attempt to determine which species of animal provided them, their size and shape show that they were left behind by some combination of mountain sheep, deer, and pronghorn. All of these animals will use caves as shelter.

Figure 8-9 shows the abundance of these pellets through time, expressed as the number of pellets per liter of Homestead Cave sediment by stratum. Clearly, fewer of them were being deposited during the early Holocene than later and the late Holocene strata tend to contain more of them than do any of the earlier strata. Broughton and his colleagues argue that these pellets reflect the abundance of artiodactyls on the local landscape through time. They also suggest that the Homestead Cave area during the early Holocene supported very few of them.

Broughton and his colleagues used a computer-based model to reconstruct multiple aspects of past environments in the Homestead Cave area during the past 14,000 years and found that the Homestead artiodactyl pellets decreased in abundance as temperature differences between winter and summer increased and as summer precipitation decreased. This result matched their prediction that seasonality extremes should work to the disadvantage of artiodactyl populations in the region.

Of course, doing this requires that we have some faith in the results of the model itself. As they note, to have confidence in those results, the model itself needs to be tested with past environmental data from the region to which it is being applied. They very much attempted to do this but were only partially successful in the attempt. As a result, we do not know exactly how accurate their model is for the Bonneville Basin.

Second, their analysis assumes that the frequency of artiodactyl pellets in Homestead Cave really is monitoring the local frequency of these animals. While that might seem to be a reasonable assumption to make, other things can explain the pattern they found. Least likely, perhaps, is the availability of other sheltering places. If other caves became less available as shelter while Homestead remained accessible, that could account for the differing pellet abundances

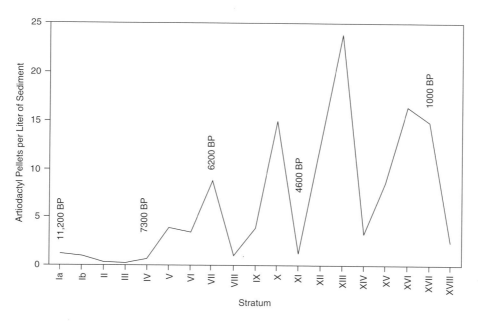

FIGURE 8-9 Changing abundance of artiodactyl fecal pellets through time at Homestead Cave (after J. M. Hunt, Rhode, and Madsen 2000).

through time. In fact, the extinction of cave bears (*Ursus spelaeus*) in France has been argued to have occurred at least in part because the increased human use of caves kept the bears from the wintering places essential to their survival. So if people were using other caves in the Homestead area during the late Holocene more frequently than they were earlier on, this could help explain the increasing numbers of fecal pellets at Homestead during the late Holocene.

In addition, if, as Broughton and his colleagues note, artiodactyls take refuge in caves during times of intense summer heat or winter cold and seasonal extremes were less during the early than during the late Holocene, then perhaps artiodactyls were quite abundant on the landscape between about 10,000 and 8,000 years ago but simply didn't need to retreat to shelters as often. In this view, it would be the late Holocene (lots of pellets), not the early Holocene (few pellets) that provided the climatic challenge.

So we need more data than just the fecal pellet record from Homestead Cave to support their argument. The Broughton team suggests a series of sites that provide just such data. At Danger, Camels Back, and Hogup caves, they note, the number of artiodactyl teeth and bones in late Holocene strata greatly outnumber those in earlier ones, which is decidedly true, although Hogup barely reaches back to the early Holocene. Further afield, work by Arianne Pinson in the Great Basin of Oregon also shows fewer artiodactyls in early Holocene archaeological sites than in late Holocene ones.

The increase in artiodactyl bones and teeth at the cave sites, however, appears to be correlated with increasing use of those sites by people. The more such use, of course, the more the opportunities for the remains of artiodactyls to be brought back to the sites involved, exactly opposite the case for the pellets. If this is what was happening, the artiodactyl record from the cave sites might simply reflect the ways in which these sites were being used by people—the greater the use, the greater the number of artiodactyl remains.

If this is correct, early Holocene sites that do show significant use of people should also show significant numbers of artiodactyl specimens. One site, Bonneville Estates Rockshelter, seems to show exactly this.

Excavated under the direction of archaeologist Ted Goebel, Bonneville Estates is located in the Lead Mine Hills about thirty miles south of Danger Cave on the western edge of the Bonneville Basin and very close to Blue Lake (figures 5-7 and 8-10). The site is extremely important not just because of the care with which it was excavated but also because it contains a rich early Holocene archaeological record. Analyses of the material from this site are still underway but Bryan Hockett's research on the animal remains shows that the early Holocene strata of this site are dominated by the remains of artiodactyls that had been introduced by people. Given this, it just might be that the artiodactyl contents of caves in the Bonneville Basin are primarily telling us about the degree to which people used those sites, not the abundance of those animals on the surrounding landscape.

The work on artiodactyl abundances in the Great Basin that Broughton and his colleagues have done is intriguing. There are many ways whereby their analysis may be problematic, but if they are correct, the early Holocene Great Basin was, on average, cool and moist but was also marked by extreme swings in temperature from winter to summer, with severe winters and hot, dry summers. What, I now ask, do other paleoclimatic analyses say about all this?

FIGURE 8-10 Bonneville Estates Rockshelter (at the end of the road) in the Lead Mine Hills, Nevada.

Early Holocene Climates

There are, in fact, strong reasons, derived from our understanding of the changing relationships between the earth and sun, to think that seasonal extremes were greater during the early Holocene than they have been since that time. Our understanding of those relationships shows that during the period of time centering on about 9,600 radiocarbon years ago, the earth's axis was tilted almost 1°F more than it is today, while the closest approach of the earth to the sun (perihelion) occurred in July, not in January as it does today. Everything else equal, the impact of both these things would be to make Northern Hemisphere summers warmer and winters colder than they are now.

"Everything else equal" is an important part of this conclusion, however, since local conditions may or may not conform to predictions meant to pertain to the earth as a whole. In addition, the models derived from our understanding of such things as changes in the tilt of the earth's axis and of the nature of the earth's orbit itself tell us about the amount and distribution of solar energy reaching the earth but not about precipitation. As a result, the question becomes what the Great Basin in particular was like during the early Holocene, how relatively warm and wet it was. The only way to answer these questions is with evidence taken directly from the past.

How wet it was seems easy. Almost no matter where we have looked in the Great Basin, from Homestead Cave and the Old River Bed on the east to the Mojave River drainage, Las Vegas Valley, and the Pintwater Range on the south, to the Lahontan Basin and the Connley Caves on the north, we have seen places unequivocally wetter than those places have been since that time.

Multiple mechanisms could have made those places wetter. Perhaps winter storms were intensified, dropping large amounts of snow in the mountains and water in the lowlands, with the lakes, springs, and marshes recharged to such a degree that no summer precipitation was needed to keep them going. Perhaps the Great Basin received far more summertime precipitation than it does now. Perhaps some combination of both these things was happening. And, just as important, perhaps different parts of the Great Basin had distinctly different climates during the early Holocene.

I begin with the southern Great Basin, including the Mojave Desert, but to do this, I have to begin slightly farther south.

During the late Wisconsin and early Holocene, woodlands that consisted of various combinations of pine, oak, and juniper were widespread in the Sonoran Desert of the American Southwest. These woodlands were gone by 8,000 years ago; in fact, most of them were gone soon after 9,000 years ago. Paleobotanists Geoff Spaulding and Lisa Graumlich have observed that the elimination of woodland in the Sonoran Desert appears to have somewhat later than it occurred in the Mojave Desert to the north. This seems to be the case even though many of the Sonoran Desert sites that document the existence of this woodland are not only farther south than the Mojave Desert sites that seem to have lost their trees somewhat earlier but also lower in elevation.

Spaulding and Graumlich argue that this difference, as well as other aspects of the late Wisconsin and early Holocene vegetational record for this region, is best explained by invoking strengthened monsoonal incursions between about 12,000 and 8,000 years ago. During this interval, they argue, summer temperatures were about 2°F to 4°F warmer than they are now, and winter temperatures about 2°F cooler. In addition, they suggest, the strengthened monsoonal system brought greatly increased rainfall to the Sonoran Desert during the summer months. Since the source of that rainfall must have been either the Gulf of Mexico or the Gulf of California and Pacific Ocean, the Sonoran Desert received more of it than did the Mojave Desert. Because the Sonoran Desert was closer to the source of the moisture, low-elevation woodlands lasted longer here than they lasted in the Mojave Desert. In this view, a good deal of early Holocene Mojave Desert precipitation fell during the summer months.

Paleobotanist Tom Van Devender has disagreed strongly with this interpretation. His analysis of the packrat midden record from the western Sonoran desert, immediately adjacent to the Mojave Desert, suggests to him that early Holocene summer temperatures here were at most modestly warmer than those of today and may have even been cooler. One might argue that this could have resulted from greater cloud cover due to increased monsoonal storms but Van Devender also finds little evidence for reliable summer rainfall in the western Sonoran Desert and calls instead for increased winter rains. If correct, this would suggest that Spaulding and Graumlich must be wrong.

Direct evidence from the Mojave Desert, however, supports their view, data provided by Hope Jahren and her colleagues. These scientists examined the ^{18}O content (see chapter 3) of the same netleaf hackberry nutlets from Pintwater Cave that show that this area, north of Las Vegas, must have been distinctly wetter during the early Holocene than it has been since that time. Their isotopic data, they argue, are best explained by the existence of significant summer monsoonal precipitation in this area at that time.

In chapter 5, I discuss both the Pacific frontal systems that bring winter rains to the Great Basin and the monsoonal systems that bring summer rains to the Southwest and parts of the Great Basin. Today, most of the precipitation that falls in the Pintwater Cave area is brought by Pacific storms, not monsoonal ones. Because of the different sources and trajectories of these storm systems, the waters they bring to this area have very different oxygen isotope contents (that from the Pacific is isotopically far lighter—less ^{18}O—than that derived from the monsoons). Because of this difference, waters brought by these different systems can be distinguished from one another by their isotopic content. The ^{18}O content of the early Holocene hackberry nutlets from Pintwater Cave shows that the plants which produced them were utilizing water brought by summer monsoonal storm systems arriving from the south.

This conclusion is consistent with the rainfall patterns of the areas in which netleaf hackberry is found today. Bob Thompson and his colleagues have examined the modern relationship between a wide range of North American plant species and various aspects of climate. Their results show that netleaf hackberry tends to occur where more precipitation falls in the summer than in the winter. The opposite may also be true, but this happens less often.

Other evidence is in accord. Adrian Harvey, Peter Wigand, and Steve Wells have examined the history of late Pleistocene and early Holocene alluvial fan development in two very different parts of the Great Basin: the edge of the Stillwater Range in the Carson Sink of western Nevada and the Pleistocene Lake Mojave area of eastern California. They discovered that the early Holocene fans in this part of the Mojave Desert were often deposited as the result of massive debris flows. These were then cut by early Holocene streams issuing from the surrounding mountains and which formed their own fans. They found nothing like this in the Stillwater area and concluded that the climatic conditions that led to fan development in the two areas must have been quite different. It is extremely unlikely, they argue, that Pacific frontal storms could have provided the wet hillsides and intense downpours needed to cause debris flows of the magnitude they discovered. They suggest that strong summer monsoonal systems that reached northward into the Mojave Desert area, but not as far north as the Carson Sink, must account for the distinctly different alluvial fan histories they found in these two areas. In line with all this, studies of the abundances of the tiny organisms known as foraminifera in cores from the Gulf of Mexico have shown that monsoonal precipitation in the Southwest was much more significant during the early Holocene than it has been during recent times.

Because it is summer heat that generates monsoons, these studies suggest that the Mojave Desert, and probably the immediately adjacent floristic Great Basin, was marked by warm and wet summers. Such summers would be consistent with the early Holocene lakes in the basin of Pleistocene Lake Mojave. They would also be consistent with the early Holocene black mats in Las Vegas Valley since those mats mark times of high water tables and organic productivity. Depending on how far north these monsoonal storms penetrated, a climate system of this sort might even explain the streams and marshes in the Old River Bed, with summer precipitation deposited in the Sevier Basin then flowing northward (chapter 5). Summer monsoons might even fit with the presence of early Holocene pinyon pine here but not to the north, because Arizona singleleaf pinyon (*Pinus monophylla fallax*: see chapter 2) is associated with summer rains.

So far so good, but some things don't quite seem to fit. In suggesting the intensified incursions of monsoonal storms into the Mojave Desert during the early Holocene, Spaulding and Graumlich noted that steppe shrubs can be expected to be common in areas dominated by winter rain. Sagebrush, *Artemisia tridentata*, is one such shrub, and the work done by Bob Thompson and his colleagues on the relationships between plant species and climate shows that while this species can be found where summer precipitation exceeds the winter kind, this is a relatively rare occurrence. It is, however, often found where winter precipitation outstrips that which falls in the summer. It appears that steppe shrubs, including sagebrush, are abundant where winter precipitation is dominant, just as Spaulding and Graumlich suggested.

We have already seen that sagebrush was quite abundant in parts of the Mojave Desert during the early Holocene: the Las Vegas Valley is one such area. This in turn suggests that winter rains must have been important here. And, of course, while such things as the early Holocene lakes in the basin of Pleistocene Lake Mojave and the black mats in the Las Vegas Valley are consistent with summer rains, they are also consistent with winter precipitation. Indeed, some interpretations of Intermittent Lake III in the Mojave Desert explain that lake in terms of high levels of winter precipitation derived from the Pacific. The presence of netleaf hackberry is consistent with winter rains, since it does grow in places dominated by rains that fall in this season.

The same can be said about singleleaf pinyon. While the Arizona ("*fallax*") variety is associated with summer rains, the California ("*californiarum*") subspecies is associated with winter ones (chapter 2). What we really need to know, but don't, is what kind of singleleaf pinyon was found in the Mojave Desert at this time. This is no fault of the scientists who have developed the paleobotanical records for this region. Until the very recent demonstration that different needle types are associated with different climatic regimes, it was enough to identify a pinyon needle from a packrat midden as coming from a single-needled variety. Now that we know that the different kinds of these needles may tell us something about the seasonal distribution of rainfall, it would be extremely helpful to go back and reanalyze them all.

So we are faced with evidence of both wet winters, as provided by sagebrush in the Las Vegas Valley, and wet summers, as provided by oxygen isotopes in the Pintwater Cave hackberry nutlets. Alongside this, we have evidence that can be explained either way, from Intermittent Lake III to the Las Vegas Valley black mats and neatleaf hackberry in areas where it does not now grow. We can even include the alluvial fan histories in eastern California. While wet slopes and heavy rains were needed to produce these particular kinds of fans, those rains could have come in the winter.

This raises the possibility that rainfall in the Mojave Desert during the early Holocene was far more evenly distributed than it is today. This is not the only place where this might have been the case. Take, for instance, Montezuma Well, a lake on the edge of the Sonoran Desert in central Arizona. Owen Davis and David Shafer analyzed the pollen sequence provided by the sediments of this lake and discovered that, from the end of the Pleistocene to about 8,000 years ago, this sequence is marked by very high abundances of pine, certainly pinyon, and juniper. To them, this suggests cool winter temperatures. During this same interval, sagebrush pollen is more common than at any other time in this 11,000 year-long sequence. That, in turn, suggests winter precipitation. On the other hand, oak pollen is also common during the early Holocene, especially between 9,500 and 8,000 years ago. From this, Davis and Shafer infer the existence of significant summer rains, since the oaks of this region today are found in areas of high summer rainfall. And their reconstructions of temperature and precipitation suggest that this part of central Arizona was, in general, cooler and wetter than what has followed. This interpretation is, in fact, quite consistent with Van Devender's interpretation of the nature of western Sonoran Desert climates during the early Holocene.

The early Holocene small-mammal histories available for the Mojave Desert do not help us that much. The most important of these have been provided by Bryan Hockett's work at Pintwater Cave (see chapter 7). The sediments of this site provided the remains of northern pocket gophers as late as 9,000 years ago, bushy-tailed woodrats as late as 8,300 years ago, and pikas as late as 7,350 years ago. Taken at face value, these records suggest cool early Holocene summers here, since none of these animals now lives in areas whose summers are nearly as hot as those in the Las Vegas area. On the other hand, the samples are not large and because none of the specimens has been directly dated, it is possible that they are not of the same age as the sediments in which they were found. Direct dates on these specimens would, as a result, be of tremendous value. As it stands, though, the Pintwater Cave small mammals suggest cool summers during the early Holocene in this part of the world.

Given current data, the best interpretation of early Holocene Mojave Desert climates is that precipitation was more evenly distributed throughout the year than it is now. Temperature is a harder call. If the mammal record is correct, summers were cooler, but if there were intensified summer monsoons, as many paleobotanists have concluded, it was likely warmer. Either way, it was certainly wetter.

Things get better in the floristic Great Basin to the north. In the Bonneville Basin, the combination of the early Holocene hackberry seed record from Homestead Cave and the small mammal faunas from Homestead and Camels Back caves provide secure indications of the nature of climates here at that time. Hackberry growing in abundance in the immediate vicinity of Homestead Cave during the early Holocene requires that there have been sufficient moisture to support these plants. The presence of pygmy rabbits requires rich sagebrush habitats in the vicinity, confirmed by the fact that Ord's kangaroo rat is far more abundant than the shadscale-loving chisel-toothed kangaroo rat during this interval. That summers must have been fairly cool is implied by the combination of yellow-bellied marmots and very abundant bushy-tailed woodrats. All this forms a consistent picture of substantial winter rains falling on a relatively cool landscape, cooler and wetter than has marked the northern Bonneville Basin at any time since.

Bob Thompson's work on the climatic implications of relatively low-elevation limber pines in the east-central Great Basin during the early Holocene has much the same implication. He has pointed out that the optimal temperature for photosynthesis in limber pine seedlings is around 59°F. Coupling that fact with the abundance of limber pine macrofossils in lower elevation packrat middens from the east-central Great Basin, he estimates that summer temperatures in this region were some 7°F to 9°F cooler than they are now. Although pertaining to much higher elevations, that figure is fully consistent with the small mammal histories from Homestead and Camels Back caves and with the vegetation histories they imply. It is also consistent with the existence of populations of relatively low-elevation pikas at this time: those animals do not tolerate hot summers.

To the west, in Ruby Valley, the presence of sagebrush steppe until about 7,700 years ago is matched by the data from Hidden Cave in the Carson Sink, where the abundance of sagebrush pollen does not fall to modern levels until about 6,800 years ago. Given what we know about the distribution of sagebrush, these records seem to suggest wintertime precipitation. In fact, there is no suggestion of significant summertime, monsoonal precipitation at all in this part of the Great Basin, nor is there any compelling indication of very hot summers. The relatively low-elevation existence of such shrubs as wild rose and current in the Painted Hills, for instance, is inconsistent with the notion that summers here were hot and dry.

I stress that all of the information we have for the early Holocene of the Great Basin indicates that this was a period of warming and drying. Desert vegetation expanded, conifers moved upslope, mammals characteristic of cool, wet habitats declined in abundance (insofar as we have records for them), and so on. However, it was in general cooler and moister compared to what was to come later. With one exception, there is no indication that summers were anywhere dramatically hot and dry. While we struggle with trying to determine the seasonality of rainfall, evidence from the Mojave Desert might best be interpreted as suggesting that it was more evenly distributed over the year than it is now, presumably with winter precipitation derived from Pacific westerlies and summer, monsoonal rains associated with warm summers. In the northern parts of the Great Basin, little evidence suggests substantial summer rainfall, though such things as the Homestead Cave hackberry seeds certainly show that there was enough to support this water-dependent species.

What stands out in all of this is the reconstruction provided by Tom Minckley and his colleagues of the nature of the early Holocene on the far northwestern edge of the Great Basin, in the Warner Mountains and the western flank of the Chewaucan River drainage. As I noted earlier, their reconstruction suggests precipitation levels from one-half to one-third of what falls here now, coupled with temperatures of the warmest month from 1.8°F to 5.4°F greater than today.

This reconstruction is at odds with almost everything else we have seen to this point. It is not, however, at odds with a variety of other analyses, extending back several decades, from other areas on the western edge of the Great Basin. Work by Scott Anderson, Owen Davis, Bob Thompson, and many others has shown that early Holocene lake levels in the Sierra Nevada were far lower than they are today and that some lakes even dried during this interval. At the same time, high-elevation plant communities in the Sierra Nevada were quite different from what they are now, with a wide series of shrubs indicative of aridity abundant in what is now forest. Analysis of stratified sediments from meadows in this region shows that the treeline was higher, that such shrubs as sagebrush were far more common, and that erosion, presumably due to lightly vegetated slopes, was far more common than it was in later times. The conclusion reached by the scientists who have done this work, that summers in this area were marked by warm temperatures and far lower precipitation compared to those that mark the region today, is very similar to that reached by Minckley and his colleagues.

There is no reason to doubt any of these conclusions. They do, however, raise a very important question. Why would the early Holocene of the Sierra Nevada and the northwestern edge of the Great Basin be warm and dry when nearly all the early Holocene paleoclimatic evidence from the Great Basin to the immediate east shows that this period of time was the wettest that the Holocene has to offer? We have no answer to this question.

The Middle Holocene (7,500 to 4,500 Years Ago)

Ernst Antevs and the Altithermal

Scientists working on the late Pleistocene and Holocene history of the Great Basin today have a tremendous advantage over those who worked here sixty or seventy years ago. There are better roads and better vehicles, making it far easier to get around. There are more people working in any given discipline, so there are more questions being asked and more information generated to answer them. There are far better analytical tools, from intricate statistical methods to isotope analyses and the extraction and identification of ancient DNA. As a result, not only is far more known about the history of, say, plants or animals, but the results from one set of investigations—on plant history, for instance—can be used to better understand the results from a different set of investigations—on mammal history, for example. In addition, we have access to an array of dating techniques that can be used to determine the absolute ages of a diverse set of materials, from seeds to fine-grained sediments.

These modern advantages make the accomplishments of Ernst Antevs all the more remarkable. We have encountered him before, in chapter 5, where I discussed his insight that the expansion of glaciers in the American north likely caused the displacement of storm tracks southward. Born in Sweden in 1888, Antevs studied under famed Swedish geologist Gerard de Geer, earning his Ph.D. in geology from the University of Stockholm in 1917. De Geer had pioneered the study of varved deposits, and Antevs followed de Geer in his interest in such deposits.

Lakes that receive their water from streams issuing from melting glaciers are often marked by sediments that show alternating layers of coarse and fine material. These couplets are formed as a result of the seasonal influx of sediments into the lake. During the warm months of the year, when suspended sediment is being introduced into the lake, it is primarily the heavier fraction that is deposited. During the winter, when the lake freezes over, the finer particles settle. Each pair of coarse and fine layers produced in this way is a varve; each represents a single year of lake history. In addition, because of increased glacial melt during warm years, the varves that make up any given sequence of lake deposits have different thicknesses. As long as it can be assumed that the temperature changes driving varve thickness are not strictly local, the changing thicknesses allow varves from different lakes to be correlated, much as tree rings can be correlated on the basis of their changing thicknesses. It follows from all of this that counting varves and patching together sequences from varved lakes provide a means of extracting an absolute chronology for those lakes. De Geer did just that. Ultimately, he compiled a varve chronology for Sweden that extended back nearly 17,000 years. Independent dating has now shown that that the Holocene portion of that chronology is extremely accurate.

If the age of each varve in a lake can be accurately determined, it follows that the age of anything incorporated in varved sediments, including pollen, can also be obtained. Pollen analysts in northern Europe were quick to take advantage of this fact. They were also quick to make use of the fact that the major changes they saw in the pollen content of these dated sediments could be correlated with similar, but undated, changes they were finding in the rich pollen and plant macrofossil sequences in Scandinavian peat bogs. By the early 1900s, it had been demonstrated that the Pleistocene had ended by 9,000 years ago or so, and that the Scandinavian middle Holocene had been warmer than earlier and later Holocene times.

In 1920, de Geer came to North America to study varved deposits in New England. Antevs came with him, but when de Geer returned, Antevs stayed, remaining here for the rest of his life. He became an American citizen in 1939, settling in Globe, Arizona.

During his early years in North America, Antevs studied varved deposits in New England and Canada, but he also started to work in the Great Basin; his first trip to this region came in 1922. By the mid-1930s, most of his publications began to deal not with varve chronologies and the northeast, but with the Pleistocene and Holocene history of the Arid West.

Antevs, like de Geer, believed that major temperature changes are "simultaneous over the entire northern or

TABLE 8-3
Ernst Antevs's Neothermal Sequence
Dates, derived from the Scandinavian varve chronology,
are in calendar years.

Temperature Age	Age (years ago)	Climates
Anathermal	9,000–7,000	Warm, Moist
Altithermal	7,000–4,500	Hot, Dry
Medithermal	4,500–Present	Cooler, Moister

SOURCE: From Antevs 1948.

southern hemisphere, and, perhaps, the entire world" (1931:1). From the very beginning of his work in North America, this belief led Antevs to assume that the broad outline of Holocene temperature history in North America was the same as it had been in Sweden. As early as 1925, he noted that the Holocene temperature maximum in Sweden appeared to fall between about 7,000 and 4,000 years ago, and that "the post-glacial maximum of temperature was fairly certainly contemporaneous in both continents" (1925:65).

Antevs was soon able to call on evidence from the Arid West to support this argument. In 1914, the geologist Walton Van Winkle had measured the salt content of Abert and Summer lakes in south-central Oregon, and had found it to be lower than it should have been had these lakes been in continuous existence since the Pleistocene. Van Winkle calculated that the rebirth of these lakes had occurred about 4,000 years ago. Likewise, Hoyt S. Gale had measured the salt content of Owens Lake, publishing his results in 1915. That content, Gale argued, was also consistent with 4,000 years of accumulation. Gale did not think that Owens Lake was dry prior to 4,000 years ago, instead speculating that this was the time that Owens Lake stopped overflowing and thus stopped flushing itself of salts. Antevs, however, argued that Gale's results were better interpreted to mean that Owens Lake had dried before this time and had been reborn at about 4,000 years ago, just as Abert and Summer lakes had been argued to do. These dates, of course, matched the varve-derived chronology for the end of the postglacial warm period in Sweden.

During the 1940s, other data on the Holocene climatic history of the West became available, including the results of Henry P. Hansen's pioneering pollen analyses in the region. In 1948, Antevs synthesized all of this information in an extremely powerful way.

What we now term the Holocene, Antevs called the "Neothermal," meaning the "new warm age." He argued that the Neothermal had begun, and the Pleistocene had ended, 9,000 years ago (because this and the other dates Antevs used were derived from the Scandinavian varve chronologies, they are in calendar years). He also divided the Neothermal into three "temperature ages." The first of these, the Anathermal, he characterized as having had temperatures that were at first like those of today, but that then became warmer; the Anathermal, he argued, dated to between 9,000 and 7,000 years ago. Although the focus of his subdivisions of the Holocene was on temperature, not moisture, he also argued that the Anathermal was wetter than the Great Basin has been during historic times. His second temperature age was the Altithermal, which he characterized as having been distinctly warmer and drier than the present, and which he argued fell between 7,000 and 4,500 years ago. The last of his ages was the Medithermal, 4,500 years ago to the present, marked by conditions much like those today (table 8-3).

As the years passed, Antevs made minor adjustments to the dates of his three-part Holocene. In 1955, for instance, he placed the end of the Pleistocene at 10,000 years ago, the beginning of the Altithermal at 7,500 years ago, and the end of this hot and dry interval at 4,500 years ago. The general nature of his Neothermal sequence, however, remained the same. A wetter early Holocene (the Anathermal) had been followed by a hot and dry middle Holocene (the Altithermal), which was in turn followed by a later Holocene (the Medithermal) marked by conditions far more like those found here now.

Proposed in detail in 1948, Antevs's Neothermal model was widely accepted by archaeologists and geologists alike. Not only did the evidence that Antevs provided to support his climatic arguments appear convincing, but his three-part sequence was associated with real dates. An archaeologist working with Holocene deposits that were fairly old and that appeared to reflect moister conditions than those of today could assign them to the Anathermal, and then argue that they dated to between 10,000 and 7,500 years ago. Deposits that seemed to reflect hot and dry times could be assigned to the Altithermal and dated to between about 7,500 and 4,500 years ago. Until radiocarbon dating came into routine use, age assignments on this basis were standard in the Arid West, especially in the Great Basin. As a result, the literature dealing with the Holocene of this area that appeared between about 1950 and the mid-1970s is often hard to follow without knowledge of Antevs's sequence.

It is important to recognize that this sequence was essentially based on two things. First, it was based on the assumption that major changes in temperature could be used to correlate depositional sequences in far-flung regions—in, for instance, Oregon and Sweden. Second, it was also based on the Scandinavian varve chronology.

Thus, when Antevs began to develop his Holocene climatic history of the Great Basin in some detail, he pointed out that Van Winkle had estimated that Abert and Summer lakes had been reborn about 4,000 years ago. However, he dismissed Gale's suggestion that Owens Lake had decreased in size 4,000 years ago, substituting in its place the argument that it was dry prior to this time. And, although the 4,000 year dates were estimates made by Gale and Van Winkle, Antevs's conclusion was that "the warm Postpluvial age some 5500–2000 B.C. also was dry" (1938:191) in the Desert West. The 2000 B.C. date coincided with Van Winkle's estimate, but the 5500 B.C. date had no secure local basis: it was drawn from the Scandinavian varve dates.

As Dave Rhode has pointed out, the advent of modern chronometric techniques—and most importantly, radiocarbon dating—has thoroughly supplanted the use of Antevs's sequence as a means of dating the past. Nonetheless, the term that Antevs gave to the most famous part of his sequence, the "Altithermal," remains in occasional use among Great Basin scholars and is in common use in other parts of the West, with much the same meaning—both chronological and climatic—that Antevs gave it.

To what degree, I now ask, was his reconstruction of middle Holocene climatic history in the Great Basin correct?

Elevated Treelines in the White Mountains

Subalpine conifers can respond to higher temperatures in many ways. As Connie Millar and other tree scientists have shown, subalpine forests and woodlands may become more densely occupied by trees while subalpine meadows and areas that were once persistent snowfields may be colonized by conifers. These trees may grow more quickly than they had before and subalpine conifers that survived in krummholz form may develop vertical shoots ("flags") that modify that form. Once-healthy trees may be attacked by diseases that can spread quickly across more tightly packed individuals. If the increase in temperature is accompanied by a decrease in soil moisture, disease organisms may also gain the upper hand because the trees are stressed by drought. An increase in winter temperatures can even lead to an increase in the rate at which disease organisms reproduce. All these things are, in fact, happening in western North America today.

Subalpine conifers can also undergo changes in elevation either up or down, with species of trees responding in an individualistic way to the climatic challenges thrown at them (chapter 5). Some may follow cold-air drainages downslope. Others may move upslope, resulting in treelines higher than they were before.

This last response—elevated treelines—was documented in one of the most elegant and chronologically well-controlled analyses of the nature of middle Holocene climates in the Great Basin. This work was done by Val LaMarche, then at the University of Arizona's famous Laboratory of Tree-Ring Research. LaMarche had become intrigued by his observation of dead trees well above treeline on both the White Mountains of eastern California and the Snake Range of eastern Nevada. He realized that at some time in the past, climatic conditions had to have been such as to allow trees to exist above where they can exist now. The

FIGURE 8-11 Campito Mountain in the White Mountains, California. Live bristlecone pines can be seen on the slopes; dead bristlecones extend nearly to the top of the mountain.

problem was to know when this was, and what those climatic conditions were.

LaMarche's detailed analysis of this situation was directed toward the dead bristlecone pines above today's treeline on Sheep and Campito mountains in the southern White Mountains. On Sheep Mountain, LaMarche found that dead trees extend as much as 500 feet above modern treeline (which falls here at about 11,500 feet). He also found that they extend nearly to the top of Campito Mountain, some 330 feet above current treeline (here at about 11,200 feet; see figure 8-11).

Using a combination of radiocarbon and tree-ring dating, LaMarche provided ages for nearly 140 dead, above-treeline trees on these two mountains. Because the innermost and outermost parts of these trees were not preserved, the dates he obtained did not tell him the precise times that the trees had become established or had died. Because the remains of trees that might have grown at even higher elevations might have decayed completely, he could not know that his uppermost dead trees also marked the uppermost limit of past treelines. What he did establish, however, was that treeline was at or above the elevation of the dead trees during the years that his dates indicated the trees were alive.

On Sheep Mountain, LaMarche's oldest tree dated to about 6,500 years ago. From that time, until about 4,200 years ago, treeline remained nearly 500 feet higher than it is now, after which it dropped sharply. On Campito Mountain, middle Holocene treelines were also distinctly higher—from about 260 feet to 360 feet higher—than they are now, but these treelines remained high until about 2,500 years ago. Putting both records together, and assuming that warmer summer temperatures are critical in allowing the upslope movement of treeline, LaMarche concluded that the period of time between about 7,400 and 4,000 years ago was marked by relatively high summer temperatures. He also observed that the period from 5,000 to 4,000 years ago was especially favorable for bristlecone pine growth; this, he suggested, reflects both warm summer temperatures and fairly high precipitation.

Although LaMarche did not attempt to estimate precipitation values from his middle Holocene trees, he did use his Sheep Mountain data to estimate the difference between recent July temperatures and those 4,500 years ago, when treeline was at or above 11,975 feet. His calculations suggested that at this time, July temperatures were 3.5°F warmer than they have been during the past few centuries. This, he observed, was strongly consistent with Antevs's notion of an Altithermal.

The Ruby Marshes Disappear

As I have discussed, Bob Thompson's analysis of sediment cores from the Ruby Marshes documents that a freshwater lake, deeper than is found here today, existed in this basin from about 10,400 years ago to about 8,000 years ago. Thompson also found that sagebrush pollen declined dramatically, while that of the cheno-ams (including saltbush) increased at about 7,700 years ago. This change Thompson interpreted as reflecting an expansion of shadscale vegetation at the expense of sagebrush.

More than that happened at about this time, however. Using the eleven radiocarbon dates he had obtained for the Holocene sections of his cores, plus the date provided by Mazama ash, which he pegged at 6,800 years ago, Thompson was able to calculate the rate at which sediments accumulated in the Ruby Marshes during much of the Holocene. Since we now know that the climactic eruption of Mount Mazama occurred at about 6,730 years ago, I have recalculated his sediment rates and use the new numbers here but the differences are tiny. Between 9,920 and 6,730 years ago, it took an average of 24 years for a centimeter (0.4 inch) of sediment to accumulate. Between 4,420 and 1,350 years ago, the same amount of material took 32 years to be deposited. But between 6,730 and 4,410 years ago, each centimeter required 87 years to be laid down.

There are only two convincing ways to explain why it took nearly a century to accomplish during the middle Holocene what took only a few decades to accomplish during the early and late Holocene. Either deposition rates slowed to a crawl during this interval or the deposits laid down during this time were eroded. No matter which of these situations pertain (and both may have happened), the implications are much the same. During the middle Holocene, the Ruby Marshes were extremely shallow and may at times have dried completely. The increased rate of sedimentation that begins at about 4,500 years ago suggests that the marshes deepened at about this time.

Thompson's careful work in Ruby Valley shows that the expansion of shadscale vegetation here was accompanied, or was perhaps soon followed by a significant decline in water levels. For the next several thousand years, those levels were markedly low. Not until about 4,500 years ago did they deepen. As Thompson notes, his data on vegetation and marsh history are consistent. Together, they suggest a warm and/or dry middle Holocene. This again looks a lot like Antevs's Altithermal.

Desert Shrubs and Marsh History in the Steens Mountain Region

Southeastern Oregon's Steens Mountain rises to an elevation of 9,733 feet, separating the Alvord Desert from Catlow Valley. Although there is a stand of grand fir on the western flank of this range, Steens Mountain is distinct from most other massive Great Basin ranges in that it lacks a zone of either montane or subalpine conifers. This fact makes Steens an excellent setting for studying the history of steppe vegetation at higher altitudes. Peter Mehringer has taken advantage of this in an important way.

Mehringer cored two Steens Mountain lakes, Fish Lake, at an elevation of 7,380 feet, and Wildhorse Lake, at 8,480 feet (figure 8-12). His analysis of the sediments he retrieved not

FIGURE 8-12 Wildhorse Lake, Steens Mountain, Oregon.

only documents that the Fish Lake Basin was free of glacial ice by 13,000 years ago, and the Wildhorse Lake cirque by 9,400 years ago, but also demonstrates that these sediments share six separate volcanic ashes deposited between about 7,000 and 1,000 years ago. Those ashes, coupled with the radiocarbon dates he obtained, allowed Mehringer to correlate the depositional records of these two lakes with exceptional precision. Armed with these tightly correlated sequences, he took a close look at the changing ratios of sagebrush to grass pollen through time as recorded in the sediments of both lakes.

He found that during the past 13,000 years, sagebrush pollen was, on the average, about six times more abundant than grass pollen in the Fish Lake sediments. By itself, of course, this figure has little meaning, and Mehringer calculated it simply to standardize the deviations around it through time (figure 8-13). Doing that showed that between 13,000 and about 8,500 years ago, the sagebrush-grass ratio remained below its Fish Lake average. By 8,700 years ago, however, it was on the rise. Shortly before 8,000 years ago, it climbed above its average value and remained there until about 5,500 years ago. For the next several hundred years, the ratio vacillated up and down; not until about 4,700 years ago did the ratio fall convincingly beneath its average value and remain there for any length of time.

Wildhorse Lake, some 1,100 feet higher than Fish Lake, showed a very similar phenomenon, though not at exactly the same time. Here, the ratio of sagebrush to grass pollen exceeded its Wildhorse Lake average at about 7,200 years ago, and remained above that average until shortly after 4,000 years ago. While the dates differ, the phenomenon is identical.

Mehringer argues that the middle Holocene episode of increased sagebrush abundance at Fish Lake indicates reduced effective moisture between about 8,700 and 4,700 years ago. The movement of sagebrush upslope, he suggests, was temperature-controlled, and accounts for the temporal offset in the Fish Lake and Wildhorse Lake records. It is also worth noting that Wildhorse Lake is in a very deep, sheltered cirque and that winter snow accumulation in this setting may also have allowed sagebrush to remain dominant here longer than at the lower, less sheltered Fish Lake setting. Mehringer argues convincingly that both records together reflect a lengthy period of relatively higher temperature and reduced snowpack.

At the same time that Mehringer was focusing on Fish and Wildhorse lakes, his student, Peter Wigand, was focusing his work on Diamond Pond, just off the northwest slope of Steens Mountain. This small (300-foot-diameter) lake occupies a volcanic explosion crater in the southern Harney Basin's Diamond Crater volcanic complex. At an elevation of 4,150 feet, it also sits near the local boundary between the shadscale and sagebrush plant communities. Mehringer and Wigand cored this lake together and retrieved a sequence of sediments nearly 50 feet deep. Wigand identified four volcanic ashes in these sediments and obtained eleven radiocarbon dates, giving him precise chronological control over the last 5,500 years of deposition in the lake. Although he was unable to obtain absolute dates for earlier sediments in his sample, the tight control that he had over deposition rates throughout most of that sample allowed him to estimate that his entire record extended back about 6,000 years.

It would have been nice if the beginning of the middle Holocene had been represented in the Diamond Pond deposits but it was not. When the pollen record begins, however, it is dominated by greasewood and other cheno-ams, these latter presumably consisting largely of *Atriplex*.

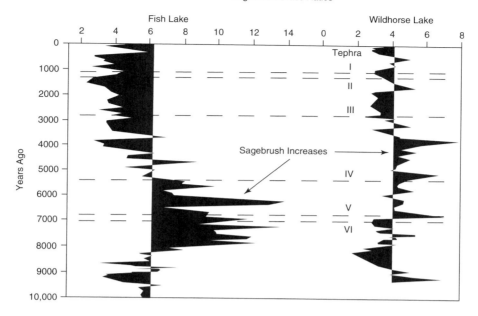

FIGURE 8-13 Ratio of sagebrush to grass pollen through time at Fish and Wildhorse lakes in Steens Mountain, Oregon (after Mehringer 1985).

From 6,000 to 5,400 years ago, greasewood pollen comprised up to 75 percent of the pollen samples as a whole. Wigand reasons that during this time, greasewood and *Atriplex* covered the lower parts of Harney Basin and that saline soils were widespread in the area.

Wigand also discovered that macrofossils from aquatic plants were virtually absent from sediments that accumulated between 6,000 and 4,500 years ago. This fact implies that, during those years, water within the crater was not deep enough to support such plants. He suggests that the water table was likely so low during this interval that marshes in the Harney Basin as a whole were greatly reduced in size, and may even have been dry for much of the year. Those marshes, I note, include those that are now found at Harney Lake and at Malheur Lake and that form the heart of the Malheur National Wildlife Refuge, just as the Ruby Marshes form the heart of Ruby Lake National Wildlife Refuge. The results of geologist Dan Dugas's work in the Malheur Basin itself match this conclusion: he found no evidence for lake deposits here between about 7,400 and 5,000 years ago.

The middle Holocene drought indicated by Diamond Pond began to end by about 5,300 years ago, when the abundance of sagebrush increased sharply in Wigand's pollen record. By 5,000 years ago, the dominance of greasewood in these sediments had ended, and sagebrush, *Atriplex*, and grasses had taken over. By the same time, macrofossils from aquatic plants had become abundant in the lake's sediments. A permanent pond had formed, and while that pond fluctuated in size through the following 5,000 years, no episode of drying comparable to that which occurred during the middle Holocene happened again.

The records extracted by Wigand and Mehringer from Steens Mountain and the Harney Basin are consistent with one another. They are also strikingly similar, as regards both lake and vegetation history, to the results obtained by Thompson in Ruby Valley, some 275 miles to the southeast, and to the lake history extracted from the Malheur Basin by Dugas. All bespeak a middle Holocene marked by relatively high temperatures and relatively low effective moisture.

The Humboldt River Dries, Pyramid Lake Shrinks, and Trees Grow beneath Lake Tahoe

Our middle Holocene record from the Lahontan Basin is extraordinarily revealing. I begin with the packrat midden record, not because it tells us so much about this period of time but because it tells us so little. In fact, there is vanishingly little information available from these middens during the middle Holocene in this area. To take one example, the exquisite packrat midden record built by Peter Wigand and Cheryl Nowak from the Virginia Mountains, just to the west of Pyramid Lake, begins at about 30,000 years ago and continues to modern times. The chronology for this record is based on forty-three radiocarbon dates. Of those dates, only two fall during the middle Holocene, both at about 7,000 years ago. In their Painted Hills midden sequence, there are eight Holocene-aged radiocarbon dates, none of which fall between 8,560 and 4,070 years ago. The implication is that whatever the climate was like during those years, it was not to the liking of the bushy-tailed woodrats that were creating middens before and after this interval. Hot and dry is the obvious choice.

At Hidden Cave, the sagebrush vegetation that marked the early Holocene is replaced by vegetation dominated by *Atriplex* and greasewood during the middle Holocene. That shift begins shortly after 7,500 years ago (figure 8-3), then continues almost but not quite unabated until about 3,500 years ago. No surprise, then, that ichthyologist Gerald Smith has shown that species of lake-dwelling fish in the deposits of Hidden Cave declined significantly in abundance at about 6,900 years ago. That decline, he suggested, reflects shrinking lakes in the Carson Desert. The decline also roughly coincides with the evidence provided by Ken Adams and his colleagues for shrinking lakes in the Lahontan Basin as a whole, as I discussed earlier.

Carson Lake is normally fed by the Carson River, which originates in the Sierra Nevada. Not far to the north, Humboldt Lake is fed by the Humboldt River, which begins in northeastern Nevada. Jerry Miller and his colleagues have analyzed the sediments that accumulated along the middle reaches of this river, near Battle Mountain, producing a sequence that extends back into the late Pleistocene. Their analysis documents that, during the early Holocene, the river was depositing organically rich muds, suggesting continually wet conditions as well as a biotically rich floodplain. All this came to an end, however, by the time Mazama ash was deposited, some 6,730 years ago. From that time until about 4,800 years ago, a gap in the depositional record is broken only briefly at about 5,500 years ago. That gap implies that the Humboldt had stopped flowing, or close to it, during this lengthy interval. Humboldt Lake, as a result, must have been greatly reduced, if it existed at all, at this time.

This possibility matches a pollen sequence developed many years ago by Roger Byrne and his colleagues from Leonard Rockshelter. This site is on the northern slope of the West Humboldt Range, at an elevation of 4,175 feet, overlooking the Humboldt River just east of the Humboldt Sink. When it was excavated for archaeological purposes in 1950, sediment samples from the site were saved. It was these samples that Byrne and his colleagues studied some thirty years later. They found a three-part pollen sequence, with two episodes of pine-pollen dominance separated by an interval during which cheno-am, presumably *Atriplex*, pollen dominated. The period of cheno-am dominance occurred as wind-blown silts were being deposited in the rockshelter. In 1955, Antevs himself had assigned these silts to the Altithermal.

Byrne, Busby, and Heizer concluded that both the introduction of wind-blown silts and the high frequencies of cheno-am pollen in those silts reflect the middle Holocene desiccation of Humboldt Sink and the colonization of the dry lake bed by *Atriplex*. All of these studies fit together perfectly.

Working in a very different part of the Lahontan Basin, Larry Benson and his colleagues extracted a sediment core from Pyramid Lake that covers the past 7,600 years or so. Their analysis of those sediments showed that the lake became significantly shallower after about 7,200 years ago and suffered from what they see as "a period of intense aridity" (Benson, Kashgarian, et al. 2002:677) between about 6,800 and 4,850 years ago. Basing their arguments on the oxygen isotope content of the calcium carbonate (to be precise, the form of calcium carbonate known as aragonite) in their cores, Benson and his team suggested that autumn temperatures in the Pyramid Lake area between during the middle Holocene must have been somewhere between 5.4°F and 13.5°F warmer than they have been, on average, during the past 2,600 years or so (autumn because this is the time of year when the aragonite they analyzed would have been precipitated from the lake). Paleobotanist Scott Mensing and his colleagues analyzed the pollen from these cores and showed that cheno-ams expanded dramatically at the expense of sagebrush at the same time Benson and his colleagues call for dramatically lower lake levels. Their data show the greatest expansion of cheno-ams, including such plants as shadscale and greasewood, between 6,650 and 5,500 years ago, a period they interpret as the driest in their record. The abundance of sagebrush increases on the landscape after that time, with significantly wetter climates beginning around 4,400 years ago.

Pyramid Lake is primarily fed by the Truckee River, with the river receiving about a third of its water from Lake Tahoe. Lake Tahoe, in its turn, overflows into the Truckee as long as the water level reaches above its rim at 6,223 feet, though artificial dams can now keep the level above this natural spill point. When the lake falls beneath this level, the river loses an important source of its water and, as a result, Pyramid Lake must decline.

That means that a history of Lake Tahoe water levels has much to tell us about the history of Pyramid Lake. While we do not have anything close to a detailed understanding of the ups and downs of Lake Tahoe during the Holocene, we do have compelling information on when it must have been low enough, long enough to have had a serious impact on Pyramid Lake levels.

In the 1970s, Jonathan Davis, Bob Elston, and Gail Townsend observed a series of phenomena along the south shore of Lake Tahoe, including valleys now beneath water, that suggested to them that Lake Tahoe had undergone a major episode of shallowing, perhaps 20 feet or more, that must have prevented the lake from overflowing into the Truckee River. Using archaeological evidence in the vicinity of the lake, they argued that this shallowing had occurred between about 7,000 and 3,500 years ago.

They appear to have been correct. The information that shows this comes from a series of tree stumps rooted well beneath the current level of the lake. The existence of these stumps has been known for some time, but when they emerged during a time of low lake levels in 1961, samples were taken for radiocarbon dating that showed some of them to be between 4,200 and 4,800 years old. More recently, Susan Lindström obtained eleven more dates for these drowned trees. Larry Benson then added three more, making a total of seventeen. The full set of dates ranges from 5,510 to 4,250 years ago. The stumps involved are substantial, up to about 10 feet tall and 3.5 feet across; tree-ring counts show that some had lived for at least 150 years. The deepest of them was growing at what is now a depth of 6,210 feet, or about 13 feet beneath the lake's natural spill point. Lindström has noted that additional, unstudied, examples seem to occur much deeper than this.

Larry Benson and Bob Thompson once worried that tectonic activity may have lowered the drowned trees to the position they now occupy. Davis, Elston, and Thompson recognized this as a potential explanation for their data but could find no evidence for it, nor have more recent investigators. All now agree that these trees reflect a period of time when Lake Tahoe had fallen well beneath its modern levels.

This work shows that prior to about 4,300 years ago, the surface of Lake Tahoe was significantly beneath the level at which the trees were rooted and that the trees were killed by rising water at about this time. Given that the trees grew well beneath the point at which the waters of Lake Tahoe spill into the Truckee River, it is also evident that during at least the interval indicated by the radiocarbon dates, Pyramid Lake was deprived of a very significant source of water. Since there appear to be trees even deeper than the ones that have been dated, it is likely that the middle Holocene decline of Lake Tahoe beneath the level at which it feeds the Truckee River began much earlier than 5,510 years ago, the earliest date in the current sequence.

During a lengthy part of the middle Holocene, then, Lake Tahoe was significantly lower than it is today, not reaching modern levels until about 4,300 years ago. This, it turns out, was also a period of frequent fires. We know this because of work done at Lily Lake, a few miles to the west of Lake Tahoe itself. Matthew Beaty and Alan Taylor extracted a sediment core from the bottom of this lake and used the changing frequencies of charcoal particles in that core to reconstruct a 14,000-year Lake Tahoe-area fire history. That history shows that during the past 14,000 years, fires were never more common than they were between about 7,200 and 4,400 years ago, with the peak in fire frequency falling about 5,700 years ago.

Obviously, it is not likely to be coincidental that the ages of Lake Tahoe's drowned trees, at 5,510 to 4,250 years ago, significantly overlaps the "period of intense aridity" that Benson and his colleagues defined from Pyramid Lake 6,800 and 4,850 years ago. Nor is it likely to be coincidental that the most frequent fires in the Lake Tahoe area during the entire Holocene occurred during the same interval.

Moisture-Dependent Small Mammals Decline in the Bonneville Basin

Just as in the Lahontan Basin, there are virtually no packrat midden data for the lower elevations of the Bonneville Basin during the middle Holocene. And just as in the Lahontan Basin, this is probably because the bushy-tailed woodrats who accumulate these middens were not there either.

The Homestead Cave small mammal faunal record shows this in almost startling clarity. Bushy-tailed woodrats simply disappeared from this area sometime between about 8,300 and 8,000 years ago and stayed disappeared through the middle Holocene and beyond (figure 7-11).

Other small mammals behaved in generally similar ways. Pygmy rabbits, so dependent on big sagebrush for habitat, disappeared at about the same time as bushy-tailed woodrats, as did yellow-bellied marmots, though neither species has returned (figures 7-4 and 7-8). Great Basin pocket mice, now found in parts of the Great Basin with heavier ground cover and higher winter precipitation, were common in the Homestead Cave area during the early Holocene. They were lost here around 8,000 years ago, only to make their way back during the late Holocene (figure 8-6). Ord's kangaroo rats, often common in sagebrush habitats and by far the most common member of the genus during the late Pleistocene and early Holocene at Homestead, remained in the area during the middle Holocene but its numbers were dwarfed by those of the shadscale-loving chisel-toothed kangaroo rat (figure 8-14). Western harvest mice, today most common in well-watered parts of the Great Basin, also hung on during the middle Holocene, albeit in much diminished numbers, then managed something of a comeback once the middle Holocene had ended (figure 8-5). Other small mammals that today depend on landscapes either cool or moist or both behaved in the same ways. All either became locally extinct and never returned (marmots, pygmy rabbits) or became locally extinct but managed to recolonize during the late Holocene (bushy-tailed woodrats, Great Basin pocket mice), or managed to hang on in greatly reduced numbers (Ord's kangaroo rats, western harvest mice) until the middle Holocene had ended.

This is not some Homestead Cave-specific response to middle Holocene climates in this area. To the south, in the Old River Bed area of the Bonneville Basin, the work done by Dave Schmitt, David Madsen, and Karen Lupo at Camels Back Cave shows that very similar things happened here. Although the samples from this cave are much smaller than those from Homestead, bushy-tailed woodrats, pygmy rabbits, western harvest mice, and Great Basin pocket mice all disappear from here as the middle Holocene begins.

Louderback and Rhode's analysis of the pollen from Blue Lake record matches the histories of these small mammals (figure 8-2). To them, the Blue Lake data suggest that the wetlands here had dried by around 7,200 years ago, the endpoint in the early Holocene drying process in this area. Between about 7,600 and 5,700 years ago, their record is marked by very low proportions of pine and sagebrush coupled with high amounts of pollen from plants well adapted to warm and dry contexts, including chenopods (presumably saltbush). The latter, in fact, are never more abundant here than they were during this lengthy interval. At the same time, the rate at which pollen accumulated in the sediments of Blue Lake dropped to very low levels. The local aridity indicated by all of this seems to end at about 5,700 years ago. At that time, the abundance of sagebrush

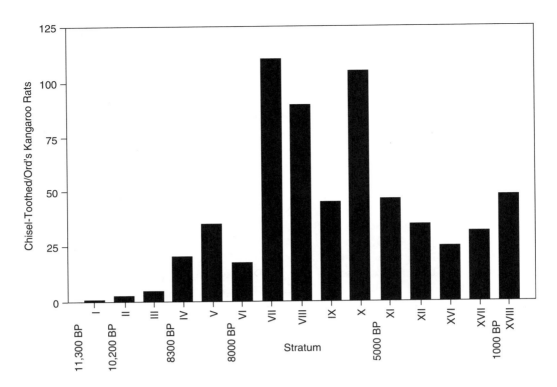

FIGURE 8-14 Changing ratios of chisel-toothed and Ord's kangaroo rats through time at Homestead Cave, Bonneville Basin, Utah (after Grayson 2000a).

pollen increases substantially, suggesting an increase in the amount of moisture on the landscape.

Very similar evidence comes from Mosquito Willie's spring, three miles to the south. Paleobotanist Chris Kiahtipes analyzed a pollen core from this site that spans the past 8,000 years, finding that the early middle Holocene here was extremely dry, with precipitation falling mainly in the winter. What he didn't find were sediments dating to between the time Mazama ash fell here (6,730 years ago) and about 3,900 years ago. The implication is either that the spring dried entirely during that time or that it dried long enough for erosion to remove whatever deposits had been there (or both). Either way, the implications for middle Holocene climates here are obvious.

Across the Bonneville Basin, the plant history extracted from Snowbird Bog by David Madsen and Don Currey (figure 8-1) strongly suggests that the period from 8,000 to 6,000 years ago here was both warm and dry, while that from 6,000 to 5,200 years ago, warm and moist.

How about the remnants of Lake Bonneville itself? We have already seen that the streams and marshes in the Old River Bed area had dried around 8,800 years ago or so. Figuring out what happened in the Great Salt Lake area, however, has proved to be a challenge. Quite some time ago, geologist Don Currey noticed that aerial photographs taken of the Great Salt Lake revealed that the lake floor was marked by the presence of what are known as desiccation polygons. These are versions of the cracked and patterned surface of drying mud, except that they are huge, some 50 to 330 feet across. The only thing they could represent is a time when the bottom of the lake lay exposed and drying. Currey suggested that these had formed during the middle Holocene.

Subsequent work by Stuart Murchison has shown that there was a Great Salt Lake during at least some of the middle Holocene, with the lake apparently spiking to 4,211 feet, 11 feet above its average historic elevation, between about 7,650 and 7,100 years ago, then again at about 5,900 years ago. Between those two times, water levels in the Great Salt Lake Basin seem to have plummeted. It is during this interval, Murchison suggests, that the huge desiccation polygons formed. Whether the entire basin dried at that time is not known, since the distribution of the polygons is not known, but it does appear that the Great Salt Lake was much reduced in scale, if not completely dry, during the heart of the middle Holocene.

This is not to say that we know a lot about the Great Salt Lake during this period. Not only is the distribution of the polygons unknown, but Murchison's important sequence is based on a very small number of radiocarbon dates. Enough is known, however, to show that the lake fell well beneath anything known historically during this interval.

Owens Lake Dries Up

All the middle Holocene histories I have discussed to this point are from within the floristic Great Basin. Owens Lake is not, but it is a lake and as such has a middle Holocene history that is better compared with the lakes I have just talked about than with the largely terrestrial records from the Mojave Desert I discuss below. That history will sound familiar since work by two separate research teams, one led by Larry Benson and the other by George Smith, have shown that it is not just Los Angeles that can cause Owens Lake to dry completely or nearly so.

Applying the tools discussed in chapter 5—the total amount of inorganic carbon and oxygen isotope ratios—to the sediments of Owens Lake, Benson and his colleagues found that the period between 6,900 and 3,000 years ago was generally drier than anything that came before or after. They could not, however, figure out how dry it became between about 5,700 and 3,600 years ago because that interval was not represented in their core. The reason, they observed, had to be that either the lake had dried entirely or had shrunk to the point that it was beneath the level of the site they had cored. Either way, it had fallen dramatically during these years. This particular result—a middle Holocene depositional gap in Owens Lake sediments—has been replicated by George Smith and his colleagues. The core that they took, from very near one of the deepest parts of the Owens Lake playa, provided fifteen radiocarbon dates that fall within the Holocene. Not one of them falls between 8,300 and 5,100 years ago—a result, they suggest, of greatly decreased lake levels during this time. Whether or not the lake dried entirely during these years is not known, but if it didn't, it surely came close to it.

A Quick Summary

All of these studies, as well as a number of others that I have not mentioned, suggest very much the same thing. An interval that began sometime between 8,000 and 7,000 years ago, and that ended sometime between 5,000 and 4,000 years ago, was far warmer or drier, or both, than what came before or after. While there is no question that this was the case, however, there is also no question that parts of the Great Basin were reasonably well watered during parts of the middle Holocene, as I discuss shortly. That middle Holocene climates were varied in both time and space should not be surprising given the evidence for it that we have already seen—the sawtooth nature of Peter Mehringer's middle Holocene pollen ratios from Fish and Wildhorse Lakes (figure 8-13), for instance, or the ups and downs of middle Holocene kangaroo rat ratios at Homestead Cave (figure 8-14). I return to this variability below.

The Mojave Desert

Earlier in this chapter I described the work done by Jay Quade and his colleagues on the late Pleistocene and Holocene black mats of southern Nevada, those highly organic, dark-colored deposits that reflect high water tables. I noted that these date to between 11,800 and 6,300 years ago and that

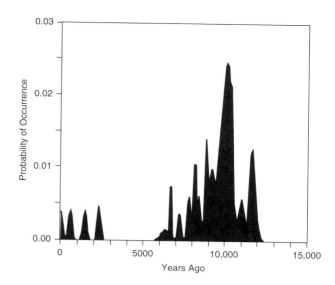

FIGURE 8-15 Changing frequencies of black mats through time in southern Nevada (after Quade et al. 1998). The higher the peak in the graph, the greater the probability that a black mat dates to that time.

they did not start forming again until about 2,300 years ago (figure 8-15). Spring discharge seems to have virtually stopped during this lengthy interval, which Quade and his colleagues refer to using Antevs's term, the Altithermal.

This dry episode matches the results that Peter Mehringer and John Sheppard obtained from their study of the sediments of Little Lake. This small lake lies in eastern California along the channel that carried water from Owens Lake to China Lake when Owens Lake was high enough to overflow in this direction (chapter 5). Their hope was to extract a core that reached well into the Pleistocene. Unfortunately, they hit sediments that they could not penetrate and so had to be content with a history that covers the past 5,000 years and thus only the tail end of the middle Holocene. While they were disappointed with the short length of that record, they were able to show that between 5,000 and 3,000 years ago, this area did not contain a lake at all but was instead marked by a salt grass meadow, marsh, and ponds. Something akin to today's Little Lake did not appear until after this time. While Mehringer and Sheppard were very careful to point out that this sequence might have been caused by faulting in this tectonically active area (remember the 1872 earthquake that I mentioned in chapter 2), their dry interval falls within the period of time when black mats were not forming in southern Nevada.

Then there is the history of creosote bush. As I have discussed, it appears that creosote bush was absent from significant parts of the Mojave Desert during the early Holocene. In fact, as Peter Koehler and his colleagues have pointed out, this plant does not begin to arrive in the central Mojave Desert until the middle Holocene: by 6,900 years ago in Silurian Valley, 6,000 years ago in the Granite Mountains, and even later in Eureka Valley. It was the middle Holocene that saw the lower elevations of the Mojave Desert began to fill with this classic Mojave Desert species.

In addition, Peter Mehringer's pollen profiles from Tule Springs, in the Las Vegas Valley, show a dramatic middle Holocene increase in the abundance of pollen very likely to have been produced by white bursage. The middens analyzed by Koehler and his colleagues from the central Mojave Desert show a very similar thing. Here, bursage does not appear in a routine way until the middle Holocene. Creosote bush and white bursage, of course, are now dominant plants in the Mojave.

One might surmise that if the Mojave Desert was warmer during the middle Holocene, there may well have been enhanced summer rainfall as a result of strengthened monsoonal storms. Paul Martin suggested exactly this for the Sonoran Desert in 1963. In Martin's view, enhanced monsoons should have made the middle Holocene in that area warm and wet, not warm and dry. Indeed, Tom Van Devender subsequently argued that his Sonoran Desert midden data indicate just such a phenomenon. Though not all have agreed, the study of foraminifera from Gulf of Mexico cores that I mentioned earlier does suggest that the greatest development of monsoonal storms in the Southwest during the past 10,000 years occurred during the middle Holocene. In addition, studies of the sediments in a core taken from Dry Lake on the Pacific side of the San Bernardino Mountains, south of the Mojave Desert, show increased monsoonal storms between 8,100 and 6,600 years ago compared to those that marked the region after this time.

If monsoonal circulation strengthened during the middle Holocene, it is possible that at least parts of the Mojave Desert would have been the recipient of increased summer rainfall. Geoff Spaulding was the first to tackle this issue head-on.

He based his arguments on a series of four middle Holocene midden samples from the McCullough Range, near the California border in far southern Nevada, some sixty miles north of the northern boundary of the Sonoran Desert. Those four samples date to 6,800, 6,480, 5,510, and 5,060 years ago, and come from an elevation of 3,430 feet. To determine whether this region was arid between 7,000 and 5,000 years ago, he compared the abundances of climatically sensitive plants in his middle Holocene middens with those in two late Holocene, and one modern, midden from the same location.

Each comparison he did suggested that the middle Holocene here was, in fact, arid. Creosote bush, for instance, was more abundant in his middle Holocene samples than in either the late Holocene or the modern ones. The same was true for desert spruce (*Peucephyllum schottii*), a shrub generally confined to low-elevation, dry settings. Spaulding also took a series of twelve species of plants represented in his middens and divided them into those that would be favored by aridity and higher winter temperatures ("thermophiles"), and those that would be favored by increased moisture availability ("mesophiles"). He then tracked the abundances of these two groups of plants through time. He found that the thermophiles peaked in abundance in his two earliest middle Holocene samples and that the mesophiles were rare in all of those samples, especially when compared to their abundances in the later ones. He concluded that conditions more arid than today characterized the McCullough Range between about 6,800 and 5,060 years ago, with the period between 5,500 and 5,060 years ago having been slightly less arid than what had come before. Spaulding's analysis thus suggests that the bulk of the middle Holocene here was warm and dry.

Peter Koehler and his colleagues note that some plants in their central Mojave Desert midden samples that date to between 4,900 and 4,400 years ago may suggest both increased temperatures and increased moisture. This joins with midden evidence provided by Spaulding for possible increased summer moisture between 5,200 and 3,500 years ago. All these midden samples fall at the tail end of the middle Holocene and match Spaulding's suggestion that this interval was somewhat wetter here than what had come before. Prior to those years, the plant evidence we have suggests that the Mojave desert was warm and dry.

In short, while the evidence is good for strengthened monsoonal incursions into the Mojave Desert during the early Holocene (in particular, the oxygen isotopes from those Pintwater Cave hackberry seeds), no such evidence exists, as of yet, for the middle Holocene. This area seems to have been substantially warmer and drier during the middle Holocene than it is now, though it is certainly possible that it was so warm that enhanced summer monsoonal precipitation had little impact on plant communities here.

Was Antevs Right?

A diverse variety of studies show that the Great Basin during the middle Holocene was marked by high temperatures or low precipitation or, most likely, both. It is tempting, as a result, to conclude that Antevs was right.

It is fairly easy, many decades after the fact, to provide a list of problems with Antevs's Neothermal sequence for the Great Basin. The period of time he assigned to the Anathermal was not marked by temperatures that were at first like today's and then grew warmer. Instead, both plants and mammals suggest that within the Great Basin, this interval was in general distinctly cooler than it is now.

Recent studies do show that Anathermal-aged lakes and marshes existed in Great Basin valleys that no longer contain them, just as Antevs suggested. Antevs, however, supported his initial arguments on this score with an interpretation of Summer Lake history that was, for reasons not of his own making, wildly wrong. At the time he developed his model, Mazama ash was thought to be present deep within the Summer Lake stratigraphic sequence, implying a lake some 90 feet deeper than modern Summer Lake at the time the ash fell. Antevs estimated that the climactic eruption of Mount Mazama occurred between 8,500 and 9,000 calendar years ago, within the range of then-current estimates for this event. He then used this date to support his argument for a wetter Anathermal. Not until 1966 was it realized that the critical Summer Lake ash had been misidentified and that it was Pleistocene in age. Antevs also depended very heavily on the salt chronologies derived from Abert, Summer, and Owens lakes to support his argument that the Altithermal had ended about 4,000 years ago. The extraction of time from the salt content of lakes, however, is now known to present virtually insurmountable difficulties.

The Altithermal, as Antevs painted it, did not really exist. He defined this interval rigidly, as a 3,000 year period during which the Great Basin was unrelentingly hot and dry: the "Long Drought," as he called it. That, in fact, was not the case. Dates for the onset of increased aridity vary from place to place; dates for the onset of less arid conditions likewise vary. As I have discussed, detailed studies of this interval do not show unrelenting aridity across 3,000 years, but instead suggest high variability within a more arid period of time.

The Fort Rock Basin provides another example of such variability. The southern end of this basin contains a number of dry lake basins that once received overflow from Silver Lake a few miles to the southwest. In the 1980s, Peter Mehringer and Bill Cannon excavated part of a sand dune next to one of these playas, at a site that came to be called Locality III. At the base of this dune, they found a small hearth that contained fish bones (tui chub, *Gila atraria*) and that dated to 9,400 years ago. Sediments above this dated to 7,200 years ago; sediments higher still contained Mazama ash (6,730 years old). Armed with the knowledge that this dune seemed to contain important archaeological materials as well as a sequence of deposits that might provide information on the nature of changing environments here during the Holocene, the University of Oregon conducted a larger series of excavations at this spot. They obtained a large series of radiocarbon dates, extending back to 10,200 years ago and that included much of the middle Holocene. Michael Droz analyzed these sediments and showed that the playa contained a small lake around 7,000 years ago, prior to the climactic eruption of Mount Mazama. The water made this location attractive to people who left behind a series of artifacts, including projectile points and shell beads. The same deposits also contain egg shells and mollusks, another indication of the fact that the playa contained water at about this time. The overlying stratum contains large amounts of Mazama ash; the stratum above it, which is not dated but holds artifacts of middle Holocene age, contains evidence that the adjacent playa was alternatively dry and wet.

Steens Mountain, with Mehringer's Wildhorse and Fish Lakes, is not far away. Long before the Locality III work was completed, Mehringer and Cannon suggested it was not coincidental that both Steens Mountain and Fort Rock Basin provide evidence for "short, sharp climatic shifts" (1994:321) during the middle Holocene. All that we now know supports this suggestion.

So Antevs's monolithically hot and dry Altithermal did not exist, even though something like it did. This is not a novel conclusion. In 1964, Alan Bryan and Ruth Gruhn observed that Holocene climates in the Arid West, including the middle Holocene, could be expected to vary through time and across space. The nature of these climates and local biotic responses to them, they argued, had to be discovered through detailed empirical investigations in particular places. They were, of course, right, as the development of our knowledge of middle Holocene climates and landscapes in the Great Basin has shown.

It is, however, a cheap shot to observe that recent studies, assisted by technology and analytical methods developed after Antevs died in 1974, show him to have been wide of the mark. His Neothermal model for the Great Basin began to be developed in detail more than sixty years ago, and it wasn't that far off. It also served to focus research on particular issues, places, and times then poorly known. His varve chronologies in northeastern North America have stood the test of time, and it was Antevs who first suggested that the expansion of glacial ice in northern North America might have shifted the jet stream southward (chapter 5). Had Antevs done only this, he would have accomplished a tremendous amount. In addition to this, however, he was also one of the pioneers in the use of geological approaches to analyze the contexts of North American archaeological sites. His contributions in this realm were of immense value when he made them and many now follow the path he pioneered. Strange, then, that while the Geological Society of America lauded his work in a lengthy obituary, the Society for American Archaeology never noticed his passing in the pages of its journal.

Pinyon Pine Arrives

Singleleaf pinyon finally arrives in the floristic Great Basin during the middle Holocene. The evidence for this seems indisputable and is provided by dated pinyon pine macrofossils and by sharp increases in the abundance of pine pollen in areas today in or near pinyon-juniper woodland.

Table 8-4 shows the dates for the macrofossils, all but one of which are from packrat middens (the exception is from Gatecliff Shelter, where the earliest pinyon pine comes from a hearth in this archaeological site). Figure 8-16 shows how the dates are arrayed across space.

There are some remarkable things about these dates and their locations. First, there are all those Pleistocene-aged dates for pinyon pine in the Mojave Desert, dates that I discuss in chapter 6. Second, the only evidence for singleleaf pinyon that predates the middle Holocene within the floristic Great Basin comes from the northwestern edge of the White Mountains, at the southwestern edge of this region. Here, Steven Jennings and Deborah Elliot-Fisk recovered pinyon pine macrofossils from a packrat midden sample that dated to 8,790 years ago. Beyond that geographically

TABLE 8-4
Latest Pleistocene and Earliest Holocene Dates for Pinyon Pine Plant Macrofossils in the Great Basin
(ND = no data)

Location	Elevation (feet)	Age (years ago)	Reference
Western Great Basin			
Painted Hills, Virginia Mountains, NV	ca. 4,750	ca. 300 calendar years[a]	Wigand 2008
Hidden Valley, Virginia Range, NV	4,860	400 ± 100	Wigand and Nowak 1992; Wigand 2002
Newton Creek, Virginia Range, NV	ND	710 ± 80	Wigand and Nowak 1992; Wigand 2002
Silver City, Virginia Range, NV	5,200	1,030 ± 90	Wigand and Nowak 1992; Nowak et al. 1994b; Wigand 2002
Navy/Army Depot, Walker Valley, NV	5,500	1,170 ± 50	Wigand and Nowak 1992; Nowak et al. 1994b; Wigand 2002
Stillwater Range, NV	ND	1,250 ± 80	Wigand and Nowak 1992; Wigand 2002
Slinkard Valley, CA	5,920	1,430 ± 80[b]	Wigand and Nowak 1992; Nowak et al. 1994b; Wigand 2002
Sherwin Summit, CA	ND	3,610 ± 70	Wigand 2002
Bodie Hills, CA	7,800	4,980 ± 80 (rare)	Halford 1998
Bodie Hills, CA	7,800	4,060 ± 110 (common)	Halford 1998
Papoose Flat, Inyo Mountains, CA	8,560	7,880 ± 60	Wigand 2002
Papoose Flat, Inyo Mountains, CA	8,560	4,690 ± 60	Wigand 2002
Falls Canyon, White Mountains, CA	6,005	8,790 ± 110	Jennings and Elliot-Fisk 1993
Eleana Range, NV	5,940	10,620 ± 120	Spaulding 1985
Alabama Hills	ND	11,450 ± 80	Wigand 2003, pers. comm.
Granite Mountain, CA	4,330	11,470 ± 70	Koehler et al. 2005
No Name Basin, CA	4,000	11,910 ± 90	Koehler et al. 2005
Forty Mile Canyon, NV	4,300	12,870 ± 50	Spaulding 1994
Robber's Roost, Scodie Mountains, CA	3,985	12,870 ± 400	McCarten and Van Devender 1988
Haystack Mountain, Owens Lake, CA	3,790	17,680 ± 150	Koehler and Anderson 1994
Skeleton Hills, NV	3,035	17,940 ± 600	Spaulding 1990; Lanner and Van Devender 1998; NAPMD
Specter Range, NV	3,905	18,740 ± 150	Spaulding 1985
Central and Eastern Great Basin			
Gatecliff Shelter, Toquima Range, NV	7,610	5,095 ± 100	Davis et al. 1983; Rhode and Thomas 1983; Thompson and Hattori 1983
Marblehead Mine, Toano Range, NV	5,905	5,960 ± 90	Rhode and Madsen 1998; Rhode 2000
Council Hall Cave, Snake Range, NV	6,695	6,120 ± 80	Thompson 1984
Valley View, Schell Creek Range, NV	7,710	6,250 ± 150	Thompson 1984
Meadow Valley Wash, NV	4,430	6,590 ± 130	Madsen 1986; NAPMD
Danger Cave, Silver Island Range, NV	4,310	6,710 ± 70	Rhode and Madsen 1998
Southern Pahranagat Range, NV	ND	8,810 ± 60	Wigand 2002

[a] Based on a tree-ring date from one of the few pinyon pines in this area.
[b] Wigand (2002) gives this date as 1,490 ± 90; Wigand and Nowak (1992), Nowak et al. (1994b), and the North American Packrat Midden Database (NAPMD) give the date as 1430 ± 80.

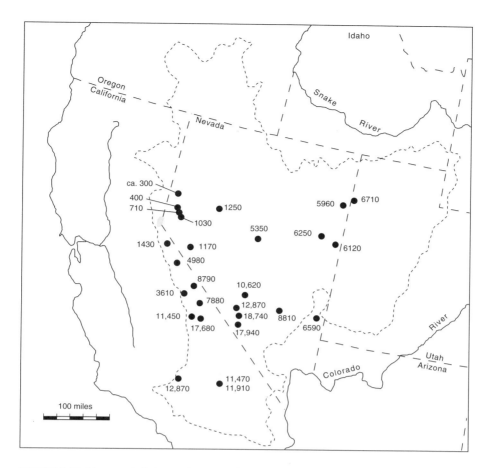

FIGURE 8-16 Movement of pinyon pine across the Great Basin during the Holocene.

marginal example, packrat middens from the floristic Great Basin that are older than about 6,700 years lack any suggestion of singleleaf pinyon.

Then there are the earliest dates from along the Nevada-Utah border. All of these dates fall between about 6,000 and 6,700 years ago. In Meadow Valley Wash, southeastern Nevada, David Madsen identified pinyon pine macrofossils from a packrat midden sample dated to 6,590 years ago; earlier midden samples had no pinyon. To the north, Bob Thompson's midden records from the Schell Creek and Snake ranges lack pinyon older than 6,300 years old. In the Schell Creek Range, his earliest records come from midden samples that date to 6,250 years ago. In the Snake Range, his earliest pinyon comes from a Smith Creek Valley midden that dates to 6,120 years ago. Today, all of these middens are in pinyon-juniper woodland. The northernmost trustworthy record for early pinyon pine in this area comes from Danger Cave. Identified by Dave Rhode and David Madsen, this material falls at 6,710 years ago. Pinyon pine appears in older deposits at the site, but without dating them directly, it is not possible to know whether these fragments are older than the 6,710-year-old specimens, or whether they represent younger material that worked its way down into older deposits. At both Blue Lake and Mosquito Willie's, pine pollen increases substantially at between 7,500 and 7,000 years ago, quite consistent with the first appearance of pinyon pine a few hundred years later at Danger Cave.

The earliest pinyon pine date from central Nevada is from Gatecliff Shelter. Surrounded by pinyon-juniper woodland in central Nevada's Toquima Range, Gatecliff, which was excavated with meticulous care by Dave Thomas, sits at an elevation of 7,610 feet, some 800 feet above the valley floor. The oldest deposits in the site were laid down more than 7,000 years ago, but the earliest directly dated pinyon comes from a hearth dated to 5,095 years ago. This, though, is most certainly not the earliest pinyon pine here, since older archaeological features at the site also contain it; the best estimate for the earliest pinyon here is that it falls at about 5,350 years. Packrat middens from Mill Canyon, in which Gatecliff sits, are in agreement. Bob Thompson and Gene Hattori analyzed these middens, and found that the three earliest samples in their series, which date to between 9,500 and 9,000 years ago, lack pinyon. The next-youngest sample in the series, which dates to 4,790 years ago, contains it.

Thompson and Bob Kautz also analyzed the pollen from the Gatecliff sediments. They found that the earliest deposits in the site, which date to between about 7,300 and 6,000 years ago, were dominated by sagebrush and cheno-am pollen. Between 6,000 and 5,600 years ago, however, both pinyon and juniper increased dramatically in abundance. A few hundred years later, the first hearths were built in Gatecliff, and they contained the remains of both pinyon and juniper. The hearths, middens, and pollen provide a convincing picture: pinyon pines arrived in the Gatecliff area soon after 6,000 years ago.

In all of these cases, from Meadow Valley Wash to Danger Cave and the Toquima Range, packrat middens that predate 6,710 years ago lack pinyon pine. The appearance of this tree in these areas after this time is not simply reflecting some preservational bias in our record. Even though the sample of middens is fairly small, the results are consistent. It appears that singleleaf pinyon entered at least the central parts of the Great Basin, presumably from the south, sometime after 7,000 years ago.

Things are very different in the western Great Basin. Here, pinyon began its movement northward much later in time. North of the northern White Mountains, with its 8,800-year-old pinyon, the earliest known pinyon pine is from the Bodie Hills, not far north of Mono Lake, and it dates to 4,980 years ago (and doesn't become common until much later). North of here, no pinyon pine is known prior to 2,000 years ago.

The difference is stark. Bob Thompson's Valley View midden, in the Schell Creek Range, has pinyon pine at 6,350 years ago. Peter Wigand's Hidden Valley samples, which go back 9,800 years, show no trace of it until 400 years ago. This is the case even though both sites are at almost exactly the same latitude.

So while pinyon pine began its movement northward during the middle Holocene, the speed with which it moved into the western and more central parts of the Great Basin differed dramatically. In the more central area, the movement seems to have been fast. In the western parts of the region, the movement seems to have been very slow. The twin challenges are to explain why pinyon began moving northward in the first place, and why it did so at such very different rates in these two areas.

The movement of pinyon pine across space depends on the movement of its seeds and that movement took time. Pinyon seeds lack the adaptations needed for wind-blown dispersal: they are far too large to be scattered by the wind and they lack the light, membranous projections, or wings, that help the seeds of some other species of pine disperse in this way. As a result, a different dispersal mechanism is needed.

That mechanism is well known. A group of birds belonging to the family Corvidae (the crows, jays, magpies, and their close relatives) collect, eat, and cache pinyon seeds, as well as the seeds of other pines. Throughout much of the Great Basin, the most important members of this group are Clark's Nutcracker (*Nucifraga columbiana*), Pinyon Jay (*Gymnorhinus cyanocephalus*), Scrub Jay (*Aphelocoma coerulescens*), and Steller's Jay (*Cyanocitta stelleri*). Scrub Jays, as Ron Lanner has observed, are not equipped to tear open closed cones, they cannot carry many seeds at a time, and those they do carry are generally not carried very far. Steller's Jays can carry and cache many more seeds but these are not common birds in much of the Great Basin. This leaves Clark's Nutcracker and the Pinyon Jay as the most important winged dispersers of pinyon seeds, at least today.

These birds are critical to the dispersal of the seeds. Ornithologists Stephen Vander Wall and Russell Balda, for instance, have calculated that a flock of 150 Clark's Nutcrackers cached approximately 4 million (!) Colorado pinyon seeds in a single autumn in an area they studied in north-central Arizona, moving those seeds as far as fourteen miles to their final destination. Indeed, pinyon pines have a variety of adaptations—from large seed size to cones that hold the seeds in such a way that they are easily visible to birds—that encourage bird-assisted dispersal.

Assuming that birds were the prime mechanism involved in the spread of pinyon pine across the floristic Great Basin, the advent of environmental conditions conducive to the establishment of pinyon populations would not translate to the rapid movement of these trees across the region. Instead, stands of pinyon would have become established slowly, as birds introduced seeds into new areas, seedlings sprouted, trees grew and produced cones (a process that might take several decades), and the seeds they produced were then dispersed, a few miles at a time, by new generations of birds. Indeed, Ron Lanner has calculated that pinyon might move across Great Basin uplands at a rate of between eight and twelve miles per century by this process. If birds were the major vector, pinyon pines would take several thousand years to spread across the Great Basin. If climatic change at the beginning of the middle Holocene, say 8,000 years ago, introduced conditions amenable to the establishment of pinyon pine in the Great Basin, then those trees might not appear in places like the Virginia Range until thousands of years later.

I mentioned above that corvids are the most important flying dispersers of pinyon seeds. There is another obvious possibility, one that walks rather than flies. As Peter Mehringer has pointed out, people may well have played an important role in moving pinyon across the landscape. In chapter 2, I note how important pinyon nuts were to the native peoples of the Great Basin. These nuts are nutritious, easily stored, and easily transported, so it is not surprising that the ethnographic and historic literature is full of examples of early historic Native Americans moving pinyon seeds across the landscape, involving distances of up to some fifty miles. Obviously, there is no reason to think that they did not do so in the past as well.

Danger Cave almost certainly provides an example. While this archaeological site contains pinyon nuts and cone fragments, they are not particularly abundant, no pinyon pine needles have been identified from the deposits of the site, and the nearest stand of pinyons is now more than ten miles away. Dave Rhode and David Madsen argue that these small amounts of pinyon most likely represent food items brought to the site by people. There are, of course, other options, the most obvious of which is that they were introduced by rodents. It is also true that owl pellets can be full of pinyon nut fragments, either because the owls had eaten pinyon-loving rodents (squirrels love them), or, far less likely since it has never been observed, because the owls themselves were eating the nuts. Carnivores will also eat pinyon nuts; I have seen coyote scat full of pinyon nut hulls, and coyotes can use caves as latrines. The Danger Cave pinyon nuts, however, have none of the tell-tale signs of having been eaten by rodents or carnivores or regurgitated by owls, making human introduction the most obvious explanation.

So there are two obvious ways that pinyon seeds, and hence pinyon trees, can be dispersed across the landscape: by birds or by people. The bird-driven process can be expected to be quite slow, but dispersal by such highly mobile people as the hunters and gatherers of the Great Basin can be quite fast. Two dispersal processes, one slow, one potentially fast. Two Great Basin pinyon histories, one slow, one fast. Putting all this together, it seems possible that the speed with which pinyon seems to have dispersed in the more central parts of the Great Basin is best explained by people carrying, and dropping, pinyon seeds as they moved across the landscape, and that the far slower process in the western Great Basin was primarily driven by bird dispersal. There is no reason, of course, to think that birds played no role in the east, and people no role in the west, but the very different nature of the two dispersal histories suggests that an important part of the explanation may reside in who was playing the major role in the dispersing.

It is, though, easy to think of other possibilities. Perhaps pinyon entered the Silver Island Range (where Danger Cave is located), and other nearby mountains, not from the south, but from the north. Perhaps a colonization route along the eastern margin of the Great Basin carried them to the mountains of northern Utah, then west through the mountains along the Utah-Idaho border, ultimately reaching the Silver Island Range and other nearby mountain masses. The problem with this is that while singleleaf pinyon is found in mountains to the west of the Wasatch Range, it is not found in the Wasatch Range itself and dispersing northward from ranges within the Bonneville Basin would require crossing a formidable biogeographic barrier.

If pinyon did move from south to north in all cases, then, as Cheryl Nowak and her colleagues have pointed out, the tree might have had an easier time of it toward the east than toward the west. This is because eastern Nevada is marked by a series of massive north-south trending valleys and mountains that would have provided a ready corridor for pinyon expansion. Although there are higher-elevation dispersal corridors in the west, they are not as continuous as the ones available to the east, and this difference might have slowed the northward movement of the trees in this area. This in itself seems unlikely to have caused the great discrepancy in arrival times at a given latitude, but Nowak and her coworkers suggest that climatic differences may have played a role as well, with more frequent cycles of warm, then cold, snaps in the late winter and spring associated with Pacific westerlies, damaging trees tricked by the warm spells into beginning their spring growth, only to be severely damaged by a subsequent cold snap.

It is also possible that pinyon pine actually existed in undetected, isolated pockets in the central, but not the western, Great Basin during the late Pleistocene and then expanded outward when conditions more amenable to their existence returned. If so, the discordant arrival times that mark the western and more central parts of the Great Basin are to be expected. If we are really desperate, we could even invoke the history of the cattle egret here (chapter 6), and wonder whether a flock of Clark's Nutcrackers was blown off course 7,000 years ago or so and ended up seeding the mountains of eastern Nevada with pinyon.

No matter what might ultimately explain the difference in the speed with which pinyon moved northward in the Great Basin beginning in the middle Holocene, something must have happened to trigger this movement in the first place. Birds and people, after all, were here to disperse pinyon seeds long before the dispersal actually happened. That something, all agree, must have been climate. Once we figure out what aspects of climate change allowed pinyons to do this in the first place, we might better understand why pinyons arrive earlier in the east than in the west.

We have yet to figure that out but a wide variety of paleobotanists have speculated that the spur was provided by a middle Holocene increase in summer temperatures and summer precipitation, perhaps combined with a northern shift in the average path of winter storms systems derived from the Pacific.

These suggestions began to be made before it was realized that pinyon did not begin to fill in the western part of its distribution until much later than it did toward the east (figure 8-16). Now that we know that this was the case, it is worth pointing out that the parts of the Great Basin that saw the arrival of this tree during the middle Holocene are exactly those parts most impacted today by summer monsoonal storms (see figure 5-2).

The argument that enhanced monsoons, associated with increased summer temperatures, triggered the movement of pinyon northward might sound inconsistent with the lack of compelling evidence for greatly enhanced monsoonal precipitation in the floristic Great Basin during the middle Holocene that I discussed above. Thanks to the work by Ken Cole and his colleagues, however, it is clear that it is not.

In chapter 2, I note that these scientists have shown that Great Basin singleleaf pinyon (*Pinus monophylla monophylla*) is now found in areas that tend to be marked by relatively dry summers and wet winters. By analyzing both precipitation and temperature patterns in the areas in which this tree is now found, they discovered that singleleaf pinyon is found in areas that receive a maximum of 1.5 inches of precipitation in the summer. Above this, all other things equal, it is Colorado pinyon, not the singleleaf variety, that would be expected to occur in the heart of the Great Basin.

If middle Holocene pinyons followed the same rules, and enhanced summer precipitation allowed the spread of pinyon northward, it couldn't have been all that enhanced. This, of course, fits with the lack of evidence for greatly strengthened monsoons in the floristic Great Basin during the middle Holocene.

How about winter precipitation? Here we are on shakier ground, since singleleaf pinyon now occurs in areas in which winter precipitation varies greatly, from less than an inch to nearly four inches. These figures, though, are twice the amount required by Colorado pinyon.

The standing hypothesis, then, is that singleleaf pinyon began its middle Holocene march northward in response to warmer summers and associated increases in summer precipitation. Ken Cole's work shows that summer rains could not have been dramatically greater than what they are now, or we might have Colorado, not singleleaf, pinyon widespread in the floristic Great Basin. If this is the case, the fact that pinyon arrived far earlier in the more easterly parts of the Great Basin than in the west could readily be explained by the fact that it is the east, not the west, that receives the lion's share of the precipitation provided by these weather systems.

As Cole and his coworkers have noted, and as people have long suspected, there is no reason to think that the northward movement of pinyon in the northwestern Great Basin has ended. Given that the birds that play such a major role in dispersing this tree extend far to the north of the tree itself, the following few hundred years should tell.

More on Middle Holocene Small Mammals

I have already discussed the impact that middle Holocene climates had on Great Basin small mammals, from the local extinction of yellow-bellied marmots, bushy-tailed woodrats, and pygmy rabbits in the areas surrounding Homestead and Camels Back caves, to the increase in the elevations of pika populations. Here, I want to provide a little more information on some of these animals.

I have already mentioned Dave Thomas's careful excavations at Gatecliff Shelter (figure 8-8), now surrounded by pinyon-juniper woodland at an elevation of 7,610 feet in central Nevada's Toquima Range, and the rich small mammal fauna that this site contains. In fact, Gatecliff provides the most detailed middle and late Holocene small mammal record available from the heart of the Great Basin.

Today, pikas are found in the upper reaches of the Toquima Range, the lowest known modern populations existing at an elevation of about 8,700 feet, some 1,100 feet above Gatecliff. Even though Gatecliff is now well beneath the lowermost Toquima Range pikas, however, there were fifty-eight pika specimens in the sediments of the site. Of these, fifty-five were deposited prior to 5,100 years ago. During the past 5,100 years, while 9,500 bones and teeth of other small mammals were introduced into the site, only three pika specimens joined them. This decline probably records the passing of the lower elevational limits of pikas above the level of the site itself, roughly coincident with the arrival of pinyon pine in the area.

Gatecliff thus seems to show the movement of pikas upslope in the Toquima Range. Equally interesting, it also documents that a small, high-elevation mammal that no longer occurs in the Great Basin occupied the Toquima Range until nearly the end of the middle Holocene. In chapter 7, I note that heather voles, which are primarily animals of coniferous forests, are today found at high elevations on the far eastern and far western fringes of the Great Basin but not within the Great Basin itself. I also noted that Smith Creek Cave shows that these animals were living in the heart of the Great Basin during the late Pleistocene. Gatecliff, however, provided heather vole specimens tightly dated to 5,300 years ago, documenting that this small rodent survived until at least that time in the Toquima Range.

There is no reason to think heather voles were living in the immediate vicinity of the site at this time. Instead, the Gatecliff specimens were probably transported here, as part of a complete vole, from much higher elevations by a raptor, probably an owl. There may also be nothing particularly meaningful about the fact that the Gatecliff specimens were deposited around 5,300 years ago. Heather voles may have lasted much longer than this on the Toquima Range, and even if they didn't, the last heather voles here may have succumbed to something as simple as random fluctuations in population numbers.

Without question, however, pygmy rabbits did respond to middle Holocene climatic change, as I have already discussed. Within the Great Basin, these sagebrush-dependent rabbits appear to have undergone two major prehistoric decreases in abundance. The first of these decreases occurred as the Pleistocene came to an end. The second, so well illustrated at Homestead Cave, happened as the heat and aridity of the middle Holocene set in. Indeed, Homestead was hardly the first place where the negative impact of middle Holocene heat and aridity on pygmy rabbit populations was detected. In 1972, archaeologist Bob Butler pointed out a dramatic decrease in the abundance of these animals at about 6,700 years ago at the Wasden site on the Snake River Plain in southern Idaho, and the same phenomenon is now reasonably well known from a number of areas within the floristic Great Basin.

These changes seem to be a response to the climate-driven decrease in abundance of sagebrush habitat on the landscape, driven by the challenging nature of middle Holocene climates. Gatecliff Shelter, however, shows that the expansion of pinyon pine itself may have caused a decline in the abundances of pygmy rabbits on the landscape.

In chapter 2, I note that Dwight Billings observed, long before we knew anything about the history of Great Basin pinyons, that it appeared as if the Great Basin's pinyon-juniper woodland had been superimposed on a once-continuous sagebrush-grass zone. With several decades of detailed work on the history of pinyon pines behind us, we now know how deeply insightful this observation was.

Given that this was the case, as pinyon pine expanded across the Great Basin and combined with Utah juniper to form pinyon-juniper woodlands, the extent of sagebrush-grass habitat must have been vastly reduced. Given that sagebrush provides the habitat required by pygmy rabbits, it follows that the formation of pinyon-juniper woodland would also have caused a decline in the abundance of these animals. Indeed, ecologists Eveline Larrucea and Peter Brussard have found that even the presence of a few pinyon pines on a site can lead to the absence of pygmy rabbits, perhaps in part by providing roosts for avian predators and cover for terrestrial ones. Since the formation of the woodlands happened at different times at different places, we can also expect this particular cause of pygmy rabbit decline to have happened at different times at different places.

Gatecliff seems to show this process in action as well. The analysis of pollen from the sediments of this site done by Bob Thompson and Robert Kautz documented a sharp increase in both pinyon and juniper pollen shortly after 6,000 years ago with, as I have discussed, the first pinyon pine macrofossils appearing not long after. Certainly not coincidentally, the relative abundances of pygmy rabbit specimens at Gatecliff dropped dramatically between about 5,350 and 5,000 years ago (figures 8-17 and 8-18). Within a few hundred years of the deposition of the first pinyon pine macrofossil at Gatecliff, pygmy rabbit relative abundances had declined by some 80 percent. They never recovered.

The Gatecliff sequence thus seems to show that the establishment of pinyon-juniper woodland in this part of the Toquima Range was tightly associated with the decline of pygmy rabbits. As other detailed sequences of this sort become available from the Great Basin, it is reasonable to suspect that similar declines will be discovered, with the timing of the decline dependent on the timing of the local formation of pinyon-juniper woodland. This process likely occurred wherever woodland replaced lower-elevation sagebrush-grass vegetation in the Great Basin, no matter what the composition of that woodland.

In short, there appear to have been at least three distinct types of decline in pygmy rabbit abundance in the prehistoric Great Basin, all associated with the loss of sagebrush habitat. Two of these seem to have been roughly synchronous across the Great Basin: one at the end of the Pleistocene, and one associated with onset of the middle Holocene. The third decline, so far shown only at Gatecliff, can be expected to have occurred as pinyon-juniper woodland developed within the floristic Great Basin.

The Late Holocene (The Last 4,500 Years)

It is natural to wonder when the Great Basin came to look as it looks today. That question, however, has only one meaningful answer: "today." Geographer Garry Rodgers, for

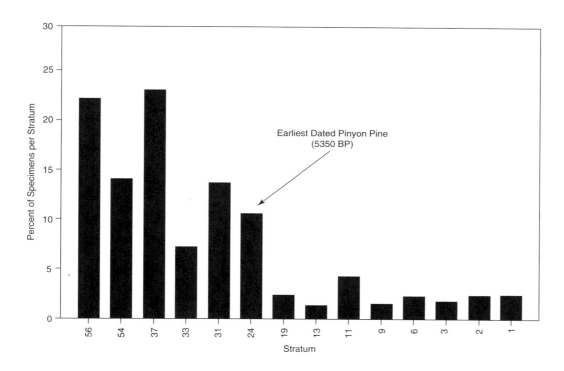

FIGURE 8-17 Changing abundance of pygmy rabbits through time at Gatecliff Shelter (after Grayson 2006a).

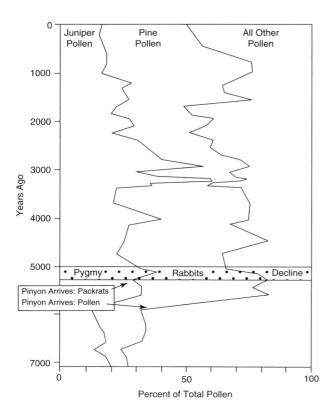

FIGURE 8-18 Relationship between the development of pinyon-juniper woodland and the decline of pygmy rabbits based on the Gatecliff Shelter pollen and packrat midden records. (Pollen data from R. S. Thompson and Kautz 1983.)

instance, has carefully matched and analyzed a large series of photographs taken from the same camera location in a diverse set of areas within the Bonneville Basin. Most of the earliest photos in these matched pairs were taken between 1900 and 1915; the second set was taken during the late 1970s. Virtually all of these pairs show dramatic changes in vegetation across the six or so decades involved. Indeed, even matched photographs taken a decade apart show significant vegetational change. Similarly, railroad buff and historian Larry Hersh has provided matched photographs of the Central Pacific Railroad route across Nevada taken first in 1868 and then again in 1997. While the vegetational differences are less pronounced than those provided by Rodgers, the implications are much the same. "Today" is the only meaningful answer to the question of when the Great Basin came to look as it now looks. But if we make the question more general and simply ask when the Great Basin came to look pretty much what it has looked like during the past few centuries, a rough answer becomes possible.

Certainly, the early Holocene cannot provide the answer. Low-elevation conifers, the absence of pinyon pine in the floristic Great Basin, expanded marshes, mammals now confined to cool and moist habitats then found in low-elevation settings: none of this suggests a Great Basin landscape in any way "modern." The generally hot and dry climatic conditions that marked much of the middle Holocene were distinctly different from what they are today, so the answer cannot lie here either. If forced to pick a time, I would choose some 4,500 years ago, when the middle Holocene came to an end. I would then hedge by noting that the answer to this question would differ in different parts of the Great Basin and would depend on whether we had plants, vertebrates, bodies of water, or something else in mind, or all of these things together. And I would also point out that there have been episodes of very significant climate change during the late Holocene, the three most important of which I discuss below.

Effective Moisture Increases

The paleoenvironmental information available for the past 5,000 years or so of Great Basin history suggests that conditions cooler and moister than those of the middle Holocene, but not as cool and moist of those of the early Holocene, were established here at roughly 4,500 years ago. The onset of these conditions defines the transition from the middle to the late Holocene.

Thompson's Ruby Valley sequence provides an excellent example. The aridity of the middle Holocene ended here around 4,700 years, when the Ruby Marshes were reborn. At the same time, cheno-am (probably *Atriplex*) pollen began to decline, and sagebrush pollen to increase, in the sediments that Thompson analyzed. The rebirth of the Ruby Marshes, coupled with the expansion of sagebrush steppe at the expense of shadscale vegetation, suggests, as Thompson notes, a return to cooler and/or moister conditions. In fact, the coolest and/or moistest conditions of the past 7,000 years seem to have been those of the past 500 years, when the marshes became even deeper and sagebrush pollen reached its highest frequencies since the early Holocene. It was, it appears, at about this time that the vegetation of Ruby Valley came to look much as it does now.

To the north and west, Mehringer's Fish and Wildhorse lake sequences carry similar implications (see figure 8-13). At lower elevation Fish Lake, the sagebrush/grass ratios, which had become elevated shortly before 8,000 years ago, first fell beneath their average at around 5,500 years ago, then spiked upward, only to fall again at about 4,700 years ago. This time, however, they fell permanently. At higher-elevation Wildhorse Lake, the sequence is much the same, but the resurgence of grass in the pollen record occurs about 1,000 years later than it occurs at Fish Lake, perhaps because Wildhorse Lake lies so much farther upslope. As Mehringer notes, lower temperatures and increased snow accumulation would seem to account for the late Holocene grass resurgence here.

To the immediate north, Diamond Pond deepened by about 5,400 years ago; a few hundred years later, sagebrush, *Atriplex*, and grasses began to replace the greasewood-dominated vegetation that marked this area during the middle Holocene. By 4,400 years ago, Diamond Pond had risen significantly. As Wigand observes, the aridity of the middle Holocene had given way to greater effective moisture by this time.

Other areas follow suit. In the Lahontan Basin, the abundance of sagebrush and grass relative to that of *Atriplex* and greasewood in the pollen from the sediments of Hidden Cave begins to climb shortly after 4,000 years ago (figure 8-3). The last drowned tree beneath the waters of Lake Tahoe dates to 4,250 years ago, marking a deepening lake here. Downstream, Pyramid Lake had began to freshen by around 3,500 years ago. To the south, Owens Lake began to refill at around 3,600 years ago. In the White Mountains, bristlecone pine treeline dropped sharply after about 4,200 years ago on Sheep Mountain, though this decline occurred later on Campito Mountain. In the Bonneville Basin, small mammals whose numbers had been greatly depressed, sometimes to the point of local extinction, begin their reappearance—Great Basin pocket mice, Ord's kangaroo rats, and western harvest mice, for instance. In the Wasatch Range to the east, the pollen sequence suggests that the area surrounding Snowbird Bog became cooler and moister at around 5,200 years ago. The Great Salt Lake seems to have begun a climb toward its historic level sometime after 4,500 years ago, although this early late Holocene increase in water depth is not well controlled chronologically and the lake has had very significant ups and downs during the past 4,000 years.

In southern Nevada, black mats do not begin to reform until around 2,300 years ago, perhaps reflecting how long it took for groundwater to be recharged here. In the Mojave Desert of southeastern California, however, increased spring discharge as early as 5,200 years ago is marked by a black mat in the Salt Spring Basin, just south of the Amargosa River's U-turn toward Death Valley. And, in the nearby Lake Mojave Basin, there is evidence for a brief resurgence of lake levels shortly after 5,000 years ago, following some 3,000 years of what appears to have been uninterrupted dry lake basins in this area. In the central Mojave Desert, packrat midden analyses show that most components of the vegetation that marks this area today had arrived by about 4,800 years ago, suggesting effective precipitation levels more similar to those that now mark the region.

Almost no matter where we look or what we look at, the extreme, but variable, heat and aridity that marked the middle Holocene during the Great Basin ended after about 4,500 years ago, sometimes earlier, sometimes later. In a general sense, this was the first time that the Great Basin began to look pretty much as it looks today. If we were to be dropped into a Great Basin valley at that time, we would not react with as much surprise as we would have reacted had we been dropped down here at, say, 9,000 or 6,000 years ago. Again, though, this does not mean that we would recognize what we saw as identical to what is there now. There was no pinyon pine in the northwestern part of its modern distribution, for instance, and, as Peter Koehler and his colleagues have observed, truly modern plant communities in parts of the central Mojave Desert did not develop until after 700 years ago.

In addition, as I mentioned above, late Holocene climates and environments were highly variable. Our knowledge of this variability is stronger than it is for earlier times for a very simple reason. Late Holocene paleoenvironmental records of virtually all sorts are easier to come by, and often easier to date, than those from earlier times.

Work by Peter Wigand and Peter Mehringer shows the kind of precision that can be obtained for this period of time. While the greater effective moisture that marks the late Holocene became evident in Diamond Pond by about 4,400 years ago, Wigand's work here showed that episodes of drought affected this area at about 2,900, 700, and 500 years ago, and that episodes of deeper water occurred four times during the past 4,000 years. These wetter episodes centered on 3,700, 2,500, 900, and 200 years ago, with Diamond Pond attaining its deepest late Holocene levels at about 3,700 years ago. Insofar as these droughts and periods of higher water levels reflect the regional water table, they probably reflect the state of the Malheur marshes as well.

Wigand also showed that the pollen of both juniper and grass increased at Diamond Pond between about 3,750 and 2,050 years ago, with these plants spreading into habitats then dominated by sagebrush and *Atriplex*. Although the species of juniper involved could not be told from the pollen itself, it must have been western juniper, the same species that occurs here today. This is evident not only from western juniper macrofossils in the Diamond Pond sediments but also from macrofossils in twenty-four nearby packrat middens. In work of impressive precision, Mehringer and Wigand have shown that fluctuations in frequencies of juniper pollen in the sediments of Diamond Pond are matched closely by the frequencies of both juniper pollen and western juniper macrofossils in the packrat middens.

The rise in juniper and grass pollen between 3,800 and 2,100 years ago coincides closely with intervals of increased water levels within Diamond Pond (from 3,750 to 3,450, and from 2,800 to 2,050, years ago, interrupted by a brief episode of drought at about 2,900 years ago). The spread of juniper and grass into sagebrush and shadscale vegetation, Wigand suggests, reflects an increase in effective moisture during this interval, matching the more general fact that western juniper increased here as effective moisture increased, and retreated during times of drought.

Rather than review particular detailed sequences from various parts of the Great Basin, however, I want to focus on what appear to be three of the most significant episodes of climate, and related environmental, change that have marked the Great Basin during the past 4,500 years or so.

Drought 2,500 to 1,900 Years Ago?

I have mentioned the work by Jerry Miller and his colleagues that documents that the middle reaches of the Humboldt River were dry, or nearly so, between about 6,700 and 5,500 years ago. That same work also suggests that an extremely arid interval occurred here between about 2,500 and 1,900 years ago. The evidence for this is provided by valley sediments of this age that were eroded from surrounding hillsides and contain multiple thin lenses of charcoal. Miller and his coworkers interpret these sediments to indicate that drought had removed hillside vegetation, making the landscape more vulnerable to both fire and erosion. They have also shown that similar events seem to have occurred elsewhere in central Nevada at this time.

A drought of this length and magnitude would, of course, be expected to have its impact elsewhere as well, and evidence suggests that this was the case. Scott Mensing's work at Pyramid Lake, for instance, detected a substantial peak in cheno-am pollen between about 2,500 and 2,000 years ago, one almost as intense as the one that occurred here during the middle Holocene. He interpreted this to reflect a significant lowering of Pyramid Lake, one that exposed greater amounts of lake edge for colonization by these plants. During roughly the same interval, oxygen isotopes and the levels of inorganic carbon in the Pyramid Lake deposits analyzed by Benson and his colleagues led them to suggest that Pyramid Lake must have been relatively shallow at that time. To the south, Scott Stine's work on the late Holocene history of Mono Lake detected a very significant lowstand here which he dated to around 2,000 years ago; Owen Davis found the same lowstand, but suggested that it dates to around 2,400 years ago. Either way, it falls within the interval of the drought called for by Miller and his colleagues, though it appears to have been fairly short-lived.

Well to the east, in the Bonneville Basin, there is evidence that the highest Holocene stand reached by the Great Salt Lake occurred between 3,400 and 2,000 years ago, with the early end pinned down by a date of 3,440 years ago. During this interval, Great Salt Lake rose to an elevation of 4,221 feet—21 feet higher than its average historic level, 9 feet higher than its historic highstand, and about 6 feet higher than the thresholds that separate the Great Salt Lake Basin from the Great Salt Lake Desert to the west (see chapter 5 and figure 5-8). As a result, the lake overflowed into the Great Salt Lake Desert, covering some 4,200 square miles with water. David Madsen and his colleagues suggest that the lake had fallen from this level by 2,400 years ago and the fish data from Homestead Cave suggest that that is certainly possible. Here, the number of fish specimens reaches an all-time low in stratum XV; recent work done by Rebecca Terry shows that this stratum accumulated between 2,500 and 2,000 years ago. This, too, is consistent with the drought inferred by Miller and his colleagues from their work in central Nevada.

All this suggests that there may have been a significant drought in at least parts of the Great Basin during these centuries. How extensive such a drought might have been, though, is not known. Peter Wigand's work at Lead Lake in the Carson Sink, for instance, suggests a shallow marsh in this area between about 2,300 and 2,000 years ago, matching the evidence from Diamond Pond for deeper water between about 2,800 and 2,050 years ago. The same wet episode seems to occur at Lower Pahranagat Lake far to the south, roughly the same period of time that black mats begin to reform in southern Nevada. While there is no reason to think that drought in parts of the northern Great Basin must or even should be matched by drought in the south, the scope and magnitude of this recently detected event has yet to be determined.

The Medieval Climatic Anomaly (900 to 1350 A.D.)

In 1965, the famed British climatologist Hubert H. Lamb gathered together a broad variety of data suggesting that there had been "a notably warm climate in many parts of the world" (1965:13) focusing on 1000 to 1200 A.D. The evidence was impressive, including vineyards in Britain where they did not exist before or after, changing distributions of trees in New Zealand, and decreasing sea ice in the northern Atlantic. Lamb referred to this interval as the Medieval Warm Epoch. Today, scientists refer to this same episode either as the Medieval Warm Period or as the Medieval Climatic Anomaly. This latter term was introduced by Scott Stine, who observed that in some places, changes in precipitation seemed far more significant than changes in temperature. As a result, he felt that a more neutral term was needed, one that made no assumptions about the precise nature of climatic change during these centuries. I follow his suggestion here.

In the same year that Lamb published his important synthesis of Medieval climate, civil engineer and water resources expert Sidney T. Harding published an analysis of the recent history of western Great Basin lakes. Among many other things, Harding noted the existence of drowned trees along the edges of Mono Lake that had become exposed as the result of the diversion of its waters by Los Angeles. He also observed that one of these trees, rooted beneath an elevation of 6,400 feet, some 20 feet below the lake's pre-diversion level, had been radiocarbon dated to 1030 A.D.

It was Stine, however, who put these trees on the global map, obtaining a lengthy series of radiocarbon dates not only for them but also for drowned shrubs and grasses. Those dates showed that Mono Lake had been significantly lower on two separate occasions during medieval times. The first of these had, from tree-ring counts, lasted from about 900 to 1110 A.D. and had seen Mono Lake some 50 feet lower than its pre-diversion level. A brief period of higher water levels separated this arid episode from the next one, which ran from before 1200 to about 1350 A.D. and marked a time when the lake fell close to the same level as it had earlier. Later work has confirmed Stine's results here, while suggesting that his two periods of extreme aridity may have fallen at 900 to 1010 A.D. and 1175 to 1275 A.D.

Stine also dated drowned trees in the West Walker River canyon (figure 8-19). These also suggested two periods of lowered water levels, one somewhere between 930 and 1130 A.D., the other between 1200 and 1340 A.D. These, of course, match the Mono Lake droughts almost perfectly.

FIGURE 8-19 Jeffrey pines emerging from the West Walker River, California.

Much the same is true for Owens Lake, which has shrubs rooted some 10 feet above the playa floor, or about 40 feet beneath the high water mark attained in 1872. One of these shrubs dated to 1020 A.D.

These analyses do not stand alone. Basing her analyses on tree-rings from foxtail pine and western juniper from the eastern crest of the Sierra Nevada west of Owens Valley, Lisa Graumlich has shown that the period from 1100 to 1375 A.D. was marked by persistently, though variably, high temperatures; the warmest 20- and 50-year periods in her 1,000-year-long record fall between about 1150 and 1200 A.D., with temperature for part of this time averaging about 1.2°F higher than modern. Graumlich and Malcolm Hughes have shown that in the White Mountains, droughts centered on 924 and 1299 A.D. were among the eight most intense droughts that appear in their 8,000-year sequence. As at Mono Lake and the Walker River gorge, these droughts, Steven Leavitt has shown, were separated by a significant wet interval, this one well dated to between 1080 and 1129 A.D.

Detailed work by Connie Millar and her colleagues at Whitewing Mountain and San Joaquin Ridge along the crest of Sierra Nevada above the headwaters of the Owens River has identical implications. They were drawn to this area by the presence of substantial amounts of deadwood on these now treeless summits, some 650 feet above the elevation of woodland in this area today. The trees turned out to include five species of pine, plus mountain hemlock. Dendrochronological work showed that these trees had thrived here from 815 to 1350 A.D. Analyses of tree-ring widths allowed Millar's research team to estimate that trees grew here at a time when annual minimum temperatures were about 5.8°F, and annual maximum temperatures about 4.4°F, higher than they are today, with both winters and summers warmer than they are now. Annual precipitation, on the other hand, was only about one inch lower than occurs here today. The very stark cut-off date for this high-elevation woodland, at 1350 A.D., does not indicate an equally stark end to warm temperatures here. In fact, the trees were doing just fine up to the point of their demise; conditions inimical to high-elevation trees here, their work demonstrated, did not begin until the following year, 1351 A.D. Rather than being killed by abrupt climate change, they were killed by a volcanic eruption, from Glass Creek, not far to the northeast. As the saying goes, if it ain't one thing, it's another.

To the north, John Kleppe has documented drowned trees rooted 120 feet beneath the surface of Fallen Leaf Lake, near South Lake Tahoe, California. Radiocarbon dates for these trees, some of which seem to have survived for about two hundred years, suggest that they were submerged at about 1215 A.D. Similarly, analyses of cores from Walker Lake provided Fasong Yuan and his colleagues with evidence for basin-wide drought between 1000 and 1360 A.D. These dates coincide closely with Stine's tree stump evidence showing that the West Walker River canyon was dry or nearly so at this time. The earliest of these two episodes also coincides with a date of 1025 A.D. for a tree rooted some 130 feet below the modern natural level of Walker Lake. At Pyramid Lake, this same general time range shows three episodes of drought separated by two wet intervals; the two latest dry periods correlate with those detected by Stine at Mono Lake. Something similar seems to happen to the Great Salt Lake, where a wet interval roughly dated to about 1,000 years ago is wedged between drier times.

Numerous other studies have identical implications. Although the analyses I have discussed here are primarily from the western Great Basin, the drought that marked this period is fully evident elsewhere. Stephen Gray and his colleagues have built a tree-ring record that covers the past 1,226 years in the Uinta Basin of northeastern Utah. During that long time span, the interval from 1250 to 1288 A.D. was the second-longest dry period to have occurred, with the year 1270 A.D. the driest of them all. In the southwestern United States, significant droughts centered on 1090, 1150, and 1280 A.D., triggering substantial migrations by the Puebloan peoples of that area. In fact, archaeologist Terry Jones and his colleagues have argued that this period saw cultural disruptions throughout much of the western United States. In the central Great Plains, drought led to a loss of wetlands and the activation of sand dunes—the Nebraska Sand Hills—between 1000 and 1300 A.D., an episode of aridity that has not been matched since. More generally, a detailed analysis of a large number of tree ring chronologies from the western half of North America, incorporating data from northern Mexico in the south to southern Canada in the north, shows that the four most intense droughts of the past 1,200 years all occurred between 900 and 1300 A.D., centering on 936, 1034, 1150, and 1253 A.D. Even though it does not appear that the Medieval Climatic Anomaly was a global affair, it was most certainly a phenomenon that affected much of western North America.

The Little Ice Age (1400 to 1900 A.D.)

The Medieval Warm Period ended shortly after 1300 A.D., introducing a period of time in western North America generally, though not always, cooler and less arid. This period is routinely referred to as the Little Ice Age, a term first used by geologist François Matthes in 1939 to refer to the rebirth of glaciers in the Sierra Nevada and elsewhere during the late Holocene.

Soon after introducing this term, Matthes suggested that the recent glaciers in the Sierra Nevada probably dated to between about 1600 and 1900 A.D. We now know that glaciers on the east face of the Sierra Nevada, between at least Lake Tahoe and the latitude of Owens Lake, expanded shortly after 700 years ago and that they did not retreat until early in the twentieth century. With equilibrium line altitudes (chapter 5) about two hundred feet lower than they are today, these glaciers, referred to as the Matthes glaciation, represent the greatest expansion of glacial ice in the Sierra Nevada to have occurred during the past 10,000 years and perhaps more.

A wide variety of other evidence falls in line. Lisa Graumlich's tree-ring work on the east slope of the southern Sierra Nevada shows that the high temperatures of the Medieval Warm Period were followed by cold temperatures that endured from about 1450 to 1850 A.D., hitting a minimum in the early 1600s. In the White Mountains, Val LaMarche documented an extended cold period between 1400 and 1800 A.D., marked by decreases in bristlecone pine tree-ring width and by a lowered treeline.

This conclusion has been confirmed in many ways, but one of the more novel pieces of supporting evidence was provided by Xiahong Feng and Samuel Epstein. These scientists examined the ratio of deuterium and hydrogen in the cellulose of tree-rings from the White Mountains. The logic behind the analysis they did is very similar to that which drives much of the work on oxygen isotopes (chapter 2). Deuterium is a stable isotope of hydrogen but has an atomic weight of about twice that of hydrogen itself. Both forms combine with oxygen to form water, but the water molecules formed with deuterium are heavier than those formed from hydrogen—hence the name heavy water for this form. Because of the difference in weight, water composed of deuterium and oxygen condenses more readily when it rains as well as evaporating more slowly. As a result, deuterium/hydrogen ratios permanently embedded in the cellulose of trees (called nonexchangeable hydrogen) can tell us about past temperatures. Using this logic on White Mountain bristlecones, Feng and Epstein showed not only that the middle Holocene was a time of great warmth in this area but also that rapid cooling began here about 1600 A.D. and peaked between 1700 and 1900 A.D.

It is not just tree-ring work that documents the existence of a generally cooler and wetter Great Basin between 1400 and 1900 A.D. An ephemeral lake formed in the Mojave Basin at about 1500 A.D., apparently fed by flooding events in the headwaters of the Mojave River. At Ash Meadows, Peter Mehringer has shown that marshes began to form about 400 years ago, after a hiatus of about 2,000 years. At Diamond Pond, the ratio of juniper to grasses climbs dramatically after about 500 years ago, suggesting either wetter climates in general or increased winter rainfall. Much the same phenomenon—an expansion of juniper woodland—occurs at Lower Pahranagat Lake at about the same time. At Ruby Lake, marshes apparently deepened during the past 500 years, coincident with an increased in the abundance of sagebrush, suggesting to Bob Thompson that this time may have seen the coolest and/or wettest conditions since the early Holocene. And, in the Bonneville Basin, the Great Salt Lake rose yet again, apparently between about 300 and 400 years ago, to reach an elevation of 4,217 feet and flood the Great Salt Lake desert to the west. Working to the south, Jack Oviatt has shown that Sevier Lake also expanded during this interval.

Then there is the curious case of the reappearing bushy-tailed woodrats at Homestead Cave (figure 7-10). As I discuss in chapter 7, these animals disappeared from the deposits of Homestead Cave around 8,300 years ago. Late Holocene stratum XVII, however, provided nine specimens of them. Today, the Homestead Cave area (figures 5-7 and 5-11) hardly seems like an appropriate spot for these animals, which prefer fairly cool settings. The specimens, though, were real enough and it seemed unlikely that they represented the dinner of a raptor that had flown them in from afar and even less likely that they got here by stratigraphic disturbance from deposits thousands of years older. All of this led us to set live traps in the cave itself, and to the discovery that the animals are living in Homestead Cave today. The stratum XVII specimens seem to represent a bushy-tailed woodrat recolonization event. Exactly when that occurred we do not know. There is only a single date, of 1,020 years ago, for this stratum and no dates for the stratum above it. If I had to guess, though, I would guess that it was during the Little Ice Age, at the same time as the Great Salt Lake and Sevier Lake were expanding.

More generally, analyses of a broad series of tree-ring records from the western United States that cover the period from 1600 to 1982 A.D. show that 1601 was the coldest year during that span, with summers a full 4°F colder than the summer average for this entire period. While there was a great deal of climatic variability during this period, the 1600s and the period from 1870 to 1930 A.D. were marked by cool summers, with the decade from 1610 to 1620 A.D. the coldest of them all.

Most generally, the Little Ice Age was decidedly not confined to the western United States, having impacted much of the Northern Hemisphere as well significant parts of its southern counterpart. Glaciers expanded in places as far-flung as the Andes, New Zealand, the Alps, the Pyrenees, and the Urals. The southern Atlantic Ocean cooled, sea ice increased in the waters surrounding New Zealand, harvests declined or crops failed entirely in Great Britain and Europe. Cod declined, and fisheries failed, off the Faroe Islands, Greenland, much of Norway, and elsewhere in the north Atlantic. In Europe, herds of livestock dwindled and milk production decreased among the survivors, wine production shrank and stored grain rotted. In the Mediterranean, frosts damaged olive crops and floods destroyed cultivated lands. In China, the Asian monsoon weakened dramatically, with the weakest monsoons correlating with the fall of the Yuan and Ming dynasties. The list is remarkably long.

The general causes of the Little Ice Age seem to lie in some combination of decreases in the amount of solar energy reaching the earth on the one hand, and increased volcanic activity on the other. The correlations between temperature and received solar energy over the past 1,200 years or so (including the Medieval Climatic Anomaly) are strong. Those correlations, however, do not seem to have been driven simply by differing amounts of energy reaching the surface of the earth from the sun, but also by changes in atmospheric chemistry (including the abundance of ozone) and perhaps even changes in the ocean's circulation. How fairly minor changes in solar activity could have caused all this remains something of a mystery, but the correlations are real enough.

While strong, these correlations are far from perfect and some scientists have observed that there are also stark relationships between Little Ice Age cold spikes and volcanic eruptions. Geologist Tom Crowley and his colleagues, for instance, have observed that the Little Ice Age was a period of substantial volcanic activity. Sixteen significant eruptions occurred between 1630 and 1850 A.D., each of which is associated with a cooling event. To take perhaps the most famous example, April 1815 saw the eruption of Tambora, on the Indonesian Island of Sumbawa. One of the two largest eruptions known to have occurred during the past 2,000 years, the Tambora eruption caused the immediate deaths of some 71,000 people on Sumbawa and the nearby island of Lompok alone. It also led to the famous "Year Without a Summer" in 1816. This disastrous year saw summer temperatures as much as 5.4°F below average in eastern North America and in central and western Europe. It also saw widespread crop failures, food riots, and outbreaks of disease. The prime mechanisms here, and in other such events, are well understood. Massive volcanic eruptions eject huge amounts of particulate matter and sulfurous gases—sulfur dioxide and hydrogen sulfide—into the atmosphere ("volcanic aerosols"). The fine particles reflect incoming solar energy, decreasing the amount that reaches the earth's surface. The sulfurous gases are injected high into the atmosphere where they intercept and absorb solar energy, causing the upper levels of the atmosphere to become warmer but the surface of the earth to become cooler.[6] As Tambora shows, the spectacular sunsets associated with such events are not worth the price.

The Walker Lake–Carson Sink Conundrum

In chapter 5, I point out that the history of Walker Lake is complicated by the fact that the Walker River, which today sustains it, can switch direction and flow not south into Walker Lake but north into the Carson Desert. That possibility makes the history of Walker Lake potentially very confusing, a potential fully realized during the late Holocene.

As we now understand that history, the Walker Lake Basin filled quickly between 4,800 and 4,700 years ago. The speed of that filling suggests that the cause was not climatic, but instead reflects the fact that the Walker River now flowed south. From about 4,700 to 2,800 years ago, the lake was perhaps about as high as it would be now were its waters not drawn for other purposes, The lake then shallowed, and remained low—to a depth of less than 3 feet—until 2,100 years ago. It then deepened, peaking at about 1,600 years ago, declined sharply to a minimum between 1,400 and 1,000 years ago, then underwent a series of ups and down during latest prehistoric times—with lows at about 870 and 250 years ago separated by a peak at 760 years ago (see figure 8-20).

There are two obvious ways to account for this complex history. First, these multiple ups and downs may have been driven by climate, with the peaks reflecting relatively moist periods and the troughs dry ones. The second possibility, one pointed to by Larry Benson and his colleagues and by Ken Adams, is that some of these troughs—in particular the ones between about 2,700 and 2,000 years ago, 1,500 and 1,000 years ago, and from 500 to 300 years ago—may be due to diversion of the Walker River into the Carson Desert.

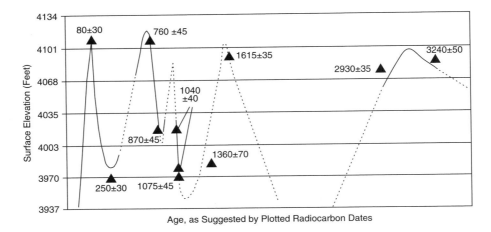

FIGURE 8-20 Changing levels of Walker Lake during the late Holocene (after K. D. Adams 2007).

We don't know which of these explanations is correct, but if the Walker River flowed into the Carson Desert during these times, there are two obvious places to look for evidence of it. The first is Adrian Valley, through which the river would have flowed. The second is the Carson Desert, into which it would have flowed.

I start with the Carson Desert, since that is the place we know most about. Obviously, when the Walker River flowed north, it must have raised water levels in the Carson Sink. Those levels should have fallen when the river flowed south to feed the Walker Lake Basin. We have two independent sets of information related to this issue. First, recent research has been done on the development of late Holocene lakes in the Carson Sink. Second, archaeological work has been done on a series of sites from low-elevation settings within this basin. The two together should tell us something about when deep lakes were, and were not, here. By so doing, they should also tell us something about when the Walker River flowed north.

The lake history has been built primarily by Ken Adams, whose work has detected two very late Holocene lakes in the Carson Sink. The earliest of these formed at 1,510 years ago, reached an elevation of 3,930 feet, and covered 1,000 square miles of the basin with water. The second dates to 810 years ago, reached an elevation of 3,950 feet, and produced a lake 1,160 square miles in extent. Both of these dates roughly coincide with periods when Walker Lake was either low or declining, and so might reflect the arrival of Walker River waters here. In addition, geographer Guy King has obtained a radiocarbon date of 710 years ago on mollusk shells from an elevation of 3,900 feet in the Carson Desert which he thinks indicates a 40 feet-deep lake here at this time. These dates overlap one another once the possible statistical errors are taken into account, so they could well refer to one and the same thing.

The archaeological dates available for low elevations in the Carson Sink, however, suggest a very different story. These dates come from sites in and around Stillwater Marsh, which forms the heart of the Stillwater National Wildlife Refuge. Two different research teams, one led by Bob Elston and the other by Bob Kelly, have conducted exacting work on these very low-elevation sites. Since the sites were full of waterbirds, fish, and freshwater molluscs, we know that when people were living here, there were marshes or shallow lakes nearby. However, we can also be certain that the ground on which those people were living was dry.

This is where the radiocarbon dates obtained from these sites become important, since they tell us when lakes in the Carson Sink were low enough to allow people to settle here. Those dates, along with the elevations of the sites, the ages and elevations of the hypothesized Carson Sink lakes, and the dates for the suggested diversions of the Walker River into the Carson Sink, are shown in table 8-5. Because we are concerned with fairly narrow periods of time, I have included the error terms for the radiocarbon dates in this table (see "Radiocarbon Dating" in chapter 3).

These dates may be telling us quite a bit about lake history in the area. If we look first at the 2,700-to-2,100-year interval, when Walker Lake was shallow and the Walker River is thought to have been flowing into the Carson Sink, it is hard to miss the fact that people were living at low elevations here at that time. Perhaps, though, we can rescue this argument by pointing out the 400-year gap in the dates (between 2,680 and 2,265 years ago), so perhaps this is the period of time when the river was flowing northward and Walker Lake was reaching a minimum.

The absence of sites that fall between 1,860 and 1,390 years ago, the interval that contains the lake called for by Adams at 1,510 years go, fits the Walker River picture well, but the lake supposed to have existed at about 810 years ago overlaps dates from a series of sites that fall between 800 and 870 years ago. These sites are at an elevation of about 3,870 feet, or 80 feet beneath the lake that Adams suggests existed here at that time. In addition, a series of sites date

TABLE 8-5
Stillwater Marsh Archaeological Dates (in Italics) Compared with Dates for Late Holocene Lakes in the Carson Sink and the Hypothesized Diversions of Walker River

Site	Elevation (feet)	Age (years ago)	Reference	Hypothesized Walker River Diversions
CH1050-3	*3870*	*290 ± 80*	*Larsen and Kelly 1995*	
				Walker River diverted 500–300 years ago (Adams 2007)
CH1070-4	*3,870*	*660 ± 30*	*Larsen and Kelly 1995*	
Carson Lake	3,900	710 ± 70	King 1996	
CH1048	*3,870*	*800 ± 90*	*Raven and Elston 1988*	
Carson Lake	3,950	810 ± 70	Adams 2003	
L72-200	*3,870*	*820 ± 70*	*Larsen and Kelly 1995*	
CH1062	*3,870*	*830 ± 80*	*Kelly 2001*	
CH1048	*3,870*	*870 ± 70*	*Raven and Elston 1988*	Walker River diverted 1,360–1,000 years ago (Adams 2007)
CH1052	*3,870*	*1,040 ± 60*	*Raven and Elston 1988*	
CH1159-4	*3,870*	*1,080 ± 50*	*Larsen and Kelly 1995*	
CH1062	*3,870*	*1,100 ± 120*	*Kelly 2001*	
CH1044-2	*3,870*	*1,140 ± 80*	*Larsen and Kelly 1995*	
CH1068	*No Data*	*1,320 ± 100*	*Kelly 2001*	
CH1173	*3,870*	*1,350 ± 70*	*Larsen and Kelly 1995*	
CH1062	*3,870*	*1,390 ± 80*	*Kelly 2001*	
Carson Lake	3,920	1,510 ± 40	Adams 2003	
CH1055	*3,870*	*1,860 ± 70*	*Raven and Elston 1988*	
CH1052	*3,870*	*2,150 ± 90*	*Raven and Elston 1988*	Walker River diverted 2,700–2,100 BP (Benson, Meyers, and Spencer 1991)
CHC1159-200	*3,870*	*2,265 ± 70*	*Raven and Elston 1988*	
CH1052	*3,870*	*2,680 ± 160*	*Raven and Elston 1988*	
CH1052	*3,870*	*2,690 ± 70*	*Raven and Elston 1988*	
CH1062	*3,870*	*2,940 ± 70*	*Kelly 2001*	
CH1052	*3,870*	*3,190 ± 70*	*Raven and Elston 1988*	
CH1052	*3,870*	*3,290 ± 90*	*Raven and Elston 1988*	

to between 1,360 and 1,040 years ago, a time that has also been suggested to mark the diversion of the Walker River northward.

As a whole, the archaeological record from the Stillwater area does not match the hypothesized diversions of the Walker River in a compelling way, nor does it match the 700- to 800-year-old Carson Sink lake called for by Adams and King.

Then there is Lead Lake, the shallow lake cored by Peter Wigand that I mentioned above. This lake is also in the Stillwater Marsh area, at an elevation of 3,810 feet. It provided a sequence covering the past 2,300 years or so and shows no evidence of having been covered by a substantial body of water during that interval.

If you find all of this confusing, you are not alone. While effortful matching of dates and events in one area with those in another can bring parts of these sequences in line (as I did above, when I observed the 400-year gap between 2,680 and 2,265 years ago), our knowledge of the late Holocene history of these two areas simply does not match up well if we are trying to use the history of Walker River diversions to do the matching.

Some important caveats have to be stressed, though. First, the ages of the late Holocene lakes called for by Adams in the Carson Sink are not well controlled chronologically: he was able to obtain only one radiocarbon date each for these two lakes, though King's work may add a second for the more recent of the two. Second, although the archaeological chronology is far better—twenty-one dates for the past 3,300 years—the gaps that it shows may still be the result of inadequate sampling of the archaeological record. If we had more sites and more dates from them, those gaps might be filled in. As it stands, the late Holocene water history of this area is mainly confusing.

There is, of course, that other option—that climate, not the Walker River, explains the history of Walker Lake. Mono Lake, located about one hundred miles to the south of the Carson Sink, suggest exactly this possibility. Although I have talked about Scott Stine's history of Mono Lake during the Medieval Climatic Anomaly, he has also shown that this lake has had a series of at least six significant high, and intervening low, stands during the late Holocene. At the time his record begins, some 3,500 years ago, the surface of Mono Lake was at an elevation of about 6,499 feet, about 82 feet above the level it had prior to the diversion of its waters to southern California. The lake reached its lowest stand of the past 3,500 years between about 2,000 and 1,800 years ago, coinciding with the tail end of the drought called for by Jerry Miller and his colleagues. At that time, its surface stood at an elevation of about 6,368 feet, some 50 feet lower than its pre-diversion level.

Stine argues that the fluctuations of Mono Lake during the past few thousand years reflect climatic change. He even notes that increases in the levels of Mono Lake during this interval are correlated with fluctuations in sunspot activity during the past 2,000 years. Benson and his colleagues observe that there are general similarities in the histories of Mono and Walker lakes during the late Holocene: both were relatively high at 3,500 years ago and during the past few hundred years, and both were low at about 2,000 and at about 1,000 years ago. Platt Bradbury and his colleagues suggest that at least some of these similarities are climatically driven. If they are right, the history of the Walker River during these times may not matter, and the discrepancies between the history of Walker Lake and the history of lakes in the Carson Sink become far less important.

I mentioned above that there are two obvious ways to learn more about possible Walker Lake diversions through Adrian Valley. The first of these involves learning more about water history in the Carson Desert. We have just seen how far this doesn't get us. The second, though, involves looking directly at Adrian Valley, the channel that the Walker River would have used to flow northward.

Guy King has done all of the recent work here and he sides firmly with the climate-change camp. His geological research suggests that any diversions of the Walker River through this valley during the Holocene must have been of very short duration and might, at best, have augmented lakes that existed in the Carson Desert for climatic reasons, as opposed to creating them in the first place.

This, of course, still doesn't solve all of our problems. Even if the ups and downs of Walker Lake and of lakes in the Carson Desert were climatically driven, with the Walker River and Adrian Valley playing only a minor role, we simply can't have lakes in the Carson Sink above the heads of people who were living there at the same time. As it stands now, we have no answers to the late Holocene questions posed by these conflicting histories.

Great Basin Bison

When people think of bison, they think of the Great Plains, not the Great Basin, but these animals did call parts of the Great Basin home during the Holocene.

Where they may have been found historically depends on what kinds of records we are willing to accept. If we stick to eyewitness written accounts, we would conclude that the only part of the Great Basin where bison were found in early historic times was in northern Utah. On the other hand, Native American oral accounts also place the animals in northeastern California, south-central and eastern Oregon, and northeastern Nevada. The eastern Oregon and northeastern Nevada accounts are matched by discoveries of bison remains lying on the surface of the ground in these areas, and the same may be true for northeastern California. If we accept these records, then bison were found across the northern Great Basin.

The problem with accepting these oral histories, though, is that there is no way to assess the time to which they refer. The apparently corroborating records provided by bison remains on, or sticking out of, the ground do not help that much, since they may be ancient. As a result, and barring the discovery of additional early historic eyewitness accounts, it is to the archaeological and paleontological records that we must turn if we want to understand the distribution of bison in the Great Basin during the Holocene.

Table 8-6 provides a list of all prehistoric Holocene occurrences of bison from the Great Basin, or, more accurately, all of them of which I am aware, along with the dates associated with them. The distribution of all specimens with secure dates is shown in figure 8-21. The sites within the large ellipse are too numerous to be labeled separately, but dates for these sites are indicated in table 8-6. Unless otherwise indicated, the sites in this ellipse fall within the time span of the archaeological cultural tradition known as Fremont, which dates to between about 1,600 and 600 years ago and which I discuss in the following chapter. The four small ellipses indicate the areas in which oral tradition and surficial discoveries of bison remains have been taken by some to indicate the historic presence of these animals (an ellipse in south-central Oregon is nearly hidden by dated sites; the figure does not show the known historic distribution of bison in Utah).

This figure shows that bison were once widespread across the eastern and northern parts of the Great Basin. It also shows, however, that this distribution is primarily a late Holocene phenomenon. Nearly all of the records (87 percent of them, to be precise) than can be placed within a general chronological framework are late Holocene in age. And of those, nearly all that are precisely dated fall within the past 1,600 years (figures 8-22 and 8-23). This distribution also suggests that bison were not to be found in the southwestern Great Basin during the past 10,000 years, even though late Pleistocene bison are known from the Mojave Desert.

So the archaeological and paleontological data suggest that bison were widespread in much of the northern and eastern Great Basin during the very late Holocene but were far less common before then. It also suggests that bison may not have existed in the Great Basin outside of these areas after about 3,000 years ago.

TABLE 8-6
Prehistoric Bison in the Great Basin

The "Age" column places sites, or strata within sites, within chronological subdivisions of the Holocene. EH = early Holocene, 10,000 yrs.–7,500 yrs. BP; MH = middle Holocene, 7,500–4,500 yrs. BP; LH = late Holocene, 4,500 yrs. BP–latest prehistoric; NP = cannot be placed in the tripartite sequence.

The "Date" column provides uncalibrated radiocarbon dates; sites indicated as "F" in this column lack radiocarbon dates but are associated with the Fremont archaeological tradition, dated to 1600–600 yrs. BP.

NISP = number of identified specimens; MNI = minimum number of individuals. MNI values are labeled as such in column 4 and are given only when NISP values are unavailable (ND = no data; SK = skull). Cave and rockshelter sites are in italics; all others are open sites.

Site	Analytical Unit	Age	NISP or MNI	Date (years ago)	Reference
26Eu1320, Little Boulder Basin, NV	Assemblage	LH	1	170	Schroedl 1995
42Bo73, Great Salt Lake Basin, UT	Assemblage	LH	11	1,150–1,090	Lupo and Schmitt 1997; Coltrain and Stafford 1999; Simms 1999
42Bo1072, Great Salt Lake Basin, UT	Assemblage	LH	ND		Lambert and Simms 2003
42Sl197, Great Salt Lake Basin, UT	Assemblage	LH	40	1,380–1,130	Lupo and Schmitt 1997; Coltrain and Stafford 1999; Simms 1999
42Sl285, Great Salt Lake Basin, UT	Assemblage	LH	ND	F	Coltrain and Leavitt 2002
42Wb42, Great Salt Lake Basin, UT	Assemblage	LH	4	F and <600	Fawcett and Simms 1993; Lupo 1993
42Wb185, Great Salt Lake Basin, UT	Assemblage	LH	5	1,430–560	Lupo and Schmitt 1997; Coltrain and Stafford 1999; Simms 1999
42Wb304, Great Salt Lake Basin, UT	Assemblage	LH	ND	1000	Lupo and Schmitt 1997; Coltrain and Stafford 1999; Simms 1999
42Wb317, Great Salt Lake Basin, UT	Assemblage	LH	6	1,015–540	Lupo and Schmitt 1997; Coltrain and Stafford 1999; Simms 1999
42Wb331, Great Salt Lake Basin, UT	Assemblage	LH	2	<600	Fawcett and Simms 1993; Lupo 1993
Baker Village, Snake Valley, NV	Assemblage	LH	137	980–680	Wilde 1992; Hockett 1998
Bear River 1, Great Salt Lake Basin, UT	Assemblage	LH	1,798	1,065	Aikens 1966; Lupo and Schmitt 1997
Bear River 2, Great Salt Lake Basin, UT	Assemblage	LH	1,220	995	Madsen and Rowe 1988; Lupo and Schmitt 1997
Bear River 3, Great Salt Lake Basin, UT	Assemblage	LH	632	1,450	Shields and Dalley 1978; Lupo and Schmitt 1997
Beatty Springs, Goose Creek Mountains, UT	Assemblage	LH	2 (MNI)	2,350	Dalley 1976
Bonneville Estates, Lead Mine Hills, NV	Assemblage	MH	ND		Rhode et al. 2005; B. Hockett, pers. comm.
Bonneville Estates, Lead Mine Hills, NV	Assemblage	LH	ND		Rhode et al. 2005; B. Hockett, pers. comm.
Bronco Charlie Cave, Ruby Mountains, NV	Surface	NP	1		Spiess 1974
Camels Back Cave, Camels Back Ridge, UT	Stratum XVIIc	LH	3	790	Schmitt and Lupo 2005
Camels Back Cave, Camels Back Ridge, UT	Stratum XIV	LH	1	3,630–3,160	Schmitt and Lupo 2005

(continued)

TABLE 8-6 (CONTINUED)

Site	Analytical Unit	Age	NISP or MNI	Date (years ago)	Reference
Camels Back Cave, Camels Back Ridge, UT	Stratum IV	MH	1	7,350	Schmitt and Lupo 2005
Cathedral Gorge, NV	Assemblage	LH	208	810–450	Johnson et al. 2005
Catlow Cave, Catlow Valley, OR	Direct Dates	LH	2	440–405	Wilde 1985; Stutte 2004
Catlow Cave, Catlow Valley, OR	Stratum I	NP	24		Cressman 1942; Wilde 1985
Catlow Cave, Catlow Valley, OR	Stratum II	NP	4		Cressman 1942; Wilde 1985
Catlow Cave, Catlow Valley, OR	Stratum III	NP	1		Cressman 1942; Wilde 1985
Connley Cave 3, Fort Rock Basin, OR	Stratum 1	LH	1	3,080	Bedwell 1973; Grayson 1979; Jenkins, Aikens, and Cannon 2002; D. L. Jenkins, pers. comm.
Connley Cave 4, Fort Rock Basin, OR	Stratum 3	NP	8		Bedwell 1973; Grayson 1979; Jenkins, Aikens, and Cannon 2002; D. L. Jenkins, pers. comm.
Connley Cave 4, Fort Rock Basin, OR	Stratum 4	NP	19		Bedwell 1973; Grayson 1979; Jenkins, Aikens, and Cannon 2002; D. L. Jenkins, pers. comm.
Connley Cave 5, Fort Rock Basin, OR	Stratum 3	NP	12		Bedwell 1973; Grayson 1979; Jenkins, Aikens, and Cannon 2002; D. L. Jenkins, pers. comm.
Danger Cave, Silver Island Range, UT	Stratum DV	LH	8	3,950–0	Jennings 1957; Grayson 1988; Rhode and Madsen 1998; Rhode, Madsen, and Jones 2006
Danger Cave, Silver Island Range, UT	Stratum DIII	MH	1		Jennings 1957; Grayson 1988; Rhode and Madsen 1998; Rhode, Madsen, and Jones 2006
Danger Cave, Silver Island Range, UT	Stratum DII	EH	2	10,080–7,920	Jennings 1957; Grayson 1988; Rhode and Madsen 1998; Rhode, Madsen, and Jones 2006
Dirty Shame Rockshelter, Owyhee Plateau, OR	Zone 4	MH	4	6,845	Aikens, Cole, and Stuckenrath 1977; Grayson 1977b
Dirty Shame Rockshelter, Owyhee Plateau, OR	Zone 5	NP	6	7,925–7,850	Aikens, Cole, and Stuckenrath 1977; Grayson 1977b
Dry Creek Ranch, Simpson Park Mountains, NV	Assemblage	LH	1 (SK)		Hall 1961; Jefferson, McDonald, and Livingston 2004
Ephraim, Sanpete Valley, UT	Assemblage	LH	ND	F	Gillin 1941
Five Finger Ridge, Clear Creek Canyon, UT	Assemblage	LH	32	840–650	Talbot et al. 2000; Fisher 2010
Fort Rock Cave, Fort Rock Basin, OR	Stratum 1	NP	5		Grayson 1979
Garrison, Snake Valley, UT	Assemblage	LH	ND	F	Taylor 1954
Gatecliff Shelter, Toquima Range, NV	Stratum 1, Horizon 2	LH	1	650	Grayson 1983; Thomas 1983b
Gatecliff Shelter, Toquima Range, NV	Strata 3–5, Horizon 5	LH	1	3,200–1,250	Grayson 1983; Thomas 1983b
Gatecliff Shelter, Toquima Range, NV	Strata 3–5, Horizon 6	LH	1	3,200–1,250	Grayson 1983; Thomas 1983b

TABLE 8-6 (CONTINUED)

Site	Analytical Unit	Age	NISP or MNI	Date (years ago)	Reference
Gilbert Peak, Uinta Mountains, UT	Assemblage	LH	1 (SK)	150	Cannon 2004
Goshen Island South, Utah Lake, UT	Assemblage	LH	8	<600	J. C. Janetski, pers. comm.
Hanging Rock Shelter, Hanging Rock Canyon, NV	Organic-Yellow	NP	2		Grayson and Parmalee 1988
Harney Dune, Harney Basin, OR	Direct Date	LH	1 (SK)	250	Raymond 1994; Stutte 2004
Helmet Crawl Cave, American Fork Canyon, UT	Assemblage	NP	ND		Miller 2002
Heron Springs, Utah Lake, UT	Assemblage	LH	13	650–440	Billat 1985; Janetski 1990; Lupo and Schmitt 1997
Hidden Cave, Eetza Mountain, NV	No Provenience	NP	1		Grayson 1985
Hinckley Mounds, Utah Lake Basin, UT	Assemblage	LH	ND	1,265–920	Green 1961; Lupo and Schmitt 1997
Hogup Cave, Hogup Mountain, UT	Stratum 16	LH	2 (MNI)	500–250	Aikens 1970; Durrant 1970
Hogup Cave, Hogup Mountain, UT	Stratum 14	LH	2 (MNI)	1,210–620	Aikens 1970; Durrant 1970
Hogup Cave, Hogup Mountain, UT	Stratum 13	LH	1 (MNI)	F	Aikens 1970; Durrant 1970
Hogup Cave, Hogup Mountain, UT	Stratum 12	NP	4 (MNI)		Aikens 1970; Durrant 1970
Hogup Cave, Hogup Mountain, UT	Stratum 10	NP	1 (MNI)		Aikens 1970; Durrant 1970
Hogup Cave, Hogup Mountain, UT	Stratum 9	NP	2 (MNI)		Aikens 1970; Durrant 1970
Hogup Cave, Hogup Mountain, UT	Stratum 8	NP	3 (MNI)		Aikens 1970; Durrant 1970
Hogup Cave, Hogup Mountain, UT	Stratum 7	MH	2 (MNI)	6,190	Aikens 1970; Durrant 1970
Hogup Cave, Hogup Mountain, UT	Stratum 6	MH	1 (MNI)	6,400–5,960	Aikens 1970; Durrant 1970
Hogup Cave, Hogup Mountain, UT	Stratum 5	MH	2 (MNI)	7,250–5,795	Aikens 1970; Durrant 1970
Hogup Cave, Hogup Mountain, UT	Stratum 4	EH	1 (MNI)	7,815	Aikens 1970; Durrant 1970
James Creek Shelter, Marys Mountain, NV	F2	LH	2	750	Budy and Katzer 1990; Grayson 1990
James Creek Shelter, Marys Mountain, NV	Horizon III	LH	15	1,250–750	Budy and Katzer 1990; Grayson 1990
James Creek Shelter, Marys Mountain, NV	Horizon II-KX	LH	5	MIXED: 1,240–240	Budy and Katzer 1990; Grayson 1990
Juke Box Cave, Silver Island Range, UT	Assemblage	NP	ND		Jennings 1957
Juniper Lake, Alvord Desert, OR	Direct Date	LH	4	370–150	Stutte 2004
Kachina Cave, Smith Creek Canyon, NV	Stratum 4	LH	ND	1,350	Miller 1979; Tuohy 1979; Mead, Thompson, and Van Devender 1982

(*continued*)

TABLE 8-6 (CONTINUED)

Site	Analytical Unit	Age	NISP or MNI	Date (years ago)	Reference
King's Dog, Surprise Valley, CA	KIV (Alkali Phase)	LH	2	1,330	O'Connell 1975; James 1983; O'Connell and Inoway 1994; J. F. O'Connell, pers. comm.
King's Dog, Surprise Valley, CA	KII (Bare Creek Phase)	LH	3	3,010–2,690	O'Connell 1975; James 1983; O'Connell and Inoway 1994; J. F. O'Connell, pers. comm.
King's Dog, Surprise Valley, CA	KI (Menlo Phase)	MH	>14	5,640	O'Connell 1975; James 1983; O'Connell and Inoway 1994; J. F. O'Connell, pers. comm.
Knoll, Great Salt Lake Basin, UT	Assemblage	LH	54	640	Fry and Dalley 1979; Lupo and Schmitt 1997
Levee, Great Salt Lake Basin, UT	Assemblage	LH	624	1,250–710	Fry and Dalley 1979; Lupo and Schmitt 1997
Lost Dune, Blitzen Valley, OR	Assemblage	LH	1 (SK)[a]	450–260	Lyons and Mehringer 1996; Lyons and Cummings 2002; Stutte 2004
Malheur Lake, OR	Direct Dates	LH	30 (SK)[b]	400–250	Stutte 2004
Marysvale, Sevier River Valley, UT	Assemblage	LH	ND	F	Gillin 1941
Median Village, Parowan Valley, UT	Assemblage	LH	2	1,050–990	Marwitt 1970
Mosquito Willie, western Bonneville Desert, UT	Stratum 3, Test Unit 6	LH	1	>2,280	Arkush 1998; Janetski 2004
Nawthis, Fishlake Plateau, UT	Assemblage	LH	9	1,075–790	Sharp 1992
Nightfire Island Lower Klamath Lake, CA	Stratum 9	LH	12	4,190–3,200	Sampson 1985
Nightfire Island Lower Klamath Lake, CA	Stratum 8	LH	5	4,190–4,055	Sampson 1985
Nightfire Island Lower Klamath Lake, CA	Stratum 6	LH	9	4,950–4,350	Sampson 1985
Nightfire Island Lower Klamath Lake, CA	Stratum 5	LH	2	4,630–4,075	Sampson 1985
Nightfire Island, Lower Klamath Lake, CA	Stratum 4	NP	2	ca. 4,950	Sampson 1985
O'Malley Shelter, Clover Valley, NV	Unit 1	MH	2	7,100–6,520	Fowler, Madsen, and Hattori 1973
Oranjeboom Cave, Goshute Mountains, NV	Assemblage	LH	6	1,220–1,060	Buck et al. 2002
Orbit Inn, Great Salt Lake Basin, UT	Assemblage	LH	29	570–300	Lupo and Schmitt 1997; Coltrain and Stafford 1999; Simms 1999
Paisley Cave 2, Summer Lake Basin, OR	Direct Date	LH	1	845	Stutte 2004
Parowan Canyon, UT	Assemblage	NP	1 (SK)		Presnall 1938
Peninsula, Warner Valley, OR	Structure 1	LH	1	625–240	Moore 1995; Eiselt 1998
Pharo Village, Scipio Valley, UT	Assemblage	LH	18	760–690	Marwitt 1968; Lupo and Schmitt 1997

TABLE 8-6 (CONTINUED)

Site	Analytical Unit	Age	NISP or MNI	Date (years ago)	Reference
Porcupine Cave, Uinta Mountains, UT	Bridge Junction	NP	1		Haman 1963; Heaton 1988
Promontory Cave, Salt Lake Basin, UT	Assemblage	LH	20	1,310–840	Aikens 1966; Marwitt 1970; Lupo and Schmitt 1997
Pyramid Lake Fishway 1016, Pyramid Lake Basin, NV	Assemblage	LH	1	3,015	Clark and Anderson 1979
Roaring Springs Cave, Catlow Valley, OR	Assemblage	NP	1		Cressman 1942; Wilde 1985
Roaring Springs Cave, Catlow Valley, OR	Direct Dates	LH	2	480–410	Wilde 1985; Stutte 2004
Rock Springs, Curlew Valley, ID	Bone Bed 1	LH	120	Historic	Arkush 2002
Rock Springs, Curlew Valley, ID	Bone Bed 2	LH	212	290	Arkush 2002
Rock Springs, Curlew Valley, ID	Bone Bed 3	LH	224	370	Arkush 2002
Rock Springs, Curlew Valley, ID	Bone Bed 4	LH	112	730	Arkush 2002
Rock Springs, Curlew Valley, ID	Bone Bed 5	LH	213		Arkush 2002
Rock Springs, Curlew Valley, ID	Bone Bed 6	LH	52		Arkush 2002
Rock Springs, Curlew Valley, ID	Bone Bed 7	LH	12	840	Arkush 2002
Sandy Beach, Utah Lake, UT	Assemblage	LH	1	510–450	Billat 1985; Janetski 1990; Lupo and Schmitt 1997
Skull Creek Dunes, Catlow Valley, OR	Direct Date	LH	6	450	Wilde 1985, Stutte 2004
Smoking Pipe, Utah Lake Basin, UT	Midden	LH	1,831	890–350	Billat 1985; Janetski 1990; Lupo and Schmitt 1997
Snake Rock Village, Castle Valley, UT	Assemblage	LH	1	F	Aikens 1967
Spotten Cave, Goshen Valley, UT	Zone 3	LH	1 (MNI)	1,310–730	Cook 1980; Lupo and Schmitt 1997
Spotten Cave, Goshen Valley, UT	Zone 5	LH	1 (MNI)		Cook 1980; Lupo and Schmitt 1997
Susie Creek, Adobe Range, NV	Assemblage	LH	1 (SK)	950	Van Vuren and Dietz 1993
Swallow Shelter, Goose Creek Mountain, UT	Stratum 10	LH	3 (MNI)	to 600	Dalley 1976
Swallow Shelter, Goose Creek Mountain, UT	Stratum 9	LH	1 (MNI)	1,120	Dalley 1976
Swallow Shelter, Goose Creek Mountain, UT	Stratum 8	LH	1 (MNI)		Dalley 1976
Swallow Shelter, Goose Creek Mountain, UT	Stratum 6	LH	1 (MNI)		Dalley 1976
Swallow Shelter, Goose Creek Mountain, UT	Stratum 5	LH	1 (MNI)	2,630	Dalley 1976

(continued)

TABLE 8-6 (CONTINUED)

Site	Analytical Unit	Age	NISP or MNI	Date (years ago)	Reference
Swallow Shelter, Goose Creek Mountain, UT	Stratum 4	LH	2 (MNI)	2,850	Dalley 1976
Swallow Shelter, Goose Creek Mountain, UT	Stratum 3	LH	1 (MNI)	3,500	Dalley 1976
Tooele, Great Salt Lake Basin, UT	Assemblage	LH	ND	F	Gillin 1941
Tosawihi Quarry (26Ek3092), Antelope Creek, NV	Assemblage	NP	1		Elston and Raven 1992; Schmitt 1992
Tosawihi Quarry (26Ek3200), Antelope Creek, NV	Assemblage	NP	1		Elston and Raven 1992; Schmitt 1992
Trego Hot Springs (26Pe118), Black Rock Desert, NV	Assemblage	LH	217	3,040–1,120	Dansie 1980; Seck 1980
Wallman Bison, Black Rock Desert, NV	Assemblage	NP	ca. 20		Dansie, Davis, and Stafford 1988
Warren, Great Salt Lake Basin, UT	Assemblage	LH	75	1,180	Lupo and Schmitt 1997; Coltrain and Stafford 1999; Simms 1999
Wells Dump, Humboldt River Basin, NV	Assemblage	LH	ND	750	Hockett and Morgenstein 2003
Wells NE, Humboldt River Basin, NV	Assemblage	NP	1 (SK)		Van Vuren and Dietz 1993
Willard Mounds, Great Salt Lake Basin, UT	Assemblage	LH	94	1,250–690	Lupo and Schmitt 1997; Coltrain and Stafford 1999; Simms 1999
Woodruff Bison Kill, Wyoming Basin, UT	Assemblage	LH	1,150	1,335	Shields 1978; Lupo and Schmitt 1997

^aAdditional uncounted specimens present.
^bThirty skulls have been reported from Malheur Lake, of which three have been dated (Stutte 2004).

Earlier in this chapter, I discussed the difficulties that may be involved in assessing the abundance of large mammals in the Great Basin during the early Holocene. Similar difficulties are involved in assessing the abundances of bison through time in the Great Basin. Most generally, human population densities were lower in the Great Basin during the early and middle Holocene than in later times (chapter 9), so there were fewer people to create archaeological sites with bison in them. In addition, since the large size of bison presents transportation problems for hunters and gatherers, the bony residues of bison hunting can be expected to be found most frequently in open sites. Buried open sites are rare for the early Holocene Great Basin and not much more common for the middle Holocene, becoming frequent only after that time. This, I suspect, is one of the prime reasons why the vast majority of the bison records from open sites come from late Holocene archaeological contexts. All this suggests caution. While we are not likely to get in scientific trouble by examining the changing abundances of Great Basin bison during the late Holocene, our record for these animals simply is not adequate for the earlier parts of this epoch.

THE WESTERN, NORTHERN, AND CENTRAL GREAT BASIN

As I noted, Native American oral histories and the presence of surficial bison remains suggest that four areas of the Great Basin supported bison in early historic times. Archaeological and paleontological data, however, confirm their presence in only one of these areas, the Malheur and Harney basins of south-central Oregon. Here, and thanks largely to the work of archaeologist Nicole Stutte, the presence of very late prehistoric bison has been well documented.

A date for bison from the King's Dog site in Surprise Valley, California, falls at 1,330 years ago, but Jim O'Connell, who excavated the site, thinks the two bison specimens from

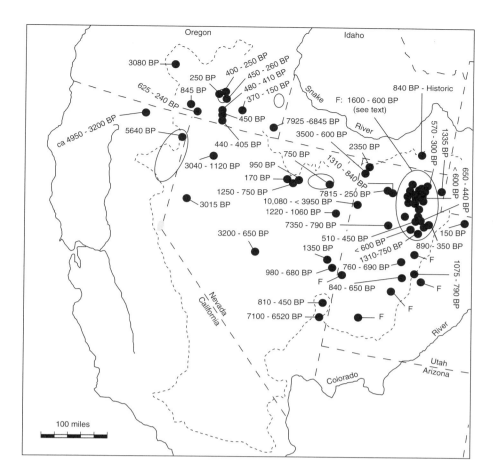

FIGURE 8-21 Distribution of dated prehistoric bison remains from the Great Basin. The ellipses in California, Oregon, and Nevada mark areas in which bison are thought to have occurred during latest prehistoric or early historic times based on oral tradition or the discovery of surficial bison remains, or both. The historic distribution of bison in Utah is not shown. (After Grayson 2006b.)

this level of the site were likely displaced upward from much earlier deposits where bison remains were far more common. Just as important, recent, intensive work on several dozen archaeological sites in this area, including locations specifically mentioned by various scholars as having supported bison historically—for instance, the Madeline Plains and Honey Lake Valley—did not provide a single specimen of this animal.

To the immediate north, in far south-central Oregon, a single specimen of bison has been reported from one of the Paisley Caves (figure 3-2). Stutte had this specimen dated and found it to be 845 years old. A single bison specimen is also known from the Peninsula site in the Warner Valley, an area mentioned by ethnographer Isabel Kelly as the scene of late prehistoric bison hunting. This specimen was associated with basketry that dates to 240 years ago. These are the only late prehistoric specimens of bison reported from this general area. Since this region has also been the focus of substantial archaeological work, it seems unlikely that the near lack of records for bison in this area reflects inadequate work by archaeologists. In nearby northwestern Nevada, only one site, on the southeastern edge of the Black Rock Desert, has provided late Holocene bison remains. While bison specimens are plentiful here, their stratigraphic position is unclear but it appears very unlikely that they are latest Holocene in age. In short, while there is no reason to question the accuracy of the oral histories suggesting the presence of bison in northeastern California and immediately adjacent Oregon and Nevada, the archaeological and paleontological records provide very little support for the suggestion that those histories refer to very recent times.

Much the same can be said for northeastern Nevada. Here, the most recent securely dated and identified bison remains date to 750 years ago. The situation for far eastern Oregon (which lies outside the Great Basin) is somewhat different. Here, there are no late Holocene records for bison at all, but this is an area that has seen relatively little archaeological and paleontological work and the absence of documented late prehistoric bison from this region may simply reflect this fact. Bison remains dating to between

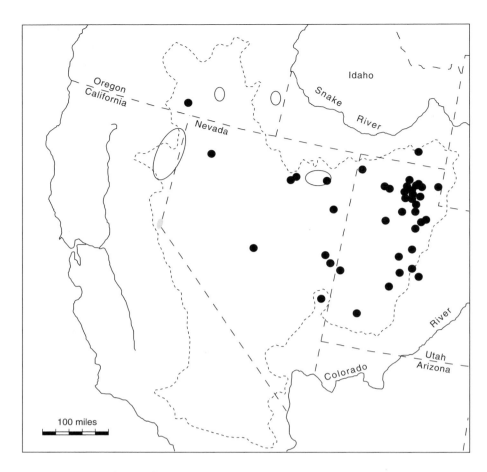

FIGURE 8-22 Distribution of Great Basin sites with bison that date to between 1,600 and 600 years ago (after Grayson 2006b).

2,000 and 150 years ago are known from the western Snake River Plain, to the northeast of the possible eastern Oregon historic records.

In short, there were, without question, bison in south-central Oregon during early historic times. Although bison may also have been in northeastern California, far eastern Oregon, and northeastern Nevada at the same time, we cannot show that this was the case.

THE EASTERN GREAT BASIN

If we were to draw a line encircling those parts of far eastern Nevada and western Utah that have failed to provide late Holocene bison remains, the results would provide a good approximation of the arid basin of Pleistocene Lake Bonneville (figures 5-5 and 8-21). Elsewhere in this area they seem to have been fairly common during this period.

Most, though not all, bison-bearing sites in this area are associated with the Fremont archaeological phenomenon and date to about 1,600 to 600 years ago (chapter 9; see figure 8-22). A wide variety of paleobotanical analyses in the eastern Great Basin and adjacent Colorado Plateau shows that this period saw increased summer temperatures, increased summer moisture, and an increased abundance of grasses. It was this set of changes that allowed the increased abundance of bison so evident in Fremont-aged sites.

This fits well with another fact. The people who lived here before the Fremont were hunters and gatherers, but the Fremont themselves were horticulturalists, growing, among other things, maize. Maize requires summer warmth, summer moisture and a lengthy growing season. Odds are that the same climatic conditions that allowed maize horticulture also led to the abundance of bison here at the same time, as I discuss in the following chapter.

I mentioned that people did not grow corn in the eastern Great Basin before Fremont times. They also did not do so afterward. And at the same time as people here were growing less and less corn, bison seem to have become less and less abundant. As Karen Lupo and Dave Schmitt observed some time ago, bison seem to have become far less common on the eastern Great Basin landscape by about six hundred years ago—about the same time as the Fremont disappeared. A broad variety of scientists have observed that this happens at about the time as summer temperature, summer precipitation, and grass abundance declines. All of this seems, in turn, to be explained by the fact that,

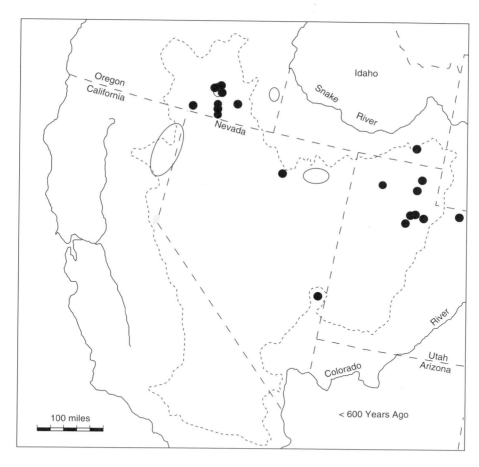

FIGURE 8-23 Distribution of Great Basin sites with bison that postdate 600 years ago (after Grayson 2006b).

during Fremont times, higher summer temperatures led to strengthened monsoonal storms here, intensified versions of the same storms that today have their greatest impact on the southern and eastern portions of the Great Basin.

Although the evidence we have is not good enough to allow us to construct a truly precise history of bison abundances in the eastern Great Basin, the decreasing abundance of bison here may have begun with the droughts of the Medieval Climatic Anomaly. As Larry Benson, Ken Peterson, and John Stein have observed, severe droughts in the Southwest at about 1150 and 1300 A.D. may have been associated with a weakened summer monsoon. The first of these droughts coincides with a decrease in maize consumption by Fremont people. The second coincides with the end of Fremont itself. All these events coincide, in turn, with a decrease in bison abundance in the eastern Great Basin. Bison were still found in the Great Salt Lake area in early historic times, but they were not nearly as abundant or as widespread as they were 1,000 years earlier, when strengthened monsoonal storms provided them with far friendlier environmental conditions.

While the history of monsoons helps us explain the history of bison in the eastern Great Basin, it cannot help with the northern Great Basin of south-central Oregon, where summer monsoonal rainfall is relatively unimportant. Indeed, the evidence we have suggests that while summer moisture may have increased in the northern Great Basin at about the same time it increased in the Bonneville area, that increase was far less pronounced in the former area than in the latter. This difference may explain why there is so little evidence for bison in this area between 1,600 and 600 years ago, compared with the great abundance of such evidence in the eastern Great Basin. Instead, the increase in the abundance of these animals here after 600 years ago (figure 8-23) may be related to the increased winter moisture and cooler temperatures that marked this region during the Little Ice Age that I discussed earlier.

No matter what the cause of this increase, however, it is not clear why late Holocene-aged bison remains are nearly absent from the far northwestern corner of the Great Basin. In spite of intensive archaeological work here during the past few decades or so, there is only a single late Holocene record for bison from this region. This is especially perplexing since no obvious biogeographic barriers prevent the movement of bison between the Malheur and Harney basins on the one hand and the Fort Rock Basin on the other.

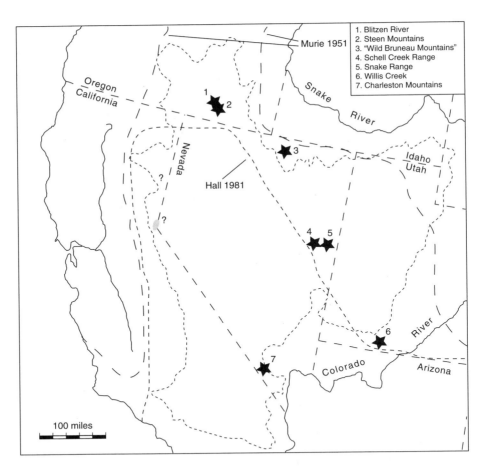

FIGURE 8-24 Early historic distribution of elk in the Great Basin according to Murie (1951) and Hall (1981). The numbered stars show the location of early historic records for elk in the Great Basin (from Bailey 1936, E. R. Hall 1946, and Murie [1951]. Unverified newspaper records from Lake Tahoe and Honey Lake Valley are indicated by question marks (Nevada Division of Wildlife 1997).

SOME BISON NEIGHBORS

As we have just seen, bison were widespread in the eastern and northern parts of the Great Basin during the late Holocene, particularly after 1,600 years ago. It is unlikely to be coincidental that archaeologists Mark Plew and Taya Sundell found exactly the same chronological pattern for the Snake River Plain of southern Idaho. They also observed that the relatively small numbers of bison specimens found in the archaeological sites of this area suggest the existence of small herds of these animals, as opposed to the huge agglomerations that characterized the Great Plains during early historic times. Much the same appears to have been the case for the eastern and northern Great Basin. And, it is certainly worth noting that the Great Basin is not the only part of the Arid West to have supported bison prehistorically. As paleontologists Jim Mead and Catherine Johnson have shown, they were also found in Arizona during, and perhaps throughout, the Holocene; they provide a record from far southwestern Arizona, just north of the Mexican border, that dates to very late prehistoric or early historic times.

A Note on Elk

Scientists who map the distribution of elk in North America follow one of two classic maps. The first of these was produced in 1951 by biologist Olaus J. Murie, whose map of the distribution of North American elk excludes nearly all of the Great Basin from elk territory. His map has been accepted by many.

Figure 8-24 shows his elk-territory boundaries for the area that included the Great Basin. It also shows the historic records for elk in this area. The real oddity in these historic records is the one for southern Nevada, in the Spring Mountains. Murie provided this record based on information he had received from famed biologist C. Hart Merriam. Although some recent authors have also mentioned this report, most simply assume it is wrong, which it almost undoubtedly is.

An elk map that is just as influential as Murie's differs in significant ways from it. This map first appeared in 1959, in an important synopsis of the mammals of North America by mammalogists E. R. Hall and K. R. Kelson. It appeared again, with minor modifications, in 1981 in the second

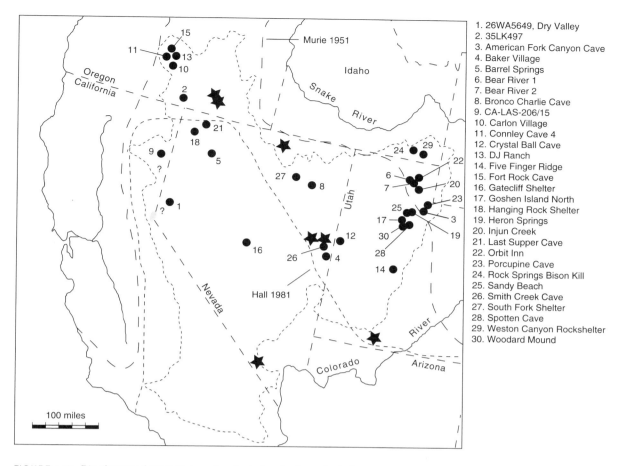

FIGURE 8-25 Distribution of Great Basin archaeological and paleontological sites that have provided elk remains. Stars indicate early historic records (see figure 8-24).

version of this synopsis by Hall alone. As figure 8-24 shows, it places elk in the northern part of the Great Basin and in nearly all of Utah. This version of elk distribution has also been followed by many later scientists.

Both maps are correct. In making his version, Murie excluded records he considered to be marginal, perhaps simply reflecting "stragglers." As a result, he essentially excluded the Great Basin from the normal territory of this animal. Hall and Kelson, on the other hand, included all records for elk, drawing the outer edges of elk territory on the nether side of those marginal extremes. This is one of those cases in which differences between scientists are based on deep and well-reasoned considerations of the information with which they were working. You can take your pick, just as later scientists have done. No matter which map these scientists have picked, all agree that elk were never common in the Great Basin, a view with which those archaeologists who have examined the issue agree.

It is the archaeological and paleontological records for this species that tell us how abundant elk were, or were not, in the Great Basin prior to the European arrival. Table 8-7 presents all of these records; the locations of the sites involved are shown in figure 8-25.

This table shows that the available chronology for the history of elk in the Great Basin is very weak. Few records can be placed in time with reasonable precision, but most of those that can be so assigned date to the late Holocene in general, and to the past 1,000 years in particular.

Table 8-7 also shows that the perception that elk have never been common in the Great Basin is correct. These animals have been reported with certainty from only twenty-one sites that can be securely placed in the Holocene Great Basin. The twenty-seven sites that report precise counts of identified bones and teeth for elk have provided only 131 such specimens—an average of only 4.9 per site, even though some of these sites record significant amounts of time. All this leaves little doubt that elk were never common in the Great Basin.

Finally, both the Murie and Hall and Kelson maps can be supported by these distributions. On the one hand, no sites with more than one elk specimen fall significantly outside of the Hall and Kelson boundary. As I said, Hall and Kelson drew their maps to include marginal records and the prehistoric record suggests that they were very successful in doing so for elk. On the other hand, if the goal of a distribution map is to indicate those places in which

TABLE 8-7

Prehistoric Great Basin Sites Containing the Remains of Elk (Sites Shown in Figure 8-25)

LW = late Wisconsin. EH = early Holocene (10,000 yrs.–7,500 yrs. BP). MH = middle Holocene (7,500–4,500 yrs. BP). LH = late Holocene (4,500 yrs. BP–latest prehistoric). ND = cannot be placed in the tripartite sequence. NP = No provenience. N = the number of identified specimens (NISP) or the minimum number of elk individuals in the assemblage (MNI). MNI are counts in bold. If no counts were provided in the original publication, the "+" simply indicates that they were present. For the meaning of "cf.", see note to table 7-1. Artifacts made from elk skeletal material are not included in the list.

Site and Excavation Level	Elevation (feet)	Age	N	Date (years ago)	References
26WA5649, Dry Valley, NV	4,660	LH	1[a]		Carpenter 2002, pers. comm.; K. R. McGuire pers. comm.
35LK497	4,325	ND	1	—	Pettigrew 1985
American Fork Canyon Cave, Wasatch Range, UT	6,200	ND	+	—	Hansen and Stokes 1941; Hall 1983; Miller 2002
Baker Village, Snake Valley, NV	5,260	LH	1 cf.	980–680	Hockett 1999; Wilde and Soper 1999
Barrel Springs, Kamma Range, NV	4,550	LH	1	—	Thomas 1972a
Bear River 1, eastern Great Salt Lake Valley, UT	4,215	LH	2[b]	1,065	Lupo and Schmitt 1997
Bear River 2, eastern Great Salt Lake Valley, UT	4,215	LH	1[c]	995	Lupo and Schmitt 1997
Bronco Charlie Cave, Ruby Mountains, NV: 210–215 cm	6,600	ND	1	—	Spiess 1974
CA-LAS-206/15, Secret Valley, CA	4,470–4,560	LH	1		Carpenter 2002, pers. comm.; K. R. McGuire pers. comm.
Carlon Village, Fort Rock Basin, OR	4,340	LH	1	1,820–590	Jenkins 1994; Wingard 2001
Connley Cave 4, Fort Rock Basin, OR: Stratum 3	4,450	ND	8	—	Grayson 1979
Connley Cave 4, Fort Rock Basin, OR: Stratum 4	4,450	ND	13	—	Grayson 1979
Connley Cave 5, Fort Rock Basin, OR: Stratum 3	4,450	ND	3	—	Grayson 1979
Connley Cave 6, Fort Rock Basin, OR: Stratum 2	4,450	LH	1	—	Grayson 1979
Crystal Ball Cave, Snake Valley, UT	5,775	LW	1 cf.[d]	—	Heaton 1985; Hockett and Dillingham 2004
DJ Ranch, Fort Rock Basin, OR, Upper Block	4,320	ND	2	—	Singer 2004
Five Finger Ridge, Clear Creek Canyon, UT	6,000	LH	5[e]	840–650	Janetski 1997; Talbot et al. 2000
Fort Rock Cave, Fort Rock Basin, OR, Stratum 1	4,445	ND	1	—	Cressman 1942; Bedwell 1973, Grayson 1979
Gatecliff Shelter, Toquima Range, NV: Stratum 1, H1	7,610	LH	1	550–470	Thomas 1983b
Goshen Island North, Utah Lake, UT	4,490	LH	2	<600	Nauta 2000; Creer, Van Dyke, and Newbold 2007; Janetski and Smith 2007
Hanging Rock Shelter, Hanging Rock Canyon, NV: Stratum 2/4	5,660	ND	2	—	Grayson and Parmalee 1988
Hanging Rock Shelter, Hanging Rock Canyon, NV: Stratum 4	5,660	ND	18	—	Grayson and Parmalee 1988
Hanging Rock Shelter, Hanging Rock Canyon, NV: Stratum 5	5,660	EH	14	—	Grayson and Parmalee 1988

TABLE 8-7 (CONTINUED)

Site and Excavation Level	Elevation (feet)	Age	N	Date (years ago)	References
Heron Springs, Utah Lake, UT	4,490	LH	6	650–440	Janetski 1990; Creer and Van Dyke 2007b; Janetski and Smith 2007
Injun Creek, eastern Great Salt Lake Valley, UT	4,210	LH	5[f]	585–345	Aikens 1966, Lupo and Schmitt 1997
Last Supper Cave, Hell Creek Canyon, NV, Midden	5,395	ND	6	—	Grayson 1988
Last Supper Cave, Hell Creek Canyon, NV, Stratum 2	5,395	ND	1	—	Grayson 1988
Last Supper Cave, Hell Creek Canyon, NV, Stratum 3	5,395	ND	1	—	Grayson 1988
Last Supper Cave, Hell Creek Canyon, NV, NP	5,395	ND	1	—	Grayson 1988
Orbit Inn, Great Salt Lake Basin, UT	4,220	LH	3	570–300	Lupo and Schmitt 1997
Porcupine Cave, Uinta Mountains, UT	9,300	ND	2	—	Haman 1963; Heaton 1988
Rock Springs Bison Kill, Curlew Valley, ID, Bone Bed 3	5,445	LH	1	370	Arkush 2002; Walker 2002
Sandy Beach, Utah Lake, UT	4,490	LH	8	510–450	Janetski 1990; Creer and Van Dyke 2007a; Janetski and Smith 2007
Smith Creek Cave, Snake Range, NV, Grey Silt et al.	6,400	ND	2	—	Miller 1979; Mead, Thompson, and Van Devender 1982
Smith Creek Cave, Snake Range, NV: Reddish-Brown	6,400	LW	>2	—	Miller 1979; Mead, Thompson, and Van Devender 1982
South Fork Shelter, South Fork, Humboldt River, NV: 30–36 in.	5,100	LH	3	—	Heizer, Baumhoff, and Clewlow 1968; Spencer et al. 1987
South Fork Shelter, South Fork, Humboldt River, NV, 48–54 in	5,100	LH	1	—	Heizer, Baumhoff, and Clewlow 1968; Spencer et al. 1987
South Fork Shelter, South Fork, Humboldt River, NV: 54-80 in.	5,100	LH	6	—	Heizer, Baumhoff, and Clewlow 1968; Spencer et al. 1987
Spotten Cave, Goshen Valley, UT: Zone 5	4,810	LH	2	—	Cook 1980
Weston Canyon Rockshelter, Weston Canyon, ID, Strata 4–5	5,215	LH	3	—	Miller 1972; Arkush 1999b
Weston Canyon Rockshelter, Weston Canyon, ID, Strata 6–8	5,215	LH	3	3,740	Miller 1972; Arkush 1999b
Weston Canyon Rockshelter, Weston Canyon, ID, Stratum 9	5,215	ND	1	—	Miller 1972; Arkush 1999b
Weston Canyon Rockshelter, Weston Canyon, ID, Strata 12–13	5,215	ND	1	—	Miller 1972; Arkush 1999b
Weston Canyon Rockshelter, Weston Canyon, ID, Strata 14–15	5,215	ND	1	—	Miller 1972; Arkush 1999b
Weston Canyon Rockshelter, Weston Canyon, ID, Strata 16–17	5,215	MH	2	7,300–7,200	Miller 1972; Arkush 1999b
Weston Canyon Rockshelter, Weston Canyon, ID, NP	5,215	ND	5	—	Miller 1972; Arkush 1999b
Woodard Mound, Utah Lake, UT	4,540	LH	4	—	Cook 1980

[a]Carpenter (2002) mentions three specimens of elk from the Tuscarora Pipeline and Alturas Transmission Line projects. Two of these were from the former project, but although Holanda (2000) notes one specimen of elk from the Alturas Project, the individual reports for this project provide no evidence for elk (McGuire 2000a, 2000b), suggesting that this might have been a typographic error carried over into Carpenter (2002).
[b]NISP given as 173 in Aikens (1966); reidentified by Lupo and Schmitt (1999).
[c]NISP given as 9 in Aikens (1967); reidentified by Lupo and Schmitt (1999).
[d]Hockett and Dillingham (2004) note that this tentatively identified specimen may pertain to the extinct genus *Navahoceros*.
[e]Fisher (2010) found no elk at this site.
[f]NISP given as 1 in Aikens (1966); reidentified by Lupo and Schmitt 1999

a species occurred more than occasionally, the Murie map meets that goal, even if the Utah boundary might be moved westward.

The Extinct Noble Marten

In 1926, mammalogist E. R. Hall described a new subspecies of American marten (*Martes americana*) on the basis of skeletal remains from two caves in northern California. Many years passed before paleontologist Elaine Anderson showed that there were consistent differences between the known remains of the marten Hall had described and the skeletons of living species of martens. In 1970, she assigned the paleontological material to the extinct species *Martes nobilis*, the noble marten. Since then, paleontologists have debated whether this form should be treated as a separate species or as a subspecies of the American marten (*Martes americana nobilis*). Most now suggest that it is probably best treated at the subspecies level.

Today, the noble marten is known from late Pleistocene sites that range from the Great Plains west through California and north to the Yukon. Two of these—Nevada's Smith Creek Cave and Snake Creek Burial Cave—are in the Great Basin. It is, however, also known from three sites from in or near the Great Basin that suggest that it may have survived here until at least 3,000 years ago or so.

Just outside of the Great Basin, Dry Creek Rockshelter, in southwestern Idaho's Boise River Valley, provided a noble marten specimen in deposits dated to about 3,300 years ago. Within the Great Basin, a single tooth was uncovered at Hidden Cave in deposits about 3,600 years old. And, at Bronco Charlie cave, on the eastern flank of the Ruby Mountains, yet another specimen came from deposits laid down between 3,500 and 1,200 years ago. All three sites together suggest that the noble marten may have existed in the Great Basin until some 3,000 years ago.

Until we learned of these specimens, it was assumed that the noble marten became extinct at the end of the Pleistocene. While it would certainly be valuable to obtain radiocarbon dates directly on all of them to make sure that they really are as young as their stratigraphic position suggests, the fact that they come from three separate sites suggests that this animal survived quite late in the Arid West. If this is correct, we are left with another obvious question: why was this animal able to survive the major environmental changes that occurred at the end of the Pleistocene and during the middle Holocene, only to become extinct as more modern conditions arrived?

Late Holocene Environmental Variability

The detailed records available for the late Holocene indicate impressive environmental variability through time and across space within the Great Basin. We do not have equally diverse records of fine-scale change for earlier Holocene times, a reflection of the fact that it is far easier to extract detailed paleoenvironmental records for more recent times than for more ancient ones. Where we do have sensitive records for the early and middle Holocene—as at Fish Lake, Diamond Pond, and Blue Lake, for instance—they also show marked variability. Earlier Holocene times were surely characterized by fine-scale change as much as the past few thousand years were so characterized. And, since these were the environments to which the prehistoric peoples of the Great Basin were adapted, we should expect to see significant spatial and temporal variability within the archaeological record as well. That record is the focus of the following chapter.

Chapter Notes

The title of this section is taken, with much respect, from a book by the same name by Paul S. Martin (1963).

Details of the late Pleistocene and Holocene history of Las Vegas Valley, including the black mat sequences for southern Nevada in general, are found in C. V. Haynes (1967), Mehringer (1967), Quade (1986, 1994), Quade and Pratt (1989), Quade et al. (1995, 1998), and Quade, Forester, and Whelan (2003); see also the notes to chapters 6 and 7. The dates I use for Las Vegas Valley Units E_1 and E_2 are from Quade et al. (1998). Quade (1986:350) suggested that the most recent early Holocene date for a black mat in the Las Vegas Valley was likely to have been contaminated by modern rootlets. C. V. Haynes (1991) argues that an episode of drought may have coincided with the extinction of at least some Pleistocene mammals. However, he has suggested to me that he hesitates to place the erosional break between Las Vegas Valley units E_1 and E_2 in this context, because he thinks it is possible that tectonic activity, not climatic change, may account for that break.

The evidence for early Holocene lakes in the Lake Mojave Basin is analyzed by W. J. Brown et al. (1990) and S. G. Wells et al. (2003); the chronology I use for these lakes is from the latter source. Quade, Forester, and Whelan (2003) discuss the relationship between Lake Mojave Intermittent Lake III and Unit E_2 in Las Vegas Valley. Late Holocene lakes in the Lake Mojave Basin are discussed by Enzel et al. (1989, 1992). References for Owens Lake history are provided in chapter 5; the key references I have used here include Bacon et al. (2006) and Orme and Orme (2008).

My discussion of late Pleistocene and early Holocene streams in the Old River Bed is based on Oviatt, Madsen, and Schmitt (2003); Schmitt et al. (2007); Madsen, Oviatt, and Young (2009); and lengthy discussions with David Madsen. K. D. Adams et al. (2008) provide a valuable synthesis of our knowledge of the early Holocene water history of the Lahontan Basin.

On the Ruby Marshes, see R. S. Thompson et al. (1990) and R. S. Thompson (1992). Huckleberry et al. (2001) present the results of their work at the Sunshine Locality in Long Valley. Four of the six dates extracted by these researchers for the early Holocene pond or wetland deposit fall between 10,060 and 9,820 years ago. They worry that because the two younger dates for this deposit (9,040 and 8,560 years ago) are conventional (that is, non-AMS) radiocarbon dates on organic sediment, they may be too young. However, they accept the conventional date on organic sediment for the overlying sands—8,120 years ago—and, in the absence of evidence suggesting that any of these dates are problematic, I accept them here.

For the Malheur Basin, see Gehr and Newman (1978), Gehr (1980), MacDowell (1992), and Dugas (1998). If you want to visit the Malheur National Wildlife Refuge, first visit www.fws.gov/malheur/. This is a spectacular place, not far from the Steens Mountain Wilderness Area. Unless you are going during the cold season, bring mosquito repellent. The Connley Caves depositional sequence is discussed in Bedwell (1973) and D. L. Jenkins, Aikens, and Cannon (2002); the former should not be read without the latter. The Connley Caves faunal data are discussed in Grayson (1979). The Lake Abert radiocarbon date I mention in the text is from Gehr (1980); Licciardi (2001) and Negrini (2002) discuss this date as well. For Walker Lake, see the references in chapter 6. The initial work on the Dietz Basin was done by Willig (1988, 1989); see also C. V. Haynes (1991). The reanalysis of the history of this basin is provided by Pinson (2008).

The information I provide on the distribution and adaptations of netleaf hackberry is from Lanner (1984); DeBolt and McCune (1995); Wang, Jahren, and Amundsen (1997); and J. M. Hunt, Rhode, and Madsen (2000); this last reference also provides a detailed discussion of the Homestead hackberry history. The Gillespie Hills midden is discussed in Rhode (2000a); for the 8,100-year date, see Wang, Jahren, and Amundsen (1997). Madsen and Currey (1979) present the Snowbird Bog pollen sequence, which has more recently been reviewed by Madsen (2000b) and Wigand and Rhode (2002). Louderback and Rhode (2009) present the Blue Lake sequence; on the Danger Cave limber pine, see Rhode and Madsen (1998) and Wigand and Rhode (2002). The Hidden Cave pollen record is discussed by Wigand and Mehringer (1985) and Wigand and Rhode (2002). See Nowak et al. (1994b) for the Painted Hills midden data.

On modern *Larrea*, see K. L. Hunter et al. (2001), Lia et al. (2001), and Hunziker and Comas (2002). Van Devender and Spaulding (1979), Spaulding (1981) and Van Devender (1990a) present the dates for the Sheep Range, Rio Grande Village, and Artillery Mountains samples I mention here. The Rampart Cave sloth dung analyses are in P. S. Martin, Sabels, and Shutler (1961) and R. M. Hansen (1978). The Tinajas Altas and Butler Mountain dates are in Van Devender (1990b); for Picacho Peak, see Cole (1986). Van Devender (1990a) discusses the strong likelihood that creosote bush was in the Chihuahuan Desert during full glacial times; see also K. L. Hunter et al. (2001). Betancourt et al. (2001) and K. L. Hunter et al. (2001) discuss the relationships between pollen and plant macrofossils in southern North America deserts, with the latter reference focusing specifically on creosote bush. The plant macrofossils from the Organ Pipe packrat middens are presented by Van Devender (1987) and O. K. Davis (1990); the pollen from these contexts is discussed by O. K. Davis and Anderson (1987), Van Devender (1988), and O. K. Davis (1990). While Van Devender (1988) is critical of certain aspects of Davis and Anderson's arguments, he does not criticize their contention that *Larrea* must have been in the Organ Pipe area long before the midden plant macrofossils suggest it was here.

References to early Holocene packrat middens in the Mojave Desert are provided in table 8-2. For the Eyrie, Owl Canyon 1, and Blue Diamond Road 3 middens, see Spaulding (1981). Spaulding (1980) argued for the possibility of glacial-age refugia for creosote bush within the Mojave Desert but supports a Holocene arrival in Spaulding (1990a, 1990b).

Table 8-2 cites the North American Packrat Midden Database (NAPMD), which is maintained jointly by the U.S. Geological Survey (USGS) and the National Oceanographic and Atmospheric Administration (NOAA). The database itself is found at http://esp.cr.usgs.gov/data/midden/. The details of packrat midden analyses can be stored here, including the exact location of the midden, the dates and identified taxa for each sample in that midden, and references to relevant publications. As a result, you can look up all packrat midden records for pinyon pine in North America that have been entered into the database, or all records for this species in Nevada, or all Nevada records in a particular elevational range. At the moment, doing the second of these searches provides a rich haul of midden records for pinyon pine. Not all midden scientists enter their data in this database (not even all who say in their publications that they are going to do so), but this is a valuable research tool nonetheless.

Spaulding (1980, 1981) provides the Basin Canyon midden data; on the Last Chance Range sequence, see Spaulding (1985). Other middens containing early, and in some cases middle, Holocene sagebrush specimens are also discussed in these references— Eureka Valley 4 (6,795 years old) and Marble Mountains 9(1) (at 8,925, 8,905, and 5,520 years ago), for instance. R. S. Thompson (1985) analyzes differences between the pollen and macrofossil content of packrat middens; see the notes to chapter 6 for other important references on the relationship between pollen and plant macrofossils in packrat middens. Koehler, Anderson, and Spaulding (2005) provide their central Mojave Desert midden data; Spaulding (1994:47) discusses the abundance of sagebrush pollen in the Fortymile Canyon middens.

Early Holocene records for neatleaf hackberry in the Mojave Desert are reviewed in Spaulding (1994) and Jahren et al. (2001); Spaulding (1994:46–47) discusses the paleoclimatic significance of these records. The Little Skull mountain record is in Forester et al. (1999); my comment that hackberry remains are common here is based on information provided by Peter Wigand in the North American Packrat Midden Database. Wigand (1997b) provides the paleobotanical record from Pintwater Cave; the general information I present on this site is from Buck et al. (1997).

On the Homestead Cave mammals, see Grayson et al. (1996), Grayson (1998, 2000a, 2000b), and Madsen (2000b). For Camels Back Cave, see Schmitt, Madsen, and Lupo (2002a, 2002b); Schmitt and Lupo (2005); and Schmitt and Madsen (2005); for Pintwater Cave, B. S. Hockett (1997, 2000) and Buck et al. (1997); for Danger Cave, Grayson (1988), and, for a discussion of stratigraphic difficulties at this site, Rhode, Madsen, and Jones (2006); for Lovelock Cave, Livingston (1988b); for Hogup Cave, Aikens (1970) and Durrant (1970); for Gatecliff, Grayson (1983) and Thomas (1983b).

Broughton et al. (2008) provide their analysis of Great Basin artiodactyl history. On pulse averaging and pulse matchers, see Kenagy and Bartholomew (1985). Gestation periods for cervids, including deer, are in Geist (1998); those for kangaroo rats, in W. T. Jones (1993). The Homestead Cave fecal pellet data are in J. M. Hunt, Rhode, and Madsen (2000). Pinson (2007) presents her data on northern Great Basin archaeological faunas. (If you want to analyze these data on your own, note that her table 10.3 has an easily corrected mistake for two of the early Holocene sites; none of the early Holocene sites she discusses contains identified artiodactyl material; see Greenspan 1993). Paleontologist Björn Kurtén

(1958) was the first to argue that cave bears (*Ursus spelaeus*) might have become extinct because of competition with people over caves. He later withdrew his support for this possibility (Kurtén 1968), but subsequent research has strongly suggested that he was correct (Grayson and Delpech 2003, Wolverton 2006). B. S. Hockett (2007) discusses the Bonneville Estates Rockshelter faunal material from early and middle Holocene strata and shows those strata are dominated by large mammal specimens. He has confirmed to me that these are from artiodactyls that could not be, or have yet to be, identified to the genus level.

On the predicted relationships between the earth and sun, and the meaning of those relationships for the earth's climate, see Kutzbach and Guetter (1986), COHMAP Members (1988), and Kutzbach et al. (1998). The 9,600 radiocarbon (or 11,000 calendar) year date that I use here for the insulation maximum is from Kutzbach et al. (1998); see also Ruter et al. (2004).

Spaulding and Graumlich (1986) argue for enhanced monsoonal rainfall in the Mojave Desert during the early Holocene. Van Devender (1990b) disagrees strongly, while Jahren et al. (2001) present evidence supporting Spaulding and Graumlich's position for the Pintwater Cave area. Harvey, Wigand, and Wells (1999) compare early Holocene alluvial fan development in the Carson Sink and the basin of Pleistocene Lake Mojave. S. G. Wells et al. (1989) suggest that Intermittent Lake III formed in response to winter storms moving eastward from the Pacific. Poore, Pavich, and Grissino-Mayer (2005) use foraminifera from Gulf of Mexico cores to reconstruct monsoonal history in the Southwest. O. K. Davis and Shafer (1992) analyze the pollen sequence from Montezuma Well. R. S. Thompson (1984, 1990) presents his reconstructions of early Holocene climates in the floristic Great Basin.

A substantial literature points to a warm and dry early Holocene in the Sierra Nevada. This literature is briefly reviewed in G. J. West et al. (2007) and includes O. K. Davis et al. (1985), R. S. Anderson (1990a), S. J. Smith and Anderson (1992), Anderson and Smith (1994), and O. K. Davis and Moratto (1988).

My biographical discussion of Ernst Antevs has been drawn from Smiley (1977), C. V. Haynes (1990), and the website given in chapter 5. Rhode (1999) provides a valuable analysis of Antevs's work in the Great Basin. For brief introductions to varves, see Lowe and Walker (1984) and Ehlers (1996). Antevs's work on the chronology of varved lakes in New England has been shown to be impressively accurate (see, for instance, Ridge and Larsen 1990). Antevs (1925) provides the scientific results of his first foray into the Great Basin; his Neothermal climatic model is developed in Antevs (1938, 1948, 1952a, 1952b, 1953a, 1953b, 1955). A full list of Antevs's publications, many of which deal with the Neothermal model, is provided by Smiley (1977) and in the website mentioned earlier. The salt chronologies used by Antevs were provided by Van Winkle (1914) and Gale (1915). The first person to suggest using this approach for estimating the age of bodies of water that had no outlet was Edmund Halley, of Halley's comet fame, in 1715.

LaMarche and Mooney (1967, 1972) discuss their work in the Snake Range and their initial work on the ages of White Mountains treelines. LaMarche (1973) presents his detailed analysis of those treelines. For the modern responses of subalpine conifers to warming temperatures, see Millar et al. (2004, 2006). References for the Ruby Marshes are provided above.

Not everyone has identified the Steens Mountain fir as grand fir; I follow Critchfield and Allenbaugh (1969) and Lanner (1983b) in doing so here. As Lanner explains, these trees appear to be hybrids between grand and white fir but are "more grand than white" (1983b:80). Mehringer (1985, 1986) discusses the results of his work at Fish Lake and Wildhorse Lake. Wigand and Rhode (2002) provide a valuable synopsis of that work. The Diamond Pond analysis is presented in Wigand (1985, 1987) and in Mehringer and Wigand (1990); see also Mehringer and Wigand (1987).

Wigand and Nowak (1992) provide the Virginia Mountain packrat midden radiocarbon dates, including those from the Painted Hills; see also the discussion in Wigand and Rhode (2002). References to Hidden Cave are provided above; see G. R. Smith (1983) on the Hidden Cave fishes. J. R. Miller, House, et al. (2004) discuss the Holocene history of the middle reaches of the Humboldt River. Benson, Kashgarian, et al. (2002) provide a detailed analysis of the Holocene history of Pyramid Lake; Mensing et al. (2004) discuss the Holocene vegetation sequence they extracted from the cores used by Benson, Kashgarian, et al. (2002).

The remarkable story of the drowned Tahoe trees can be followed in Harding (1965); Lindström (1990, 1996); Benson, Kashgarian, et al. (2002); and Mensing et al. (2004). Benson and Thompson (1987a) suggested that tectonic activity might account for the drowned trees. See Schweikert et al. (2000) for a discussion of possible explanations for these trees, including tectonic activity; like all others, these authors conclude that climatic change is by far the most likely explanation. Furgurson (1992) discusses Lindström's work briefly and provides an excellent photo of drowned tree stumps. J. O. Davis, Elston, and Townsend (1976) present their analysis of possible Lake Tahoe lake levels during the middle Holocene. The Lily Lake fire history is provided by Beaty and Taylor (2009).

References for Homestead Cave, Camels Back Cave, Blue Lake, and Snowbird Bog are provided above. Currey (1980) discusses the possibility that the Great Salt Lake desiccated the middle Holocene times; Murchison (1989a, 2000) provides the general middle Holocene Great Salt Lake history that I review here. For Mosquito Willie's, see Kiahtipes (2009). The key references on the middle Holocene desiccation of Owens Lake include Benson, Burdett, et al. (1997); Benson, Kashgarian, et al. (2002); Bischoff, Stafford, and Rubin (2007); and Bacon et al. (2006). Other studies whose results mirror the ones I discuss here include R. C. Bright (1966); Byrne, Busby, and Heizer (1979); R. S. Thompson (1984); and Mehringer (1985); for a more detailed review, see Wigand and Rhode (2002).

The Little Lake sequence is provided by Mehringer and Sheppard (1978) and the Tule Springs pollen analyses by Mehringer (1967). Wigand (2002) and Wigand and Rhode (2002) note that the mid-Holocene increase in the abundance of low-spine composite pollen in Mehringer's Tule Springs profiles likely reflects an increase of white bursage on the landscape. P. S. Martin (1963) argued for a warm and moist middle Holocene in the Sonoran Desert. Van Devender (1990b) presents his arguments for strengthened monsoons in the Sonoran Desert during the middle Holocene; Spaulding (1991) disagrees at the same time as he argues for a warm and dry middle Holocene in southwestern Nevada. Spaulding (1985:44–45) presents evidence for increased moisture in southern Nevada at the very end of the middle Holocene and at the beginning of the late Holocene. The Dry Lake sequence is provided by Bird and Kirkby (2006).

Antevs's use of Summer Lake history can be seen in the references to his work provided above; I have taken his estimate of the age of Mazama ash from Antevs (1948). See I. S. Allison (1945, 1966b) on Summer Lake ashes, and Feth (1959) on Great Basin salt chronologies. The Fort Rock Basin work is presented by Mehringer and Cannon (1994), Droz (1997), D. L. Jenkins (2000), and D. L. Jenkins, Droz, and Connolly (2004). Bryan and Gruhn (1964) provide their critique of Antevs's Neothermal model; T. A. Minckley, Bartlein, and Shinker (2004) make an identical argument. Aschmann (1958) also critiqued that model, though he essentially dismissed in its entirety the notion of an Altithermal. See C. V. Haynes (1990) for a discussion of Antevs's geological analyses of archaeological sites.

References to the pinyon pine locations discussed in the text are provided in table 8-4. On the Gatecliff pollen record, see R. S. Thompson and Kautz (1983); Thomas (1983b) provides a wealth of information on the site as a whole. The role of corvids in dispersing pinyon pine seeds has been analyzed in many places; my discussion depends on Vander Wall and Balda (1977) and Lanner (1983a, 1996); see also Madsen (1986a). By far the best general introduction to the dispersal of pines by birds is provided by Lanner (1996). That pinyon did not enter much of the Great Basin until well into the Holocene was not fully recognized until the 1980s, but Madsen and Berry (1975) had discussed this very possibility long before.

Mehringer (1986) suggested that people may have played a significant role in moving pinyon seeds across the Great Basin. Rhode (1990) provides a brief review of the distances that people are known to have transported pinyon pine in the Great Basin in early historic times; the fifty-mile figure I give comes from this source, calculated by Rhode (1990) from information provided by Steward (1938:178). If you are interested in learning more about issues relating to the human transport of such items as pinyon nuts, see Rhode (1990) and Metcalfe and Barlow (1992). Madsen and Rhode (1990) and Rhode and Madsen (1998) discuss the Danger Cave pinyon pine in detail, including the possibility that it was introduced by people. If you read the former paper, be sure to read the latter one, since it clarifies the differentiation of limber and pinyon pine macrofossils from this site. On Danger Cave in general, see J. D. Jennings (1957), Grayson (1988), Rhode et al. (2005), and Rhode, Madsen, and Jones (2006). Data on the modern distribution of pinyon pine in Utah are taken from Albee, Shultz, and Goodrich (1988).

R. S. Thompson and Hattori (1983), R. S. Thompson (1984, 1990), Wigand and Nowak (1992), Wigand (2002), and Wigand and Rhode (2002) have all suggested that increased summer rainfall may have allowed the northward expansion of singleleaf pinyon. C. L. Nowak et al. (1994b) discuss possible physiographic and climatic constraints to the dispersal of pinyon pine toward the northwestern limits of its current distribution. The rainfall figures that I have used for singleleaf and Colorado pinyon come from Cole et al. (2008), including the electronic appendix to that source.

The Gatecliff small mammals are discussed by Grayson (1981, 1983, 1987, 2006a). B. R. Butler (1972) discusses the Wasden site sequence. Larrucea and Brussard (2008b) is required reading on the relationship between pygmy rabbits and pinyon pine. On the Holocene history of pygmy rabbits in Washington State, see Lyman (1991, 2004b).

Rodgers (1982) and Hersh (2000) analyze matched photographs of Great Basin landscapes. Black mat formation in Salt Spring Basin is discussed by K. C. Anderson and Wells (2003); evidence for late Holocene lake resurgence in the Mojave Basin is provided by Tchakerian and Lancaster (2002). On the history of western juniper in the Diamond Pond area, see especially Mehringer and Wigand (1990).

Evidence for a drought between 2,500 and 1,900 years ago along the middle Humboldt River and elsewhere in central Nevada is presented by J. R. Miller et al. (2001, 2004). Stine (1990) and O. K. Davis (1999a) present the Holocene history of Mono Lake; Madsen et al. (2001) discuss the significance of the Homestead Cave Stratum XV fish remains; Terry (2008) provides additional radiocarbon dates for this stratum. The 3,440-year date I mention here is from Murchison (1989a). Wigand and Rhode (2002) present the Lower Pahranagat Lake data as well as a comparison with the Diamond Pond record.

H. H. Lamb (1965) introduced the term "Medieval Warm Epoch"; Stine (1994) substituted "Medieval Climatic Anomaly"; Kington (2008) provides a review of Lamb's remarkable career. Harding (1965) discusses the Mono Lake drowned trees; he had been alerted to these by Donald B. Lawrence, who obtained the radiocarbon date that Harding reported (Lawrence and Lawrence 1961). Details on Sidney Harding's accomplishments as a civil engineer are provided in Harding (1967), available online at http://openlibrary.org/a/OL2173117A; see pages 500–503 for his discussion of his work on recent Great Basin lake history. Stine (1990, 1994, 1998) discusses these trees and their implications in detail, along with those from the Walker River Canyon and a number of other places, including, in his 1998 paper, Owens Lake and Walker Lake; the later Mono Lake work to which I refer is by N. E. Graham and Hughes (2007). Graumlich (1993), Leavitt (1994), Graumlich and Lloyd (1996), Hughes and Graumlich (1996), and Millar et al. (2006) provide the details on the Sierran and White Mountains sequences I discuss here. Kleppe (2005) presents the Fallen Leaf Lake drowned tree data. He does not indicate whether the 1215 A.D. date he provides is a calibrated or radiocarbon date; if the latter, it is closer to 1270 A.D. in real years. See Yuan et al. (2004) for the late Holocene history of Walker Lake, and Benson and Kashgarian et al. (2002) and Mensing et al. (2004) on Pyramid Lake. Broughton (2000b); Broughton, Madsen, and Quade (2000) and Madsen et al. (2001) present evidence for increased Great Salt Lake levels about 1,000 years ago. S. T. Gray, Jackson, and Betancourt (2004) and S. T. Gray et al. (2006) present their Uinta Basin analyses; see Benson, Peterson, and Stein (2007) for the Southwest. E. R. Cook et al. (2004) reconstruct changing aridity in western North America as a whole during the past 1,200 years, observing the four intense droughts that I mention here.

There is a very substantial literature on the Medieval Climatic Anomaly (aka the Medieval Warm Period). If you are interested in learning more, including possible explanations for it, the best place to start is N. E. Graham et al. (2007). Crowley (2000); Crowley and Lowery (2000); Broecker (2001); D. D. Jones, Osborn, and Briffa (2001); R. S. Bradley, Hughes, and Diaz (2003); P. D. Jones and Mann (2004); and Woodhouse (2004) also provide excellent discussions of this interval. Mann et al. (2009) provide an important global assessment of the both the Medieval Climatic Anomaly and the Little Ice Age.

Matthes (1939) introduced the term "Little Ice Age"; Matthes (1941) provided an early analysis of the late Holocene glaciers of the Sierra Nevada, suggesting that they dated to the sixteenth through nineteenth centuries. D. H. Clark and Gillespie (1997) and A. R. Gillespie and Zehfuss (2004) provide details on the Matthes glacial episode. LaMarche (1973, 1974) discusses Little

Ice Age treelines and climates in the White Mountains. Graumlich (1993) and Lloyd and Graumlich (1997) provide the Sierran tree-ring analyses. Feng and Epstein (1994) provide the isotope analysis of White Mountains bristlecones; for more on this approach, see Augusti, Betson, and Schleucher (2008). On late Holocene lakes in the Mojave Basin, see Enzel et al. (1989, 1992); on the Ash Meadows sequence, Mehringer and Warren (1976). The story of the Homestead Cave bushy-tailed woodrats, both modern and ancient, is found in Grayson et al. (1996) and Grayson and Madsen (2000). The Diamond Pond and Lower Pahranagat Lake analyses I refer to here are in Wigand (1997a). Briffa, Jones, and Schweingruber (1992) provide the tree-ring analysis of the western United States as a whole that I refer to here. Stahle et al. (2007) also provide a very important discussion of droughts in North America, including the West, since 1300 A.D., supporting the evidence for variable Little Ice Age climates presented by Briffa, Jones, and Schweingruber (1992), as well as others.

The most authoritative and detailed introduction to the Little Ice Age is provided by J. M. Grove (2004), but Fagan (2000) provides a beautifully written introduction as well. For an equally accessible, and much shorter, introduction to the Little Ice Age and its possible causes, see A. T. Grove (2008). My discussion of the global nature, and causes, of this period draws on the two Grove references, as well as on Haigh (1994), Crowley (2000), Bard et al. (2001), Shindell et al. (2001), P. D. Jones and Mann (2004), Crowley et al. (2008), G. H. Denton and Broecker (2008), Wanner et al. (2008), Zhang et al. (2008), Licciardi et al. (2009), and Mann et al. (2009). On Tambora, see the discussion by Oppenheimer (2003).

The Walker Lake sequence is developed by Bradbury, Forester, and Thompson (1989); Benson, Meyers, and Spencer (1991); and K. D. Adams (2007). K. D. Adams (2003) discusses late Holocene lakes in the Carson Sink. The results of the Stillwater Marsh excavations are presented in Raven and Elston (1988); the Stillwater archaeological survey is discussed in Raven and Elston (1989) and Raven (1990); data on the freshwater molluscs, fish, and birds from these sites are provided by Drews (1988), Greenspan (1988), and Livingston (1988a, 1991). C. S. Larsen and Kelly (1995) and R. L. Kelly (1999, 2001) provide detailed discussions of the archaeology of this area. Wigand and Rhode (2002) provide information on the Lead Lake core. Stine (1984, 1990) presents his analysis of the late Holocene history of Mono Lake; since Stine (1990) gives his results in terms of calendar, not radiocarbon, years, his dates differ in a minor way from the ones I provide here. G. Q. King (1978, 1993, 1996) provides nearly all that we know about the water history of Adrian Valley. See R. L. Kelly (2007:68) for a different examination of the relationship between water levels and archaeology in the Carson Sink; he gets similar results but provides a more optimistic interpretation.

Late Holocene highstands of the Great Salt Lake are discussed by Currey (1987, 1990); Merola, Currey, and Ridd (1989); Murchison (1989a); Oviatt and Miller (1997); and Madsen et al. (2001); comparable highstands in the Sevier Lake Basin are discussed by Oviatt (1988).

See Grayson (2006b) for a more detailed discussion of the Holocene history of bison in the Great Basin, and Lupo (1996) and Lupo and Schmitt (1997) for important discussions of bison in the eastern Great Basin. Native American accounts of bison in this area are provided by Merriam (1926), I. T. Kelly (1932), Bailey (1936), Steward (1938), Riddell (1952, 1960), and Van Vuren and Bray (1985). For matching surficial records of bison, see Bailey (1936), Riddell (1952), Van Vuren and Bray (1985), Van Buren and Dietz (1993), and Verts and Carraway (1998). Evidence for increased summer warmth and summer precipitation in the eastern Great Basin during Fremont times is provided by Hemphill and Wigand (1995), Newman (1996), Wigand (1997a), Rhode (2000), Wigand and Rhode (2002), and Kiahtipes (2009). Salzer (2000), among many others, discusses the climates required by maize horticulture. Benson, Peterson, and Stein (2007) discuss the latest Holocene climatic history of the Southwest, including the history of monsoonal incursions here. See Plew and Sundell (2000) on Snake River Plain bison, and J. I. Mead and Johnson (2004) for the Southwest. References on the Fremont archaeological phenomenon are provided in the following chapter.

On prehistoric elk in the Great Basin, see Grayson and Fisher (2009); on prehistoric bison, Lupo (1996), Lupo and Schmitt (1997), and Grayson (2006b). Murie (1951), E. R. Hall and Kelson (1959), and E. R. Hall (1981) provide the maps I discuss here. Later scientists who have followed the Murie map include Bryant and Maser (1982) and O'Gara and Dundas (2002); those who have adopted the E. R. Hall (1981) map include Zeveloff and Collett (1988) and Verts and Carraway (1998). Although outside the area covered here, McCorquodale (1985), Dixon and Lyman (1996), and Lyman (2004a, 2004c) discuss the prehistoric distribution of elk in the Columbia Plateau, north of the Great Basin; for the Southwest, see Truett (1996). Archaeologists who have observed that elk seem to have been very scarce in the prehistoric Great Basin include Grayson (1988) and Janetski (2006).

As an aside, I note that the term "elk" is used in English-speaking Europe to refer to *Alces alces*, which Americans refer to as the moose. Because of this, North American *Cervus elaphus* has often been referred to as "wapiti." Here, I follow current formal usage (Wilson and Cole 2000, Wilson and Reeder 2005) and retain the term "elk" for this animal, just as almost everybody else does. To make things even more potentially confusing, *Cervus elaphus* also occurs in Europe, where it is referred to as red deer. Bryant and Maser (1982) and Geist (1998) provide interesting historical reviews of these terms.

Since I am on this topic, most people living in pronghorn territory refer to these animals as pronghorn. Real antelope, however, are Old World animals that belong to a different family, which doesn't usually stop me from calling a pronghorn an antelope when I see one. In case all this reminds you of the terms "bison" and "buffalo," I point out that "bison" refers to animals that belong to the genus *Bison*, "buffalo" to certain animals of the Asian genus *Bubalus* (water buffalo, for instance) or the African genus *Syncerus* (the African buffalo, *Syncerus caffer*).

On the noble marten, see E. R. Hall (1926), E. Anderson (1970), E. M. Mead and Mead (1989), R. W. Graham and Graham (1994), Grayson (1984b, 1985, 1987) and Meyers (2007). Youngman and Schueler (1991) argued that the noble marten does not differ sufficiently from the American marten to be treated as a separate species. Elaine Anderson (1994) responded that the extinct form is morphologically distinct but that it might be best to treat it as a subspecies of American marten (*Martes americana nobilis*). A statistical comparison of the skulls of the extinct form with those of living martens led paleontologist Jeff Meyers (2007) to agree with this assessment at the same time that he confirmed the distinct nature of the noble marten. Analysis of the genetics of this form would be most useful, but Meyers's work has gone far to clarify the relationships of the noble marten to other New World members of the genus.

PART FIVE

GREAT BASIN ARCHAEOLOGY

CHAPTER NINE

The Prehistoric Archaeology of the Great Basin

Pre-Clovis Sites in the Great Basin

The Clovis archaeological phenomenon provides the earliest widespread evidence we have for people in North America (chapter 4). Dating to between 11,200 and 10,800 years ago, Clovis is best known from the Plains and Southwest and is marked by very distinctive lanceolate, concave-based fluted projectile points (figure 4-3). These distinctive stone points are also famous for having frequently been found associated with the remains of extinct Pleistocene mammoths.

Claims of pre-Clovis sites have been made for most parts of North America. The Great Basin is no exception. Not only does this region contain the Calico site and Fort Rock Cave, the latter with a 13,200-year date from the very bottom (chapter 3), but it also contains a wide variety of other sites that have been claimed to be pre-Clovis in age. All but one of these has been soundly rejected by virtually all professional archaeologists.

Tule Springs, in southern Nevada, provides an excellent example of the rejected sites. Although the site was at one time argued to contain archaeological materials more than 28,000 years old, substantial excavations done in the 1960s showed conclusively that the artifacts here date to no more than about 11,000 years ago. The same excavations led to the important research (discussed in chapters 6 and 8) by Peter Mehringer and Vance Haynes, among others, on the latest Pleistocene and Holocene environments of the Las Vegas Valley.

While the Great Basin is not exceptional in having a significant number of sites that have been held by some to be pre-Clovis in age, it is alone in North America in containing one site that clearly does predate Clovis: the Paisley Caves, discussed in chapter 3, dated to about 12,350 years ago (figures 3-2 and 9-1).

Even though this is the case, it is also true that the Great Basin archaeological record is entirely silent between the time people stopped at Paisley to do what they did there and about 11,200 years ago. Then, the archaeological record begins again, and it begins with a richness that is in some important ways hard to match in other parts of North America.

The Latest Pleistocene and Early Holocene

Great Basin Fluted Point Sites

Fluted points are fairly common in the Great Basin. Some of these are fluted from base to tip and look very much like Folsom points from the Great Plains and Southwest, which date to between 10,900 and 10,200 years ago (see chapter 3). However, most Great Basin fluted points tend to look more like classic Clovis points than like anything else. As a result, the term "Clovis" was at one time routinely applied to both the points and to the sites that contain them, and sometimes still is. Doing that, however, almost automatically implies that the Great Basin examples are the same age as those from elsewhere—that is, between 11,200 and 10,800 years ago. This is a jump that many Great Basin archaeologists are unwilling to make.

There is a very good reason for that. Archaeologists Charlotte Beck and Tom Jones have analyzed a large series of Great Basin fluted points in great detail, measuring just about everything on these points that could be measured and comparing the results to classic Clovis points. Their results confirmed what many had suspected but had never shown: that many Great Basin fluted points are quite different from those that mark Clovis. They are often shorter and thinner and have a wider and deeper (more concave) base. These results suggested to them that most of these points are likely to be younger than their Clovis relatives in the Plains and Southwest, perhaps appearing late in Clovis times or even contemporary with Folsom. To mark

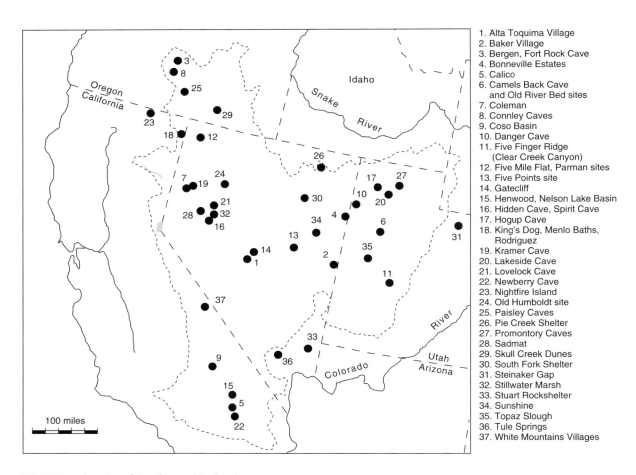

FIGURE 9.1 Location of sites discussed in this chapter.

this important difference, I simply refer to them as Great Basin fluted points (figure 9-2).

Fluted points are known from most, but not all parts of the Great Basin. In fact, it is easier to specify the major areas from which they are poorly known—the high-elevation valleys of central Nevada and the Bonneville Basin—than to list the places that have provided them in some number. This is not to say that the people who made these points did not visit these areas. William Davis and his colleagues, for instance, have described a fluted point site in the southern end of Tule Valley, Utah, a spot that might have overlooked a declining Pleistocene Lake Bonneville if this undated site had been occupied at the right time (figure 5-8). But fluted points seem to have been used, or at least discarded, much less frequently in these parts of the Great Basin than elsewhere in the region. It is not likely that the relative rarity of fluted points in these areas reflects insufficient archaeological work, since both central Nevada and the Bonneville Basin have been the focus of intensive archaeological surveys in environments at least roughly comparable to those that have provided fluted points elsewhere.

Because fluted points represent one of the earliest known artifact styles from North America, discoveries of isolated examples often become widely known. As a result, most fluted point "sites" in the Great Basin consist of isolated points and nothing more. A number of locations have, however, provided sizable numbers of these objects, along with other artifacts. Such concentrations are known, for instance, from China Lake in southeastern California, from the area of Pleistocene Lake Tonopah in the southern end of Big Smoky Valley in southern Nevada, from Long Valley in eastern Nevada, and from the Alkali Basin in south-central Oregon. When other artifacts are found so closely associated with these points that one can be fairly sure that they actually belong together, the associated items tend to be fairly nondescript stone tools that, if found on their own, would not be recognized as belonging to a fluted point tool kit. The exceptions to this statement generally involve debris from the manufacture of the points themselves: the channel flakes from producing the flutes are, for instance, distinctive. This situation parallels that for Clovis tool kits in general, as I discuss in chapter 4.

Because nearly all fluted point sites from the Great Basin have been found on the surface, little progress has been made toward dating them. Indeed, not even the six sites in which fluted points that have been found in a buried context have been able to shed much light on their age.

In the early 1940s, Elmer Smith of the University of Utah recovered two fluted points from deep within Danger Cave.

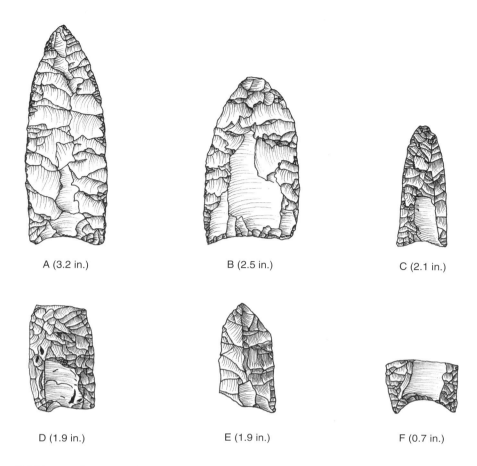

FIGURE 9.2 Great Basin fluted points: A, B from Alkali Lake Basin, Oregon (after Willig 1988); C, D, E from Fort Irwin, Mojave Desert, California (after Warren and Phagan 1988); F from Long Valley, Nevada (after Beck and Jones [2009]). The Long Valley point comes from beneath material dated to 10,320 years ago (Beck and Jones [2009]). Measurements indicate the length of the specimens. Drawings by Peggy Corson.

These points were misplaced and were unavailable to Jesse Jennings when he did his own work at this crucial site between 1949 and 1953; not until 1986 was it reported that one of the missing points had been rediscovered. Although the earliest deposits at Danger Cave were laid down about 10,300 years ago, all we know about the Danger Cave fluted points is that they came from the "lower levels" of this site.

Things are far better at the famous Sunshine Locality at the southern end of Long Valley in eastern Nevada, where Charlotte Beck and Tom Jones have done some of the best research that has ever been done on the late Pleistocene and early Holocene archaeology of the Great Basin. I will talk more about this work below and now just mention that nearly all of the artifacts they found here, both by themselves and those who worked here before them, were on the surface. Those artifacts include seventeen fluted points, only two of which match classic Clovis points in form. Of the seventeen, one (figure 9-2, F) was uncovered during excavations. This point was found beneath organic material that dates to 10,320 years ago and thus must be older than that, but how much older is not known.

In the Mojave Desert just south of the Granite Mountains (and not far northeast of the Calico site), Claude Warren and his colleagues excavated and analyzed the material from the Henwood site as part of their archaeological work on the Fort Irwin Military Reservation. Here, they found two fluted points, one on the surface, one buried, both similar to a series of classic Clovis points from the Plains and elsewhere. They also obtained a radiocarbon date of 8,470 years from material associated with the buried point. This, of course, is far younger than the known time range for classic Clovis points.

There are three obvious ways of accounting for this young date. First, it is possible that Great Basin fluted points were in use well into the early Holocene. Second, it is equally possible that this point had been picked up and transported here by later peoples. Indeed, the Henwood site also provided a series of artifacts that we know to have been in use at about 8,500 years ago. Third, it may also be that the date is wrong. A second date from the same context provided an age of 4,360 years and was rejected by the Warren and his team as being incorrect. Perhaps, then, the 8,470-year date is incorrect as well.

Then there is a fluted point from the Connley Caves, excavated by Stephen Bedwell (see chapter 8). This point came from deposits that are now known to have been disturbed but that at the same time have fairly consistent radiocarbon dates that fall between about 10,600 and 7,200 years ago. Finally, fluted point fragments were excavated from the Old Humboldt Site, along the Humboldt River north of Lovelock, Nevada, but all that can be said of the age of these specimens is that they were deposited before the climactic eruption of Mount Mazama around 6,730 years ago. A fluted point was also excavated by Ted Goebel and his colleagues from Bonneville Estates Rockshelter some thirty miles south of Danger Cave (see chapter 8 and figures 5-7 and 8-10), but this point does not help us at all, because it came from deposits that were historic in age.

In short, not a single buried site known from the Great Basin has provided fluted points in a situation that allows us to know how old they actually are. As a result, we do not know exactly when these artifacts were in use, although the Sunshine results tell us that at least some were deposited more than 10,300 years ago.

Fortunately, we can do far more than bemoan the fact that we have no buried fluted point sites in the Great Basin. Instead, we can point out that since so many of these sites, and isolated points, are found on the surface of the ground, we know far more about the distribution of fluted points than we could possibly know if all of this material were buried. Nearly all of these sites are located along the edges of the now-extinct lakes and marshes that existed in the Great Basin during the late Pleistocene and early Holocene. Archaeologist Amanda Taylor, for instance, analyzed the distribution of more than 150 Great Basin fluted points and found that 90 percent of them came from such valley-bottom settings.

This is not to say that no high-elevation fluted point sites are known from the Great Basin. Avocational archaeologist Vonn Larsen found a fluted point at 6,680 feet in central Utah's Clear Creek Canyon. While there is some chance this was brought here by a later prehistoric inhabitant of the area, the same is not likely to be the case for the two fluted points and associated Clovis-like blade (chapter 4) found by Dave Rhode and his colleagues at 7,810 feet in the Pine Grove Hills of western Nevada.

These, though, are exceptions and it is seems evident that whatever the people who were making and using these fluted points were doing, they were doing a lot of it near shallow water.

At one time, some Great Basin archaeologists argued that the makers of fluted points in this region were big-game hunters. That interpretation was derived directly from the spectacular Clovis sites of the Plains and Southwest that contain fluted points tightly associated with mammoth. As I discuss in chapter 4, most archaeologists have abandoned the notion that Clovis peoples made their living by hunting now-extinct mammals and few, if any, archaeologists think that the fluted point users of the Great Basin made their living in this way.

There are no convincing associations between fluted points and the remains of extinct Pleistocene mammals in the Great Basin. Although the lack of such associations may have led some to discard the notion that the makers of Great Basin fluted points were big game hunters, that lack probably means very little. Since virtually all of our fluted point sites come from the surface, convincing associations with mammals, large or small, are hardly to be expected. As I mentioned in chapter 7, it would be exciting, but not all that surprising, if an association of this sort were to be discovered. Given the general North American pattern, one would predict that if fluted points are found tightly associated with an extinct mammal in the Great Basin, that mammal will be a mammoth.

Instead, the demise of the big-game hunting interpretation for the makers of Great Basin fluted points follows in the wake of similar conclusions that have been reached for other parts of North America. It also follows from the fact that the distribution of fluted points in the Great Basin is so often tied to what would have been highly productive shallow-water environments. Had fluted point makers been heavily involved in the pursuit of such mammals as mammoth and horse, the distribution of the artifacts they produced would not be this restricted.

If Great Basin fluted points are latest Pleistocene or even early Holocene in age, the concentration of these sites adjacent to productive shallow-water settings makes good ecological sense. As I have discussed, our knowledge of the vegetational history of the Great Basin strongly suggests that in many parts of the Great Basin, subalpine conifers extended well down along mountain flanks until after 11,000 years ago or so. Although such subalpine conifers as limber pine do produce edible nuts, they are relatively expensive to procure and process, while the productivity of subalpine woodlands is, in general, low. The most productive environments in the latest Pleistocene Great Basin would have been provided by the plants and animals associated with shallow-water settings and in the steppe vegetation immediately above those settings, and this is precisely where fluted points are found.

Great Basin Stemmed Point Sites

In the 1930s, Elizabeth and William Campbell conducted a detailed archaeological survey of the shores of Pleistocene Lake Mojave in southeastern California (see figures 5-5 and 5-16). Deeply interested in the early peoples of arid western North America, they had come to Lake Mojave for two specific reasons. First, they thought that in order to demonstrate that a site was of truly great antiquity, they would have to show that it did not, or at least was not likely to, contain artifacts representing a jumble of different occupations that had occurred through the ages—that it was, in their words, "pure" (Campbell et al. 1937:9). That logic dictated that they look in areas that appeared to have been habitable only during remote times.

Second, they also recognized that the best case they could make for the antiquity of anything they found would stem from its geological context. If it could be shown that the context was ancient and that the area had not been suitable for human occupation since that time, it would follow that the site must be ancient as well.

Their reasoning was sound and they were successful. Scattered along the shorelines of Lake Mojave, the Campbells and their colleagues found a series of sites that contained artifacts so weathered that they often had to be turned over to be identified as such. While those artifacts included a diverse variety of fairly large core and flake tools, most distinctive among them were projectile points and a series of objects that they called "crescentic stones" and that are now simply referred to as crescents.

Nearly all of the projectile points found at these sites fell into two groups. These they called Lake Mohave and Silver Lake points (figure 9-3). Both of these styles have distinct stems with bases that are generally, but not always, rounded. In Lake Mohave points, however, the stems—the part of the point that was hafted—are distinctly longer than they are in Silver Lake points. Silver Lake points generally have stems that are less than half the length of the point, while Lake Mohave points have stems that take up half or more of the whole object.

The crescents that the Campbells found are as distinctive as the projectile points. These crescent-shaped stone objects are generally one to two and a half inches from tip to tip, and up to about one-half inch wide at their center (figure 9-3). Often carefully flaked, the front and back edges of the central portion of crescents are frequently rendered dull by grinding or very steep retouching. To some, the fact that the edges of the midsections of these implements are dulled suggests that they were hafted at this spot. In this view, the dulling was performed either to prevent the hafting material, presumably sinew, from being damaged or to remove irregularities from the edges that would increase the chances that the tool would break.

Since the geological context of their sites was critical to the Campbells, they relied on experts to interpret that context. Ernst Antevs spent several weeks with them at Lake Mojave and, in 1937, was convinced that the location of these sites, on terraces of Pleistocene Lake Mojave, implied that they were at least 15,000 years old

The Campbells assigned the material they had found to what they called the Lake Mohave Culture. With seventy years of additional research and fifty years of radiocarbon dating behind us, we now have a deeper understanding of the age of what they found, and of how it relates to other aspects of Great Basin prehistory.

In earlier chapters, I noted that Lake Mojave seems to have reached high levels between 18,400 and 16,600 years ago (Lake Mojave I) and between 13,700 and 11,400 years ago (Lake Mojave II). I also noted that a series of shallow lakes formed in this basin between 11,400 and 8,700 years ago (Intermittent Lake III). It now appears that the occupation of the sites collected and analyzed by the Campbells was contemporary with Intermittent Lake III. That this was

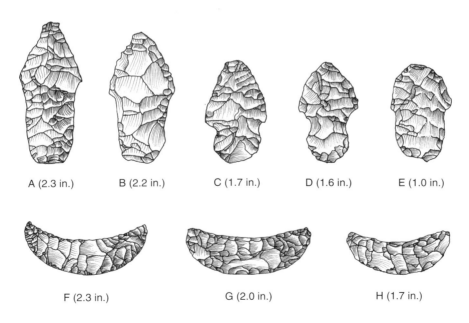

FIGURE 9.3 Lake Mohave points (A, B), Silver Lake points (C, D, E), and crescents (F, G, H). A–E and G–H from Long Valley, Nevada (after Beck and Jones 2009); F from Alkali Basin, Oregon (after Willig 1988). Measurements indicate the length (points) or width (crescents) of the specimens. Drawings by Peggy Corson.

so follows not so much from the placement of the sites on various Lake Mojave terraces, though the lowest of the artifact-bearing terraces is beneath the Lake Mojave II highstand. Instead, this conclusion follows more securely from the fact that the general kinds of distinctive artifacts that mark the Lake Mojave Culture—stemmed projectile points and crescents—are now reasonably well dated in various parts of the Great Basin. Those dates fall between about 11,200 and 7,500 years ago. Their known time range in the Mojave Desert falls between 10,100 and 8,400 years ago, but this shorter time span may not mean much, since there are not many secure radiocarbon dates for this material from the Mojave itself. Either way, the lifespan of Campbell's Lake Mojave Culture falls within the time that Intermittent Lake III was (intermittently) forming in the Lake Mojave Basin.

The Lake Mohave and Silver Lake points that characterize the Campbell's Lake Mojave Culture are not the only stemmed, and often large, points known from the Great Basin and immediately adjacent areas during the latest Pleistocene and early Holocene. In addition to these two forms, there are Cougar Mountain, Haskett, Lind Coulee, and Parman points; some would include an additional form called Windust with these as well (figure 9-4).

All of these artifacts have in common the fact that they have fairly thick stems that usually contract to a base that is rounded to square in outline; many also have a distinct shoulder that separates the stem from the blade portion of the point. Typically, the edges of the stems have been dulled by grinding, an attribute they share with fluted points. As with crescents, the grinding was probably done either to prevent damage to the sinew used to haft them or to remove irregularities that would have increased the chances of breakage.

Great Basin archaeologists recognize both the variability displayed by these early stemmed forms and that they are similar in many significant ways. As a result, they are routinely grouped together and referred to as "Great Basin Stemmed" points, though they are also referred to as belonging to the "Western Stemmed Point Tradition".

Sites that contain Great Basin stemmed points also often contain both fluted and unfluted lanceolate points with concave bases. The fluted versions are the same as the ones discussed in the previous section; the unfluted ones tend

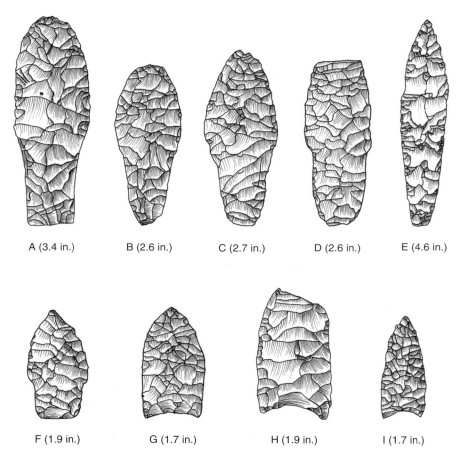

FIGURE 9.4 Cougar Mountain (A, B), Parman (C, D), Haskett (E), Windust (F), and unfluted lanceolate (G, H, I) points. A–D and F–I from Long Valley, Nevada (after Beck and Jones 2009), E from the Snake River Plain, Idaho (after B. R. Butler 1978). Measurements indicate the length of the specimens. Drawings by Peggy Corson.

to look very much like them but have bases that have been carefully thinned rather than fluted. Just like the stemmed and fluted forms, the edges of these points are usually ground. Often the fluted and unfluted forms are lumped together and referred to as "Great Basin Concave-based" points. It is tempting to think that these points differ only in whether or not they happened to have been fluted; perhaps, for instance, the unfluted examples were simply too thin to flute and so were basally thinned instead. Charlotte Beck and Tom Jones, however, have shown that this is not the case in Long Valley. Here, the fluted and unfluted forms have the same widths, lengths, and thicknesses but have very different flaking patterns.

The stemmed and concave-based points are distinctly different, even though many of them seem to have been in use at about the same time and are often found on the same sites. At the Sunshine Locality in southern Long Valley, for instance, Beck and Jones tallied about seven hundred stemmed points alongside nearly one hundred fluted and unfluted concave-based ones. Why these points co-occur so frequently is one of the more vexing questions the early Holocene archaeology of the Great Basin has provided. There are, as we shall see shortly, several possible answers, but discussing those possibilities requires that we know more about the archaeology of this time period.

In addition to the points, stemmed point sites have provided a diverse set of associated artifacts. Most distinctive among these are the crescents that I mentioned above. The function of these objects is a mystery, though there are clues that may someday lead to a solution. First, as I have mentioned, the midsections of the edges are routinely ground. Of the 245 crescents known from the Sunshine Locality, for instance, 82 percent are ground on at least one edge. This, in turn, suggests that they were hafted. Second, crescents were routinely manufactured from chert or similar stone even when other raw materials were easily available. At Sunshine, 94 percent of the crescents were of chert or chalcedony, compared to only 18 percent of the stemmed points; in the Black Rock Desert; 89 percent of the crescents are on chert or chalcedony, but only 1 percent of the stemmed points. This consistency in raw material selection suggests that whatever these objects were used for, they required high-quality, durable stone to do the job. Third, all of the wear, breakage, and resharpening on crescents is concentrated on the tips of the implements. Finally, and very much unlike the lanceolate points, crescents are rarely found in isolated contexts. When they are found, they are almost always found on sites marked by large numbers of stemmed points. Even then, they are found in abundance only in some parts of the Great Basin.

Over the years, almost everything that could be suggested for the use of these artifacts have been suggested, from amulets to transversely mounted projectile points (leading to the name "Great Basin Transverse" points). Gene Hattori and his colleagues analyzed the protein residues adhering to a sample of thirty crescents and discovered that those proteins had come from a wide variety of animals, from fish and waterfowl to rabbits or hares. However, they also recognized that the residues could represent either the kinds of animals the tools were used to procure or the animal products that had been used to haft the objects. A separate analysis, by Phil Shelley, found that some crescents have deposits of plant silicates, known as opal phytoliths, on them, suggesting that they had been used for plant processing. One could infer from these residues that the tools were hafted using materials derived from various vertebrates and then used to process plants, but the best inference would be that we need more studies of this sort on much larger samples of crescents. All we really know with any certainty is that they seem to have been hafted, either transversely or longitudinally, that they were made on durable raw material, and that the tips seem to have been the focus of use.

The stemmed point sites also provide a series of less distinctive artifact types. These include a variety of stone flake and core tools commonly interpreted as scrapers, knives, and gravers, although their functions are not well known. It is also clear that people occupying these sites brought with them raw material that had been flaked on both surfaces to produce small cores that could either be used as tools themselves or further reduced by flaking to produce whatever tool the situation called for.

Just as important, stemmed point sites at times contain small numbers of ground stone implements, presumably manos and metates. Manos and metates are the classic seed-grinding implements used by the native peoples of the Great Basin in historic times. The material to be ground—often but not necessarily seeds—was placed on the larger metate, then ground with the smaller hand stone, or mano (not coincidentally, the Spanish word for hand). Later prehistoric sites in the Great Basin often contain large numbers of these important implements but stemmed point sites routinely contain at most a small number of them. Archaeologist Mark Basgall, whose work has done much to increase our understanding of California's Mojave Desert, has noted that stemmed point sites in the north-central part of the Mojave contain an average of 3.6 milling stones per site, compared to a whopping 125.4 per site during the middle Holocene, although the numbers decline after this time. Just as important, he notes that the milling stones found on stemmed point show very low levels of wear.

This pattern is fairly typical of stemmed point sites throughout the Great Basin. If milling stones are present, they are uncommon and show very little indication of intensive use. In addition, there is no direct evidence those ground stone tools were used for plant processing—evidence that might be provided, for instance, by residue analysis—and it is well known that such tools were used historically for a wide variety of purposes. This all changes dramatically once the early Holocene comes to an end.

As I have discussed, fluted point sites are routinely located in valley bottoms, in areas that would have been marked by shallow lakes and marshes during the late Pleistocene and early Holocene. The same is true for many stemmed point sites. The Sunshine Locality, in southern Long Valley, again provides an excellent example. Work conducted by Beck, Jones, and their colleagues has shown that when the rich archaeological material at this site was accumulating, the area was marked by a permanent stream feeding a substantial marsh. Today, there is no surface water at all at this spot.

There are many other comparable examples. One of the best studied of these is provided by the work done by David Madsen and his colleagues along the Old River Bed on the Dugway Proving Ground in central Utah. That work has revealed the presence of more than seventy stemmed point sites associated with the delta of the river that once flowed south from the Sevier Lake Basin into a shallow Lake Bonneville to the north (see chapters 5 and 8). Analyses of the stone tools collected from these sites shows that they lack groundstone and fluted points but are rich in a wide variety of stemmed points. The position of these sites shows that when they were occupied, between about 10,500 and 8,800 years ago, the area was characterized by rich marshes. When the marshes disappeared, so did the people. Work in other parts of the Old River Bed delta has found exactly the same thing. These areas, today best described as hyperarid, were marked by extensive marshes toward the end of the Pleistocene and during the early Holocene—marshes that proved extremely attractive to peoples who are today recognized by the stemmed points they left behind.

The occupation of stemmed point sites in valley-bottom locations that held shallow lakes and marshes might make this use of the landscape sound very similar to that known for the users of fluted points. The latter, after all, seem to have been largely confined to this setting. In fact, one obvious, but perhaps incorrect, interpretation of the frequent discovery of fluted and stemmed points on the same surface sites is that different peoples used the same patches of landscape at different times.

The distribution of stemmed point sites, however, is far broader than valley bottoms alone and in this they differ from fluted point sites. Site after site shows these differences. Northwestern Nevada's Last Supper Cave, for instance, held a significant stemmed point occupation dated to between about 9,000 and 8,500 years ago, yet is located adjacent to a small stream in rugged country, some twenty-five miles from, and 1,200 feet higher than, the nearest valley setting that would have supported a significant lake or marsh. To the south, the Old Humboldt site is adjacent to the Humboldt River, at Rye Patch Reservoir, and contains a variety of Great Basin stemmed points deposited prior to the eruption of Mount Mazama 6,730 years ago. In central Nevada, the Five Points site sits at an elevation of 8,520 feet, just beneath the summit of Park Mountain and overlooking a series of small meadows. This site is undated but contains Silver Lake and other point styles that show its affinities well. In the Yucca Mountain area of southern Nevada, work by Paul Buck, Gregory Haynes and their colleagues has shown that stemmed point sites are most often found on the terraces of now-dry streams that drain large watersheds. In the Coso Basin of the northwestern Mojave Desert, Jelmer Eerkens and his colleagues have shown that early Holocene sites are found not only along the edges of what might then have been a lake or marsh but in a broad variety of other settings as well. Mark Basgall has shown that a similar situation exists in the north-central Mojave Desert, including the Nelson Lake Basin.

There is even a fairly high-elevation house structure known from the very edge of the northwestern Great Basin that is associated with both early Holocene dates and stemmed points. This structure is part of the Paulina Lake site, located in Newberry Crater in Oregon's Cascade Range. The site is adjacent to Paulina Creek, which is part of the Deschutes River drainage and thus not actually in the Great Basin but it doesn't miss by much. Here, Tom Connolly uncovered the remains of an oval house with a centrally located hearth. The structure was at least thirteen feet long, defined in part by charred support posts made of lodgepole pine. Dates from the posts suggest that the house was constructed around 8,500 years ago, consistent with the stemmed points (and a small number of groundstone tools) associated with this occupation. This, in fact, is the earliest well-dated dwelling known from the Great Basin or its immediate periphery.

These sites, along with many others, show that while nearly all fluted point occupations are located in valley bottoms, stemmed point sites are found both here and in a wide range of other environments, from riversides to mountain meadows. Insofar as we can judge from artifact distributions, the makers of stemmed points were utilizing a far wider range of environments than were the makers of fluted points.

Because stemmed point sites in areas that were once the scene of biologically productive marshes often have thick concentrations of artifacts—sometimes thousands of them—it is likely that they were used for substantial amounts of time. At the same time, analyses of the stone tools on these sites have shown that the people who made these tools traveled impressive distances across the landscape.

The logic behind these studies seems simple once someone else has thought of it. Imagine a group of mobile hunter-gatherers moving across the landscape. When they come to a good source of raw material for stone tools—obsidian, for instance—they stop, manufacture new tools and take away carefully prepared bifaces, pieces of raw material worked on both sides from which new tools can be made as needed. The further they get from the last source of good raw material, the more worn their tools become and the more reduced the bifaces that they brought with them. If the raw material was obsidian, we can tell from the tools themselves where the raw material was obtained in the first place (figure 9-5). This is because different obsidian sources have different chemical signatures. If we have a catalogue of the possible sources and their chemical constituents, the tools can be chemically analyzed to determine the sources

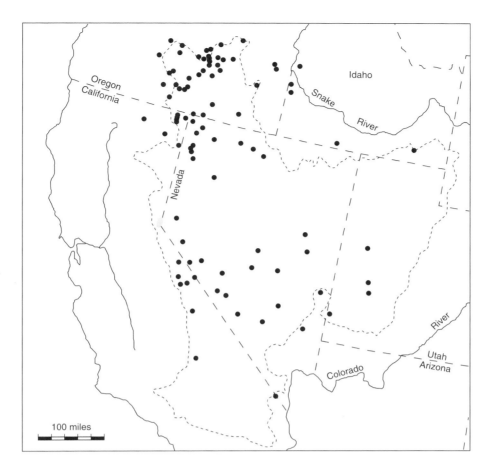

FIGURE 9.5 Known obsidian sources in and adjacent to the Great Basin (after G. T. Jones et al. 2003).

that provided the obsidian from which they were made. Putting these two things together—the degree to which a set of tools have been reduced and knowledge of the source of the stone from which they were made—allows archaeological sites to be read like a travel diary.

A number of Great Basin archaeologists have taken this approach to reading the stemmed point travel map but the most detailed effort has been made by Tom Jones and his colleagues, with geochemist Richard Hughes providing the information on raw material sources. They showed that the obsidian from stemmed point sites in adjacent Long, Butte, and Jakes valleys in eastern Nevada came from an enormous area, measuring some 280 miles north-south and nearly 100 miles east-west.

Jones and his coworkers are not alone in providing data suggesting that the users of stemmed points were remarkably mobile. Archaeologists Daron Duke and Craig Young discovered that obsidian on stemmed point sites from the northern edge of the Old River Bed, north of the area in which Madsen and his colleagues have worked, came from areas that ranged from 50 to 310 miles away. Geoff Smith's work on stemmed point sites from Five Mile Flat in northwestern Nevada shows that the obsidian here was derived from an area that measures 175 miles north-south and 115 miles east west. Kelly Graf has determined that obsidian from the Sadmat site, located in the Carson Sink, was derived from sources located between 150 miles to the north and 135 miles to the south; at the Coleman site, in the Winnemucca subbasin of Pleistocene Lake Lahontan, the obsidian came from places as far as 150 miles north and 60 miles south.

Not all results from obsidian sourcing indicate the use of such far-flung raw materials. Albert Oetting, for instance, found that all the obsidian sources used in two stemmed point sites in the Fort Rock Basin came from within forty-five miles of those sites. Most, however, replicate the kinds of results I have just discussed: the obsidian in stemmed point sites often comes from vast distances, suggesting that the people using these artifacts were themselves moving across vast distances.

Amassing all of this information allowed Jones and his coworkers to define a series of obsidian "conveyance zones" in the Great Basin, areas within which, but apparently not between which, people were moving significant amounts of obsidian across space (figure 9-6). That, in turn, suggests relatively little contact between the people living in different zones. Had they been in contact, Jones and his colleagues suggest, one might expect more evidence for raw material exchange among them.

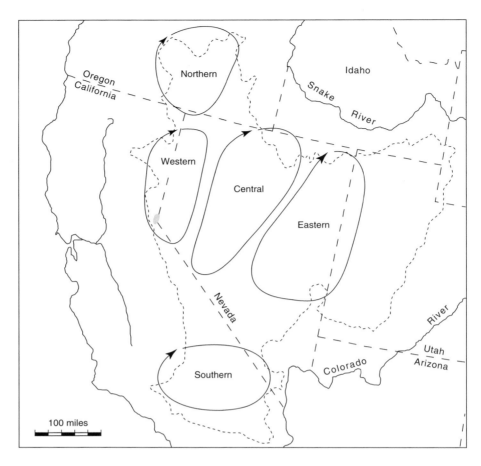

FIGURE 9.6 Latest Pleistocene and early Holocene obsidian conveyance zones in the Great Basin (after G. T. Jones et al. 2003).

The broad distances that users of stemmed points were moving across the landscape, coupled with the fact that raw materials occur only at particular points on that landscape, has important implications for why the late Pleistocene and early Holocene archaeological record looks the way it does. There are, for instance, no local obsidian sources appropriate for making stemmed points in either Long Valley or at the Old River Bed. Nonetheless, 14 percent of the Sunshine Locality stemmed points and about half of those from the Old River Bed sites on the Dugway Proving Ground are made out of this material. This means that people arrived either with these points in hand or with obsidian bifaces from which they could be made. The more the tools were used, the more they had to be resharpened until, ultimately, they wore out, were discarded, and had to be replaced with more easily available raw material, often fine-grained volcanic rock which was itself subject to the same process. One of the end points of these constraints on the availability of raw material can be seen with the points called "Dugway Stubbies". These are tiny little points, about one inch long, that are common at the Old River Bed sites, the large majority of which are of obsidian. They represent the remnants of discarded artifacts, now reworked into nubbins. Much the same thing was happening in other parts of the Great Basin, including the Mojave Desert.

The fact that the areas in which people were living were often not the same areas that provided raw materials for stone tool use may help explain the tremendous variability in the forms of stemmed points. Although Charlotte Beck has shown that the relative abundances of the different forms change as time passes, most or all of these point styles seem to have been in use for something on the order of 4,000 years. Rather than reflecting the passage of time, the variability that these points show may primarily reflect the degree to which artifact shapes changed as the objects were continually resharpened. As Tom Jones and his colleagues have noted, "were we to have equal access to points that had never been used, our typologies might be quite different" (Jones, Beck, and Kessler 2002:7).

Mark Basgall, Charlotte Beck, Tom Jones and others have also noted that even though stemmed points have traditionally been called "points," as if they were used to tip weapons, it is far more likely that they were used for a wide variety of functions. Some may have served as weapon tips but the fact that they are often asymmetrical, have rounded tips, thick cross-sections, and heavy wear on their edges suggests that they were used for a wide variety of purposes, including cutting, sawing, and chiseling. In that sense, they might better be called Stemmed Swiss Army Knives. This is very different from the situation with fluted and unfluted lanceolate concave-based points, which meet all the requirements for having been the tips of weapons used for hunting.

This might help explain why different sets of artifacts are routinely made of different raw materials. The lanceolate points are routinely, though not always, on chert, but the stemmed "points" are typically made on obsidian or fine-grained volcanic rocks, such as certain forms of basalt. Difference uses, different raw materials.

The Relationship between Fluted and Stemmed Point Sites

The lanceolate points, fluted or not, are often found on the same sites as stemmed points but appear to have had different functions. The former seem to have served most often as weapon tips, perhaps on throwing or thrusting spears or as atlatl dart points. The latter seem to have had multiple uses, perhaps only rarely serving as true points. Since the two kinds are so often found together, perhaps they are different tools in the same tool kit.

Beck and Jones have made this argument most recently and in most detail. Agreeing with a position that has long been argued by archaeologist Alan Bryan, they suggest that people using stemmed points were living in the Great Basin before fluted points were introduced, perhaps the descendants of populations who had moved inland from the Pacific coast. People using fluted points arrived in the Intermountain West later, coming from the interior of North America and encountering stemmed point users in the Columbia Plateau and on the northern edge of the Great Basin itself. Great Basin peoples then adopted fluted points as hunting implements, freeing stemmed points to become multiple-use tools. As time went on, the lanceolate points lost their flutes, just as we know happened in the Great Plains. In this view, fluted and stemmed points were made by the same people in the Great Basin, merging technologies with two very different origins.

This would account for many otherwise perplexing attributes of the late Pleistocene and early Holocene archaeology of the Great Basin. Obviously, it would explain why the two kinds of very distinctive artifacts are so often found together. Such sites have been found in the China Lake Basin, in Jakes Valley, in Railroad Valley, in Long Valley, in the southern Big Smoky Valley, in the Sevier Desert, and elsewhere. It would also explain why fluted points are found as isolated artifacts more often than stemmed points seem to be. Mark Basgall and Matt Hall, for instance, discovered that 29 percent of the fluted points known from Fort Irwin, in southeastern California's Mojave Desert, were isolated finds, compared to only 1.5 percent of the stemmed points. If stemmed points (and crescents, also rarely found as isolates) were used for tasks at sites occupied for substantial amounts of time, while fluted points were used by hunting groups away from those sites, this is exactly what we would expect to find. It would explain why lanceolate and stemmed points tend to be made from very different raw materials and why there are both fluted (earlier) and unfluted (later) lanceolate points. And, it would explain why Great Basin stemmed point sites have dates as early as the Clovis sites of the plains and Southwest (11,200 years ago) but Great Basin fluted points have yet to be shown this old. The stemmed point users were here first and adopted the technology represented by fluted points when they came into contact with people using it at some later, but unknown, date.

It may not, though, explain everything. First, detailed analyses of the steps taken to manufacture fluted and stemmed points have shown that these groups of artifacts were made in distinctly different ways. Second, a number of valley bottom settings in the Great Basin have provided concentrations of fluted points in the absence of stemmed points. These places include the basin of Pleistocene Lake Tonopah in southern Nevada and the Alkali Lake Basin of eastern Oregon. Finally, as discussed above, fluted point sites are found in a much narrower range of environments than are the stemmed point ones, with the former largely, though not entirely, confined to valley bottoms and the latter much more widely dispersed across the landscape.

These things suggest another explanation for the presence of both fluted and stemmed points in the Great Basin. Perhaps, rather than having formed parts of the same tool kit, these artifacts were used by culturally distinct groups of people who lived here, at least early on, at about the same time. Mark Basgall, among others, has made this argument.

But this argument poses problems, as well. While it is possible that fluted and stemmed points were made in different ways because they were made by culturally distinct peoples, it is also possible that these tools had to be made in different ways because they had different uses, just as has been argued for atlatl dart and arrow points in parts of eastern North America. Perhaps some sites contain only fluted points because people came to these places only to hunt. Perhaps fluted points were found in a narrower range of environments because the purposes they served were appropriate only for those places.

That brings us to a third possibility, that fluted points really did come first, as has often been assumed. Most of the earliest radiocarbon dates associated with stemmed points in the Great Basin postdate the classic Clovis sites of the Southwest and Great Plains, though not by very much. This leaves open the possibility that fluted points came first, stemmed ones later. In this view, fluted point occupations may have been largely confined to valley bottoms because these marsh-rich settings provided environments that not only were highly productive but were also separated from one another by mountains whose lower elevations, at least, were covered by relatively unproductive subalpine woodlands. Occupying a region where human population density was surely low, these people focused on the rich resources provided by the valleys themselves and made little use of the much less productive settings that lay above them. Following them in time, later Pleistocene stemmed point occupations were at first also confined to valley bottoms. As the Pleistocene ended and the subalpine treeline began to move upward, the range of environments utilized by people in the Great Basin would have expanded to include the increasingly productive uplands, accounting for the fact that stemmed point sites are distributed across a broader variety of landscapes.

In this view, fluted point occupations come first but are soon replaced by those marked by stemmed points. Valley bottom sites that contain both fluted and stemmed point sites are explained by the fact that both sets of people used very similar parts of the landscape. Fluted and stemmed points are made in different ways because they were made by different people. Stemmed point sites have a broader distribution because they were deposited across a much broader period of time, measured in thousands rather than hundreds of years, during which time the nature of the Great Basin landscape changed dramatically. If this view is correct, it seems unlikely that the technology represented by fluted points was ancestral to that used by the makers of stemmed points because the techniques used to make these artifacts diverge so much. And if that is the case, it raises the very real problem of knowing where the people who made the stemmed points came from.

These are the three prime ways in play to explain the relationship between fluted and stemmed points in the Great Basin. In the first, they represent different tools in the same tool kit, perhaps because of interactions, on the Columbia Plateau or the very northern fringes of the Great Basin, between the makers of the two different kinds of tools. In the second, they represent culturally distinct but, at least at first, contemporary peoples. In the third, they represent temporally distinct occupations, with the fluted points coming first. Each has the advantage of explaining something that the others cannot. Once we have a better chronology for all this, we will be in a better position to know which is most likely to be correct, or if even better explanations exist.

The Spirit Cave Burials

In 1940, Sydney and Georgia Wheeler were working for the Nevada State Parks Commission when they excavated a small rockshelter near Grimes Point in Nevada's Carson Desert, not far from Hidden Cave (see chapter 6). They quickly uncovered five separate sets of human remains, some of which they reburied but the rest of which were given to the Nevada State Museum. Not until 1994, when some of this material was radiocarbon dated, was it discovered that the Wheelers had removed some of the oldest human skeletal material ever found in the Americas.

The small rockshelter that provided this material is now known as Spirit Cave. The human remains they excavated included a woman who had been carefully wrapped with a very well-made woven mat. Both the mat and the woman herself dated to about 9,300 years ago. Somewhat deeper in the site was the body of an adult male, forty to forty-five years old, with preserved soft tissue (the "Spirit Cave Mummy"). He was buried with his hide moccasins and a woven breechcloth and had also been wrapped in exquisite textiles. At the time of his death, about 9,400 years ago, he was suffering from a series of undoubtedly painful abscesses in his upper and lower jaws that may have led to his demise. The other early material consisted of cremated human remains buried in woven bags, one of which has been dated to 9,040 years ago. Since the remains in both bags seem to have come from the same individual, this date must apply to both of them. The fifth burial was much younger, dating to 4,640 years ago.

While no stemmed or lanceolate points were found associated with the early burials from Spirit Cave, odds are that these people used the kinds of stone tools I have been discussing. Indeed, the Sadmat site is only some twenty-five miles to the northwest, and it is possible that the early individuals so carefully placed in Spirit Cave were the same as those who created this site. It is also likely that they visited Hidden Cave as well, since similar basketry, dated to 9,300 years ago, is known from that site. Spirit Cave makes the people involved very real.

Detailed analyses of the remains of the Spirit Cave Mummy have been conducted by a series of biological anthropologists. All found them to be distinctly different from any known modern Native Americans. The same is true for a skeleton found in 1968 at the northwest edge of Pyramid Lake that has been dated to 9,200 years ago As Heather Edgar and her colleagues have pointed out, these differences follow from the antiquity of these individuals. That people who died more than 9,000 years ago differ in marked ways from modern people is neither surprising nor an indication that they are not related to the native peoples of North America who were encountered by Europeans when they arrived in the Americas. Given the quality of the textiles associated with the Spirit Cave individuals, however, they are certainly telling us that at least some of the people who lived in the Great Basin during the early Holocene were superb weavers. Indeed, Kay Fowler and Gene Hattori have suggested that the largest textiles from Spirit Cave were probably made on a frame or loom. If so, this would represent the earliest use of such a device known from North America.

Late Pleistocene and Early Holocene Subsistence

The fact that nearly all fluted point sites, and many of the stemmed point ones, are found in settings that would have been adjacent to shallow lakes and marshes suggests that the resources provided by these bodies of water were critical to the lives of the people who occupied them. In fact, archaeologist Steve Bedwell once defined what he called the Western Pluvial Lakes Tradition to encompass the late Pleistocene and early Holocene human occupations of the western and northern Great Basin, occupations that he saw as "directed toward the exploitation of a lake environment" (1973:171). The use of this term soon expanded to include the entire Great Basin and adjacent areas.

The term, though, is no longer in use. The more we have learned of the stemmed point occupations, the more we have realized that these people utilized a wide range of environments. There is no reason to doubt that lakes and marshes were key to their livelihood in many areas. Not only can the location of so many of these sites be explained in no other way, but when these bodies of shallow water largely disappeared from the Great Basin, between about 8,300 and 7,500 years ago, so did the stemmed point sites. However, stemmed point sites are found in such a broad variety of settings that it is clear that the people who created them were not tethered to low-elevation lakes and marshes.

We have no direct information on the subsistence pursuits of the people who used the fluted and unfluted lanceolate points. If, as seems highly likely, the points were hunting implements, we have no idea what animals were being pursued with them. We are far better off with the stemmed point occupations since a number of excavated sites have provided the remains of the plants and animals that these people were eating. In many, but not all, ways, those remains suggest that late Pleistocene and early Holocene diets were similar to those that came later.

We know most about the animals. At Last Supper Cave the stemmed point occupation was associated with large numbers of the freshwater mollusc *Margaritifera falcata*, suggesting that the early occupants of this site were here at least in part to collect these bivalves. At the Old Humboldt site, Amy Dansie found that the fauna associated with the stemmed point occupation was dominated by jackrabbits but also included mollusc shells, the bones of small artiodactyls (deer, mountain sheep, or pronghorn) and even a few bison bones. In the Mojave Desert, Charles Douglas and his colleagues found that the faunas from the stemmed point occupations at Henwood and nearby sites were dominated by the bones of small artiodactyls, rabbits, and jackrabbits; the presence of burned bones from small rodents and lizards suggests that these animals may have been taken as well. Two sites in the eastern Fort Rock Basin that date to between 8,800 and 9,100 years ago had pits jammed full of jackrabbit bones. Elsewhere in the Fort Rock Basin, a hearth dated to 9,400 years ago provided the remains of small fish; coprolites from the Spirit Cave mummy were full of such fish, probably representing that person's last meal. Bryan Hockett's work at Bonneville Estates Rockshelter documents that the people who used this site between about 10,800 and 8,800 years ago were taking deer, pronghorn, and mountain sheep, as well as jackrabbits, sage grouse, and grasshoppers.

The only detailed analysis of dietary plant use for this period has been provided by Dave Rhode and Lisbeth Louderback. Their analysis of the plants from the early Holocene deposits of Bonneville Estates Rockshelter suggests that cactus pads and a wide variety of seeds, from Indian ricegrass to sunflower, had been eaten. The coprolites from the Spirit Cave mummy are also said to contain seeds, but no detailed analyses of these plants have been done.

Anthropologists who study human diet often evaluate those diets in terms of the number of different kinds of items they include. Broad diets include a wide variety of species; narrow ones, relatively few. What we know about the diets of the people who were making stemmed points suggests that they were impressively broad. In this, they are similar to the diets of the Great Basin hunters and gatherers who followed them in time.

Understanding human dietary adaptations, however, requires that we know a lot more than the number of plants and animals included in the diet, since different species provide very different returns in energy and nutrients for a given level of effort put into obtaining them. Because of this, diets that include the same number of species may reflect very different levels of dietary affluence.

In the Great Basin, seeds are the classic example of an expensive food item. This is not so much because they are so time-consuming to collect but because they need to be processed if they are to provide the energy and nutrients they contain. Among both hunters and gatherers and many agricultural peoples, that processing was routinely done by hand with stone tools; in the Great Basin, manos and metates were the primary tools involved. Grinding seeds with stone tools is a slow and laborious task, and it is this that makes them so expensive in terms of time and energy.

An example provided by archaeologist Jim O'Connell helps to show the implications of all this. O'Connell spent nearly a year living among the Alyawara in the central Australian desert. Interested in why these hunting and gathering people utilized only some of the food resources available to them, and those in very different abundances, he collected detailed information on the amount of effort needed to acquire and process the plants and animals on which the Alyawara depend for their subsistence. He also collected data on the number of calories they gained in return for these efforts. When O'Connell and his colleague Kristen Hawkes analyzed this information, they found that the decisions the Alyawara make regarding what plants to collect and what animals to hunt were well explained by the costs and benefits involved in taking the species involved. They also discovered that seeds were among the most costly of Alyawara foods.

O'Connell found that processing one kilogram (2.2 pounds) of seeds routinely took about five hours, providing about 500 to 750 kilocalories per hour. That caloric return is very low: some roots and fruits, for instance, routinely returned ten times the calories that seeds returned in a given amount of time. No wonder, then, that seeds are among the first foods to be dropped from the Alyawara diet when other foods become available, and that they are dropped before such things as lizards and insect grubs. Seeds simply rank so low in terms of their costs and benefits that they are quickly replaced when higher-ranked foods can be utilized.

What O'Connell and his colleagues found for the Alyawara, Scott Cane found for the Gugadja, hunter-gatherers of the Great Sandy Desert in north Western Australia. Here, too, seeds were costly, taking an average of five hours to be turned into a kilogram of food, and providing an average return of only 340 kilocalories an hour. As Cane noted, one woman—and women are the seed collectors and processors here, as they are among virtually all hunters and gatherers—would have to work some ten to fifteen hours a day to provide a family of five with at most half their daily caloric intake. Many other studies have shown the same thing, as has experimental work done by archaeologist Steve Simms on a wide variety of Great Basin seed plants. Small seeds are expensive and are utilized only when less expensive foods are not available.

These relationships tell us something very important about the late Pleistocene and early Holocene archaeology of the Great Basin. There are no convincing associations between ground stone tools and fluted point sites here. They are found on some stemmed point sites, but they are always rare and were never intensely used. And even though Bonneville Estates Rockshelter suggests that small seeds were used during the early Holocene, no ground stone tools have been found in the early layers of this site.

The implications are clear. The diets of the late Pleistocene and early Holocene hunters and gatherers of the Great Basin may have been incorporated many species and, in this sense, have been similar to the diets of later peoples in this area. However, the resources available to them were sufficiently more productive than those available to later peoples that they had no need to spend large amounts of time processing small seeds. That did not happen until the onset of the middle Holocene, when increasing aridity led to the declining abundance of productive resources, including those associated with shallow-water habitats. When that happened, the stemmed point tradition disappeared.

The Middle Holocene

Adapting to a Poorer World

During the middle Holocene, between about 8,300 and 4,500 years ago, the Ruby Marshes shrank dramatically and may have disappeared, trees grew beneath what is now the surface of Lake Tahoe, Pyramid Lake fell, the Humboldt River and Humboldt Lake lost much if not all of the water they had held during the early Holocene, Owens Lake and Malheur lakes may have dried entirely, bristlecone pines moved far upslope on the White Mountains, and vegetation dominated by saltbush replaced a flora dominated by sagebrush in many part of the Great Basin. The relatively rich resources supported by widespread shallow-water systems were largely gone. Pinyon pine was not to be found outside the southern Great Basin until well into the middle Holocene or beyond and so could not help take the place of what was lost to regional drying at the end of the early Holocene. Were I to choose a time during the past 10,000 years to not live in the Great Basin, this would be it.

The challenges posed to human life here at this time are captured well by Mark Basgall's counts of the changing numbers of grinding stones per site in the north-central Mojave Desert that I mentioned above. From about 4 per site during the early Holocene, the numbers climb to 125 per site in the middle Holocene, then decline once this challenging climatic episode ends. This is the case even though the two kinds of sites can be found at very similar spots on the landscape. The critical change was in the kinds of resources those spots supported. For the first time in the Great Basin, people now had to include large amounts of small seeds in their diet, paying the energetic costs demanded by processing them with stone tools and causing grinding stones to become common in Great Basin archaeological sites. They then remained so throughout the rest of prehistory, emerging as key plant processing tools during early historic times.

Some of the best information we have on human lifeways during the middle Holocene in the Great Basin comes from three sites in the Bonneville Basin: Danger Cave, Bonneville Estates Rockshelter, and Camels Back Cave. I begin with what those sites have told us.

When Danger Cave was first professionally excavated by Jesse Jennings between 1949 and 1953, it seemed to show that the significant use of seeds and grinding stones began here during the early Holocene and perhaps even during the latest Pleistocene. Jennings used his discoveries at Danger Cave to argue for a long and stable human adaptation to the desert environments offered by the Great Basin. He referred to this adaptation as the Desert Culture, marked, among other things, by the heavy use of milling stones and small seeds. This way of life, he suggested, may have extended back some 10,000 years yet was essentially the same as that encountered by Europeans when they entered the Great Basin a few hundred years ago.

As time went on, however, Danger Cave began to appear more and more enigmatic. Sites that were early Holocene in age showed no evidence for the significant use of grinding stones or for the incorporation of highly processed seeds into the diet. Danger Cave moved from being the type site for a long and relatively unchanging Desert Culture to being the only site in the Great Basin that suggested the significant use of small seeds and milling stones during the early Holocene.

Now, thanks to work by David Madsen, Dave Rhode, and others, we know that Danger Cave doesn't show that at all. Instead, what it shows is how tricky the often complex deposits of caves can be. Jennings's excavations provided the remains of a series of human coprolites from the deepest levels of Danger Cave. When analyzed, these proved to be full of the small seeds from iodinebush, a plant that grows on the salty clays found along the edges of playas. That evidence matched the presence of grinding stones from these deep deposits as well as the remains of iodinebush plants themselves, the detritus produced by processing the plants within the cave. Madsen, Rhode, and their colleagues have now shown that the oldest iodinebush-filled coprolite within Danger Cave dates to about 8,700 years ago; the oldest iodinebush remnants that resulted from plant processing, to about 8,600 years ago. And although a few grinding stones may have been deposited at the cave before this time, it is just as possible that the ones that Jennings found in the earliest deposits here were there because they had moved downward from later, milling stone–rich sediments.

Small seed use, then, begins at about 8,600 years ago at Danger Cave. Not surprisingly, this is about the same time that Homestead Cave, across the Bonneville Basin to the east (chapter 8) provides powerful evidence that the landscape was drying out in a very significant way. The same climatic changes that led to the drying of Owens Lake, Malheur Lake, and the Ruby Marshes, to decreased pollen deposition at Blue Lake and to the gap in the depositional record at Mosquito Willie's now forced people to begin the intensive use of high-cost resources that before were simply not important to them.

Other sites in the Bonneville Basin have similar implications. Bonneville Estates Rockshelter seems to have been largely abandoned as the early Holocene came to an end but people used the site sporadically during the middle Holocene. When they first returned, around 7,200 years ago, they did so to hunt artiodactyls and jackrabbits and make heavy use of small seeds, including iodinebush, processing those seeds with the earliest grinding stones found at the site. Much the same happened when they returned here at about 1,000 years later. Camels Back Cave, directly across the Bonneville Salt Flats from Bonneville Estates, was first occupied, albeit briefly, around 7,500 years ago. And whenever people stayed at the site during the middle Holocene, they did so briefly but made heavy use of grinding stones, small seeds, and jackrabbits while they were there.

People were also importing small amounts of pinyon nuts to both Danger Cave (chapter 8) and Bonneville Estates, but the overwhelming impression provided by these sites is that the peoples of the Bonneville Basin were focusing their lives on, as Dave Rhode has aptly put it, "a few scattered pockets of promise in an otherwise forbidding landscape" (2008:16).

Middle Holocene Houses

Northeastern California's Surprise Valley provided one of those scattered pockets of promise, probably because the valley receives its water primarily from streams and springs fed by the massive Warner Mountains to the immediate west. Archaeological work done here during the late 1960s by Jim O'Connell showed not only that the area saw significant human occupation during the middle Holocene but that the people who lived here at that time were building substantial houses.

That evidence comes primarily from the King's Dog site, located just west of Middle Alkali Lake adjacent to a hot spring (the site takes its name from Danny King's dog, which died after jumping into this spring). The spring provides potable water and feeds a substantial marsh, and was likely to have been key to the occupation of the site itself.

O'Connell's work in the earliest, middle Holocene, deposits of this site yielded a wide variety of artifacts: side-notched projectile points, chipped stone drills and a diverse set of other kinds of stone tools; antler wedges, bone awls and other bone tools; and large cylindrical mortars and carefully fashioned pestles.

O'Connell also excavated a series of five superimposed house floors in the deeper parts of King's Dog. Although the oldest of these house floors was not dated, the second oldest dated to 5,640 years ago. These early houses were substantial: 22 feet to 25 feet across and about 2.5 feet deep, with a central fireplace and a sloping ramp entryway. From the pattern of postholes preserved on these floors, O'Connell observed that a ring of five to seven posts must have been arranged around the central hearth within the house pits. He speculated that horizontal stringers had been attached to the tops of these and that rafters sloped down from the stringers to just beyond the edge of the house pit itself. Presumably, the rafters were covered by mats and then by earth.

Much of this reconstruction stems directly from what O'Connell discovered from his excavations here: substantial house pits, central fireplaces, and postholes arrayed around those fireplaces. He found no timbers and no mats, so the rest of his reconstruction is speculative. That he did not find these items, however, is not surprising. Had they once been there, they might easily have decayed during the past 5,600 years or so. It is just as likely, however, that they were not left behind to begin with. When houses of this sort were vacated for substantial periods during historic times, the timbers and mats were removed and carefully stored.

Not far from King's Dog, the Menlo Baths site provided the remains of a middle Holocene structure much like those at King's Dog. Also located next to a hot spring (hence the name), Menlo Baths had been the focus of uncontrolled digging by artifact collectors. As a result, much of the site had been destroyed by the time O'Connell arrived, and he was able to excavate only a small portion of the house he encountered here.

Analysis of the faunal remains from both these sites suggests that they were winter occupations. In fact, we don't have to look far to find early historic winter houses remarkably similar to those unearthed by O'Connell. Those houses were made by the Modoc, who lived some eighty miles to the northwest and spoke a language belonging to the Penutian language group, one very different from that spoken by the Northern Paiute people of Surprise Valley. The Modoc built earth-covered winter lodges above circular pits that averaged about 22 feet across and 4 feet deep. (The Klamath, to the immediate north, built similar winter houses but with pits about 2.5 feet deep; the difference in depth may reflect differences in winter temperatures.) The superstructure of these dwellings was supported by four main posts placed around a central fire. Horizontal stringers were placed at the top of these posts, and rafters angled down from the stringers to just beyond the edges of the house pit; the rafters were then covered by mats and earth. These houses were costly to construct: a typical one would take the one or more families who were to occupy it about a month to build. They repaid the expense, however, since they were spacious, warm, and durable.

Those parts of the King's Dog structures that were well represented archaeologically—the large pits, central fireplaces, and central support posts—are nearly identical to the Modoc earth lodges. Even what wasn't there matches what the Modoc did, since they were among those who carefully removed and stored the timbers and mats when they left their homes for lengthy periods. There is one exception to these similarities: most Modoc structures were entered through an opening in the roof, while the King's Dog houses were entered by a rampway that exited to the east. However, some Modoc houses did have ramp entrances, and, although the similarity may be entirely coincidental, the Modoc ramp entries would have faced east. The land of the dead lay to the west, and most Modoc houses that had steps or doorways had them facing east.

The Surprise Valley is not the only area to have provided middle Holocene houses in the Great Basin. To the north, in the Fort Rock Basin, the Bergen site provided evidence for two house structures, marked by shallow circular basins some thirteen feet across with central hearths that dated to about 5,100 years ago. Today, the area in which this site is located is quite dry, but Margaret Helzer's analysis of the plant remains from both houses showed that they were loaded with bulrush seeds, showing that water must then have been nearby. Large numbers of small fish bones in one of the structures shows exactly the same thing. These houses also had abalone and *Olivella* shells in them, suggesting an exchange system that reached from the northwestern Great Basin to the Pacific coast, from which these shells must ultimately have come.

Declining Human Population Densities?

If the middle Holocene peoples of the Great Basin were living in a forbidding landscape occasionally interrupted by productive settings provided by places like Surprise Valley, then perhaps resources were so scarce that the Great Basin was able to support fewer people at this time than either before or after. Perhaps productivity had fallen so far that parts of the Great Basin were even abandoned by people.

In fact, this is an old argument, one commonly made from the 1940s to the 1970s. The argument followed not just from certain aspects of the archaeological record but also from the wide acceptance that Antevs's Altithermal climate model, with its several thousand years of drought, had received. Indeed, Antevs himself was one of the first to make the population decrease argument, in 1948. As it became increasingly obvious that the climates of this period were more variable than had once been thought, the idea that significant parts of the Great Basin were abandoned during the middle Holocene became much less popular.

However, that the Altithermal notion is dated doesn't change the fact that while the middle Holocene was climatically variable, it was also quite arid and that shallow-water resources played a key role in the lives of hunter-gatherers in the Great Basin whenever they were available. In historic times, Great Basin population densities were highest in areas that were well watered, lowest in areas that were not. There is every reason to believe that the disappearance of the Ruby Marshes, Owens Lake, and Malheur Lake had a dramatic impact on the people who lived in those areas. Paulina Marsh, for instance, appears to have dried during the middle Holocene (chapter 8); it is probably not coincidental that the nearby Connley Caves appear to have been used by people before and after, but saw little significant use during, this period.

It is not easy to infer human population densities from archaeological data, especially for hunter-gatherers. Counting sites can be problematic since the existence of many sites might reflect the existence of small numbers of highly mobile people or many people who did not move often at all. The stemmed point sites that I have discussed provide a good example. There are many such sites in certain parts of the Great Basin, but the work of Tom Jones, Charlotte Beck, and others have shown that they appear to have been created by small numbers of very highly mobile people.

Another possible approach is to count the number of radiocarbon dates associated with archaeological material and examine how those dates are distributed through time. This approach, too, poses obvious problems. For instance, the older a deposit is, the less the chance that organic materials will have been preserved to be dated, and the less the chance that associated archaeological material will have survived intact. Geologically active periods of time may destroy significant numbers of sites and causes others to be deeply buried and so be less visible. Archaeologists may spend great amounts of effort to get radiocarbon dates for particularly interesting sites (for instance, very old ones) and less amounts of time to get dates for sites whose age is made obvious by the artifacts they contain (for instance, historic sites). A particular occupation layer for which two dates are available will count twice as much as a layer for which only one date was obtained; well-funded projects may produce more dates than those with less support, and so on. The problems are real but the guiding assumption is probably reasonably accurate, given that we recognize that it is almost always the case that more recent parts of the archaeological record will provide more dates than older ones. All other things equal, greater numbers of people on the landscape should, in general, leave behind more things to be dated than do smaller numbers of people. If middle Holocene climates led to a decrease in the number of people living in the Great Basin during that time, the frequencies of radiocarbon dates should show it.

Arguments for low human population densities during the middle Holocene in the Great Basin have been made for virtually all parts of this region. I begin with the western Great Basin and, in particular, the western Lake Lahontan Basin (figure 9-7), an area that archaeologist Robert Heizer long argued saw very low numbers of people during this time. Figure 9-8 shows the frequencies of radiocarbon-dated archaeological sites through time for this area: the higher the peak on the graph, the greater the chances that a site will date to that time.

Just as Heizer's argument requires, it is the middle Holocene that has lowest sustained troughs in the number of dated sites. Indeed, we need go back into the Pleistocene to find even lower lows.

While Heizer was making his argument for reduced middle Holocene human population densities in the western Great Basin, archaeologists working in the eastern Great Basin were busy pointing out that they saw no such phenomenon there. Excavations at Danger Cave by Jesse Jennings and by Mel Aikens at Hogup Cave led both of them to reject the notion that at least the lower elevations of the Bonneville Basin had seen declining numbers of people during Antevs's Altithermal, a position that has more recently been argued by David Madsen. Figures 9-9 and 9-10 show the frequencies of radiocarbon dates through time for this region. Figure 9-9 doesn't help very much, because the huge number of dates during the latest Holocene, almost all for the Fremont (see below), makes it hard to see the details in the earlier parts of the graph, so figure 9-10 begins at 1,500 years ago. There are several things to note about the pattern seen here. First, the middle Holocene spike shortly before 6,000 years ago is caused largely by the careful and well-dated excavations at Camels Back Cave, discussed earlier. Second, the lowest point in the radiocarbon curve during the past 10,000 years is, in fact, found during the middle Holocene. Beyond this, however, the ups and downs shown in the middle Holocene part of this graph do not differ all that much from those that come immediately before and after it.

At about the same time as Heizer and Jennings were working, Luther Cressman suggested that the climatic challenges of the middle Holocene led to decreased human use of at least the lower-lying parts of the Fort Rock Basin. Figure 9-11

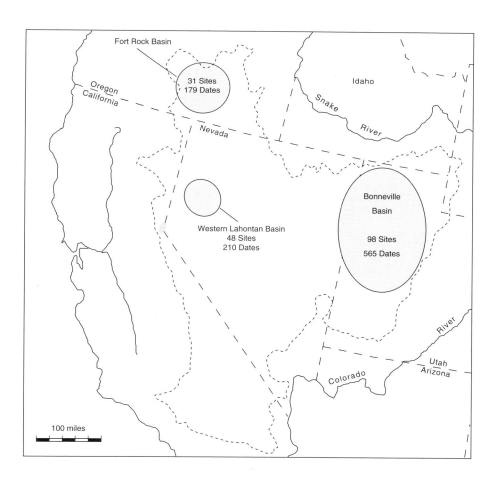

FIGURE 9.7 Areas within the Great Basin with numbers of radiocarbon-dated archaeological sites, and numbers of associated dates, displayed in figures 9-8 through 9-12.

THE PREHISTORIC ARCHAEOLOGY 305

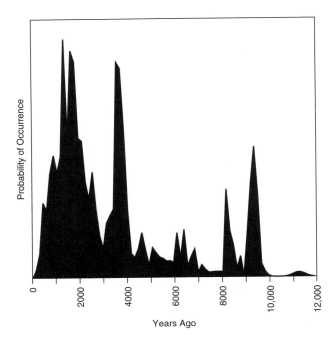

FIGURE 9.8 Changing frequencies of radiocarbon dates from archaeological sites in the western Lahontan Basin. The higher the peak in the graph, the greater the probability that a site dates to that time. (After Louderback, Grayson, and Llobera 2010.)

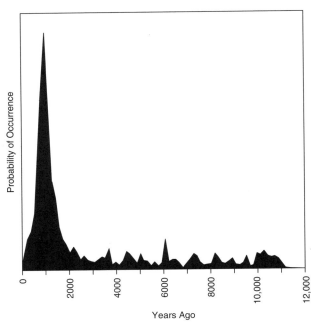

FIGURE 9.9 Changing frequencies of radiocarbon dates from archaeological sites in the Bonneville Basin. The higher the peak in the graph, the greater the probability that a site dates to that time. The huge spike in dates around 1,000 years ago is caused by the large number of dates for the Fremont phenomenon. (After Louderback, Grayson, and Llobera 2010.)

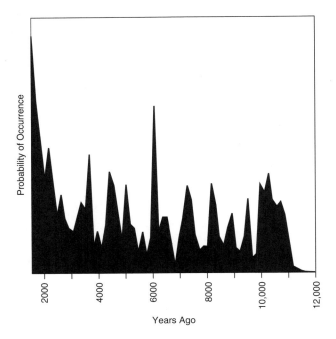

FIGURE 9.10 Changing frequencies of radiocarbon dates from archaeological sites in the Bonneville Basin prior to 1,500 years ago. The higher the peak in the graph, the greater the probability that a site dates to that time. (After Louderback, Grayson, and Llobera 2010.)

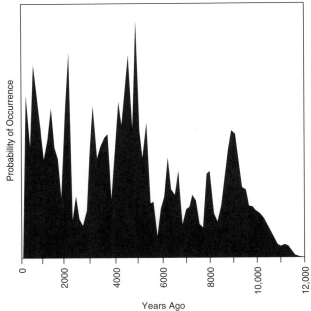

FIGURE 9.11 Changing frequencies of radiocarbon dates from archaeological sites in the Fort Rock Basin. The higher the peak in the graph, the greater the probability that a site dates to that time. (After Louderback, Grayson, and Llobera 2010.)

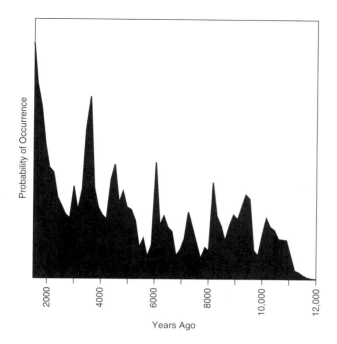

FIGURE 9.12 Changing frequencies of radiocarbon dates falling before 1,500 years ago in the western Lahontan Basin, Bonneville Basin, and Fort Rock Basin combined. The higher the peak in the graph, the greater the probability that a site dates to that time. (After Louderback, Grayson, and Llobera 2010.)

suggests that he appears to have been right. The lowest peaks in radiocarbon date counts tend to fall in the middle Holocene as do the lowest lows. And just as in the western Lahontan Basin, peaks in the early Holocene are higher than those in the middle.

Figure 9-12 shows that if we add up all the dates from all three areas, the lowest lows tend to fall within the middle Holocene. The obvious conclusion to draw from all of this is that nearly everyone who argued about the effects of the middle Holocene on human population densities in the Great Basin was, to one degree or another, correct. On the one hand, the parts of the Great Basin I have examined here were not abandoned throughout this period. On the other, there is every reason to believe that in significant parts of the Great Basin, human population densities thinned in response to the same climatic challenges that led to the widespread adaptation of small seeds as a food source, along with the grinding stones needed to process them. And, as all of the illustrations show, populations seem to have increased during parts of the middle Holocene, matching the evidence I discuss in chapter 8 for wetter interludes during this generally very arid interval.

Much the same appears to be true for the Mojave Desert, which has substantial radiocarbon evidence for middle Holocene occupations. But even here, very few sites date to the hottest and driest times of the middle Holocene. Indeed, Mark Sutton and his colleagues note that parts of the Mojave may have been abandoned during this time.

Projectile Point Chronologies

As I discussed, tallying the number of radiocarbon dates may provide trustworthy information on changing numbers of people on past landscapes through time but they can also lead us astray. This is particularly true for the Great Basin, where most archaeological sites are found on the surface of the ground, either because the surface has never been buried or because the sediments that once covered it have been eroded. Unfortunately, no known dating techniques can be used to assign precise ages to the vast majority of these surface sites. To get dates for these sites, archaeologists rely on the presence of artifact types whose forms are known to change through time.

Throughout North America, two major kinds of artifacts—pottery and projectile points—are routinely used for this purpose. In general, ceramics tend to provide North American archaeologists with their best artifactual time markers because they were often decorated in ways that changed sensitively as time passed. The highly mobile lifestyle of most Great Basin peoples, however, did not lend itself readily to the routine use of items that were cumbersome and difficult to transport securely. In addition, many of the functions for which ceramics were used elsewhere—food storage and cooking, for instance—were taken over by basketry in the Great Basin. As I discuss later, pottery did not become widespread across the Great Basin until after about six hundred years ago, and even then did not become very common in most areas.

As a result, archaeologists use projectile points as the major artifactual time-telling device for Great Basin sites. These tend to be far less sensitive temporal indicators than ceramics, but they are found throughout the area and, as far as we know, were in use during the entire span of human occupation in the Great Basin.

I have, of course, already used projectile points in just this way. Great Basin stemmed point sites from buried contexts routinely date to between about 11,200 and 7,500 years ago. It follows that surface sites that contain these points were also occupied during this period. This is a substantial time span, but it is far better than having no temporal control at all.

My concern here, however, is not with late Pleistocene and early Holocene projectile points but instead with the forms that follow them. Quite a bit of what we know about the changing human use of Great Basin landscapes during the past 7,500 years has come from using these later points to date surface sites, and to assign ages to buried sites that could be dated in no other way.

Middle and Late Holocene Great Basin Projectile Points

Following a system devised by Robert F. Heizer and his students, most Great Basin projectile point types that occurred during the past 8,000 years are given a two-part name, in which the first term refers to a site or to an area from which the type was first defined, and the second refers to some aspect of the shape of the point involved. "Cottonwood

Triangular," for instance, refers to a triangular point first defined from a site along Cottonwood Creek in Owens Valley. In addition to being named in this fashion, points that are similar to one another are routinely grouped into "series," with the various types within each series generally distinguished from one another on the basis of the morphology of the hafted end of the point.

Heizer and his students began the difficult process of making sense of the morphological and temporal variation shown by middle and late Holocene Great Basin projectile points. In a work of enduring importance, archaeologist Dave Thomas then codified these definitions and provided a key to allow their identification in the central Great Basin. The result of the work of these and other scholars has been the definition of four major series of chronologically sensitive Great Basin projectile point types in use during the past 7,500 years or so. Many other named point types exist, especially for the edges of the Great Basin, but I focus on these four series here.

Thomas's definitions are based on a large series of projectile points from the Monitor Valley, including Gatecliff Shelter (figure 8-8), and, he is careful to emphasize, are meant to apply only to this part of Nevada. As he noted, the classification he devised can be expected to become "fuzzier" the farther one goes from this part of the Great Basin.

Gatecliff Series points have triangular blades and medium-to-large stems (the hafted part of the point); the stems contract toward the base and are set off from the blade by angles that exceed 60°. There are two types in the series: Gatecliff Split Stem and Gatecliff Contracting Stem (figure 9-13).

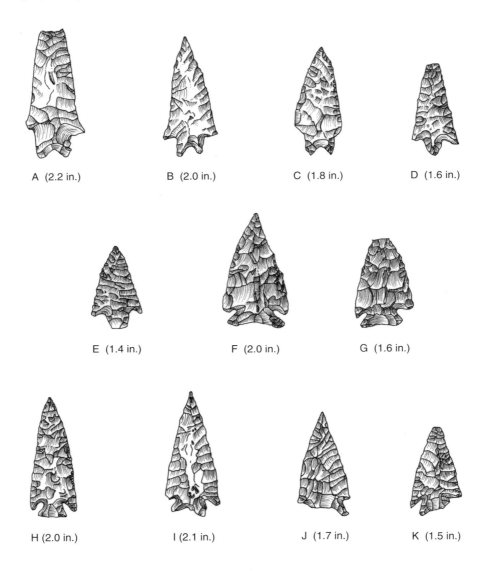

FIGURE 9.13 Gatecliff Split Stem (A, B, C), Gatecliff Contracting Stem (D, E), Elko Corner-Notched (F, G, H), and Elko Eared (I, J, K) points from Gatecliff Shelter, Toquima Range, Nevada (after Thomas 1983b). All are atlatl dart points. Measurements indicate the length of the specimens. Drawings by Peggy Corson.

Elko Series points are large and corner-notched; the angles at the notches are less than 60°, the width at the base greater than one centimeter (0.40 inch), and the weight generally greater than 1.5 grams. There are two named types: Elko Corner-Notched and Elko Eared (figure 9-13).

Rosegate Series points are small and corner-notched with triangular blades and expanding stems; the base is generally less than one centimeter wide, the weight generally less than 1.5 grams. Many archaeologists recognize two major types within the series (Rose Spring and Eastgate points), but others feel that the two are so similar to one another that they are best treated as a single type, routinely called Rosegate (figure 9-14).

Desert Series points include three named forms: Desert Side-Notched, Cottonwood Triangular, and Cottonwood Leaf-Shaped (figure 9-14). Desert Side-Notched points are small and triangular with notches fairly high on the sides and weights of no more than 1.5 grams. Cottonwood Triangular points are unnotched, thin, triangular points with straight to convex bases; they also weigh no more than 1.5 grams. Cottonwood Leaf-shaped points are identical to Cottonwood Triangular points except that they have rounded bases.

Gatecliff and Elko series points are broader at the base and heavier than Rosegate and Desert series points. They were used to tip atlatl darts, which were then propelled with an atlatl or "spear-thrower" (figure 9-15). The atlatl was not in use in the Great Basin or any nearby area during historic times, though it was used in the Arctic, along the Northwest Coast, near the mouth of the Mississippi, and in northern Mexico, including parts of Baja California. When Europeans entered the Great Basin, the native peoples of this area were armed with bows and arrows. The more gracile Rosegate and Desert series points were used as arrowpoints, with at least the latter in use during early historic times.

A good case for these different uses can be made from the points themselves. Both Dave Thomas and Michael Shott, for instance, have shown that historic period atlatl points tend to be more robust than historic arrowpoints. However, the best case can be made from the archaeological record itself. Atlatls, atlatl darts, bows, and arrows (and even more commonly, fragments of these weapons) have been found in some number in Great Basin archaeological sites. Most of the darts and arrows do not have points attached, but when they do, Gatecliff and Elko series points tip darts, while Desert series points tip arrows. In fact, Desert series points are also found on historic Great Basin arrows. To my knowledge, Rosegate points have not been found attached to arrows, but they have been found in deposits that have yielded arrow fragments.

If atlatls were not in use during historic times in the Great Basin but bows were, it follows that the atlatl dart points must be earlier than the arrow points. If the known historic arrows of the Great Basin were tipped with Desert series points, it follows that Rosegate points, which were also used to tip arrows, may have gone out of use before then.

In fact, the introduction of the bow and arrow into the Great Basin is marked by the advent of Rosegate points.

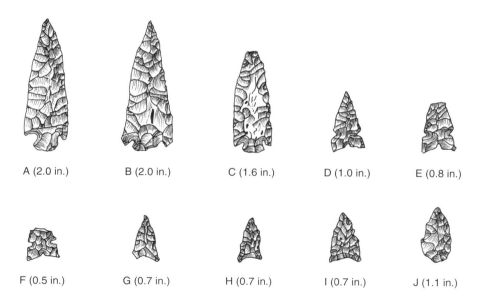

FIGURE 9.14 Rosegate (A, B, C), Desert Side-Notched (D, E, F), Cottonwood Triangular (G, H, I), and Cottonwood Leaf-Shaped (J) points. A–I from Gatecliff Shelter, Toquima Range, Nevada (after Thomas 1983b); J from Mateo's Ridge, Toiyabe Range, Nevada (after Hatoff and Thomas [1976]). Measurements indicate the length of the specimens. Drawings by Peggy Corson.

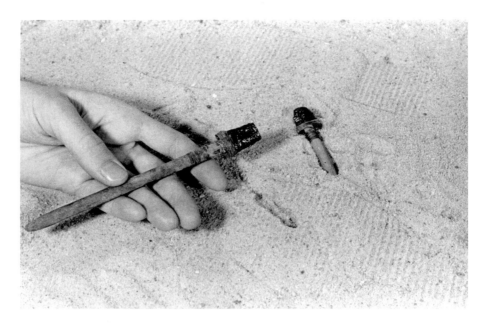

FIGURE 9.15 Projectile points hafted to atlatl foreshafts from Hidden Cave, Nevada. Photograph by Judith Silverstein, courtesy of David H. Thomas and the Division of Anthropology, American Museum of Natural History, New York.

These seem to appear first on the edges of the Great Basin, known from northeastern California as early as about 1,800 years ago, from the eastern Great Basin by about 1,700 years ago, and from the northern edge of the Mojave Desert south of Owens Lake by around 1,600 years ago. Elsewhere, they seem somewhat later. In the central Great Basin, for instance, they appear around 1,300 years ago. Desert series points begin to replace them after about 600 years ago, appearing at different times in different places.

It is perhaps not surprising that the bow and arrow seems to appear at different places at different times. Bob Bettinger and Jelmer Eerkens have shown that Rosegate points from eastern California are more variable than those from central Nevada, suggesting that people in the former area were experimenting with the form, while those in central Nevada received the bow and arrow as a package not subject to experimentation. This does not mean that the bow and arrow reached the interior of the Great Basin from California (no one has replicated the Bettinger and Eerkens analysis along the northern or eastern edge of the Great Basin), but it does provide insight into the process by which this particular set of tools became incorporated into the Great Basin tool kit.

Given that the bow and arrow was subjected to tinkering within the Great Basin and that it took some time to spread across this region, it follows that it must have coexisted with the atlatl and dart complex that it ultimately replaced. That it did so seems fairly clear, but the chronology of the atlatl dart points that they replaced is quite complicated.

I begin where the beginning is easiest, in the central Great Basin. Here, Thomas's work, confirmed by others, shows that Gatecliff series points are the earliest dart points, dating from about 5,000 to 3,500 years ago. These are followed by Elko points, in use between about 3,500 and 1,300 years ago, when they are replaced by Rosegate points. Rosegate points then begin to be replaced by Desert series points around 600 years ago, with these latter forms in use into historic times.

If things were like this everywhere in the Great Basin, we would be in good shape. We would have a discrete set of artifact types each of which marked a well-defined period of time. If we found a surface site marked only by Elko points, we would know that it was created sometime between 3,500 and 1,300 years ago. In fact, this is exactly what can be done in the central Great Basin. Unfortunately, as we move away from this area, things get messier.

First, just as Thomas warned might happen, his key, carefully constructed to apply to dart points from central Nevada, begins to lose accuracy. This is especially true along the edges of the Great Basin, where different kinds of projectile points are found, points that the key was not meant to handle. This is no big deal, since altering the key can handle this situation. Second, though, the chronology of these points also begins to break down.

The problem first became evident with Elko points. Archaeologist Rick Holmer observed that these artifacts seemed to be in use earlier in the eastern Great Basin than in central Nevada and that they lasted much later; he

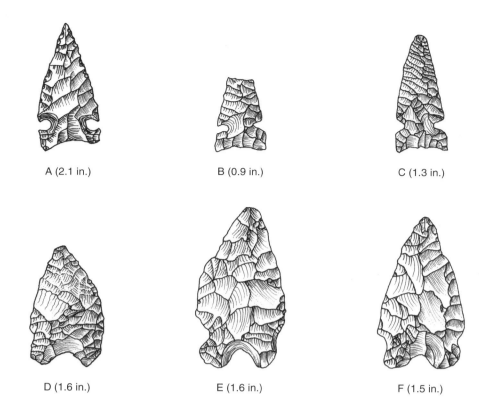

FIGURE 9.16 Northern Side-Notched (A, B, C) and Pinto (D, E, F) points. A from Surprise Valley, California (after O'Connell 1975); B–C from Nightfire Island, Lower Klamath Lake, California (after Sampson 1985); D from Long Valley, Nevada (after Beck and Jones 2009); E–F from the Stahl Site, Rose Valley, California (after Schroth 1994). Measurements indicate the length of the specimens. Drawings by Peggy Corson.

suggested a time range from as early as about 8,000 years ago to as late as about 1,000 years ago. The early use of Elko points is also known from in the northern Great Basin. Jim Wilde, for instance, found Elko points beneath Mazama ash at the Skull Creek Dunes site in Catlow Valley, just west of Steens Mountain. While no radiocarbon dates are available for this level of the site, the points must predate the deposition of the ash 6,730 years ago. When Mel Aikens first argued from his Hogup Cave data that Elko points were far earlier in the east than they were to the west, many archaeologists were reluctant to believe it. A number of sites, however, now show that he was right.

It is not just the eastern and northern Great Basin where this happens. In the Honey Lake Basin of northeastern California, Elko points date to as early as 5,000 years ago, though their ending dates seem to be about the same as they are in the central Great Basin. And it is not just Elko points that refuse to stay put in time: Gatecliff points do the same thing. In the Mojave Desert, these points seem to date to about the same time as they do in the central Great Basin, but in northeastern California and adjacent Nevada they are at times found in secure stratigraphic contexts that show that they were in use well after 3,000 years ago; indeed, some have suggested they may have lasted until 2,000 years ago or so in the northwestern Great Basin.

The obvious conclusion to draw here is that the different major forms of dart points seem to have different histories in different parts of the Great Basin. Arrow points behave the same way. Desert Side-Notched points show up at different times in different places, and Rosegate points may have remained in use in parts of the far western Great Basin until close to historic times. Rather than being able to apply the well-defined chronology available for the central Great Basin to other parts of this region, archaeologists have to build separate chronologies and separate keys for the particular areas in which they are working before they can know the degree to which the points overlap in time and precisely what those times are.

I began this discussion by noting that much of the Great Basin's archaeological record sits on the surface of the ground, that these sites cannot be dated precisely, and that, as a result, counting the number of radiocarbon dates through time might provide a misleading indicator of changing human population densities. Using time-sensitive artifacts—projectile points in the Great Basin—is an obvious way to provide an independent measure of these changing densities. Now, though, we have seen that these artifact types seem to have different chronological meaning in different parts of the Great Basin. A site with Elko points in the central Great Basin certainly dates to the late Holocene, but the same site in the eastern Great Basin might be either middle or late Holocene. That makes things tough.

It is also true that the central Great Basin chronology, built by Dave Thomas, begins 5,000 years ago, with Gatecliff points and thus does not cover the middle Holocene at all. This, as Charlotte Beck and others have pointed out, may well be a reflection of the light human use this area saw during the middle Holocene. If there are no well-dated middle Holocene sites in an area, we have no access to the kinds of projectile points that might have been used at that time.

In fact, there seem to be only two projectile point types that we know were in use throughout the middle Holocene, and one of those was in use before the middle Holocene began. These are Northern Side-Notched and Pinto points.

Northern Side-Notched points (figure 9-16) are known from the Columbia Plateau just north of the Great Basin, the northern parts of the Great Basin, and Utah's Wasatch Front. They are most abundant north of the Humboldt River, but they reach farther south than this in some areas—known, for instance, from Camels Back Cave on central Utah's Dugway Proving Ground. In general, they date to between about 7,300 and 4,500 years ago, though they might have lasted somewhat later than this in some areas. As such, they are seem to be good markers of the middle Holocene where they are found. The middle Holocene Surprise Valley houses that I discussed are characterized by these points. Similar versions provide the earliest known evidence for communal pronghorn hunting in the Great Basin.

Pinto points were defined in the 1930s, before the Berkeley system came into being. As a result, it is one of the few Great Basin point types whose name does not reflect a combination of where it was first defined (the Pinto Basin site in Joshua Tree National Monument) and what it looks like. Found primarily in the southern and eastern Great Basin, these artifacts resemble Gatecliff Split-stem points but tend to be more robust and differ from them in details on the base of the point (figure 9-16); they may have tipped spears rather than atlatl darts. For many years, Pinto points were thought to be a strong marker of the middle Holocene in the Mojave Desert but work by Mark Basgall, Adela Schroth, and others shows that they date from 9,000 to perhaps as late as 4,000 years ago. Much the same is true for the eastern Great Basin, where they seem to date to between about 9,000 and 5,500 years ago or later. Archaeologist Claude Warren has argued that these points essentially mark of the end of the stemmed point tradition and he is probably right.

Elusive Sites and Low Population Densities

These are the only abundant artifactual time markers available for the past 8,000 years or so in the Great Basin. If our concern is with understanding the nature of middle Holocene human occupation in this region, there is an obvious problem. Surface sites, and buried sites that cannot be directly dated, make up the bulk of the known Great Basin archaeological record. These sites are routinely assigned ages on the basis of the time-sensitive artifacts they contain. Unfortunately, precious few of these are known for the middle Holocene. In the more northerly parts of the Great Basin, Northern Side-Notched points fill the bill. In both the southwestern and eastern Great Basin, Pinto points come close to doing the same, but they came into being during the later parts of the early Holocene. If our goal is to use artifactual time-markers to identify archaeological sites as middle Holocene in age, only Northern Side-Notched points will do the trick

John Fagan put these points to use some years ago, following arguments that had been made by Bedwell and Cressman. Fagan reasoned that if aridity had caused places like the Fort Rock Basin to be less intensively used during the middle Holocene, people might have focused their activities on higher-elevation springs. Accordingly, he examined twelve springside archaeological sites scattered across a broad swath of southeastern Oregon, from just west of Warner Valley to just east of Steens Mountain. He found that ten of his twelve sites contained Northern Side–notched points and concluded that these were middle Holocene occupations. Given that Northern Side–notched points were in use between 7,300 and 4,500 years ago, there is no reason to doubt him.

Fagan's northern Great Basin sites are similar to many middle Holocene sites elsewhere in Great Basin in an important way. They are near dependable supplies of water, just as are such sites as Danger Cave, Hogup Cave, King's Dog, Menlo Baths, and Nightfire Island. In the Honey Lake area of northeastern California, Randall Milliken and William Hildebrandt (1997:158) have observed that middle Holocene occupations centered on the wetlands that happened to remain. In those places where the wetlands disappeared, so did the people. The Connley Caves also provide an example, occupied during the early and late Holocene but only sparsely, if at all, in between. And there are many caves whose occupations did not begin at all until the middle Holocene was over or nearly so, including Gatecliff Shelter (5,370 years ago) and Lovelock Cave (4,690 years ago). In the Mojave Desert, middle Holocene sites occur in a broad variety of settings, but sites in this area seem to disappear for something on the order of 1,000 years during the harshest times the middle Holocene had to offer.

Middle Holocene aridity seems to have had a dramatic impact on the human occupants of the Great Basin. Grinding stones became common as this interval began, reflecting the intensified use of energetically expensive seeds. In the Mojave Desert, as we have seen, grinding stones were more abundant during this interval than either before or after. Sites that had once been occupied were vacated. and in many areas human use of the landscape seems to have been heavily focused on dependable sources of water. Some areas may actually have been abandoned, but even if this were not the case, the lowest human population densities in the Great Basin during the past 10,000 years or so seem to have been reached during this lengthy interval, as figures 9-8 through 9-12 show. As these figures also show, human population densities varied during this period, just as the nature of middle Holocene environments varied.

All this has to be tempered by the fact that, outside of the area where Northern Side-Notched points occur, we are unable to date middle Holocene surface sites, and often even buried sites, securely. Nonetheless, unless the archaeological record is playing an evil trick on us, it does look as if the Great Basin during the middle Holocene was marked by relatively low human population densities and that those low densities were very likely caused by increased aridity. Ernst Antevs, Luther Cressman, Robert Heizer, and other scholars of earlier decades who took this position may have gotten some of the details wrong, but they appear to have been right about the general picture.

The Late Holocene

The Record Expands

In chapter 8, I note that if I were forced to pick a time when the Great Basin first began to look generally as it looked when Europeans first arrived, I would pick 4,500 years ago, the beginning, more or less, of the late Holocene. That conclusion would also seem to apply to the adaptations of the people who occupied this region. As with Great Basin environments, the exact date differs from place to place, but between about 5,000 and 4,000 years ago, the archaeological record begins to look very much as if it could have been created by people living in ways similar to the ways in which Great Basin native peoples lived when Europeans first encountered them.

The beginning of the late Holocene is marked by a tremendous increase in the numbers of known archaeological sites, both those dated directly and those dated by associated projectile points. In the western Lahontan Basin, this increase centers on 4,000 years ago (figure 9-8). In the Fort Rock Basin, it occurs around 4,500 years ago (figure 9-11). In the Bonneville Basin, the increase across the transition from the middle-to-late Holocene is far less dramatic (just as Aikens, Jennings, and Madsen have argued) but it is there and it, too, falls at about 4,500 years ago (figure 9-10).

Many caves were also first occupied as the middle Holocene came to an end or soon after. At the Connley Caves the occupational hiatus ends at about 4,700 years ago. The first human occupation of Newberry Cave, in the western Mojave Desert, is marked by Gatecliff Contracting Stem points; the earliest radiocarbon date here falls at 3,765 years ago. At Stuart Rockshelter, in the eastern Mojave Desert, the earliest occupation occurred at 4,050 years ago. Hidden Cave has basketry dated as early as 9,300 years ago but the earliest intense use of this site, again marked by Gatecliff series points, began at about 3,700 years ago. At Kramer Cave, the earliest intense use began at about 3,900 years ago, also marked by Gatecliff series points. Lovelock Cave seems to have been utilized soon after 4,700 years ago; South Fork Shelter, at about 4,360 years ago, Pie Creek Shelter at 4,840 years ago, Lakeside Cave at 4,710 years ago, and so on. These dates are impressively consistent and are matched by other aspects of the archaeological record. Substantial sites with structures appear along the shores of Lake Abert soon after 4,000 years ago, for instance, and the number of sites in northeastern California increases significantly once we reach the late Holocene. In the Mojave Desert, the number of sites begins, in Mark Basgall's words, "a gradual but inexorable increase" (2000b:133) at this time.

This situation is completely different from that encountered with the middle Holocene, when it can be difficult to find any sites at all. Once the late Holocene begins, sites become abundant. They are also found almost everywhere. In the eastern Great Basin, where such sites as Danger Cave and Hogup Cave appear to have been utilized throughout the middle Holocene, the onset of the late Holocene sees not only an increase in the number of known sites but also, as Mel Aikens and David Madsen have pointed out, the first significant use of upland settings. Indeed, Gatecliff series points are frequently found in the alpine zone of the mountains of the central Great Basin. Dating to between 5,000 and 3,500 years ago here, they are commonly associated with rock walls, cairns, and rings, features that appear to have been used in conjunction with hunting. Although, as I have discussed, high-elevation settings were used well before this time, the onset of the late Holocene seems to mark the first substantial use of such settings throughout the Great Basin.

The intensive use of a broad variety of environments was fully characteristic of the native peoples of the Great Basin at the time of European contact. The archaeological record suggests that such use began as the late Holocene began. In some areas, even the houses that date to early in the late Holocene are much like those known from the region historically.

In Surprise Valley, for instance, the large housepit structures that marked the middle Holocene occupation at King's Dog and Menlo Baths gave way to far less substantial structures, associated with Gatecliff series points, after 5,000 years ago. Jim O'Connell excavated forty-six of these structures at two Surprise Valley sites, King's Dog (where they are stratigraphically above the large middle Holocene houses) and Rodriguez. All of these houses were preserved as saucer-shaped depressions from ten to eighteen feet in diameter and up to about twenty inches deep, with postholes ringing the inner edge of the depression and hearths located in the center.

Unlike the situation with the larger, earlier houses, the superstructures of many of these later forms had burned and were preserved as charred remains. As a result, O'Connell was able to show that two different kinds of structures were involved. The most abundant of these was a dome-shaped wickiup whose superstructure was formed by light willow or aspen poles which were then covered by grasses, brush, and tule mats. The other form was a simple brush windscreen—a circular, unroofed frame of poles that had brush and mats tied to it. Both kinds of structures were widely used in the Great Basin during historic times.

The forty-six structures at these two sites were by no means contemporaneous. King's Dog, for instance, had fifteen of the saucer-shaped depressions stacked one on top of the other, and the radiocarbon dates indicate that these structures had been built across several thousand years. While the earliest projectile points associated with these structures are Gatecliff series, the latest are Rosegate. It is not clear why such a dramatic shift in house types from middle to late Holocene times occurred here, though it is possible that the substantial middle Holocene houses represent winter occupations, while the more lightly built late Holocene structures were for summer use.

This represents a dramatic change not just from the middle Holocene but from anything that appears to have gone on before. Fluted point sites are confined to valley bottom settings. Although some stemmed point sites are found in the uplands, most are also in valley bottoms. Middle Holocene sites are simply scarce, and many of those that we can date securely to this period are located near lakes and springs. The scarcity of such sites suggests that human population densities in the Great Basin were generally low during this period, just as they were low in water-poor areas during early historic times. Once the middle Holocene ends, however, sites become common, and they become so in diverse environmental settings. The difference is pronounced, and makes it appear as if we are seeing, for the first time, human adaptations, and human population densities, much like those of early historic times.

It is tempting to argue that this pronounced shift was related to the arrival of pinyon pine through much of the Great Basin. As I discuss in chapter 2, pinyon was, historically, an extremely important food source throughout those areas of the Great Basin that contained it, providing a resource often critical for winter survival. Pinyon, however, did not arrive in the central latitudes of the Great Basin until after about 6,500 years ago, and was probably not abundant here until well after that. Much of the middle Holocene had neither the rich valley-bottom resources that were critical to people during the late Pleistocene and early Holocene, nor the pinyon that was critical in many areas during early historic times. Scarce resources lead to scarce people.

Once pinyon arrived, we would expect human populations to respond, and perhaps to respond quickly. Gatecliff Shelter shows the kind of correlation between the arrival of this productive resource and the arrival of people that would be expected in a relatively resource-poor environment. Pinyon arrived in this part of the Toquima Range soon after 6,000 years ago (figures 8-16 and 8-18). The first human occupation at Gatecliff occurred at about 5,350 years ago, and the first hearths in this site, which date to about the same time, contain the remains of pinyon pine. That is, once pinyon arrived in the Toquima Range, it appears to have been used and Gatecliff itself was not used until it was there.

Pinyon, however, cannot account for the changes that occur in the archaeological record at the same time in many other parts of the Great Basin. Hidden Cave saw significant human use by 3,700 years ago, yet pinyon pine did not arrive in this area until after 1,500 years ago. Pinyon never reached south-central Oregon, yet the occupational hiatus at the Connley Caves ends at about 4,700 years ago and substantial sites appear along the shores of Lake Abert soon after 4,000 years ago. The phenomenon appears to match what was happening far to the south, but it happened in the absence of pinyon.

Although the arrival of pinyon certainly made a difference, the end of middle Holocene aridity appears to have been far more important. The Ruby Marshes were reborn at about 4,700 years ago, about the same time that the sagebrush-grass ratios at Fish Lake, in Steens Mountain, fell, and the Connley Caves were reoccupied. The pollen from Leonard Rockshelter strongly suggests desiccation of the Humboldt Sink during the middle Holocene (chapter 8), but when nearby Lovelock Cave was first utilized, at or soon after 4,600 years ago, the coprolite evidence shows that people were making heavy use of wetland resources. The same is true for Hidden Cave: coprolites from the earliest intense human use of this site are full of bulrush seeds, cattail pollen, and the bones of small fish. Amazingly, Dave Rhode's analysis of hormone residues in these coprolites shows that all of them came from women.

During historic times in the Great Basin, human population densities were highly correlated with the abundance of shallow water. Where shallow water was abundant, population densities were high. Where water was scarce, population densities were low. Prehistorically, when middle Holocene aridity ended, human population densities appear to have increased. In some cases, pinyon was important, but in others it was not. What counted everywhere was that it was no longer as arid as it had been.

Numic Expansion

The foodstuffs utilized by the native peoples of the Great Basin during early historic times often presented significant incongruities in both time and space. Pinyon nuts, for instance, are to be found only in the middle elevations of the mountains and only during the fall. Sedge fruits are also available only in the fall but are most abundant in low-elevation marshes. The flowering heads of cattails are found in these marshes as well, but only during the summer. Yellow-bellied marmots are generally active from late spring to early fall but they are often found only above the upper treeline. Townsend's ground squirrels can be taken during spring and summer but only beneath the lower treeline.

Such spatial and temporal incongruities presented significant challenges to Great Basin native peoples. To be sure, the severity of the challenge varied from place to place. Where resources were densely packed, generally because the area was well watered, as in Owens Valley, the challenges were lessened and population densities higher. Where resources were widely scattered, generally because the area was poorly watered, as in much of the Bonneville Basin, the challenges were more severe and population densities lower. But whether we look at the Owens Valley Paiute or the Gosiute, the solution was the same. People dispersed across the landscape to gain access to far-flung resources only seasonally available, and they stored many of these resources to level the temporal differences in resource availability.

This pattern was also in existence in the Great Basin early in the late Holocene. Although there are a number of ways to document this fact, the coprolites alone are compelling. As I mentioned, the earliest coprolites from the deeper deposits at Hidden Cave, dated to between 3,700 and 3,400 years ago, contained both cattail pollen and sedge seeds. As Peter Wigand and Peter Mehringer pointed out, either the pollen or the seeds, or both, had to have been stored. Although they may have been available in the same environment, they do not become available at the same time.

What we know of the adaptations of many Great Basin peoples some 4,000 years ago makes them seem similar to the adaptations of the peoples encountered by Europeans in early historic times. There were, of course, some substantial differences: to take but one example, the earlier peoples did not have the bow and arrow. But in many areas, the general use of the landscape seems to have been quite similar.

That the use of the landscape at 4,000 years ago and at the time of European contact was similar, however, does not have to mean that the modern Native peoples of the Great Basin are lineal descendants of those who lived here far earlier during the Holocene. Obviously, distinctly different peoples can use the same landscape in similar ways. Because language, race, and culture are totally independent phenomena—at birth, any person has the capability of speaking any language and belonging to any culture—it is at best difficult to tell from the prehistoric archaeological record how a given set of people might be related to those who were in the same area thousands of years before and impossible to determine what language those earlier people spoke. As a result, the general similarities between human adaptations during early historic times and those 4,000 years ago in the Great Basin cannot be taken to mean that these earlier peoples spoke languages closely related to those spoken by the native peoples of the region during historic times, or that the people in the Great Basin earlier in the Holocene were the direct ancestors of those encountered when Europeans arrived.

In fact, most archaeologists and anthropologists think that they were not closely related at all. Many scholars believe that the Numic speakers who occupied much of the Great Basin at the time of European contact (see chapter 2 and figure 2-16) spread across this region beginning about 1,000 years ago. This postulated late prehistoric population movement is referred to as the "Numic expansion."

Similar movements are documented to have occurred at other times and other places. To take but one of many prehistoric North American examples, the homeland of the ancestors of the people we know as the Navajo and Apache was in far northwestern North America: they are related to the Haida and Tlingit of the Pacific Northwest Coast and to many peoples in northwestern Canada, as well as to the Yurok and Wiyot of northwestern California. Their ancestors arrived in the Southwest only a few hundred years ago, by a route still unknown.

That the ancestors of the Navajo and Apache arrived in the Southwest relatively recently is not controversial among scientists. There is no other way to account for the languages spoken by these people, for their genetic affiliation, or for certain aspects of the archaeological record. That Numic speakers spread across much of the Great Basin during the past 1,000 years or so, however, has been highly controversial.

The idea is an old one. As David Madsen has pointed out, archaeologists have been suggesting things along this line since at least 1930. But the modern debate stems not from these early suggestions but from the efforts of scholars whose work focused on the nature of Uto-Aztecan languages. To see how this happened requires that we take a brief tour through some aspects of historical linguistics.

GLOTTOCHRONOLOGY

French, Italian, Spanish, and Portuguese are all descended from Latin, a fact evident from the similarities these languages show in their structure and content. For instance, significant amounts of the vocabulary in all four of these modern languages is derived from Latin. Milk in Spanish is *leche*; in French, *lait*; in Italian, *latte*; and in Portuguese, *leite*. In the ancestral Latin, it was *lactem*. Words possessing a common origin of this sort are called cognates.

The differences between cognates, as between *lait* and *leche*, are due to the slow changes that languages undergo as time passes. In addition to these kinds of changes, however, words used to convey a given concept can also be replaced by a new term, either borrowed from a different language or created anew. When, years ago, I learned French, I was taught to say "la fin de semaine" for weekend. Today, the common French term for weekend is *le weekend*, a loan word from English.

Given that related languages become increasingly different from one another as time passes, it follows that the number of cognates they share will decrease through time. This in turn suggests that the percentage of cognates in a pair of related languages might be used as a linguistic clock. The fewer the number of remaining cognates in a pair of related languages, the longer those languages have been separated. Further, if the rate of "decay" or loss in the number of cognates were known, perhaps the actual time of divergence of a language pair could be calculated.

Beginning late in the 1940s, the brilliant linguist Morris Swadesh began to develop these thoughts in detail. Ultimately, he and others produced a method for calculating the time of divergence of language pairs, a method that became known as glottochronology. Glottochronological methods other than the one developed by Swadesh exist, but only Swadesh's approach has been applied to Great Basin languages, and it is this approach to which I refer in what follows.

In Swadesh's method, languages weren't compared on the basis of their entire vocabulary, but instead on the basis of a core vocabulary containing items felt to be universal and noncultural, words referring to "things found anywhere in the world and familiar to every member of a society" (Swadesh 1952:457). At first, the core vocabulary used by Swadesh contained two hundred words; later, it was shortened to one hundred. Both lists contained such words as (in English) "see," "woman," "tongue," "hair," and "hand."

In 1950, Swadesh determined that during the past 1,000 years, English had retained about 85 percent of a set of its core vocabulary words. Soon after, linguist Robert Lees examined the histories of thirteen languages and determined that the average core vocabulary retention rate in these languages for the two-hundred-word list was 81 percent per thousand years. If it is assumed that this retention rate is both accurate and constant through time, it can be used to estimate the time of divergence of two languages from a common parent. For instance, two languages that have 66 percent cognate items in their core vocabulary can be calculated to have been separated for 1,000 years. The calculation is simple: each language will retain 81 percent of core vocabulary items after 1,000 years: $81\% \times 81\% = 66\%$.

For a variety of reasons, the estimates of the length of linguistic divergence provided by Swadesh's glottochronology are minimum estimates only. As a result, they are often referred to as "minimum centuries of divergence." If glottochronology really provides us with a "glottoclock," that clock would provide a minimum estimate of the amount of time that has passed since two languages diverged from a common parent.

Unfortunately, there is every reason to think that Swadesh's glottochronology does not work. For starters, no reasons compel us to think that there really is a culture-free vocabulary. In addition, serious difficulties are associated with both the idea and the calculation of constant retention rates. When Swadesh, Lees, and others calculated their retention rates, they did it the only way they could—by using written languages. Writing, however, stabilizes a language, and there is no reason to think that retention rates drawn from written languages have anything to do with retention rates for those that are not written. Worse, there isn't even any way to find out, since we can't talk to people who lived 1,000 or 2,000 years ago.

It is also true that retention rates should not be constant, but should instead reflect the contexts in which people live. Such things as population densities, degrees of mobility, environmental complexity, and external contacts should play a significant role in determining the length of time items are retained in a vocabulary. The greater the degree of contact among members of a linguistic group, the slower the diversification of the languages among the populations that constitute that group, and vice versa.

Even if Swadesh's glottochronology did work for unwritten languages, we would have no way of knowing it. But if these arguments are not enough, there is also the fact that applications of glottochronology to languages whose history is known do not provide encouraging results. To take but one example, John Rea's glottochronological analysis of Italian and French determined that these languages had split by 1586 A.D. As Rea observed, however, French and Italian were well differentiated by that time. The first written record of French dates to 842 A.D. and of Italian, to 960 A.D., but these written records postdate the divergence of the languages themselves. At the Council of Tours, in 813 A.D., French bishops decided that priests were to speak in the vernacular, not in Latin, "so that all may understand what is said" (Rickard 1989:18).

Many scholars obtained similar incorrect results for a diverse variety of other languages. Theresa Bynon (1977:270) pointed out that where they can be tested, many of the results of glottochronological analyses are, simply put, "absurd."

Classic glottochronological methods are now rejected by linguists for a simple reason: the assumptions on which they were based are now known to be incorrect. There are far more sophisticated methods for estimating times of language divergence, but even these have been questioned. It was, though, the classical, now-rejected approach that was applied to the languages of the Great Basin and that led to the popularity of the notion of a "Numic Expansion" about 1,000 years ago.

SYDNEY LAMB AND NUMIC TIME DEPTH

In 1955, Swadesh applied his glottochronological methods to Uto-Aztecan languages, hoping thereby to clarify the position of Nahuatl, the language of the Aztecs (see chapter 2). As part of this work, Swadesh calculated that Numic and what was then perceived to be its closest relative, Tübatulabal, had diverged a minimum of 3,500 years ago (some still think that Tübatulabal is Numic's closest relative, while others give this position to Hopi). In addition, he suggested that Ute and Tübatulabal had split at least 2,900 years ago and that Mono and Ute had diverged at least 1,900 years ago.

Swadesh's paper, published in Spanish in a Mexican journal and dealing primarily with Mexican languages, might have gone unnoticed by Great Basin archaeologists were it not for the linguist Sydney Lamb. In a tremendously influential paper on Numic languages, Lamb observed many of the facts that I recounted in chapter 2: that there are six closely related Numic languages in the Great Basin (see figure 2-16); that these fall into three groups of two languages each; that one member of each pair occupied a very small part of the southwestern Great Basin, while the second member of each pair spread over tremendous distances to the north and east; and, that dialectical variation within languages seemed greater in the southwestern Great Basin than in other areas occupied by Numic speakers.

These geographic considerations led Lamb (1958:98) to suggest an original homeland for Numic speakers:

> We have six languages in the Numic family, two belonging to each subdivision. Three of these languages are in a small cluster near the southwestern part of the Great Basin, while the other three occupy a vast area to the north and east. Remembering that the split into the three subfamilies preceded the separation of each of them into a pair of languages, one must place the linguistic center of gravity of the Numic family somewhere around Death Valley.

This conclusion, Lamb noted, is supported by the fact that Tübatulabal, then thought to be Numic's closest relative, was spoken directly across the Sierra Nevada from the area he had targeted as the Numic homeland.

Having inferred the location of this homeland, Lamb attempted to determine when Numic peoples had begun to disperse across the Great Basin. Here, he turned to Swadesh's glottochronological work. He followed Swadesh in suggesting that Numic and Tübatulabal were becoming distinct between 3,000 and 4,000 years ago and that the division of Numic into three separate languages, all then located in the southwestern Great Basin, occurred at about 2,000 years ago or shortly thereafter.

From this point, Lamb was on his own, since Swadesh had not calculated separation times within the three pairs of languages that comprise Numic. Lamb speculated that Numic speakers remained confined to the southwestern Great Basin until about 1,000 years ago, at which time "there began a great movement northward and eastward, which was to extend the domain of Numic far beyond its earlier limits" (Lamb 1958:99). This movement—the Numic expansion—saw the replacement of the people who had lived here earlier. The linguistic affiliation of these earlier people, Lamb observed, is simply unknown.

In Lamb's view, the Numic speakers encountered by Europeans in much of the Great Basin had been there for a very short period of time indeed—no longer than 1,000 years and perhaps, toward the northern and eastern edges of the region, far less. Although Lamb's estimated 1,000 year date for the Numic expansion was speculative, glottochronological data that appeared after Lamb wrote his essay supported that estimate. In 1958, linguist Kenneth Hale calculated minimum dates of under 1,000 years for the development of Northern Paiute, Shoshone, and Ute; in 1968, James Goss estimated a minimum of 900 years for the division of Kawaiisu-Ute into its component languages; in 1971, Wick Miller and his colleagues calculated a minimum of 700 years for the split between Panamint and Shoshone. The consistency is impressive.

In making his linguistic arguments, Lamb was unaware that archaeologists had made roughly similar suggestions. However, it certainly did not hurt that the general idea was in the archaeological intellectual air, nor did it hurt that Lamb's argument was by far the most coherent and compelling of them all. Soon, many Great Basin archaeologists and anthropologists had accepted Lamb's view and continue to do so.

Care, however, is required here. Lamb himself was fully aware of potential problems with his glottochronological results. In 1958, he noted that the results of glottochronology are merely "very rough approximations" that are "better than nothing at all" and that it must be used "only with great caution" (1958:98). More recently, he has suggested that the linguistic data could be consistent with the movement of Numic peoples across the Great Basin as early as 3,000 years ago. Today, even greater caution is required, given that Swadesh's glottochronology "is now largely discredited" (Gray and Atkinson 2003:436). That early glottochronological estimates provided consistent results for the divergence of Numic languages does not mean that it provided the correct ones. On the other hand, it also does not mean that the answers it provided were wrong.

CORN, BEANS, AND PROTO-UTO-AZTECAN: AN ALTERNATIVE VIEW OF NUMIC PREHISTORY

Without question, Numic peoples did expand across the Great Basin. After all, they are here now and they weren't here 20,000 years ago. The question is when this happened and, if people were not filling a landscape devoid of others, why it happened.

In the standard account, Numic peoples were confined to the southwestern corner of the Great Basin until about 1,000 years and then began to move to the north and east. The date for the beginning of it all is derived from glottochronology, while the direction of movement is derived from the fan-shaped distribution of Numic languages in the Great Basin, with the maximum diversity of those languages occurring in the southwestern Great Basin.

There are, though, other ways of explaining the distinctive geography of Numic languages in the Great Basin. In 1925, the great anthropologist Alfred Kroeber surveyed the distribution of Uto-Aztecan languages in California and the Great Basin. He found it more likely that "the semicircular fan" (1953:577) that described this distribution reflected not the direction of movement of people across space but the places where people "gradually agglomerated." Indeed, he concluded that Numic speakers in the Great Basin might have been here "from time immemorial" (1953:580). In 1977, linguist James Goss suggested that the Numic linguistic geography might be accounted for by the degree of isolation among linguistic groups across the Great Basin. Most recently, linguist Jane Hill has argued that the greater complexity of Numic languages in the southwestern Great Basin might be best explained by their nearness to speakers of other languages in California. In none of these views does the geography of Numic languages provide a key to the movement of Numic peoples in the deeper past.

Hill, in fact, completely dismisses the standard account of Numic expansion and calls instead for what she refers to as "Numic ethnogenesis *in situ*" (2002a:333). This name for her version of Numic history makes it sound like she agrees with Kroeber in thinking that the people who speak these languages have been in the Great Basin from time immemorial. This, however, is not what she thinks at all.

All Uto-Aztecan languages share a common ancestor, known as Proto-Uto-Aztecan, but Hill focuses much of her attention on a subset of these, known as Northern Uto-Aztecan—Hopi, Numic, Tübatulabal, and Takic (the Shoshone peoples of southern California). These share a more recent ancestor than Uto-Aztecan as a whole, much as Italian and French share a more recent common ancestor than does English with either Italian or French. Her analysis of the vocabulary of Northern Uto-Aztecan languages suggests to her that many of them have cognate words related to corn (or maize) horticulture. She argues that these terms have developed from a common ancestor (think *lait, latte,* and *leche*), and that they are shared with the more southerly Uto-Aztecan languages. These southern languages range deep into Mesoamerica, the source of corn itself. She concludes from this that the ancestors of today's Northern Uto-Aztecan speakers must have moved north into the Southwest after they already become corn farmers. Indeed, she argues that it was this lifestyle that allowed them to move into territory once occupied by hunters and gatherers. Since the earliest evidence for corn-growing in the Southwest dates to about 4,000 years ago, she argues that this marks the earliest time that the ancestors of Northern Uto-Aztecan speakers, including the Numic, could have arrived in the Southwest.

If possessing cognates related to corn horticulture suggests the earliest time that Uto-Aztecan speakers could have arrived in what is now the southwestern United States, then the lack of cognates for particular items among Northern Uto-Aztecans might tell us something about the minimum amount of time they have been here. In fact, Northern Uto-Aztecan languages do not share cognate terms for beans with the more southerly members of the family. Given that the earliest beans from the Southwest are now dated to about 2,500 years ago, Hill argues that the ancestors of modern Northern Uto-Aztecan people must have been in the Southwest, and have separated from their southern relatives, before this time.

All these things together suggest to Hill that the ancestors of today's Numic speakers arrived in the Southwest, perhaps the Four Corners region, between 4,000 and 3,000 years ago. She suggests that they began to move into the Great Basin about 3,000 years ago, ultimately spreading throughout this region. In the more southerly parts of the Great Basin, they continued to farm, the farming communities of the southern Paiute and Ute representing the descendants of these people. As they moved north, into territory inhospitable to corn, beans, and squash horticulture, they became hunters and gatherers. The highly developed technology that Great Basin hunters and gatherers had for gathering and processing seeds, she suggests, was developed directly from technology that had been used for cultivated plants.

This view of deeper Numic history shares some important similarities with the standard view but it is also distinctly different from it in some significant ways. The most important similarity is that they both see the modern Numic-speaking peoples of the Great Basin as relatively recent arrivals. The most important difference is that while Hill often refers to dates derived from glottochronology, her chronology for the arrival of the ancestors of modern Numic peoples in the southwestern United States is based on the reconstruction of cognate terms, or the lack thereof, coupled with radiocarbon-dated evidence for the first domesticated plants in the Southwest. It is this approach that helps make her work so insightful.

The standard view of Numic history and that provided by Hill also differ as regards the chronology of the movement of Numic peoples across the Great Basin and the explanation of the fan-shaped distribution of the languages they spoke. In the standard view, that movement occurred within the past 1,000 years in most areas, creating the fan-shaped distribution. In Hill's view, Numic peoples arrived first in the southern Great Basin, perhaps beginning around 3,000 years ago, and then later—she suggests between 2,000 and 1,000 years ago—in the north, moving in from multiple directions. The distinctive geography of Numic languages is explained not by the movement of people outward from the southwestern Great Basin, but by the relatively unreliable nature of resources found away from the Sierran front in the Great Basin and by the location of Numic peoples in the southwestern Great Basin adjacent to speakers of a diverse set of Californian languages. This setting, she suggests, led to greater linguistic diversification within the Numic languages found here.

Hill's chronology of the movement of Numic peoples, however, seems to be derived from glottochronology and so lacks the very substantial support provided by the combination of cognates and radiocarbon dates for the arrival of the ancestors of modern Numic peoples in the Southwest between about 4,000 and 3,000 years ago. That combination, though, is not necessarily inconsistent with the relatively recent arrival of Numic-speaking peoples throughout much of the Great Basin.

A NORTHERN ALTERNATIVE

There are evident weaknesses in Hill's account. As I mentioned, it is problematic that her chronology for Numic expansion appears to be based on glottochronology. In addition, she does not provide a compelling explanation as to why Numic peoples would have occupied vast portions of the Great Basin where corn horticulture was not possible and, as part of that, how they were able to replace or incorporate the people who were there before them.

There may be other problems as well. William Merrill and his colleagues have questioned nearly all aspects of Hill's reconstruction of Uto-Aztecan history and, in so doing, have forwarded a distinctly different vision of that history, and of Numic speakers among them.

Merrill's team denies that the terms for corn that are found in Northern Uto-Aztecan languages are true cognates for words that are found in Southern Uto-Aztecan, but they provide little evidence to support this rejection. More convincingly, they also reject the argument that corn was carried to the southwestern United States by immigrants from the south. Instead, they argue that corn arrived by the process known as diffusion, moving from group to group from south to north.

They make this argument for a number of cogent reasons. First, they note that the arrival of corn in the Southwest, around 4,000 years ago, is not accompanied by any of the very distinctive artifact forms that existed to the south at this time. Second, they observe that there were more domesticated plants than just corn in Mesoamerica at the time this plant arrived in the north, including bottle gourd and squash, but only corn shows up around 4,000 years ago. The arrival of these other domesticated plants is spread across the following 1,500 years. If immigrants brought corn to the Southwest, they ask, why did they not bring any of their distinctive artifacts or other important plants with them? Just as important, they note that work by geneticist Brian Kemp has shown that the Uto-Aztecan peoples of the Southwest bear little genetic similarity to those of Mesoamerica. Finally, they observe that the contexts in which early corn is found in the Southwest strongly suggests continuity with cultural traditions that had developed in place. All this, they conclude, shows that corn diffused across space, rather than having been brought by the Uto-Aztecan immigrants called for by Hill.

They also argue that Proto-Uto-Aztecan peoples—the ancestors of all Uto-Aztecan speakers—came from the Great Basin itself. In particular, they argue that their homeland was in the heart of Nevada. Following observations by linguist Kay Fowler, they observe that Northern and Southern Uto-Aztecan languages do not share cognate words for oak and pinyon. To them, this means that the people who spoke Proto-Uto-Aztecan must have come from an area that had neither of these trees. Central Nevada meets this requirement, as long as Proto-Uto-Aztecan developed, and then diverged, prior to the arrival of pinyon pine here. Given that the earliest evidence we have for pinyon pine in central Nevada falls at about 5,400 years ago if we rely on the Gatecliff plant macrofossils, or about 6,000 years ago if we rely on the Gatecliff pollen (chapter 7; figures 8-16 and 8-18), these people must have been here before this time.

Indeed, Merrill and his colleagues suggest that Proto-Uto-Aztecan peoples were in this part of the Great Basin as early as 10,000 years ago. With the beginning of middle Holocene aridity, they suggest, these people began to disperse to the south and west. Ultimately, they reached Mesoamerica, where they acquired the corn that subsequently diffused northward to reach the Southwest by 4,000 years ago or so.

Merrill's team does not address issues relating to the origins of Numic speakers *per se*, since this was not the focus of their work. If they are correct, however, some of the implications seem clear since, in their view, the ancestors of Numic peoples were in central Nevada during the early Holocene. Two things, however, strongly suggest otherwise.

First, central Nevada is famous among Great Basin archaeologists not only for a series of crucial archaeological sites, including Gatecliff, but also for the fact that, unlike many others parts of the Great Basin, it is virtually devoid of early Holocene archaeology. If Proto-Uto-Aztecan speakers were here during the early Holocene, they do not seem to have left much behind. Second, Kay Fowler's reconstruction of Proto-Uto-Aztecan terms for plants and animals showed not only that it lacked words for oak and pinyon pine but that it also seems to have had terms for turkey and agave. There have never been turkeys in central Nevada, though they are found in the Southwest, and agaves, with their distinctive spiny leaves, are limited to the southern part of the state. These were among the reasons that Fowler suggested that the origins of Uto-Aztecan lie to the south of the botanical Great Basin. Given the nature of the archaeological record for central Nevada, and the results of Fowler's linguistic work—which Merrill and his colleagues rely on heavily in other ways—there still seems little reason to doubt her conclusion.

THE ARCHAEOLOGICAL EVIDENCE

As I have mentioned several times, it is virtually impossible to assess linguistic affiliation from prehistoric archaeological data. The problems associated with attempting to do this should be clear. First, no certain artifactual markers of linguistic affiliation are known from the Great Basin. Second, although Hill's linguistic analysis suggests that Numic speakers could not have been in the Great Basin prior to their arrival in the Southwest between 3,000 and 4,000 years ago, it does not tell us when Numic peoples began to move across this region. Both her approach, and Lamb's, suggest that Numic peoples must have arrived later in the northern Great Basin than they arrived in the southern part of this region, but neither provides a compelling chronology for this movement. The reconstruction by Merrill and his colleagues would seem to have very different implications, but that reconstruction is strongly contradicted by both archaeological and linguistic evidence.

It is certainly true that the 2,000 years of Great Basin prehistory prior to the European arrival were impressively dynamic from an archaeological standpoint. To some extent, this may simply reflect the fact that more recent times are always easier to see in the archaeological record than are more ancient ones. However, there is no question that the changes are real, since they are so visible and would be even if they had occurred far earlier. In addition, they may, some think, be related to the movement of Numic speakers across the Great Basin.

I begin with one of the more remarkable discoveries to have been made in this region during the past few decades: the existence of small villages in the alpine tundra zone of several Great Basin mountain ranges.

The Alpine Villages

ALTA TOQUIMA

In 1978, Dave Thomas led a small archaeological crew to Mount Jefferson, the summit of the Toquima Range, above and south of Gatecliff Shelter. They came here because the U.S. Forest Service had reported the existence of stone structures in this area and Thomas wanted to see exactly what they were. He was expecting to discover hunting features of the sort that I have already discussed. These he found in abundance but, at an elevation of 11,000 feet, he also found something distinctly different: a complex of some three dozen stone structures adjacent to a small stand of limber pine and overlooking a small creek in an area that Thomas appropriately refers to as "on the way to nowhere" (Thomas 2010; see figure 2-13).

These structures include the stone foundations of a large number of houses as well as stone windbreaks and what appear to be storage facilities. Above and beyond the site stretches a large expanse of alpine tundra vegetation. The site soon became known as Alta Toquima Village; at the time of its discovery, it was the highest village site known from North America.

Thomas later discovered a series of much smaller sites that also contain houses in the Alta Toquima area. These have been studied in less detail, but they establish that there was more than one village in the alpine tundra zone of the Toquima Range.

In 1981 and 1983, Thomas excavated portions of Alta Toquima. This work, coupled with careful mapping of the associated surface features, has gone far toward revealing the nature of this site. The most obvious feature of Alta Toquima is the houses, which are marked by carefully constructed circles consisting of several courses of flat rocks stacked on top of one another to heights of about three feet (figures 9-17 and 9-18). These circles formed the foundation for a superstructure that has not been preserved and whose nature is not known. As at King's Dog, the superstructures may have been removed after use and stored for the future. However the superstructure was built, the houses were substantial. From wall to wall, the interiors measured about ten feet across. There were also quite a few of them: Thomas counted twenty-eight, and it is very likely that more existed here.

Because the structures sit on the surface, their existence and size were known before the excavations began. However, by excavating more than a dozen of the houses, as well as part of the rich midden deposits associated with some of them, Thomas recovered a diverse set of artifacts and a smaller sample of faunal and floral remains. He also discovered that most of the houses had one or more hearths inside. Because these hearths held well-preserved charcoal, Thomas was able to obtain a radiocarbon dates for twelve of the structures. These dates showed that about half of the

FIGURE 9.17 Stone circle footing for an Alta Toquima house.

FIGURE 9.18 Stone circle footing for another Alta Toquima house.

houses had seen their first use before 1,000 years ago, about half after that time. The earliest of them had been occupied some 1,840 years ago; the latest, during very late prehistoric (and perhaps even early historic) times. The projectile points from the houses match the radiocarbon dates: they include Desert and Cottonwood series points, but smaller numbers of Elko points are present in some structures as well.

There are three important things that these dates do not mean. First, they do not mean that the first people to use this particular spot came here about 1,840 years ago. In fact, archaeological deposits adjacent to some of the houses date back to 3,000 years ago; they are associated with Elko and Gatecliff points and were left behind by people doing something other than living in the house structures that make Alta Toquima so special. Second, the radiocarbon dates from the houses do not mean that the structures had been used continuously for nearly 2,000 years. In fact, there are gaps of up to two centuries in the radiocarbon dates. Finally, nothing about these dates, or anything else at Alta Toquima, suggests that all of the houses were ever occupied at the same time. Although there is no way to know how many were in contemporaneous use, the radiocarbon dates suggests that only a few may have been in use at once.

In addition to the points, Alta Toquima Village provided a wide variety of other artifacts, including large numbers of grinding stones and a surprising number of broken ceramic vessels. The Alta Toquima sherds—some four hundred of them—are fairly thick, with coarse temper (the inclusions that prevent the vessel from cracking when fired), and are highly variable in both the paste used to make them and in the hardness and color of the end product. The unimpressive nature of the Alta Toquima ceramics is fully characteristic of the pottery that appears across much of the Great Basin late in prehistory and that was used in this area during early historic times. While perhaps not artistically impressive, these vessels likely played an important role in food processing, as I will discuss in more detail shortly.

The small samples of plant and mammal remains that Thomas recovered provide direct information on the food resources utilized by the people who were living in these houses. Two species of mammals—yellow-bellied marmots and mountain sheep—provided the bulk of the animal remains; both species may be seen in the immediate vicinity of the site today. Marmots remain above ground only during the late spring and summer at this elevation; their presence in the deposits of Alta Toquima demonstrates that this site must have been occupied during at least those months.

The plant remains from the site were analyzed by Dave Rhode. He found that the fuel requirements of these people were met by using locally available limber pine and sagebrush but that plant foods apparently came from both local and valley-bottom settings. Foods derived from local plants include limber pine nuts, the flesh and perhaps seeds of Simpson's pediocactus (*Pediocactus simpsonii*), and small bulbs of spikerush (*Eleocharis*). On the other hand, a surprising diversity of seeds found at Alta Toquima must have been carried up from lower elevations, the most abundant of which came from bulrush (*Schoenoplectus maritimus*), almost certainly collected from marshes that lie at least 4,300 feet beneath the site.

Alta Toquima Village is a remarkable site. A good deal of labor was invested in building the houses here; this, along with the storage facilities, suggests that more than a single brief stay was involved. The artifacts document that a variety of activities was conducted: the projectile points suggest hunting; the grinding tools, plant processing. The direct evidence from the animal and plant remains shows that while local resources were utilized, some plant foods were transported here from the valley bottom. It would appear that entire family groups moved to this alpine location to utilize a diverse set of local resources and that this movement occurred during the late spring or summer. It also appears that such use began as early 1,800 years ago or so and then continued, apparently sporadically, until, or nearly until, historic times. It also fairly likely that these same people used Gatecliff Shelter, located some eighteen miles to the north and 3,400 feet lower.

In addition to the sites with house structures, Thomas found that the Toquima Range highlands hold a diverse set of hunting features: rock walls that appear to have served as parts of game drives, for instance, and stone circles that seem to have functioned as hunting blinds. It is difficult to date such features, but it is certainly meaningful that while the projectile point assemblage at Alta Toquima Village is dominated by arrowpoints, the hunting features are dominated by earlier Elko and Gatecliff atlatl dart points. To judge from the projectile points, the Alta Toquima assemblages were deposited within the past 2,000 years, while the hunting features came into use between 5,000 and 3,500 years ago.

On a global basis, there is a fairly distinct division of labor among hunters and gatherers: women gather and process plants and take small game, while men hunt large mammals. There are many interacting reasons for this. Not only may having and caring for children conflict with the physical demands of hunting, but men and women have different goals in pursuing food. To women, provisioning their families looms large. Since this is the case, they pursue food resources that provide reliable contributions to the diet—plants and small game. These have the provisioning advantage that they can remain private if the gatherer wishes. Men, on the other hand, often pursue game not just to provision their families, but also to gain various kinds of social status. As a result, they pursue large prey that may be difficult, and so less dependable, to acquire but that carry an important signal about hunting prowess. The results of this hunting are routinely shared with members of the community at large. No wonder, then, that men who successfully hunt risky prey tend to have more kids than men who do not.

The results of this division of labor are evident in the remains of the people themselves, including the prehistoric hunters and gatherers of the Great Basin. For instance, human skeletons from the Stillwater Marsh area of the Carson Sink, remains that date to within the past 2,600 years, show very high frequencies of osteoarthritis, a disorder that can be caused by mechanical stress to the joints. The skeletons of men, however, show more osteoarthritis in the hip and ankle than do those of women, while those of women show more of it in the lower back. These differences are consistent with men having traveled greater distances over challenging terrain; the lower-back difficulties experienced by women might have been caused by carrying heavy loads or intense involvement with seed processing. Going along with all this is the fact that the lower limbs of these people were impressively robust, suggesting a great deal of mobility, but those of men showed the characteristic shapes taken on by people who spend great amounts of time negotiating difficult landscapes. These differences are all consistent with men having been the hunters, women the gatherers and plant processors.

If we put all of this together with what we know of the archaeology of the Toquima Range, it is reasonable to suggest, as Thomas does, that the earliest use of the alpine tundra of this area was by male hunting groups, a use that began between 5,000 and 3,500 years ago. These are the people who constructed the hunting features and left behind the Gatecliff and Elko points found with them. Then, for some reason, family groups began visiting this area some 1,800 years ago or so, bringing plant foods from far lower elevations. While there, they hunted mammals, including mountain sheep and marmots, and gathered local plants, including cactus, spikerush, and limber pine. As time went on, this new use of the uplands seems to have become increasingly important; by the time the bow and arrow was introduced, most of the use of these uplands was based at Alta Toquima and the other small villages nearby. This was a process, not an event, but it was a process that saw dramatic changes in the way prehistoric people used what is now the heart of the Alta Toquima Wilderness Area.

WHITE MOUNTAINS

No sooner had Great Basin archaeologists taken in the surprise provided by Alta Toquima than archaeologist Bob Bettinger presented them with another. In 1982, Bettinger found a near duplicate of Alta Toquima at an elevation of 10,400 feet in the southern White Mountains of eastern California, again above treeline. This discovery spurred Bettinger to launch a detailed archaeological survey of this area. By the time he was done, he had discovered twelve village sites that contained from one to about a dozen circular houses and that fell between 10,330 feet and 12,640 feet in elevation (see figures 9-19 and 9-20).

While it stretches the concept to call a site with one house a "village," these sites are, as a whole, very similar to Alta Toquima and to the smaller settlements associated with it. The White Mountains sites have circular stone house foundations with internal hearths, storage facilities, occasional ceramics, abundant grinding tools, and hunting equipment. As at Alta Toquima, the faunas from these sites are dominated by the remains of mountain sheep and marmots, both of which are locally available. Paleobotanist Elizabeth Scharf has shown that the plant remains from at least one of

FIGURE 9.19 Site 12640 on the White Mountains. Stone circle house footings emerge from the snow in the center of the photograph, with the Sierra Nevada in the background.

FIGURE 9.20 The Pressure Drop site in the White Mountains, with a house ring in the center foreground and the Sierra Nevada in the distance.

these sites—Midway, at an elevation of 11,290 feet—include both local species (for instance, chickweed, *Cerastium*), and those transported from below (for instance, pinyon pine nuts, today available only some 1,800 feet downslope). This, too, mirrors the situation at Alta Toquima. In many ways, the White Mountains and Alta Toquima sites are peas in a pod, even though Alta Toquima is 130 miles northeast of the White Mountains villages.

Some of the White Mountains villages contain glass trade beads, showing that they were used in earliest historic times. At the other end of the time scale, the earliest radiocarbon date for these sites falls at 1,780 years ago, very close to the earliest house date at Alta Toquima. This, though, is where the chronological similarity ends. Of the twenty-one radiocarbon dates available for the White Mountains villages, twenty fall after 900 years ago. This, as Dave Thomas has

THE PREHISTORIC ARCHAEOLOGY   323

pointed out, is very different from the pattern seen in the Toquima Range, where the dates are fairly evenly arrayed between 1,840 years ago and very late prehistoric times.

I noted that the uppermost elevations of the Toquima Range are marked by the presence of both small villages, which were in use during the past 1,800 years or so and hunting features, which seem to have been the focus of earlier use. The same pattern appears in the White Mountains, but here it is somewhat better understood because many of the sites occupied during village times also hold earlier occupations. Bettinger has excavated several sites that were in use only during "previllage" times, and has analyzed the artifacts associated with the hunting features scattered across the White Mountains alpine tundra zone.

As on the Toquima Range, the high-elevation hunting features in the White Mountains are primarily associated with Elko and Gatecliff series atlatl dart points. At the excavated village sites, the occupations of the houses are dominated by Desert and Rosegate series points. The occupations that predate the houses are associated with Elko and Gatecliff points, suggesting that these earlier occupants were also the people who used the hunting features.

Bettinger has documented a series of other differences between the earlier and later occupations of these sites. Compared to the previllage occupations, the villages have significantly more plant-processing equipment, tools to make other tools (for instance, drills), and projectile points. The previllage occupations, on the other hand, are loaded with artifacts that appear to reflect the repair of hunting tools and also have large numbers of implements that could have been used to process hunted animals.

Just as Thomas argues, and for the same reasons, Bettinger argues that the earlier occupations of these sites were focused on hunting, probably by small groups of men. The villages, he suggests, were used by family groups performing a diverse set of activities. Since the faunal remains from both village and previllage occupations contain marmots, these sites were likely used between the late spring and late summer month, perhaps, paleobotanist Linah Ababneh has observed, during periods marked by above-average precipitation. Although the chronologies are quite different, with the routine occupation of the Toquima Range villages beginning distinctly earlier than those in the White Mountains, the changing patterns of landscape use are extremely similar in both areas.

The Bow, the Arrow, Ceramics, Pinyon Pine, and Changing Human Adaptations in the Great Basin

As I have discussed, the bow and arrow seem to have arrived in the Great Basin between 1,800 and 1,300 years ago, with different areas receiving this weapon system at different times. It is difficult to know what the impact of this introduction was here, but it must have been significant. Compared to the atlatl and dart, the bow and arrow provides greater range and accuracy, allows the hunter to release more shots in a given amount of time, can be used to greater effect in stealth hunting since it requires less physical movement to use, and can be far more effective in pursuing smaller prey. It also has real advantages in conflicts between people; indeed, archaeologist John Blitz has suggested that this is one, and perhaps the prime, reason, that it spread so rapidly across North America in later prehistoric times.

Bob Kelly has observed that the bow and arrow may have made individual hunting more profitable than it had been before and that it might have brought about substantial social change as well. It is, however, Bob Bettinger who has developed these potentially wide-ranging effects in most detail.

Bettinger observes that the small game taken by individual bow hunting probably did not have to be shared, unlike large mammals taken by any method. If the bow and arrow had this effect, he suggests, large mammals may have remained shared, "public," goods, while smaller prey remained entirely in the possession of the hunter who bagged them. At the same time, the effective nature of the bow and arrow as a hunting tool may have reduced any expectations that plant foods had to be shared. If this were the case, it would pay to amass greater amounts of food that could be stored since one would not have to worry that all the effort involved would simply end up feeding those who had not done the work involved. Bettinger argues that this bow and arrow-driven shift, from an emphasis on the importance of foodstuffs that were seen as a public resource to one in which a significant subset of them were seen as being private, might have dramatically altered social relationships, landscape use, and many other aspects of life in the Great Basin. As part of this, he suggests, once wide-ranging bands of hunters and gatherers might have divided themselves into smaller social groups focused around independent nuclear families able to respond quickly to changes in resource abundance, with the food they obtained largely remaining within family groups. All of this might have produced, as Bettinger (1999:74) put it, some of the "very first 'me' generations" in this region.

But is there any evidence that such a change actually happened? In fact, there is, and from multiple directions. Great Basin ceramics provide an excellent example.

There were multiple kinds of ceramics in the Great Basin during late prehistoric times, including those associated with the Fremont (see below), but the ones most important here are those in use through much of the Great Basin in areas occupied by Numic-speaking peoples.

This pottery was generally, though not always, made by building up clay coils from the bottom of the pot, smoothing the edges of the coils together, and then firing it under low temperature to produce vessels that varied in color, even on the same object. The pots are rarely decorated and everything thing we know about them suggests that their manufacture was an individual or family affair. The results have been given different names in different places, including Shoshone Ware (as at Alta Toquima), Southern Paiute Utility Ware, Owens Valley Brown Ware, Death Valley

Brown Ware, and so on. All of these ceramics, however, are highly variable and can be very difficult to tell apart. As a result, I follow archaeologist Lonnie Pippin's suggestion that they all be referred to as "Intermountain Brown Wares," although they are not all brown. With extremely rare exceptions, all secure dates for this pottery fall within the past six hundred years.

Because of the work of archaeologist Jelmer Eerkens, we now know a great deal about the uses to which these vessels were put in the southwestern Great Basin, from the northern Mojave Desert into Owens Valley. He has observed that the shape of these pots strongly suggests they were used for boiling. Their wide mouths allow heat to escape, preventing the pots from self-destructing as boiling proceeds, and their rough exteriors increase surface area and so allow the efficient transfer of heat from fire to contents. In addition, the exteriors of the pots are routinely fire-blackened. Chemical analyses that he did showed that the residues that adhere to the innards of these vessels came from seeds. The pots, he concluded, were used to cook seeds by boiling.

Pots aren't the only way to boil seeds. The same end can be achieved by putting hot rocks into a water- and seed-filled basket. In this approach, the basket needs to be attended continuously, both to make sure that it doesn't succumb to the process and to replace cooling rocks from a nearby fire. Pots, on the other hand, can cook away on their own. The savings in time and effort are obvious.

Eerkens's research is not the only work to suggest that many Intermountain Brown Ware pots were used to process seeds, but it is the most detailed. He was also the first to observe that everything about these pots seems directed toward private use. In the Owens Valley, sherds are found more often inside houses than outside, the lack of decoration suggests that they were made for personal use with no thought as to their display value, and they can be both made and used individually. Private pots for private seeds, coming into use about six hundred years ago. This, of course, is substantially after the arrival of the bow and arrow but it is what might be expected if Bettinger is right in thinking that the bow and arrow may have ultimately triggered a broad range of adaptive changes among the prehistoric peoples of the Great Basin.

Then there is the evidence for pinyon nut use in the Great Basin. Pinyon pine nuts can be gathered at two significantly different stages of their existence. One can wait until the autumn when the seeds are ripe and the cones have opened. The seeds can then be knocked out of their cones and then gathered quickly and efficiently. Taking this approach, however, involves several problems. Most obviously, there is a reason that pinyon pines display their seeds so obviously: they are meant to be taken and dispersed by birds. Because they are there for the taking, other animals consume great numbers of them. As a result, collecting seeds when the cones are open and the seeds ripe must be done at one particular time of the year and in competition with the winged and furry.

On the other hand, the cones can be harvested when they are still green and closed. That allows harvesting to be spread out over a longer period of time, allows more of the crop to be collected, and avoids losses to nonhuman competitors. This approach, however, requires far more work. Rather than simply knocking the seeds from the cones, the cones themselves have to be plucked from the trees and then roasted. The roasting opens the cones, allowing the nuts to be removed.

Both methods were used in the Great Basin during historic times, though the harvesting of green cones seems to have been somewhat more common. There is every reason to think that the harvesting of ripe seeds began as soon as pinyon trees had become abundant enough in any given area to make it worthwhile. However, the evidence is convincing that green-cone harvesting is relatively late in at least some areas.

The most compelling information of this sort comes from the southwestern Great Basin. In an area extending from the Coso Range to Deep Springs Valley and west across Owens Valley itself, the earliest evidence for substantial amounts of green-cone harvesting falls between 1,350 and 600 years ago, depending on where you are. Some of that evidence comes from radiocarbon-dated features associated with the roasting of pinyon cones. Other evidence is provided by radiocarbon dates from, or arrow points associated with, the small rock-ring features used to store the pinyon crop after it had been collected.

This evidence falls in line with the suggestion, from ceramics, of more intensive use of small seeds after the bow and arrow arrived. It also matches Bettinger's suggestion about the increased emphasis on the gathering of storable foods, particularly plants, after this new weapon system was in place.

Other things match Bettinger's suggestions, but these are among the most significant. If he is right, the implication is that the introduction of the bow and arrow ultimately led to substantial changes in how the prehistoric peoples of the Great Basin interacted with their landscape and with each other. It would be good to have comparable evidence from other parts of the Great Basin, and good to know that there are not better explanations for these phenomena, but what we know so far is consistent with Bettinger's intriguing arguments.

LATE PREHISTORIC ADAPTIVE CHANGE AND THE NUMIC EXPANSION

Early historic Native American lifeways in much of the Great Basin placed great emphasis on small seeds, pinyon nuts, and the pursuit of small game, all embedded in a small-scale social system emphasizing the nuclear family or small groups of nuclear families. The later prehistoric developments that we have just discussed, phenomena that seem to have occurred within the past 2,000 years in those places within the Great Basin where they have been sufficiently studied, make it seem that this was the time when these historic lifeways came into being. Even the earliest known use of snares, primarily used to take small mammals, dates to this period. This is precisely what Bob Bettinger has argued.

Bettinger, however, takes all this a step further. He argues that the appearance of this constellation of adaptations during later prehistoric times in the southwestern Great Basin represents the development of the core adaptations of Numic peoples. These adaptations allowed them to spread across the Great Basin, outcompeting the people who had been there before them, just as the classic Numic expansion hypothesis holds.

In particular, Bettinger distinguishes between the lifeways of the people who were here just prior to Numic speakers (the "pre-Numic peoples" of the Great Basin) and those of the Numic peoples themselves. In particular, he argues that pre-Numic peoples traveled widely in search of such high-quality resources as mountain sheep. Numic peoples, on the other hand, he sees as having been far more dependent on lower-quality resources, including small mammals and seeds, that demanded less time to find but more time and effort to process. Though Numic peoples would take high-quality resources when they encountered them, pre-Numic peoples simply did not utilize lower-quality foods to the same extent as did those who replaced them. Indeed, he refers to pre-Numic groups as "travelers," people who emphasized large-scale movements to track down the resources on which they depended. Numic groups he refers to as "processors," people who were organized into smaller groups focused on nuclear families, traveled far less and put more energy into amassing and processing smaller packages of resources. He argues that the development of the Numic adaptation in the southwestern Great Basin—the very development I have just reviewed—was due to population growth in this area, growth that fostered the development of more effective ways of extracting calories from resources that had been used less intensively before. In fact, he sees the appearance of the White Mountains villages as one indication of this development.

Bettinger argues that these adaptive differences gave Numic peoples the ability to respond more effectively to bursts of resource availability and to extract more calories per unit of territory. This, in turn, allowed higher population densities than those that characterized pre-Numic peoples. As a result, "pre-Numic peoples were powerless to stay the invasive spread of the more costly, but less spatially demanding, Numic adaptation" (1991:674). Movement outward from the southwestern Great Basin ultimately created the fan-shaped distribution of Numic languages that many have seen as indicating the direction of movement of these people across space.

Many criticisms of Bettinger's explanation of Numic expansion have been raised over the years. It is easy to observe, for instance, that there is a deep history of small seed use in the Great Basin, as the analyses of dietary remains from such sites as Bonneville Estates Rockshelter and Danger Cave, discussed earlier in this chapter, show. In addition, ample evidence shows that there were places in the Great Basin prior to the time of the hypothesized Numic expansion where people were spending significant amounts of time in one spot—substantial settlements in Surprise Valley and the Lake Abert Basin provide examples. And, since Bettinger sees the advent of the alpine villages as a key marker of this adaptation, the routine use of Alta Toquima Village before the routine use of the White Mountains villages does not match the time-line his view requires, nor does the fact that, as Elizabeth Scharf has shown, small seeds were being heavily used in the White Mountains prior to the establishment of the villages there.

It is unlikely to be coincidental that the two sets of high-elevation villages that have been excavated in the Great Basin are near low-elevation areas that supported some of the highest early historic human population densities reported from the Great Basin: the Owens and Reese River valleys. This suggests that high-elevation villages may have occurred in response to increasing low-elevation population densities and attendant pressures on at least lowland resources. If that occurred first in the Reese River Valley, it would follow that a distinctly earlier, more intensive, use of the alpine zone would occur in this area before it occurred in the White Mountains. Whether this actually happened, though, is not known.

I am not questioning Bettinger's explanation of the development of the lifeways of Numic peoples as they existed in early historic times. No other account explains as much as his does and no other account has as much archaeological support. But it is one thing to provide a powerful explanation of how modern Numic lifestyles came into being and quite another to demonstrate that this was related to the spread of a people across the landscape. As should be clear from Jane Hill's work, it appears likely that Numic peoples arrived in the Great Basin sometime after 4,000 years ago or so. When that might have been, however, we do not know, unless we want to believe the dates provided by a method that has been thoroughly rejected by linguists. Until we have some secure way of detecting when Numic speakers arrived in any given part of the Great Basin, it will remain fully possible that if Bettinger is right about the processes that saw the development of Numic lifeways, those processes occurred within populations of Numic speakers themselves.

ARTIFACT MARKERS AND THE NUMIC EXPANSION

Many archaeologists have argued that the arrival of Numic peoples in various parts of the Great Basin can be detected through specific artifact types, regardless of the adaptive significance of those implements. In particular, they have argued that Desert Side-Notched projectile points, ceramics, and certain kinds of basketry mark the appearance of Numic speakers. Archaeologist Steve Simms, however, has correctly observed that specific artifact types cannot be counted on to identify the moment of arrival of Numic peoples, that "whether it is pottery or arrowheads, each of these arrives in places for its own reasons, and the artifacts lay across cultures in different ways" (2008:254).

The strongest arguments for a Numic-specific artifact marker have been made for basketry, with the most strenuous efforts in this realm made by textile expert Jim Adovasio. He argues that at about 1,000 years ago, roughly the time called for by the classic Numic hypothesis, a dramatic change in basketry styles occurs in all those parts of the Great Basin historically occupied by Numic peoples, with the exception of the area occupied by the Fremont (see below). He and his colleagues even find that "only those who are never permitted outdoors without their keepers" (1994:122) could accept an explanation for this that does not involve population replacement at this time—that is, the arrival of Numic speakers.

Others, though, tread more gently here. There are two reasons for this. Most obvious are the difficulties associated with inferring linguistic affiliation from prehistoric material objects. A case in point is provided by Gatecliff Shelter. Here, a basketry fragment that came from a stratum deposited at about 1300 A.D. is virtually identical to basketry used by modern Numic peoples. However, when this specimen was dated directly, it was found to be 3,200 years old. Although this fragment may, in fact, have been made by a Numic speaker, it is some 2,500 years earlier than called for by the classic Numic expansion hypothesis.

In addition to the difficulties involved in associating material culture with languages, only in two areas in the Great Basin do we know enough about prehistoric basketry to be able to make the case that population replacement may be related to changing human populations. One of these areas saw the fluorescence of the Fremont, discussed in the following section. The other is centered on the Lahontan Basin, from the Carson Sink northwest to Pyramid Lake.

Here, as a wide variety of scholars have noted, a very distinctive set of baskets, trays, hats, and other items was in use from about 4,000 to as recently as 600 years ago. These items, often referred to as belonging to the Lovelock Culture (after Lovelock Cave), are so different from those made by the Northern Paiute people who live here now that it is, just as Adovasio says, hard to believe that this can be explained by local people simply giving up one approach to basketry and adopting another. Detailed analysis of a rich set of basketry from this time period by Ed Jolie and Ruth Burgett has shown that they are not only very distinct from those made by the historic Numic residents of this area but also very similar to those made by the Maidu-speaking people of California.

There are three Maidu languages, spoken in an area to the northwest and west of the Washoe and extending across the Sierra Nevada into western California. These are Penutian languages, allied not to the Uto-Aztecan languages spoken by Numic peoples, but to the languages spoken by the Klamath, Modoc, Nez Perce, and others. If we were to infer linguistic affiliation from basketry, we would have to guess that before six hundred years ago or so, this part of the Lahontan Basin was occupied not by Numic peoples but by Penutians and perhaps the ancestors of the modern Maidu.

This fits well with a much earlier argument made by archaeologist Gene Hattori. He pointed out that cave sites in the Winnemucca Lake Basin contain a broad series of artifacts far more at home in the territory of prehistoric Penutian speakers in California than they are in the historic Lahontan Basin. And all this fits well with Northern Paiute stories recounting attacks on enemies who lived both along the Humboldt River and at Pyramid Lake "many hundred years ago" (Sarah Winnemucca Hopkins 1883:73). It also fits well with the timing of the Medieval Warm Period and the significant droughts in this part of the world that I discuss in chapter 8. Indeed, Larry Benson and his colleagues argue that it was the aridity of this period of time that saw the end of the "Lovelock Culture" and that allowed Numic peoples, so well adapted to living in arid environments, to replace them.

Very similar arguments have been made for the broad swath of the Great Basin that falls within south-central Oregon. More than seventy years ago, Luther Cressman reported that basketry that he had retrieved from southeastern Oregon, between Warner Valley and Steens Mountain, bore little resemblance to that made by the modern Numic peoples of the area but was instead similar to that made by the Klamath and the Modoc. Klamath tradition, he noted, maintained that they had once occupied southern Oregon as far east as Steens Mountain, matching the implications he derived from the prehistoric basketry he had found. He amplified these arguments a few years later, armed with the richer body of data that had resulted from all the fieldwork he had done in this area.

More recent work continues to suggest that Cressman may have been correct. Albert Oetting, for instance, has shown that the area surrounding Lake Abert was occupied, from about 4,000 to less than 1,000 years ago, by people living in villages composed of houses constructed, Klamath-style, in substantial excavated pits, some of which were associated with distinctive artifacts that match those known ethnographically from the Klamath-Modoc area. He suggests, as did Cressman, that the people who created these sites may have been the ancestors of the modern Klamath who were then replaced, in very late prehistoric times, by Numic speakers. To the west, in the Warner Valley, Sunday Eiselt has reported the presence of a house, dated to between about 650 and 240 years ago, and probably closer to the latter, that is very similar to those used historically by the Klamath. Some of the artifacts associated with this structure would not be out of place in an historic Klamath setting. Other sites have similar implications.

All of this suggests that during late prehistoric times, Penutian speakers may have lived as far east as Steens Mountain in the far northern Great Basin (the Klamath) and as far east as the Carson Sink in the Lahontan area (Maidu speakers). Of the two, however, the Lahontan situation is the more compelling, based primarily on substantial and well-dated textile assemblages that have been compared in tremendous detail with those made by people in surrounding areas—and, as we will soon see, by genetic data as well. The northern Great Basin situation is intriguing, and the argument that close relatives of the modern Klamath once lived as far east as Steens Mountain may be correct, but the archaeological record hardly requires this conclusion. Here, as Eiselt correctly pointed out, you can "choose your truth" (1997:119).

THE FREMONT COMPLEX

One of the most intriguing episodes in the human prehistory of the Great Basin occurred between about 400 A.D. and 1350 A.D., when much of the eastern Great Basin and adjacent northern Colorado Plateau came to support people whose lifeways differed significantly from those who were there before and after (figure 9-21). These people not only manufactured well-made, thin-walled gray pottery but also grew corn, beans, and squash and often lived in sizable villages. The villages were routinely located along perennial streams near arable land and include a wide variety of substantial structures: pithouses, above-ground adobe dwellings and storage facilities of either adobe or stone, and carefully made storage pits. These sites may also contain very distinctive basketry, moccasins, metates, unfired clay figurines, and ornaments, the last of these at times made from imported turquoise and Pacific Coast shell. Associated with all of this is a very distinctive rock art style, one that mirrors designs found on both pottery and the figurines. Because sites of this sort were first described from along south-central Utah's Fremont River, the people who made them are referred to as the "Fremont."

Those familiar with the archaeology of the Southwest might be struck by the fact that the use of corn, beans, and squash, of well-made gray ceramics, and of above-ground adobe and masonry structures sounds very southwestern. In fact, the similarities go even further. For instance, some Fremont peoples constructed irrigation ditches, much as was done in the Southwest. In addition, many of the pithouses on Fremont sites have constructional features that tie them to the Southwest. As a result of all this, there is no doubt that the source of some of the traits that make the Fremont distinctive lies to the immediate south.

It does not necessarily follow, however, that Fremont peoples themselves came from that area. Almost everything

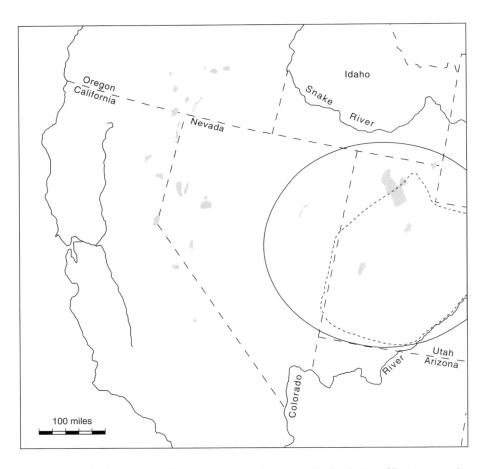

FIGURE 9.21 The distribution of Fremont. The solid ellipse marks the distribution of Fremont ceramics (after Madsen and Schmitt 2005); the hatched line marks the core Fremont area (after Janetski 2008).

we know about these people suggests that their way of life developed locally, even if some of the items they incorporated into that way of life came from elsewhere. Many of the characteristics that came together to define what we call the Fremont were present in this area well before the Fremont lifeway came into being. Jim Wilde and Deborah Newman, for instance, have shown that corn was being grown in the Sevier River Valley of central Utah by 200 B.C., long before Fremont existed, and detailed studies of eastern Great Basin basketry by Jim Adovasio document that this distinctive tradition developed from local antecedents.

In fact, one of the most interesting things about the Fremont is that the things that tend to characterize it fell into place in a piecemeal fashion. The basketry developed locally. The corn came from the south more than 2,000 years ago but did not become widespread until centuries later. The ceramics came after this, with the earliest securely dated examples dating to sometime after 400 A.D. But not until after 800 A.D. does the full-blown version of Fremont come into being, marked by all those things that I listed above. It is because Fremont grew in this piecemeal fashion that different archaeologists will give slightly different dates for its beginnings.

The Steinaker Gap site, in northeastern Utah's Uinta Basin (and outside of the Great Basin: see figure 9-1) provides one of many examples of the piecemeal way in which the Fremont complex came into being. The earliest occupation at this site dates to about 250 A.D. and is marked by lightly built houses that suggest a summer occupation. The people who lived in those houses lacked ceramics and appear to have used both the atlatl and bow and arrow, since projectile points associated with both weapon systems were found. They also, however, grew corn—about half their diet seems to have been provided by this plant—while data from a nearby, contemporary site shows that they grew squash as well. In addition to all that, they had also dug at least two lengthy ditches that probably transported water to the fields in which they grew their crops.

Archaeologists do not argue about the dates that should be assigned to the onset of the Fremont phenomenon, since all recognize the way in which it grew and that slicing up a continuum of this sort is bound to be arbitrary. They do, however, argue about what the term "Fremont" actually means. On the one hand, some, like Fremont expert Joel Janetski and his colleagues, correctly point out that there are commonalities to many Fremont sites—those things I list above—and that those commonalities mean Fremont is a distinctive archaeological phenomenon. On the other hand, others, like Fremont experts David Madsen and Steve Simms, correctly observe that the list of traits meant to define Fremont sites are differentially distributed through time and across space, and that some of the most diagnostic of them—the basketry and moccasins, for instance—are so rarely preserved that they are not much help. Not even the nearly ubiquitous Fremont ceramics help here since they are far more widely distributed than any of the other possible Fremont markers (figure 9-21). This is not because they were traded outside of the core Fremont region but because hunters and gatherers on the edge of that region—the Fremont Fringe as Madsen and Dave Schmitt have called it—adopted this ceramic style and began making them locally.

In fact, the term I used to head this section—the Fremont Complex—was introduced by Madsen and Simms to denote a highly variable set of archaeological sites grouped under a single name for analytical convenience. But even though they come close to denying the existence of something called Fremont, they call it Fremont nonetheless.

The debate over the nature of Fremont stems from the fact that Fremont sites, and Fremont lifeways, were extremely variable. The sites range from large villages with substantial structures to small, isolated locations that contain one or two wickiups and associated living debris. Occupations in caves, as at Hogup Cave and Camels Back Cave in the Bonneville Basin, and large numbers of open-air sites consist of little more than scattered lithics and gray sherds. The architecture varies from north to south, from east to west, and through time. Some Fremont sites contain substantial amounts of domesticated plants, while others provide none at all. "The key to understanding the Fremont," Madsen (1989:63) has said, "is variation," and he is right.

Steve Simms has suggested that at least some of the variability shown by Fremont sites may reflect the fact that Fremont peoples practiced very different lifestyles. Some may have lived year-round in large horticultural villages while making short-term forays into the surrounding environment for particular kinds of resources. Some may have been far more mobile, moving widely across the landscape during certain times of the year when not tied down by the demands of their crops, or for multiple years during times when plant crops were not productive. And, some may have been full-time hunter-gatherers, utilizing the same general areas as people more dependent on domesticated plants and interacting with them. Indeed, Madsen and Simms (1998:323) suggest that there may have been people "who could have been all of these things at one time or another in their lives."

The large villages containing multiple houses, substantial storage facilities, significant amounts of domesticates, carefully crafted metates, and, at times, large central structures that seem to have served specialized purposes, are relatively easy to interpret as having been left behind by people who were heavily invested in maize horticulture. Significant numbers of sites are, however, far harder to interpret. Steve Simms and Marilyn Isgreen, for instance, excavated a site in the Sevier Desert that consisted of two circular structures that had been used at slightly different times. At least one of these had been a brush-covered wickiup some eight feet across. The site, occupied at about 1100 A.D., also provided Fremont ceramics and a single corn cob. People had returned to this single spot in the Sevier Desert on several occasions, but they had done so not to grow corn but, apparently, to gather seeds and hunt small mammals. Were they full time hunters and gatherers who had obtained corn from people living in one of the village sites? Or were they people from one of those villages who had come here on several occasions to augment their diet with wild foods? We do not know. But given the date, the corn, and the ceramics, there is no reason not to call them "Fremont."

Fortunately, some very compelling information shows just how variable the diets of Fremont people were. That information comes from the analysis of the bones of Fremont peoples themselves.

When the Great Salt Lake retreated from the very high elevations it reached during the late 1980s (chapter 5), it was discovered that the erosive action of waves, ice, and wind had uncovered a rich assemblage of archaeological sites, including numerous burials, along the eastern edge of the lake. Because the combination of continued erosion and vandalism was quickly leading to the destruction of this material, the State of Utah decided to fund archaeological work here, with the support of the Northwestern Band of the Shoshoni Nation. The remains of eighty-five people were carefully and efficiently removed and have now been reburied elsewhere.

Between the excavations and the reburials, however, and thanks to the permission of the Native Americans involved, Simms and his colleagues were able to show that these burials had been made between 700 and 1300 A.D. and were thus Fremont in age. In addition, archaeological chemist Joan Coltrain and her colleagues took advantage of the fact that all of our skeletons carry within them chemical signatures of our diets and that analyses of these signatures can provide information on what people were eating. They discovered that the earliest people buried here were incorporating ever-increasing amounts of corn in their diet. This was no surprise, given what we know about the history of Fremont in general. However, she also discovered that between 850 and 1150 A.D., the diets of these people were extremely varied, ranging from those who had consumed large amounts of corn to those who had consumed very little of it if any, with males consuming far more corn than females. Then, at about 1150 A.D., corn seems to have dropped out of the diet entirely.

Earlier, Coltrain had been able to analyze the bone chemistry of a small series of individuals from four different Fremont Village sites from south of the Great Salt Lake area. All of those individuals had depended heavily on corn for their subsistence. She had also shown that corn had provided about half the diet for the people who had lived at the Steinaker Gap site that I discussed above. Now, she had shown that in the Great Salt Lake area, during the height of the Fremont phenomenon, different people had very different diets. This is precisely the kind of variability that archaeologists have come to expect from the Fremont, even if we do not understand it yet.

It is not possible to know how many people lived in the Fremont area when it was at its peak. It is, though, possible to estimate how many people lived at single sites, by counting the number of houses that seem to have been occupied at any one time and multiplying that by the number of people that probably lived in them. A good example is provided by Five Finger Ridge, in south-central Utah's Clear Creek Canyon. Archaeologists Richard Talbot and Lane Richens excavated this site with great care and, because it was going to be destroyed by activities related to highway construction, in its entirety. They found a total of thirty-seven pithouses, nineteen adobe-walled surface structures that served for food storage, and a wide variety of other features. Joel Janetski and his colleagues estimate that as the thirteenth century came to an end, at least twelve of the pithouses were in simultaneous use, suggesting a population of at least sixty, and perhaps as many as a hundred, people. Across the state, on the Nevada side of the Utah-Nevada border not far from Great Basin National Park, Baker Village may have held as many as fifty people at or soon after 1200 A.D.

Most Fremont sites are far smaller than these; many have just two or three pithouses, each of which may have held a single family. But people were numerous enough that interpersonal relationships were not always smooth. At several of the larger sites, human bones have been found that had been butchered and burned. While it is hard to know why they had been treated this way, the fact that some of the people involved had died violent deaths leaves little doubt that whatever the reasons behind this, they were not friendly ones. As Madsen (1989:64) has eloquently observed, Fremont people "laughed, lusted, and lied . . . fought, feasted, and feared the unknown. They worked and they played and they raised as many children as they could." They were, in other words, very much like ourselves.

And it all ended, but just as the beginning was piecemeal, so was the end. In the Great Salt Lake area, Coltrain's work shows that corn dropped out of the diet at about 1150 A.D., marking the last horticulture in this part of the Great Basin. Some horticulture continued until after 1400 A.D. on the far northeastern edge of the Fremont region, but in nearly all places it is gone by about 1350 A.D. Fremont, whatever it was, disappeared.

Probably no one thing can account for both the beginnings and the ends of Fremont. However, if we can account for the adoption and loss of corn, beans, and squash horticulture, we can at least understand the context that allowed all of this to happen.

Given that nearly all of the world's peoples are now dependent on domesticated plants for their survival, it is tempting to think that if you can grow your own, you must be better off than if you cannot. Archaeologist Renee Barlow, however, observes that growing corn with simple hand tools is not all that productive when compared to the returns that can be gotten from many wild foods. Planting corn with the use of a digging stick and without any labor-intensive field preparation, as was done in early historic times in parts of the far southern Great Basin, can provide yields on the order of 1,500 kilocalories per hour, roughly comparable to the yields provided by pine nuts, beneath those that can be provided by cattail pollen, and far below those provided by many vertebrates. Barlow also shows that additional labor invested in horticulture does not reap a particularly substantial gain. As a result, she suggests that agriculture should be adopted only at times when highly productive resources have become scarce on the landscape and should be dropped when such resources are abundant. In this view, farming was adopted in the Fremont area when things were bad and abandoned when they were good.

These are interesting arguments, but the truth is that not many wild plant resources in the Great Basin can actually outdo the returns that might be gotten from farming. Large mammals certainly do that, but those were likely pursued by men. As Barlow notes, at least the initial adoption of horticulture was probably in the female realm. In addition, while there is evidence that hunting during Fremont times reduced the abundance of large mammals on the landscape, no evidence suggests that the beginnings of horticulture here were associated with declining abundances of productive wild resources or that the abandonment of horticulture was associated with their increased availability. In fact the kinds of climate change that seem to account for the adoption of horticulture by Fremont people in the first place, and that I am about to discuss, would have increased, not decreased, the abundance of wild foods on the landscape. While Barlow's arguments help account for the decisions made by Fremont peoples after horticulture had arrived, they do not account for why all this started or for why it stopped.

There is, on the other hand, compelling evidence that climate change both fostered the adoption of domesticates and forced their loss. In chapter 8, I noted that the Fremont phenomenon coincided with a great increase in the abundance of bison in the eastern Great Basin—one of the wild food resources that became more common during Fremont times—and that this increase was associated with increases in summer temperatures and summer rainfall. Both of these were, in turn, associated with the strengthened northward incursions of monsoonal storm systems. Corn began its lengthy career as a tropical grass and, as such, requires summer warmth, summer moisture, and a long growing season. All these things were provided by the kinds of climates that appear in the Fremont region at about the time the Fremont begins and that disappeared at about the same time the Fremont way of life disappeared.

All this fits together to form a tight package. Fremont horticulture was allowed by climate change. When summer rains decreased and growing seasons shortened, bison populations shrank dramatically, horticulture on the Fremont scale became impossible, and the Fremont phenomenon ended. In fact, the decline and disappearance of Fremont seems to coincide tightly both with the loss of the monsoons and with a series of significant droughts that impacted both the Southwest and Fremont country at about 1050 A.D., 1150 A.D., and shortly before 1300 A.D.—the Medieval droughts that I discuss in chapter 8. The first of these matches a decrease in the number of Fremont sites, the second the loss of corn horticulture in the Great Salt Lake area, and the third the end of Fremont itself. It was climate change that doomed the Fremont.

But if climate change doomed the Fremont way of life, what happened to the people themselves? After all, when Europeans arrived in what was once Fremont territory, they encountered Numic peoples who, with the exception of some groups in far southern Utah, did not grow corn, beans, and squash (explained by the altered climatic conditions), but who also used basketry distinctly different from that which marked the Fremont, used ceramics not particularly similar to the bulk of Fremont wares, and even used footwear having little in common with Fremont moccasins. Did Fremont peoples change their lifestyles this sharply, or did Numic peoples, in one way or another, replace them? And if Numic peoples did replace them, might this have been part of the Numic expansion that we have been discussing here?

ANCIENT HUMAN DNA, THE DISAPPEARANCE OF THE FREMONT, AND NUMIC HISTORY

We cannot yet answer all of these questions about the Fremont, but recent research using ancient DNA preserved in human bone has shown us how these questions will likely be answered. At the same time, that research has started to clarify many aspects of Great Basin prehistory that have long been confusing and at times contentious.

The bulk of this work has focused on what is known as mitochondrial DNA (mtDNA), genetic material located in small, self-contained structures found in cells outside the cell nucleus, which has its own DNA. Mitochondrial DNA offers some very convenient things to those interested in past human population. First, it is abundant. Second, compared to nuclear DNA, it changes, or mutates, relatively rapidly through time, meaning that it can be a fairly sensitive indicator of relatedness at the population level. Third, its composition and structure is very well understood. Fourth, mtDNA is passed along solely through the mother. Fifth, it does not recombine, or trade genetic material, with other DNA, or does so only rarely. This makes interpretations of its composition straightforward compared with nuclear DNA, which does recombine. Finally, compared to ancient nuclear DNA, it is relatively easy to extract from bone.

Native American mtDNA falls into five major lineages, or haplogroups, all of which are also found in south-central Asia and all of which were certainly brought by people who came across the Bering Land Bridge (chapter 3). Different Native American groups are marked by different frequencies of these five haplogroups. Because this is the case, mtDNA offers the opportunity to trace the relationships among modern populations of Native Americans and between those populations and populations in the past.

In only two major settings has this approach been applied in the Great Basin. The first of these involves the Stillwater

Marsh burials discussed earlier in this chapter, along with a separate set of prehistoric individuals found in the Pyramid Lake area. The second involves the Great Salt Lake Fremont burials.

I begin with the results of the analysis of the individuals from the Stillwater and Pyramid Lake areas. Frederika Kaestle and her colleagues analyzed these remains, dated to between about 300 and 6,000 years ago, and found that, as a group, they were genetically distinct from the Northern Paiute peoples who occupy the area today but very similar to Penutian speakers who now occupy California. This, of course, matches what Ed Jolie, Ruth Burgett, Gene Hattori, and others have argued from the archaeological record—that it appears that Penutian peoples had once occupied the western Lahontan Basin, only to be replaced by Numic speakers. If all this is correct, the western Lahontan Basin was, until relatively recent times, occupied by Penutian peoples who were then replaced by the Numic peoples, presumably Northern Paiute, who live here today.

However, there are some caveats here. First, and least important, recall that mtDNA is inherited only through the mother. As a result, these results are telling us about the history of women, not necessarily the history of everyone. That said, it seems highly unlikely that Penutian women simply left on their own, leaving the men behind. Women worldwide might find this reasonable, but the women who left would have had to be replaced by Numic-speaking women to create the genetic picture now found in this region. The possibility of that having happened seems very remote.

The second caveat involves the nature of the prehistoric "population" that Kaestle and her colleagues have analyzed. Since it is the frequencies of the five haplogroups that distinguish between populations, not the haplogroup to which any given individual belongs, Kaestle and her colleagues had to create a "population" by combining individuals that ranged across some 5,700 years. As a result, the composite pattern does not tell us the genetic makeup of people who lived on the landscape at any one point in time. Instead, it tells us the composite makeup of people who lived here across nearly 6,000 years.

Finally, and most important of all, recent analyses have shown if the earlier Stillwater populations were sufficiently small and isolated, the genetic differences they show are not large enough to remove them as possible ancestors of the modern Numic peoples of the area. Unfortunately, as things stand now, we have no way of assessing either how large the earlier populations were, or how much genetic material they may have shared with others. The analysis of much larger samples of prehistoric peoples from a much more restricted span of time coupled with more detailed genetic studies are needed to address this issue.

The Fremont research has different implications. Dennis O'Rourke and his colleagues analyzed the mtDNA from more than forty of the Fremont burials exposed by the Great Salt Lake floods I discussed earlier. They found these individuals to have been very different genetically from the prehistoric western Lahontan Basin populations studied by Kaestle and her colleagues. They also found the Great Salt Lake Fremont people to have been closely related to both the prehistoric Anasazi of the Southwest and to contemporary Puebloan populations (the descendants of the Anasazi). The match with the Anasazi was not perfect, or even close to it, but it was close enough to suggest that the people who left behind both the Anasazi and the Fremont archaeological remains are likely to have shared a common ancestor.

O'Rourke and his colleagues suggest that the origins of the Fremont are to be sought in the Southwest. In a general sense, of course, this must be the case, since such things as corns, beans, and squash horticulture and aspects of Fremont architecture have long been recognized to be Southwestern in origin. But while their work might suggest that the Fremont came into being as a result of people moving into the eastern Great Basin from the Southwest, it does not require it. It is possible that there were genetically continuous populations of people from the eastern Great Basin into the Southwest, people who shared genes and ideas and who gave rise to both the Anasazi and the Fremont. That would account for both the archaeology and the genetics and, at the moment, seems like the most likely possibility.

Unfortunately, the work by O'Rourke and his colleagues does not tell us about the relationship between the Fremont and the people who came after them. As they observe, we simply do not know enough about the genetics of the people who followed the Fremont to be able to assess this relationship in any meaningful way.

CLOSING THOUGHTS ON THE NUMIC EXPANSION

As I mentioned earlier, there is no question that the Native peoples who now occupy the Great Basin must have come from elsewhere, since, if we go back far enough in time, there was no one here at all. In that sense, a "Numic Expansion" must have occurred, and the remaining question is when.

Jane Hill's work suggests that it was after 4,000 to 3,000 years ago. Work by geneticist Jason Eshleman on burials from the Central Valley of California indicates that the Uto-Aztecan speakers of western California may have been in this area between 4,000 and 2,000 years ago, suggesting that they may have been in at least parts of the Great Basin by this time as well. The most secure information we have, provided by the combination of archaeology and genetics, comes close to demonstrating that the western Lahontan Basin was occupied by Penutian speakers until quite late in prehistoric times, a conclusion that matches Northern Paiute oral tradition. Even though this is the case, the genetic analyses may suggest, but they certainly do not demonstrate, that a population replacement occurred here. To the north, archaeology suggests that Penutian speakers may also have occupied the far northern Great Basin of south-central Oregon into late prehistoric times. This, however, is a conclusion that has no genetic support at all.

The stark changes in basketry technology and other artifact forms that mark the end of the Fremont suggests that human populations changed here as well. If this happened, and it looks likely that it did, what happened here may have been extremely complex.

This is because of a phenomenon known as the "Promontory Culture," first described by Julian Steward decades ago on the basis of his excavations at the Promontory caves on the northern edge of Great Salt Lake. Steward observed that some of the artifacts he had found seemed quite similar to those known from far northern North America, suggesting that they may represent "remains left by one of the Athapascan-speaking tribes" (Steward 1937a:87). Joel Janetski has pointed out that this archaeological phenomenon, known from Utah Lake north through the Great Salt Lake area, seems to date to between about 1300 and 1600 A.D. He agrees with Steward in thinking that the people who made these artifacts may have been Athapaskan speakers, since the artifact similarities point to those parts of the American subarctic where these people live. As I have discussed, the northern Athapaskan people are closely related, both linguistically and genetically, to the Navajo and Apache of the Southwest. Everything we know about Southwestern archaeology shows that they are fairly recent arrivals there. As a result, archaeologists have long looked for traces of their movements southward and it could be that the "Promontory Culture" reflects part of that movement. The distinctive Promontory artifacts are gone by about 1600 A.D., replaced by artifacts, and a lifestyle, that looks just like that of the Ute and Shoshone people who were encountered by Europeans when they entered the area.

It is tempting to speculate that the ancestors of the famous Navajo code talkers of World War II may have once lived along the shores of the Great Salt Lake. That would, however, be premature, since it is far from obvious that the post-Fremont "Promontory Culture" was, in fact, a product of Athapaskan people and that they, in turn, were replaced by Numic peoples. All of this is in need of genetic support. Either way, however, the archaeological record suggests that the Numic speakers who lived in the Fremont area when Europeans arrived, and who live here now, are relatively recent arrivals as well.

Elsewhere, things are far less clear. The arrival of such things as particular kinds of pottery and arrowpoints are simply not trustworthy markers of ethnicity. The fact that, for instance, Desert Side-Notched arrow points arrive late across the Great Basin and coincide with a hypothetical chronology for an equally late arrival of Numic peoples is not compelling: those artifacts, or ones very much like them, are far more widespread than the people themselves. It is the combination of detailed analyses of artifacts and human genetics that are helping to clarify the situation in the western Lahontan Basin, and it is likely to be this combination that helps clarify this situation elsewhere.

No matter when Numic peoples arrived in any given part of the Great Basin, however, nearly all of this land was their land when Europeans arrived.

Chapter Notes

The best place to start for a detailed overview of Great Basin prehistory is Simms (2008). C. S. Fowler and Fowler (2008) contains a broad set of excellent and well-illustrated papers on Great Basin prehistory by a wide variety of experts. Beck and Jones (1997) provide an important review of the early Holocene archaeology of the Great Basin. (Archaeologist Tom Jones is the great-great-nephew of General Jesse Lee Reno, who was killed in the Civil War and for whom Reno, Nevada, is named.) R. L. Kelly (1997) does the same for the middle and late Holocene, as do Madsen, Oviatt, and Schmitt (2005) for the prehistory of the Bonneville Basin. T. L. Jones and Klar (2007) covers parts of the western Great Basin and is very much worth the read. Harrington and Simpson (1961) argued that the Tule Springs site was pre-Clovis in age; for the refutation, see Wormington and Ellis (1967). References to the Paisley Caves are provided in chapter 3.

Important discussions of Great Basin fluted points are provided by Beck and Jones (1997, 2007, 2009, 2010a, 2010b). Those publications also provide references to all of the relevant literature. W. E. Davis, Sack, and Shearin (1996) describe the Tule Valley fluted point site. E. L. Davis (1978a, 1978b) describes the China Lake fluted point sites. On Lake Tonopah, see E. W. C. Campbell et al. (1937), E. W. C. Campbell and Campbell (1940), and, especially, Pendleton (1979), which includes an excellent analysis of a sample of the Tonopah fluted points as well as Elizabeth Campbell's field notes on the early work at this locale. On Long Valley and the Sunshine Locality within it, see Beck and Jones (2009). There is a lengthy and fairly complex literature on the fluted point sites in the Alkali Basin; Pinson (2008) is the place to start.

J. D. Jennings (1957) and Holmer (1986) discuss the fluted points from Danger Cave; the chronology of this site has been greatly clarified by Rhode, Madsen, and Jones (2006), and I follow their suggested chronology in this chapter. Jennings (1994) provides more personal insights into the Danger Cave excavations and on his very important career as a whole. The Sunshine Locality is described in detail by Huckleberry et al. (2001) and Beck and Jones (2009). The Henwood site is discussed by Warren and Phagan (1988) and Douglas, Jenkins, and Warren (1988). The Connley Caves fluted point is analyzed by Beck et al. (2004). On the Old Humboldt site, see J. O. Davis and Rusco (1987); on the Bonneville Estates fluted point, Goebel (2007). Fort Rock Cave was once thought to have provided a fluted point; J. L. Fagan (1975) showed that this is not the case.

A. Taylor (2002) analyzes the geographic distribution of fluted points in the Great Basin. This important paper is not published, but a brief synopsis of it is (A. Taylor 2003). V. Larsen (1999) describes the Clear Creek Canyon fluted point. The initial discovery of a fluted point at 7,810 feet in the Pine Grove Hills was reported by Rhode (1987); he has told me about the later discoveries, along with a reevaluation of the elevation of the site. Many of the general points I discuss here about Great Basin fluted points and their makers were made many years ago by Heizer and Baumhoff (1970) and Wilke, King, and Bettinger (1974).

E. W. C. Campbell et al. (1937) present the results of their work at Pleistocene Lake Mojave; Antevs's 15,000-year estimate appears in his contribution to this volume. They spelled Lake Mojave with an *h* (as many have); as a result, contemporary archaeologists often refer to "Lake Mohave points," since that was the first official spelling. Warren and De Costa (1964) and Ore and Warren (1971) discuss the dating of the Lake Mojave artifacts within the constraints provided by what was then known of the history of the lake; Warren and Ore (1978) discovered buried artifacts at Lake Mojave, obtained a single radiocarbon date on a mollusc shell, and argued that these materials had been deposited from at least 10,270 years ago to about 8,000 years ago. The history of Pleistocene Lake Mojave is discussed in chapters 5 and 8. Warren (1967) remains a valuable clarification of early work done on Lake Mojave materials. The best recent synthesis of this material is by Sutton et al. (2007).

My comment that the lowest artifact-bearing terrace at Lake Mojave is beneath the Lake Mojave II highstand is based on the fact that, with the exception of eroded sites, the lowest artifacts found by the Campbells were at an elevation of 937 feet (285.6 meters). Lake Mojave shoreline A, which marks the highstands of both Lake Mojave I and II times, is higher, sitting at 943 feet (287.5 meters; see S. G. Wells et al. 2003).

My discussion of the chronology of stemmed point occupations follows Beck and Jones (1997, 2010a, 2010b). Tuohy and Layton (1977) suggested the name "Great Basin Stemmed" for these projectile points. Bryan (1980, 1988), Musil (1988), Titmus and Woods (1991), and Beck and Jones (2009) discuss the hafting of these points. Beck and Jones (1997, 2009) provide excellent descriptive overviews of Great Basin stemmed points, along with references to other relevant literature. I have followed their usage closely. Warren (1991) provides a key to the early points of the Mojave Desert.

On crescents, see Tadlock (1966), Clewlow (1968), B. R. Butler (1970), J. O. Davis (1978:57–60), Beck and Jones (1990, 1997, 2009), Amick (1995, 1999), and Jones, Beck, and Kessler (2002). Analyses of residues on crescents are provided by Hattori, Shelley, and Tuohy (1976); Shelley (1976); and Hattori, Newman, and Tuohy (1990). Renewed research on these residues, using newly available technologies, would be valuable.

Grinding stones have been found on stemmed point sites in many parts of the Great Basin, including the Alkali Basin (J. L. Fagan 1988, Willig 1988, Pinson 2004), the Fort Rock Basin (Oetting 1993), the Bonneville Basin (Simms and Isgreen 1984; Arkush and Pitblado 2000), the Nevada Test Site (Buck et al. 1998), and the Mojave Desert of southeastern California (E. W. C. Campbell et al. 1937; Basgall 1993, 2000b; Sutton et al. 2007). The changing ratios of ground stone through time in the north-central Mojave are from Basgall (2000b). It is important to realize that some of these artifacts may have been deposited by later peoples, though this is clearly not the case in many instances. And it is always possible, as Simms (1983b) pointed out, that later peoples removed ground stone that was once there.

The remarkable record of late Pleistocene and early Holocene archaeological sites along the Old River Bed on the Dugway Proving Ground is discussed by Jones, Beck, and Kessler (2002); Oviatt, Madsen, and Schmitt (2003); and Schmitt et al. (2007).

Arkush and Pitblado (2000) and Duke and Young (2007) describe the early Holocene archaeology of nearby areas.

Last Supper Cave is discussed by Layton (1970) and Grayson (1988); the Old Humboldt Site by Dansie (1987) and J. O. Davis and Rusco (1987); the Five Points Site by Price and Johnston (1988); the Yucca Mountain area by Buck et al. (1998) and G. M. Haynes (2004). Eerkens et al. (2007) describe the early Holocene archaeology of the Coso Basin; see Basgall (1993, 1994, 2000a, 2000b) for the Nelson Lake Basin and the north-central Mojave Desert as a whole. Connolly (1999) presents a thorough description of the Paulina Lake Site.

R. E. Hughes (1998) provides an excellent brief introduction to obsidian sourcing. G. T. Jones et al. (2003) is the place to start for the translation of lithic artifacts into ancient travel maps in the Great Basin; Beck et al. (2002) and Kessler, Beck, and Jones (2009) apply similar logic on a smaller scale. Duke and Young (2007) discuss obsidian transport from Old River Bed stemmed point sites. The Five Mile Flat sites (also called the Parman sites; these are the type sites for the Parman stemmed point style) were originally discussed in detail by Layton (1970, 1979); G. M. Smith (2006, 2007) provides a detailed analysis of the stone tools from this location, as well as of the sources of the raw materials on which those tools are made and the implications of all this for human movement across the landscape. Graf (2001, 2002) provides similar information for the Sadmat and Coleman sites; see Warren and Ranere (1968) and Tuohy (1970, 1974, 1981, 1988) for earlier discussions of these locations. Oetting (1993) provides information on the Fort Rock Basin (Buffalo Flat) sites. Dugway Stubbies are defined in G. T. Jones, Beck, and Kessler (2002) and discussed and illustrated in Schmitt et al. (2007).

There is a substantial literature on the possible uses of stemmed points. The best starting place is Beck and Jones (2009); other important discussions are provided by Beck and Jones (1990, 1993, 1997, 2007), Basgall and Hall (1991), Basgall (1993), and G. T. Jones and Beck (1999).

Bryan (1977, 1980, 1988) has long argued that the users of stemmed points preceded users of fluted points in the Great Basin; Tuohy (1974) argued that fluted and stemmed points belong to the same tool kit. Beck and Jones (1997, 2009, 2010a, 2010b) have examined these arguments in impressive detail. Basgall and Hall (1991) provide data on isolated fluted and stemmed points from Fort Irwin. For sites that contain both fluted and stemmed points, see E. L. Davis (1978b) on China Lake, Price and Johnston (1988) on Jakes Valley, Zancanella (1988) on Railroad Valley, Hutchinson (1988) and Beck and Jones (2009) on Long Valley, Pendleton (1979) and Tuohy (1988) on southern Big Smoky Valley, and Simms and Lindsay (1989) on the Sevier Desert. Comparative analyses of the production of stemmed and fluted points have been conducted by J. L. Fagan (1988), Pendleton (1979), Bryan (1988), Warren and Phagan (1988), and Beck and Jones (2008, 2009, 2010a, 2010b). Pendleton (1979) and J. L. Fagan (1988) describe sites in the Pleistocene Lake Tonopah and Alkali Lake basins, respectively, in which fluted points are found in the absence of stemmed ones. Sutton et al. (2007) note that such sites exist in the Mojave Desert as well. Basgall and Hall (1991) and Basgall (1993) suggest that different sets of people are more likely to be represented by these artifact types. The earliest radiocarbon dates for stemmed points in the Intermountain West are listed and discussed in Beck and Jones (2009, 2010a).

Nassany and Pyle (1999) discuss the different ways in which dart and arrow points were manufactured in central Arkansas.

The Spirit Cave material was first published in Wheeler and Wheeler (1969). Wheeler (1997) provides a slightly different version of this paper, without the very moving photographs and without Georgia Wheeler as a coauthor. Dansie (1997) recounts the history of the work here and on other, nearby human burials; Tuohy and Dansie (1997) present relevant radiocarbon dates and a discussion of associated artifacts. Edgar (1997) and Jantz and Owsley (1997) analyze the pathologies shown by the Spirit Cave Mummy. The remarkable textiles from Spirit Cave are analyzed by C. S. Fowler, Hattori, and Dansie (2000) and C. S. Fowler and Hattori (2009); C. S. Fowler and Hattori (2008) provide a more general discussion of these objects, along with excellent photographs of some of them. Jantz and Owsley (1997) and Steele and Powell (2002), among others, examine the similarities and differences between the Spirit Cave Mummy and contemporary peoples. The Fallon Paiute-Shoshoni Tribe (www.fpst.org/index.php) has requested that the Spirit Cave burials be returned to them under the terms of the Native American Graves Protection and Repatriation Act (http://www.nps.gov/history/nagpra/). The legal dispute over the final disposition of the Spirit Cave remains has yet to be resolved (see www.narf.org/nill/bulletins/dct/unreported/fallonpaiute.pdf and Nevada State Museum Anthropology Program 2007); I should note that I served as a consultant to the tribe in this request. Edgar et al. (2007) present a superb overview of the Spirit Cave material and the analyses that have been done of it; it is the place to start if you are interested in a fair and balanced analysis of the situation.

See Bedwell and Cressman (1971) and Bedwell (1973) for the definition of the Western Pluvial Lakes Tradition, and Hester (1973) for the expanded version of this concept. All agree that, as valuable as the concept once was, it has lost its descriptive and explanatory value.

References to the faunal remains from Last Supper Cave, Old Humboldt, and Henwood are provided above. Greenspan (1993) and Oetting (1993) provide details on faunal remains from early Holocene sites in the eastern Fort Rock Basin (Buffalo Flat); for elsewhere in the Fort Rock Basin, see Greenspan (1994) and Mehringer and Cannon (1994). Eiselt (1997b) identifies the fish remains from the Spirit Cave Mummy coprolites; Napton (1997) mentions that these contain seeds as well. B. S. Hockett (2007) discusses the faunal remains from Bonneville Estates Rockshelter; Rhode and Louderback (2007) discuss the plant remains from this site. Madsen (1999, 2007) provides valuable discussions of early Holocene human adaptations in the Great Basin in general, as do Elston and Zeanah (2002).

On the costs of seeds, see O'Connell and Hawkes (1981, 1984); O'Connell, Latz, and Barnett (1983); Simms (1984); Cane (1987); R. L. Kelly (1995); and Barlow and Metcalfe (1999). O'Connell, Jones, and Simms (1982) were the first to suggest that the middle Holocene increase in grinding stones could be explained by a cost-benefit analysis; I have simply followed their lead here.

On Danger Cave, see J. D. Jennings (1957); Rhode et al. (2005); and Rhode, Madsen, and Jones (2006); on Bonneville Estates Rockshelter, Rhode et al. (2005); Goebel (2007); Graf (2007); Rhode and Louderback (2007); and Rhode (2008); on Camels Back Cave, Schmitt, Madsen, and Lupo (2002a, 2002b); Schmitt and Madsen (2005); and Rhode (2008).

In addition to O'Connell (1971, 1975), see O'Connell and Hayward (1972), O'Connell and Ericson (1974), and O'Connell and Inoway (1994) on the Surprise Valley sites. Descriptions of Modoc and Klamath winter earth lodges are provided by Ray (1963) and Stern (1966). Sampson (1985) provides a detailed analysis of the Nightfire Island site. Helzer (2001, 2004) provides details on the Bergen site.

Heizer (1951), Baumhoff and Heizer (1965), Cressman (1977, 1986), and many others have seen the middle Holocene Great Basin as markedly inhospitable; they were, of course, working long before many of the detailed paleoclimatic records I have discussed in chapter 8 became available. J. D. Jennings (1957), Aikens (1970), and Madsen (2002), among others, have disagreed with that argument. Surovell and Brantingham (2007) provide an excellent discussion of the use of radiocarbon frequency distributions to infer past human population numbers, including the very real pitfalls involved. The radiocarbon curves presented in figures 9-8 through 9-12 were developed by Louderback, Grayson, and Llobera (2010). No such curves have been developed for the Mojave Desert, but Sutton et al. (2007) present a substantial list of middle Holocene radiocarbon dates from archaeological sites here and observe that parts of this region may have been abandoned during this interval.

The relationship between shallow-water resources and human population densities may be seen in Steward (1938); see Thomas (1972b) for a discussion of this issue. C. S. Fowler (1990a, 1990b, 1992), Janetski (1991), R. L. Kelly (1997, 2001), Madsen (2002, 2007), and Janetski and Smith (2007) discuss marsh- and lake-oriented adaptations in the Great Basin.

Because of their potential chronological and adaptive significance, there is a huge literature on Great Basin projectile points in addition to the stemmed and fluted points discussed earlier. Key references to begin with include Hester (1973), Thomas (1981), and Holmer (1986). On the distribution of the atlatl during historic times in North America, see Driver and Massey (1957). Gatecliff and Elko series points hafted to atlatl darts are known from, among other places, Kramer Cave (Hattori 1982) and the NC site in southern Nevada (Tuohy 1982), respectively; Desert series points hafted on arrows are known from Tommy Tucker Cave (Fenenga and Riddell 1949) and Gypsum Cave (Harrington 1933); Desert series points on historic Great Basin arrows are illustrated in D. D. Fowler and Matley (1979). Rosegate points have been found associated with arrow fragments at James Creek Shelter (Elston and Budy 1990). Thomas (1978), Schott (1997), and Ames, Fuld, and Davis (2010) discuss the morphological differences between arrow and dart points.

In addition to Thomas (1981) and Holmer (1986), my discussion of the morphology and chronology of Great Basin projectile points has depended heavily on S. J. Vaughan and Warren (1987); O'Connell and Inoway (1994); Beck (1995); B. S. Hockett (1995); Delacorte (1997); McGuire (1997, 2000a); Schroth (1999); Yohe (1998, 2000); Bettinger and Eerkens (1999); Arkush and Pitblado (2000); Basgall and Hall (2000); Hildebrandt and King (2000, 2002); G. T. Jones et al. (2003); Oviatt, Madsen, and Schmitt (2003); McGuire, Delacorte, and Carpenter (2004); Elston (2005); Schmitt et al. (2007); and Thomas (2009). Largaespada (2006) presents a key specifically built for the northern Great Basin. Flenniken and Wilke (1989) argued that the variability in Great Basin atlatl dart points does not reflect time, but instead different stages in the rejuvenation of broken points. Although their argument has been rejected, it led to an increased understanding of variability in these artifact types. As Zeanah and Elston (2001) note, an important idea does not have to be correct to lead to productive research.

Hockett and Murphy (2009) provides evidence for middle Holocene communal pronghorn hunts in the Great Basin. Fagan (1974) presents the results of his work at spring sites in southeastern Oregon. Milliken and Hildebrandt (1997) and Hildebrandt and King (2000) review the middle Holocene archaeology of northeastern California. Other sites that saw their first human occupation at or just after the end of the middle Holocene include Amy's Shelter (Gruhn 1979), Newark Cave (D. D. Fowler 1968), Pie Creek Shelter (McGuire, Delacorte, and Carpenter 2004), Remnant Cave (Berry 1976), and Swallow Shelter (Dalley 1976). On the middle Holocene in the Mojave Desert, see Basgall (2000b) and Sutton et al. (2007).

On Newberry Cave, see C. A. Davis and Smith (1981) and C. A. Davis, Taylor, and Smith (1981); on Stuart Rockshelter, D. Shutler, Shutler, and Griffith (1960); on Hidden Cave, Thomas (1985), Rhode, Adams, and Elston (2000), Rhode (2003), and C. S. Fowler and Hattori (2009); on Kramer Cave, Hattori (1982); on Lovelock Cave, Heizer and Napton (1970) and Livingston (1988b); on South Fork Shelter, Heizer, Baumhoff, and Clewlow (1968) and L. Spencer et al. (1987); on Pie Creek Shelter, McGuire, Delacorte, and Carpenter (2004); and, on Lakeside Cave, Madsen and Kirkman (1988). On the increased use of upland settings in the Bonneville Basin, see Aikens and Madsen (1986) and Madsen (1982, 2002); on Gatecliff series points at high elevation, Thomas (1988). The Hidden Cave coprolites are discussed by Roust (1967), Thomas (1985, chapter 27), Wigand and Mehringer (1985), Butler and Schroeder (1998), and Rhode (2003); the date I use for the earliest intense occupation at Hidden Cave follows Rhode (2003). The Lovelock Cave coprolite data are in Ambro (1967), Cowan (1967), and Napton and Heizer (1970).

Goddard (1996) and L. Campbell (1997) provide excellent introductions to Native American languages, as does Golla (2007) on the linguistic prehistory of California, including the Uto-Aztecan speakers at issue here. Malhi et al. (2008) review the genetic relationships between southern and northern Athapaskan speakers. Madsen (1994) discusses archaeological precursors to the linguistically derived Numic expansion hypothesis. My discussion of glottochronological work relevant to this hypothesis is based on Swadesh (1950, 1952, 1959), Lees (1953), Gudschinsky (1956), C. F. Hockett (1958), Hymes (1960), and Bynon (1977). Rea (1958, 1973) provides his telling analyses of the validity of glottochronology. My discussion of the history of French is based on Rickard (1989); the 960 A.D. date for written Italian comes from Embleton (1991).

Swadesh (1956) presented his glottochronological analysis of Uto-Aztecan. In one of the classic papers in Great Basin anthropology, S. M. Lamb (1958) placed this work in the Numic context. He told me that, when he wrote this seminal paper, he was unaware of archaeological speculations about relatively recent movements of Numic peoples in the Arid West. On the relationship of Numic to Tübatulabal, see Ruhlen (1991a), Silver and Miller (1997), and Golla (2007). The other glottochronological assessments I discuss were provided by Hale (1958), J. H. Goss (1968), and W. R. Miller, Tanner and Foley (1971). Lamb's statement that the Numic expansion might have occurred as early as 3,000 years ago is in Thomas (1994). J. H. Goss (1977) rejected Lamb's Numic expansion hypothesis. For recent applications of more sophisticated versions of glottochronology, see R. D. Gray and Atkinson (2003) and R. D. Gray, Drummond, and Greenhill (2009). If you are interested in learning more about glottochronology, including current versions thereof, Renfrew, McMahon, and Trask (2000) and McMahon and McMahon (2005) are the best places to start. Madsen and Rhode (1994) is required reading for issues related to Numic expansion, though it predates Jane Hill's work.

For many years, and joined by others (e.g., Aikens and Witherspoon 1986, Holmer 1990, Aikens 1994), I strongly questioned the Numic expansion hypothesis (Grayson 1993, 1994). My objections were simple: there was no scientific reason to accept the dates provided by glottochronology and thus no reason to accept the hypothesis. Jane Hill's research began to change this situation dramatically: see Hill (2001a, 2001b, 2002a, 2002b, 2006, 2008). My comment that her dates for the spread of Numic across the Great Basin seem derived from glottochronology are based on her discussion in Hill (2001b:927). Merrill et al. (2009) question her conclusions and provide their version of Uto-Aztecan prehistory; on the genetics of southwestern and Mesoamerican Uto-Aztecan peoples, see Kemp (2006), Kohler et al. (2008), and Merrill et al. (2009). D. D. Fowler (1983) provides her critically important analysis of cognate words for plants and animals in Uto-Aztecan languages. See Steadman (1980) and Bocheński and Campbell (2006) for discussions of the fossil record for turkeys in North America, and Munro (2006) and Speller et al. (2010) for turkeys in arid western North America. Ingram (2008) reviews the distribution of agaves in Nevada.

Bettinger (2008) synthesizes high-elevation archaeological sites in the Great Basin. The archaeology of the Toquima Range is presented by Thomas (1982, 1994) and Thomas and Pendleton (2009); the plant and animal remains from Alta Toquima Village, by Rhode (2009) and Grayson and Canaday (2009). Richard Adams has discovered a similar site, though with many more structures, at 10,700 feet in Wyoming's Wind River Range and reports that others exist nearby (R. Adams 2009, Wingerson 2009). See R. L. Kelly (1995), Bird (1999), Hawkes and Bird (2002), E. A. Smith (2004), and Bleige Bird and Bird (2008) for discussions of the sexual division of labor among hunters and gatherers. The implications of the Stillwater human skeletons for prehistoric mobility are presented by C. S. Larsen, Ruff, and Kelly (1995).

Key references on the archaeology of the White Mountains include Bettinger and Oglesby (1985), Bettinger (1991a, 1991b), Grayson (1991), and Thomas (1994); additional information on the faunas from these sites is provided by Grayson and Millar (2008) and Grayson and Canaday (2009). Scharf (1992, 2009) provides information on the plant remains from Midway. Ababneh (2008) analyzes the possible relationship between precipitation and the use of the White Mountains villages. Steward (1941) noted the existence of very high–elevation houses in the White Mountains constructed in ways similar to those discovered by Bettinger, though the peoples Steward interviewed for his ethnography of the Owens Valley Paiute did not mention them (Steward 1933).

See R. L. Kelly (1997) and Blitz (1988) on the bow and arrow, and Bettinger (1999a, 1999b) for his insights into the possible impact of the introduction of this weapon system into the Great Basin, including the ultimate impact on public-versus-private goods. See Madsen (1986b), Pippin (1986), and the papers in Griset (1986) and Mack (1990) for important earlier discussions of Great Basin ceramics, and Rhode (1994) and Feathers and Rhode (1998) for dates for Intermountain Brown Wares. My brief synthesis of Eerken's work on Intermountain Brown Wares has depended heavily on Eerkens (2003a, 2003b, 2003c, 2004, 2005) and Eerkens, Neff, and Glascock (1999, 2002a, 2002b).

Excellent overviews of the uses of pinyon by the native peoples of the Great Basin are provided by Madsen (1986a), Janetski (1999b), and C. S. Fowler and Rhode (2006). Bettinger (1976, 1977) first suggested the late appearance of intensive pinyon use in the Owens Valley area. Madsen (1986a) made the argument for the Great Basin as a whole and suggested that the bow and arrow might have had something to do with this. My discussion of the late appearance of intensive pinyon use in the Owens Valley area is based on Bettinger (1989); Delacorte (1990); Eerkens, King, and Wohlgemuth (2004); and Hildebrandt and Ruby (2006). For owls feeding on pine nuts, see Rhode and Madsen (1998). Janetski (1979) discusses prehistoric and ethnographic snares in the Great Basin.

Bettinger first developed his views on Numic expansion in Bettinger and Baumhoff (1982, 1983), Bettinger (1991a), and Young and Bettinger (1992); these must be read in conjunction with his later papers on the origins of Numic adaptations, cited above. Simms (1983a), Aikens and Witherspoon (1986), Shaul (1986), Grayson (1993, 1994), Musil (1995), and Thomas (1994), among others, critiqued those initial but highly influential arguments. Canaday (1997) provides an important analysis of the alpine archaeology of the Toiyabe Range; Steward (1938) provides population density estimates for the early historic Great Basin.

Adovasio has published widely on the prehistoric textiles of the Great Basin: in addition to his work on the Fremont, cited below, good places to start include Adovasio (1986a, 1986b), Adovasio and Andrews (1983), and Adovasio and Peddler (1994). Thomas et al. (2008) provide the date for the Gatecliff basketry fragment I discuss here. They give the calibrated date; Gene Hattori was kind enough to provide me with the radiocarbon date (3,210 ± 50 years BP). Adovasio and Andrews (1983) originally described the basketry from Gatecliff.

Ed Jolie and Ruth Burgett's arguments for the Maidu affinities of Lovelock Culture textiles are focused on the remarkable materials recovered from Charlie Brown Cave, in the Winnemucca Basin, many of which are on display at the Nevada State Museum: see Burgett (2004), Jolie (2004), and Jolie and Jolie (2008). This last reference contains superb photographs of these stunning objects. Fowler (1994b) provides a well-reasoned discussion of the relationship between Numic material culture and the Numic expansion, including textiles and pottery. Tuohy and Hattori (1994b) provide a good introduction to Lovelock Wickerware, one of the key forms marking the Lovelock Culture. Hattori (1982) discusses the possible Penutian affiliations of the Winnemucca Basin archaeological record. The possible consequences of late Holocene climate change and the demise of the Lovelock Culture is explored in Benson, Berry, et al. (2007); Benson, Hattori, et al. (2006) demonstrate that Lovelock Culture textiles are locally made.

Cressman (1936, 1942) first used archaeological data to suggest that Klamath peoples once lived as far east as Steens Mountain. Pettigrew (1985) and Oetting (1989, 1990) continued the argument; see also O'Neill et al. (2006). Eiselt (1997a, 1998) discusses the Warner Valley situation. The "other sites" to which I refer here include Carlon Village, at the south end of Silver Lake, Oregon (Wingard 2001).

Four general introductions to Fremont are very much worth reading: Madsen (1989), Janetski (1998, 2008), and Simms (2008, chapter 5). Madsen and Simms (1998) and Talbot, Baker, and Janetski (2005) are more technical but repay the effort. Talbot (2000a, 2000b) are required reading on Fremont architecture and settlement patterns. Adovasio (1975, 1979, 1980, 1986a, 1986b; Adovasio, Pedler, and Illingworth 2002, 2008) has examined Fremont basketry in tremendous detail. Janetski (2002) analyzes Fremont trade, both internal and external. Metcalfe and Larrabee (1985), among others, provide evidence for Fremont irrigation. The chronology of the Fremont phenomenon is discussed in detail by Massimino and Metcalfe (1999), Spangler (2000b), and Benson and Berry et al. (2007).

The earliest dates for corn in the Fremont area are from the Elsinore burial site, near the town of the same name in central Utah's Sevier River Valley. The site—a bell-shaped burial pit that contained more than two hundred corn cobs and cob fragments as well as a human burial—was found as a result of highway-related construction. Had it not been for the quick action of Utah Department of Transportation supervisor Mike Munroe, it, and the earliest known corn from Utah, would have been destroyed before it could have been studied (Wilde, Newman, and Godfrey 1986; Wilde and Newman 1989; and Wilde and Tasa 1991). The Steinaker Gap site is discussed in detail by Talbot and Richens (1996, 2004). Barlow (2006) and Madsen and Schmitt (2005) provide recent synopses of the chronological development of Fremont. Adovasio (1979, 1986a, Adovasio, Pedler, and Illington 2002, 2008); Janetski (1998, 2002), Talbot, Baker, and Janetski (2005), and others are comfortable with the notion of Fremont as a distinct, and distinctive, cultural entity. Madsen (1979, 1989), Madsen and Simms (1998), and others are less so.

B. S. Hockett and Morgenstern (2003) document the manufacture of Fremont ceramics at a site near Elko, Nevada, far outside the core Fremont zone; the term "Fremont Fringe" is from Madsen and Schmitt (2005:18). Simms (1986; see also Simms 1990) began the current argument over the nature of Fremont subsistence in the Great Basin; Madsen and Simms (1998), Madsen and Schmitt (2005), and Talbot, Baker, and Janetski (2005) are among those who have continued it. On large Fremont structures that seem to have special purposes, see B. S. Hockett (1998), Janetski (1998), Wilde and Soper (1999), and the additional references on Five Finger Ridge provided below. The small Sevier Desert Fremont site discussed in this chapter is Topaz Slough, reported by Simms and Isgreen (1984) and Simms (1986); S. J. Smith (1994) reports a similar site in Skull Valley. Spangler (2000a) presents strong evidence from northeastern Utah's Tavaputs Plateau for Fremont peoples who switched between horticulture and hunting and gathering, perhaps on a seasonal basis.

The very important sites exposed as a result of the high levels reached by the Great Salt Lake in the 1980s are discussed in detail by Simms, Loveland, and Stuart (1991); Fawcett and Simms (1993); Simms (1999); Simms and Raymond (1999); and other papers in Hemphill and Larsen (1999). Coltrain (1993, 1996), Coltrain and Stafford (1999), and Coltrain and Leavitt (2002) present the bone chemistry analyses discussed here.

Five Finger Ridge is the only Fremont village site to have been excavated in its entirety. For that we have multiple people to thank. First, there was local resident Marvin Magleby, who pointed out the existence of this site and thus prevented it from being destroyed by activities related to highway construction. Second, there was the Paiute Indian Tribe of Utah, which worked hard to ensure that the entire site would be excavated and which assisted with those excavations. Third, Joel Janetski and his Brigham Young University colleagues organized and led the work and published the results in an important series of monographs. Janetski (1998) provides a superb synthesis of the Clear Creek

Canyon project, including Five Finger Ridge, and lists the monographs that resulted from it; the ones to begin with are Janetski et al. (2000) and Talbot et al. (2000). On Baker Village, see B. S. Hockett (1998) and Wilde and Soper (1999). Novak and Kollmann (1999) discuss the evidence for violent interactions among people shown from Fremont skeletons; that evidence has now expanded beyond the four sites they discuss.

Baker Village is located just north of the town of Baker, Nevada (the gateway to Great Basin National Park), and is easy to find: follow the signs for the Bureau of Land Management's Baker Archaeological Site down a well-maintained dirt road to the picnic area. The trail through the site begins here and is supported by an informative guidebook; replicas of the surface structures duplicate some of what was found during excavation.

The materials from both Five Finger Ridge and the Elsinore Burial Site are housed at Fremont Indian State Park, just off I-70 near Ridgefield, Utah (http://stateparks.utah.gov/parks/fremont). The park's museum has award-winning interpretive exhibits of Fremont artifacts and lifestyles, including material from both Five Finger Ridge and Elsinore (and a great bookstore for those interested in Fremont archaeology). Five Finger Ridge was located just across the highway from the park headquarters; the park's grounds contain impressive displays of Fremont rock art, some of which are easily accessible by well-maintained trails. You can also try your hand at using an atlatl. Camping is available nearby, but the campground fills at times, so you might want to reserve a place in advance (motels are available in Ridgefield).

Barlow (2002, 2006) develops her ideas on the conditions under which people in the Fremont area should have adopted, and abandoned, horticulture. Janetski (1997) examines the possible impact of Fremont hunters on the animals they were hunting; Ugan (2005) offers a nuanced view of the factors that might drive changing abundances of mammal remains in Fremont archaeological sites. Simms (1984, 1987) provides estimates of energetic return rates from various Great Basin plants and animals; see also R. L. Kelly (1995).

Evidence for increased monsoonal incursions in the eastern Great Basin and adjacent Colorado Plateau is reviewed by Wigand and Rhode (2002) and Grayson (2006b); see also the notes for chapter 8. Increased summer rainfall should also have boosted the abundance of Colorado pinyon in the eastern Great Basin and adjacent Colorado Plateau (Cole et al. 2008), which does not match the predictions of the Barlow model. Benson, Berry, et al. (2007) discuss the eleventh-, twelfth-, and thirteenth-century droughts and relate them to the demise of the Fremont. Kloor (2007) provides a very readable synthesis of the Fremont disappearance; Benson (2008) does the same for the 1150 and 1300 A.D. droughts.

The key references on the genetics of the prehistoric peoples of the Great Basin are Kaestle, Lorenz, and Smith (1999); O'Rourke, Parr, and Carlyle (1999); Carlyle et al. (2000); O'Rourke, Hayes, and Carlyle (2000a, 2000b); Kaestle and Smith (2001); Eshleman (2002); Eshleman and Smith (2007); and Cabana, Hunley, and Kaestle (2008). Eshleman, Malhi, and Smith (2003) and O'Rourke, Hayes, and Carlyle (2000a) provide excellent introductions to the analysis of mitochondrial DNA in North America.

D. G. Smith et al. (2000) and Malhi et al. (2003) observe that the distribution of haplogroups among Uto-Aztecan speakers is not consistent with Hill's argument that Uto-Aztecan people moved northward as part of the spread of agriculture. As a result, they suggest, on the basis of other genetic markers, that if this movement occurred, it may have primarily involved males, a process that J. N. Hill (2002b) sees as quite possible. Janetski (1994) and Janetski and Smith (2007) provide excellent introductions to the perplexing Promontory phenomenon.

If you are in or near Carson City, I strongly recommend a visit to the Nevada State Museum (600 North Carson Street; 775-687-4810; http://nevadaculture.org/museums/). This museum has long supported important archaeological and ethnographic research and has world-class collections and exhibits. Among other things, you can see Northern Paiute artist Mike Williams's superb reproduction of one of the 2,100-year-old Lovelock Cave duck decoys (the eleven originals, discovered in 1924, are in the collections of the National Museum of the American Indian in Washington, DC), a series of exquisite baskets by famed Washoe basket weaver Dat So La Lee, and many of the Charlie Brown Cave baskets. There is also a reproduction of a Columbian mammoth from the Black Rock Desert (the bones themselves are too fragile to display but are in the museum's collections). The museum, in the former Carson City Mint, is one of Nevada's gems.

Other Great Basin museums also have notable collections of Great Basin basketry on display. The Mono County Museum, in Bridgeport, California, is one of them. Housed in what was originally the Bridgeport Elementary School building, built in 1880 and in use as such until 1964, this museum has a superb display of early-twentieth-century Paiute and Washoe baskets, many from the Ella M. Cain collection. There is also a significant collection of items representing the early Euro-American pioneers of the area, but it is the basketry that sets this museum apart (www.monocomuseum.org).

PART SIX

CONCLUSIONS

CHAPTER TEN

The Great Basin Today and Tomorrow

It has become routine to end regional natural histories of the Americas with discussions of all the disastrous things that have happened since the arrival of Europeans, and of the dire consequences that will emerge if we do not change our ways. The warnings are warranted. The nature of the human world is driven, for good and bad, by population increase. Technological advances and environmental degradation follow as a result.

The technological advances follow in part from the greater number of innovators found in larger populations. They also follow from the increasing specialization that is made possible by greater numbers of people working in a context in which knowledge has been made rapidly cumulative by the printing press. Environmental degradation follows both from the technological advances and from the sheer impact that so many people have on the landscape.

It is also true, however, that many of the concerns expressed by those worried about the fate of natural landscapes in the Americas tend to forget that it is not always easy to know what a "natural" landscape really is. Many Americans—natural history writers included—have a highly romanticized view of Native Americans, one in which Indians are presumed to have lived in peace and harmony with the natural world. In this view, North America prior to the arrival of Europeans was pristine; North America after that arrival, increasingly despoiled.

That view is wrong in many ways. The prehistoric peoples of the Americas are, after all, people, and had significant impact on their environment. The hunters and gatherers of the Great Basin were no exception. As I mention in Chapter 2, a communal pronghorn hunt in a Great Basin valley might not be repeated for a decade, since it took that long for the animals to recover. In many places, people set the landscape on fire to increase the efficiency of hunting and the productivity of seed plants. In the White Mountains, the importance of mountain sheep in the diet fell steadily from the earliest occupations some 4,500 years ago to the latest use of the small villages that I describe in chapter 9. This steady decrease cannot be explained by climate change but instead appears to reflect the impact of human hunting on local mountain sheep populations. The landscape that Europeans encountered when they first arrived in the Great Basin was not pristine, if by "pristine" we mean "devoid of significant human influence."

Nonetheless, when Europeans arrived in the Great Basin, a vast portion of this region would have met the definition of wilderness provided in the Wilderness Protection Act of 1964, as an area that "generally appears to have been affected primarily by the forces of nature." That, of course, was not to last long. Some of the subsequent impact is obvious. Elko, Reno, Las Vegas, and Salt Lake City provide examples; the Newlands Project in the Truckee and Carson river basins and the diversion of the waters of Mono Lake and Owens Valley are others. Indeed, if the Southern Nevada Water Authority has its way and gains permission to draw down groundwater from a broad swath of Nevada, from Railroad Valley on the east to Snake Valley on the west, we might quickly learn what a severe version of middle Holocene aridity looks like. In comparison to these proposals, the damage done to Owens Valley by the Los Angeles Department of Water and Power may come to look benign.

However, many changes have occurred in Great Basin ecosystems as a result of the European arrival that are far less evident than these. I give one example here, chosen for the simplicity of its complexity.

Deer, Cougars, Porcupines, and Cattle

Early European explorers and settlers routinely noted the scarcity of deer across much of the Great Basin. As James Moffit (1934:53) noted for the area surrounding Lower

Klamath Lake, in the late nineteenth century "one could ride for a day without seeing a deer in regions where similar excursions would today reveal many of these animals." Today, however, deer are abundant here.

That deer have undergone a tremendous increase in abundance in the Great Basin during the past century or so has been known for many years. Archaeological and paleontological work undertaken during the past few decades has expanded this picture by showing that, with the exception of the far eastern Great Basin (some Fremont sites are loaded with them), deer were also uncommon during later prehistoric times. The faunas from the White Mountains sites, for instance, held but four specimens of deer, compared to more than five hundred of mountain sheep. In fact, there were even more specimens of pronghorn here (seven) than of deer, even though pronghorn are hardly animals of the alpine tundra. Many other sites tell the same story: deer are now far more abundant in the Great Basin than they were during the preceding 10,000 years.

Zoologists Joel Berger and John Wehausen have observed one of the apparent side-effects of this population explosion. To judge from archaeological and paleontological faunas, mountain lions or cougars (*Felis concolor*) were also rare in the Great Basin during late prehistoric times. The virtual absence of cougars from late prehistoric faunas does not reflect a general scarcity of carnivores from these sites: bobcats (*Lynx rufus*), coyotes (*Canis latrans*), and badgers (*Taxidea taxus*) are routinely found in Holocene faunas, and even such historically rare carnivores as bears (*Ursus americanus*) and wolves (*Canis lupus*) are found as well. The remains of cougars, though extremely rare in prehistoric sites, are frequently seen in the Great Basin today. It appears that cougars are now far more abundant in this region than they were during late prehistoric times, and perhaps even far more abundant here than they have been since the end of the Pleistocene.

It might be argued that the increased abundance of deer during the past century simply reflects the removal of Native American hunting pressure on these animals. Deer were scarce in early historic times, this argument would go, because Indian hunting kept their numbers down; deer populations rebounded when that hunting declined.

There is nothing conceptually wrong with that argument. Some years ago, anthropologist Harold Hickerson showed that adjacent but hostile groups of Chippewa and Sioux in western Wisconsin and central Minnesota had virtually eliminated deer from their respective territories. Those territories, however, were separated by a buffer zone controlled by neither group but coveted by both. If either Sioux or Chippewa entered that zone, they risked armed confrontation with their enemies. Freed from human pressure, deer abounded here.

A similar argument has been made for the Snake River Plain and the Columbia Plateau to the immediate north of the Great Basin, but this approach will not work in the Great Basin itself. Archaeological faunas show that deer were rarely taken prehistorically in any number. The implication is that these animals were uncommon long before the European arrival and that their rarity has little or nothing to do with Native American hunting. As a result, their increase in historic times can have nothing to do with the removal of Native American hunting pressure.

It is far more likely that the modern abundance of these animals in the Great Basin has resulted from complex interactions between plants, domestic livestock, and the deer themselves. Berger and Wehausen, among others, note that the introduction of cattle, sheep, and horses into the Great Basin led to the removal of grasses and to the spread of plant species favorable to deer. In response to this massive habitat alteration, deer expanded tremendously in number. Berger and Wehausen also observe that it is not likely to be coincidental that cougars have increased in number at the same time deer have increased. Deer provide prime prey for these large carnivores, and Berger and Wehausen argue that cougars have become more abundant because their prey has become so.

If cougars are now common in the Great Basin because deer have become common, then what might happen when, for whatever reason, deer become less abundant? It turns out that they do the obvious and switch to other prey. Richard Sweitzer and his colleagues have shown that when deer populations shrink in this region, not only does mountain lion predation on domestic livestock increase, but porcupines also become a prime target for these cats. In the Granite Range of northwestern Nevada, just north of Gerlach, porcupines were virtually eliminated by cougars as drought followed by a severe winter led to a steep decline in the number of deer. Anecdotal evidence suggests that cougars have had the same impact on porcupines elsewhere in the Great Basin.

So the increased abundance of deer in the Great Basin follows from the introduction of exotic livestock. The increased abundance of cougars follows from the increased abundance of deer, to the point that there now appear to be more cougars here than at any time during the previous 10,000 years. When deer decline, porcupines are among the native animals that pay the price. All this because of the introduction of domestic herbivores.

More Lessons from the Past

Were it not for the archaeological and paleontological records, we could not be certain that the current abundance of either cougars or deer in the Great Basin is a very recent phenomenon. In fact, our understanding of the way the Great Basin has come to be permeates our understanding of the way it is now and of what it may be like in the future.

Pikas provide a prime example. It was biologist Jim Brown who first suggested, on the basis of modern distributions, that pikas must have occupied Great Basin mountains during the Pleistocene, after which they became differentially extinct across those mountains. It was biologist Erik Beever and his colleagues who first observed that many populations of these animals that existed during historic times have recently become extinct. The record from the past provided the deeper temporal context for these observations. During the past 8,000 years, that record shows, increasing temperatures have driven pikas upslope nearly half a mile in the Great Basin (figure 8-4). Odds are that future increases in temperature—global warming—will drive them higher still and, because there is not much higher left for them on Great Basin mountains, cause their extinction here. Indeed, in just the past eighty years or so, the lower elevational limits of pikas in Yellowstone National Park, just beyond the eastern edge of the Great Basin, have increased by about five hundred feet, an increase that seems associated with rising minimum temperatures.

Likewise, observations of pygmy rabbits in the field showed how dependent they are on healthy stands of tall sagebrush. Records from the deeper past came later, showing that whenever sagebrush has dwindled on the landscape, so have these diminutive rabbits. Homestead Cave may show this best, but other sites show it as well. On the other hand, it was the record from an archaeological site, Gatecliff Shelter on the Toquima Range, that first showed that the expansion of pinyon-juniper woodland takes a toll on these animals as well, presumably by decreasing the areal expanse of sagebrush habitat. Work by Eveline Larrucea and Peter Brussard then showed that the same process is happening now. They surveyed areas in Nevada and California where pygmy rabbits had been reported between 1877 and 1946 and found that fifteen of those sites now have pinyon-juniper woodland on them. Of those fifteen, only one still has pygmy rabbits. They also found that the average elevation of the pygmy rabbit sites they surveyed is 515 feet higher than the sites occupied historically. Because appropriate habitat appears to exist at lower elevations, they attribute this increase to climatic warming during the past century. If these processes—pinyon-juniper expansion and warming—continue in the future, there is no need to guess what will happen to pygmy rabbit populations in much of the Great Basin.

Bison provide a different kind of example. These animals were never very common in the Great Basin, but their numbers increased and decreased at different times and different places in this region, with the processes behind this changing abundance now reasonably apparent. Their numbers grew in the eastern Great Basin, in Fremont territory, as summer precipitation and summer length increased, then decreased as the same suite of climate changes that did in the Fremont did in the bison as well. In post-Fremont times—that is, after about 1300 A.D.—it was the northern Great Basin that supported the greatest numbers of these animals, again in response to climate change. While the bison data were provided by the archaeological and paleontological records, the climate changes were detected primarily through detailed analyses of plants from packrat middens, of pollen from various settings, and of tree rings—all standard tools in the scientific arsenal of paleoecologists who work in the Arid West.

Given all this hard-earned knowledge, it seems odd to learn that there is a proposal afoot to "reintroduce" bison to state parks in southern Nevada. This, of course, would not be a reintroduction in any meaningful sense of the term, since bison have not roamed here for centuries, and certainly not in any number under current climatic conditions. Unless future climate change provides more appropriate living conditions for them, putting then here makes little historical sense. Far better to turn the idea over to Oregon's Malheur National Wildlife Refuge, where the term "reintroduction" could be accurately used. Unless of course, the goal is an economic one, akin to the 1967 and subsequent introductions of mountain goats to the Wasatch Range, or of elk to many places in Nevada, neither of which can be justified by the known histories of these animals.

And what might the Great Basin's past tell us about its future if predictions about global warming are correct? Here, unfortunately, things are murkier as regards both the past and the present. As I have discussed, the hottest and driest prolonged spell the Great Basin has seen during the past 10,000 years was during the middle Holocene, between roughly 7,500 and 4,500 years ago. What we don't really know, however, is the exact combination of decreased precipitation and increased temperatures, either annually or by season, that caused this generally arid interval. Even secure temperature estimates are hard to come by, though the estimate of a temperature increase between 5.4°F and 13.5°F for the Pyramid Lake area (for autumn) does encompass estimates for Great Basin temperatures by the end of the current century.

Those estimates for the future, based on multiple versions of computerized climate models that successfully account for many aspects of modern climate, are quite consistent. They suggest an increase in average annual temperature of about 7°F by the year 2100 across much of the Great Basin, with winter increases of about 4.5°F to 7.2°F and summer increases of about 7.2°F. The murkiness here lies in the fact that predictions for precipitation tend to give conflicting results. Most, however, predict that both summer and winter precipitation will decrease throughout the Great Basin. The result would be a dramatic decrease in surface waters, and the models predict that as well.

Because we are not certain of the magnitude of temperature and precipitation changes, or their seasonal distribution, during the middle Holocene in the Great Basin, and because some models give conflicting predictions about the nature of precipitation by the end of the current century, it is not possible to know how good an analog the middle Holocene is for what the near future may bring us in the Arid West. If it is a good analog, and if the predicted climate changes last long enough, all those things I discuss in chapter 8 may await us sooner than we might wish, from steep declines in the levels of Great Basin marshes and lakes to significantly altered distributions of plants and local but possibly permanent (think pikas) extinctions of small mammals.

Many have attempted to predict the general future of plants and animals in the Desert West under conditions of global warming, but few have attempted to predict the futures of particular species. For most vertebrates, this follows from a lack of sufficiently detailed knowledge of their ecological requirements, though some real progress has been made in this realm in recent years. Things are different for plants, though, whose habitat requirements are far better known.

Paleoecologists Sarah Shafer, Pat Bartlein, and Bob Thompson have taken advantage of this fact to model the distribution of a broad set of plant species in western North America by the end of this century. Their results suggest that a globally warmed Great Basin might, in fact, have little in common with what happened here during the middle Holocene. Most notable from our perspective are the predicted responses of four species: big sagebrush, creosote bush, Joshua tree, and saguaro cactus. In their models, big sagebrush is lost from a broad swath of the Great Basin while expanding its distribution northwards. At the same time, creosote bush and Joshua tree move into the western and central Great Basin, while saguaro cactus spreads into the Mojave. Creosote bush and Joshua tree are now plants of the Mojave Desert; saguaro is one of the iconic species of the Sonoran Desert. If these predictions are correct, then, by the end of this century, significant hunks of the Great Basin will be covered by plants now characteristic of the Mojave Desert, and this latter area will have plants now characteristic of the Sonoran Desert.

Nothing of this sort happened during the middle Holocene, but nothing of this sort might happen even if global warming proceeds as predicted. As Shafer and her colleagues carefully stress, their model is just that—a simplified version of reality. It does not take into account such things as the impact of increased carbon dioxide on the plants themselves, the substrates on which plants grow, or competitive interactions among plants. It also does not, because it cannot, take into account inaccuracies in the climate models themselves. But the important thing is that two different ways of interpreting what the future might be like given global warming—one that uses the middle Holocene as an analog and one that uses current knowledge alone to predict the future—provide very different results. That said, neither is particularly comforting, and neither holds much hope for an ever-expanding Las Vegas.

The Great Basin Today

Our intuitive understanding of the meaning of time depth seems to fade by powers of ten. It is easy to intuit the meaning of a year or a decade ago, somewhat harder for a century ago, harder still for a thousand years ago, and perhaps impossible for 10,000 or 100,000 years ago. Predicting the near-term future is easy for astrophysicists but hard for geologists and biologists. That's why we can land people on the moon and have tables of future eclipses but are unable to predict earthquakes or the fate of pygmy rabbits with any precision. The present is a lot easier.

And at present, the Great Basin contains some of the most spectacular country the earth has to offer, much of it immediately accessible to the public. Death Valley and Great Basin national parks and their nearby wilderness areas are obvious examples, but there are many places in the Great Basin of astonishing beauty that are known to only a relative few and that get fewer visitors still. There are so many of these that picking out some of them does a disservice to the others. Among official federal wilderness areas, there are the Arc Dome (Toiyabe Range), Alta Toquima (Toquima Range), Cedar Mountains, Deseret Peak (Stansbury Range), Inyo Mountains, Jarbidge, Ruby Mountains, Steens Mountain, and White Mountains wilderness areas. Birder (and sometimes mosquito) heaven is to be found at the Bear River, Fish Springs, Malheur, Pahranagat, Ruby Lake, and Stillwater national wildlife refuges, among other such places. There are places where some of the endangered fishes that I discuss in chapter 7 can easily be seen—Death Valley National Park, of course, but also Nevada's Ash Meadows and Moapa Valley national wildlife refuges. All this, without even mentioning the numerous state parks that dot the Great Basin: Utah's Fremont Indian State Park, for instance, or California's Mono Lake Tufa State Natural Reserve. All of these, and many others, provide the best of all possible introductions to the places I have discussed here. And maybe, if I am fortunate, you will take this book with you when you visit.

Chapter Notes

Eisenstein (1979) evaluates the profound impact of the printing press on western history. On the use of fire as a management tool by Native Americans in the Great Basin, see I. T. Kelly (1932), Steward (1933, 1941), O. C. Stewart (1941), Irwin (1980), C. S. Fowler (2000), C. S. Fowler and Rhode (2006), and Janetski (2006). The decline of mountain sheep through time in the White Mountains is shown in Grayson (2001). The story of the Newlands Project is provided by Townley (1977); see chapter 5 on water diversions from Mono and Owens lakes. The Southern Nevada Water Authority (2009) discusses its plans for accessing groundwater from a huge area of central and eastern Nevada, using language reminiscent of the slick pamphlets once used by the Los Angeles Department of Water and Power (1989a, 1989b) to convince readers of the beneficence of its intent. Deacon et al. (2007) provide a more realistic assessment of the ultimate impact of the Southern Nevada Water Authority's plans; see also Greenwald and Bradley (2008) and Patten, Rouse, and Stromberg (2008). The website of the Great Basin Water Network (www.greatbasinwater.net/) provides an effective way to monitor the issues involved here.

On the abundance of deer in the prehistoric Great Basin, including Fremont sites, see Janetski (1997, 2006) and Lupo and Schmitt (1997). Berger and Wehausen (1991) discuss the recent history of deer and cougars in the Great Basin; the porcupine story is told by Sweitzer, Jenkins, and Berger (1997). For the relationship between cougars and mountain sheep in the eastern Sierra Nevada and Mojave Desert, see Wehausen (1996) and, for New Mexico (where mountain lions help support themselves by preying on cattle), Rominger et al. (2004). Berger (1986) evaluates the relationship between horses, cattle, and deer in the Great Basin; Hickerson (1965) analyzes the role of boundary zones in maintaining mammalian populations. The impact of Native Americans on large mammals in the Snake River Plain and Columbia Plateau in early historic times is assessed by P. S. Martin and Szuter (1999a, 1999b, 2004), Lyman and Wolverton (2002), and Laliberte and Ripple (2003). The recent history of pikas in Yellowstone National Park is recounted by Moritz et al. (2008).

My discussion of the Great Basin under conditions of global warming is based on Hayhoe et al. (2004), Christensen et al. (2007), Pachauri and Reisinger (2007), and Bates et al. (2008). General predictions of plant and animal responses to global warming that incorporate the Great Basin include McKenney et al. (2007) and Lawler et al. (2009). The website associated with the former reference provides specific predictions of future distributions for many North American trees and shrubs (www.planthardiness.gc.ca/ph_futurehabitat.pl?lang=en). Martínez-Mayer, Peterson, and Hargrove (2004), among others, use mathematical models of the ecological requirements of a selected set of mammal species to predict, often successfully, the distributions of those species in the past. Shafer, Bartlein, and Thompson (2001) provide the predictions of Great Basin plant histories that I discuss here.

For easy access to the wilderness areas of the Great Basin, see www.wilderness.net; for National Wildlife Refuges, Butcher (2003) or www.fws.gov/refuges/.

APPENDIX 1

Relationship between Radiocarbon and Calendar Years for the Past 25,000 Radiocarbon Years

Conversions were done with Calib Version 6.0. (http://calib.qub.ac.uk/calib/), using a standard deviation of 50 years and the IntCal09 calibration dataset. The calendar year is the median age provided by Calib for each radiocarbon age.

Radiocarbon Years	Calendar Years
100	119
200	181
300	382
400	449
500	531
600	601
700	658
800	722
900	824
1,000	913
1,100	1,012
1,200	1,127
1,300	1,230
1,400	1,316
1,500	1,388
1,600	1,481
1,700	1,610
1,800	1,732
1,900	1,843
2,000	1,954
2,100	2,075
2,200	2,224
2,300	2,313

Radiocarbon Years	Calendar Years
2,400	2,450
2,500	2,578
2,600	2,731
2,700	2,810
2,800	2,905
2,900	3,044
3,000	3,195
3,100	3,320
3,200	3,424
3,300	3,529
3,400	3,650
3,500	3,773
3,600	3,909
3,700	4,040
3,800	4,192
3,900	4,331
4,000	4,478
4,100	4,626
4,200	4,726
4,300	4,871
4,400	4,981
4,500	5,158
4,600	5,318
4,700	5,426
4,800	5,521
4,900	5,636
5,000	5,736

Radiocarbon Years	Calendar Years	Radiocarbon Years	Calendar Years
5,100	5,823	9,700	11,137
5,200	5,963	9,800	11,221
5,300	6,084	9,900	11,302
5,400	6,211	10,000	11,478
5,500	6,302	10,100	11,700
5,600	6,375	10,200	11,902
5,700	6,489	10,300	12,095
5,800	6,599	10,400	12,273
5,900	6,722	10,500	12,466
6,000	6,841	10,600	12,559
6,100	6,977	10,700	12,622
6,200	7,095	10,800	12,678
6,300	7,226	10,900	12,767
6,400	7,337	11,000	12,867
6,500	7,413	11,100	12,993
6,600	7,497	11,200	13,107
6,700	7,571	11,300	13,197
6,800	7,640	11,400	13,271
6,900	7,735	11,500	13,352
7,000	7,837	11,600	13,434
7,100	7,932	11,700	13,551
7,200	8,012	11,800	13,652
7,300	8,105	11,900	13,759
7,400	8,243	12,000	13,856
7,500	8,323	12,100	13,944
7,600	8,403	12,200	14,050
7,700	8,487	12,300	14,224
7,800	8,576	12,400	14,451
7,900	8,727	12,500	14,657
8,000	8,868	12,600	14,878
8,100	9,041	12,700	15,053
8,200	9,162	12,800	15,211
8,300	9,317	12,900	15,401
8,400	9,433	13,000	15,617
8,500	9,504	13,100	15,869
8,600	9,561	13,200	16,115
8,700	9,653	13,300	16,376
8,800	9,834	13,400	16,574
8,900	10,029	13,500	16,683
9,000	10,186	13,600	16,763
9,100	10,252	13,700	16,836
9,200	10,361	13,800	16,901
9,300	10,498	13,900	16,964
9,400	10,631	14,000	17,043
9,500	10,786	14,100	17,139
9,600	10,940	14,200	17,274

Radiocarbon Years	Calendar Years	Radiocarbon Years	Calendar Years
14,300	17,386	18,900	22,477
14,400	17,520	19,000	22,590
14,500	17,654	19,100	22,758
14,600	17,769	19,200	22,885
14,700	17,882	19,300	23,010
14,800	18,000	19,400	23,113
14,900	18,279	19,500	23,317
15,000	18,250	19,600	23,452
15,100	18,243	19,700	23,567
15,200	18,497	19,800	23,663
15,300	18,588	19,900	23,761
15,400	18,640	20,000	23,888
15,500	18,707	20,100	24,036
15,600	18,762	20,200	24,122
15,700	18,829	20,300	24,228
15,800	18,950	20,400	24,343
15,900	19,113	20,500	24,453
16,000	19,153	20,600	24,596
16,100	19,255	20,700	24,691
16,200	19,371	20,800	24,790
16,300	19,454	20,900	24,916
16,400	19,513	21,000	25,042
16,500	19,606	21,100	25,170
16,600	19,737	21,200	25,299
16,700	19,835	21,300	25,422
16,800	19,980	21,400	25,611
16,900	20,097	21,500	25,750
17,000	20,206	21,600	25,898
17,100	20,298	21,700	26,028
17,200	20,394	21,800	26,127
17,300	20,518	21,900	26,287
17,400	20,739	22,000	26,442
17,500	20,826	22,100	26,505
17,600	20,993	22,200	26,640
17,700	21,174	22,300	26,844
17,800	21,301	22,400	27,121
17,900	21,382	22,500	27,281
18,000	21,454	22,600	27,321
18,100	21,561	22,700	27,411
18,200	21,730	22,800	27,640
18,300	21,844	22,900	27,777
18,400	21,967	23,000	27,867
18,500	22,141	23,100	27,950
18,600	22,239	23,200	28,045
18,700	22,316	23,300	28,141
18,800	22,387	23,400	28,214

Radiocarbon Years	Calendar Years	Radiocarbon Years	Calendar Years
23,500	28,270	24,300	29,164
23,600	28,342	24,400	29,309
23,700	28,439	24,500	29,389
23,800	28,543	24,600	29,449
23,900	28,663	24,700	29,514
24,000	28,804	24,800	29,595
24,100	28,920	24,900	29,727
24,200	29,023	25,000	29,883

APPENDIX 2

Concordance of Common and Scientific Plant Names

In each entry, the scientific name follows that used by the U.S. Department of Agriculture Natural Resources Conservation Service (www.plants.usda.gov/); then, vice versa. Common names for plants are not standardized. Those I use here are provided by the USDA or by the *Flora of North America* (www.efloras.org/); in certain cases, I have also included names that have long been in common use in the Great Basin but that do not appear in either of these references.

Common Names (Scientific Name)

Alkali sacaton (*Sporobolus airoides*)

Antelope bitterbrush (*Purshia tridentata*)

Arrowweed (*Pluchea sericea*)

Bailey's greasewood (*Sarcobatus baileyi*)

Baltic rush (*Juncus balticus*)

Beaked sedge (*Carex rostrata*)

Beavertail pricklypear (*Opuntia basilaris*)

Big greasewood (*Sarcobatus vermiculatus*)

Big sagebrush (*Artemisia tridentata*)

Bigtooth maple (*Acer grandidentatum*)

Black sagebrush (*Artemisia nova*)

Blackbrush (*Coleogyne ramosissima*)

Blue spruce (*Picea pungens*)

Brittlebush (*Encelia farinosa*)

Broadleaf cattail (*Typha latifolia*)

Bud sagebrush (*Picrothamnus desertorum*)

Bulrush (*Schoenoplectus*)

Cattail (*Typha*)

Cheatgrass (*Bromus tectorum*)

Chia (*Salvia columbariae*)

Chickweed (*Cerastium*)

Colorado pinyon (*Pinus edulis*)

Common juniper (*Juniperus communis*)

Cosmopolitian bulrush (*Schoenoplectus maritimus*)

Creosote bush (*Larrea tridentata*)

Curl-leaf mountain mahogany (*Cercocarpus ledifolius*)

Currant (*Ribes*)

Desert almond (*Prunus fasciculata*)

Desert holly (*Atriplex hymenelytra*)

Desert peach (*Prunus andersonii*)

Desert saltbush, Cattle saltbush (*Atriplex polycarpa*)

Desert snowberry (*Symphoricarpos longiflorus*)

Desert spruce, Schott's pygmycedar (*Peucephyllum schottii*)

Douglas-fir (*Pseudotsuga menziesii*)

Engelmann spruce (*Picea engelmannii*)

Fourwing saltbush (*Atriplex canescens*)

Foxtail pine (*Pinus balfouriana*)

Fremont cottonwood (*Populus fremontii*)

Gambel oak (*Quercus gambelii*)

Giant sequoia (*Sequoiadendron giganteum*)

Gooseberry currant (*Ribes montigenum*)

Grand fir (*Abies grandis*)

Great Basin bristlecone pine (*Pinus longaeva*)

Great Basin wildrye (*Leymus cinereus*)

Green rabbitbrush (*Ericameria teretifolia*)

Hardstem bulrush, tule (*Schoenoplectus acutus*)

Honey mesquite (*Prosopis glandulosa*)

Horsebrush (*Tetradymia*)
Incense cedar (*Calocedrus decurrens*)
Indian ricegrass (*Achnatherum hymenoides*)
Iodinebush, pickleweed (*Allenrolfea occidentalis*)
Jeffrey pine (*Pinus jeffreyi*)
Jointfir (*Ephedra*)
Joshua tree (*Yucca brevifolia*)
Limber pine (*Pinus flexilis*)
Little sagebrush (*Artemisia arbuscula*)
Lodgepole pine (*Pinus contorta*)
Mormon tea (*Ephedra viridis*)
Mountain hemlock (*Tsuga mertensiana*)
Mojave seablite (*Sueada moquinii*)
Netleaf hackberry (*Celtis reticulata*)
Nevada jointfir (*Ephedra nevadensis*)
Nuttall's saltbush (*Atriplex nuttallii*)
Ocean spray (*Holodiscus*)
Parry's saltbush (*Atriplex parryi*)
Ponderosa pine (*Pinus ponderosa*)
Pricklypear (*Opuntia*)
Quaking aspen (*Populus tremuloides*)
Rabbitbrush (*Chrysothamnus*)
Red fir (*Abies magnifica*)
Rocky Mountain juniper (*Juniperus scopulorum*)
Rubber rabbitbrush (*Ericameria nauseosa*)
Sagebrush (*Artemisia*)
Saltbush (*Atriplex*)
Saltgrass (*Distichlis spicata*)
Screwbean mesquite (*Prosopis pubescens*)
Shadscale (*Atriplex confertifolia*)
Shrubby cinquefoil (*Dasiphora fruticosa*)
Silver buffaloberry (*Shepherdia argentea*)
Silverweed cinquefoil (*Argentina anserina*)
Simpson's pediocactus (*Pediocactus simpsonii*)
Singleleaf pinyon (*Pinus monophylla*)
Smokebush, Nevada dalea (*Psorothamnus polydenius*)
Snowberry (*Symphoricarpos*)
Spikerush (*Eleocharis*)
Spineless horsebrush (*Tetradymia canescens*)
Stansbury cliffrose (*Purshia stansburiana*)
Subalpine fir (*Abies lasiocarpa*)
Sugar pine (*Pinus lambertiana*)
Tamarisk (*Tamarix*)
Teddybear cholla (*Cylindropuntia bigelovii*)

Torrey saltbush (*Atriplex torreyi*)
Utah agave (*Agave utahensis*)
Utah juniper (*Juniperus osteosperma*)
Washoe pine (*Pinus washoensis*)
Western juniper (*Juniperus occidentalis*)
Western white pine (*Pinus monticola*)
Whipple yucca (*Hesperoyucca whipplei*)
White bursage, burrobush (*Ambrosia dumosa*)
White fir (*Abies concolor**)
Whitebark pine (*Pinus albicaulis*)
Winterfat (*Krascheninnikovia lanata*)
Woods' rose (*Rosa woodsii*)
Yellow rabbitbrush (*Chrysothamnus viscidiflorus*)

Scientific Name (Common Name)

Abies concolor* (white fir)
Abies grandis (grand fir)
Abies lasiocarpa (subalpine fir)
Abies magnifica (red fir)
Acer grandidentatum (bigtooth maple)
Achnatherum hymenoides (Indian ricegrass)
Agave utahensis (Utah agave)
Allenrolfea occidentalis (iodinebush, pickleweed)
Ambrosia dumosa (white bursage, burrobush)
Argentina anserina (silverweed cinquefoil)
Artemisia (sagebrush)
Artemisia arbuscula (little sagebrush)
Artemisia nova (black sagebrush)
Artemisia tridentata (big sagebrush)
Atriplex (saltbush)
Atriplex canescens (fourwing saltbush)
Atriplex confertifolia (shadscale)
Atriplex hymenelytra (desert holly)
Atriplex nuttallii (Nuttall's saltbush)
Atriplex parryi (Parry's saltbush)
Atriplex polycarpa (desert saltbush, cattle saltbush)
Atriplex torreyi (torrey saltbush)
Bromus tectorum (cheatgrass)
Calocedrus decurrens (incense cedar)
Carex rostrata (beaked sedge)
Celtis reticulata (netleaf hackberry)
Cerastium (chickweed)
Cercocarpus ledifolius (curl-leaf mountain mahogany)

Chrysothamnus (rabbitbrush)

Chrysothamnus viscidiflorus (yellow rabbitbrush)

Coleogyne ramosissima (blackbrush)

Cylindropuntia bigelovii (teddybear cholla)

Dasiphora fruticosa (shrubby cinquefoil)

Distichlis spicata (saltgrass)

Eleocharis (spikerush)

Encelia farinosa (brittlebush)

Ephedra (jointfir)

Ephedra nevadensis (Nevada jointfir)

Ephedra viridis (Mormon tea)

Ericameria nauseosa (rubber rabbitbrush)

Ericameria teretifolia (green rabbitbrush)

Hesperoyucca whipplei (whipple yucca)

Holodiscus sp. (ocean spray)

Juncus balticus (Baltic rush)

Juniperus communis (common juniper)

Juniperus occidentalis (western juniper)

Juniperus osteosperma (Utah juniper)

Juniperus scopulorum (Rocky Mountain juniper)

Krascheninnikovia lanata (winterfat)

Larrea tridentata (creosote bush)

Leymus cinereus (Great Basin wildrye)

Opuntia (pricklypear)

Opuntia basilaris (beavertail pricklypear)

Pediocactus simpsonii (Simpson's pediocactus)

Peucephyllum schottii (desert spruce, Schott's pygmycedar)

Picea engelmannii (Engelmann spruce)

Picea pungens (blue spruce)

Picrothamnus desertorum (bud sagebrush)

Pinus albicaulis (whitebark pine)

Pinus balfouriana (foxtail pine)

Pinus contorta (lodgepole pine)

Pinus edulis (Colorado pinyon)

Pinus flexilis (limber pine)

Pinus jeffreyi (Jeffrey pine)

Pinus lambertiana (sugar pine)

Pinus longaeva (Great Basin bristlecone pine)

Pinus monophylla (singleleaf pinyon)

Pinus monticola (western white pine)

Pinus ponderosa (ponderosa pine)

Pinus washoensis (Washoe pine)

Pluchea sericea (arrowweed)

Populus fremontii (Fremont cottonwood)

Populus tremuloides (quaking aspen)

Prosopis glandulosa (honey mesquite)

Prosopis pubescens (screwbean mesquite)

Prunus andersonii (desert peach)

Prunus fasciculata (desert almond)

Pseudotsuga menziesii (Douglas-fir)

Psorothamnus polydenius (smokebush, Nevada dalea)

Purshia stansburiana (Stansbury cliffrose)

Purshia tridentata (antelope bitterbrush)

Quercus gambelii (gambel oak)

Ribes (currant)

Ribes montigenum (gooseberry currant)

Rosa woodsii (Woods' rose)

Salvia columbariae (chia)

Sarcobatus baileyi (Bailey's greasewood)

Sarcobatus vermiculatus (big greasewood)

Schoenoplectus (bulrush)

Schoenoplectus acutus (hardstem bulrush, tule)

Schoenoplectus maritimus (cosmopolitan bulrush)

Sequoiadendron giganteum (sequoia)

Shepherdia argentea (silver buffaloberry)

Sporobolus airoides (alkali sacaton)

Sueada moquinii (Mojave seablite)

Symphoricarpos (snowberry)

Symphoricarpos longiflorus (desert snowberry)

Tamarix (tamarisk)

Tetradymia (horsebrush)

Tetradymia canescens (spineless horsebrush)

Tsuga mertensiana (mountain hemlock)

Typha (cattail)

Typha latifolia (broadleaf cattail)

Yucca brevifolia (Joshua tree)

*The *Flora of North America* places Sierran populations of white fir in a different species, *Abies lowiana*. However, the U.S. Department of Agriculture includes these populations as a subspecies of *A. concolor*. In addition, paleobotanists have identified all ancient white fir specimens as *Abies concolor*, and this is the usage I follow here (see Forester et al. 1999).

REFERENCES

Ababneh, L. 2008. Bristlecone pine paleoclimatic model for archaeological patterns in the White Mountain [sic] of California. *Quaternary International* 188:59–78.

Adams, D. B. 1979. The cheetah: Native American. *Science* 205:1155–1158.

Adams, D. K., and A. C. Comrie. 1997. The North American monsoon. *Bulletin of the American Meteorological Society* 78:2197–2213.

Adams, K. D. 2003. Age and paleoclimatic significance of late Holocene lakes in the Carson Sink, NV, USA. *Quaternary Research* 60:294–306.

Adams, K. D. 2007. Late Holocene sedimentary environments and lake-level fluctuations at Walker Lake, Nevada, USA. *Geological Society of America Bulletin* 119:126–139.

Adams, K. D. 2010. Lake levels and sedimentary environments during deposition of the Trego Hot Springs and Wono tephras in the Lake Lahontan Basin, Nevada, USA. *Quaternary Research* 73:118–129.

Adams, K. D., and S. G. Wesnousky. 1998. Shoreline processes and the age of the Lake Lahontan highstand in the Jessup embayment, Nevada. *Bulletin of the Geological Society of America* 110:1318–1332.

Adams, K. D., and S. G. Wesnousky. 1999. The Lake Lahontan highstand: Age, surficial characteristics, soil development, and regional shoreline correlation. *Geomorphology* 30:357–392.

Adams, K. D., S. G. Wesnousky, and B. G. Bills. 1999. Isostatic rebound, active faulting, and potential geomorphic effects in the Lake Lahontan basin, Nevada and California. *Geological Society of American Bulletin* 111:1739–1756.

Adams, K. D., T. Goebel, K. Graf, G. S. Smith, A. J. Camp, R. W. Briggs, and D. Rhode. 2008. Late Pleistocene and early Holocene lake-level fluctuations in the Lahontan Basin, Nevada: Implications for the distribution of archaeological sites. *Geoarchaeology* 23:608–643.

Adams, R. 2009. Archaeology at the High Rise Village Site (48FR5891): Domestic life at 10,700 Feet, Wind River Range, Wyoming. Paper presented at the 66th Plains Anthropological Conference, Laramie, WY.

Adovasio, J. M. 1975. Fremont basketry. *Tebiwa* 17(2):67–76.

Adovasio, J. M. 1979. Comment by Adovasio. *American Antiquity* 44:723–731.

Adovasio, J. M. 1980. Fremont: An artifactual perspective. In D. B. Madsen, ed., *Fremont Perspectives*, pp. 35–40. State of Utah Division of State History Antiquities Section Selected Papers 7(16).

Adovasio, J. M. 1986a. Artifacts and ethnicity: Basketry as an indicator of territoriality and population movements in the prehistoric Great Basin. In C. J. Condie and D. D. Fowler, eds., *Anthropology of the Desert West: Essays in Honor of Jesse D. Jennings*, pp. 43–88. University of Utah Anthropological Papers 110.

Adovasio, J. M. 1986b. Prehistoric basketry. In W. L. d'Azevedo, ed., *Great Basin, volume 11*, of *Handbook of North American Indians*, pp. 194–205. Smithsonian Institution Press, Washington, DC.

Adovasio, J. M., and R. L. Andrews. 1983. Material culture of Gatecliff Shelter: Basketry, cordage, and miscellaneous fiber constructions. In D. H. Thomas, *The Archaeology of Monitor Valley*, vol. 2, *Gatecliff Shelter*, pp. 279–289. American Museum of Natural History Anthropological Papers 59(1).

Adovasio, J. M., and J. Page. 2002. *The First Americans: In Pursuit of Archaeology's Greatest Mystery*. Random House, New York.

Adovasio, J. M., and D. R. Pedler. 1994. A tisket, a tasket: Looking at the Numic speakers through the "lens" of a basket. In D. B. Madsen and D. Rhode, eds., *Across the West: Human Population Movement and the Expansion of the Numa*, pp. 114–123. University of Utah Press, Salt Lake City.

Adovasio, J. M., and D. R. Pedler. 2005. A long view of deep time at Meadowcroft Rockshelter. In R. Bonnichsen, B. T. Lepper, D. Stanford, and M. R. Waters, eds., *Paleoamerican Origins: Beyond Clovis*, pp. 23–28. Center for the Study of the First Americans; Texas A&M University Press, College Station.

Adovasio, J. M., D. R. Pedler, and J. S. Illingworth. 2002. Fremont basketry. *Utah Archaeology* 15:5–26.

Adovasio, J. M., D. R. Pedler, and J. S. Illingworth. 2008. Fremont basketry. In C. S. Fowler and D. D. Fowler, eds., *The Great Basin: People and Place in Ancient Times*, pp. 124–127. School for Advanced Research Press, Santa Fe, NM.

Adovasio, J. M., J. Donahue, R. C. Carlisle, K. Cushman, R. Stuckenrath, and P. Wiegman. 1984. Meadowcroft Rockshelter and the Pleistocene/Holocene transition in southwestern

Pennsylvania. In H. H. Genoways and M. R. Dawson, eds., *Contributions in Quaternary Vertebrate Paleontology: A Volume in Memorial to John E. Guilday*, pp. 347–369. Carnegie Museum of Natural History Special Publication 8.

Adovasio, J. M., R. L. Andrews, and J. S. Illingworth. 2009. Netting, net hunting, and human adaptation in the eastern Great Basin. In B. Hockett, ed., *Past, Present and Future Issues in Great Basin Archaeology: Papers in Honor of Don D. Fowler*, pp. 84–102. Bureau of Land Management Cultural Resource Series 20, Reno, NV.

Ager, T. A. 2003. Late Quaternary vegetation and climate history of the central Bering Land Bridge from St. Michael Island, western Alaska. *Quaternary Research* 60:10–32.

Aikens, C. M. 1966. *Fremont-Promontory-Plains Relationships in Northern Utah*. University of Utah Anthropological Papers 82.

Aikens, C. M. 1967. *Excavations at Snake Rock Village and the Bear River No. 2 Site*. University of Utah Anthropological Papers 87.

Aikens, C. M. 1970. *Hogup Cave*. University of Utah Anthropological Papers 93.

Aikens, C. M. 1994. Adaptive strategies and environmental change in the Great Basin and its peripheries as determinants in the migrations of Numic-speaking peoples. In D. B. Madsen and D. Rhode, eds., *Across the West: Human Population Movement and the Expansion of the Numa*, pp. 35–43. University of Utah Press, Salt Lake City.

Aikens, C. M., and D. B. Madsen. 1986. Prehistory of the eastern area. In W. L. d'Azevedo, ed., *Great Basin, Handbook of North American Indians, volume 11* pp. 149–160. Smithsonian Institution Press, Washington, DC.

Aikens, C. M., and Y. T. Witherspoon. 1986. Great Basin Numic prehistory: Linguistics, archaeology, and environment. In C. J. Condie and D. D. Fowler, eds., *Anthropology of the Desert West: Essays in Honor of Jesse D. Jennings*, pp. 7–20. University of Utah Anthropological Papers 110.

Aikens, C. M., D. L. Cole, and R. Stuckenrath. 1977. Excavations at Dirty Shame Rockshelter, southeastern Oregon. *Tebiwa* 4:1–29.

Albee, B. J., L. M. Shultz, and S. Goodrich. 1988. *Atlas of the Vascular Plants of Utah*. Utah Museum of Natural History Occasional Paper 7.

Alcorn, G. 1988. *The Birds of Nevada*. Fairview West, Fallon, NV.

Alder, W. 2002. The National Weather Service, weather across Utah in the 1980s, and its effects on Great Salt Lake. In J. W. Gwynn, ed., *Great Salt Lake: An Overview of Change*, pp. 295–301. Utah Department of Natural Resources Special Publication. Salt Lake City.

Al-Fenadi, Y. 2007. Hottest temperature record on the world, El Azisia, Libya. www.wmo.ch/pages/mediacentre/news_members/documents/Libya.pdf.

Allen, B. D., and R. Y. Anderson. 2000. A continuous, high-resolution record of late Pleistocene climate variability from the Estancia basin, New Mexico. *Geological Society of America Bulletin* 112:1444–1458.

Allen, J., and J. F. O'Connell. 2008. Getting from Sunda to Sahul. In G. Clark, F. Leach, and S. O'Connor, eds., *Islands of Inquiry: Colonization, Seafaring, and the Archaeology of Maritime Landscapes*, pp. 31–46. ANU E Press, Australian National University, Canberra.

Alley, R. B. 2000. The Younger Dryas cold interval as viewed from central Greenland. *Quaternary Science Reviews* 19:213–116.

Alley, R. B., D. A. Meese, C. A. Shuman, A. J. Gow, K. C. Taylor, P. M. Grootes, J. W. C. White, M. Ram, E. D. Waddington, P. A. Mayewski, and G. A. Zielinski. 1993. Abrupt increase in Greenland snow accumulation at the end of the Younger Dryas event. *Nature* 362:527–529.

Allison, I. S. 1945. Pumice beds at Summer Lake, Oregon. *Geological Society of America Bulletin* 56:789–807.

Allison, I. S. 1966a. *Fossil Lake, Oregon: Its Geology and Fossil Faunas*. Oregon State Monographs, Studies in Geology 9.

Allison, I. S. 1966b. Pumice at Summer Lake, Oregon: A correction. *Geological Society of America Bulletin* 77:329–330.

Allison, I. S. 1979. *Pluvial Fort Rock Lake, Lake County, Oregon*. State of Oregon Department of Geology and Mineral Industries Special Paper 7.

Allison, I. S. 1982. Geology of Pluvial Lake Chewaucan, Lake County, Oregon. *Oregon State Monographs, Studies in Geology* 11.

Allison, N. 1988. Lehner Ranch site: Officially on the map. *Mammoth Trumpet* 4(4):3.

Ambro, R. D. 1967. Dietary-technological-ecological aspects of Lovelock Cave coprolites. *University of California Archaeological Survey Reports* 70:37–48.

Ames, K. M., K. A. Fuld, and S. Davis. 2010. Dart and arrow points on the Columbia Plateau of western North America. *American Antiquity* 75:287–325.

Amick, D. S. 1995. Raw material selection patterns among Paleoindian tools from the Black Rock Desert, Nevada. *Current Research in the Pleistocene* 12:55–57.

Amick, D. S. 1999. Using lithic artifacts to explain past behavior. In C. Beck, ed., *Models for the Millennium: Great Basin Anthropology Today*, pp. 161–170. University of Utah Press, Salt Lake City.

Anderson, D. E. 1998. Late Quaternary Paleohydrology, Lacustrine Stratigraphy, Fluvial Geomorphology, and Modern Hydroclimatology of the Amargosa River/Death Valley Hydrologic System, California and Nevada. Ph.D. dissertation, University of California, Davis.

Anderson, D. E., and S. G. Wells. 2003. Latest Pleistocene lake highstands in Death Valley, California. In Y. Enzel, S. G. Wells, and N. Lancaster, eds., *Paleoenvironments and Paleohydrology of the Mojave and Southern Great Basin Deserts*, pp. 115–128. Geological Society of America Special Paper 368.

Anderson, E. 1970. *Quaternary Evolution of the Genus* Martes (Carnivora, Mustelidae). Acta Zoologica Fennica 130.

Anderson, E. 1994. Evolution, prehistoric distribution, and systematics of *Martes*. In S. W. Buskirk, A. S. Harestad, M. G. Raphael, and R. A. Powell, eds., *Martens, Sables, and Fishers: Biology and Conservation*, pp. 13–25. Cornell University Press, Ithaca.

Anderson, K. C., and S. G. Wells. 2003. Latest Quaternary paleohydrology of Silurian Lake and Salt Spring basin, Silurian Valley, California. In Y. Enzel, S. G. Wells, and N. Lancaster, eds., *Paleoenvironments and Paleohydrology of the Mojave and Southern Great Basin Deserts*, pp. 129–141. Geological Society of America Special Paper 368.

Anderson, M. E., and J. E. Deacon. 2001. Population size of Devils Hole pupfish (*Cyprinodon diabolis*) correlates with water level. *Copeia* 2001:224–228.

Anderson, P. M., and A. V. Lozhkin. 2001. The Stage 3 interstadial complex (Karginskii/middle Wisconsinan interval) of Beringia: Variations in paleoenvironments and implications for paleoclimatic interpretations. *Quaternary Science Reviews* 20:93–125.

Anderson, P. M., M. E. Edwards, and L. B. Brubaker. 2004. Results and paleoclimate implications of 35 years of paleoecological research in Alaska. In A. R. Gillespie, S. C. Porter, and B. F.

Atwater, eds., *The Quaternary Period in the United States*, pp. 427–440. Developments in Quaternary Science 1. Elsevier, Amsterdam.

Anderson, P. M., A. V. Lozhkin, and L. B. Brubaker. 2002. Implications of a 24,000-yr palynological record for a Younger Dryas cooling and for boreal forest development in northeastern Siberia. *Quaternary Research* 57:325–333.

Anderson, R. S., 1990a. Holocene forest development and paleoclimates within the central Sierra Nevada, California. *Journal of Ecology* 78:470–489.

Anderson, R. S. 1990b. Modern pollen rain within and adjacent to two giant sequoia (*Sequoiadendron giganteum*) groves, Yosemite and Sequoia National Parks, California. *Canadian Journal of Forest Research* 20:1289–1305.

Anderson, R. S., and S. J. Smith. 1994. Paleoclimatic interpretations of meadow sediment and pollen stratigraphies from California. *Geology* 22:723–726.

Anderson, R. Y., B. D. Allen, and K. M. Menking. 2002. Geomorphic expression of abrupt climate change in southwestern North America at the glacial termination. *Quaternary Research* 57:371–381.

Andrews, J. T. 1987. The Late Wisconsin glaciation and deglaciation of the Laurentide Ice Sheet. In W. F. Ruddiman and H. E. Wright. Jr., eds., *North America and Adjacent Oceans during the Last Deglaciation*, pp. 13–37. The Geology of North America, vol. K-3. Geological Society of America, Boulder, CO.

Antevs, E. 1925. On the Pleistocene history of the Great Basin. *Carnegie Institute of Washington Publication* 352:51–114.

Antevs, E. 1931. *Late-glacial Correlations and Ice Recession in Manitoba*. Canada Geological Survey Memoir 168.

Antevs, E. 1938. Postpluvial climatic variations in the Southwest. *Bulletin of the American Meteorological Society* 19: 190–193.

Antevs, E. 1948. Climatic changes and pre-white man. In *The Great Basin, with Emphasis on Postglacial Times*, pp. 167–191. University of Utah Bulletin 38(20).

Antevs, E. 1952a. Cenozoic climates of the Great Basin. *Geologische Rundschau* 40:96–109.

Antevs, E. 1952b. Climatic history and the antiquity of man in California. *University of California Archaeological Survey Reports* 16:23–31.

Antevs, E. 1953a. Geochronology of the deglacial and Neothermal ages. *Journal of Geology* 61:195–230.

Antevs, E. 1953b. The Postpluvial or Neothermal. *University of California Archaeological Survey Reports* 22:9–23.

Antevs, E. 1955. Geologic-climatic dating in the West. *American Antiquity* 20:317–335.

Anyonge, W., and C. Roman. 2006. New body mass estimates for *Canis dirus*, the extinct Pleistocene dire wolf. *Journal of Vertebrate Paleontology* 26:209–212.

Arkush, B. S. 1986. Aboriginal exploitation of pronghorn in the Great Basin. *Journal of Ethnobiology* 6:239–255.

Arkush, B. S. 1995. *The Archaeology of CA-MNO-2122: A Study of Pre-Contact and Post-Contact Lifeways among the Mono Basin Paiute*. University of California Publications: Anthropological Records 31.

Arkush, B. S. 1998. *Archaeological Investigations at Mosquito Willie Rockshelter and Lower Lead Mine Hills Cave, Great Salt Lake Desert, Utah*. Coyote Press, Salinas, CA.

Arkush, B. S. 1999a. Numic pronghorn exploitation: A reassessment of Stewardian-derived models of big-game hunting in the Great Basin. In R. O. Clemmer, L. D. Myers, and M. E. Rudden, eds., *Julian Steward and the Great Basin: The Making of an Anthropologist*, pp. 35–52. University of Utah Press, Salt Lake City.

Arkush, B. S. 1999b. Recent small-scale excavations at Weston Canyon Rockshelter in southeastern Idaho. *Tebiwa* 27(1): 1–64.

Arkush, B. S. 2002. *Archaeology of the Rock Springs Site: A Multicomponent Bison Kill and Processing Camp in Curlew Valley, Southeastern Idaho*. Boise State University Monographs in Archaeology 1.

Arkush, B. S., and B. L. Pitblado. 2000. Paleoarchaic surface assemblages in the Great Salt Lake Desert, northwestern Utah. *Journal of California and Great Basin Anthropology* 22:12–42.

Arno, S. F., and R. P. Hammersly. 1984. *Timberlines: Mountain and Arctic Forest Frontiers*. The Mountaineers, Seattle.

Arnow, T. 1980. Water budget and water surface fluctuations of Great Salt Lake. In J. W. Gwynn, ed., *Great Salt Lake: A Scientific, Historical, and Economic Overview*, pp. 255–261. Utah Geological and Mineral Survey Bulletin 116.

Arnow, T. 1984. *Water-Level and Water-Quality Changes in Great Salt Lake, Utah, 1847–1983*. U.S. Geological Survey Circular 913.

Arnow, T., and D. Stephens. 1990. *Hydrologic Characteristics of the Great Salt Lake, Utah, 1847–1986*. U.S. Geological Survey Water-Supply Paper 2332.

Arroyo-Cabrales, J., and O. J. Polaco. 2003. Caves and the Pleistocene vertebrate paleontology of Mexico. In B. W. Schubert, J. I. Mead, and R. W. Graham, eds., *Ice Age Cave Faunas of North America*, pp. 273–291. Indiana University Press, Bloomington.

Aschmann, H. H. 1958. Great Basin climates in relation to human occupance. *University of California Archaeological Survey Reports* 42:23–40.

Atwood, G. 2002. Storm-related flooding hazards, coastal processes, and shoreline evidence of Great Salt Lake. In J. W. Gwynn, ed., *Great Salt Lake: An Overview of Change*, pp. 43–53. Utah Department of Natural Resources Special Publication, Salt Lake City.

Augusti, A., T. R. Betson, and J. Schleucher. 2008. Deriving correlated climate and physiological signals from deuterium isotopomers in tree rings. *Chemical Geology* 252:1–8.

Austin, G. T., and D. D. Murphy. 1987. Zoogeography of Great Basin butterflies: Patterns of distribution and differentiation. *Great Basin Naturalist* 47:186–201.

Austin, L. H. 2002. Problems and management alternatives related to the selection and construction of the West Desert Pumping Project. In J. W. Gwynn, ed., *Great Salt Lake: An Overview of Change*, pp. 303–312. Utah Department of Natural Resources Special Publication, Salt Lake City.

Bacon, S. N., R. M. Burke, S. K. Pezzopane, and A. S. Jayko. 2006. Last glacial maximum and Holocene lake levels of Owens Lake, eastern California, USA. *Quaternary Science Reviews* 25:1264–1282.

Bader, N. E. 2000. Pollen analysis of Death Valley sediments deposited between 166 and 114 ka. *Palynology* 24:49–61.

Bagley, M. 1988. A sensitive-plant monitoring study on the Eureka Dunes, Inyo County, California. In C. A. Hall, Jr., and V. Doyle-Jones, eds., *Plant Biology of Eastern California*, pp. 223–243. Natural History of the White-Inyo Range Symposium Volume 2. White Mountain Research Station, University of California, Los Angeles.

Bagley, W. 2002. *Blood of the Prophets: Brigham Young and the Massacre at Mountain Meadows*. University of Oklahoma Press, Norman.

Bailey, R. G. 1994. *Descriptions of the Ecoregions of the United States,* 2nd ed. USDA Forest Service Miscellaneous Publication 1391.

Bailey, R. G. 1995. *Ecoregions of North America.* USDA Forest Service, Washington, DC.

Bailey, R. G. 1998. *Ecoregions: The Ecosystem Geography of the Oceans and Continents.* Springer-Verlag, New York.

Bailey, V. 1936. *The Mammals and Life Zones of Oregon.* North American Fauna 55.

Baker, A. J., L. J. Huynen, O. Haddrat, C. D. Millar, and D. M. Lambert. 2005. Reconstructing the tempo and mode of evolution in an extinct clade of birds with ancient DNA: The giant moas of New Zealand. *Proceedings of the National Academy of Sciences* 102:8257–8262.

Baker, V. R. 2002. High-energy megafloods: Planetary settings and sedimentary dynamics. In I. P. Martini, V. R. Baker, and G. Garzón, eds., *Floods and Megaflood Processes and Deposits: Recent and Ancient Examples,* pp. 3–15. International Association of Sedimentologists Special Publication 32.

Baker, V. R. 2008. The Spokane Flood debates: Historical background and philosophical perspective. In R. H. Grapes, D. Oldroyd, and A. Grigelis, eds., *History of Geomorphology and Quaternary Geology,* pp. 33–50. Geological Society Special Publication 101.

Baker, V. R., G. Benito, and A. N. Rudoy. 1993. Paleohydrology of late Pleistocene superflooding, Altay Mountains, Siberia. *Science* 259:348–350.

Bamforth, D. B. 1988. *Ecology and Human Organization on the Great Plains.* Plenum Press, New York.

Barbour, M. G. 1988. Mojave Desert scrub vegetation. In M. G. Barbour and J. Major, eds., *Terrestrial Vegetation of California,* pp. 835–867. California Native Plant Society Publication 9.

Bard, E. G., G. Raisbeck, F. Yiou, and J. Jouzel. 2001. Solar irradiance during the last 1200 years based on cosmogenic nuclides. *Tellus* 52B:985–992.

Barendregt, R. W., and A. Duk-Rodkin. 2004. Chronology and extent of late Cenozoic ice sheets in North America: A magnetostratigraphic assessment. In J. Ehlers and P. L Gibbard, eds., *Quaternary Glaciations: Extent and Chronology,* part 2: *North America,* pp. 1–8. Developments in Quaternary Science 2. Elsevier, Amsterdam.

Barlow, K. R. 2002. Predicting maize agriculture among the Fremont: An economic comparison of farming and foraging in the America Southwest. *American Antiquity* 67:65–88.

Barlow, K. R. 2006. A formal model for predicting agriculture among the Fremont. In D. J. Kennett and B. Winterhalder, eds., *Behavioral Ecology and the Transition to Agriculture,* pp. 87–102. University of California Press, Berkeley.

Barlow, K. R., and D. Metcalfe. 1999. Plant utility indices: Two Great Basin examples. *Journal of Archaeological Science* 23:351–372.

Barnes, C. T. 1927. *Utah Mammals.* Bulletin of the University of Utah 17(12).

Barnosky, A. D., P. L. Koch, R. S. Feranec, S. L. Wing, and A. B. Shabel. 2004. Assessing the causes of late Pleistocene extinctions on continents. *Science* 306:70–75.

Basgall, M. E. 1993. Early Holocene prehistory of the north-central Mojave Desert. Ph.D. dissertation, University of California, Davis.

Basgall, M. E. 1994. Deception Knoll (CA-SBR-5047): An early Holocene encampment in the north-central Mohave Desert. In R. E. Reynolds, ed., *Off Limits in the Mojave Desert,* pp. 61–70. San Bernardino County Museum Association Special Publication 94-1.

Basgall, M. E. 2000a. Patterns of Toolstone use in late-Pleistocene/early-Holocene assemblages of the Mojave Desert. *Current Research in the Pleistocene* 17:4–6.

Basgall, M. E. 2000b. The structure of archaeological landscapes in the north-central Mojave Desert. In J. S. Schneider, R. M. Yohe II, and J. K. Gardner, eds., *Archaeological Passages: A Volume in Honor of Claude Nelson Warren,* pp. 123–138. Western Center for Archaeology and Paleontology Publications in Archaeology 1. Hemet, CA.

Basgall, M. E., and M. C. Hall. 1991. Relationships between fluted and stemmed points in the Mojave Desert. *Current Research in the Pleistocene* 8:61–64.

Basgall, M. E., and M. C. Hall. 2000. Morphological and temporal variation in bifurcate-stemmed dart points of the western Great Basin. *Journal of California and Great Basin Anthropology* 22:237–276.

Bates, B., Z. W. Kundzewicz, S. Wu, and J. Palutikof. 2008. *Climate Change and Water.* Technical Paper of the Intergovernmental Panel on Climate Change, IPCC Secretariat, Geneva.

Baugh, T. M., and J. E. Deacon. 1983. The most endangered pupfish. *Freshwater and Marine Aquarium* 6(6):22–26, 78–79.

Baumhoff, M. A., and R. F. Heizer. 1965. Postglacial climate and archaeology in the desert West. In H. E. Wright, Jr. and D. G. Frey, eds., *The Quaternary of the United States,* pp. 697–708. Princeton University Press, Princeton, NJ.

Beanland, S., and M. C. Clark. 1994. *The Owens Valley Fault Zone, Eastern California, and Surface Faulting Associated with the 1872 Earthquake.* U.S. Geological Survey Bulletin 1982.

Beaty, R. M., and A. H. Taylor. 2009. A 14000 year sedimentary charcoal record of fire from the northern Sierra Nevada, Lake Tahoe Basin, California, USA. *The Holocene* 19:347–358.

Beck, C. 1995. Functional attributes and the differential persistence of Great Basin dart forms. *Journal of California and Great Basin Anthropology* 17:222–243.

Beck, C., and G. T. Jones. 1990. Toolstone selection and lithic technology in early Great Basin prehistory. *Journal of Field Archaeology* 17:283–299.

Beck, C., and G. T. Jones. 1993. The multipurpose function of Great Basin stemmed series points. *Current Research in the Pleistocene* 10:52–54.

Beck, C., and G. T. Jones. 1997. The Terminal Pleistocene/Early Holocene archaeology of the Great Basin. *Journal of World Prehistory* 11:161–236.

Beck, C., and G. T. Jones. 2007. Early Paleoarchaic point morphology and chronology. In K. E. Graf and D. N. Schmitt, eds., *Paleoindian or Paleoarchaic: Great Basin Human Ecology at the Pleistocene-Holocene Transition,* pp. 23–41. University of Utah Press, Salt Lake City.

Beck, C., and G. T. Jones. 2008. The Clovis-last hypothesis: Investigating early lithic technology in the Intermountain West. Paper presented at the Great Basin Anthropological Conference, Portland, OR.

Beck, C., and G. T. Jones. 2009. *The Archaeology of the Eastern Nevada Paleoarchaic, part 1: The Sunshine Locality.* University of Utah Anthropological Papers 126.

Beck, C., and G. T. Jones. 2010a. Clovis and Western Stemmed: Population migration and the meeting of two technologies in the Intermountain West. *American Antiquity* 75:81–116.

Beck, C., and G. T. Jones. 2010b. The Clovis-last hypothesis: Investigating early lithic technology in the Intermountain West. In D. Rhode, ed., *Meetings at the Margins: Prehistoric*

Cultural Interactions in the Intermountain West. University of Utah Press, Salt Lake City, in press.

Beck, C., G. T. Jones, D. L. Jenkins, C. E. Skinner, and J. J. Thatcher. 2004. Fluted or basally-thinned? Re-examination of a lanceolate point from the Connley Caves in the Fort Rock Basin. In D. L. Jenkins, T. J. Connolly, and C. M. Aikens, eds., *Early and Middle Holocene Archaeology of the Northern Great Basin,* pp. 282–294. University of Oregon Anthropological Papers 62.

Beck, C., A. K. Taylor, G. T. Jones, C. M. Fadem, C. R. Cook, and S. A. Millward. 2002. Rocks are heavy: Transport costs and Paleoarchaic quarry behavior in the Great Basin. *Journal of Anthropological Archaeology* 21:481–507.

Bedinger, M. S., J. R. Harril, and J. M. Thomas. 1984. *Maps Showing Ground-Water Units and Withdrawal, Basin and Range Province, Nevada.* U.S. Geological Survey Water-Resources Investigations Report 83-4119-A.

Bedwell, S. F. 1973. *Fort Rock Basin: Prehistory and Environment.* University of Oregon Books, Eugene.

Bedwell, S. F., and L. S. Cressman. 1971. Fort Rock report: Prehistory and environment of the pluvial Fort Rock Lake area of south-central Oregon. In C. M. Aikens, ed., *Great Basin Anthropological Conference 1970: Selected Papers,* pp. 1–26. University of Oregon Anthropological Papers 1.

Beever, E. A., P. F. Brussard, and J. Berger 2003. Patterns of apparent extinction among isolated populations of pikas (*Ochotona princeps*) in the Great Basin. *Journal of Mammalogy* 84:37–54.

Beever, E. A., J. L. Wilkening, D. E. Mc Ivor, S. S. Weber, and P. F. Brussard. 2008. American pikas (*Ochotona princeps*) in northwestern Nevada: A newly discovered population at a low-elevation site. *Western North American Naturalist* 68:8–14.

Behle, W. H. 1978. Avian biogeography of the Great Basin and Intermountain region. In K. T. Harper and J. L. Reveal, eds., *Intermountain Biogeography: A Symposium,* pp. 55–80. Great Basin Naturalist Memoirs 2.

Beiswenger, J. M. 1991. Late Quaternary vegetational history of Grays Lake, Idaho. *Ecological Monographs* 61:165–182.

Belk, M. C., and H. D. Smith. 1990. *Ammospermophilus leucurus.* Mammalian Species 368.

Bell, C. J. 1993. A late Pleistocene mammalian fauna from Cathedral Cave, White Pine County, Nevada. *Journal of Vertebrate Paleontology* 13(3), Supplement:26A.

Bell, C. J. 1995. A middle Pleistocene (Irvingtonian) microtine rodent fauna from White Pine County, Nevada, and its implications for microtine rodent biochronology. *Journal of Vertebrate Paleontology* 15(3), Supplement:18A.

Bell, C. J., and C. N. Jass. 2004. Arvicoline rodents from Kokoweef Cave, Ivanpah Mountains, San Bernardino County, California. *Bulletin of the Southern California Academy of Science* 103:1–11.

Bell, C. J., and J. I. Mead. 1998. Late Pleistocene microtine rodents from Snake Creek Burial Cave, White Pine County, Nevada. *Great Basin Naturalist* 58:82–86.

Belnap J. 2003. Biological soil crusts and wind erosion. In J. Belnap and O. L. Lange, eds. *Biological Soil Crusts: Structure, Function, and Management.* Revised 2nd printing, pp. 339–347. Ecological Studies Series 150. Springer-Verlag, Berlin.

Belnap, J. 2006. The potential roles of biological soil crusts in dryland hydrologic cycles. *Hydrological Processes* 20:3159–3178.

Belnap J., and D. Eldridge. 2003. Disturbance and recovery of biological soil crusts. In J. Belnap and O. L. Lange, eds. *Biological Soil Crusts: Structure, Function, and Management.* Revised 2nd printing, pp. 363–383. Ecological Studies Series 150, Springer-Verlag, Berlin.

Belnap, J., S. L. Phillips, and T. Troxler. 2006. Soil lichen and moss cover and species richness can be highly dynamic: The effects of invasion by the annual exotic grass *Bromus tectorum* and the effects of climate on biological soil crusts. *Applied Soil Ecology* 32:63–76.

Belnap J., J. H. Kaltenecker, R. Rosentreter, J. Williams, S. Leonard, and D. Eldridge. 2001. *Biological Soil Crusts: Ecology and Management.* U.S. Department of the Interior, Bureau of Land Management Technical Reference 1730-2.

Benson, L., and R. A. Darrow. 1981. *Trees and Shrubs of the Southwestern Deserts.* University of Arizona Press, Tucson.

Benson, L. V. 1988. *Preliminary Paleolimnologic Data for the Walker Lake Subbasin, California and Nevada.* U.S. Geological Survey Water-Resources Investigations Report 87-4258.

Benson, L. V. 2004a. *The Tufas of Pyramid Lake.* U.S. Geological Survey Circular 1267.

Benson, L. V. 2004b. Western lakes. In A. R. Gillespie, S. C. Porter, and B. F. Atwater, eds., *The Quaternary Period in the United States,* pp. 185–204. Developments in Quaternary Science 1. Elsevier, Amsterdam.

Benson, L. V. 2008. Impact of drought on prehistoric western Native Americans. In Jones, J., ed., *California Drought: An Update,* pp. 28–35. Department of Water Resources, Sacramento.

Benson, L. V., and H. Klieforth. 1989. Stable isotopes in precipitation and ground water in the Yucca Mountain region, southern Nevada: Paleoclimatic implications. In D. H. Peterson, ed., Aspects of climate variability in the Pacific and western Americas. *American Geophysical Union Geophysical Monograph* 55:41–49.

Benson, L. V., and M. D. Mifflin. 1986. *Reconnaissance Bathymetry of Basins Occupied by Pleistocene Lake Lahontan, Nevada and California.* U.S. Geological Survey Water-Resources Investigation Report 85-4262.

Benson, L. V., and F. L. Paillet. 1989. The use of total lake-surface area as an indicator of climatic change. *Quaternary Research* 32:262–275.

Benson, L. V., and Z. Peterman. 1995. Carbonate deposition, Pyramid Lake subbasin, Nevada: 3. The use of ^{87}Sr values in carbonate deposits (tufas) to determine the hydrologic state of paleolake systems. *Palaeogeography, Palaeoclimatology, Palaeoecology* 119:201–213.

Benson, L. V., and R. S. Thompson. 1987a. Lake-level variation in the Lahontan Basin for the past 50,000 years. *Quaternary Research* 28:69–85.

Benson, L. V., and R. S. Thompson. 1987b. The physical record of lakes in the Great Basin. In W. F. Ruddiman and H. E. Wright, Jr., eds., *North America and Adjacent Oceans during the Last Deglaciation,* pp. 241–260. The Geology of North America, Volume K-3. Geological Society of America, Boulder, CO.

Benson, L. V., M. Kashgarian, and M. Rubin. 1995. Carbonate deposition, Pyramid Lake subbasin, Nevada, 2: Lake levels and polar jet stream positions reconstructed from radiocarbon ages and elevations of carbonates (tufas) deposited in the Lahontan Basin. *Palaeogeography, Palaeoclimatology, Palaeoecology* 117:1–30.

Benson, L. V., P. A. Meyers, and R. J. Spencer. 1991. Change in the size of Walker Lake during the past 5000 years. *Palaeogeography, Palaeoclimatology, Palaeoecology* 81:189–214.

Benson, L. V., K. Peterson, and J. Stein. 2007. Anasazi (Pre-Columbian Native American) migrants during the middle-12th and late-13th centuries: Were they drought induced? *Climatic Change* 83:187–213.

Benson, L. V., L. D. White, and R. Rye. 1996. Carbonate deposition, Pyramid Lake subbasin, Nevada: 4. Comparison of the stable isotope values of carbonate deposits (tufas) and the Lahontan lake-level record. *Palaeogeography, Palaeoclimatology, Palaeoecology* 122:45–76.

Benson, L. V., J. Burdett, S. Lund, M. Kashgarian, and S. Mensing. 1997. Nearly synchronous climate change in the Northern Hemisphere during the last glacial termination. *Nature* 388:263–265.

Benson, L. V., E. M. Hattori, H. E Taylor, S. R. Poulson, and E. A. Jolie. 2006. Isotope sourcing of prehistoric willow and tule textiles recovered from Great Basin rock shelters and caves: Proof of concept. *Journal of Archaeological Science* 33:1588–1599.

Benson, L. V., S. Lund, R. Negrini, B. Linsley, and M. Zic. 2004. Response of North American Great Basin lakes to Dansgaard-Oeschger oscillations. *Quaternary Science Reviews* 23:193–206.

Benson, L. V., H. M. May, R. C. Antweiler, and T. I. Brinton. 1998. Continuous lake-sediment records of glaciation in the Sierra Nevada between 52,600 and 12,500 ^{14}C yr B.P. *Quaternary Research* 50:113–127.

Benson, L. V., J. P. Smoot, M. Kashgarian, A. Sarna-Wojcicki, and J. W. Burden. 1997. Radiocarbon age and environments of deposition of the Wono and Trego Hot Springs tephra layers in the Pyramid Lake subbasin, Nevada. *Quaternary Research* 47:251–260.

Benson, L. V., M. S. Berry, E. A. Jolie, J. D. Spangler, D. W. Stahle, and E. M. Hattori. 2007. Possible impacts of early-11th-, middle-12th-, and late-13th-century droughts on western Native Americans and the Mississippian Cahokians. *Quaternary Science Reviews* 26:336–350.

Benson, L. V., J. W. Burdett, M. Kashgarian, S. P. Lund, F. M. Phillips, and R. O. Rye. 1996. Climatic and hydrologic oscillations in the Owens Lake Basin and adjacent Sierra Nevada, California. *Science* 274:746–749.

Benson, L. V., J. Liddicoat, J. Smoot, A. Sarna-Wojcicki, R. Negrini, and S. Lund. 2003. Age of the Mono Lake excursion and associated tephra. *Quaternary Science Reviews* 22:135–140.

Benson, L. V., S. P. Lund, J. W. Burdett, M. Kashgarian, T. P. Rose, J. P. Smoot, and M. Schwartz. 1998. Correlation of late-Pleistocene lake-level oscillations in Mono Lake, California, with North Atlantic climate events. *Quaternary Research* 49:1–10.

Benson, L. V., D. R. Currey, R. I. Dorn, K. R. Lajoie, C. G. Oviatt, S. W. Robinson, G. I. Smith, and S. Stine. 1990. Chronology of expansion and contraction of four Great Basin lake systems during the past 35,000 years. *Palaeogeography, Palaeoclimatology, Palaeoecology* 78:241–286.

Benson, L. V., M. Kashgarian, R. Rye, S. Lund, F. Paillet, J. Smoot, J., C. Kester, S. Mensing, D. Meko, and S. Lindström. 2002. Holocene multidecadal and multicentennial droughts affecting northern California and Nevada. *Quaternary Science Reviews* 21:659–682.

Berger, J. 1986. *Wild Horses of the Great Basin.* University of Chicago Press, Chicago.

Berger, J., and J. D. Wehausen. 1991. Consequences of a mammalian predator-prey disequilibrium in the Great Basin desert. *Conservation Biology* 5:244–248.

Berry, M. S. 1976. Remnant Cave. In G. F. Dalley, ed., *Swallow Shelter and Associated Sites*, pp. 115–127. University of Utah Anthropological Papers 96.

Betancourt, J. L., T. R. Van Devender, and P. S. Martin. 1990a. Introduction. In J. L. Betancourt, T. R. Van Devender, and P. S. Martin, eds., *Packrat Middens: The Last 40,000 Years of Biotic Change*, pp. 2–11. University of Arizona Press, Tucson.

Betancourt, J. L., T. R. Van Devender, and P. S. Martin, eds. 1990b. *Packrat Middens: The Last 40,000 Years of Biotic Change.* University of Arizona Press, Tucson.

Betancourt, J. L., K. A. Rylander, C. Peñalba, and J. L. McVickar. 2001. Late Quaternary vegetation history of Rough Canyon, south-central New Mexico, USA. *Palaeogeography, Palaeoclimatology, Palaeoecology* 165:71–95.

Bettinger, R. L. 1976. The development of pinyon exploitation in central eastern California. *Journal of California and Great Basin Prehistory* 3:81–95.

Bettinger, R. L. 1977. Aboriginal human ecology in Owens Valley: Prehistoric change in the Great Basin. *American Antiquity* 42:3–17.

Bettinger, R. L. 1989. *The Archaeology of Pinyon House, Two Eagles, and Crater Middens: Three Residential Sites in Owens Valley, Eastern California.* American Museum of Natural History Anthropological Papers 67.

Bettinger, R. L. 1991a. Aboriginal occupation at high altitude: Alpine villages in the White Mountains of eastern California. *American Anthropologist* 93:656–679.

Bettinger, R. L. 1991b. Native land use: Archaeology and anthropology. In C. A. Hall, Jr., ed., *Natural History of the White-Inyo Range*, pp. 463–486. University of California Press, Berkeley.

Bettinger, R. L. 1999a. From traveler to processor: Regional trajectories of hunter-gatherer sedentism in the Inyo-Mono region, California. In B. R. Billman and G. M. Feinman, eds., *Settlement Pattern Studies in the Americas: Fifty Years since Virú*, pp. 39–55. Smithsonian Institution Press, Washington, DC.

Bettinger, R. L. 1999b. What happened in the Medithermal. In C. Beck, ed., *Models for the Millennium: Great Basin Anthropology Today*, pp. 62–74. University of Utah Press, Salt Lake City.

Bettinger, R. L. 2008. High altitude sites in the Great Basin. In C. S. Fowler and D. D. Fowler, eds., *The Great Basin: People and Place in Ancient Times*, pp. 86–93. School for Advanced Research Press, Santa Fe, NM.

Bettinger, R. L., and M. A. Baumhoff. 1982. The Numic spread: Great Basin cultures in competition. *American Antiquity* 47:485–503.

Bettinger, R. L., and M. A. Baumhoff. 1983. Return rates and intensity of resource use in Numic and Prenumic adaptive strategies. *American Antiquity* 48:830–834.

Bettinger, R. L., and J. Eerkens. 1999. Point typologies, cultural transmission, and the spread of bow-and-arrow technology in the prehistoric Great Basin. *American Antiquity* 64:231–242.

Bettinger, R. L., and R. Oglesby. 1985. Lichen dating of alpine villages in the White Mountains, California. *Journal of California and Great Basin Anthropology* 7:202–224.

Bevis, K. A. 1995. Reconstruction of Late Pleistocene paleoclimatic characteristics in the Great Basin and adjacent areas. Ph.D. dissertation, Oregon State University, Corvallis.

Bevis, K. A. 1999. A brief history of glaciation at Steens Mountain, southeast Oregon. In C. Narwold, organizer, *Quaternary Geology of the Northern Quinn River and Alvord Valleys, Southeastern Oregon. 1999 Friends of the Pleistocene Field Trip, Pacific Cell, September 24–26, 1999*, Article A10, pp. 1–5. Geology Department, Humboldt State University, Arcata, CA.

Billat, S. E. 1985. A study of Fremont subsistence at the Smoking Pipe site. M.A. thesis. Brigham Young University, Provo, UT.

Billings, W. D. 1945. The plant associations of the Carson Desert region. *Butler University Botanical Studies* 7:89–123.

Billings, W. D. 1950. Vegetation and plant growth as affected by chemically altered rocks in the western Great Basin. *Ecology* 31:62–74.

Billings, W. D. 1951. Vegetational zonation in the Great Basin of western North America. In *Les Bases Ecologiques de la Régénération de la Végétation des Zones arides*, pp. 101–122. International Union of Biological Sciences, Series B, No. 9.

Billings, W. D. 1954. Temperature inversions in the Pinyon-Juniper Zone of a Nevada mountain range. *Butler University Botanical Studies* 11:112–118.

Billings, W. D. 1978. Alpine phytogeography across the Great Basin. In K. T. Harper and J. L. Reveal, eds., *Intermountain Biogeography: A Symposium*, pp. 105–117. Great Basin Naturalist Memoirs 2.

Billings, W. D. 1988. Alpine vegetation. In M. G. Barbour and W. D. Billings, eds., *North American Terrestrial Vegetation*, pp. 391–420. California Native Plant Society Publication 9.

Billings, W. D. 1990. The mountain forests of North America and their environments. In C. B. Osmond, L. F. Pitelka, and G. M. Hidy, eds., *Plant Biology of the Basin and Range*, pp. 47–86. Springer-Verlag, Berlin.

Bills, B. G., T. J. Wambeam, and D. R. Currey. 2002. Geodynamics of Lake Bonneville. In J. W. Gwynn, ed., *Great Salt Lake: An Overview of Change*, pp. 7–33. Utah Department of Natural Resources Special Publication, Salt Lake City.

Bills, T. M., and H. G. McDonald. 1998. Fauna from late-Pleistocene sediments of the Sheriden Cave Site (33WY252), Wyandot County, Ohio. *Current Research in the Pleistocene* 15:101–103.

Bird, B. W., and M. E. Kirkby. 2006. An alpine lacustrine record of early Holocene North American monsoon dynamics from Dry Lake, southern California (USA). *Journal of Paleolimnology* 35:179–192.

Bird, R. 1999. Cooperation and conflict: The behavioral ecology of the sexual division of labor. *Evolutionary Anthropology* 8:65–75.

BirdLife International. 2008. Species Factsheet: *Gymnogyps californianus*. www.birdlife.org.

Birks, H. J. B., and H. H. Birks. 2008. Biological responses to rapid climate change at the Younger Dryas–Holocene transition at Kråkenes, western Norway. *The Holocene* 18:19–30.

Bischoff, J. L., and K. Cummins. 2001. Wisconsin glaciation of the Sierra Nevada (79,000–15,000 yr B.P.) as recorded by rock flour in sediments of Owens Lake, California. *Quaternary Research* 55:14–24.

Bischoff, J. L., T. W. Stafford, Jr., and M. Rubin. 1997. A time-depth scale for Owens Lake sediments of core OL-92: Radiocarbon dates and constant mass-accumulation rate. In G. I. Smith and J. L. Bischoff, eds., *An 800,000-year Paleoclimatic Record from Core OL-92, Owens Lake, Southeast California*, pp. 91–98. Geological Society of America Special Paper 317.

Bischoff, J. L., R. J. Shlemon, T. L. Ku, R. J. Rosenbaum, and F. E. Budinger, Jr. 1981. Uranium-series and soil-geomorphic dating of the Calico archaeological site, California. *Geology* 9:576–582.

Blackwelder, E. 1934. Supplementary notes on Pleistocene glaciation in the Great Basin. *Journal of the Washington Academy of Sciences* 24:212–222.

Blake, E. R. 1977. *Manual of Neotropical birds*, vol. 1: *Spheniscidae (Penguins) to Laridae (Gulls and allies)*. University of Chicago Press, Chicago.

Bleige Bird, R., and D. W. Bird. 2008. Why women hunt: Risk and contemporary foraging in a Western Desert Aboriginal community. *Current Anthropology* 49:655–693.

Blitz, J. H. 1988. Adoption of the bow and arrow in prehistoric North America. *North American Archaeologist* 9:123–145.

Bocheński, Z. M., and K. E. Campbell, Jr. 2006. *The Extinct California Turkey,* Meleagris californica, *from Rancho La Brea: Comparative Osteology and Systematics*. Natural History Museum of Los Angeles County Contributions in Science 509.

Booth, D. B., K. G. Troost, J. J. Clague, and R. B. Waitt. 2004. The Cordilleran Ice Sheet. In A. R. Gillespie, S. C. Porter, and B. F. Atwater, eds., *The Quaternary Period in the United States*, pp. 17–44. Developments in Quaternary Science 1. Elsevier, Amsterdam.

Bouchard, D. P., D. S. Kaufman, A. Hochberg, and J. Quade. 1998. Quaternary history of the Thatcher Basin, Idaho, reconstructed form the $^{87}Sr/^{86}Sr$ and amino acid composition of lacustrine fossils: Implications for the diversion of the Bear River into the Bonneville Basin. *Palaeogeography, Palaeoclimatology, Palaeoecology* 14:95–114.

Bowers, R., and R. M. Burke. 2005. Preliminary research on the relative age estimates of shorelines associated with pluvial Lake Madeline, Lassen County, northeastern California. *Geological Society of America Abstracts with Programs* 37(4):43.

Boxall, B. 2008. Number of Devil's Hole pupfish increasing. *Los Angeles Times*, October 14. http://latimesblogs.latimes.com/unleashed/2008/10/number-of-devil.html.

Bradbury, J. P. 1987. Late Holocene diatom paleolimnology of Walker Lake, Nevada. *Archiv für Hydrobiologie,* Supplement 79, *Monographische Beitrage* 1:1–27.

Bradbury, J. P., R. M. Forester, and R. S. Thompson. 1989. Late Quaternary paleolimnology of Walker Lake, Nevada. *Journal of Paleolimnology* 1:249–267.

Bradley, B. A., and E. Fleishman. 2008. Relationships between expanding pinyon-juniper cover and topography in the central Great Basin, Nevada. *Journal of Biogeography* 35:951–964.

Bradley, B., and D. Stanford. 2004. The North Atlantic ice-edge corridor: A possible Paleolithic route to the New World. *World Archaeology* 36(4):459–478.

Bradley, R. S. 2008. The Younger Dryas and the sea of ancient ice. *Quaternary Research* 70:1–10.

Bradley, R. S., M. K. Hughes, and H. F. Diaz. 2003. Climate in medieval time. *Science* 302:404–405.

Briffa, K. R., P. D. Jones, and F. H. Schweingruber. 1992. Tree-ring density reconstructions of summer temperature patterns across western North America since 1600. *Journal of Climate* 5:735–754.

Briggs, R. W., S. G. Wesnousky, and K. D. Adams. 2005. Late Pleistocene and late Holocene lake highstands in the Pyramid Lake subbasin of Lake Lahontan, Nevada, USA. *Quaternary Research* 64:257–263.

Brigham-Grette, J., A. V. Lozhkin, P. M. Anderson, and O. Y. Glushkova. 2004. Paleoenvironmental conditions in western Beringia before and during the Last Glacial Maximum. In D. B. Madsen, ed., *Entering America: Northeast Asia and Beringia before the Last Glacial Maximum*, pp. 29–61. University of Utah Press, Salt Lake City.

Bright, J., D. S. Kaufman, R. M. Forester, and W. E. Dean. 2006. A continuous 250,000 yr record of oxygen and carbon isotopes in ostracode and bulk-sediment carbonate from Bear Lake, Utah-Idaho. *Quaternary Science Reviews* 25:2258–2270.

Bright, R. C. 1966. Pollen and seed stratigraphy of Swan Lake, southeastern Idaho. *Tebiwa* 9(2):1–47.

Broecker, W. S. 2001. Was the Medieval Warm Period global? *Science* 291:1497–1499.

Broecker, W. S. 2006. Was the Younger Dryas triggered by a flood? *Science* 312:1146–1148.

Broecker, W. S., G. H. Denton, R. L. Edwards, H. Cheng, R. B. Alley, and A. E. Putnam. 2010. Putting the Younger Dryas into context. *Quaternary Science Reviews* 29:1078–1081.

Brooks, G. R. 1989. *The Southwest Expedition of Jedediah S. Smith: His Personal Account for the Journey to California, 1826–1827*. University of Nebraska Press, Lincoln.

Brooks, J. 1962. *The Mountains Meadows Massacre*. University of Oklahoma Press, Norman.

Brooks, S. T., M. B. Haldeman, and R. H. Brooks. 1988. *Osteological Analyses of the Stillwater Skeletal Series, Stillwater Marsh, Churchill Count, Nevada*. U.S. Department of the Interior, U.S. Fish and Wildlife Service, Region 1, Cultural Resource Series 2.

Broughton, J. M. 2000a. Terminal Pleistocene fish remains from Homestead Cave, Utah, and implications of fish biogeography in the Bonneville Basin. *Copeia* 2000:645–656.

Broughton, J. M. 2000b. The Homestead Cave icthyofauna. In D. B. Madsen, *Late Quaternary paleoecology in the Bonneville Basin*, pp. 103–122. Utah Geological Survey Bulletin 130.

Broughton, J. M., D. B. Madsen, and J. Quade. 2000. Fish remains from Homestead Cave and lake levels of the past 13,000 years in the Bonneville Basin. *Quaternary Research* 53:392–401.

Broughton, J. M., D. A. Byers, R. A. Bryson, W. Eckerle, and D. B. Madsen. 2008. Did climatic seasonality control late Quaternary artiodactyl densities in western North America? *Quaternary Science Reviews* 27:1916–1937.

Broughton, J. M., V. I. Cannon, S. Arnold, R. J. Bogiatto, and K. Dalton. 2004. The taphonomy of owl-deposited fish remains and the origin of the Homestead Cave ichthyofauna. *Journal of Taphonomy* 4:69–95.

Brown, J. H. 1971. Mammals on mountaintops: Nonequilibrium insular biogeography. *American Naturalist*, 105:467–478.

Brown, J. H. 1978. The theory of insular biogeography and the distribution of boreal birds and mammals. In K. T. Harper and J. L. Reveal, eds., *Intermountain biogeography: A symposium*, pp. 209–227. Great Basin Naturalist Memoirs 2.

Brown, J. H. 2004. Concluding remarks. In M. V. Lomolino and L. R. Heaney, eds., *Frontiers of Biogeography: New Directions in the Geography of Nature*, pp. 361–368. Sinauer Associates, Sunderland, MA.

Brown, L. 1971. *African Birds of Prey*. Houghton Mifflin, Boston. MA.

Brown, W. J., S. G. Wells, Y. Enzel, R. Y. Anderson, and L. D. McFadden. 1990. The late Quaternary history of pluvial Lake Mojave–Silver Lake and Soda Lake basins, California. In R. E. Reynolds, S. G. Wells, and R. H. Brady III, eds., *At the End of the Mojave: Quaternary Studies in the Eastern Mojave Desert*, pp. 55–72. San Bernardino County Museum Association, Redlands, CA.

Brubaker, L. B., P. M. Anderson, M. E. Edwards, and A. V. Loshkin. 2005. Beringia as a glacial refugium for boreal trees and shrubs: New perspectives from mapped pollen data. *Journal of Biogeography* 32:833–848.

Bryan, A. L. 1977. Developmental stages and technological traditions. In W. S. Newman and B. Salwen, eds., *Amerinds and their Paleoenvironments in Northeastern North America*, pp. 255–368. Annals of the New York Academy of Sciences 288.

Bryan, A. L. 1979. Council Hall Cave. In D. R. Tuohy and D. L. Rendall, eds., *The Archaeology of Smith Creek Canyon, Eastern Nevada*, pp. 254–271. Nevada State Museum Anthropological Papers 17.

Bryan, A. L. 1980. The stemmed point tradition: An early technological tradition in western North America. In L. B. Harten, C. N. Warren, and D. R. Tuohy, eds., *Anthropological Papers in Memory of Earl H. Swanson, Jr.*, pp. 77–107. Idaho Museum of Natural History, Pocatello, ID.

Bryan, A. L. 1988. The relationship of the stemmed point and the fluted point traditions in the Great Basin. In J. A. Willig, C. M. Aikens, and J. L. Fagan, eds., *Early Human Occupation in Far Western North America: The Clovis-Archaic Interface*, pp. 53–74. Nevada State Museum Anthropological Papers 21.

Bryan, A. L., and R. Gruhn. 1964. Problems relating to the Neothermal climate sequence. *American Antiquity* 29:307–315.

Bryant, L. D., and C. Maser. 1982. Classification and distribution. In Thomas, J. W., and D. E. Toweill, eds., *Elk of North America: Ecology and Management*, pp. 1–59. Stackpole Books, Harrisburg, PA.

Buchanan, B., M. Collard, and K. Edinborough. 2008. Paleoindian demography and the extraterrestrial impact hypothesis. *Proceedings of the National Academy of Sciences* 105:11651–11654.

Buck, P. E., W. T. Hartwell, G. Haynes, and D. Rhode. 1998. *Archaeological Investigations at Two Early Holocene Sites near Yucca Mountain, Nye County, Nevada*. Topics in Yucca Mountain Archaeology 2. Desert Research Institute, Las Vegas, NV.

Buck, P. E., B. Hockett, F. Nials, and P. E. Wigand. 1997. *Prehistory and Paleoenvironment of Pintwater Cave, Nevada: Results of Field Work during the 1996 Season*. Project Number OS-005071, Nellis Air Force Base, Nevada.

Buck, P. E., B. Hockett, K. Graf, T. Goebel, G. Griego, L. Perry, and E. Dillingham. 2002. Oranjeboom Cave: A single component Eastgate site in northeastern Nevada. *Utah Archaeology* 15:99–112.

Buckland, W. 1823. *Reliquiae diluvianae; or, observations on the organic remains contained in caves, fissures, and diluvial gravel, and on other geological phenomena, attesting the action of an universal deluge*. John Murray, London.

Budinger, F. E., Jr. 1983. The Calico Early Man Site, San Bernardino, California. *California Geology* 36:75–82.

Budy, E. E., and K. L. Katzer. 1990. Chronology of cultural occupation. In R. G. Elston and E. E. Budy, eds., *The Archaeology of James Creek Shelter*, pp. 47–55. University of Utah Anthropological Papers 115.

Bunch, T. E., A. West, R. B. Firestone, J. P. Kennett, J. H. Wittke, C. R. Kinzie, and W. S. Wolbach. 2010. Geochemical data reported by Paquay et al. do not refute Younger Dryas impact event. *Proceedings of the National Academy of Sciences* 107:E58.

Burgett, R. B. 2004. Coiled basketry designs from Charlie Brown Cave, Western Nevada. M.A. thesis, University of Nevada, Reno.

Burney, D. A., L. P. Burney, L. R. Godfrey, W. L. Jungers, S. M. Goodman, H. T. Wright, A. J. T. Jull. 2004. A chronology for late prehistoric Madagascar. *Journal of Human Evolution* 47:25–63.

Bursik, M. I., and A. R. Gillespie. 1993. Late Pleistocene glaciation of Mono Basin, California. *Quaternary Research* 39:24–35.

Butcher, R. D. 2003. *America's National Wildlife Refuges*. Roberts Rinehart Publishers, Lanham, MD.

Butler, B. R. 1970. A surface collection from Coyote Flat, southeastern Oregon. *Tebiwa* 13(1):34–57.

Butler, B. R. 1972. The Holocene or postglacial ecological crisis on the eastern Snake River Plain. *Tebiwa* 15(1):49–63.

Butler, B. R., 1978. *A Guide to Understanding Utah Archaeology (Third edition): The Upper Snake and Salmon River Country*. The Idaho Museum of Natural History, Pocatello, ID.

Butler, V. L., and R. A. Schroeder. 1998. Do digestive processes leave diagnostic traces on fish bones? *Journal of Archaeological Science* 25:957–971.

Bynon, T. 1977. *Historical Linguistics*. Cambridge University Press, London.

Byrne, R., C. Busby, and R. F. Heizer. 1979. The Altithermal revisited: Pollen evidence from the Leonard Rockshelter. *Journal of California and Great Basin Anthropology* 1:280–294.

Cabana, G. S., K. Hunley, and F. A. Kaestle. 2008. Population replacement or continuity? A novel computer simulation approach and its application to the Numic expansion (western Great Basin, USA). *American Journal of Physical Anthropology* 135:438–447.

Campbell, E. W. C., and W. H. Campbell. 1940. A Folsom complex in the Great Basin. *The Masterkey* 14(1):7–11.

Campbell, E. W. C., W. H. Campbell, E. Antevs, C. A. Amsden, J. A. Barbieri, and F. D. Bode. 1937. *The Archaeology of Pleistocene Lake Mohave: A Symposium*. Southwest Museum Papers 11.

Campbell, K. E., Jr. 1995. Additional specimens of the Giant Teratorn, *Argentavis magnificens*, from Argentina (Aves: Teratornithidae). *Courier Forschungsinstitut Senckenberg* 181:199–201.

Campbell, K. E., Jr. 2004. A new species of late Pleistocene lapwing from Rancho La Brea, California. *The Condor* 104:170–174.

Campbell, K. E., Jr., and E. Tonni. 1981. Preliminary observations on the paleobiology and evolution of teratorns (Aves: Teratornithidae). *Journal of Vertebrate Paleontology* 1:265–272.

Campbell, K. E., Jr., and E. Tonni. 1983. Size and locomotion in teratorns (Aves: Teratornithidae). *The Auk* 100:390–403.

Campbell, K. E., Jr., E. Scott, and K. B. Springer. 1999. A new genus for the Incredible Teratorn (Aves: Teratornithidae). In P. Wellnhofer, C. Mourer-Chauviré, D. W. Steadman, and L. D. Martin, eds., *Avian Paleontology at the Close of the 20th Century: Proceedings of the 4th International Meeting of the Society of Avian Paleontology and Evolution, Washington, D. C., 4–7 June 1996*, pp. 169–175. Smithsonian Contributions to Paleobiology 89.

Campbell, L. 1997. *American Indian Languages: The Historical Linguistics of Native America*. Oxford University Press, Oxford, UK.

Canaday, T. W. 1997. Prehistoric Alpine Hunting Patterns in the Great Basin. Ph.D. dissertation, University of Washington, Seattle.

Cane, S. 1987. Australian Aboriginal subsistence in the Western Desert. *Human Ecology* 15:391–434.

Cannon, K. P. 2004. *The Analysis of a Late Holocene Bison Skull from the Ashley National Forest*. Vernon, UT: Ashley National Forest Report AS-04-1010, Ashley National Forest, US Forest Service.

Cannon, M. D., and D. J. Meltzer. 2004. Early Paleoindian foraging: Examining the faunal evidence for large mammal specialization and regional variability in prey choice. *Quaternary Science Reviews* 23:1955–1987.

Carey, H. V., and J. D. Wehausen. 1991. Mammals. In C. A. Hall, Jr., ed., *Natural History of the White-Inyo Range*, pp. 437–560. University of California Press, Berkeley.

Carling, P. A., A. D. Kirkbride, S. Parnachov, P. S. Borodavko, and G. W. Berger. 2002. Late Quaternary catastrophic flooding in the Altai Mountains of south-central Siberia: A synoptic overview and an introduction to flood deposit sedimentology. In I. P. Martini, V. R. Baker, and G. Garzón, eds., *Floods and Megaflood Processes and Deposits: Recent and Ancient Examples*, pp. 17–35. International Association of Sedimentologists Special Publication 32.

Carlyle, S. W., R. L. Parr, M. G. Hayes, and D. H. O'Rourke. 2000. Context of material lineages in the greater Southwest. *American Journal of Physical Anthropology* 113:85–101.

Carpenter, K. L. 2002. Reversing the trend: Late Holocene subsistence change in northeastern California. In McGuire, K. R., editor, *Boundary Lands: Archaeological Investigations along the California–Great Basin Interface* pp. 49–59. Nevada State Museum Anthropological Papers 24.

Carter, D. T., L. L. Ely, J. E. O'Connor, and C. R. Fenton. 2006. Late Pleistocene outburst flooding from pluvial Lake Alvord into the Owyhee River, Oregon. *Geomorphology* 75:346–367.

Carter, H. L. 1968. *"Dear Old Kit": The Historical Kit Carson, with a New Edition of the Carson Memoirs*. University of Oklahoma Press, Norman.

Caskey, S. J., and S. G. Wesnousky. 1997. Static stress changes and earthquake triggering during the 1954 Fairview Peak and Dixie Valley earthquakes, central Nevada. *Bulletin of the Seismological Society of America* 87:521–527.

Caskey, S. J., J. W. Bell, A. R. Ramelli, and S. G. Wesnousky. 2004. Historic surface faulting and paleoseismicity in the area of the 1954 Rainbow Mountain–Stillwater earthquake sequence, central Nevada. *Bulletin of the Seismological Society of America* 94:1255–1275.

Caskey, S. J., J. W. Bell, B. D. Slemmons, and A. R. Ramelli. 2000. Historical surface faulting and paleoseismology of the central Nevada seismic belt. In D. R. Lageson, S. G. Peters, and M. M. Lahren, eds., *Great Basin and Sierra Nevada: Geological Society of America Field Guide* 2, pp. 23–44.

Caskey, S. J., S. G. Wesnousky, P. Zhang, and D. B. Slemmons. 1996. Surface faulting of the 1954 Fairview Peak (M_s 7.2) and Dixie Valley (M_s 6.8) earthquakes, central Nevada. *Bulletin of the Seismological Society of America* 86:761–787.

Caveney, S., D. A. Charlet, H. Freitag, M. Maier-Stolte, and A. N. Starratt. 2001. New observations on the secondary chemistry of world *Ephedra* (Ephedraceae). *American Journal of Botany* 88:1199–1208.

Chaffin, T. 2002. *Pathfinder: John Charles Frémont and the Course of American Empire*. Hill and Wang, New York.

Chalfant, W. A. 1975. *The Story of Inyo*. Chalfant Press, Bishop, CA.

Chamberlain, C. P., J. R. Waldbauer, K. Fox-Dobbs, S. D. Newsome, P. L. Koch, D. R. Smith, M. E. Church, S. D. Chamberlain, K. J. Sorenson, and R. Risebrough. 2005. Pleistocene to recent dietary shifts in California condors. *Proceedings of the National Academy of Sciences* 102:16707–16711.

Chandler, R. M., and L. D. Martin. 1991. A record for caracara from the late Pleistocene of Nebraska. *Current Research in the Pleistocene* 8:87–88.

Channell, J. E. T. 2006. Late Brunhes polarity excursions (Mono Lake, Laschamp, Iceland Basin and Pringle Falls) recorded at ODP Site 919 (Irminger Basin). *Earth and Planetary Science Letters* 244:378–393.

Charlet, D. A. 1991. Relationships of the Great Basin alpine flora: A quantitative analysis. M.A. thesis, University of Nevada, Reno.

Charlet, D. A. 1996. *Atlas of Nevada Conifers: A Phytogeographic Reference*. University of Nevada Press, Reno.

Charlet, D. A. 2007. Distribution patterns of Great Basin conifers: Implications of extinction and immigration. *Aliso* 24:31–61.

Charlet, D. A. 2008. Shah-Kan-Daw: Anthropogenic simplication of semi-arid vegetation structure. In S. G. Kitchen, R. L. Pendleton, T. A. Monaco, and J. Vernon, compilers, *Proceedings: Shrublands under Fire; Disturbance and Recovery in a Changing*

World, pp. 5–24. U.S. Department of Agriculture Proceedings RMRS-P-52.

Charlet, D. A. 2009a. *Atlas of Nevada Vegetation. Volume 1. Mountains*. Unpublished Manuscript.

Charlet, D. A. 2009b. Floristics of the Mt. Jefferson alpine florula. In D. H. Thomas and L. S. A. Pendleton, *The Archaeology of Monitor Valley: 4. Alta Toquima and the Mt. Jefferson Complex*. American Museum of Natural History Anthropological Papers, in press.

Chatterjee, S., R. J. Templin, and K. E. Campbell, Jr. 2007. The aerodynamics of *Argentavis*, the world's largest flying bird from the Miocene of Argentina. *Proceedings of the National Academy of Sciences* 104:12398–12403.

Chavez, A, and T. Warner. 1995. *The Domínguez-Escalante Journal: Their Expedition through Colorado, Utah, Arizona, and New Mexico in 1776*. University of Utah Press, Salt Lake City.

Christensen, J. H., B. Hewitson, A. Busuioc, A. Chen, X. Gao, I. Held, R. Jones, R. K. Kolli, W-T. Kwon, R. Laprise, V. Magaña Rueda, L. Mearns, C. G. Menéndez, J. Räisänen, A. Rinke, A. Sarr, and P. Whetton. 2007. Regional projections. In S. Solomon, D. Qin, M. Manning, Z. Chen, M. Marquis, K. B. Averyt, M. Tignor, and H. L. Miller, eds., *Climate Change 2007: The Physical Science Basis. Contributions of Working Group I to the Fourth Assessment Report of the Intergovernmental Panel on Climate Change*, pp. 847–940. Cambridge University Press, Cambridge, UK.

Clague J. J., and T. S. James. 2002. History and isostatic effects of the last ice sheet in southern British Columbia. *Quaternary Science Reviews* 21:71–87.

Clague, J. J., R. W. Mathewes, and T. A. Ager. 2004. Environments of northwestern North America before the Last Glacial Maximum. In D. B. Madsen, ed., *Entering America: Northeast Asia and Beringia before the Last Glacial Maximum*, pp. 63–94. University of Utah Press, Salt Lake City.

Clague, J. J., R. Barendregt, R. J. Enkin, and F. F. Foit, Jr. 2003. Paleomagnetic and tephra evidence for tens of Missoula floods in southern Washington. *Geology* 31:247–250.

Clark, D. H., and A. R. Gillespie. 1997. Timing and significance of late-glacial and Holocene cirque glaciation in the Sierra Nevada, California. *Quaternary International* 38/39:21–38.

Clark, P. U., and A. C. Mix. 2002. Ice sheets and sea level of the Last Glacial Maximum. *Quaternary Science Reviews* 21:1–7.

Clark, P. U., A. S. Dyke, J. D. Shakun, A. E. Carlson, J. Clark, B. Wohlfarth, J. X. Mitrovica, S. W. Hostetler, and A. M. McCabe. 2009. The Last Glacial Maximum. *Science* 325:710–714.

Clements, F. E. 1936. Nature and structure of the climax. *Journal of Ecology* 24:252–284.

Clemmer, R. O., L. D. Myers, and M. E. Rudden, eds. 1999. *Julian Steward and the Great Basin: The Making of an Anthropologist*. University of Utah Press, Salt Lake City.

Clewlow, C. W., Jr. 1968. Surface archaeology of the Black Rock Desert, Nevada. *University of California Archaeological Survey Reports* 73:1–94.

Cline, G. G. 1988. *Exploring the Great Basin*. University of Nevada Press, Reno.

Coates, P. 2006. *American Perceptions of Immigrant and Invasive Species*. University of California Press, Berkeley.

Coats, R. R. 1964. *Geology of the Jarbidge Quadrangle, Nevada-Idaho*. U.S. Geological Survey Bulletin 1141-M.

Coats, R. R., R. C. Green, and L. D. Cress. 1977. *Mineral Resources of the Jarbidge Wilderness and Adjacent Areas, Elko County, Nevada*. U.S. Geological Survey Bulletin 1439.

Cohen, A. S., M. R. Palacios-Fest. R. M. Negrini, P. E. Wigand, and D. B. Erbes. 2000. A paleoclimate record for the past 250,000 years for Summer Lake, Oregon, USA: II. Sedimentology, paleontology and geochemistry. *Journal of Paleolimnology* 24:151–182.

COHMAP Members. 1988. Climatic changes of the last 18,000 years: Observations and model simulations. *Science* 241: 1043–1052.

Cole, K. L. 1986. The lower Colorado River Valley: A Pleistocene Desert. *Quaternary Research* 25:392–400.

Cole, K. L., J. Fisher, S. T. Arundel, J. Cannella, and S. Swift. 2008. Geographical and climatic limits of needle types of one- and two-needled pinyon pines. *Journal of Biogeography* 35:257–269.

Collard, M., B. Buchanan, and K. Edinborough. 2008. Reply to Anderson et al., Jones, Kennett and West, and Kennett et al.: Further evidence against the extraterrestrial impact hypothesis. *Proceedings of the National Academy of Sciences* 105: E112–E114.

Colman, S. M., D. S. Kaufman, J. Bright, C. Heil, J. W. King, W. E. Dean, J. G. Rosenbaum, R. M. Forester, J. L. Bischoff, M. Perkins, and J. P. McGeehin. 2006. Age model for a continuous, ca. 250 ka Quaternary lacustrine record from Bear Lake, Utah-Idaho. *Quaternary Science Reviews* 25:2271–2282.

Coltrain, J. B. 1993. Fremont corn agriculture: A pilot stable carbon isotope study. *Utah Archaeology* 6:49–56.

Coltrain, J. B. 1996. Stable carbon and radioisotope analyses. In R. K. Talbot and L. D. Richens, *Steinaker Gap: An Early Fremont Homestead*, pp. 115–122 Brigham Young University Museum of Peoples and Cultures Occasional Papers 4.

Coltrain, J. B., and S. W. Leavitt. 2002. Climate and diet in Fremont prehistory: Economic variability and abandonment of maize agriculture in the Great Salt Lake basin. *American Antiquity* 67:453–485.

Coltrain, J. B., and T. W. Stafford, Jr. 1999. Stable carbon isotopes and Great Salt Lake wetlands diet. In B. E. Hemphill, and C. S. Larsen, eds., *Prehistoric Lifeways in the Great Basin Wetlands: Bioarchaeological Reconstruction and Interpretation*, pp. 55–83. University of Utah Press, Salt Lake City.

Conaway, J. S. 2000. Hydrogeology and paleohydrology in the Williamson River Basin, Klamath County, Oregon. M.S. thesis, Portland State University, Portland, OR.

Connolly, T. J. 1999. *Newberry Crater: A Ten-thousand-year Record of Human Occupation and Environmental Change in the Basin-Plateau Borderlands*. University of Utah Anthropological Paper 121.

Cook, C. W. 1980. Faunal analysis from five Utah Valley Sites: A test of a subsistence model from the Sevier Fremont area. M.A. thesis. Brigham Young University, Provo, UT.

Cook, E. R., C. A. Woodhouse, C. M. Eakin, D. M. Meko, and D. W. Stahle. 2004. Long-term aridity changes in the western United States. *Science* 306:1015–1018.

Cope, E. D. 1889. The Silver Lake of Oregon and its region. *American Naturalist* 23:970–982.

Cowan, R. A. 1967. Lake-margin ecologic exploitation in the Great Basin as demonstrated by an analysis of coprolites from Lovelock Cave, Nevada. *University of California Archaeological Survey Reports* 70:21–36.

Cox, B. F., J. W. Hillhouse, and L. A. Owen. 2003. Pliocene and Pleistocene evolution of the Mojave River, and associated tectonic development of the Transverse Ranges and Mojave Desert, based on borehole stratigraphy studies and mapping of landforms and sediments near Victorville, California. In Y. Enzel, S. G. Wells, and N. Lancaster, eds., *Paleoenvironments and Paleohydrology of the Mojave and Southern Great Basin Deserts*, pp. 1–42. Geological Society of America Special Paper 368.

Cox, L. 2004. *Swimming to Antarctica: Tales of a Long-Distance Swimmer*. Harcourt, Orlando.

Creer, S., R. Van Dyke, and B. Newbold. 2007. Unmodified faunal remains from Goshen Island North. In J. C. Janetski and G. C. Smith., *Hunter-Gatherer Archaeology in Utah Valley*, pp. 301–310. Brigham Young University Museum of People and Cultures Occasional Papers 12.

Creer, S., and R. Van Dyke. 2007a. Unmodified faunal remains. In J. C. Janetski and G. C. Smith., *Hunter-Gatherer Archaeology in Utah Valley*, pp. 217–225. Brigham Young University Museum of People and Cultures Occasional Papers 12.

Creer, S., and R. Van Dyke. 2007b. Vertebrate faunal remains. In J. C. Janetski and G. C. Smith., *Hunter-Gatherer Archaeology in Utah Valley*, pp. 156–166. Brigham Young University Museum of People and Cultures Occasional Papers 12.

Crégut-Bonnoure, E. 1996. Ordre des carnivores. In C. Guérin and M. Patou-Mathis, eds., *Les Grands Mammifères Plio-Pléistocènes d'Europe*, pp. 155–230. Masson, Paris.

Cressman, L. S. 1936. *Archaeological Survey of the Guano Valley Region in Southeastern Oregon*. University of Oregon Monographs, Studies in Anthropology 1.

Cressman, L. S. 1937. *Petroglyphs of Oregon*. University of Oregon Monographs, Studies in Anthropology 2.

Cressman, L. S. 1942. *Archaeological Researches in the Northern Great Basin*. Carnegie Institute of Washington Publication 538.

Cressman, L. S. 1951. Western prehistory in the light of carbon 14 dating. *Southwestern Journal of Anthropology* 7:289–313.

Cressman, L. S. 1964. *The Sandal and the Cave*. Beaver Books, Portland, OR.

Cressman, L. S. 1966. Man in association with extinct fauna in the Great Basin. *American Antiquity* 31:866–867.

Cressman, L. S. 1977. *Prehistory of the Far West: Homes of Vanished Peoples*. University of Utah Press, Salt Lake City.

Cressman, L. S. 1986. Prehistory of the northern area. In W. L. d'Azevedo, ed., *Great Basin, Handbook of North American Indians, Volume 11*, pp. 120–126. Smithsonian Institution Press, Washington, DC.

Cressman, L. S. 1988. *A Golden Journey: Memoirs of an Archaeologist*. University of Utah Press, Salt Lake City.

Cressman, L. S., H. Williams, and A. Krieger. 1940. *Early Man in Oregon: Archaeological Studies in the Northern Great Basin*. University of Oregon Monographs, Studies in Anthropology 2.

Critchfield, W. B., and G. L. Allenbaugh. 1969. The distribution of Pinaceae in and near northern Nevada. *Madroño* 19:12–26.

Cronquist, A., A. H. Holmgren, N. H. Holmgren, and J. L. Reveal. 1972. *Intermountain Flora: Vascular Plants of the Intermountain West, U.S.A.*, Volume 1. Hafner, New York.

Crosby, A. W. 1986. *Ecological Imperialism: The Biological Expansion of Europe, 900–1900*. Cambridge University Press, Cambridge, UK.

Crowley, T. J. 2000. Causes of climate change over the past 1000 years. *Science* 289:270–277.

Crowley, T. J., and T. S. Lowery. 2000. How warm was the Medieval Warm Period? *Ambio* 29:51–54.

Crowley, T. J., G. Zielinski, B. Vinther, R. Udisti, H. Kreutz, J. Cole-Dai, and E. Castellano. 2008. Volcanism and the Little Ice Age. *PAGES News* 16(2):22–23.

Crum, S. J. 1994. *The Road on Which We Came: A History of the Western Shoshone*. University of Utah Press, Salt Lake City.

Crum, S. J. 1999. Julian Steward's vision of the Great Basin: A critique and response. In R. O. Clemmer, L. D. Myers, and M. E. Rudden, eds., *Julian Steward and the Great Basin: The Making of an Anthropologist*, pp. 117–127. University of Utah Press, Salt Lake City.

Csuti, B., A. J. Kimerling, T. A. O'Neil, M. M. Shaugnessy, E. P Gaines, and M. M. P. Huso. 1997. *Atlas of Oregon Wildlife*. Oregon State University Press, Corvallis.

Curran, H. 1982. *Fearful Crossing: The Central Overland Trail through Nevada*. Great Basin Press, Reno.

Currey, D. R. 1980. Coastal geomorphology of Great Salt Lake and vicinity. In J. W. Gwynn, ed., *Great Salt Lake: A Scientific, Historical, and Economic Overview*, pp. 69–82. Utah Geological and Mineral Survey Bulletin 116.

Currey, D. R. 1983. *Lake Bonneville: Selected Features of Relevance to Neotectonic Analysis*. U.S. Geological Survey Open-File Report 82–1070.

Currey, D. R. 1987. Great Salt Lake levels: Holocene geomorphic development and hydrographic history. In *Third Annual Landsat Workshop*, pp. 127–132. Laboratory for Terrestrial Physics, National Aeronautics and Space Administration Goddard Flight Center, Greenbelt, MD.

Currey, D. R. 1988. Isochronism of final Pleistocene lakes in the Great Salt Lake and Carson Desert region of the Great Basin. *Programs and Abstracts of the Tenth Biennial Meeting of the American Quaternary Association*, p. 117.

Currey, D. R. 1990. Quaternary paleolakes in the evolution of semidesert basins, with special emphasis on Lake Bonneville and the Great Basin, U.S.A. *Palaeogeography, Palaeoclimatology, Palaeoecology* 76:189–214.

Currey, D. R., and C. G. Oviatt. 1985. Durations, average rates, and probable causes of Bonneville expansion, stillstands, and contractions during the last deep-lake cycle, 32,000 to 10,000 years ago. In P. A. Kay and H. F. Diaz, eds., *Problems of and Prospects for Predicting Great Salt Lake Levels*, pp. 9–24. Center for Public Affairs and Administration, University of Utah, Salt Lake City.

Currey, D. R., G. Atwood, and D. R. Mabey. 1984. *Major Levels of Great Salt Lake and Lake Bonneville*. Utah Geological and Mineral Survey Map 73.

Cushman, K. A. 1982. Floral remains from Meadowcroft Rockshelter, Washington County, southwestern Pennsylvania. In R. C. Carlisle and J. M. Adovasio, eds., *Meadowcroft: Collected Papers on the Archaeology of Meadowcroft Rockshelter and the Cross Creek Drainage*, pp. 207–220. OCLC 10395512.

Cutting, L. 2007. Status of restoration in the Mono Basin. *Mono Lake Newsletter* (Spring):5–9.

Czaplewski, N. J., J. I. Mead, C. J. Bell, W. D. Peachey, and T-L. Ku. 1999. Pagapo Springs Cave revisited, Part II: Vertebrate paleofauna. *Oklahoma Museum of Natural History Occasional Papers* 5:1–41.

Dahlgren, R. A., J. H. Richards, and Z. Yu. 1997. Soil and groundwater chemistry and vegetation distribution in a desert playa, Owens Lake, California. *Arid Soil Research and Rehabilitation* 11:221–244.

Dalley, G. F. 1976. *Swallow Shelter and Associated Sites*. University of Utah Anthropological Papers 96.

Dalquest, W. W., and F. B. Stangl, Jr. 1984. Late Pleistocene and early Recent mammals from Fowlkes Cave, southern Culberson County, Texas. In H. H. Genoways and M. R. Dawson, M.

R., eds., *Contributions in Quaternary Vertebrate Paleontology: A Volume in Memorial to John E. Guilday*, pp. 432–455. Carnegie Museum of Natural History Special Publications 8.

Dansie, A. J. 1980. Notes on 26Pe118 fauna. In S. M. Seck, The archaeology of Trego Hot Springs: 26Pe118, pp. 196–209. M.A. thesis, University of Nevada, Reno.

Dansie, A. J. 1987. Archaeofauna from 26PE670. In M. K. Rusco and J. O. Davis, eds., *Studies in Archaeology, Geology and Paleontology at Rye Patch Reservoir, Pershing County, Nevada*, pp. 69–73. Nevada State Museum Anthropological Papers 20.

Dansie, A. J. 1997. Early Holocene burials in Nevada: Overview of localities, research, and legal issues. *Nevada Historical Society Quarterly* 40:4–14.

Dansie, A. J., and W. J. Jerrems. 2005. More bits and pieces: A new look at Lahontan Basin chronology and human occupation. In R. Bonnichsen, B. T. Lepper, D. Stanford, and M. R. Waters, eds., *Paleoamerican Origins: Beyond Clovis*, pp. 51–80. Center for the Study of the First Americans; Texas A & M University Press, College Station.

Dansie, A. J., J. O. Davis, and T. W. Stafford, Jr. 1988. The Wizards Beach Recession: Farmdalian (25,500 yr B.P.) vertebrate fossils co-occur with early Holocene artifacts. In J. A. Willig, C. M. Aikens, and J. L. Fagan, eds., *Early Human Occupation in Far Western North America: The Clovis-Archaic Interface*, pp. 153–200. Nevada State Museum Anthropological Papers 21.

Daulton, T. L., N. Pinter, and A. C. Scott. 2010. No evidence of nanodiamonds in Younger-Dryas sediments to support an impact event. *Proceedings of the National Academy of Sciences* 107:16043–16047.

Davis, C. A., and G. A. Smith. 1981. *Newberry Cave*. San Bernardino County Museum Association, Redlands, CA.

Davis, C. A., R. E. Taylor, and G. A. Smith. 1981. New radiocarbon determinations from Newberry Cave. *Journal of California and Great Basin Anthropology* 3:144–147.

Davis, E. L. 1978a. Associations of people and a Rancholabrean fauna at China Lake, California. In A. L. Bryan, ed., *Early Man in America from a Circum-Pacific Perspective*, pp. 183–217. Department of Anthropology, University of Alberta Occasional Papers 1.

Davis, E. L., ed. 1978b. *The Ancient Californians: Rancholabrean Hunters of the Mojave Lakes Country*. Natural History Museum of Los Angeles County Science Series 29.

Davis, J. O. 1978. *Quaternary Tephrochronology of the Lake Lahontan Area, Nevada and California*. Nevada Archaeological Survey Research Paper 7.

Davis, J. O. 1982. Bits and pieces: The last 35,000 years in the Lahontan area. In D. B. Madsen and J. F. O'Connell, eds., *Man and Environment in the Great Basin*, pp. 53–75. Society for American Archaeology Papers 2.

Davis, J. O. 1983. Level of Lake Lahontan during deposition of the Trego Hot Springs tephra about 23,400 years ago. *Quaternary Research* 19:312–324.

Davis, J. O. 1985a. Correlation of late Quaternary tephra layers in a long pluvial sequence near Summer Lake, Oregon. *Quaternary Research* 23:38–53.

Davis, J. O. 1985b. Sediments and geological setting of Hidden Cave. In D. H. Thomas, ed., *The Archaeology of Hidden Cave, Nevada*, pp. 80–103. Anthropological Papers of the American Museum of Natural History 61(1).

Davis, J. O. 1987. Geology at Rye Patch. In M. K. Rusco and J. O. Davis, eds., *Studies in Archaeology, Geology and Paleontology at Rye Patch Reservoir, Pershing County, Nevada*, pp. 9–22. Nevada State Museum Anthropological Papers 20.

Davis, J. O., and M. K. Rusco. 1987. The Old Humboldt Site—26Pe670. In M. K. Rusco and J. O. Davis, eds., *Studies in Archaeology, Geology and Paleontology at Rye Patch Reservoir, Pershing County, Nevada*, pp. 41–73. Nevada State Museum Anthropological Papers 20.

Davis, J. O., R. Elston, and G. Townsend. 1976. Coastal geomorphology of the south shore of Lake Tahoe. In R. Elston, ed., *Holocene Environmental Change in the Great Basin*, pp. 40–65. Nevada Archaeological Survey Research Paper 6.

Davis, J. O., W. N. Melhorn, D. T. Trexler, and D. H. Thomas. 1983. Geology of Gatecliff Shelter: Physical stratigraphy. In D. H. Thomas, *The Archaeology of Monitor Valley. 2. Gatecliff Shelter*, pp. 39–63. American Museum of Natural History Anthropological Papers 59(1).

Davis, O. K. 1987. Spores of the dung fungus *Sporormiella*: Increased abundance in historic sediments and before Pleistocene megafaunal extinction. *Quaternary Research* 28: 290–294.

Davis, O. K. 1990. Caves as sources of biotic remains in arid western North America. *Palaeogeography, Palaeoclimatology, Palaeoecology* 76:331–348.

Davis, O. K. 1998. Palynological evidence for vegetation cycles in a 1.5 million year pollen record from the Great Salt Lake, Utah, U.S.A. *Palaeogeography, Palaeoclimatology, Palaeoecology* 138:175–185.

Davis, O. K. 1999a. Pollen analysis of a late-glacial and Holocene sediment core from Mono Lake, Mono County, California. *Quaternary Research* 52:243–249.

Davis, O. K. 1999b. Pollen analysis of Tulare Lake, California: Great Basin–like vegetation in central California during the full-glacial and early Holocene. *Review of Paleobotany and Palynology* 107:249–257.

Davis, O. K. 2001. Palynology: An important tool for discovering historic ecosystems. In D. Egan and E. A. Howell, eds., *The Historical Ecology Handbook: A Restorationist's Guide to Reference Ecosystems*, pp. 229–255. Island Press, Washington, DC.

Davis, O. K. 2002. Late Neogene environmental history of the northern Bonneville Basin: A review of palynological studies. In R. Hershler, D. B. Madsen, and D. R. Currey, eds., *Great Basin Aquatic Systems History*, pp. 295–307. Smithsonian Contributions to the Earth Sciences 33.

Davis, O. K., and R. S. Anderson. 1987. Pollen in packrat (*Neotoma*) middens: Pollen transport and the relationship of pollen to vegetation. *Palynology* 11:185–198.

Davis, O. K., and M. J. Moratto. 1988. Evidence for a warm dry early Holocene in the western Sierra Nevada of California: Pollen and plant macrofossil analysis of Dinkey and Exchequer meadows. *Madroño* 35:132–149.

Davis, O. K., and D. S. Shafer. 1992. A Holocene climatic record for the Sonoran Desert from pollen analysis of Montezuma Well, Arizona, USA. *Palaeogeography, Palaeoclimatology, Palaeoecology* 92:107–119.

Davis, O. K., and D. S. Shafer. 2006. *Sporormiella* fungal spores, a palynological means of detecting herbivore density. *Palaeogeography, Palaeoclimatology, Palaeoecology* 237:40–50.

Davis, O. K., R. S. Anderson, P. L. Fall, M. K. O'Rourke, and R. S. Thompson. 1985. Palynological evidence for early Holocene aridity in the southern Sierra Nevada, California. *Quaternary Research* 24:322–332.

Davis, R., and J. R. Callahan. 1992. Post-Pleistocene dispersal in the Mexican vole (*Microtus mexicanus*): An example of an apparent trend in the distribution of southwestern mammals. *Great Basin Naturalist* 52, 262–268.

Davis, R., and C. Dunford. 1987. An example of contemporary colonization of montane islands by small nonflying mammals in the American southwest. *The American Naturalist* 129, 398–406.

Davis, W. E., D. Sack, and N. Shearin. 1996. The Hell'N Moriah Clovis site. *Utah Archaeology* 9:55–70.

Davis, W. H. 1929. *Seventy-five Years in California*. John Howell, San Francisco.

D'Azevedo, W. L., ed. 1986. *Great Basin, Handbook of North American Indians, volume 11*. Smithsonian Institution Press, Washington, DC.

Deacon, J. E., and M. S. Deacon. 1979. Research on endangered fishes in the National Parks with special emphasis on the Devils Hole pupfish. In R. M. Linn, ed., *First Conference on Scientific Research in National Parks*, pp. 9–19. National Park Service Transactions and Proceeding Series 5.

Deacon, J. E., and C. D. Williams. 1991. Ash Meadows and the legacy of the Devils Hole pupfish. In W. L. Minckley and J. E. Deacon, eds., *Battle Against Extinction: Native Fish Management in the American West*, pp. 69–91. University of Arizona Press, Tucson.

Deacon, J. E., A. E. Williams, C. D. Williams, and J. E. Williams. 2007. Fueling population growth in Las Vegas: How large-scale groundwater withdrawal could burn regional biodiversity. *BioScience* 57:688–698.

Dean, W. E., W. A. Wurtsbaugh, and V. A. Lamarra. 2009. Climatic and limnologic setting of Bear Lake, Utah and Idaho. In J. G. Rosenbaum and D. S. Kaufman, eds., *Paleoenvironments of Bear Lake, Utah and Idaho, and its Catchment*, pp. 1–14. Geological Society of America Special Paper 450.

DeBoer, W. R., 2001. Of dice and women: gambling and exchange in native North America. *Journal of Archaeological Method and Theory* 8:215–268.

DeBolt, A. M., and B. McCune. 1995. Ecology of *Celtis reticulata* in Idaho. *Great Basin Naturalist* 55:237–248.

DeDecker, M. 1984. *Flora of the Northern Mojave Desert, California*. California Native Plant Society Special Publication 7.

DeDecker, M. 1991. Shrubs and flowering plants. In C. A. Hall, Jr., ed., *Natural History of the White-Inyo Range*, pp. 108–241. University of California Press, Berkeley.

Delacorte, M. G. 1990. The prehistory of Deep Springs Valley, Eastern California: Adaptive variation in the western Great Basin. Ph.D. dissertation, University of California, Davis.

Delacorte, M. G. 1997. *Culture Change along the Eastern Sierra Nevada/Cascade Front. Volume 1: History of Investigations and Summary of Findings*. Coyote Press, Salinas, CA.

Delorme, L. D., and S. C. Zoltai. 1984. Distribution of an Arctic ostracode fauna in space and time. *Quaternary Research* 21:65–74.

Delpech, F. 1983. *Les Faunes du Paléolithique Supérieur dans le Sud-Ouest de la France*. Cahiers du Quaternaire 6.

Delpech, F. 1999. Biomasse d'ongulés au Paléolithique et inférences démographiques. *Paléo* 11:19–42.

DeLucia, E. H., and W. H. Schlesinger. 1990. Ecophysiology of Great Basin and Sierra Nevada vegetation on contrasting soils. In C. B. Osmond, L. F. Pitelka, and G. M. Hidy, eds., *Plant Biology of the Basin and Range*, pp. 143–178. Springer-Verlag, Berlin.

DeLucia, E. H., and W. H. Schlesinger. 1991. Resource-use efficiency and drought tolerance in adjacent Great Basin and Sierran plants. *Ecology* 72:51–58.

DeLucia, E. H., W. H. Schlesinger, and D. W. Billings. 1988. Water relations and the maintenance of Sierran conifers on hydrothermally altered rock. *Ecology* 69:303–311.

DeLucia, E. H., W. H. Schlesinger, and D. W. Billings. 1989. Edaphic limitations to growth and photosynthesis in Sierran and Great Basin vegetation. *Oecologia* 78:184–190.

DeMay, I. S. 1941a. Pleistocene bird life of the Carpinteria asphalt, California. *Carnegie Institute of Washington Publication* 530:61–66.

DeMay, I. S. 1941b. Quaternary bird life of the McKittrick asphalt, California. *Carnegie Institute of Washington Publication* 530:35–60.

Denton, G. H., and W. S. Broecker. 2008. Wobbly ocean conveyor circulation during the Holocene? *Quaternary Science Reviews* 27:1939–1950.

Denton, G. H., R. F. Anderson, J. R. Toggweiler, R. L. Edwards, J. M. Schaefer, and A. E. Putnam. 2010. The last glacial termination. *Science* 328:1652–1656.

Denton, S. 2003. *American Massacre: The Tragedy at Mountain Meadows, September 1857*. Alfred A. Knopf, New York.

DePolo, D. M., and C. M. dePolo. 1999. *Earthquakes in Nevada, 1852–1998*. Nevada Bureau of Mines and Geology Map 119.

Dial, K. P., and N. J. Czaplewski. 1990. Do woodrat middens accurately represent the animals' environments and diets? The Woodhouse Mesa Study. In J. L. Betancourt, T. R. Van Devender, and P. S. Martin, eds., *Packrat Middens: The Last 40,000 Years of Biotic Change*, pp. 43–58. University of Arizona Press, Tucson.

Diamond, J. M. 1992. *The Third Chimpanzee: The Evolution and Future of the Human Animal*. HarperCollins, New York.

Dicken, S. N. 1980. Pluvial Lake Modoc, Klamath County, Oregon, and Modoc and Siskyou Counties, California. *Oregon Geology* 42:179–187.

Digonnet, M. 2007. *Hiking Death Valley: A Guide to Its Natural Wonders and Mining Past*. M. Digonnet, Palo Alto, CA.

Dillehay, T. D. 1989. *Monte Verde: A Late Pleistocene Settlement in Chile*, vol. 1, *Paleoenvironment and Site Context*. Smithsonian Institution Press, Washington, DC.

Dillehay, T. D. 1997. *Monte Verde: A Late Pleistocene Settlement in Chile*, vol. 2, *The Archaeological Context and Interpretation*. Smithsonian Institution Press, Washington, DC.

Dillehay, T. D., C. Ramírez, M. Pino, M. B. Collins, J. Rossen, and J. D. Pino-Navarro. 2008. Monte Verde: Seaweed, food, medicine, and the peopling of South America. *Science* 320:784–786.

Dively-White, D. V. 1989. Late Quaternary zoogeography of the Toano Range based on Mad Chipmunk Cave, Elko County, Nevada. *Journal of Vertebrate Paleontology* 9(3), 19A.

Dively-White, D. V. 1990. Mad Chipmunk Cave faunule: A late Pleistocene-late Holocene record for the north-central Great Basin. *Journal of Vertebrate Paleontology* 10(3), 20A–21A.

Division of Water Resources, State of Nevada. 1972. Water resources and inter-basin flows. In *Water for Nevada: Hydrologic Atlas*, Map S-13. Department of Water Resources, State of Nevada, Carson City.

Dixon, S. L., and R. L. Lyman. 1996. On the Holocene history of elk (*Cervus elaphus*) in eastern Washington. *Northwest Science* 70:262–273.

Dobler, F. C., and K. R. Dixon. 1990. The pygmy rabbit. In J. A. Chapman and J. E. C. Flux, eds., *Rabbits, Hares, and Pikas: Status Survey and Conservation Action Plan*, pp. 111–115. IUCN, Gland.

Dohrenwend, J. C. 1984. Nivation landforms in the western Great Basin and their paleoclimatic significance. *Quaternary Research* 22:275–288.

Douglas, C. L., D. L. Jenkins, and C. N. Warren. 1988. Spatial and temporal variability in faunal remains from four Lake Mojave-Pinto period sites in the Mojave Desert. In J. A. Willig, C. M. Aikens, and J. L. Fagan, eds., *Early Human Occupation in Far Western North America: The Clovis-Archaic Interface*, pp. 131–144. Nevada State Museum Anthropological Papers 21.

Downs, T., H. Howard, T. Clements, and G. A. Smith. 1959. *Quaternary Animals from Schuiling Cave in the Mojave Desert, California*. Los Angeles City Museum Contributions in Science 29.

Drews, M. P. 1988. Freshwater molluscs. In C. Raven and R. G. Elston, eds., *Preliminary Investigations in Stillwater Marsh: Human Prehistory and Geoarchaeology*, pp. 328–339. U.S. Department of the Interior, U.S. Fish and Wildlife Service, Region 1, Cultural Resource Series 1.

Driver, H. E., and W. C. Massey 1957. *Comparative Studies of North American Indians*. Transactions of the American Philosophical Society 47(2).

Droz, M. S. 1997. Geomorphic and climatic history of Holocene channel, playas, and lunettes in the Fort Rock Basin, Lake County, Oregon. M.A. thesis, University of Oregon, Eugene.

Dugas, D. P. 1998. Late Quaternary variations in the level of Paleo–Lake Malheur, eastern Oregon. *Quaternary Research* 50:276–282.

Duke, D. G., and D. C. Young. 2007. Episodic permanence in Paleoarchaic basin selection and settlement. In K. E. Graf and D. N. Schmitt, eds., *Paleoindian or Paleoarchaic: Great Basin Human Ecology at the Pleistocene-Holocene Transition*, pp. 123–138. University of Utah Press, Salt Lake City.

Duk-Rodkin, A., R. W. Barendregt, D. G. Froese, F. Weber, R. Enkin, I. R. Smith, G. D. Zazula, P. Waters, and R. Klassen. 2004. Timing and extent of Plio-Pleistocene glaciations in northwestern Canada and east-central Alaska. In J. Ehlers and P. L Gibbard, eds., *Quaternary Glaciations: Extent and Chronology, Part 2: North America*, pp. 313–346. Developments in Quaternary Science 2. Elsevier, Amsterdam.

Dundas, R. G. 1999. Quaternary records of the dire wolf, *Canis dirus*, in North and South America. *Boreas* 28:375–385.

Dunlay, T. 2000. *Kit Carson and the Indians*. University of Nebraska Press, Lincoln.

Durrant, S. D. 1952. *Mammals of Utah: Taxonomy and Distribution*. University of Kansas Publications, Museum of Natural History 6.

Durrant, S. D. 1970. Faunal remains as indicators of Neothermal climates at Hogup Cave. In C. M. Aikens, *Hogup Cave*, pp. 241–245. University of Utah Anthropological Papers 93.

Dyke, A. S. 2004. An outline of North American deglaciation with emphasis on central and northern Canada. In J. Ehlers and P. L. Gibbard, eds., *Quaternary Glaciations: Extent and Chronology*, Part 2, *North America*, pp. 373–424. Developments in Quaternary Science 2. Elsevier, Amsterdam.

Dyke, A. S., D. Giroux, and L. Robertson. 2004. *Paleovegetation Maps of Northern North America*. Geological Survey of Canada Open File 4682.

Dyke, A. S., A. Moore, and L. Robertson. 2003. *Deglaciation of North America*. Geological Survey of Canada Open File 1574.

Dyke, A. S., J. T. Andrews, P. U. Clark, J. H. England, G. H. Miler, J. Shaw, and J. J. Veillette. 2002. The Laurentide and Innuitian Ice Sheets during the Last Glacial Maximum. *Quaternary Science Reviews* 21:9–31.

Echelle, A. A. 2008. The western North American pupfish clade (Cyprinodontidae: *Cyprinodon*): Mitochondrial DNA divergence and drainage history. In M. C. Reheis, R. Hershler, and D. M. Miller, eds., *Late Cenozoic Drainage History of the Southwestern Great Basin and Lower Colorado River Region: Geologic and Biotic Perspectives*, pp. 27–38. Geological Society of America Special Paper 439.

Echelle, A. A., and A. F. Echelle. 1993. Allozyme perspective on mitochondrial DNA variation and evolution of the Death Valley pupfishes (Cyprinodontidae: *Cyprinodon*). *Copeia* 1993:275–287.

Echelle, A. A., E. W. Carson, A. F. Echelle, R. A. Van Den Bussche, T. E. Dowling, and A. Meyer. 2005. Historical biogeography of the New-World pupfish genus *Cyprinodon* (Teleostei: Cyprinodontidae). *Copeia* 2005:320–339.

Echlin, D. R., P. J. Wilke, and L. E. Dawson. 1981. Ord Shelter. *Journal of California and Great Basin Anthropology* 3, 49–68.

Edgar, H. J. H. 1997. Paleopathology of the Wizards Beach Man (AHUR 2023) and the Spirit Cave mummy (AHUR 2064). *Nevada Historical Society Quarterly* 40:57–61.

Edgar, H. J. H., E. A Jolie, J. F. Powell, and J. E. Watkins. 2007. Contextual issues in Paleoindian repatriation: Spirit Cave Man as a case study. *Journal of Social Archaeology* 7:101–122.

Edwards, M. E., L. B. Brubaker, A. V. Lozhkin, and P. M. Anderson. 2005. Structurally novel biomes: A response to past warming in Beringia. *Ecology* 86:1696–1703.

Eerkens, J. W. 2003a. Residential mobility and pottery use in the western Great Basin. *Current Anthropology* 44:728–738.

Eerkens, J. W. 2003b. Sedentism, storage, and the intensification of small seeds: Prehistoric developments in Owens Valley, California. *North American Archaeologist* 24:281–309.

Eerkens, J. W. 2003c. Towards a chronology of brownware pottery in the Western Great Basin: a case study from Owens Valley. *North American Archaeologist* 24:1–27.

Eerkens, J. W. 2004. Privatization, small-seed intensification, and the origins of pottery in the western Great Basin. *American Antiquity* 69:653–670.

Eerkens, J. W. 2005. GC-MS analysis and fatty acid ratios of archaeological potsherds from the western Great Basin of North America. *Archaeometry* 47:83–102.

Eerkens, J. W., J. King, and E. Wohlgemuth. 2004. The prehistoric development of intensive green-cone piñon processing in eastern California. *Journal of Field Archaeology* 29:17–27.

Eerkens, J. W., H. Neff, and M. Glascock. 1999. Early pottery from Sun'gava and implications for the development of ceramics in Owens Valley. *Journal of California and Great Basin Anthropology* 21:275–285.

Eerkens, J. W., H. Neff, and M. D. Glascock. 2002a. Ceramic production among small-scale and mobile hunters and gatherers: A case study from the southwestern Great Basin. *Journal of Anthropological Archaeology* 21:200–229.

Eerkens, J. W., H. Neff, and M. D. Glascock. 2002b. Typologies and classification of Great Basin pottery: A new look at Death Valley brownwares. In D. M. Glowacki and H. Neff, eds., *Ceramic Production and Circulation in the Greater Southwest: Source Determination by INAA and Complementary Mineralogical Investigations*, pp. 140–151. Cotsen Institute of Archaeology, Monograph 44.

Eerkens, J. W., J. S. Rosenthal, D. C. Young, and J. King. 2007. Early Holocene landscape archaeology in the Coso Basin, northwestern Mojave Desert, California. *North American Archaeologist* 28:87–112.

Egan, F. 1985. *Frémont: Explorer for a Restless Nation*. University of Nevada Press, Reno.

Ehlers, J. 1996. *Quaternary and Glacial Geology*. Wiley, Chichester.

Eiselt, B. S. 1997a. *Defining Ethnicity in Warner Valley: An Analysis of House and Home*. University of Nevada, Reno, Department of Anthropology Technical Report 97-2.

Eiselt, B. S. 1997b. Fish remains from the Spirit Cave paleofecal material: 9,400 year old evidence for Great Basin utilization of small fishes. *Nevada Historical Society Quarterly* 40:117–139.

Eiselt, B. S. 1998. *Household Activity and Marsh Utilization in the Archaeological Record of Warner Valley: The Peninsula Site*. University of Nevada, Reno, Department of Anthropology Technical Report 98-2.

Eisenstein, E. L. 1979. *The Printing Press as an Agent of Change*. Cambridge University Press, Cambridge.

Elftman, H. O. 1931. *Pleistocene Mammals of Fossil Lake, Oregon*. American Museum Novitates 481.

Elias, S. A., S. K. Short, and H. H. Birks. 1997. Late Wisconsin environments of the Bering Land Bridge. *Palaeogeography, Palaeoclimatology, and Palaeoecology* 136:293–308.

Elias, S. A., S. K. Short, C. H. Nelson, and H. H. Birks. 1996. Life and times of the Bering Land Bridge. *Nature* 382:60–63.

Elliot-Fisk, D. L. 1987. Glacial geomorphology of the White Mountains, California and Nevada: Establishment of a glacial chronology. *Physical Geography* 8:299–323.

Elston, R. G. 2005. Flaked- and battered-stone artifacts. In D. N. Schmitt and D. B. Madsen, *Camels Back Cave*, pp. 92–119. University of Utah Anthropological Papers 125.

Elston, R. G., and E. E. Budy, eds. 1990. *The Archaeology of James Creek Shelter*. University of Utah Anthropological Papers 115.

Elston, R. G., and C. Raven. 1992. *Archaeological Investigations at Tosawihi, A Great Basin Quarry. Part I: The Periphery*. Intermountain Research, Silver City, NV.

Elston. R. G., and D. W. Zeanah. 2002. Thinking outside the box: A new perspective on diet breadth and sexual division of labor in the prearchaic Great Basin. *World Archaeology* 34:103–130.

Elston, R. G., K. L. Katzer, and D. R. Currey. 1988. Chronological summary. In C. Raven and R. G. Elston, eds., *Preliminary Investigations in Stillwater Marsh: Human Prehistory and Geoarchaeology*, pp. 373–384. U.S. Department of the Interior, U.S. Fish and Wildlife Service, Region 1, Cultural Resource Series 1.

Embleton, S. 1991. Mathematical methods of genetic classification. In S. M. Lamb and E. D. Mitchell, eds., *Sprung from Some Common Source*, pp. 365–388. Stanford University Press, Stanford, CA.

Emslie, S. D. 1985. The late Pleistocene (Rancholabrean) avifauna of Little Box Elder Cave, Wyoming. *Contributions to Geology, University of Wyoming* 23:63–82.

Emslie, S. D. 1987. Age and diet of fossil California Condors in Grand Canyon, Arizona. *Science* 237:768–770.

Emslie, S. D. 1988. The fossil history and phylogenetic relationships of condors (Ciconiiformes: Vulturidae) in the New World. *Journal of Vertebrate Paleontology* 8:212–228.

Emslie, S. D. 1990. Additional ^{14}C dates of fossil California Condors. *National Geographic Research* 6:134–135.

Emslie, S. D. 1998. *Avian Community, Climate, and Sea-level Changes in the Plio-Pleistocene of the Florida Peninsula*. Ornithological Monographs 50.

Emslie, S. D., and N. J. Czaplewski. 1985. A new record of giant short-faced bear, *Arctodus simus*, from western North America with a re-evaluation of its paleobiology. *Natural History Museum of Los Angeles County Contributions in Science* 731:1–12.

Emslie, S. D., and N. J. Czaplewski. 1999. Two new fossil eagles from the late Pliocene (late Blancan) of Florida and Arizona and their biogeographic implications. In P. Wellnhofer, C. Mourer-Chauviré, D. W. Steadman, and L. D. Martin, eds., *Avian Paleontology at the Close of the 20th Century: Proceedings of the 4th International Meeting of the Society of Avian Paleontology and Evolution, Washington, D.C., 4–7 June 1996*, pp. 185–198. Smithsonian Contributions to Paleobiology 89.

Emslie, S. D., and T. H. Heaton. 1987. The late Pleistocene avifauna of Crystal Ball Cave, Utah. *Journal of the Arizona-Nevada Academy of Science* 21:53–60.

England, J., N. Atkinson, J. Bednarski, A. S. Dyke, D. A. Hodgson, and C. Ó. Cofaigh. 2006. The Innuitian Ice Sheet: Configuration, dynamics, and chronology. *Quaternary Science Reviews* 25:689–703.

Enzel, Y., S. G. Wells, and N. Lancaster. 2003a. Late Pleistocene lakes along the Mojave River, southeast California. In Y. Enzel, S. G. Wells, and N. Lancaster, eds., *Paleoenvironments and Paleohydrology of the Mojave and Southern Great Basin Deserts*, pp. 61–77. Geological Society of America Special Paper 368.

Enzel, Y., S. G. Wells, and N. Lancaster, eds. 2003b. *Paleoenvironments and Paleohydrology of the Mojave and Southern Great Basin Deserts*. Geological Society of America Special Paper 368.

Enzel, Y., D. R. Cayan, R. Y. Anderson, and S. G. Wells. 1989. Atmospheric circulation during Holocene lake stands in the Mojave Desert: Evidence of regional climatic change. *Nature* 341:44–47.

Enzel, Y., W. J. Brown, R. Y. Anderson, L. D. McFadden, and S. G. Wells. 1992. Short-duration Holocene lakes in the Mojave River drainage basin, southern California. *Quaternary Research* 38:60–73.

Erlandson, J. M. 2002. Anatomically modern humans, maritime voyaging, and the Pleistocene colonization of the Americas. In N. G. Jablonski, ed., *The First Americans: The Pleistocene Colonization of the Americas*, pp. 59–92. Memoirs of the California Academy of Science 27.

Eshleman, J. A. 2002. Mitochondrial DNA and prehistoric population movements in western North America. Ph.D. dissertation, University of California, Davis.

Eshleman, J. A., and D. G. Smith. 2007. Prehistoric mitochondrial DNA and population movements. In T. L. Jones and K. A. Klar, eds., *California Prehistory: Colonization, Culture, and Complexity*, pp. 291–298. AltaMira Press, Lanham, MD.

Eshleman, J. A., R. S. Malhi, and D. G. Smith. 2003. Mitochondrial DNA studies of Native Americans: Conceptions and misconceptions of the population prehistory of the Americas. *Evolutionary Anthropology* 12:7–18.

Everett, D. E., and F. E. Rush. 1967. *A Brief Appraisal of the Water Resources of the Walker Lake Area, Mineral, Lyon, and Churchill Counties, Nevada*. Nevada Department of Conservation and Natural Resources Water Resources-Reconnaissance Series Report 40.

Ewing, R. C., and F. N. von Hippel. 2009. Nuclear waste management in the United States: Starting over. *Science* 325:151–152.

Fagan, B. 2000. *The Little Ice Age*. Basic Books, New York.

Fagan, J. L. 1974. *Altithermal Occupation of Spring Sites in the Northern Great Basin*. University of Oregon Anthropological Papers 6.

Fagan, J. L. 1975. A supposed fluted point from Fort Rock Cave, an error of identification, and its consequences. *American Antiquity* 40:356–357.

Fagan, J. L. 1988. Clovis and western Pluvial Lakes Tradition lithic technologies at the Dietz site in south-central Oregon. In J. A. Willig, C. M. Aikens, and J. L. Fagan, eds., *Early Human*

Occupation in Far Western North America: The Clovis-Archaic Interface, pp. 389–416. Nevada State Museum Anthropological Papers 21.

Fairbanks, R. G., R. A. Mortlock, T.-C. Chiu, L. Cao, A. Kaplan, T. P. Guilderson, T. W. Fairbanks, and A. L. Bloom. 2005. Marine radiocarbon calibration curve spanning 0 to 50,000 Years B.P. based on paired ^{230}Th/^{234}U/^{238}U and ^{14}C dates on pristine corals. *Quaternary Science Reviews* 24:1781–1796.

Faith, J. T., and T. A. Surovell. 2009. Synchronous extinction of North America's Pleistocene mammals. *Proceedings of the National Academy of Sciences* 106:20641–20645.

Farlow, J. O., and J. McClain. 1996. A spectacular specimen of the elk-moose *Cervalces scotti* from Noble County, Indiana, U.S.A. In K. M. Stewart and K. L. Seymour, eds., *Palaeoecology and Palaeoenvironments of Late Cenozoic Mammals: Tributes to the Career of C. S. (Rufus) Churcher*, pp. 322–330. University of Toronto Press, Toronto.

FAUNMAP Working Group. 1994. *FAUNMAP: A Database Documenting Late Quaternary Distributions of Mammal Species in the United States*. Illinois State Museum Scientific Papers 25.

Fawcett, W. B., and S. R. Simms. 1993. *Archaeological Test Excavations in the Great Salt Lake Wetlands and Associated Analyses, Weber and Box Elder Counties, Utah*. Utah State University Anthropological Papers 14.

Feathers, J. 2003. Use of luminescence dating in archaeology. *Measurement Science and Technology* 14:1493–1509.

Feathers, J., and D. Rhode. 1998. Luminescence dating of protohistoric pottery from the Great Basin. *Geoarchaeology* 13:287–308.

Fedje, D. W., and H. Josenhans. 2000. Drowned forests and archaeology on the continental shelf of British Columbia, Canada. *Geology* 28:99–102.

Fedje, D. W., Q. Mackie, E. J. Dixon, and T. H. Heaton. 2004. Late Wisconsin environments and archaeological visibility on the northern Northwest Coast. In D. B. Madsen, ed., *Entering America: Northeast Asia and Beringia before the Last Glacial Maximum*, pp. 97–138. University of Utah Press, Salt Lake City.

Fenenga, F., and F. A. Riddell. 1949. Excavation of Tommy Tucker Cave, Lassen County, California. *American Antiquity* 14:203–214.

Feng, X., and S. Epstein. 1994. Climatic implications of an 8000-year hydrogen isotope time series from bristlecone pine trees. *Science* 265:1079–1080.

Ferring, C. R. 2001. *The Archaeology and Paleoecology of the Aubrey Clovis Site (41DN479), Denton County, Texas*. Center for Environmental Archaeology, Department of Geography, University of North Texas, Denton.

Feth, J. H. 1959. Re-evaluation of the salt chronology of several Great Basin lakes: A discussion. *Geological Society of America Bulletin* 70:637–640.

Finley, R. B., Jr. 1990. Woodrat ecology and behavior and the interpretation of paleomiddens. In J. L. Betancourt, T. R. Van Devender, and P. S. Martin, eds., *Packrat Middens: The Last 40,000 Years of Biotic Change*, pp. 28–42. University of Arizona Press, Tucson.

Firestone, R. B., Jr., and W. Topping. 2001. Terrestrial evidence of a nuclear catastrophe in Paleoindian times. *The Mammoth Trumpet* 16(2):9–16.

Firestone, R. B., et al. (26 authors). 2007. Evidence for an extraterrestrial impact 12,900 years ago that contributed to the megafaunal extinctions and the Younger Dryas cooling. *Proceedings of the National Academy of Sciences* 104:16016–16021.

Fisher, D. C. 1984. Mastodon butchery by North American Paleo-Indians. *Nature* 308:271–272.

Fisher, F. I. 1945. Locomotion in the fossil vulture *Teratornis*. *American Midland Naturalist* 33:725–742.

Fisher, J. L. 2010. Costly signaling and changing faunal abundances at Five Finger Ridge, Utah. Ph.D. dissertation, University of Washington, Seattle.

Fitzpatrick, J. A., and J. L. Bischoff. 1993. *Uranium-series Dates on Sediments of the High Shoreline of Panamint Valley, California*. U.S. Geological Survey Open-File Report 93-232.

Fladmark, K. R. 1979. Routes: Alternative migration corridors for early man in North America. *American Antiquity* 44:55–69.

Fleishman, E., D. D. Murphy, and G. T. Austin. 1999. Butterflies of the Toquima Range, Nevada: Distribution, natural history, and comparison to the Toiyabe Range. *Great Basin Naturalist* 59:50–62.

Fleishman, E., G. T. Austin, and D. D. Murphy. 2001. Biogeography of Great Basin butterflies: Revisiting patterns, paradigms, and climate change scenarios. *Biological Journal of the Linnean Society* 74:501–515.

Flenniken, J. J., and P. J. Wilke. 1989. Typology, technology, and chronology of Great Basin dart points. *American Anthropologist* 91:149–158.

Floyd, C. H., D. H. Van Vuren, and B. May. 2005. Marmots on Great Basin mountaintops: Using genetics to test a biogeographic paradigm. *Ecology* 86:2145–2153.

Floyd, T., C. S. Elphick, G. Chisholm, K. Mack, R. G. Elston, E. M. Ammon, and J. D. Boone. 2007. *Atlas of the Breeding Birds of Nevada*. University of Nevada Press, Reno.

Forester, R. M. 1987. Late Quaternary paleoclimatic records from lacustrine ostracodes. In W. F. Ruddiman and H. E. Wright, Jr., eds., *North America and Adjacent Oceans during the Last Deglaciation*, pp. 261–276. The Geology of North America, Volume K-3. Geological Society of America, Boulder. CO.

Forester, R. M., J. P. Bradbury, C. Carter, A. B. Elvidge-Tuma, M. L. Hemphill, S. C. Lundstrom, S. A. Mahan, B. D. Marshall, L. A. Neymark, J. B. Paces, S. E. Sharpe, J. F. Whelan, and P. E. Wigand. 1999. *The Climatic and Hydrologic History of Southern Nevada during the Late Quaternary*. U.S. Geological Survey Open-File Report 98–635.

Fowler, C. S. 1982. Settlement patterns and subsistence systems in the Great Basin: The ethnographic record. In D. B. Madsen and J. F. O'Connell, eds., *Man and Environment in the Great Basin*, pp. 121–138. Society for American Archaeology Papers 2.

Fowler, C. S. 1983. Some lexical clues to Uto-Aztecan prehistory. *International Journal of American Linguistics* 49:224–257.

Fowler, C. S. 1986. Subsistence. In W. L. d'Azevedo, ed., *Great Basin, Handbook of North American Indians, Volume 11*, pp. 64–97. Smithsonian Institution Press, Washington, DC.

Fowler, C. S., ed. 1989. *Willard Z. Park's Ethnographic Notes on the Northern Paiute of western Nevada, 1933–1944*. University of Utah Anthropological Papers 114.

Fowler, C. S. 1990a. Ethnographic perspectives on marsh-based cultures in western Nevada. In J. C. Janetski and D. B. Madsen, eds., *Wetland Adaptations in the Great Basin*, pp. 17–32. Brigham Young University Museum of Peoples and Cultures, Occasional Papers 1.

Fowler, C. S. 1990b. *Tule Technology. Northern Paiute Uses of Marsh Resources in Western Nevada*. Smithsonian Folklife Studies 6.

Fowler, C. S. 1992. *In the Shadow of Fox Peak: An Ethnography of the Cattail-Eater Northern Paiute People of Stillwater Marsh*. U.S. Department of the Interior, U.S. Fish and Wildlife Service, Region 1, Cultural Resource Series 5.

Fowler, C. S. 1994a. Beginning to understand: Twenty-eight years of fieldwork in the Great Basin of western North America. In D. D. Fowler and D. L. Hardesty, eds., *Others Knowing Others: Perspectives on Ethnographic Careers*, pp. 145–166. Smithsonian Institution Press, Washington, DC.

Fowler, C. S. 1994b. Material culture and the proposed Numic expansion. In D. B. Madsen and D. Rhode, eds., *Across the West: Human Population Movement and the Expansion of the Numa*, pp. 103–113. University of Utah Press, Salt Lake City.

Fowler, C. S. 2000. "We Live by Them": Native knowledge of biodiversity in the Great Basin of western North America. In P. E. Minnis and W. J. Elisens, eds., *Biodiversity and Native America*, pp. 99–132. University of Oklahoma Press, Norman, OK.

Fowler, C. S., and D. D. Fowler, eds. 2008. *The Great Basin: People and Place in Ancient Times*. School for Advanced Research Press, Santa Fe, NM.

Fowler, C. S., and E. M. Hattori. 2008. The Great Basin's oldest textiles. In C. S. Fowler and D. D. Fowler, eds., *The Great Basin: People and Place in Ancient Times*, pp. 60–67. School for Advanced Research Press, Santa Fe, NM.

Fowler, C. S., and E. M. Hattori. 2009. Recent advances in Great Basin textile research. In B. Hockett, ed., *Past, Present and Future Issues in Great Basin Archaeology: Papers in Honor of Don D. Fowler*, pp. 103–124. Bureau of Land Management Cultural Resource Series 20, Reno, NV.

Fowler, C. S., and D. E. Rhode. 2006. Great Basin Plants. In D. J. Stanford, B. D. Smith, D. H. Ubelaker, and E. J. E. Szathmáry, eds., *Environment, Origins, and Population, Handbook of North American Indians, Volume 3*, pp. 331–350. Smithsonian Institution Press, Washington, DC.

Fowler, C. S., and N. P. Walter. 1985. Harvesting Pandora moth larvae with the Owens Valley Paiute. *Journal of California and Great Basin Anthropology* 7:155–165.

Fowler, C. S., E. M. Hattori, and A. J. Dansie. 2000. Ancient matting from Spirit Cave, Nevada: Technical implications. In P. B. Drooker and L. D. Webster, eds., *Beyond Cloth and Cordage: Archaeological Textile Research in the Americas*, pp. 119–139. University of Utah Press, Salt Lake City.

Fowler, D. D. 1968. *The Archaeology of Newark Cave, White Pine County, Nevada*. Desert Research Institute Technical Report Series S-H, Social Sciences and Humanities, Publication 3.

Fowler, D. D. 1973. Biographical Sketch of S. M. Wheeler. In D. D. Fowler, ed., *The Archaeology of Etna Cave, Lincoln County, Nevada*, p. 7. Desert Research Institute Publications in the Social Sciences 7.

Fowler, D. D., and J. F. Matley. 1979. *Material Culture of the Numa: The John Wesley Powell Collection, 1867–1880*. Smithsonian Contributions to Anthropology 26.

Fowler, D. D., D. B. Madsen, and E. M Hattori. 1973. *Prehistory of Southeastern Nevada*. Desert Research Institute Publications in the Social Sciences 6.

Fox-Dobbs, K., T. S. Stidham, G. J. Bowen, and S. D. Emslie. 2006. Dietary controls on extinction versus survival among avian megafauna in the late Pleistocene. *Geology* 34:685–688.

Francaviglia, R. V. 2005. *Mapping and Imagination in the Great Basin*. University of Nevada Press, Reno.

Franzwa, G. M. 1999. *Maps of the California Trail*. Patrice Press, Tucson.

Frémont, J. C. 1845. *Report of the Exploring Expedition to the Rocky Mountains in the Year 1842, and to Oregon and California in the Years 1843–1844*. Goles and Seaton, Washington, DC.

Frémont, J. C. 1887. *Memoirs of My Life*. Belford, Clarke, New York.

Friedel, D. E. 1993. Chronology and climatic controls of late Quaternary lake-level fluctuations in Chewaucan, Fort Rock, and Alkali basins, south-central Oregon. Ph.D. dissertation, University of Oregon, Eugene.

Friedel, D. E. 1994. Paleolake shorelines and lake level chronology of the Fort Rock Basin, Oregon. In C. M. Aikens and D. L. Jenkins, eds., *Archaeological Research in the Northern Great Basin: Fort Rock Archaeology since Cressman*, pp. 21–40. University of Oregon Anthropological Papers 50.

Friedel, D. E. 2001. Pleistocene Lake Chewaucan: Two short pieces on hydrological connections and lake-level oscillations. In R. Negrini, S. Pezzopane, and T. Badger, eds., *Quaternary Studies near Summer Lake, Oregon: Friends of the Pleistocene Ninth Annual Pacific Northwest Cell Field Trip September 28–30, 2001*, pp. DF1–DF3.

Frison, G. C. 2000. A ^{14}C Date on a late-Pleistocene *Camelops* at the Casper-Hell Gap Site, Wyoming. *Current Research in the Pleistocene* 17:28–29.

Frison, G., and B. Bradley. 1999. *The Fenn Cache: Clovis Weapons and Tools*. One Horse Land and Cattle Company, Santa Fe, NM.

Frison, G., and D. J. Stanford. 1982. *The Agate Basin Site: A Record of the Paleoindian Occupation of the Northwestern High Plains*. Academic Press, New York.

Froese, R., and D. Pauly, eds. 2008. *FishBase*. World Wide Web electronic publication. www.fishbase.org, version (06/2008).

Fry, G. F., and G. F. Dalley. 1979. *The Levee Site and the Knoll Site*. University of Utah Anthropological Papers 100.

Fullerton, D. S., C. A. Bush, and J. N. Pennell. 2003. *Map of Surficial Deposits and Materials in the Eastern and Central United States (East of 102° West Longitude)*. U.S. Geological Survey Geologic Investigations Series I-2789.

Furgurson, E. B. 1992. Lake Tahoe: Playing for high stakes. *National Geographic* 181:112–132.

Gale, H. S. 1915. *Salines in the Owens, Searles, and Panamint Basins, Southeastern California*. U.S. Geological Survey Bulletin 580:251–323.

Garcia, A. F., and M. Stokes. 2006. Late Pleistocene highstand and recession of a small, high-altitude pluvial lake, Jakes Valley, central Great Basin, USA. *Quaternary Research* 65:179–186.

Gehr, K. D. 1980. Late Pleistocene and recent archaeology and geomorphology of the south shore of Harney Lake, Oregon. M.A. thesis, Portland State University, Portland.

Gehr, K. D., and T. M. Newman. 1978. Preliminary note on the late Pleistocene geomorphology and archaeology of the Harney Basin, Oregon. *The Ore Bin* 40:165–170.

Geist, V. 1998. *Deer of the World: Their Evolution, Behavior, and Ecology*. Stackpole Books, Mechanicsburg.

Genoways, H. H., and J. H. Brown, eds. 1993. *Biology of the Heteromyidae*. American Society of Mammalogists Special Publication 10.

Gilbert, B. 1983. *Westering Man: The Life of Joseph Walker*. University of Oklahoma Press, Norman.

Gilbert, G. K. 1890. *Lake Bonneville*. U.S. Geological Survey Monograph 1.

Gilbert, M. T. P., D. L. Jenkins, A. Götherstrom, N. Naveran, J. J. Sanchez, M. Hofreiter, P. F. Thomsen, J. Binladen, T. F. G. Higham, R. M. Yohe II, R. Parr, L. S. Cummings, and E. Willerslev. 2008. DNA from pre-Clovis human coprolites in Oregon. North America. *Science* 320:786–789.

Gilbert, M. T. P., D. L. Jenkins, T. F. G. Higham, M. Rasmussen, H. Malmström, E. M. Svensson, J. J. Sanchez, L. S. Cummings,

R. M. Yohe II, M. Hofreiter, A. Götherström, and E. Willerslev. 2009. Response to Poinar et al. on "DNA from pre-Clovis human coprolites in Oregon, North America." *Science* 325:148-b.

Gill, F., M. Wright, and D. Donsker. 2008. *IOC World Bird Names* (version 1.6). http://www.worldbirdnames.org/.

Gill, J. L., J. W. Williams, S. T. Jackson, K. B. Lininger, and G. S. Robinson. 2009. Pleistocene megafaunal collapse, novel plant communities, and enhanced fire regimes in North America. *Science* 326:1100–1103.

Gillespie, A. R., and P. H. Zehfuss. 2004. Glaciations of the Sierra Nevada. In J. Ehlers and P. L Gibbard, eds., *Quaternary Glaciations: Extent and Chronology. Part II: North America*, pp. 51–62. Developments in Quaternary Science 2. Elsevier, Amsterdam.

Gillespie, W. B. 1984. Late Quaternary small vertebrates from Chaco Canyon, northwest New Mexico. *New Mexico Geology* 6:16.

Gillette, D. D., and D. B. Madsen. 1992. The short-faced bear *Arctodus simus* from the late Quaternary in the Wasatch Mountains of central Utah. *Journal of Vertebrate Paleontology* 12:107–112.

Gillette, D. D., and D. B. Madsen. 1993. The Columbian mammoth, *Mammuthus columbi*, from the Wasatch Mountains of central Utah. *Journal of Paleontology* 67:669–680.

Gillette, D. D., and W. E. Miller. 1999. Catalogue of new Pleistocene mammalian sites and recovered fossils from Utah. In D. D. Gillette, ed., *Vertebrate Paleontology in Utah*, pp. 523–530. Utah Geological Survey Miscellaneous Publication 99-1.

Gillette, D. D., H. G. McDonald, and M. C. Hayden. 1999. The first record of Jefferson's ground sloth, *Megalonyx jeffersonii*, in Utah (Pleistocene, Rancholabrean Land Mammal Age). In D. D. Gillette, ed., *Vertebrate Paleontology in Utah*, pp. 509–521. Utah Geological Survey Miscellaneous Publication 99-1.

Gillin, J. 1941. *Archaeological Excavations in Central Utah*. Papers of the Peabody Museum of American Archaeology and Ethnology 17(2).

Gleason, H. A. 1926. The individualistic concept of the plant association. *Bulletin of the Torrey Botanical Club* 53:7–26.

Gleason, H. A., and A. Cronquist. 1964. *The Natural Geography of Plants*. Columbia University Press, New York.

Gobalet, K. W., and R. M. Negrini. 1992. Evidence for endemism in fossil tui chub, *Gila bicolor*, from Pleistocene Lake Chewaucan, Oregon. *Copeia* 1992:539–544.

Goddard, I., ed. 1996. Languages. *Handbook of North American Indians, Volume 17*. Smithsonian Institution, Washington, DC.

Godfrey, W. E. 1966. *The Birds of Canada*. National Museums of Canada Bulletin 203, Biological Series 73.

Godsey, H. S., D. R. Currey, and M. A. Chan. 2005. New evidence for an extended occupation of the Provo shoreline and implications for regional climate change, Pleistocene Lake Bonneville, Utah, USA. *Quaternary Research* 63:212–223.

Godsey, H. S., G. Atwood, E. Lips, D. M. Miller, M. Milligan, and C. G. Oviatt. 2005. Don R. Currey Memorial Field Trip to the shores of Pleistocene Lake Bonneville. In J. Pederson and C. M. Dehler, eds., *Interior Western United States*, pp. 419–448. Geological Society of America Field Guide 6.

Goebel, T. 2004. The search for a Clovis progenitor in Subarctic Siberia. In D. B. Madsen, ed., *Entering America: Northeast Asia and Beringia before the Last Glacial Maximum*, pp. 311–356. University of Utah Press, Salt Lake City. UT.

Goebel, T. 2007. Pre-Archaic and early Archaic technological activities at Bonneville Estates Rockshelter: A first look at the lithic artifact record. In K. E. Graf and D. N. Schmitt, eds., *Paleoindian or Paleoarchaic: Great Basin Human Ecology at the Pleistocene-Holocene Transition*, pp. 156–184. University of Utah Press, Salt Lake City.

Goebel, T., R. Powers, and B. Bigelow. 1991. The Nenana complex of Alaska and Clovis origins. In R. Bonnichsen and K. Turnmire, eds., *Clovis: Origins and Adaptations*, pp. 49–79. Center for the Study of the First Americans, Oregon State University, Corvallis.

Goebel, T., M. R. Waters, and E. H. O'Rourke. 2008. The late Pleistocene dispersal of modern humans in the Americas. *Science* 319:1497–1502.

Goldberg, P., and T. L. Arpin. 1999. Micromorphological analysis of sediments from Meadowcroft Rockshelter, Pennsylvania: Implications for radiocarbon dating. *Journal of Field Archaeology* 26:325–342.

Goldberg, P., F. Berna, and R. I. Macphail. 2009. Comment on "DNA from pre-Clovis human coprolites in Oregon, North America." *Science* 325:148-c.

Golla, V. 2007. Linguistic prehistory. In T. L. Jones and K. A. Klar, eds., *California Prehistory: Colonization, Culture, and Complexity*, pp. 71–82. AltaMira Press, Lanham, MD.

Gongora, J, N. J. Rawlence, V. A. Mobegis, H. Jianlin, J. A. Alcalde, J. T. Matus, O. Hanotte, C. Moran, J. J. Austin, S. Ulm, A. J. Anderson, G. Larson, and A. Cooper. 2008. Indo-European and Asian origins for Chilean and Pacific chickens revealed by mtDNA. *Proceedings of the National Academy of Sciences* 105:10308–10313.

Goodwin, H. T., and R. E. Reynolds. 1989a. Late Quaternary Sciuridae from Kokoweef Cave, San Bernardino County, California. *Bulletin of the Southern California Academy of Sciences* 88:21–32.

Goodwin, H. T., and R. E. Reynolds. 1989b. Late Quaternary Sciuridae from low elevations in the Mojave Desert, California. *Southwestern Naturalist* 34:506–512.

Gordon, R. G., Jr., ed. 2005. *Ethnologue: Languages of the World*. Fifteenth edition. SIL International, Dallas. http://www.ethnologue.com/.

Goss, J. H. 1968. Culture-historical inference from Utaztecan linguistic evidence. In E. H. Swanson, ed., *Utaztecan Prehistory*, pp. 1–42. Occasional Papers of the Idaho State University Museum 22.

Goss, J. H. 1977. Linguistic tools for the Great Basin prehistorian. In D. D. Fowler, ed., *Models and Great Basin Prehistory: A Symposium*, pp. 49–70. Desert Research Institute Publications in the Social Sciences 12.

Goss, J. A. 1999. The Yamparika—Shoshones, Comanches, or Utes—or Does it Matter? In R. O. Clemmer, L. D. Myers, and M. E. Rudden, eds., *Julian Steward and the Great Basin: The Making of an Anthropologist*, pp. 74–84. University of Utah Press, Salt Lake City.

Graf, K. E. 2001. Paleoindian technological provisioning in the western Great Basin. M.A. thesis, University of Nevada, Reno.

Graf, K. E. 2002. Paleoindian procurement and mobility in the western Great Basin. *Current Research in the Pleistocene* 19:87–89.

Graf, K. E. 2007. Stratigraphy and chronology of the Pleistocene to Holocene transition at Bonneville Estates Rockshelter, eastern Great Basin. In K. E. Graf and D. N. Schmitt, eds., *Paleoindian or Paleoarchaic: Great Basin Human Ecology at the Pleistocene-Holocene Transition*, pp. 82–104. University of Utah Press, Salt Lake City.

Graf, K. E. 2009. "The good, the bad, the ugly": Evaluating the radiocarbon chronology of the middle and late Upper

Paleolithic in the Enisei River Valley, south-central Siberia. *Journal of Archaeological Science* 36:694–707.

Graham, B. 2003. *The Crossing of the Sierra Nevada in the Winter of 1843–1844*. B. Graham, Sacramento.

Graham, N. E., and M. K. Hughes. 2007. Reconstructing the Mediaeval low stands of Mono Lake, Sierra Nevada, California, USA. *The Holocene* 17:1197–1210.

Graham, N. E., M. K. Hughes, C. M. Ammann, K. M. Cobb, M. P. Hoerling, D. J. Kennett, J. P. Kennett, B. Rein, L. Stott, P. E. Wigand, and T. Xu. 2007. Tropical-Pacific–mid-latitude teleconnections in medieval times. *Climatic Change* 83:241–285.

Graham, R. W. 1985. Diversity and community structure of the late Pleistocene mammal fauna of North America. *Acta Zoologica Fennica* 170:181–192.

Graham, R. W. 1988. The role of climatic change in the design of biological reserves: The paleoecological perspective for conservation biology. *Conservation Biology* 2:391–394.

Graham, R. W. 1992. Late Pleistocene faunal changes as a guide to understanding effects of greenhouse warming on the mammalian fauna of North America. In R. L. Peters and T. W. Lovejoy, eds., *Global Warming and Biological Diversity*, pp. 76–87. Yale University Press, New Haven, CT.

Graham, R. W., and M. A. Graham. 1994. Late Quaternary distribution of *Martes* in North America. In S. W. Buskirk, A. S. Harestad, M. G. Raphael, and R. A. Powell, eds., *Martens, Sables, and Fishers: Biology and Conservation*, pp. 26–58. Cornell University Press, Ithaca, NY.

Graham, R. W., and E. L. Lundelius, Jr. 1984. Coevolutionary disequilibrium and Pleistocene extinctions. In P. S. Martin and R. G. Klein, eds., *Quaternary Extinctions: A Prehistoric Revolution*, pp. 223–249. University of Arizona Press, Tucson.

Graham, R. W., C. V. Haynes, D. Johnson, and M. Kay. 1981. Kimmswick: A Clovis-mastodon association in eastern Missouri. *Science* 213:1115–1117.

Graumlich, L. J. 1993. A 1000-yr record of temperature and precipitation in the Sierra Nevada. *Quaternary Research* 39:249–255.

Graumlich, L. J., and A. H. Lloyd. 1996. Dendroclimatic, ecological and geomorphological evidence for long-term climatic change in the Sierra Nevada, U.S.A. In J. S. Dean, D. M. Meko, and T. W. Swetnam, eds., *Tree Rings, Environment and Humanity*, pp. 51–59. Radiocarbon, Department of Geosciences, University of Arizona, Tucson.

Gray, R. D., and Q. D. Atkinson. 2003. Language-tree divergence times support the Anatolian theory of Indo-European origin. *Nature* 426:435–439.

Gray, R. D., A. J. Drummond, and S. J. Greenhill. 2009. Language phylogenies reveal expansion pulses and pauses in Pacific settlement. *Science* 323:479–530.

Gray, S. T., J. L Betancourt, S. T. Jackson, and R. G. Eddy. 2006. Role of multidecadal climate variability in a range extension of pinyon pine. *Ecology* 87:1124–1130.

Gray, S. T., S. T. Jackson, and J. L. Betancourt. 2004. Tree-ring based reconstructions of interannual to decadal scale precipitation variability for northeastern Utah since 1226 A.D. *Journal of the American Water Resources Association* 40:947–960.

Grayson, D. K. 1977a. On the Holocene history of some Great Basin lagomorphs. *Journal of Mammalogy*, 58:507–513.

Grayson, D. K. 1977b. Paleoclimatic implications of the Dirty Shame Rockshelter mammalian fauna. *Tebiwa* 9:1–26.

Grayson, D. K. 1977c. Pleistocene avifaunas and the overkill hypothesis. *Science* 195:691–693.

Grayson, D. K. 1979. Mt. Mazama, climatic change, and Fort Rock Basin archaeofaunas. In P. D. Sheets and D. K. Grayson, eds., *Volcanic Activity and Human Ecology*, pp. 427–458. Academic Press, New York.

Grayson, D. K. 1981. A mid-Holocene record for the heather vole, *Phenacomys* cf. *intermedius*, in the central Great Basin, and its biogeographic significance. *Journal of Mammalogy* 62:115–121.

Grayson, D. K. 1982. Toward a history of Great Basin mammals during the past 15,000 Years. In D. B. Madsen and J. F. O'Connell, eds., *Man and Environment in the Great Basin*, pp. 82–101. Society for American Archaeology Papers 2.

Grayson, D. K. 1983. The paleontology of Gatecliff Shelter: small mammals. In D. H. Thomas, *The Archaeology of Monitor Valley. 2. Gatecliff Shelter*, pp. 98–126. American Museum of Natural History Anthropological Papers 59(1).

Grayson, D. K. 1984a. Nineteenth century explanations of Pleistocene extinctions: A review and analysis. In P. S. Martin and R. G. Klein, eds., *Quaternary Extinctions: A Prehistoric Revolution*, pp. 5–39. University of Arizona Press, Tucson.

Grayson, D. K. 1984b. The time of extinction and nature of adaptation of the noble marten, *Martes nobilis*. In H. H. Genoways and M. R. Dawson, eds., *Contributions in Quaternary Vertebrate Paleontology: A Volume in Memorial to John E. Guilday*, pp. 233–240. Carnegie Museum of Natural History Special Publication 8.

Grayson, D. K. 1985. The paleontology of Hidden Cave: Birds and mammals. In D. H. Thomas, ed., *The Archaeology of Hidden Cave, Nevada*, pp. 125–161. American Museum of Natural History Anthropological Papers 61(1).

Grayson, D. K. 1986. Eoliths, archaeological ambiguity, and the generation of "middle-range" research. In D. J. Meltzer, D. D. Fowler, and J. A. Sabloff, eds., *American Archaeology: Past and Present*, pp. 77–133. Smithsonian Institution Press, Washington, D.C.

Grayson, D. K. 1987. The biogeographic history of small mammals in the Great Basin: Observations on the last 20,000 years. *Journal of Mammalogy* 68:359–375.

Grayson, D. K. 1988. *Danger Cave, Last Supper Cave, Hanging Rock Shelter: The Faunas*. American Museum of Natural History Anthropological Papers 66(1).

Grayson, D. K. 1990. The James Creek Shelter mammals. In R. G. Elston and E. E. Budy, eds., *The Archaeology of James Creek Shelter*, pp. 87–98. University of Utah Anthropological Papers 115.

Grayson, D. K. 1991. Alpine faunas from the White Mountains, California: Adaptive change in the late prehistoric Great Basin? *Journal of Archaeological Science* 18:483–506.

Grayson, D. K. 1993. *The Deserts' Past: A Natural Prehistory of the Great Basin*. Smithsonian Institution Press, Washington, DC.

Grayson, D. K. 1994. Chronology, glottochronology, and Numic expansion. In D. B. Madsen and D. Rhode, eds., *Across the West: Human Population Movement and the Expansion of the Numa*, pp. 20–23. University of Utah Press, Salt Lake City.

Grayson, D. K. 1998. Moisture history and small mammal community richness during the latest Pleistocene and Holocene, northern Bonneville Basin, Utah. *Quaternary Research* 49:330–334.

Grayson, D. K. 2000a. Mammalian responses to middle Holocene climatic change in the Great Basin of the western United States. *Journal of Biogeography* 27, 181–192.

Grayson, D. K. 2000b. The Homestead Cave mammals. In. D. B. Madsen, *Late Quaternary paleoecology in the Bonneville Basin*, pp. 67–89. Utah Geological Survey Bulletin 130.

Grayson, D. K. 2001. The archaeological record of human impacts on animal populations. *Journal of World Prehistory* 15:1–68.

Grayson, D. K. 2004. Monte Verde, field archaeology, and the human colonization of the Americas. In D. B. Madsen, ed., *Entering America: Northeast Asia and Beringia before the Last Glacial Maximum*, pp. 289–297. University of Utah Press, Salt Lake City.

Grayson, D. K. 2005. A brief history of Great Basin pikas. *Journal of Biogeography* 32:2101–2111.

Grayson, D. K. 2006a. Brief histories of some Great Basin mammals: Extinctions, extirpations, and abundance histories. *Quaternary Science Reviews* 25:2964–2991.

Grayson, D. K. 2006b. Holocene bison in the Great Basin, western USA. *The Holocene* 16:913–925.

Grayson, D. K. 2006c. Ice age extinctions. *Quarterly Review of Biology* 81:259–264.

Grayson, D. K. 2006d. Late Pleistocene faunal extinctions. In D. J. Stanford, B. D. Smith, D. H. Ubelaker, and E. J. E. Szathmáry, eds., *Environment, Origins, and Population, Handbook of North American Indians, Volume 3*, pp. 208–218. Smithsonian Institution Press, Washington, DC.

Grayson, D. K. 2007. Deciphering North American Pleistocene extinctions. *Journal of Anthropological Research* 63:185–214.

Grayson, D. K. 2008. Holocene underkill. *Proceedings of the National Academy of Sciences* 105:4077–4078.

Grayson, D. K., and T. W. Canaday. 2009. The Alta Toquima vertebrates. In D. H. Thomas and L. S. A. Pendleton, *The Archaeology of Monitor Valley: 4. Alta Toquima and the Mt. Jefferson Tablelands Complex*. American Museum of Natural History Anthropological Papers, in press.

Grayson, D. K., and F. Delpech. 2003. Ungulates and the Middle-to-Upper Paleolithic transition at Grotte XVI (Dordogne, France). *Journal of Archaeological Science* 30:1633–1648.

Grayson, D. K., and F. Delpech. 2006. Was there increasing dietary specialization across the Middle-to-Upper Paleolithic transition in France? In N. J. Conard, ed., *When Neanderthals and Modern Humans Met*, pp. 377–417. Tübingen Publications in Prehistory, Kerns Verlag, Tübingen.

Grayson, D. K., and J. L. Fisher. 2009. Holocene elk (*Cervus elaphus*) in the Great Basin. In B. Hockett, ed., *Past, Present and Future Issues in Great Basin Archaeology: Papers in Honor of Don D. Fowler*, pp. 67–83. Bureau of Land Management Cultural Resource Series 20, Reno, NV.

Grayson, D. K., and S. D. Livingston. 1989. High-elevation records for *Neotoma cinerea* in the White Mountains, California. *Great Basin Naturalist* 49:392–395.

Grayson, D. K., and S. D. Livingston. 1993. Missing mammals on Great Basin mountains: Holocene extinctions and inadequate knowledge. *Conservation Biology* 7:527–532.

Grayson, D. K., and D. B. Madsen. 2000. Biogeographic implications of recent low-elevation recolonization by *Neotoma cinerea* in the Great Basin. *Journal of Mammalogy* 81:1100–1105.

Grayson, D. K., and D. J. Meltzer. 2002. The human colonization of North America, Clovis hunting and large mammal extinction. *Journal of World Prehistory* 16:313–359.

Grayson, D. K., and D. J. Meltzer. 2003. A requiem for North American overkill. *Journal of Archaeological Science* 30: 585–593.

Grayson, D. K., and C. I. Millar. 2008. Prehistoric human influence on the abundance and distribution of deadwood in alpine landscapes. *Perspectives in Plant Ecology, Evolution and Systematics* 10:101–108.

Grayson, D. K., and P. W. Parmalee. 1988. Hanging Rock Shelter. In D. K. Grayson, *Danger Cave, Last Supper Cave, Hanging Rock Shelter: The Faunas*, pp. 105–115. American Museum of Natural History Anthropological Papers 66(1).

Grayson, D. K., S. D. Livingston, E. Rickart, and M. W. Shaver. 1996. The biogeographic significance of low elevation records for *Neotoma cinerea* from the northern Bonneville Basin, Utah. *Great Basin Naturalist* 56:191–196.

Green, D. F. 1961. *Archaeological Investigations at the G. M. Hinckley Farm Site, Utah County, Utah, 1956–1960*. Brigham Young University Press, Provo, UT.

Green, J. S., and J. T. Flinders. 1980. *Brachylagus idahoensis*. *Mammalian Species* 125.

Greenberg, J. H. 1987. *Language in the Americas*. Stanford University Press, Stanford, CA.

Greenspan, R. L. 1988. Fish remains. In C. Raven and R. G. Elston, eds., *Preliminary Investigations in Stillwater Marsh: Human Prehistory and Geoarchaeology*, pp. 313–327. U.S. Department of the Interior, U.S. Fish and Wildlife Service, Region 1, Cultural Resource Series 1.

Greenspan, R. L. 1993. Analysis of the Buffalo Flat vertebrate fauna remains. In A. C. Oetting, *The Archaeology of Buffalo Flat: Cultural Resources Investigations for the CONUS OTH-B Buffalo Flat Radar Transmitter Site, Christmas Lake Valley, Oregon*, pp. 613–634. Heritage Research Associates Report 151, Eugene, OR.

Greenspan, R. L. 1994. Archaeological fish remains in the Fort Rock Basin. In C. M. Aikens and D. L. Jenkins, eds., *Archaeological Research in the Northern Great Basin: Fort Rock Archaeology since Cressman*, pp. 485–504. University of Oregon Archaeological Papers 50.

Greenwald, D. N., and C. Bradley. 2008. Assessing protection for imperiled species of Nevada, U.S.A.: Are species slipping through the cracks of existing protections? *Biodiversity and Conservation* 17:2951–2960.

Grinnell, J. 1919. The English Sparrow has arrived in Death Valley: An experiment in nature. *American Naturalist* 53:468–472.

Griset, S., ed. 1986. *Pottery of the Great Basin and Adjacent Areas*. University of Utah Anthropological Papers 111.

Grove, A. T. 2008. A brief consideration of climate forcing factors in view of the Holocene glacier record. *Global and Planetary Change* 66:141–147.

Grove, J. M. 2004. *Little Ice Ages: Ancient and Modern*. Routledge, London.

Gruhn, R. 1979. Excavation in Amy's Shelter, eastern Nevada. In D. R. Tuohy and D. L. Rendall, eds., *The Archaeology of Smith Creek Canyon, Eastern Nevada*, pp. 90–161. Nevada State Museum Anthropological Papers 17.

Gudschinsky, S. 1956. The ABCs of lexicostatistics (glottochronology). *Word* 12:175–210.

Guérin, C., and M. Patou-Mathis, eds. 1996. *Les grands mammifères Plio-Pléistocènes d'Europe*. Masson, Paris.

Guilday, J., and P. W. Parmalee. 1982. Vertebrate faunal remains from Meadowcroft Rockshelter, Washington County, Pennsylvania: Summary and interpretation. In R. C. Carlisle and J. M. Adovasio, eds., *Meadowcroft: Collected Papers on the Archaeology of Meadowcroft Rockshelter and the Cross Creek Drainage*, pp. 163–174. OCLC 10395512.

Guilday, J. E., P. W. Parmalee, and R. W. Wilson. 1980. Vertebrate faunal remains from Meadowcroft Rockshelter (36WH297), Washington County, Pennsylvania. Unpublished manuscript.

Guilday, J. E., H. W. Hamilton, E. Anderson, and P. W. Parmalee. 1978. *The Baker Bluff Cave Deposit and the Late Pleistocene Faunal Gradient*. Bulletin of the Carnegie Museum of Natural History 11.

Guthrie, R. D. 1984. Mosaics, allelochemics, and nutrients: An ecological theory of late Pleistocene megafaunal extinctions. In P. S. Martin and R. G. Klein, eds., *Quaternary Extinctions: A Prehistoric Revolution*, pp. 259–298. University of Arizona Press, Tucson.

Guthrie, R. D. 2001. Origin and causes of the mammoth steppe: A story of cloud cover, woolly mammoth tooth pits, buckles, and inside-out Beringia. *Quaternary Science Reviews* 20:549–574.

Guthrie, R. D. 2003. Rapid body size decline in Alaskan Pleistocene horses before extinction. *Nature* 426:169–171.

Guthrie, R. D. 2004. Radiocarbon evidence of mid-Holocene mammoths stranded on an Alaskan Bering Sea Island. *Nature* 429:746–749.

Guthrie, R. D. 2006. New carbon dates link climatic change with human colonization and Pleistocene extinctions. *Nature* 441:207–209.

Gwynn, J. W., ed. 1980. *Great Salt Lake: A Scientific, Historical, and Economic Overview*. Utah Geological and Mineral Survey Bulletin 116.

Gwynn, J. W., ed. 2002. *Great Salt Lake: An Overview of Change*. Utah Department of Natural Resources Special Publication, Salt Lake City.

Hafner, D. J. 1993. North American pika (*Ochotona princeps*) as a late Quaternary biogeographic indicator species. *Quaternary Research* 39:373–380.

Hafner, D. J. 1994. Pikas and permafrost: Post-Wisconsin zoogeography of *Ochotona* in the southern Rocky Mountains, U.S.A. *Arctic and Alpine Research* 26:375–382.

Hafner, J. C., E. Reddington, and M. T. Craig. 2006. Kangaroo mice (*Microdipodops megacephalus*) of the Mono Basin: Phylogeography of a peripheral isolate. *Journal of Mammalogy* 87:1204–1217.

Haigh, J. D. 1994. The role of stratospheric ozone in modulating the solar radiative forcing of the atmosphere. *Nature* 370:544–546.

Haile, J., D. G. Froese, R. D. E. MacPhee, R. G. Roberts, L. J. Arnold, A. V. Reyes, M. Rasmussen, R. Nielsen, B. W. Brook, S. Robinson, M. Demuro, M. T. P. Gilbert, K. Munch, J. J. Austin, A. Cooper, I. Barnes, P. Möller, and E. Willerslev. 2009. Ancient DNA reveals late survival of mammoth and horse in interior Alaska. *Proceedings of the National Academy of Sciences* 106:22363–22368.

Hale, K. 1958. Internal diversity in Uto-Aztecan: I. *International Journal of American Linguistics* 24:101–107.

Halford, F. K. 1998. Archaeology and environment on the Dry Lakes Plateau, Bodie Hills, California: Hunter-gatherer coping strategies for Holocene environmental variability. M.A. thesis, University of Nevada, Reno.

Hall, E. R. 1926. A new marten from the Pleistocene cave deposits of California. *Journal of Mammalogy* 7:127–139.

Hall, E. R. 1946. *Mammals of Nevada*. University of California Press, Berkeley.

Hall, E. R. 1961. *Bison bison* in Nevada. *Journal of Mammalogy* 42:279–280.

Hall, E. R. 1981. *Mammals of North America*. Second edition. Wiley and Sons, New York.

Hall, E. R., and K. R. Kelson. 1959. *Mammals of North America*. Ronald Press, New York.

Hall, M. J. 1983. *A Reassessment of American Fork Cave (42UT135), Utah County, Utah*. Brigham Young University Department of Anthropology Technical Series 83–51.

Hallett, D. J., L. V. Hills, and J. J. Clague. 1997. New accelerator mass spectrometry radiocarbon ages for the Mazama tephra layer from Kootenay National Park, British Columbia, Canada. *Canadian Journal of Earth Sciences* 34:1202–1209.

Halley, E. 1715. A short account of the cause of the saltness of the ocean, and of the several lakes that emit no rivers; with a proposal, by help thereof, to discover the age of the world. *Philosophical Transactions of the Royal Society of London* 29:296–300.

Halsey, J. H. 1953. Geology of parts of the Bridgeport, California and Wellington, Nevada quadrangles. Ph.D. dissertation, University of California, Berkeley.

Haman, J. F. 1963. *Porcupine Cave, Summit County, Utah. Salt Lake City: Salt Lake Grotto*. National Speleological Society Technical Note 1.

Hamrick, J. L., A. F. Schnabel, and P. V. Wells. 1994. Distribution and genetic diversity within and among populations of Great Basin conifers. In K. T. Harper, L. L. St. Clair, K. H. Thorne, and W. M. Hess, eds., *Natural History of the Colorado Plateau and Great Basin*, pp. 146–161. University Press of Colorado, Niwot.

Hansen, G. H., and W. L. Stokes. 1941. An ancient cave in American Fork Canyon. *Utah Academy of Sciences, Arts, and Letters Proceedings* 18:27–37.

Hansen, H. P. 1947. Postglacial vegetation in the Northern Great Basin. *American Journal of Botany* 34:164–171.

Hansen, R. M. 1978. Shasta ground sloth food habits, Rampart Cave, Arizona. *Paleobiology* 4:302–319.

Harding, S. T. 1965. *Recent Variations in the Water Supply of the Western Great Basin*. Water Resources Center Archives Series 16, University of California, Berkeley.

Harding, S. T. 1967. *A Life in Western Water Development*, edited by G. J. Giefer. Statewide Water Resources Center, University of California Library, in cooperation with the Regional Oral History Office, Bancroft Library, Berkeley (http://openlibrary.org/a/OL2173117A).

Harper, K. T., D. L. Freeman, W. K. Ostler, and L. G. Klikoff. 1978. The flora of Great Basin mountain ranges: Diversity, sources, and dispersal ecology. In K. T. Harper, and J. L. Reveal, eds., *Intermountain Biogeography: A Symposium*, pp. 81–104. Great Basin Naturalist Memoirs 2.

Harril, J. R., J. S. Gates, and J. M. Thomas. 1988. *Major Ground-Water Flow Systems in the Great Basin Region of Nevada, Utah, and Adjacent States: Regional Aquifer Systems of the Great Basin*. U.S. Geological Survey Hydrologic Investigations Atlas HA-694-C.

Harrington, M. R. 1933. *Gypsum Cave, Nevada*. Southwest Museum Papers 8.

Harrington, M. R. 1945. Bug sugar. *The Masterkey* 19:95–96.

Harrington, M. R., and R. D. Simpson. 1961. *Tule Springs, Nevada with other Evidences of Pleistocene Man in North America*. Southwest Museum Paper 18.

Harris, A. H. 1970. The Dry Cave mammalian fauna and late pluvial conditions in southeastern New Mexico. *The Texas Journal of Science* 22:3–27.

Harris, A. H. 1977. Wisconsin Age environments in the northern Chihuahuan Desert: evidence from higher vertebrates. In R. H. Wauer and D. H. Riskind, eds., *Transactions of the Symposium on the Chihuahuan Desert Region, United States and Mexico*, pp. 23–52. National Park Service Transactions and Proceedings Series 3.

Harris, A. H. 1984. *Neotoma* in the late Pleistocene of New Mexico and Chihuahua. In H. H. Genoways and M. R. Dawson, eds., *Contributions in Quaternary Vertebrate Paleontology: A Volume in Memorial to John E. Guilday*, pp. 164–178. Carnegie Museum of Natural History Special Publications 8.

Harris, A. H. 1985. *Late Pleistocene Vertebrate Paleoecology of the West*. University of Texas Press, Austin.

Harris, A. H. 1987. Reconstruction of mid-Wisconsin environments in southern New Mexico. *National Geographic Research* 3(2):142–151.

Harris, A. H. 1990. Fossil evidence bearing on southwestern mammalian biogeography. *Journal of Mammalogy* 71:219–229.

Harris, A. H. 1993a. Quaternary vertebrates of New Mexico. *New Mexico Museum of Natural History and Science Bulletin* 2:179–197.

Harris, A. H. 1993b. Wisconsinan pre-pleniglacial biotic change in southeastern New Mexico. *Quaternary Research* 40:127–133.

Harris, A. H. 2003. The Pleistocene vertebrate fauna from Pendejo Cave. In R. S. MacNeish, ed., *Pendejo Cave*, pp. 37–65. University of New Mexico Press, Albuquerque.

Harris, A. H., and J. S. Findley. 1964. Pleistocene-Recent fauna of the Isleta Caves, Bernalillo County, New Mexico. *American Journal of Science* 262:114–120.

Hart, W. S., J. Quade, D. B. Madsen, D. S. Kaufman, and C. G. Oviatt. 2004. The ^{87}Sr/^{86}Sr ratios of lacustrine carbonates and lake-level history of the Bonneville paleolake system. *Geological Society of America Bulletin* 16:1107–1119.

Harvey, A. M., P. E. Wigand, and S. G. Wells. 1999. Response of alluvial fan systems to the late Pleistocene to early Holocene climatic transition: Contrasts between the margins of pluvial lakes Lahontan and Mojave, Nevada and California, USA. *Catena* 36:255–281.

Hatoff, B. W., and D. H. Thomas. 1976. Mateo's Ridge Site. In D. H. Thomas and R. L. Bettinger, *Prehistoric Piñon Ecotone Settlements in the Upper Reese River Valley, Central Nevada*, pp. 276–312. American Museum of Natural History Anthropological Papers 53(3).

Hattori, E. M. 1982. *The Archaeology of Falcon Hill Cave, Winnemucca Lake, Washoe County, Nevada*. Nevada State Museum Anthropological Papers 18.

Hattori, E. M., M. E. Newman, and D. R. Tuohy. 1990. Blood residue analysis of Great Basin crescents. Paper presented at the Great Basin Anthropological Conference, Reno, NV.

Hattori, E. M., P. H. Shelley, and D. R. Tuohy. 1976. Preliminary report on a techno-functional examination of flaked stone artifacts. Paper presented at the Great Basin Anthropological Conference, Las Vegas, NV.

Haury, E. W. 1953. Artifacts with mammoth remains, Naco, Arizona: Discovery of the Naco mammoth and the associated projectile points. *American Antiquity* 19:1–14.

Haury, E., E. B. Sayles, and W. W. Wasley. 1959. The Lehner Mammoth Site, southeastern Arizona. *American Antiquity* 25:2–30.

Hawkes, K., and R. Bliege Bird. 2002. Showing off, handicap signalling, and the evolution of men's work. *Evolutionary Anthropology* 11:58–67.

Hayhoe, K., D. Cayan, C. B. Field, P. C. Frumhoff, E. P. Maurer, N. L. Miller, S. C. Moser, S. H. Schneider, K. N. Cahill, E. E. Cleland, L. Dale, R. Drapek, R. M. Hanemann, L. S. Kalkstein, J. Lenihan, C. K. Lunch, R. P. Neilson, S. C. Sheridan, and J. H. Verville. 2004. Emissions pathways, climate change, and impacts on California. *Proceedings of the National Academy of Sciences* 101:12422–12427.

Haynes, C. V., Jr. 1967. Quaternary geology of the Tule Springs area, Clark County, Nevada. In H. M. Wormington and D. Ellis, eds., *Pleistocene Studies in Southern Nevada*, pp. 15–104. Nevada State Museum Anthropological Papers 13.

Haynes, C. V., Jr. 1969. The earliest Americans. *Science* 166:709–715.

Haynes, C. V., Jr. 1971. Time, environment, and early man. *Arctic Anthropology* 8(2):3–14.

Haynes, C. V., Jr. 1973. The Calico Site: Artifacts or geofacts? *Science* 181:305–310.

Haynes, C. V., Jr. 1990. The Antevs-Bryan years and the legacy for Paleoindian geochronology. In L. F Laporte, ed., *Establishment of a Geologic Framework for Paleoanthropology*, pp. 55–58. Geological Society of America Special Paper 242.

Haynes, C. V., Jr. 1991. Geoarchaeological and paleohydrological evidence for a Clovis-age drought in North America and its bearing on extinction. *Quaternary Research* 35:438–450.

Haynes, C. V., Jr. 1992. Contributions of radiocarbon dating to the geochronology of the peopling of the New World. In R. E. Taylor, A. Long, and R. Kra, eds., *Radiocarbon After Four Decades*, pp. 255–374. Springer-Verlag, New York.

Haynes, C. V., Jr. 2005. Clovis, pre-Clovis, climate change, and extinction. In R. Bonnichsen, B. T. Lepper, D. Stanford, and M. R. Waters, eds., *Paleoamerican Origins: Beyond Clovis*, pp. 113–132. Center for the Study of the First Americans, Texas A&M University, College Station, TX.

Haynes. C. V., Jr. 2008. Younger Dryas "black mats" and the Rancholabrean termination in North America. *Proceedings of the National Academy of Sciences* 105:6520–6525.

Haynes, C. V., Jr., and B. B. Huckell, eds. 2007. *Murray Springs: A Clovis Site with Multiple Activity Areas in the San Pedro Valley, Arizona*. University of Arizona Anthropological Papers 27.

Haynes, C. V., Jr., J. Boerner, K. Domanik, D. Lauretta, J. Ballenger, and J. Goreva. 2010. The Murray Springs Clovis site, Pleistocene extinction, and the question of extraterrestrial impact. *Proceedings of the National Academy of Sciences* 107:4010–4015.

Haynes, G. 2002. *The Early Settlement of North America: The Clovis Era*. Cambridge University Press, Cambridge, UK.

Haynes, G., D. G. Anderson, C. R. Ferring, S. J. Fiedel, D. K. Grayson, C. V. Haynes, Jr., V. T. Holliday, B. B. Huckell, M. Kornfeld, D. J. Meltzer, J. Morrow, T. Surovell, N. M. Waguespack, P. Wigand, and R. M. Yohe, II. 2007. Comment on "Redefining the Age of Clovis: Implications for the Peopling of the Americas." *Science* 317:320.

Haynes, G. M. 2004. Assemblage richness and composition at Paleoarchaic sites near Yucca Mountain, Nevada. Ph.D. dissertation, University of Nevada, Reno.

Hayssen, V. 1991. *Dipodomys microps*. Mammalian Species 389.

Hayward, C. L., C. Cottam, A. M. Woodbury, and H. H. Frost. 1976. Birds of Utah. Great Basin Naturalist Memoirs 1.

Heald, W. F. 1956. An active glacier in Nevada. *American Alpine Journal* 10(1):164–167.

Heaton, T. H. 1985. Quaternary paleontology and paleoecology of Crystal Ball Cave, Millard County, Utah: With emphasis on mammals and description of a new species of fossil skunk. *Great Basin Naturalist* 45:337–390.

Heaton, T. H. 1987. Initial investigation of vertebrate remains from Snake Creek Burial Cave, White Pine County, Nevada. *Current Research in the Pleistocene* 4:107–109.

Heaton, T. H. 1988. Bears and man at Porcupine Cave, western Uinta Mountains, Utah. *Current Research in the Pleistocene* 5:71–73.

Heckmann, R. A., C. W. Thompson, and D. A. White. 1981. Fishes of Utah Lake. *Great Basin Naturalist Memoirs* 5:107–127.

Heizer, R. F. 1945. Honey-dew "sugar" in western North America. *The Masterkey* 19:140–145.

Heizer, R. F. 1951. Preliminary report on the Leonard Rockshelter site, Pershing County, Nevada. *American Antiquity* 17:89–98.

Heizer, R. F., and M. A. Baumhoff. 1970. Big game hunters in the Great Basin: A critical review of the evidence. *University of California Archaeological Research Facility Report* 7:1–12.

Heizer, R. F., and L. K. Napton. 1970. *Archaeology and the Prehistoric Great Basin Lacustrine Subsistence Regime as seen from Lovelock Cave, Nevada*. University of California Archaeological Research Facility Contributions 10.

Heizer, R. H., M. A. Baumhoff and C. W. Clewlow, Jr. 1968. Archaeology of South Fork Shelter (NV-El-11), Elko County, Nevada. *University of California Archaeological Survey Reports* 71:1–58.

Helzer, M. M. 2001. Paleoethnobotany and household archaeology at the Bergen site: A middle Holocene occupation in the Fort Rock Basin, Oregon. Ph.D. dissertation, University of Oregon, Eugene.

Helzer, M. M. 2004. Archaeological investigations at the Bergen Site: Middle Holocene lakeside occupations near Fort Rock, Oregon. In D. L. Jenkins, T. J. Connolly, and C. M. Aikens, eds., *Early and Middle Holocene Archaeology of the Northern Great Basin*, pp. 77–94. University of Oregon Anthropological Papers 62.

Hemphill, B. E., and C. S. Larsen. 1999. *Prehistoric Lifeways in the Great Basin Wetlands: Bioarchaeological Reconstruction and Interpretation*. University of Utah Press, Salt Lake City.

Hemphill, M. L., and P. E. Wigand. 1995. A detailed 2,000-year late Holocene pollen record from Lower Pahranagat Lake, southern Nevada, USA. In W. J. Waugh, K. L. Petersen, P. E. Wigand, B. D. Louthan, and R. D. Walker, eds., *Climate Change in the Four Corners and Adjacent Regions: Implications for Environmental Restoration and Land-use Planning*, pp. 41–49. National Technical Information Service: US Department of Energy CONF-94093250, Springfield, VA.

Hemphill-Haley, M., D. Lindberg, and M. Reheis. 1999. Lake Alvord and Lake Coyote; A hypothesized flood. In C. Narwold, organizer, *Quaternary Geology of the Northern Quinn River and Alvord Valleys, Southeastern Oregon. 1999 Friends of the Pleistocene Field Trip, Pacific Cell, September 24–26, 1999*, Article A2, pp. 1–7. Geology Department, Humboldt State University, Arcata, CA.

Herr, P. 1987. *Jessie Benton Frémont*. University of Oklahoma Press, Norman.

Herr, P., and M. L. Spence. 1993. *The Letters of Jessie Benton Frémont*. University of Illinois Press, Urbana.

Herschy, R. 2001. The world's maximum observed floods. In Á. Snorrason, H. P. Finnsdóttir, and M. E. Moss, eds., *The Extremes of the Extremes: Extraordinary Floods*, pp. 355–360. International Association of Hydrological Sciences Publication 271.

Herschy, R., compiler. 2003. *World Catalogue of Maximum Observed Floods*. International Association of Hydrological Sciences Publication 284.

Hersh, L. 2000. *The Central Pacific Railroad across Nevada 1868 & 1997: Photographic Comparatives*. Lawrence K. Hersh, Fernley, NV.

Hershler, R., and D. W. Sada. 2002. Biogeography of Great Basin aquatic snails of the genus *Pyrgulopsis*. In R. Hershler, D. B. Madsen, and D. R. Currey, eds., *Great Basin Aquatic Systems History*, pp. 255–276. Smithsonian Contributions to the Earth Sciences 33.

Hershler, R., and H.-P. Liu. 2008a. Ancient vicariance and recent dispersal of springsnails (Hydrobiidae: *Pyrgulopsis*) in the Death Valley system, California-Nevada. In M. C. Reheis, R. Hershler, and D. M. Miller, eds., *Late Cenozoic Drainage History of the Southwestern Great Basin and Lower Colorado River Region: Geologic and Biotic Perspectives*, pp. 91–101. Geological Society of America Special Paper 439.

Hershler, R., and H.-P. Liu. 2008b. Phylogenetic relationships of assimineid gastropods of the Death Valley–lower Colorado River region: Relicts of the late Neogene marine incursion? *Journal of Biogeography* 35:1816–1825.

Hershler, R., D. B. Madsen, and D. R. Currey, eds. 2002. *Great Basin Aquatic Systems History*. Smithsonian Contributions to the Earth Sciences 33.

Hertel, F. 1994. Diversity in body size and feeding morphology within past and present vulture assemblages. *Ecology* 75:1074–1084.

Hertel, F. 1995. Ecomorphological indicators of feeding behavior in recent and fossil raptors. *The Auk* 112:890–903.

Hester, T. R. 1973. *Chronological Ordering of Great Basin Prehistory*. Contributions of the University of California Archaeological Research Facility 17.

Heusser, C. J. 1960. *Late-Pleistocene Environments of North Pacific North America*. American Geographical Society Special Publication 35.

Hickerson, H. 1965. The Virginia deer and intertribal buffer zones in the upper Mississippi valley. In A. Leeds and A. P. Vayda, eds., *Man, Culture, and Animals*, pp. 43–66. American Association for the Advancement of Science Publication 78.

Hidy, G. M., and H. E. Klieforth. 1990. Atmospheric processes and the climates of the basin and range. In C. B. Osmond, L. F. Pitelka, and G. M. Hidy, eds., *Plant Biology of the Basin and Range*, pp. 17–46. Springer-Verlag, Berlin.

Hiebert, R. D., and J. L. Hamrick. 1983. Patterns and levels of genetic variation in Great Basin bristlecone pine, *Pinus longaeva*. *Evolution* 37:302–310.

Hildebrandt, W. R., and J. H. King. 2000. Projectile point variability along the northern California–Great Basin interface: Results from the Tuscarora-Alturas projects. In K. R. McGuire, ed., *Archaeological Investigations along the California–Great Basin Interface: The Alturas Transmission Line Project. Volume 1. Prehistoric Archaeological Studies: The Pit River Uplands, Madeline Plains, Honey Lake and Secret Valley, and Sierra Front Project Segments*, pp. 221–252. Coyote Press, Salinas, CA.

Hildebrandt, W. R., and J. H. King. 2002. Projectile point variability along the northern California–Great Basin interface: Results from the Tuscarora-Alturas projects. In K. R. McGuire, ed., *Boundary Lands: Archaeological Investigations along the California–Great Basin Interface*, pp. 5–28. Nevada State Museum Anthropological Papers 24.

Hildebrandt, W. R., and A. Ruby. 2006. Prehistoric pinyon exploitation in the southwestern Great Basin: A view from the Coso Range. *Journal of California and Great Basin Anthropology* 26:11–31.

Hill, J. N. 2001a. Languages on the land: Toward an anthropological dialectology. In J. Terrell, ed., *Archaeology, Languages, and History: Essays on Culture and Ethnicity*, pp. 257–282. Bergin and Garvey, Westport, CT.

Hill, J. N. 2001b. Proto-Uto-Aztecan: A community of cultivators in central Mexico? *American Anthropologist* 103:913–914.

Hill, J. N. 2002a. Proto-Uto-Aztecan cultivation and the northern devolution. In P. Bellwood and C. Renfrew, eds., *Examining the Farming/Language Dispersal Hypothesis*, pp. 331–340. McDonald Institute Monographs, McDonald Institute for Archaeological Research, University of Cambridge, Cambridge, UK.

Hill, J. N. 2002b. Toward a linguistic prehistory of the Southwest: "Aztec-Tanoan" and the arrival of maize cultivation. *Journal of Anthropological Research* 58:457–475.

Hill, J. N. 2006. *Uto-Aztecan hunter-gatherers: Language change in the Takic spread and Numic spread compared.* Abstract for Historical Linguistics and Hunter-Gatherer Populations in Global Perspective, 2006. Available at http://lingweb.eva.mpg.de/HunterGathererWorkshop2006/Abstracts.pdf.

Hill, J. N. 2008. Northern Uto-Aztecan and Kiowa-Tanoan: Evidence of contact between the proto-languages. *International Journal of American Linguistics* 74:155–188.

Hill, W. E. 1986. *The Oregon Trail: Yesterday and Today.* Caxton, Caldwell, ID.

Hockett, B. S. 1995. Chronology of Elko series and split stemmed points from northeastern Nevada. *Journal of California and Great Basin Anthropology* 17:41–53.

Hockett, B. S. 1997. Faunal Remains. In P. Buck, B. Hockett, F. Nials, and P. Wigand, *Prehistory and Paleoenvironment of Pintwater Cave, Nevada: Results of Field Work during the 1996 Season*, pp. 63–71. Project Number OS-005071, Nellis Air Force Base, Nevada.

Hockett, B. S. 1998. Sociopolitical meaning of faunal remains from Baker Village. *American Antiquity* 63:289–302.

Hockett, B. S. 1999. Unmodified faunal remains. In J. Wilde and R. A. Soper, *Baker Village. Report of Excavations, 1990–1994*, pp. 147–150. Brigham Young University Museum of Peoples and Cultures Technical Series 99–12.

Hockett, B. S. 2000. Paleobiogeographic changes at the Pleistocene-Holocene boundary near Pintwater Cave, southern Nevada. *Quaternary Research* 53:263–269.

Hockett, B. S. 2007. Nutritional ecology of late Pleistocene to middle Holocene subsistence in the Great Basin. In K. E. Graf and D. N. Schmitt, eds., *Paleoindian or Paleoarchaic: Great Basin Human Ecology at the Pleistocene/Holocene Transition*, pp. 204–230. University of Utah Press, Salt Lake City.

Hockett, B. S., and E. Dillingham. 2004. *Paleontological Investigations at Mineral Hill Cave.* Contributions to the Study of Cultural Resources Technical Report 18. U.S. Department of the Interior, Bureau of Land Management, Reno, NV.

Hockett, B. S., and M. Morgenstein. 2003. Ceramic production, Fremont foragers, and the late Archaic prehistory of the north-central Great Basin. *Utah Archaeology* 16:1–36.

Hockett, B. S., and T. W. Murphy. 2009. Antiquity of communal pronghorn hunting in the north-central Great Basin. *American Antiquity* 74:708–734.

Hockett, B. S., B. Brothers, and L. Seymour. 1997. *The Spring Creek Mastodon from Discovery to Exhibit.* Northeastern Nevada Historical Society, Elko.

Hockett, C. F. 1958. *A Course in Modern Linguistics.* Macmillan, New York.

Hoffecker, J. J., and S. A. Elias. 2007. *Human Ecology of Beringia.* Columbia University Press, New York.

Hoffmeister, D. F. 1986. *Mammals of Arizona.* University of Arizona Press; Arizona Fish and Game Department, Tucson.

Holanda, K. L. 2000. Reversing the trend: Late Holocene subsistence change in northeastern California. In K. R. McGuire, ed., *Archaeological Investigations along the California–Great Basin Interface: The Alturas Transmission Line Project. Volume 1. Prehistoric Archaeological Studies: The Pit River Uplands, Madeline Plains, Honey Lake and Secret Valley, and Sierra Front Project Segments*, pp. 283–297. Coyote Press, Salinas, CA.

Holden, C. 2008. Stone axes from the deep. *Science* 320:159.

Holland, D. 2005. *On Location in Lone Pine.* Second edition. Holland House, Santa Clarita, CA.

Holliday, V. T. 2009. Geoarchaeology and the search for the first Americans. *Catena* 78:310–322.

Holliday, V. T., and D. J. Meltzer. 2010. The 12.9ka impact hypothesis and North American Paleoindians. *Current Anthropology* 51:575–607.

Holman, J. A. 1961. Osteology of living and fossil New World quails (Aves, Galliformes). *Bulletin of the Florida State Museum, Biological Sciences* 6(2):131–233.

Holmer, R. N. 1986. Common projectile points of the Intermountain West. In C. J. Condie and D. D. Fowler, eds., *Anthropology of the Desert West: Essays in Honor of Jesse D. Jennings*, pp. 89–116. University of Utah Anthropological Papers 110.

Holmer, R. N. 1990. Prehistory of the Northern Shoshone. In E. S. Lohse and R. N. Holmer, eds., *Fort Hall and the Shoshone-Bannock*, pp. 41–59. Idaho State University Press, Pocatello, ID.

Hooke, R. Le B. 1972. Geomorphic evidence for late-Wisconsin and Holocene tectonic deformation, Death Valley, California. *Geological Society of America Bulletin* 83:2073–2098.

Hooke, R. LeB. 2002. Is there any evidence of Mega-Lake Manly in the eastern Mojave Desert during oxygen isotope stage 5e/6? *Quaternary Research* 57:177–179.

Hooke, R. LeB. 2005. Where was the southern end of Lake Manly during the Blackwelder stand? In Reheis, M. C., ed., *Geologic and Biotic Perspectives on Late Cenozoic Drainage History of the Southwestern Great Basin and Lower Colorado River Region: Conference Abstracts*, pp. 3–4. U.S. Geological Survey Open-File Report 2005–1004.

Hopkins, S. W. 1883 [1969]. *Life among the Paiutes: Their Wrongs and Claims.* Cupples, Upham, and Co. and G. P. Putnam's Sons, New York; reprinted by Sierra Media, Inc., Bishop.

Hostetler, S. W., and L. V. Benson. 1990. Paleoclimatic implications of the high stand of Lake Lahontan derived from models of evaporation and lake level. *Climate Dynamics* 4:207–217.

Hostetler, S. W., and P. U. Clark. 1997. Climate controls of western U.S. glaciation at the Last Glacial Maximum. *Quaternary Science Reviews* 16:505–511.

Hostetler, S. W., F. Giorgi, G. T. Bates, and P. J. Bartlein. 1994. Lake-atmosphere feedbacks associated with paleolakes Bonneville and Lahontan. *Science* 263:665–668.

Houghton, J. G. 1969. *Characteristics of Rainfall in the Great Basin.* Desert Research Institute, Reno, NV.

Houghton, J. G., C. M. Sakamoto, and R. O. Gifford. 1975. *Nevada's Weather and Climate.* Nevada Bureau of Mines and Geology Special Publication 2.

Houston, D. C. 1975. Ecological isolation of African scavenging birds. *Ardea* 63:55–64.

Howard, H. 1932. *Eagles and Eagle-like Vultures of the Pleistocene of Rancho La Brea.* Carnegie Institute of Washington Publication 429.

Howard, H. 1935. A new species of eagle from a Quaternary Cave deposit in eastern Nevada. *The Condor* 37:206–209.

Howard, H. 1946. A Review of the Pleistocene Birds of Fossil Lake, Oregon. *Carnegie Institute of Washington Publication* 551:141–195.

Howard, H. 1952. The prehistoric avifauna of Smith Creek Cave, Nevada, with a description of a new gigantic raptor. *Bulletin of the Southern California Academy of Sciences* 51:50–54.

Howard, H. 1955. *Fossil Birds from Manix Lake, California.* U.S. Geological Survey Professional Paper 264-J.

Howard, H. 1962. *A Comparison of Avian Assemblages from Individual Pits at Rancho La Brea, California.* Los Angeles County Museum Contributions in Science 58.

Howard, H. 1964a. *A New Species of the "Pigmy Goose," Anabernicula, from the Oregon Pleistocene, with a Discussion of the Genus.* American Museum Novitates 2200.

Howard, H. 1964b. Fossil Anseriformes. In J. Delacour, *The Waterfowl of the World*, Volume 4, pp. 233–326. Country Life, London.

Howard, H. 1971. Quaternary avian remains from Dark Canyon Cave, New Mexico. *The Condor* 73:237–240.

Howard, H. 1972. The incredible teratorn again. *The Condor* 74:341–344.

Howard, H., and A. H. Miller. 1933. Bird remains from cave deposits in New Mexico. *The Condor* 35:15–18.

Howard, H., and A. H. Miller. 1939. The avifauna associated with human remains at Rancho La Brea, California. *Carnegie Institute of Washington Publication* 514:41–48.

Hubbs, C. L., and R. R. Miller. 1948. The zoological evidence: Correlation between fish distribution and hydrographic history in the desert basins of western United States. *Bulletin of the University of Utah* 38(20), *Biological Series* 10(7):17–166.

Hubbs, C. L., R. R. Miller, and L. C. Hubbs. 1974. *Hydrographic History and Relict Fishes of the North-Central Great Basin.* Memoirs of the California Academy of Sciences 7.

Huckleberry, G., C. Beck, G. T. Jones, A. Holmes, M. Cannon, S. D. Livingston, and J. M. Broughton. 2001. Terminal Pleistocene/Early Holocene environmental change at the Sunshine Locality, north-central Nevada, U.S.A. *Quaternary Research* 55:303–312.

Hughes, M. K., and L. J. Graumlich. 1996. Multimillennial dendroclimatic studies from the western United States. In P. D. Jones, R. S. Bradley, and J. Jouzel, eds., *Climatic variations and forcing mechanisms of the last 2000 years*, pp. 109–124. NATO ASI Series 141.

Hughes, R. E. 1998 On reliability, validity and scale in obsidian sourcing research. In A. F. Ramenofsky and A. Steffen, eds., *Unit Issues in Archaeology: Measuring Time, Space, and Material*, pp. 103–114. University of Utah Press, Salt Lake City.

Hunt, C. B. 1966. *Plant Ecology of Death Valley, California*. U.S. Geological Survey Professional Paper 509.

Hunt, C. B. 1967. *Physiography of the United States*. W. H. Freeman, San Francisco.

Hunt, C. B. 1975. *Death Valley: Geology, Ecology, Archaeology*. University of California Press, Berkeley.

Hunt, C. B., and D. R. Mabey. 1966. *Stratigraphy and Structure, Death Valley, California*. U.S. Geological Survey Professional Paper 494-A.

Hunt, C. B., T. W. Robinson, W. A. Bowles, and A. L. Washburn. 1966. *Hydrologic Basin, Death Valley, California*. U.S. Geological Survey Professional Paper 494-B.

Hunt, J. M., D. Rhode, and D. B. Madsen. 2000. Homestead Cave flora and non-vertebrate fauna. In D. B. Madsen, *Late Quaternary Paleoecology in the Bonneville Basin*, pp. 47–58. Utah Geological Survey Bulletin 130.

Hunter, J. S., S. M. Durant, and T. M. Caro. 2006. Patterns of scavenger arrival at cheetah kills in Serengeti National Park, Tanzania. *African Journal of Ecology* 45:275–281.

Hunter, K. L., J. L. Betancourt, B. R. Riddle, T. R. Van Devender, K. L. Cole, and W. G. Spaulding. 2001. Ploidy race distributions since the Last Glacial Maximum in the North American desert shrub, *Larrea tridentata*. *Global Ecology and Biogeography* 10:521–533.

Hunziker, J. H., and C. Comas. 2002. *Larrea* interspecific hybrids revisited (Zygophyllaceae). *Darwiniana* 40:33–38.

Hurlbert, R. C., and J. J. Becker. 2001. Reptilia 3: Birds. In R. C. Hurlbert, ed., *The Fossil Vertebrates of Florida*, pp. 152–165. University Press of Florida, Gainesville. FL.

Hutchinson, P. W. 1988. The prehistoric dwellers at Lake Hubbs. In J. A. Willig, C. M. Aikens, and J. L. Fagan, eds., *Early Human Occupation in Far Western North America: The Clovis-Archaic Interface*, pp. 303–318. Nevada State Museum Anthropological Papers 21.

Hymes, D. H. 1960. Lexicostatistics so far. *Current Anthropology* 1:3–44.

Ingram, S. 2008. *Cacti, Agaves, and Yuccas of California and Nevada*. Cachuma Press, Los Olivos.

Irving, W. N., and C. R. Harington. 1973. Upper Pleistocene radiocarbon-dated artifacts from the northern Yukon. *Science* 179:335–340.

Irwin, C. N., ed. 1980. *The Shoshoni Indians of Inyo County, California: The Kerr Manuscript*. Ballena Press Publications in Archaeology, Ethnology, and History 15.

Ives, R. L. 1946. Glaciation in the desert ranges, Utah. *Journal of Geology* 54:335.

Jackson, D., and M. L. Spence, eds. 1970. *The Expeditions of John Charles Frémont. Volume 1: Travels from 1838 to 1844*. University of Illinois Press, Urbana.

Jackson, D., and M. L. Spence, eds. 1973. *The Expeditions of John Charles Frémont. Volume 2: The Bear Flag Revolt and the Court-Martial*. University of Illinois Press, Urbana. IL.

Jackson, L. E., Jr., and A. Duk-Rodkin. 1996. Quaternary geology of the ice-free corridor: Glacial controls on the peopling of the New World. In T. Akazawa and E. J. E. Szathmáry, eds., *Prehistoric Mongoloid Dispersals*, pp. 214–227. Oxford University Press, Oxford.

Jackson, L. E., Jr., and M. C. Wilson. 2004. The ice-free corridor revisited. *Geotimes* 49(2):16–19.

Jackson, R. H., and D. J. Stevens. 1981. Physical and cultural environment of Utah Lake and adjacent areas. *Great Basin Naturalist Memoirs* 5:3–23.

Jacobsen, W. H., Jr. 1986. Washoe language. In W. L. d'Azevedo, ed., *Great Basin, Handbook of North American Indians, Volume 11*, pp. 107–112. Smithsonian Institution Press, Washington, DC.

Jahren, A. H., R. Amundson, C. Kendall, and P. Wigand. 2001. Paleoclimatic reconstruction using the correlation in $\delta^{18}O$ of hackberry carbonate and environmental water, North America. *Quaternary Research* 56:253–263.

James, S. R. 1983. Surprise Valley settlement and subsistence: A critical review of the faunal evidence. *Journal of California and Great Basin Anthropology* 5:156–175.

James, S. R. 2004. Mineral Hill Cave avifauna. In B. S. Hockett, and E. Dillingham, *Paleontological Investigations at Mineral Hill Cave*, pp. 136–142. Contributions to the Study of Cultural Resources Technical Report 18. U.S. Department of the Interior Bureau of Land Management, Reno, NV.

Janetski, J. C. 1979. Implications of snare bundles in the Great Basin and Southwest. *Journal of California and Great Basin Anthropology* 1:306–321.

Janetski, J. C. 1986. The Great Basin Lacustrine subsistence pattern: Insights from Utah Valley. In C. J. Condie and D. D. Fowler, eds., *Anthropology of the Desert West: Essays in Honor of Jesse D. Jennings*, pp. 145–167. University of Utah Anthropological Papers 110.

Janetski, J. C. 1990. Wetlands in Utah Valley prehistory. In J. C. Janetski and D. B. Madsen, eds., *Wetland Adaptations in the*

Great Basin, pp. 233–257. Brigham Young University Museum of Peoples and Cultures Occasional Papers 1.

Janetski, J. C. 1991. *The Utes of Utah Lake*. University of Utah Anthropological Papers 116.

Janetski, J. C. 1994. Recent transitions in the eastern Great Basin: The archaeological record. In D. B. Madsen and D. Rhode, eds., *Across the West: Human Population Movement and the Expansion of the Numa*, pp. 157–178. University of Utah Press, Salt Lake City.

Janetski, J. C. 1997. Fremont hunting and resource intensification in the eastern Great Basin. *Journal of Archaeological Science* 24:1075–1088.

Janetski, J. C. 1998. *Archaeology of Clear Creek Canyon*. Museum of Peoples and Cultures, Brigham Young University, Provo, UT.

Janetski, J. C. 1999a. Julian Steward and Utah archaeology. In R. O. Clemmer, L. D. Myers, and M. E. Rudden, eds., *Julian Steward and the Great Basin: The Making of an Anthropologist*, pp. 19–34. University of Utah Press, Salt Lake City.

Janetski, J. C. 1999b. Role of pinyon-juniper woodlands in aboriginal societies of the Desert West. In S. B. Monsen and R. Stevens, compilers, *Proceedings: Ecology and Management of Pinyon-Juniper Communities within the Interior West*, pp. 249–253. USDA Forest Service Proceedings RMRS-P-9.

Janetski, J. C. 2002. Trade in Fremont society: Contexts and contrasts. *Journal of Anthropological Archaeology* 21:344–370.

Janetski, J. C. 2004. *2003 Test Excavations at Mosquito Willie (42TO137)*. Brigham Young University Museum of Peoples and Cultures Technical Series 04-12.

Janetski, J. C. 2006. Great Basin animals. In D. J. Stanford, B. D. Smith, D. H. Ubelaker, and E. J. E. Szathmáry, eds., *Environment, Origins, and Population, Handbook of North American Indians, Volume 3*, pp. 351–364. Smithsonian Institution Press, Washington, DC.

Janetski, J. C. 2007. Hunter-gatherer strategies in Utah Valley. In J. C. Janetski and G. C. Smith, *Hunter-Gatherer Archaeology in Utah Valley*, pp. 317–340. Brigham Young University Museum of Peoples and Cultures Occasional Papers 12.

Janetski, J. C. 2008. The enigmatic Fremont. In C. S. Fowler and D. D. Fowler, eds., *The Great Basin: People and Place in Ancient Times*, pp. 105–115. School for Advanced Research Press, Santa Fe, NM.

Janetski, J. C., and D. B. Madsen, eds. 1990. *Wetland Adaptations in the Great Basin*. Brigham Young University Museum of Peoples and Cultures, Occasional Papers 1.

Janetski, J. C., and G. C. Smith. 2007. *Hunter-Gatherer Archaeology in Utah Valley*. Brigham Young University Museum of Peoples and Cultures Occasional Papers 12.

Janetski, J. C., R. K. Talbot, D. E. Newman, L. D. Richens, and J. D. Wilde. 2000. *Clear Creek Canyon Archaeological Project: Results and Synthesis*. Brigham Young University Museum of Peoples and Cultures Occasional Papers 7.

Jantz, R. L., and D. W. Owsley. 1997. Pathology, taphonomy, and cranial morphometrics of the Spirit Cave mummy. *Nevada Historical Society Quarterly* 40:62–84.

Jass, C. N. 2007. New perspectives on Pleistocene biochronology and biotic change in the east-central Great Basin: An examination of the vertebrate fauna from Cathedral Cave, Nevada. Ph.D. dissertation, University of Texas, Austin.

Jayko, A. S. 2005. Late Quaternary denudation, Death and Panamint Valleys, eastern California. *Earth-Science Reviews* 73:271–289.

Jayko, A. S., and S. N. Bacon. 2008. Late Quaternary MIS 6–8 shoreline features of pluvial Owens Lake, Owens Valley, eastern California. In M. C. Reheis, R. Hershler, and D. M. Miller, eds., *Late Cenozoic Drainage History of the Southwestern Great Basin and Lower Colorado River Region: Geologic and Biotic Perspectives*, pp. 185–206. Geological Society of America Special Paper 439.

Jayko, A. S., R. M. Forester, D. S. Kaufmann, F. M. Phillips, J. C. Yount, J. McGeehin, and S. A. Mahan. 2008. Late Pleistocene lakes and wetlands, Panamint Valley, Inyo County, California. In M. C. Reheis, R. Hershler, and D. M. Miller, eds., *Late Cenozoic Drainage History of the Southwestern Great Basin and Lower Colorado River Region: Geologic and Biotic Perspectives*, pp. 151–184. Geological Society of America Special Paper 439.

Jefferson, G. T. 1982. Late Pleistocene vertebrates from a Mormon Mountain Cave in southern Nevada. *Bulletin of the Southern California Academy of Sciences* 81:121–127.

Jefferson, G. T. 1985. *Review of the Late Pleistocene Avifauna from Lake Manix, Central Mojave Desert, California*. Natural History Museum of Los Angeles County Contributions in Science 362.

Jefferson, G. T. 1987. The Camp Cady Local Fauna: Paleoenvironment of the Lake Manix Basin. *San Bernardino County Museum Association Quarterly* 34 (3–4):3–35.

Jefferson, G. T. 1991. Rancholabrean age vertebrates from the southeastern Mojave Desert, California. In R. E. Reynolds, ed., *Crossing the Borders: Quaternary Studies in Eastern California and Southwestern Nevada*, pp. 163–176. San Bernardino County Museum Association, Redlands, CA.

Jefferson, G. T. 2003. Stratigraphy and paleontology of the middle to late Pleistocene Manix Formation and paleoenvironments of the central Mojave River, southern California. In Y. Enzel, S. G. Wells, and N. Lancaster, eds., *Paleoenvironments and Paleohydrology of the Mojave and Southern Great Basin Deserts*, pp. 43–60. Geological Society of America Special Paper 368.

Jefferson, G. T., and A. E. Tejada-Flores. 1993. The late Pleistocene record of *Homotherium* (Felidae: Machairodontinae) in the southwestern United States. *PaleoBios* 15:37–46.

Jefferson, G. T., H. G. McDonald, and S. D. Livingston. 2004. *Catalogue of Late Quaternary and Holocene Fossil Vertebrates from Nevada*. Nevada State Museum Occasional Papers 6.

Jefferson, G. T., H. G. McDonald, W. A. Akerstein, and S. J. Miller. 2002. Catalogue of late Pleistocene and Holocene fossil vertebrates from Idaho. In W. A. Akerstein, M. E. Thompson, D. J. Meldrum, R. A. Rapp, and H. G. McDonald, eds., *And Whereas . . . Papers on the Vertebrate Paleontology of Idaho Honoring John A. White, Volume 2*, pp. 157–192. Idaho Museum of Natural History Occasional Paper 37.

Jefferson, G. T., W. E. Miller, M. E. Nelson, and J. H. Madsen, Jr. 1994. *Catalogue of Late Quaternary Vertebrates from Utah*. Natural History Museum of Los Angeles County Technical Reports 9.

Jehl, J. R., Jr. 1967. Pleistocene birds from Fossil Lake, Oregon. *The Condor* 69:24–27.

Jenkins, D. S. 1994. Archaeological investigations at three wetlands sites in the Silver Lake area of the Fort Rock Basin. In C. M. Aikens and D. L. Jenkins, eds., *Archaeological Researches in the Northern Great Basin: Fort Rock Archaeology Since Cressman*, pp. 213–258. University of Oregon Anthropological Papers 50.

Jenkins, D. L. 2000. Early to middle Holocene cultural transitions in the Northern Great Basin of Oregon: The view from Fort Rock. In J. S. Schneider, R. M. Yohe II, and J. K. Gardner, eds., *Archaeological Passages: a Volume in Honor of Claude Nelson Warren*, pp. 69–109. Western Center for Archaeology and Paleontology Publications in Archaeology 1.

Jenkins, D. L. 2007. Distribution and dating of cultural and paleontological remains at the Paisley Five Mile Point caves in the

Jenkins, D. L., C. M. Aikens, and W. J. Cannon. In K. E. Graf and D. N. Schmitt, eds., *Paleoindian or Paleoarchaic: Great Basin Human Ecology at the Pleistocene/Holocene Transition*, pp. 57–81. University of Utah Press, Salt Lake City.

Jenkins, D. L., C. M. Aikens, and W. J. Cannon. 2002. *Reinvestigation of the Connley Caves (35LK50): A Pivotal Early Holocene Site in the Fort Rock Basin of South-central Oregon.* Manuscript on file at the University of Oregon Museum of Natural History, Eugene.

Jenkins, D. L., M. S. Droz, and T. J. Connolly. 2004. Geoarchaeology of wetland settings in the Fort Rock Basin, south-central Oregon. In D. L. Jenkins, T. J. Connolly, and C. M. Aikens, eds., *Early and Middle Holocene Archaeology of the Northern Great Basin*, pp. 13–52. University of Oregon Anthropological Papers 62.

Jennings, J. D. 1957. *Danger Cave*. University of Utah Anthropological Papers 27.

Jennings, J. D. 1986. Prehistory: Introduction. In W. L. d'Azevedo, ed., *Great Basin, Handbook of North American Indians, Volume 11*, pp. 113–119. Smithsonian Institution Press, Washington, DC.

Jennings, J. D. 1994. *Accidental Archaeologist: Memoirs of Jesse D. Jennings*. University of Utah Press, Salt Lake City.

Jennings, S. A. 1996. Analysis of pollen contained in middens from the White Mountains and volcanic tableland of eastern California. *Palynology* 20:5–13.

Jennings, S. A., and D. L. Elliott-Fisk. 1993. Packrat midden evidence of late Quaternary vegetation change in the White Mountains, California-Nevada. *Quaternary Research* 39:214–221.

Jewell, P. W. 2008. Morphology and paleoclimatic significance of Pleistocene Lake Bonneville spits. *Quaternary Research* 68:421–430.

Johnson, D. M., R. R. Petersen, D. R. Lycan, J. W. Sweet, M. Neuhaus, and A. L. Schaedel. 1985. *Atlas of Oregon Lakes*. Oregon State University, Corvallis.

Johnson, L., and J. Johnson. 1987. *Escape from Death Valley*. University of Nevada Press, Reno.

Johnson, W. G., S. E. Sharpe, T. F. Bullard, and K. Lupo. 2005. Characterizing a first occurrence of bison deposits in southeastern Nevada. *Western North American Naturalist* 65:24–35.

Johnston, W. A. 1933. Quaternary geology of North America in relation to the migration of man. In D. Jenness, ed., *The American Aborigines: Their Origin and Antiquity*, pp. 11–45. University of Toronto Press, Toronto.

Jolie, E. A. 2004. Coiled basketry from Charlie Brown Cave, western Nevada. M.A. thesis, University of Nevada, Reno.

Jolie, E. A., and R. B. Jolie. 2008. Hats, baskets, and trays from Charlie Brown Cave. In C. S. Fowler and D. D. Fowler, eds., *The Great Basin: People and Place in Ancient Times*, pp. 74–77. School for Advanced Research Press, Santa Fe, NM.

Jones, G. T., and B. Beck. 1999. Paleoarchaic archaeology in the Great Basin. In C. Beck, ed., *Models for the Millennium: Great Basin Anthropology Today*, pp. 83–95. University of Utah Press, Salt Lake City.

Jones, G. T., C. Beck, and R. A. Kessler. 2002. *Analysis of Artifacts from the Old River Bed Delta, Dugway Proving Ground, Utah.* Report prepared for and submitted to David B. Madsen in fulfillment of the contract agreement with the University and Community College System of Nevada, signed on June 6, 2002.

Jones, G. T., C. Beck, E. E. Jones, and R. E. Hughes. 2003. Lithic source use and Paleoarchaic foraging territories in the Great Basin. *American Antiquity* 68:5–38.

Jones, P. D., and M. E. Mann. 2004. Climate over past millennia. *Reviews of Geophysics* 42, RG2002, doi: 10.1029/2003RG000143.

Jones, P. D., T. J. Osborn, and K. R. Briffa. 2001. The evolution of climate over the last millennium. *Science* 292:662–667.

Jones, T. L, and K. A. Klar, eds. 2007. *California Prehistory: Colonization, Culture, and Complexity*, pp. 11–34. AltaMira Press, Lanham.

Jones, T. L., J. F. Porcasi, J. M. Erlandson, H. Dallas, Jr., T. A. Wake, and R. Schwaderer, 2008. The protracted Holocene extinction of California's flightless sea duck (*Chendytes lawi*) and its implications for the Pleistocene overkill hypothesis. *Proceedings of the National Academy of Sciences* 105:4105–4108.

Jones, T. L., G. M. Brown, L. M. Raab, J. L. McVickar, W. G. Spaulding, D. J. Kennett, A. York, and P. L. Walker. 1999. Environmental imperatives reconsidered. *Current Anthropology* 40:137–170.

Jones, V. H. 1945. The use of honey-dew as food by Indians. *The Masterkey* 19:145–149.

Jones, W. T. 1993. The social systems of heteromyid rodents. In H. H. Genoways and J. H. Brown, eds., *Biology of the Heteromyidae*, pp. 575–595. American Society of Mammalogists Special Publication 10.

Jørgensen, S., J. L. Hamrick, and P. V. Wells. 2002. Regional patterns of genetic diversity in *Pinus flexilis* (Pinaceae) reveal complex species history. *American Journal of Botany* 89:792–800.

Kaestle, F. A., and D. G. Smith. 2001. Ancient mitochondrial DNA evidence for prehistoric population movement: The Numic expansion. *American Journal of Physical Anthropology* 115:1–12.

Kaestle, F. A., J. G. Lorenz, and D. G. Smith. 1999. Molecular genetics and the Numic expansion: A molecular investigation of the prehistoric inhabitants of Stillwater Marsh. In B. E. Hemphill, and C. S. Larsen, eds., *Prehistoric Lifeways in the Great Basin Wetlands: Bioarchaeological Reconstruction and Interpretation*, pp. 167–183. University of Utah Press, Salt Lake City.

Kaliser, B. N. 1989. *Water-Related Geologic Problems of 1983: Utah Occurrences by County.* Utah Geological and Mineral Survey Miscellaneous Publication 89-4.

Kaliser, B. N., and J. E. Slosson. 1988. *Geologic Consequences of the 1983 Wet Year in Utah.* Utah Geological and Mineral Survey Miscellaneous Publication 88-3.

Kaufman, D. S. 2003. Amino acid paleothermometry of Quaternary ostracodes from the Bonneville Basin, Utah. *Quaternary Science Reviews* 22:899–914.

Kaufman, D. S., S. L. Forman, and J. Bright. 2001. Age of the Cutler Dam alloformation (Late Pleistocene), Bonneville Basin, Utah. *Quaternary Research* 56:322–334.

Kaufman, D. S., S. C. Porter, and A. R. Gillespie. 2004. Quaternary glaciation in Alaska, the Pacific Northwest, and Hawaii. In A. R. Gillespie, S. C. Porter and B. F. Atwater, eds., *The Quaternary Period in the United States*. pp. 77–103. Developments in Quaternary Science 1. Elsevier, Amsterdam.

Kaufman, D. S., J. Bright, W. E. Dean, J. G. Rosenbaum, K. Moser, R. S. Anderson, S. M. Colman, C. W. Heil, Jr., G. Jiménez-Moreno, M. C. Reheis, and K. R. Simmons. 2009. A quarter-million years of paleoenvironmental change at Bear Lake, Utah and Idaho. In J. G. Rosenbaum and D. S. Kaufman, eds., *Paleoenvironments of Bear Lake, Utah and Idaho, and its Catchment*, pp. 311–351. Geological Society of America Special Paper 450.

Kavanagh, T. W. 2001. Comanche. In R. J. DeMaille, ed., *Plains, Handbook of North American Indians, Volume 13*, pp. 886–906. Smithsonian Institution, Washington, DC.

Keigwin, L. D., J. P. Donnelly, M. S. Cook, N. W. Driscoll, and J. Brigham-Grette. 2006. Rapid sea-level rise and Holocene climate in the Chuckchi Sea. *Geology* 34:861–864.

Kelly, I. T. 1932. Ethnography of the Surprise Valley Paiute. *University of California Publications in American Anthropology and Archaeology* 31(3):67–210.

Kelly, R. L. 1995. *The Foraging Spectrum: Diversity in Hunter-Gatherer Lifeways*. Smithsonian Institution Press, Washington, DC.

Kelly, R. L. 1997. Late Holocene Great Basin prehistory. *Journal of World Archaeology* 11:1–49.

Kelly, R. L. 1999. Theoretical and archaeological insights into foraging strategies among the prehistoric inhabitants of the Stillwater Marsh wetlands. In B. E. Hemphill and C. S. Larsen, eds., *Prehistoric Lifeways in the Great Basin Wetlands: Bioarchaeological Reconstruction and Interpretation*, pp. 117–150. University of Utah Press, Salt Lake City.

Kelly, R. L. 2001. *Prehistory of the Carson Desert and Stillwater Mountains: Environment, Mobility, and Subsistence in a Great Basin Wetland*. University of Utah Anthropological Papers 123.

Kelly, R. L. 2007. *Mustang Shelter: Test excavations of a Rockshelter in the Stillwater Mountains, Western Nevada*. Bureau of Land Management Cultural Resource Series 18, Reno, NV.

Kelly, R. L., and E. Hattori. 1985. Present environment and history. In D. H. Thomas, ed., *The Archaeology of Hidden Cave, Nevada*, pp. 39–46. American Museum of Natural History Anthropological Papers 61(1).

Kemp, B. M. 2006. Mesoamerica and Southwest prehistory and the entrance of humans into the Americas: Mitochondrial DNA evidence. Ph.D. dissertation, University of California, Davis.

Kenagy, G. J. 1972. Saltbush leaves: excision of hypersaline tissues by a kangaroo rat. *Science* 178:1094–1096.

Kenagy, G. J. 1973. Adaptations for leaf eating in the Great Basin kangaroo rat, *Dipodomys microps*. *Oecologia* 12:383–412.

Kenagy, G. J., and G. A. Bartholomew. 1985. Seasonal reproductive patterns in five coexisting California desert rodent species. *Ecological Monographs* 55:371–397.

Kennett, D. J., J. P. Kennett, A. West, C. Mercer, S. S. Que Hee, L. Bement, T. E. Bunch, M. Sellers, and W, S, Wolbach. 2009a. Nanodiamonds in the Younger Dryas boundary sediment layer. *Science* 323:94.

Kennett, D. J., J. P. Kennett, A. West, G. J. West, T. E. Bunch, B. J. Culleton, J. M. Erlandson, S. S. Que Hee, J. R. Johnson, C. Mercer, F. Shen, M. Sellers, T. W. Stafford, Jr., A. Stich, J. C. Weaver, J. H. Wittke, and W. S. Wolbach. 2009b. Shock-synthesized hexagonal diamonds in Younger Dryas boundary sediments. *Proceedings of the National Academy of Sciences* 106:2519–2524.

Kerns, V. 1999. Learning the Land. In R. O. Clemmer, L. D. Myers, and M. E. Rudden, eds., *Julian Steward and the Great Basin: The Making of an Anthropologist*, pp. 1–18. University of Utah Press, Salt Lake City.

Kerns, V. 2003. *Scenes from the High Desert: Julian Steward's Life and Theory*. University of Illinois Press, Urbana.

Kerr, R. 2008. Experts find no evidence for a mammoth-killer impact. *Science* 319:1331–1332.

Kerr, R. 2009. Did the mammoth slayer leave a diamond calling card? *Science* 323:26.

Kerr, R. 2010. Mammoth-killer impact flunks out. *Science* 329:1140–1141.

Kessler, R. A., C. Beck, and G. T. Jones. 2009. Trash: The structure of Great Basin Paleoarchaic debitage assemblages in western North America. In B. Adams and R. S. Blades, eds., *Lithic Materials and Paleolithic Societies*, pp. 144–159. Blackwell Publishing, Chichester, UK.

Kiahtipes, C. A. 2009. Fire in the desert: Holocene paleoenvironments in the Bonneville Basin. M.A. thesis, Washington State University, Pullman.

Kidder, A. V., Guernsey., S. J. 1919. *Archaeological Explorations in Northeastern Arizona*. Bulletin of American Ethnology 65.

King, G. Q. 1978. The late Quaternary history of Adrian Valley, Lyon County, Nevada. M.S. thesis, University of Utah, Salt Lake City.

King, G. Q. 1993. Late Quaternary history of the lower Walker River and its implications for the Lahontan paleolake system. *Physical Geography* 14:81–96.

King, G. Q. 1996. Geomorphology of a dry valley: Adrian Pass, Lahontan Basin, Nevada. *Association of Pacific Coast Geographers Yearbook* 58:89–114.

King, G. Q. 2003. *The Forty Mile Desert Emigrant Trail: Its Natural and Human History*. Camp Nevada Monographs 11.

King, T. J., Jr. 1976. Late Pleistocene-early Holocene history of coniferous woodlands in the Lucerne Valley region, Mohave Desert, California. *Great Basin Naturalist* 36:227–238.

Kington, H. H. 2008. Hubert H. Lamb: A review of his life and work. *Weather* 63:187–189.

Kirby, M. E., and J. T. Andrews. 1999. Mid-Wisconsin Laurentide Ice Sheet growth and decay: Implications for Heinrich events 3 and 4. *Paleoceanography* 14:211–233.

Kirch, P. V. 2000. *On the Road of the Winds: An Archaeological History of the Pacific Islands before European Contact*. University of California Press, Berkeley.

Klein, R. G. 2009. *The Human Career: Human Biological and Cultural Origins*. Third edition. University of Chicago Press, Chicago, IL.

Kleppe, J. A. 2005. A study of ancient trees rooted 36.5 m (120′) below the surface level of Fallen Leaf Lake, California. *Journal of the Nevada Water Resources Association* 2:29–40.

Klinger, R. E. 2001. Lacustrine deposition of Lake Manly or springs near Titus Canyon? In M. N. Machette, M. L. Johnson, and J. L. Slate, eds., *Quaternary and Late Pliocene Geology of the Death Valley Region: Recent Observations on Tectonics, Stratigraphy, and Lake Cycles (Guidebook for the 2001 Pacific Cell—Friends of the Pleistocene Field Trip)*, pp. A38–A40. U.S. Geological Survey Open-File Report 01-51.

Kloor, K. 2007. The vanishing Fremont. *Science* 318:1540–1543.

Knapp, P. A. 1996. Cheatgrass (*Bromus tectorum L*) dominance in the Great Basin Desert. *Global Environmental Change* 6:37–52.

Knott, J. R. 1997. An early to middle Pleistocene pluvial Death Valley lake, Mormon Point, central Death Valley, California. In R. E. Reynolds and J. Reynolds, eds., *Death Valley: The Amargosa Route*, pp. 85–88. San Bernardino County Museum Association Quarterly 44(2).

Knott, J. R., J. C. Tinsley III, and S. G. Wells. 2004. Reply to Hooke (2004). *Quaternary Research* 61:344–347.

Knott, J. R., M. N. Machette, R. E. Klinger, A. M. Sarna-Wojcicki, J. C. Liddicoat, J. C. Tinsley III, B. T. David, and V. M. Ebbs. 2008. Reconstructing late Pliocene to middle Pleistocene Death Valley lakes and river systems as a test of pupfish (Cyprinodontidae) dispersal hypotheses. In M. C. Reheis, R. Hershler, and D. M. Miller, eds., *Late Cenozoic Drainage History of the Southwestern Great Basin and Lower Colorado River Region: Geologic and Biotic Perspectives*, pp. 1–26. Geological Society of America Special Paper 439.

Koch, P. L., and A. D. Barnosky. 2006. Late Quaternary extinctions: State of the debate. *Annual Review of Ecology, Evolution, and Systematics* 37:215–250.

Koehler, P. A., and R. S. Anderson. 1994. Full-glacial shoreline vegetation during the maximum highstand at Owens Lake, California. *Great Basin Naturalist* 54:142–149.

Koehler, P. A., and R. S. Anderson. 1995. Thirty thousand years of vegetation changes in the Alabama Hills, Owens Valley, California. *Quaternary Research* 43:238–248.

Koehler, P. A., R. S. Anderson, and W. G. Spaulding. 2005. Development of vegetation in the central Mojave Desert of California during the Quaternary. *Palaeogeography, Palaeoclimatology, Palaeoecology* 215:297–311.

Kohler, T. A., M. P. Glaude, J.-P. Bocquet-Appel, and B. M. Kemp. 2008. The Neolithic demographic transition in the U. S. Southwest. *American Antiquity* 73:645–669.

Kooyman, B., M. E. Newman, C. Cluney, M. Lobb, S. Tolman, P. McNeil and L. Hills. 2001. Identification of horse exploitation by Clovis hunters based on protein analysis. *American Antiquity* 66:686–691.

Krider, P. R. 1998. Paleoclimatic significance of late Quaternary lacustrine and alluvial stratigraphy, Animas Valley, New Mexico. *Quaternary Research* 50:283–289.

Kroeber, A. L. 1953. *Handbook of the Indians of California*. California Book Company, Berkeley.

Kropf, M., J. I. Mead, and R. S. Anderson. 2007. Dung, diet, and the paleoenvironment of the extinct shrub-ox (*Euceratherium collinum*) on the Colorado Plateau, USA. *Quaternary Research* 67:143–151.

Kruuk, H. 1967. Competition for food between vultures in East Africa. *Ardea* 55:171–193.

Ku, T.-L., S. Luo, T. Lowenstein, J. Li, and R. J. Spencer. 1998. U-series chronology of lacustrine deposits in Death Valley, California. *Quaternary Research* 50:261–275.

Kuehn, S. C., and F. F. Foit. 2001. Updated tephra stratigraphy at Summer Lake, Oregon, a sub-basin of pluvial Lake Chewaucan. In R. Negrini, S. Pezzopane, and T. Badger, eds., *Quaternary Studies near Summer Lake, Oregon: Guidebook for the Friends of the Pleistocene Ninth Annual Pacific Northwest Cell Field Trip*, pp. SK1–SK9.

Kuehn, S. C., and F. F. Foit. 2006. Correlation of widespread Holocene and Pleistocene tephra layers from Newberry Volcano, Oregon, USA, using glass composition and numerical analysis. *Quaternary International* 148:113–137.

Kurtén, B. 1958. *Life and Death of the Pleistocene Cave Bear*. Acta Zoologica Fennica 95.

Kurtén, B. 1968. *The Cave Bear Story*. Columbia University Press, New York.

Kurtén, B., and E. Anderson. 1980. *Pleistocene Mammals of North America*. Columbia University Press, New York.

Kurzius, M. A. 1981. *Vegetation and Flora of the Grapevine Mountains, Death Valley National Monument, California-Nevada*. Cooperative National Park Resources Studies Unit, University of Nevada, Las Vegas, Contribution CPSU/UNLV 017/06.

Kutzbach, J. E. 1987. Model simulations of the climatic patterns during the deglaciation of North America. In W. F. Ruddiman and H. E. Wright Jr., eds., *North America and Adjacent Oceans during the Last Deglaciation*, pp. 425–446. The Geology of North America, Volume K-3. Geological Society of America, Boulder, CO.

Kutzbach, J. E., and P. J. Guetter. 1986. The influence of changing orbital parameters and surface boundary conditions on climate simulations for the past 18,000 years. *Journal of the Atmospheric Sciences* 43:1726–1759.

Kutzbach, J., R. Gallimore, S. Harrison, P. Behling, R. Selin, and F. Laarif. 1998. Climate and biome simulations for the past 21,000 years. *Quaternary Science Reviews* 17:473–506.

Kuzmin, Y. V. 2010. Extinction of the woolly mammoth (*Mammuthus primigenius*) and woolly rhinoceros (*Coelodonta antiquitatis*) in Eurasia: Review of chronological and environmental issues. *Boreas* 39:247–261.

La Rivers, I. 1962. *Fishes and Fisheries of Nevada*. Nevada State Fish and Game Commission, Carson City.

Laabs, B. J. C., and D. S. Kaufman. 2003. Quaternary highstands in Bear Lake Valley, Utah and Idaho. *Geological Society of America Bulletin* 115:463–478.

Laabs, B. J. C., M. A. Plummer, and D. M. Mickelson. 2006. Climate during the Last Glacial Maximum in the Wasatch and southern Uinta Mountains inferred from glacier modeling. *Geomorphology* 75:300–317.

Laabs, B. J. C., J. S. Munroe, J. G. Rosenbaum, K. A. Refsnider, D. M. Mickelson, B. S. Singer, and M. W. Caffee. 2007. Chronology of the Last Glacial Maximum in the Upper Bear River Basin, Utah. *Arctic, Antarctic, and Alpine Research* 39:537–548.

Lajoie, K. R. 1968. Late Quaternary stratigraphy and geologic history of Mono Basin, eastern California. Ph.D. dissertation, University of California, Berkeley.

Lajoie, K. R., and S. W. Robinson. 1982. Late Quaternary glaciolacustrine chronology Mono Basin, California. *Geological Society of America Abstracts with Programs* 14:179.

Laliberte, A. S., and W. J. Ripple. 2003. Wildlife encounters by Lewis and Clark: A spatial analysis of interactions between native Americans and wildlife. *BioScience* 53:994–1002.

LaMarche, V. C., Jr. 1965. Distribution of Pleistocene glaciers in the White Mountains of California and Nevada. *U.S. Geological Survey Professional Paper* 525-C: C144–C146.

LaMarche, V. C., Jr. 1973. Holocene climatic variations inferred from treeline fluctuations in the White Mountains, California. *Quaternary Research* 3:632–660.

LaMarche, V. C., Jr. 1974. Paleoclimatic inferences from long tree-ring records. *Science* 183:1043–1048.

LaMarche, V. C., Jr., and H. A. Mooney. 1967. Altithermal timberline advance in western United States. *Nature* 213:980–982.

LaMarche, V. C., Jr., and H. A. Mooney. 1972. Recent climatic change and development of the bristlecone pine (*P. longaeva* Bailey) krummholz zone, Mt. Washington, Nevada. *Arctic and Alpine Research* 4:61–72.

Lamarra, V., C. Liff, and J. Carter. 1986. Hydrology of Bear Lake Basin and its impact on the trophic state of Bear Lake, Utah-Idaho. *Great Basin Naturalist* 46:690–705.

Lamb, H. H. 1965. The early Medieval Warm Epoch and its sequel. *Palaeogeography, Palaeoclimatology, Palaeoecology* 1:13–37.

Lamb, S. M. 1958. Linguistic prehistory in the Great Basin. *International Journal of American Linguistics* 29:95–100.

Lambeck, K., Y. Yoyoyama, and T. Purcell. 2002. Into and out of the Last Glacial Maximum: Sea-level change during Oxygen Isotope States 3 and 2. *Quaternary Science Reviews* 21:343–360.

Lambert, P. M., and S. R. Simms. 2003. *Archaeology and Analysis of Prehistoric Human Remains near Willard Bay, Utah*. Utah State University Contributions to Anthropology 33.

Lange, A. L. 1956. Woodchuck remains in northern Arizona caves. *Journal of Mammalogy* 37, 289–291.

Lanner, R. M. 1981. *The Piñon Pine: A Cultural and Natural History*. University of Nevada Press, Reno.

Lanner, R. M. 1983a. The expansion of singleaf piñon in the Great Basin. In D. H. Thomas, *The Archaeology of Monitor*

Valley. 2. *Gatecliff Shelter*, pp. 167–171. American Museum of Natural History Anthropological Papers 59(1).

Lanner, R. M. 1983b. *Trees of the Great Basin*. University of Nevada Press, Reno.

Lanner, R. M. 1996. *Made for Each Other: A Symbiosis of Birds and Pines*. Oxford University Press, New York.

Lanner, R. M. 1999. *Conifers of California*. Cachuma Press. Los Olivos, CA.

Lanner, R. M. 2007. *The Bristlecone Book: A Natural History of the World's Oldest Trees*. Mountain Press, Missoula, MT.

Lanner, R. M., and T. R. Van Devender. 1998. The recent history of pinyon pines in the American Southwest. In D. M. Richardson, ed., *Ecology and Biogeography of Pinus*, pp. 171–182. Cambridge University Press, Cambridge, UK.

Lao, Y., and L. Benson. 1988. Uranium-series age estimates and paleoclimatic significance of Pleistocene tufas from the Lahontan Basin, California and Nevada. *Quaternary Research* 30:165–176.

Largaespada, T. D. 2006. Significant points in time: A typology and chronology of middle and late Holocene projectile points from the northern Great Basin. In B. L. O'Neill, ed., *Beads, Points, and Pit Houses: A Northern Great Basin Miscellany*, pp. 69–92. University of Oregon Anthropological Papers 66.

Larrucea, E. S., and P. F. Brussard. 2008a. Habitat selection and current distribution of the pygmy rabbit in Nevada and California, USA. *Journal of Mammalogy* 89:691–699.

Larrucea, E. S., and P. F. Brussard. 2008b. Shift in location of pygmy rabbit *(Brachylagus idahoensis)* habitat in response to changing environments. *Journal of Arid Environments* 72:1636–1643.

Larsen, C. S., and R. L. Kelly. 1995. *Bioarchaeology of the Stillwater Marsh: Prehistoric Human Adaptation in the Western Great Basin*. American Museum of Natural History Anthropological Papers 77.

Larsen, C. S., C. B. Ruff, and R. L. Kelly. 1995. Structural analysis of the Stillwater postcranial remains: Behavioral implications of articular joint pathology and long bone diaphyseal morphology. In C. S. Larsen and R. L. Kelly, *Bioarchaeology of the Stillwater Marsh: Prehistoric Human Adaptation in the Western Great Basin*, pp. 107–133. American Museum of Natural History Anthropological Papers 77.

Larsen, V. 1990. A fluted point from Clear Creek Canyon, central Utah. *Utah Archaeologist* 3:133–136.

Lawler, J. J., S. L. Shafer, D. White, P. Kareiva, E. P. Maurer, A. R. Blaustein, and P. J. Bartlein. 2009. Projected climate-induced faunal change in the Western Hemisphere. *Ecology* 90:588–597.

Lawlor, T. E. 1998. Biogeography of Great Basin mammals: Paradigm lost? *Journal of Mammalogy* 79:1111–1130.

Lawrence, D. B., and E. G. Lawrence. 1961. Response of enclosed lakes to current glaciopluvial climatic conditions in middle latitude western North America. *Annals of the New York Academy of Sciences* 95:341–350.

Layton, T. N. 1970. High rock archaeology: An interpretation of the prehistory of the northwestern Great Basin. Ph.D. dissertation, Harvard University, Cambridge, MA.

Layton, T. N. 1979. Archaeology and paleo-ecology of pluvial Lake Parman, northwestern Great Basin. *Journal of New World Archaeology* 3(3):41–46.

Leavitt, S. W. 1994. Major wet interval in White Mountains Medieval Warm Period evidenced in $\delta^{13}C$ of bristlecone pine tree rings. *Climatic Change* 26:299–307.

Lees, R. B. 1953. The basis of glottochronology. *Language* 29:113–127.

Lemke, P., J. Ren, R. B. Alley, I. Allison, G. Flato, Y. Fujii, G. Kaser, P. Mote, R. H. Thomas, and T. Zhang. 2007. Observations: Changes in snow, ice, and frozen ground. In S. Solomon, D. Qin, M. Manning, M. Marquis, K. B. Averyt, M. M. B. Tignor, H. L. Miller, and Z. Chen, eds., *Climate Change 2007: The Physical Science Basis. Contributions of Working Group 1 to the Fourth Assessment Report of the Intergovernmental Panel on Climate Change*, pp. 337–384. Cambridge University Press, Cambridge.

Lemons, D. R., M. R. Milligan, and M. A Chan. 1996. Paleoclimatic implications of late Pleistocene sediment yield rates for the Bonneville Basin, northern Utah. *Paleogeography, Palaeoclimatology, Palaeoecology* 123:147–159.

Lewis, E. A. 1992. *The Frémont Cannon: High Up and Far Back*. Second revised edition. Western Trails Press, Penn Valley, CA.

Li, H.-C., C.-F. You, T-L. Ku, X.-M. Xu, H. P. Buchheim, N-J. Wan, R-M. Wang, and M-L. Shen. 2008a. Isotopic and geochemical evidence of palaeoclimate changes in Salton Basin, California, during the past 20 kyr: 2. $^{87}Sr/^{86}Sr$ ratio in lake tufa as an indicator of connection between Colorado River and Salton Basin. *Palaeogeography, Palaeoclimatology, Palaeoecology* 259:198–212.

Li, H.-C, X.-M. Xu, T.-L. Ku, C.-F. You, H. P. Buchheim, and R. Peters. 2008b. Isotopic and geochemical evidence of palaeoclimate changes in Salton Basin, California, during the past 20 kyr: 1. $\delta^{18}O$ and $\delta^{13}C$ records in lake tufa deposits. *Palaeogeography, Palaeoclimatology, Palaeoecology* 259:182–197.

Li, J., T. K. Lowenstein, and I. R. Blackburn. 1997. Responses of evaporate mineralogy to inflow water sources and climate during the past 100 k.y. in Death Valley, California. *Geological Society of America Bulletin* 109:1361–1371.

Li, J., T. K. Lowenstein, C. B. Brown, T.-L. Ku, and S. Luo. 1996. A 100 ka record of water tables and paleoclimates from salt cores, Death Valley, California. *Palaeogeography, Palaeoclimatology, Palaeoecology* 123:179–203.

Lia, V. V., A. A. Confalonieri, C. I. Comas, and J. H. Hunziker. 2001. Molecular phylogeny of *Larrea* and its allies (Zygophyllaceae): Reticulate evolution and the probable time of creosote bush arrival to North America. *Molecular Phylogenetics and Evolution* 21:309–320.

Lian, O. B., and R. G. Roberts. 2006. Dating the Quaternary: Progress in luminescence dating of sediments. *Quaternary Science Reviews* 25:2449–2468.

Licciardi, J. M. 2001. Chronology of latest Pleistocene lake-level fluctuations in the pluvial Lake Chewaucan basin, Oregon, USA. *Journal of Quaternary Science* 16:545–553.

Licciardi, J. M., P. U. Clark, E. J. Brook, D. Elmore, and P. Sharma. 2004. Variable responses of western U. S. glaciers during the last deglaciation. *Geology* 32:81–84.

Licciardi, J. M., J. M. Schaefer, J. R. Taggart, and D. C. Lund. 2009. Holocene glacier fluctuations in the Peruvian Andes indicate northern climate linkages. *Science* 325:1677–1679.

Liddicoat, J. C. 1996. Mono Lake excursion in the Lahontan Basin, Nevada. *Geophysical Journal International* 125:630–635.

Liddicoat, J., and R. Coe. 2008. Mono Lake excursion in the U. S. Great Basin. *Geophysical Research Abstracts* 10, EGU2008-A-01350.

Lillquist, K. D. 1994. Late Quaternary Lake Franklin: Lacustrine chronology, coastal geomorphology, and hydro-isostatic deflection in Ruby Valley and northern Butte Valley, Nevada. Ph.D. dissertation, University of Utah, Salt Lake City.

Lin, J. C., W. S. Broecker, S. R. Hemming, I. Hajdas, R. F. Anderson, G. I. Smith, M. Kelley, and G. Bonani. 1998. A reassessment of U-Th and ^{14}C ages for late-glacial high frequency hydrological events at Searles Lake, California. *Quaternary Research* 49:11–23.

Lindberg, D. N. 1999. A synopsis of late Pleistocene shorelines and faulting, Tule Springs rim to Mickey Basin, Alvord Desert, Harney County, Oregon. In C. Narwold, organizer, *Quaternary Geology of the northern Quinn River and Alvord valleys, southeastern Oregon. 1999 Friends of the Pleistocene Field Trip, Pacific Cell, September 24–26, 1999*, Article A3, pp. 1–13. Geology Department, Humboldt State University, Arcata, CA.

Lindskov, K. L. 1984. *Floods of May to June 1983 along the Northern Wasatch Front, Salt Lake City to North Ogden, Utah*. Utah Geological and Mineral Survey Water-Resources Bulletin 24.

Lindström, S. 1990. Submerged tree stumps as indicators of mid-Holocene aridity in the Lake Tahoe region. *Journal of California and Great Basin Anthropology* 12:146–157.

Lindström, S. 1996. Lake Tahoe case study: Lake levels. In *Sierra Nevada Ecosystem Project: Final Report to Congress, Addendum*, pp. 265–268. Centers for Water and Wildland Resources, University of California, Davis.

Lingenfelter, R. E. 1986. *Death Valley and the Amargosa: A Land of Illusion*. University of California Press, Berkeley.

Lips, E. W., D. W. Marchetti, and J. C. Gosse. 2005. Revised chronology of late Pleistocene glaciers, Wasatch Mountains, Utah. *Geological Society of America Abstracts with Programs* 37(7):41.

Litwin, R. J., D. P. Adam, N. O. Fredericksen, and W. B. Woolfenden. 1997. An 800,000-year pollen record for Owens Lake, California: Preliminary analyses. In G. I. Smith and J. L. Bischoff, eds., *An 800,000-year Paleoclimatic Record from Core OL-92, Owens Lake, Southeast California*. Geological Society of America Special Paper 317.

Livingston, S. D. 1986. Archaeology of the Humboldt Lakebed Site. *Journal of California and Great Basin Anthropology* 8(1):99–115.

Livingston, S. D. 1988a. Avian fauna. In C. Raven and R. G. Elston, eds., *Preliminary Investigations in Stillwater Marsh: Human Prehistory and Geoarchaeology*, pp. 292–311. U.S. Department of the Interior, U.S. Fish and Wildlife Service, Region 1, Cultural Resource Series 1.

Livingston, S. D. 1988b. The avian and mammalian remains from Lovelock Cave and the Humboldt lakebed site. Ph.D. dissertation, University of Washington, Seattle.

Livingston, S. D. 1991. Aboriginal utilization of birds in the western Great Basin. In J. R. Purdue, W. E. Klippel, and B. W. Styles, eds., *Beamers, Bobwhites, and Blue-Points: Tributes to the Career of Paul W. Parmalee*, pp. 341–357. Illinois State Museum Scientific Papers 23.

Livingston, S. D. 1992a. Mammals of Serendipity Cave and the Roberts Mountains. Paper presented at the Great Basin Anthropological Conference, Boise, ID.

Livingston, S. D. 1992b. The DeLong Mammoth Locality, Black Rock Desert, Nevada. *Current Research in the Pleistocene* 8:94–97.

Livingston, S. D. 2000. The Homestead Cave avifauna. In Madsen, D. B., *Late Quaternary Paleoecology in the Bonneville Basin*, pp. 91–102. Utah Geological Survey Bulletin 130.

Lloyd, A. H., and L. J. Graumlich. 1997. Holocene dynamics of treeline forests in the Sierra Nevada. *Ecology* 78:1199–2010.

Lockett, H. C., and L. L. Hargrave. 1953. *Woodchuck Cave: a Basketmaker II Site in Tsegi Canyon, Arizona*. Museum of Northern Arizona Bulletin 26.

Logan, L. E. 1981. The mammalian fossils of Muskox Cave, Eddy County, New Mexico. *Proceedings of the Eighth International Congress of Speleology* 21:159–160.

Logan, L. E. 1983. Paleoecological implications of the mammalian fauna of Lower Sloth Cave, Guadalupe Mountains, Texas. *National Speleological Society Bulletin* 45:3–11.

Logan, L. E., and C. C. Black. 1979. The Quaternary vertebrate fauna of Upper Sloth Cave, Guadalupe Mountains, Texas. In H. H. Genoways and R. J. Baker, R. J., eds., *Biological Investigations in the Guadalupe Mountains National Park, Texas*, pp. 141–158. National Park Service Proceedings and Transactions 5.

Lomolino, M. V., and R. Davis. 1997. Biogeographic scale and biodiversity of mountain forest mammals of western North America. *Global Ecology and Biogeography Letters* 6:57–76.

Lomolino, M. V., B. R. Riddle, and J. H. Brown. 2006. *Biogeography*. Third edition. Sinauer Associates, Sunderland, MA.

Long, A., and P. S. Martin. 1974. Death of North American sloths. *Science* 186:638–640.

Long, J. L. 2003. *Introduced Mammals of the World*. CSIRO Publishing, Collingwood, Australia.

Loope, L. L. 1969. Subalpine and alpine vegetation of northeastern Nevada. Ph.D. dissertation, Duke University, Durham, NC.

Los Angeles Department of Water and Power. 1989a. *Along the Owens River*. Los Angeles Department of Water and Power, Los Angeles.

Los Angeles Department of Water and Power. 1989b. *Los Angeles Aqueduct System*. Los Angeles Department of Water and Power, Los Angeles.

Louderback, L., and D. Rhode. 2009. 15,000 years of vegetation change in the Bonneville Basin: The Blue Lake record. *Quaternary Science Reviews* 28:308–326.

Louderback, L. A., D. K. Grayson, and M. Llobera. 2010. Mid-Holocene environments and human history in the Great Basin. *The Holocene*, in press.

Lowe, J. J., and M. J. C. Walker. 1984. *Reconstructing Quaternary Environments*. Longman, London.

Lowenstein, T. K. 2002. Pleistocene lakes and paleoclimates (0 to 200 ka) in Death Valley, California. In R. Hershler, D. B. Madsen, and D. R. Currey, eds., *Great Basin Aquatic Systems History*, pp. 109–120. Smithsonian Contributions to the Earth Sciences 33.

Lowie, R. H. 1939. Ethnographic notes on the Washo. *University of California Publications in American Archaeology and Ethnology* 36:301–352.

Lu, J., G. Sun, S. G., McNulty, and D. M. Amatya. 2005. A comparison of six potential evapotranspiration methods for regional use in the southeastern United States. *Journal of the American Water Resources Association* 41:621–633.

Lubinski, P. M. 1999. The communal pronghorn hunt: A review of the ethnographic and archaeological evidence. *Journal of California and Great Basin Anthropology* 21:158–181.

Lund, E. H., and E. Bentley. 1976. Steens Mountain, Oregon. *The Ore Bin* 38(4):51–66.

Lund, S. P., J. C. Liddicoat, K. R. Lajoie, T. L. Henyey, and S. W. Robinson. 1988. Paleomagnetic evidence for long-term (10^4 year) memory and periodic behavior in the earth's core dynamo process. *Geophysical Research Letters* 15:1101–1104.

Lund, S., J. S. Stoner, J. E. T. Channell, and G. Acton. 2006. A summary of Brunhes paleomagnetic field variability recorded in Ocean Drilling Program cores. *Physics of the Earth and Planetary Interiors* 156:194–204.

Lund, W. R. 2005. Utah Quaternary Fault Parameters Working Group: Critical review of Utah paleoseismic-trenching data and consensus recurrence-interval and vertical slip-rate estimates for Utah's Quaternary faults. In W. R. Lund, ed., *Proceedings Volume: Basin and Range Province Seismic Hazards Summit II* (no pagination). Utah Geological Survey Miscellaneous Publication 05-2.

Lundelius, E. L., Jr. 1979. Post-Pleistocene mammals from Pratt Cave and their environmental significance. In H. H. Genoways and R. J. Baker, eds., *Biological Investigations in the Guadalupe Mountains National Park, Texas*, pp. 239–258. National Park Service Proceedings and Transactions 5.

Lundelius, E. L., Jr., R. W. Graham, E. Anderson, J. Guilday, J. A. Holman, D. W. Steadman, and S. D. Webb. 1983. Terrestrial vertebrate faunas. In S. C. Porter, ed., *Late Quaternary Environments of the United States, Volume 1, The Late Pleistocene*, pp. 311–353. University of Minnesota Press, Minneapolis, MN.

Lupo, K. D. 1993. Small faunal assemblages from the Willard Bay sites. In W. B. Fawcett and S. R. Simms, *Archaeological Test Excavations in the Great Salt Lake Wetlands and Associated Analysis*, pp. 197–215. Utah State University Contributions to Anthropology 14.

Lupo, K. D. 1996. The historical occurrence and demise of bison in northern Utah. *Utah Historical Quarterly* 62:168–180.

Lupo, K. D., and D. N. Schmitt. 1997. On late Holocene variability in bison populations in the northeastern Great Basin. *Journal of California and Great Basin Anthropology* 19:50–69.

Lyman, R. L. 1991. Late Quaternary biogeography of the pygmy rabbit (*Brachylagus idahoensis*) in eastern Washington. *Journal of Mammalogy* 72:110–117.

Lyman, R. L. 2004a. Aboriginal overkill in the Intermountain West of North America. *Human Nature* 15:169–208.

Lyman, R. L. 2004b. Biogeographic and conservation implications of late Quaternary pygmy rabbits (*Brachylagus idahoensis*) in eastern Washington. *Western North American Naturalist* 64:1–6.

Lyman, R. L. 2004c. Prehistoric biogeography, abundance, and phenotypic plasticity of elk (*Cervus elaphus*) in Washington state. In R. L. Lyman and K. P. Cannon, eds., *Zooarchaeology and Conservation Biology*, pp. 136–163. University of Utah Press, Salt Lake City.

Lyman, R. L., and S. Wolverton. 2002. The late prehistoric-early historic game sink in the northwestern United States. *Conservation Biology* 16:73–85.

Lyons, W. H., and M. L. Cummings. 2002. Sources of sandstone artifacts and pottery from Lost Dune, a late prehistoric site in Harney County, southeastern Oregon, U.S.A. *Geoarchaeology* 17:717–748.

Lyons, W. H., and P. J. Mehringer, Jr. 1996. *Archaeology of the Lost Dune Site (35HA792), Blitzen Valley, Harney County, Oregon: A Report of Excavations by the 1995 WSU Field School*. Manuscript on file, Bureau of Land Management, Burns, OR.

Mabey, D. R. 1986. Notes on the historic high level of Great Salt Lake. *Utah Geological and Mineral Survey Notes* 20(2):13–15.

Mabey, D. R. 1987. The end of the wet cycle. *Utah Geological and Mineral Survey Notes* 21(2–3):8–9.

MacDowell, P. F. 1992. An overview of Harney Basin geomorphic history, climate and hydrology. In C. Raven and R. G. Elston, eds., *Land and Life at Malheur Lake: Preliminary Geomorphological and Archaeological Investigations*, pp. 13–34. U.S. Department of the Interior, U.S. Fish and Wildlife Service, Region 1, Cultural Resource Series 8.

Machette, M. N. 2005. Summary of the late Quaternary tectonics of the Basin and Range Province in Nevada, eastern California, and Utah. In W. R. Lund, ed., *Proceedings Volume: Basin and Range Province Seismic Hazards Summit II* (unpaginated). Utah Geological Survey Miscellaneous Publication 05-2.

Machette, M. N., R. E. Klinger, and J. R. Knott. 2001. Questions about Lake Manly's age, extent, and source. In M. N. Machette, M. L. Johnson, and J. L. Slate, eds., *Quaternary and Late Pliocene Geology of the Death Valley Region: Recent Observations on Tectonics, Stratigraphy, and Lake Cycles (Guidebook for the 2001 Pacific Cell—Friends of the Pleistocene Field Trip)*, pp. G143-G149. U.S. Geological Survey Open-File Report 01-51.

Mack, J. M., ed. 1990. *Hunter-gatherer Pottery from the Far West*. Nevada State Museum Anthropological Papers 23.

Mack, R. N., and J. N. Thompson. 1982. Evolution in steppe with few large, hooved mammals. *American Naturalist* 119:757–773.

MacMahon, J. A. 2000. Warm deserts. In M. G. Barbour and W. D. Billings, eds., *North American Terrestrial Vegetation*, pp. 285–322. Cambridge University Press, Cambridge, UK.

MacPhee, R. D. E., and P. A. Marx. 1997. The 40,000 year plague: Humans, hyperdisease, and first-contact extinctions. In S. M. Goodman and B. D. Patterson, eds., *Natural Change and Human Impact in Madagascar*. pp. 169–217. Smithsonian Institution Press, Washington, DC.

MacPhee, R. D. E., M. A. Iturralde-Vinent, and O. Jiménez Vázquez. 2007. Prehistoric sloth extinctions in Cuba: Implications of the new "last" appearance date. *Caribbean Journal of Science* 43:94–98.

Madsen, D. B. 1972. Paleoecological investigations in Meadow Valley Wash, Nevada. In D. D. Fowler, ed., *Great Basin Cultural Ecology: A Symposium*, pp. 57–66. Desert Research Institute Publications in the Social Sciences 8.

Madsen, D. B. 1976. Pluvial–post-pluvial vegetation changes in the southeastern Great Basin. In R. Elston, ed., *Holocene Environmental Change in the Great Basin*, pp. 104–119. Nevada Archaeological Survey Paper 6.

Madsen, D. B. 1979. The Fremont and the Sevier: Defining prehistoric agriculturalists north of the Anasazi. *American Antiquity* 44:711–722.

Madsen, D. B. 1982. Get it where the gettin's good: A variable model of Great Basin subsistence and settlement based on data from the eastern Great Basin. In D. B. Madsen and J. F. O'Connell, eds., *Man and Environment in the Great Basin*, pp. 207–226. Society for American Archaeology Papers 2.

Madsen, D. B. 1986a. Great Basin nuts: A short treatise on the distribution, productivity, and prehistoric use of pinyon. In C. J. Condie and D. D. Fowler, eds., *Anthropology of the Desert West: Essays in Honor of Jesse D. Jennings*, pp. 21–41. University of Utah Anthropological Papers 110.

Madsen, D. B. 1986b. Prehistoric ceramics. In W. L. d'Azevedo, ed., *Great Basin, Handbook of North American Indians, Volume 11*, pp. 206–214. Smithsonian Institution Press, Washington, DC.

Madsen, D. B. 1989. *Exploring the Fremont*. University of Utah Occasional Publication 8.

Madsen, D. B. 1994. Mesa Verde and Sleeping Ute Mountain: The geographical and chronological dimensions of the Numic Expansion. In D. B. Madsen and D. Rhode, eds., *Across the West: Human Population Movement and the Expansion of the Numa*, pp. 24–31. University of Utah Press, Salt Lake City.

Madsen, D. B. 1999. Environmental change during the Pleistocene-Holocene transition and its possible impact on human populations. In C. Beck, ed., *Models for the Millennium: Great Basin Anthropology Today*, pp. 75–82. University of Utah Press, Salt Lake City.

Madsen, D. B. 2000a. A high-elevation Allerød: Younger Dryas megafauna from the west-central Rocky Mountains. In D. B. Madsen and M. D. Metcalf, eds., *Intermountain Archaeology*, pp. 100–115. University of Utah Anthropological Papers 122.

Madsen, D. B. 2000b. *Late Quaternary Paleoecology in the Bonneville Basin*. Utah Geological Survey Bulletin 130.

Madsen, D. B. 2002. Great Basin peoples and late Quaternary aquatic history. In R. Hershler, D. B. Madsen, and D. R. Currey, eds., *Great Basin Aquatic Systems History*, pp. 387–405. Smithsonian Contributions to the Earth Sciences 33.

Madsen, D. B. 2007. The Paleoarchaic to Archaic transition in the Great Basin. In K. E. Graf and D. N. Schmitt, eds., *Paleoindian or Paleoarchaic: Great Basin Human Ecology at the Pleistocene-Holocene Transition*, pp. 3–20. University of Utah Press, Salt Lake City.

Madsen, D. B., and M. S. Berry. 1975. A reassessment of northeastern Great Basin prehistory. *American Antiquity* 40: 391–405.

Madsen, D. B., and D. R. Currey. 1979. Late Quaternary glacial and vegetation changes, Little Cottonwood Canyon area, Wasatch Mountains, Utah. *Quaternary Research* 12:254–270.

Madsen, D. B., and J. E. Kirkman. 1988. Hunting hoppers. *American Antiquity* 53:593–604.

Madsen, D. B., and D. Rhode. 1990. Early Holocene pinyon (*Pinus monophylla*) in the northeastern Great Basin. *Quaternary Research* 33:94–101.

Madsen, D. B., and D. Rhode, eds. 1994. *Across the West: Human Population Movement and the Expansion of the Numa*. University of Utah Press, Salt Lake City.

Madsen, D. B., and D. Rowe. 1988. Utah radiocarbon dates I: pre-1970 dates. *Utah Archaeology* 1:52–56.

Madsen, D. B., and D. N. Schmitt. 2005. *Buzz-Cut Dune and Fremont Foraging at the Margin of Agriculture*. University of Utah Anthropological Papers 124.

Madsen, D. B., and S. R. Simms. 1998. The Fremont Complex: A behavioral perspective. *Journal of World Prehistory* 12:255–336.

Madsen, D. B., C. G. Oviatt, and D. N. Schmitt. 2005. A geomorphic, environmental, and cultural history of the Camels Back Cave Region. In D. N Schmitt and D. B. Madsen, *Camels Back Cave*, pp. 20–45. University of Utah Anthropological Papers 125.

Madsen, D. B., C. G. Oviatt, and D. C. Young. 2009. Old River Bed geomorphology and chronology. In D. B. Madsen, D. Schmitt, and D. Page, *The Paleoarchaic Occupation of the Old River Bed*. Unpublished manuscript.

Madsen, D. B., D. Rhode, D. K Grayson, J. M. Broughton, S. D. Livingston, J. Hunt, J. Quade, D. N. Schmitt, and M. W. Shaver III. 2001. Late Quaternary environmental change in the Bonneville Basin, western USA. *Palaeogeography, Palaeoclimatology, Palaeoecology* 167:243–271.

Maher, L. J., Jr. 1969. *Ephedra* pollen in sediments of the Great Lakes Region. *Ecology* 45:391–395.

Malhi, R. S., H. M. Mortensen, J. A. Eshleman, B. M. Kemp, J. G. Lorenz, F. A. Kaestle, J. R. Johnson, C. Gorodezky, and D. G. Smith. 2003. Native American mtDNA prehistory in the American Southwest. *American Journal of Physical Anthropology* 120:108–124.

Malhi, R. S., A. Gonzalez-Oliver, K. B. Schroeder, B. M. Kemp, J. A. Greenberg, S. Z. Dobrowski, D. G. Smith, A. Resendez, T. Karafet, M. Hammer, S. Zegura, and T. Brovko. 2008. Distribution of Y chromosomes among native North Americans: A study of Athapaskan population history. *American Journal of Physical Anthropology* 137:412–424.

Mandryk, C. A. S., H. Josenhans, D. W. Fedje, and R. W. Mathewes. 2001. Late Quaternary paleoenvironments of northwestern North America: Implications for inland versus coastal migration routes. *Quaternary Science Reviews* 20:301–314.

Manley, W. F. 2002. *Postglacial Flooding of the Bering Land Bridge: A Geospatial Animation*. INSTAAR, University of Colorado, v1, http://instaar.colorado.edu/QGISL/bering_land_bridge/.

Manly, W. L. 1894. *Death Valley in '49*. Pacific Tree and Vine Co., San Jose, CA (reprinted by Chalfant Press, Bishop, CA, 1977).

Mann, M. E., Z. Zhang, S. Rutherford, R. S. Bradley, M. K. Hughes, D. Shindell, C. Ammann, G. Falugevi, and F. Ni. 2009. Global signatures and dynamical origins of the Little Ice Age and Medieval Climate Anomaly. *Science* 326:1256–1260.

Marcus, L. F., and R. Berger. 1984. The significance of radiocarbon dates for Rancho La Brea. In P. S. Martin and R. G. Klein, eds., *Quaternary Extinctions: A Prehistoric Revolution*, pp. 159–183. University of Arizona Press, Tucson.

Marlon, J. R., P. J. Bartlein, M. K. Walsh, S. P. Harrison, K. J. Brown, M. E. Edwards, P. E. Higuera, M. J. Power, R. S. Anderson, C. Briles, A. Brunelle, C. Carcaillet, M. Daniels, F. S. Hu, M. Lavoie, C. Long, T. Minckley, P. J. H. Richard, A. C. Scott, D. S. Shafer, W. Tinner, C. E. Umbanhowar Jr., and C. Whitlock. 2009. Wildfire responses to abrupt climate change in North America. *Proceedings of the National Academy of Sciences* 106:2519–2524.

Marr, J. W. 1977. The development of movement of tree islands near the upper limit of tree growth in the southern Rocky Mountains. *Ecology* 58:1159–1164.

Marshall, S. J., T. S. James, and G. K. C. Clarke. 2002. North American ice sheet reconstructions at the Last Glacial Maximum. *Quaternary Science Reviews* 21:175–192.

Martin, J. E. 2006. Vertebrate remains from the upper portion of the Pleistocene Fossil Lake section and their relationship to the assemblage at the Paisley 5 Mile Point Caves in south-central Oregon. Paper presented at the Great Basin Anthropological Conference, Las Vegas, NV.

Martin, J. E., D. Patrick, A. J. Kihm, F. F. Foit, Jr., and D. E. Grandstaff. 2005. Lithostratigraphy, tephrochronology, and rare earth element geochemistry of fossils at the classical Pleistocene Fossil Lake area, south central Oregon. *Journal of Geology* 113:139–155.

Martin, P. S. 1963. *The Last 10,000 Years: A Fossil Pollen Record of the American Southwest*. University of Arizona Press, Tucson.

Martin, P. S. 1984. Prehistoric overkill: The global model. In P. S. Martin and R. G. Klein, eds., *Quaternary Extinctions: A Prehistoric Revolution*, pp. 354–403. University of Arizona Press, Tucson.

Martin, P. S. 2005. *Twilight of the Mammoths: Ice Age Extinctions and the Rewilding of North America*. University of California Press, Berkeley.

Martin, P. S., and D. W. Steadman. 1999. Prehistoric extinctions on islands and continents. In R. D. E. MacPhee, ed., *Extinctions in Near Time*, pp. 17–52. Kluwer Academic/Plenum, New York.

Martin, P. S., and C. R. Szuter. 1999a. Megafauna of the Columbia Basin, 1800–1840: Lewis and Clark in a game sink. In D. D. Goble and P. W. Hirt, eds., *Northwest Lands, Northwest Peoples*, pp. 188–204. University of Washington Press, Seattle.

Martin, P. S., and C. R. Szuter. 1999b. War zones and game sinks in Lewis and Clark's west. *Conservation Biology* 13:36–45.

Martin, P. S., and C. R. Szuter. 2004. Revising the "Wild West": big game meets the ultimate keystone species. In C. L. Redman, S. R. James, P. R. Fish, and J. D. Rogers, eds., *The Archaeology of Global Change: The Impacts of Humans on Their Environment*, pp. 63–88. Smithsonian Institution, Washington, DC.

Martin, P. S., B. E. Sabels, and D. Shutler. 1961. Rampart Cave coprolite and ecology of the Shasta ground sloth. *American Journal of Science* 259:102–127.

Martin, P. S., R. S. Thompson, and A. Long. 1985. Shasta ground sloth extinction: A test of the blitzkrieg model. In J. I. Mead and D. J. Meltzer, eds., *Environments and Extinctions: Man in Late Glacial North America*, pp. 5–14. Center for the Study of Early Man, University of Maine, Orono, ME.

Martin, P. Si., G. I. Quimby, and D. Collier. 1947. *Indians before Columbus*. University of Chicago Press, Chicago.

Martínez-Meyer, E., A. T. Peterson, and W. W. Hargrove. 2004. Ecological niches as stable distributional constraints on mammal species, with implications for Pleistocene extinctions and climate change projections for biodiversity. *Global Ecology and Biogeography* 13:305–314.

Marwitt, J. P. 1968. *Pharo Village*. University of Utah Anthropological Papers 91.

Marwitt, J. P. 1970. *Median Village and Fremont Culture Regional Variation*. University of Utah Anthropological Papers 95.

Mason, J. A., J. B. Swinehart, R. J. Goble, and D. B. Loope. 2004. Late-Holocene dune activity linked to hydrological drought, Nebraska Sand Hills, USA. *The Holocene* 14:209–217.

Mason, J. L., and K. L. Kipp, Jr. 2002. Investigation of hydrology and solute transport, Bonneville Salt Flats, northwestern Utah. In J. W. Gwynn, ed., *Great Salt Lake: An Overview of Change*, pp. 423–432. Utah Department of Natural Resources Special Publication, Salt Lake City.

Massimino, J., and D. Metcalfe. 1999. New form for the Formative. *Utah Archaeology* 12:1–16.

Matsubara, Y., and A. D. Howard. 2009. A spatially explicit model of runoff, evaporation, and lake extent: Application to modern and late Pleistocene lakes in the Great Basin region, western United States. *Water Resources Research* 45: W06425, doi:10.1029/2007WR005953.

Matthes, F. E. 1939. Report of Committee on Glaciers. *American Geophysical Union Transactions* 1939:518–523.

Matthes, F. E. 1941. Rebirth of the glaciers of the Sierra Nevada during late Post-Pleistocene times. *Geological Society of America Bulletin* 52:2030.

Mawby, J. E. 1967. Fossil vertebrates of the Tule Springs Site, Nevada. In H. M. Wormington and D. Ellis, eds., *Pleistocene studies in Southern Nevada*, pp. 105–129. Nevada State Museum Anthropological Papers 13.

McAllister, J. A., and R. S. Hoffman. 1988. *Phenacomys intermedius*. *Mammalian Species* 305.

McCain, E. B., and J. L Childs. 2008. Evidence of resident jaguars (*Panthera onca*) in the southwestern United States and the implications for conservation. *Journal of Mammalogy* 89:1–10.

McCorquodale, S. M. 1985. Archaeological evidence of elk in the Columbia Basin. *Northwest Science* 59:192–197.

McDonald, H. G. 1996. Biogeography and paleoecology of ground sloths in California, Arizona, and Nevada. *SBCMA Quarterly* 43(1):61–65.

McDonald, H. G. 2002. *Platygonus compressus* from Franklin County, Idaho, and a review of the genus in Idaho. In W. A. Akersten, M. E. Thompson, D. J. Meldrum, R. A. Rapp, and H. G. McDonald, eds., *And Whereas . . . Papers on the Vertebrate Paleontology of Idaho Honoring John A. White, Volume 2*, pp. 141–149. Idaho Museum of Natural History Occasional Paper 37.

McDonald, H. G. 2003. Sloth remains from North American caves and associated karst features. In B. Schubert, J. I. Mead, and R. W. Graham, eds., *Ice Age Cave Faunas of North America*, pp. 1–16. Indiana University Press, Bloomington.

McDonald, H. G., and E. L. Lundelius, Jr. 2009. The giant ground sloth *Eremotherium laurillardi* (Xenarthra, Megatheriidae) in Texas. In L. B. Albright, ed., *Papers on Geology, Vertebrate Paleontology, and Biostratigraphy in Honor of Michael O. Woodburne*, pp. 407–421. Museum of Northern Arizona Bulletin 65.

McDonald, H. G., W. E. Miller, and T. H. Morris. 2001. Taphonomy and significance of Jefferson's ground sloth (Xenarthra: Megalonychidae) from Utah. *Western North American Naturalist* 61(1): 64–77.

McDonald, K. A., and J. H. Brown. 1992. Using montane mammals to model extinctions due to global change. *Conservation Biology* 6:409–415.

McGuire, K. R. 1997. *Culture Change along the Eastern Sierra Nevada/Cascade Front. Volume 4: Secret Valley*. Coyote Press, Salinas.

McGuire, K. R., ed. 2000a. *Archaeological Investigations along the California–Great Basin Interface: The Alturas Transmission Line Project. Volume 1. Prehistoric Archaeological Studies: The Pit River Uplands, Madeline Plains, Honey Lake and Secret Valley, and Sierra Front Project Segments*. Coyote Press, Salinas.

McGuire, K. R., ed. 2000b. *Archaeological Investigations along the California–Great Basin Interface: The Alturas Transmission Line Project. Volume 1. Prehistoric Archaeological Studies: The Pit River Uplands, Madeline Plains, Honey Lake and Secret Valley, and Sierra Front Project Segments. Site Reports and Data Appendices*. Coyote Press, Salinas.

McGuire, K. R., M. G. Delacorte, and K. Carpenter. 2004. *Archaeological Excavations at Pie Creek and Tule Valley Shelters, Elko County, Nevada*. Nevada State Museum Anthropological Papers 25.

McKenney, D. W., J. H. Pedlar, K. Lawrence, K. Campbell, and M. F. Hutchinson. 2007. Potential impacts of climate change on the distribution of North American trees. *BioScience* 57:939–948.

McLane, A. R. 1978. *Silent Cordilleras: The Mountain Ranges of Nevada*. Camp Nevada Monographs 4.

McMahon, A., and R. McMahon. 2005. *Language Classification by Numbers*. Oxford University Press, Oxford, UK.

McManus, J. F., R. Francois, J-M. Gherardi, L. D. Keigwin, and S. Brown-Leger. 2004. Collapse and rapid resumption of Atlantic meridional circulation linked to deglacial climate changes. *Nature* 428:834–837.

Mead, E. M., and J. I. Mead. 1989. Snake Creek Burial Cave and a review of the Quaternary mustelids of the Great Basin. *Great Basin Naturalist* 49:143–154.

Mead, J. I. 1983. Harrington's extinct mountain goat (*Oreamnos harringtoni*) and its environment in the Grand Canyon, Arizona. Ph.D. dissertation, University of Arizona, Tucson.

Mead, J. I. 1987. Quaternary records of pika, *Ochotona*, in North America. *Boreas* 16:165–171.

Mead, J. I., and L. D. Agenbroad. 1992. Isotope dating of Pleistocene dung deposits from the Colorado Plateau, Arizona and Utah. *Radiocarbon* 34:1019.

Mead, J. I., and C. J. Bell. 1994. Late Pleistocene and Holocene herpetofaunas of the Great Basin and Colorado Plateau. In K. T. Harper, L. L. St. Clair, K. H. Thorne, and W. M. Hess, eds., *Natural History of the Colorado Plateau and Great Basin*, pp. 255–275. University Press of Colorado, Niwot.

Mead, J. I., and F. Grady. 1996. *Ochotona* (Lagomorpha) from late Quaternary cave deposits in eastern North America. *Quaternary Research* 45:93–101.

Mead, J. I., and C. B. Johnson. 2004. *Bison* and *Bos* from protohistoric and historic localities in the San Rafeal Valley, Arizona. *Kiva* 70:183–193.

Mead, J. I., and M. C. Lawler. 1994. Skull, mandible, and metapodials of the extinct Harrington's mountain goat (*Oreamnos harringtoni*). *Journal of Vertebrate Paleontology* 14:562–576.

Mead, J. I., and L. K. Murray. 1991. Late Pleistocene vertebrates from the Potosi Mountain packrat midden, Spring Range, Nevada. In R. E. Reynolds, ed., *Crossing the Borders: Quaternary Studies in Eastern California and Southwestern Nevada*, pp. 124–126. San Bernardino County Museum Association, Redlands.

Mead, J. I., and W. G. Spaulding. 1995 Pika (*Ochotona*) and paleoecological reconstructions of the Intermountain West, Nevada and Utah. In D. W. Steadman and J. I. Mead, eds., *Late Quaternary Environments and Deep History: A Tribute to Paul S. Martin*, pp. 165–186. The Mammoth Site of Hot Springs, South Dakota Scientific Paper 3.

Mead, J. I., C. J. Bell, and L. K. Murray. 1992. *Mictomys borealis* (northern bog lemming) and the Wisconsin paleoecology of the east-central Great Basin. *Quaternary Research* 37:229–238.

Mead, J. I., R. S. Thompson, and T. R. Van Devender. 1982. Late Wisconsinan and Holocene fauna from Smith Creek Canyon, Snake Range, Nevada. *Transactions of the San Diego Society of Natural History* 20:1–26.

Mead, J. I., L. D. Agenbroad, A. M. Phillips, III, and L. T. Middleton. 1987. Extinct mountain goat (*Oreamnos harringtoni*) in southeastern Utah. *Quaternary Research* 27:323–331.

Mead, J. I., P. S. Martin, R. C. Euler, A. Long, A. J. T. Jull, L. S. Toolin, D. J. Donahue, and T. W. Linick. 1986. Extinction of Harington's mountain goat. *Proceedings of the National Academy of Sciences* 83:836–839.

Mee, A. M., B. A. Rideout, J. A. Hamber, J. N. Todd, G. Austin, M. Clark, and M. P. Wallace. 2007. Junk ingestion and nestling mortality in a reintroduced population of California Condors *Gymnogyps californianus*. *Bird Conservation International* 17:119–130.

Meek, N. 1999. New discoveries about the late Wisconsinan history of the Mojave River system. In R. E. Reynolds and J. Reynolds, eds., *Tracks along the Mojave*, pp. 113–117. San Bernardino County Museum Association Quarterly 46(3).

Meek, N. 2000. The late Wisconsinan history of the Afton Canyon area, Mojave Desert, California. In R. E. Reynolds and J. Reynolds, eds., *Empty Basins, Vanished Lakes*, pp. 32–34. San Bernardino County Museum Association Quarterly 47(2).

Meek, N. 2004. Mojave River history from an upstream perspective. In R. E. Reynolds, ed., *Breaking Up: The 2004 Desert Symposium Field Trip and Abstracts*, pp. 41–49. University of California Fullerton, Desert Studies Consortium, Fullerton.

Mehringer, P. J., Jr. 1965. Late Pleistocene vegetation in the Mojave Desert of southern Nevada. *Journal of the Arizona Academy of Science* 3:172–188.

Mehringer, P. J., Jr. 1967. Pollen analysis of the Tule Springs Site, Nevada. In H. M. Wormington and D. Ellis, eds., *Pleistocene Studies in Southern Nevada*, pp. 130–200. Nevada State Museum Anthropological Papers 13.

Mehringer, P. J., Jr. 1985. Late-Quaternary pollen records from the interior Pacific Northwest and northern Great Basin of the United States. In V. M. Bryant, Jr., and R. G. Holloway, eds., *Pollen Records of Late-Quaternary North American Sediments*, pp. 167–190. American Association of Stratigraphic Palynologists, Dallas.

Mehringer, P. J., Jr. 1986. Prehistoric environments. In W. L. d'Azevedo, ed., *Great Basin, Handbook of North American Indians, Volume 11*, pp. 31–50. Smithsonian Institution Press, Washington, DC.

Mehringer, P. J., Jr., and W. J. Cannon. 1994. Volcaniclastic dunes of the Fort Rock Valley, Oregon: Stratigraphy, chronology, and archaeology. In C. M. Aikens and D. L. Jenkins, eds., *Archaeological Research in the Northern Great Basin: Fort Rock Archaeology since Cressman*, pp. 283–327. University of Oregon Archaeological Papers 50.

Mehringer, P. J., Jr., and C. W. Ferguson. 1969. Pluvial occurrence of bristlecone pine (*Pinus aristata*) in a Mojave Desert mountain range. *Journal of the Arizona Academy of Science* 5:284–292.

Mehringer, P. J., Jr., and C. V. Haynes, Jr. 1965. The pollen evidence for the environment of early man and extinct animals at the Lehner Mammoth Site, southeastern Arizona. *American Antiquity* 31:17–23.

Mehringer, P. J., Jr., and J. C. Sheppard. 1978. Holocene history of Little Lake, California. In E. L. Davis, ed., *The Ancient Californians: Rancholabrean Hunters of the Mojave Lakes Country*, pp. 153–166. Natural History Museum of Los Angeles County Science Series 29.

Mehringer, P. J., Jr., and C. N. Warren. 1976. Marsh, dune, and archaeological chronology, Ash Meadows, Amargosa Desert, Nevada. In R. Elston, ed., *Holocene environmental change in the Great Basin*, pp. 120–151. Nevada Archaeological Survey Paper 6.

Mehringer, P. J., Jr., and P. E. Wigand. 1987. Western juniper in the Holocene. In R. L. Everett, compiler, *Proceedings of the Pinyon-Juniper Conference, Reno, Nevada, January 13–16, 1986*, pp. 109–119. U.S. Forest Service Intermountain Research Station, General Technical Report INT-215.

Mehringer, P. J., Jr., and P. E. Wigand. 1990. Comparison of late Holocene environments from woodrat middens and pollen: Diamond Craters, Oregon. In J. L. Betancourt, T. R. Van Devender, and P. S. Martin, eds., *Packrat Middens: The Last 40,000 Years of Biotic Change*. University of Arizona Press, Tucson.

Meltzer, D. J. 2004. Peopling of North America, in A. R. Gillespie, S. C. Porter and B. F. Atwater, eds., *The Quaternary Period in the United States*, pp. 539–563. Developments in Quaternary Science 1. Elsevier, Amsterdam.

Meltzer, D. J. 2006. *Folsom: New Archaeological Investigations of a Classic Paleoindian Bison Kill*. University of California Press, Berkeley.

Meltzer. D. J. 2009. *First Peoples in a New World: Colonizing Ice Age America*. University of California Press, Berkeley, CA.

Meltzer, D. J., D. K. Grayson, G. Ardila, A. W. Barker, D. F. Dincauze, C. V. Haynes, F. Mena, L. Núñez, and D. J. Stanford. 1997. On the Pleistocene antiquity of Monte Verde, southern Chile. *American Antiquity* 62:659–993.

Menges, C. M. 2008. Multistage late Cenozoic evolution of the Amargosa River drainage, southwestern Nevada and eastern California. In M. C. Reheis, R. Hershler, and D. M. Miller, eds., *Late Cenozoic Drainage History of the Southwestern Great Basin and Lower Colorado River Region: Geologic and Biotic Perspectives*, pp. 39–90. Geological Society of America Special Paper 439.

Menking, K. M., R. Y. Anderson, N. G. Shafike, K. H. Syed, and B. D. Allen. 2004. Wetter or colder during the Last Glacial Maximum? Revisiting the pluvial lake question in southwestern North America. *Quaternary Research* 62:280–288.

Mensing, S. A. 2001. Late-glacial and early Holocene vegetation and climate change near Owens Lake, eastern California. *Quaternary Research* 55:57–65.

Mensing, S. A., L. V. Benson, M. Kashgarian, and S. Lund. 2004. A Holocene pollen record of persistent droughts from Pyramid Lake, Nevada, USA. *Quaternary Research* 62:29–38.

Merola, J. A., D. R. Currey, and M. K. Ridd. 1989. Thematic mapper laser profile resolution of Holocene lake limit, Great Salt Lake Desert, Utah. *Remote Sensing of Environment* 28:233–244.

Merriam, C. H. 1926. The buffalo in northeastern California. *Journal of Mammalogy* 7:211–214.

Merrill, W. L., R. J. Hard, J. B. Mabry, G. J. Fritz, K. R. Adams, J. R. Roney, and A. C. MacWilliams. 2009. The diffusion of maize to the southwestern United States and its impact. *Proceedings of the National Academy of Sciences* 106:21019–21026.

Messing, H. J. 1986. A late Pleistocene-Holocene fauna from Chihuahua, Mexico. *Southwestern Naturalist* 31:277–288.

Metcalfe, D., and K. R. Barlow. 1992. A model for exploring the optimal trade-off between field processing and transport. *American Anthropologist* 94:340–356.

Metcalfe, D., and L. V. Larrabee. 1985. Fremont irrigation: Evidence from Gooseberry Valley, central Utah. *Journal of California and Great Basin Anthropology* 7(2):244–254.

Meyers, J. I. 2007. Basicranial analysis of *Martes* and the extinct *Martes nobilis* (Carnivora: Mustelidae) using geometric morphometrics. M.S. thesis, Northern Arizona University, Flagstaff.

Mickelson, D. M., and P. M. Colgan. 2004. The southern Laurentide Ice Sheet. In A. R. Gillespie, S. C. Porter, and B. F. Atwater, eds., *The Quaternary Period in the United States*, pp. 1–17. Developments in Quaternary Science 1. Elsevier, Amsterdam.

Mifflin, M. D., and M. M. Wheat. 1979. *Pluvial Lakes and Estimated Pluvial Climates of Nevada*. Nevada Bureau of Mines and Geology Bulletin 94.

Millar, C. I. 1996. Tertiary vegetation history. In *Sierra Nevada Ecosystem Project: Final Report to Congress, Vol. II, Assessments and Scientific Basis for Management Options*, pp. 71–122. University of California Centers for Water and Wildland Resources, Davis. CA.

Millar, C. I. 1998. Early evolution of pines. In D. M. Richardson, ed., *Ecology and Biogeography of Pinus*, pp. 69–90. Cambridge University Press, Cambridge, UK.

Millar, C. I., J. C. King, R. D. Westfall, H. A. Alden, and D. L. Delany. 2006. Late Holocene forest dynamics, volcanism, and climate change at Whitewing Mountain and San Joaquin Ridge, Mono County, Sierra Nevada, CA, USA. *Quaternary Research* 66:273–287.

Millar, C. I., R. D. Westfall, D. L. Delany, J. C. King, and L. J. Graumlich. 2004. Response of subalpine conifers in the Sierra Nevada, California, U.S.A., to 20th-century warming and decadal climate variability. *Arctic, Antarctic, and Alpine Research* 36: 181–200.

Miller, A. H. 1932. An extinct icterid from Shelter Cave, New Mexico. *The Auk* 49:38–41.

Miller, A. H. 1947. A new genus of icterid from Rancho La Brea. *The Condor* 49:22–24.

Miller, B. A. 2006. The phylogeography of *Prosopium* in western North America. M.A. thesis, Brigham Young University, Provo. UT.

Miller, J. R., D. Germanoski, K. Waltman, R. Tausch, and J. Chambers. 2001. Influence of late Holocene hillslope processes and landforms on modern channel dynamics in upland watersheds of central Nevada. *Geomorphology* 38:373–391.

Miller, J. R., K. House, D. Germanoski, R. J. Tausch, and J. C. Chambers. 2004. Fluvial geomorphic responses to Holocene climate change. In J. C. Chambers and J. R. Miller, eds., *Great Basin Riparian Ecosystems: Ecology, Management, and Restoration*, pp. 49–87. Island Press, Washington, DC.

Miller, L. 1931. The California Condor in Nevada. *The Condor* 33:32.

Miller, L., and I. S. DeMay. 1942. The fossil birds of California. *University of California Publications in Zoology* 47(4):47–142.

Miller, R. F., and J. A. Rose. 1995. Historic expansion of *Juniperus occidentalis* (western juniper) in southeastern Oregon. *Great Basin Naturalist* 55:37–45.

Miller, R. F., and J. A. Rose. 1999. Fire history and western juniper encroachment in sagebrush steppe. *Journal of Range Management* 52:550–559.

Miller, R. F., and R. J. Tausch. 2001. The role of fire in juniper and pinyon woodlands: A descriptive analysis. In K. E. M. Galley and T. P. Wilson, eds., *Proceedings of the Invasive Species Workshop: The Role of Fire in the Control and Spread of Invasive Species. Fire Conference 2000: The First National Congress on Fire Ecology, Prevention, and Management*. Tall Timbers Research Station Miscellaneous Publication 11.

Miller, R. F., T. J. Svejcar, and J. A. Rose. 2000. Impacts of western juniper on plant community composition and structure. *Journal of Range Management* 53:574–585.

Miller, R. F., F. A. Branson, I. S. McQueen, and C. T. Snyder. 1982. Water relations in soils as related to plant communities in Ruby Valley, Nevada. *Journal of Range Management* 35:462–468.

Miller, R. F., J. D. Bates, T. J. Svejcar, F. B. Pierson, and L. E. Eddleman. 2005. *Biology, Ecology, and Management of Western Juniper* (Juniperus occidentalis). Oregon State University Agricultural Experiment Station Technical Bulletin 152.

Miller, R. R. 1945. Four new species of fossil cyprinodont fishes from eastern California. *Journal of the Washington Academy of Sciences* 35:315–321.

Miller, R. R. 1948. *The Cyprinodont Fishes of the Death Valley System of Eastern California and Southwestern Nevada*. University of Michigan Museum of Zoology Miscellaneous Publications 68.

Miller, R. R. 1981. Coevolution of deserts and pupfishes (genus *Cyprinodon*) in the American Southwest. In R. J. Naiman and D. L. Soltz, eds., *Fishes in North American Deserts*, pp. 39–94. Wiley, New York.

Miller, R. R., J. D. Williams, and J. E. Williams. 1989. Extinctions of North American fishes during the past century. *Fisheries* 14(6):22–38.

Miller, S. J. 1972. Weston Canyon rockshelter: Big-game hunting in southeastern Idaho. M.A. thesis, Idaho State University, Pocatello.

Miller, S. J., 1979. The archaeological fauna of four sites in Smith Creek Canyon. In D. R. Tuohy and D. L. Rendall, eds., *The Archaeology of Smith Creek Canyon, Eastern Nevada*, pp. 272–331. Nevada State Museum Anthropological Papers 17.

Miller, W. E. 1976. Late Pleistocene vertebrates of the Silver Creek Local Fauna from north central Utah. *Great Basin Naturalist* 36:387–424.

Miller, W. E. 1987. *Mammut americanum*, Utah's first record of the American mastodon. *Journal of Paleontology* 61:168–183.

Miller, W. E. 2002. Quaternary vertebrates of the northeastern Bonneville Basin and vicinity of Utah. In J. W. Gwynn, ed., *Great Salt Lake: An Overview of Change*, pp. 54–69. Utah Department of Natural Resources Special Publication, Salt Lake City.

Miller, W. R. 1983. Uto-Aztecan languages. In A. Ortiz, ed., *Southwest, Handbook of North American Indians, Volume 10*, pp. 113–124. Smithsonian Institution Press, Washington, DC.

Miller, W. R. 1984. The classification of Uto-Aztecan languages based on lexical evidence. *International Journal of American Linguistics* 50:1–24.

Miller, W. R. 1986. Numic languages. In W. L. d'Azevedo, ed., *Great Basin, Handbook of North American Indians, Volume 11*, pp. 98–106. Smithsonian Institution Press, Washington, DC.

Miller, W. R., J. L. Tanner, and L. P. Foley. 1971. A lexicostatistic study of Shoshoni dialects. *Anthropological Linguistics* 13:142–164.

Milliken, R., and W. R. Hildebrandt. 1997. *Culture Change along the Eastern Sierra Nevada/Cascade Front. Volume V: Honey Lake Basin*. Coyote Press, Salinas.

Milne, G. A., J. X. Mitrovica, and D. P. Schrag. 2002. Estimating past continental ice volume from sea-level data. *Quaternary Science Reviews* 21:361–376.

Minckley, T. A., P. J. Bartlein, and J. J. Shinker. 2004. Paleoecological response to climate change in the Great Basin since the Last Glacial Maximum. In D. L. Jenkins, T. J. Connolly, and C. M. Aikens, eds., *Early and Middle Holocene Archaeology of the Northern Great Basin*, pp. 21–30. University of Oregon Anthropological Papers 62.

Minckley, T. A., C. Whitlock, and P. J. Bartlein. 2007. Vegetation, fire, and climate history of the northwestern Great Basin during the last 14,000 years. *Quaternary Science Reviews* 26: 2167–2184.

Minckley, W. L., and J. E. Deacon, eds. 1991. *Battle Against Extinction: Native Fish Management in the American West*. University of Arizona Press, Tucson.

Minckley, W. L., and P. C. Marsh. 2009. *Inland Fishes of the Greater Southwest*. University of Arizona Press, Tucson.

Minnich, R. A. 2007. Climate, paleoclimate, and paleovegetation. In M. G. Barbour, R. Keeler-Wolf, and A. S. Schoenherr, eds., *Terrestrial Vegetation of California*. Third edition, pp. 43–70. University of California Press, Berkeley.

Mitton, J. B., B. R. Kreiser, and R. G. Latta. 2000. Glacial refugia of limber pine (*Pinus flexilis* James) inferred from the population structure of mitochondrial DNA. *Molecular Ecology* 9:91–97.

Mock, C. J. 1996. Climatic controls and spatial variations of precipitation in the western United States. *Journal of Climate* 9:1111–1125.

Moffit, J. 1934. Mule deer study program. *California Fish and Game* 20:52–66.

Montenegro, Á., R. Hetherington, M. Eby, and A. J. Weaver. 2006. Modelling pre-historic transoceanic crossings into the Americas. *Quaternary Science Reviews* 25:1323–1338.

Moore, M. 1995. Analysis of fauna from sites in Warner Valley, Oregon. M.A. thesis, University of Nevada, Reno.

Moore, P. D., J. A. Webb, and M. E. Collinson. 1991. *Pollen Analysis*. Second edition. Blackwell, Oxford, UK.

Morgan, D. L. 1964. *Jedediah Smith and the Opening of the West*. University of Nebraska Press, Lincoln.

Morgan, D. L. 1985. *The Humboldt: Highroad of the West*. University of Nebraska Press, Lincoln.

Morgan, G. S. 2002. Late Rancholabrean mammals from southernmost Florida, and the Neotropical influence in Florida Pleistocene faunas. In R. J. Emry, ed., *Cenozoic Mammals of Land and Sea: Tributes to the Career of Clayton E. Ray*, pp. 15–38. Smithsonian Contributions to Paleobiology 93.

Moritz, C., J. L. Patton, C. J. Conroy, J. L. Parra, G. C. White, and S. R. Beissinger. 2008. Impact of a century of climate change on small-mammal communities in Yosemite National Park, USA. *Science* 322:261–264.

Morlan, R. E. 1987. The Pleistocene archaeology of Beringia. In M. H. Nitecki and D. V. Nitecki, eds., *The Evolution of Human Hunting*. Plenum, New York.

Morrison, J. L. 1996. Crested Caracara (*Caracara cheriway*). In A. Poole, ed., *The Birds of North America Online*. Ithaca: Cornell Lab of Ornithology. http://bna.birds.cornell.edu/bna/species/249.

Morrison, R. B. 1964. *Lake Lahontan: Geology of Southern Carson Desert, Nevada*. U.S. Geological Survey Professional Paper 401.

Morrison, R. B. 1991. Quaternary stratigraphic, hydrologic, and climatic history of the Great Basin, with emphasis on Lakes Lahontan, Bonneville, and Tecopa. In R. B. Morrison, ed., *Quaternary Nonglacial Geology: Conterminous U.S.*, pp. 283–320. The Geology of North America, Volume K-2. Geological Society of America, Boulder, CO.

Morrison, R. B. 1999. Lake Tecopa: Quaternary geology of Tecopa Valley, California, a multimillion-year record and its relevance to the proposed nuclear-waste repository at Yucca Mountain, Nevada. In L. A. Wright and B. W. Troxel, eds., *Cenozoic Basins of the Death Valley Region*. Geological Society of America Special Paper 333.

Morrison, R. B., and M. D. Mifflin. 2000. Lake Tecopa and its environs: 2.5 million years of exposed history relevant to climate, groundwater, and erosion issues at the proposed nuclear waste repository at Yucca Mountain, Nevada. In D. R. Lageson, S. G. Peters, and M. M. Lahren, eds., *Great Basin and Sierra Nevada: Geological Society of America Field Guide* 2:355–382.

Mozingo, H. N. 1987. *Shrubs of the Great Basin*. University of Nevada Press, Reno.

Muniz, M. 1998. Preliminary results of excavations and analysis of Little River Rapids: A prehistoric inundated site in North Florida. *Current Research in the Pleistocene* 15:48–49.

Munro, N. 2006. The role of the turkey in the Southwest. In D. J. Stanford, B. D. Smith, D. H. Ubelaker, and E. J. E. Szathmáry, eds., *Environment, Origins, and Population, Handbook of North American Indians, Volume 3*, pp. 463–470. Smithsonian Institution Press, Washington, DC.

Munroe, J. S., and D. M. Mickelson. 2002. Last Glacial Maximum equilibrium-line altitudes and paleoclimate, northern Uinta Mountains, Utah, U.S.A. *Journal of Glaciology* 48:257–266.

Munroe, J. S., B. J. C. Laabs, J. D. Shakun, B. S. Singer, D. M. Mickelson, K. A. Refsnider, and M. W. Caffee. 2006. Latest Pleistocene advance of alpine glaciers in the southwestern Uinta Mountains, Utah, USA: Evidence for the influence of local moisture sources. *Geology* 34:841–844.

Murchison, S. B. 1989a. Fluctuation history of Great Salt Lake, Utah, during the last 13,000 Years. Ph.D. dissertation and Limneotectonics Laboratory Technical Report 89-2, University of Utah, Salt Lake City.

Murchison, S. B. 1989b. Utah chub (*Gila atraria*) from the latest Pleistocene Gilbert shoreline, west of Corrine, Utah. *Great Basin Naturalist* 49:131–133.

Murchison, S. B., and W. E. Mulvey. 2000. Late Pleistocene and Holocene shoreline stratigraphy on Antelope Island, Davis County, Utah. In J. K. King and G. C. Willis, eds., *Geology of Antelope Island*, pp. 77–83. Utah Geological Survey Miscellaneous Publications 00-1.

Murie, O. J. 1951. *The Elk of North America*. Stackpole, Harrisburg, PA; Wildlife Management Institute, Washington, DC.

Murray, L. K., C. J. Bell, M. T. Dolan, and J. I. Mead. 2005. Late Pleistocene fauna from the southern Colorado Plateau, Navajo County, Arizona. *Southwestern Naturalist* 50:363–374.

Murton, J. B., M. D. Bateman, S. R. Dallimore, J. T. Teller, and Z. Yang. 2010. Identification of Younger Dryas outburst flood path from Lake Agassiz to the Arctic Ocean. *Nature* 464: 740–743.

Musil, R. R. 1988. Functional efficiency and technological change: A hafting tradition model for prehistoric North America. In J. A. Willig, C. M. Aikens, and J. L. Fagan, eds., *Early Human Occupation in Far Western North America: The Clovis-Archaic Interface*, pp. 373–389. Nevada State Museum Anthropological Papers 21.

Musil, R. R. 1995. *Adaptive Transitions and Environmental Change in the Northern Great Basin: A View from Diamond Swamp*. University of Oregon Anthropological Papers 51.

Nachlinger, J., K. Sochi, P. Comer, G. Kittel, and D. Dorfman. 2001. *Great Basin: An Ecoregion-Based Conservation Blueprint*. The Nature Conservancy, Reno, NV.

Naiman, R. J. 1981. An ecosystem overview: Desert fishes and their habitats. In R. J. Naiman and D. L. Soltz, eds., *Fishes in North American Deserts*, pp. 493–531. Wiley, New York.

Napton, L. K. 1997. The Spirit Cave mummy: Coprolite investigations. *Nevada Historical Society Quarterly* 40:97–104.

Napton, L. K., and R. F. Heizer. 1970. Analysis of human coprolites from archaeological contexts, with primary reference to Lovelock Cave, Nevada. *University of California Archaeological Research Facility Contributions* 10:87–130.

Nassaney, M. S., and K. Pyle. 1999. The adoption of the bow and arrow in eastern North America: A view from central Arkansas. *American Antiquity* 64:243–263.

Nauta, L. T. 2000. Utah Chub size utilization at Goshen Island. In D. B. Madsen and M. D. Metcalf, eds., Intermountain archaeology. *University of Utah Anthropological Papers* 122:148–156.

Negrini, R. 2002. Pluvial lake sizes in the northwestern Great Basin throughout the Quaternary Period. In R. Hershler, D. B. Madsen, and D. R. Currey, eds., *Great Basin Aquatic Systems History*, pp. 11–52. Smithsonian Contributions to the Earth Sciences 33.

Negrini, R. M., and J. O. Davis. 1992. Dating late Pleistocene pluvial events and tephras by correlating paleomagnetic secular variation records from the western Great Basin. *Quaternary Research* 38:46–59.

Negrini, R. M., D. B. Erbes, A. P. Roberts, K. L Verosub, A. M. Sarna-Wojcicki, and C. E. Meyer. 1994. Repeating waveform initiated by a 180–190ka geomagnetic excursion in western North America: Implications for field behavior during polarity transitions and subsequent secular variation. *Journal of Geophysical Research* 99 (B12) 24:105–124.

Negrini, R. M., D. B. Erbes, K. Faber, A. M. Herrera, A. P. Roberts, A. S. Cohen, P. E. Wigand, and F. F. Foit, Jr. 2000. A paleoclimate record for the past 250,000 years for Summer Lake, Oregon, USA: I. Chronology and magnetic proxies for lake level. *Journal of Paleolimnology* 24:125–149.

Nelson, D. E., R. E. Morlan, J. S. Vogel, J. R. Southon, and C. R. Harington. 1986. New dates on northern Yukon artifacts: Holocene not Upper Pleistocene. *Science* 232:749–751.

Nelson, E. W. 1983. *The Eskimo about Bering Strait*. Smithsonian Institution Press, Washington, DC.

Nelson, M. E., and J. H. Madsen, Jr. 1978. Late Pleistocene musk oxen from Utah. *Kansas Academy of Science Transactions* 81:277–295.

Nelson, M. E., and J. H. Madsen, Jr. 1980. A summary of Pleistocene, fossil vertebrate localities in the northern Bonneville Basin of Utah. In J. W. Gwynn, ed., *Great Salt Lake: A Scientific, Historical, and Economic Overview*, pp. 97–114. Utah Geological and Mineral Survey Bulletin 116.

Nevada Division of Wildlife. 1997. *Nevada Elk Species Management Plan*. Nevada Division of Wildlife, Reno.

Nevada Fish and Wildlife Office; Region 8, California and Nevada. 2008. *Devils Hole Pupfish*. www.fws.gov/Nevada/protected_species/fish/species/dhp/dhp.html

Nevada State Museum Anthropology Program. 2007. Spirit Cave Update—January 2007. *Nevada State Museum Newsletter* 7(1):3.

Newman, D. E. 1996. Pollen and macrofossil analyses. In R. K. Talbot and L. D. Richens, *Steinaker Gap: An Early Fremont Homestead*, pp. 123–147. Brigham Young University Museum of Peoples and Cultures Occasional Papers 4.

Nicolson, M. 1990. Henry Allan Gleason and the individualistic hypothesis: The structure of a botanist's career. *The Botanical Review* 56:91–161.

Nogués-Bravo, D., J. Rodríguez, J. Hortal, P. Batra, and M. B. Araújo. 2008. Climate change, humans, and the extinction of the woolly mammoth. *PLoS Biology* 6:685–692. doi:10.1371/journal.pbio.0060079.

Norris, R. M. 1988. Eureka Valley sand dunes. In C. A. Hall, Jr., and V. Doyle-Jones, eds. *Plant Biology of Eastern California*, pp. 207–211. Natural History of the White-Inyo Range Symposium Volume 2. White Mountain Research Station, University of California, Los Angeles.

Novak, S. A. 2008. *House of Mourning: A Biocultural History of the Mountain Meadows Massacre*. University of Utah Press, Salt Lake City.

Novak, S. A., and D. D. Kollmann. 1999. Perimortem processing of human remains among the Great Basin Fremont. *International Journal of Osteoarchaeology* 10:65–75.

Nowak, C. L., R. S. Nowak, R. J. Tausch, and P. E. Wigand. 1994a. A 30 000 year record of vegetation dynamics at a semi-arid locale in the Great Basin. *Journal of Vegetation Science* 5:579–590.

Nowak, C. L., R. S. Nowak, R. J. Tausch, and P. E. Wigand. 1994b. Tree and shrub dynamics in northwestern Great Basin woodland and shrub steppe during the Late-Pleistocene and Holocene. *American Journal of Botany* 81:265–277.

Nowak, R. M. 1999. *Walker's Mammals of the World*. Sixth edition. Johns Hopkins University Press, Baltimore.

Nowak, R. N. 1979. *North American Quaternary* Canis. Museum of Natural History, University of Kansas Monograph 6.

O'Connell, J. F. 1971. The archaeology and cultural ecology of Surprise Valley, northeast California. Ph.D. dissertation, University of California, Berkeley.

O'Connell, J. F. 1975. *The Prehistory of Surprise Valley*. Ballena Press Anthropological Papers 4.

O'Connell, J. F., and J. Allen. 2004. Dating the colonization of Sahul (Pleistocene Australia–New Guinea): A review of recent research. *Journal of Archaeological Science* 31:835–853.

O'Connell, J. F., and J. E. Ericson. 1974. Earth lodges to wickiups: A long sequence of domestic structures from the northern Great Basin. *Nevada Archaeological Survey Research Paper* 5:43–61.

O'Connell, J. F., and K. Hawkes. 1981. Alyawara plant use and optimal foraging theory. In B. Winterhalder and E. A. Smith, eds., *Hunter-Gatherer Foraging Strategies: Ethnographic and Archaeological Analyses*, pp. 99–125. University of Chicago, Chicago, IL.

O'Connell, J. F., and K. Hawkes. 1984. Food choice and foraging sites among the Alyawara. *Journal of Anthropological Research* 40:504–535.

O'Connell, J. F., and P. S. Hayward. 1972. Altithermal and Medithermal human adaptations in Surprise Valley, northeast California. In D. D. Fowler, ed., *Great Basin Cultural Ecology: A Symposium*, pp. 25–42. Desert Research Institute Publications in the Social Sciences 8.

O'Connell, J. F., and C. M. Inoway. 1994. Surprise Valley projectile points and their chronological implications. *Journal of California and Great Basin Anthropology* 16:162–198.

O'Connell, J. F., J. Allen, and K. Hawkes. 2009. Pleistocene Sahul and the Origins of Seafaring. In A. Anderson, J. Barrett, and K. Boyle, eds., *The Global Origins and Development of Seafaring*. McDonald Institute for Archaeological Research, Cambridge University, Cambridge.

O'Connell, J. F., K. T. Jones, and S. R. Simms. 1982. Some thoughts on prehistoric archaeology in the Great Basin. In D. B. Madsen and J. F. O'Connell, eds., *Man and Environment in the Great Basin*, pp. 227–240. Society for American Archaeology Papers 2.

O'Connell, J. F., P. K. Latz, and P. Barnett. 1983. Traditional and modern plant use among the Alyawara of central Australia. *Economic Botany* 37:80–109.

O'Connor, J. E. 1993. *Hydrology, Hydraulics, and Geomorphology of the Bonneville Flood*. Geological Society of America Special Paper 274.

O'Connor, J. E., and V. R. Baker. 1992. Magnitudes and implications of peak discharges from glacial Lake Missoula. *Geological Society of America Bulletin* 104:267–279.

O'Farrell, M. J., and A. R. Blaustein. 1974. *Microdipodops megacephalus*. *Mammalian Species* 46.

O'Gara, B. W., and R. G. Dundas. 2002. Distribution: Past and present. In D. E. Toweill., and J. W. Thomas, eds., *North American Elk: Ecology and Management*, pp. 67–119. Smithsonian Institution Press, Washington, DC.

O'Neill, B. L., D. L. Jenkins, C. M. Hodges, P. W. O'Grady, and T. J. Connolly. 2006. Housepits in the Chewaucan Marsh: Investigations at the Gravelly Ford Bridge Site. In B. L. O'Neill, ed., *Beads, Points, and Pit Houses: A Northern Great Basin Miscellany*, pp. 93–136. University of Oregon Anthropological Papers 66.

O'Rourke, D. H., M. G. Hayes, and S. W. Carlyle. 2000a. Ancient DNA studies in physical anthropology. *Annual Review of Anthropology* 29:217–242.

O'Rourke, D. H., M. G. Hayes, and S. W. Carlyle. 2000b. Spatial and temporal stability of mtDNA haplogroup frequencies in native North America. *Human Biology* 72:15–34.

O'Rourke, D. H., R. L. Parr, and S. W. Carlyle. 1999. Molecular genetic variation in prehistoric inhabitants of the eastern Great Basin. In B. E. Hemphill, and C. S. Larsen, eds., *Prehistoric lifeways in the Great Basin Wetlands: Bioarchaeological Reconstruction and Interpretation*, pp. 84–102. University of Utah Press, Salt Lake City.

O'Rourke, M. K. 1991. Pollen in packrat middens: The contribution of filtration. *Grana* 30:337–341.

O'Sullivan, P. B., M. Morwood, D. Hobbs, F. Aziz, Suminto, M. Situmorang, A. Raza, and R. Maas. 2001. Archaeological implications of the geology and chronology of the Soa basin, Flores, Indonesia. *Geology* 29:607–610.

Oetting, A. C. 1989. *Villages and Wetlands Adaptations in the Northern Great Basin: Chronology and Land Use in the Lake Abert–Chewaucan Marsh Basin, Lake County, Oregon*. University of Oregon Anthropological Papers 41.

Oetting, A. C. 1990. Aboriginal settlement in the Lake Abert–Chewaucan Marsh Basin, Lake County, Oregon. In J. C. Janetski and D. B. Madsen, eds., *Wetland Adaptations in the Great Basin*, pp. 183–206. Brigham Young University Museum of Peoples and Cultures, Occasional Papers 1.

Oetting, A. C. 1993. *The Archaeology of Buffalo Flat: Cultural Resources Investigations for the CONUS OTH-B Buffalo Flat Radar Transmitter Site, Christmas Lake Valley, Oregon*. Heritage Research Associates Report 151, Eugene, OR.

Oetting, A. C. 1994. Early Holocene rabbit drives and prehistoric land use patterns on Buffalo Flat, Christmas Lake Valley, Oregon. In C. M. Aikens and D. L. Jenkins, eds., *Archaeological Research in the Northern Great Basin: Fort Rock Archaeology since Cressman*, pp. 155–169. University of Oregon Archaeological Papers 50.

Office of Management and Budget. 2009. *A New Era of Responsibility: Renewing America's Promise*. U.S. Government Printing Office, Washington, DC.

Olson, S. L. 2007. The "Walking Eagle" *Wetmoregyps daggetti* Miller: A scaled-up version of the Savanna Hawk (*Buteogallus meridionalis*). *Ornithological Monographs* 63:100–114.

Omernik. J. M. 1995. Ecoregions: A spatial framework for environmental management. In W. S. Davis and T. P. Simon, eds., *Biological Assessment and Criteria: Tools for Water Resource Planning and Decision Making*, pp. 49–62. Lewis Publishers, Boca Raton, FL.

Omernik, J. M. 1999. *Level III Ecoregions of the Continental United States*. National Health and Environmental Effects Research Laboratory, U.S. Environmental Protection Agency. Corvallis, OR.

Oppenheimer, C. 2003. Climatic, environmental and human consequences of the largest known historic eruption: Tambora volcano (Indonesia) 1815. *Progress in Physical Geography* 27:230–259.

Ore, H. T., and C. N. Warren. 1971. Late Pleistocene–Early Holocene geomorphic history of Lake Mojave, California. *Geological Society of America Bulletin* 82:2553–2562.

Orme, A. R. 2008a. Lake Thompson, Mojave Desert, California: The late Pleistocene lake system and its Holocene desiccation. In M. C. Reheis, R. Hershler, and D. M. Miller, eds., *Late Cenozoic Drainage History of the Southwestern Great Basin and Lower Colorado River Region: Geologic and Biotic Perspectives*, pp. 261–278. Geological Society of America Special Paper 439.

Orme, A. R. 2008b. Pleistocene pluvial lakes of the American West: A short history of research. In R. H. Grapes, D. Oldroyd, and A. Grigelis, eds., *History of Geomorphology and Quaternary Geology*, pp. 51–78. Geological Society Special Publication 101.

Orme, A. R., and A. J. Orme. 2008. Late Pleistocene shorelines of Owens Lake, California, and their hydroclimatic and tectonic implications. In M. C. Reheis, R. Hershler, and D. M. Miller, eds., *Late Cenozoic Drainage History of the Southwestern Great Basin and Lower Colorado River Region: Geologic and Biotic Perspectives*, pp. 207–225. Geological Society of America Special Paper 439.

Orr, P. C. 1969. *Felis trumani*, a new radiocarbon dated cat skull from Crypt Cave, Nevada. *Bulletin of the Santa Barbara Museum of Natural History Department of Geology* 2:1–8.

Osborn, G. 1989. Glacial deposits and tephra in the Toiyabe Range, Nevada, U.S.A. *Arctic and Alpine Research* 21:256–267.

Osborn, G. 2004. Great Basin of the western United States. In J. Ehlers and P. L. Gibbard, eds., *Quaternary Glaciations: Extent and Chronology, Part II*, pp. 63–68. Elsevier, Amsterdam.

Osborn, G., and K. Bevis. 2001. Glaciation in the Great Basin of the western United States. *Quaternary Science Reviews* 20:1377–1410.

Osborn, J., M. Lachniet, and M. Saines. 2008. Interpretation of Pleistocene glaciation in the Spring Mountains of Nevada: Pros and cons. In E. M. Duebendorfer and E. I. Smith, eds., *Field Guide to Plutons, Volcanoes, Faults, Reefs, Dinosaurs, and Possible Glaciation in Selected Areas of Arizona, California, and Nevada*, pp. 153–172. Geological Society of America Field Guide 11.

Osmond, C. B., L. F. Pitelka, and G. M. Hidy, eds. 1990. *Plant Biology of the Basin and Range.* Springer-Verlag, Berlin.

Oviatt, C. G. 1988. Late Pleistocene and Holocene lake fluctuations in the Sevier Lake Basin, Utah, USA. *Journal of Paleolimnology* 1:9–21.

Oviatt, C. G. 1989. *Quaternary Geology of Part of the Sevier Desert, Millard County, Utah.* Utah Geological and Mineral Survey Special Studies 70.

Oviatt, C. G. 1997. Lake Bonneville fluctuations and global climate change. *Geology* 25:155–158.

Oviatt, C. G., and D. M. Miller. 1997. New explorations along the northern shores of Lake Bonneville. In P. K. Link and B. J. Kowallis, eds., *Mesozoic to Recent Geology of Utah*, pp. 345–371. Brigham Young University Geology Studies 42(II).

Oviatt, C. G., and R. S. Thompson. 2002. Recent developments in the study of Lake Bonneville since 1980. In J. W. Gwynn, ed., *Great Salt Lake: An Overview of Change*, pp. 1–6. Utah Department of Natural Resources Special Publication, Salt Lake City.

Oviatt, C. G., D. R. Currey, and D. M. Miller. 1990. Age and paleoclimatic significance of the Stansbury shoreline of Lake Bonneville, northeastern Great Basin. *Quaternary Research* 33:291–305.

Oviatt, C. G., D. B. Madsen, and D. N. Schmitt. 2003. Late Pleistocene and early Holocene rivers and wetlands in the Bonneville Basin of western North America. *Quaternary Research* 60:200–210.

Oviatt, C. G., W. D. McCoy, and R. G. Reider. 1987. Evidence for a shallow early or middle Wisconsin-age lake in the Bonneville Basin, Utah. *Quaternary Research* 27:248–262.

Oviatt, C. G., R. S. Thompson, D. S. Kaufman, J. Bright, and R. M. Forester. 1999. Reinterpretation of the Burmester Core, Bonneville Basin, Utah. *Quaternary Research* 52:180–184.

Oviatt, C. G., D. M. Miller, J. P. McGeehin, C. Zachary, and S. Mahan. 2005. The Younger Dryas phase of Great Salt Lake, Utah, USA. *Palaeogeography, Palaeoclimatology, Palaeoecology* 219:263–284.

Palacios-Fest, M. R., A. D. Cohen, J. Ruiz, and B. Blank. 1993. Comparative paleoclimatic interpretations from nonmarine ostracodes using faunal assemblages, trace elements shell chemistry and stable isotope data. In P. K. Swart, K. C. Lohmann, J. McKenzie, and S. Savin, eds., *Climate Change in Continental Isotopic Records*, pp. 179–190. American Geophysical Union Geophysical Monograph 78.

Pachauri, R. K., and A. Reisinger, eds. 2007. *Climate Change 2007: Synthesis Report. Contributions of Working Groups I, II, and III to the Fourth Assessment Report of the Intergovernmental Panel on Climate Change.* IPCC, Geneva.

Palmer, R. S., ed. 1962. *Handbook of North American Birds. Volume 1: Loons through Flamingos.* Yale University Press, New Haven.

Paquay, F. S., S. Goderis, G. Ravizza, F. Vanhaeck, M. Boyd, T. A. Surovell, V. T. Holliday, C. V. Haynes, Jr., and P. Claeys. 2009. Absence of geochemical evidence for an impact event at the Bølling-Allerød/Younger Dryas transition. *Proceedings of the National Academy of Sciences* 106:21505–21510.

Paquay, F. S., S. Goderis, G. Ravizza, and P. Claeys. 2010. Reply to Bunch et al.: Younger Dryas impact proponents challenge new platinum group elements and osmium data unsupportive of their hypothesis. *Proceedings of the National Academy of Sciences* 107:E59–E60.

Patrickson, S. J., D. Sack, A. R. Brunelle, and K. A. Moser. 2010. Late Pleistocene to early Holocene lake level and paleoclimate insights from Stansbury Island, Bonneville Basin, Utah. *Quaternary Research* 73:237–246.

Patten, D. T., L. Rouse, and J. Stromberg. 2008. Vegetation dynamics of Great Basin springs: Potential effects of groundwater withdrawal. In L. E. Stevens and V. J. Meretsky., eds., *Aridland Springs in North America: Ecology and Conservation*, pp. 279–289. University of Arizona Press, Tucson.

Patterson, L. W. 1983. Criteria for determining the attributes of man-made lithics. *Journal of Field Archaeology* 10:297–307.

Pavlik, B. 2008. *The California Deserts: An Ecological Discovery.* University of California Press, Berkeley.

Pavlov, P., J. I. Svendsen, and S. Indrelid. 2001. Human presence in the European Arctic nearly 40,000 years ago. *Nature* 413:64–67.

Pendleton, L. S. A. 1979. Lithic technology in early Nevada assemblages. M.A. thesis, California State University, Long Beach.

Pendleton, L. S. A., and D. H. Thomas. 1983. *The Fort Sage Drift Fence, Washoe County, Nevada.* American Museum of Natural History Anthropological Papers 58(2).

Perry, J. P., Jr. 1991. *The Pines of Mexico and Central America.* Timber Press, Portland. OR.

Peterson, P. M. 1984. *Flora and Physiognomy of the Cottonwood Mountains, Death Valley National Monument, California.* Cooperative National Park Resources Studies Unit, University of Nevada, Las Vegas.

Pettigrew, R. M. 1985. *Archaeological investigations on the East Shore of Lake Abert, Lake County, Oregon.* University of Oregon Anthropological Papers 32.

Phillips, F. M. 2008. Geological and hydrological history of the paleo–Owens River drainage since the late Miocene. In M. C. Reheis, R. Hershler, and D. M. Miller, eds., *Late Cenozoic Drainage History of the Southwestern Great Basin and Lower Colorado River Region: Geologic and Biotic Perspectives*, pp. 115–150. Geological Society of America Special Paper 439.

Phillips, F. M., A. R. Campbell, C. Kruger, P. Johnson, R. Roberts, and E. Keyes. 1992. *A Reconstruction of the Response of the Water Balance in Western United States Lake Basins to Climatic Change.* New Mexico Water Resources Research Institute Technical Report 269.

Phillips, F. M., A. R. Campbell, G. I. Smith, and J. L. Bischoff. 1994. Interstadial climatic cycles: A link between western North America and Greenland? *Geology* 22:1115–1118.

Phillips, F. M., M. G. Zreda, L. V. Benson, M. A. Plummer, D. Elmore, and P. Sharma. 1996. Chronology for fluctuations in late Pleistocene Sierra Nevada glaciers and lakes. *Science* 274:749–751.

Phillips, K. N., and A. S. Van Denburgh. 1971. *Hydrology and Geochemistry of Abert, Summer, and Goose Lakes, and other Closed Basin Lakes in South Central Oregon.* U.S. Geological Survey Professional Paper 502-B.

Piegat, J. J. 1980. Glacial geology of central Nevada. M.S. thesis, Purdue University, Lafayette, IN.

Pierce, K. L. 2004. Pleistocene glaciation in the Rocky Mountains. In A. R. Gillespie, S. C. Porter and B. F. Atwater, eds., *The Quaternary Period in the United States*, pp. 63–76. Developments in Quaternary Science 1. Elsevier, Amsterdam.

Pinkoski, M. 2008. Julian Steward, American anthropology, and colonialism. *Histories of Anthropology Annual* 4:172–204.

Pinson, A. O. 2004. Of lakeshores and dry basin floors: A regional perspective on the early Holocene record of environmental change and human adaptation at the Tucker site. In D. L. Jenkins, T. J. Connolly, and C. M. Aikens, eds., *Early and Middle Holocene Archaeology of the Northern Great Basin*, pp. 52–76. University of Oregon Anthropological Papers 62.

Pinson, A. O. 2007. Artiodactyl use and adaptive discontinuity across the Paleoarchaic/Archaic transition in the northern Great Basin. In K. E. Graf and D. N. Schmitt, eds., *Paleoindian or Paleoarchaic: Great Basin Human Ecology at the Pleistocene/Holocene Transition*, pp. 187–203. University of Utah Press, Salt Lake City.

Pinson, A. O. 2008. Geoarchaeological context of Clovis and Western Stemmed Tradition sites in Dietz Basin, Lake County, Oregon. *Geoarchaeology* 23:63–106.

Pinter, N., and S. E. Ishman. 2008a. Impacts, mega-tsunami, and other extraordinary claims. *GSA Today* 18:37–38.

Pinter, N., and S. E. Ishman. 2008b. Impacts, mega-tsunami, and other extraordinary claims. *GSA Today* 18:e11–e14.

Pippin, L. C. 1986. Intermountain Brown Wares: An assessment. In S. Griset, ed., *Pottery of the Great Basin and Adjacent Areas*, pp. 9–21. University of Utah Anthropological Papers 111.

Pister, E. P. 1981. The conservation of desert fishes. In R. J. Naiman and D. L. Soltz, eds., *Fishes in North American Deserts*, pp. 493–531. Wiley, New York.

Pitulko, V. V., P. A. Nikolsky, E. Yu. Girya, A. E. Basilyan, V. E. Tumskoy, S. A. Koulakov, S. N. Astakhov, E. Yu. Pavlova, and M. A. Anisomov. 2004. The Yana RHS Site: Humans in the Arctic before the Last Glacial Maximum. *Science* 303:52–56.

Plew, M. G., and Sundell, T. 2000. The archaeological occurrence of bison on the Snake River Plain. *North American Archaeologist* 21:119–137.

Poinar, H., S. Fiedel, C. E. King, A. M. Devault, K. Bos, M. Kuch, and R. Debruyne. 2009. Comment on "DNA from pre-Clovis human coprolites in Oregon, North America." *Science* 325:148.

Polhemus, D. A., and J. T. Polhemus. 2002. Basin and ranges: The biogeography of aquatic true bugs (Insecta: Heteroptera) in the Great Basin. In R. Hershler, D. B. Madsen, and D. R. Currey, eds., *Great Basin Aquatic Systems History*, pp. 235–254. Smithsonian Contributions to the Earth Sciences 33.

Poore, R. Z., M. J. Pavich, and H. D. Grissino-Mayer. 2005. Record of the North American southwest monsoon from Gulf of Mexico sediment cores. *Geology* 33:209–212.

Porter, S. C. 1977. Present and past glaciation threshold in the Cascade Range, Washington, U.S.A.: Topographic and climatic controls, and paleoclimatic implications. *Journal of Glaciology* 18:101–116.

Porter, S. C., K. L. Pierce, and T. D. Hamilton. 1983. Late Wisconsin mountain glaciation in the western United States. In S. C. Porter, ed., *Late-Quaternary Environments of the United States*. Volume 1: The Late Pleistocene, pp. 71–111. University of Minnesota Press, Minneapolis, MN.

Powledge, T. M. 2006. What is the Hobbit? *PLoS Biology* 4(12): e440 doi:10.1371/journal.pbio.0040440.

Prather, M. 2008. Owens Lake is returning to WILDlife. *Mono Lake Newsletter* Winter–Spring 2008:13.

Presnall, C. C. 1938. Evidences of bison in southwestern Utah. *Journal of Mammalogy* 19:111–112.

Preuss, C. 1958. *Exploring with Frémont: The Private Diaries of Charles Preuss, Cartographer for John C. Frémont on His First, Second, and Fourth Expeditions to the Far West*. Translated and edited by E. G. and E. K. Gudde. University of Oklahoma Press, Norman.

Price, B. A., and S. E. Johnston. 1988. A model of late Pleistocene and early Holocene adaptation in eastern Nevada. In J. A. Willig, C. M. Aikens, and J. L. Fagan, eds., *Early Human Occupation in Far Western North America: The Clovis-Archaic Interface*, pp. 231–250. Nevada State Museum Anthropological Papers 21.

Price, R. A., A. Liston, and S. H. Strauss. 1998. Phylogeny and systematics of *Pinus*. In D. M. Richardson, ed., *Ecology and Biogeography of Pinus*, pp. 49–68. Cambridge University Press, Cambridge, UK.

Putnam, W. C. 1949. Quaternary geology of the June Lake district, California. *Geological Society of America Bulletin* 60:1281–1302.

Putnam, W. C. 1950. Moraine and shoreline relationships at Mono Lake, California. *Geological Society of America Bulletin* 61:115–122.

Quade, J. 1986. Late Quaternary environmental changes in the Upper Las Vegas Valley, Nevada. *Quaternary Research* 26:340–357.

Quade, J. 1994. Spring deposits and late Pleistocene groundwater levels in southern Nevada. In *High Level Radioactive Waste Management: Proceedings of the Fifth Annual International Conference, Las Vegas, Nevada, May 22–26, 1994*, 4:2530–2537. American Nuclear Society, La Grange Park, Illinois, and American Society of Civil Engineers, New York.

Quade, J., and W. L. Pratt. 1989. Late Wisconsin groundwater discharge environments of the southwestern Indian Springs Valley, southern Nevada. *Quaternary Research* 31:351–370.

Quade, J., R. M. Forester, and J. E. Whelan. 2003. Late Quaternary paleohydrologic and paleotemperature change in southern Nevada. In Y. Enzel, S. G. Wells, and N. Lancaster, eds. *Paleoenvironments and Paleohydrology of the Mojave and Southern Great Basin Deserts*, pp. 165–168. Geological Society of America Special Paper 368.

Quade, J., R. M. Forester, W. L. Pratt, and C. Carter. 1998. Black mats, spring-fed streams, and late-glacial-age recharge in the southern Great Basin. *Quaternary Research* 49:129–148.

Quade, J., M. D. Mifflin, W. L. Pratt, W. McCoy, and L. Burckle. 1995. Fossil spring deposits in the southern Great Basin and their implications for changes in water-table levels near Yucca Mountain, Nevada, during Quaternary time. *Geological Society of America Bulletin* 107:213–230.

Quaife, M. M., ed. 1966. *Kit Carson's Autobiography*. University of Nebraska Press, Lincoln.

Quaife, M. M., ed. 1978. *Adventures of a Mountain Man: The Narrative of Zenas Leonard*. University of Nebraska Press, Lincoln.

Ramenofsky, A. F. 1987. *Vectors of Death*. University of New Mexico Press, Albuquerque.

Ramis, D., and P. Bover. 2001. A review of the evidence for domestication of *Myotragus balearicus* Bate 1909 (Artiodactyla, Caprinae) in the Balearic Islands. *Journal of Archaeological Science* 28:265–282.

Raper, D., and M. Bush. 2009. A test of *Sporormiella* representation as a predictor of megaherbivore presence and abundance. *Quaternary Research* 71:490–496.

Rasmussen, M., L. S. Cummings, M. T. P. Gilbert, V. Bryant, C. Smith, D. L. Jenkins, and E. Willerslev. 2009. Response to comment by Goldberg et al. on "DNA from pre-Clovis human coprolites in Oregon, North America." *Science* 325:148-d.

Raven, C. 1990. *Prehistoric Human Geography in the Carson Desert. Part II: Archaeological Field Tests of Model Predictions*. U.S. Department of the Interior, U.S. Fish and Wildlife Service, Region 1, Cultural Resource Series 4.

Raven, C., and R. G. Elston, eds. 1988. *Preliminary Investigations in Stillwater Marsh. Human Prehistory and Geoarchaeology*. U.S. Department of the Interior, U.S. Fish and Wildlife Service, Region 1, Cultural Resource Series 1.

Raven, C., and R. G. Elston. 1989. *Prehistoric Human Geography in the Carson Desert. Part I: A Predictive Model of Land-Use in the Stillwater Wildlife Management Area*. U.S. Department of the Interior, U.S. Fish and Wildlife Service, Region 1, Cultural Resource Series 3.

Raven, C., and R. G. Elston. 1990. *Prehistoric Human Geography in the Carson Desert. Part II: Archaeological Field Tests of Model Predictions*. U.S. Department of the Interior, U.S. Fish and Wildlife Service, Region 1, Cultural Resource Series 4.

Ray, V. F. 1963. *Primitive Pragmatists: The Modoc Indians of Northern California*. University of Washington Press, Seattle.

Raymond, A. W. 1994. *The surface archaeology of Harney Dune (35Ha718), Malheur National Wildlife Refuge, Oregon*. U.S. Department of the Interior, U.S. Fish and Wildlife Service Region 1, Cultural Resource Series 9.

Rea, J. A. 1958. Concerning the validity of lexicostatistics. *International Journal of American Linguistics* 24:145–150.

Rea, J. A. 1973. The Romance data of the pilot studies for glottochronology. *Current Trends in Linguistics* 11:355–367.

Redmond, B. G., and K. B. Tankersley. 2005. Evidence of early Paleoindian bone modification and use at the Sheriden Cave site (33WY252), Wyandot County, Ohio. *American Antiquity* 70:503–526.

Redwine, K. L. 2003a. Paleoclimate implications of the middle to late Quaternary pluvial history of Newark Valley, east-central Nevada. In D. J. Easterbook, ed., *Quaternary Geology of the United States: INQUA 2003 Field Guide*, pp. 184–191. Desert Research Institute, Reno, NV.

Redwine, K. L. 2003b. The Quaternary pluvial history and paleoclimate implications of Newark Valley, east-central Nevada; derived from mapping and interpretation of surficial units and geomorphic features. M.S. thesis, Humboldt State University, Arcata, CA.

Refsnider, K. A., B. J. C. Laabs, M. A. Plummer, D. M. Mickelson, B. S. Singer, and M. W. Caffee. 2008. Last Glacial Maximum climate inferences from cosmogenic dating and glacier modeling of the western Uinta ice field, Uinta Mountains, Utah. *Quaternary Research* 69:130–144.

Reheis, M. C. 1994. Comments on "Soil Development Parameters in the absence of a chronosequence in a glaciated basin of the White Mountains, California-Nevada,' by T. W. Swanson, D. L. Elliot-Fisk, and R. J. Southard. *Quaternary Research* 41:245–249.

Reheis, M. C. 1997. Dust deposition downwind of Owens (dry) Lake, 1991–1994: Preliminary findings. *Journal of Geophysical Research* 102, D22:25999–26008.

Reheis, M. C. 1999a. *Extent of Pleistocene Lakes in the Western Great Basin*. U.S. Geological Survey Miscellaneous Field Studies Map MF-2323.

Reheis, M. C. 1999b. Highest pluvial-lake shorelines and Pleistocene climate of the western Great Basin. *Quaternary Research* 52:196–205.

Reheis, M. C., and R. Morrison. 1997. High, old pluvial lakes of western Nevada. In P. K. Link and B. J. Kowallis, eds., *Proterozoic to Recent Stratigraphy, Tectonics, and Volcanology, Utah, Nevada, Southern Idaho and Central Mexico*, pp. 459–492. Brigham Young University Geology Studies 42.

Reheis, M. C., and J. L. Redwine. 2008. Lake Manix shorelines and Afton Canyon terraces: Implications for incision of Afton Canyon. In M. C. Reheis, R. Hershler, and D. M. Miller, eds., *Late Cenozoic Drainage History of the Southwestern Great Basin and Lower Colorado River Region: Geologic and Biotic Perspectives*, pp. 227–259. Geological Society of America Special Paper 439.

Reheis, M C., B. J. C. Laabs, and D. S. Kaufman. 2009. Geology and geomorphology of Bear Lake Valley and upper Bear River, Utah and Idaho. In J. G. Rosenbaum and D. S. Kaufman, eds., *Paleoenvironments of Bear Lake, Utah and Idaho, and its Catchment*, pp. 15–48. Geological Society of America Special Paper 450.

Reheis, M. C., D. M. Miller, and J. L. Redwine. 2007. *Quaternary Stratigraphy, Drainage-basin Development, and Geomorphology of the Lake Manix Basin, Mohave Desert*. U.S. Geological Survey Open-File Report 2007-1281.

Reheis, M. C., S. Stine, and A. D. Sarna-Wojcicki. 2002. Drainage reversals in Mono Basin during the late Pliocene and Pleistocene. *Geological Society of America Bulletin* 114:991–1006.

Reheis, M. C., J. L Slate, A. M. Sarna-Wojcicki, and C. E. Meyer. 1993. A late Pliocene to middle Pleistocene pluvial lake in Fish Lake Valley, Nevada and California. *Geological Society of America Bulletin* 105:953–967.

Reheis, M. C., A. M. Sarna-Wojcicki, R. L. Reynolds, C. A Repenning, and M. D. Mifflin. 2002. Pliocene to middle Pleistocene lakes in the western Great Basin: Ages and connections. In R. Hershler, D. B. Madsen, and D. R. Currey, eds., *Great Basin Aquatic Systems History*, pp. 53–108. Smithsonian Contributions to the Earth Sciences 33:53–108.

Reheis, M. C., J. Redwine, K. Adams, S. Stine, K. Parker, R. Negrini, R. Burke, G. Kurth, J. McGeehin, J. Paces, F. Phillips, A. Sarna-Wojcicki, and J. Smoot. 2008. Pliocene to Holocene lakes in the Great Basin: New perspectives on paleoclimate, landscape dynamics, tectonics, and paleodistribution of aquatic species. In M. C. Reheis, R. Hershler, and D. M. Miller, eds., *Late Cenozoic Drainage History of the Southwestern Great Basin and Lower Colorado River Region: Geologic and Biotic Perspectives*, pp. 155–194. Geological Society of America Special Paper 439.

Reis, G. 2008. Mono Lake drops 2.4 feet since 2006 high point; Meromixis ends. *Mono Lake Newsletter* Winter–Spring 2008:14.

Reisner, M. 1993. *Cadillac Desert: The American West and Its Disappearing Water*. Revised edition. Penguin Books, New York.

Renfrew, C., A. McMahon, and L. Trask, eds. 2000. *Time Depth in Historical Linguistics*. The McDonald Institute for Archaeological Research, Cambridge.

Rennie, D. P. 1987. Late Pleistocene alpine glacial deposits in the Pine Forest Range, Nevada. M.S. thesis, University of Nevada, Reno.

Reumer, J. W., L. Rook, K. Van der Borg, K. Post, D. Mol, and J. De Vos. 2003. Late Pleistocene survival of the saber-toothed cat *Homotherium* in northwestern Europe. *Journal of Vertebrate Paleontology* 23:260–262.

Reveal, J. L. 1979. Biogeography of the intermountain region: A speculative appraisal. *Mentzelia* 4:1–92.

Reveal, J. L., and J. L. Reveal. 1985. The missing Fremont cannon: An ecological solution? *Madroño* 32:106–117.

Reynolds, R. E., R. L. Reynolds, and C. J. Bell. 1991. The Devil Peak sloth. In R. E. Reynolds, ed., *Crossing the Borders: Quaternary Studies in Eastern California and Southwestern Nevada*, pp. 115–116. San Bernardino County Museum Association, Redlands, CA.

Reynolds, R. E., R. L. Reynolds, C. J. Bell, and B. Pitzer. 1991b. Vertebrate remains from Antelope Cave, Mescal Range, San Bernardino County, California. In R. E. Reynolds, ed., *Crossing the Borders: Quaternary Studies in Eastern California and southwestern Nevada*, pp. 107–109. San Bernardino County Museum Association, Redlands, CA.

Reynolds, R. E., R. L. Reynolds, C. J. Bell, N. J. Czaplewski, H. T. Goodwin, J. I. Mead, and B. Roth. 1991a. The Kokoweef Cave faunal assemblage. In R. E. Reynolds, ed., *Crossing the Borders: Quaternary Studies in Eastern California and Southwestern Nevada*, pp. 97–103. San Bernardino County Museum Association, Redlands, CA.

Rhode, D. 1987. The mountains and the lake: Prehistoric lacustrine-upland settlement relationships in the Walker watershed, western Nevada. Ph.D. dissertation, University of Washington, Seattle.

Rhode, D. 1990. On transportation costs of Great Basin resources: An assessment of the Jones-Madsen model. *Current Anthropology* 31:413–419.

Rhode, D. 1994. Direct dating of brown ware ceramics using thermoluminescence and its relation to the Numic spread. In D. B. Madsen and D. Rhode, eds., *Across the West: Human Population Movement and the Expansion of the Numa*, pp. 124–130. University of Utah Press, Salt Lake City.

Rhode, D. 1999. The role of paleoecology in the development of Great Basin archaeology, and vice versa. In C. Beck, ed., *Models for the Millennium: Great Basin Anthropology Today*, pp. 29–49. University of Utah Press, Salt Lake City.

Rhode, D. 2000a. Holocene vegetation history in the Bonneville Basin. In Madsen, D. B., *Late Quaternary Paleoecology in the Bonneville Basin*, pp. 149–163. Utah Geological Survey Bulletin 130.

Rhode, D. 2000b. Middle and late Wisconsin vegetation in the Bonneville Basin. In Madsen, D. B., *Late Quaternary Paleoecology in the Bonneville Basin*, pp. 137–147. Utah Geological Survey Bulletin 130.

Rhode, D. 2001. Packrat middens as a tool for reconstructing historic ecosystems. In D. Egan and E. A. Howell, eds., *The Historical Ecology Handbook: A Restorationist's Guide to Reference Ecosystems*, pp. 257–293. Island Press, Washington, DC.

Rhode, D. 2002. *Native Plants of Southern Nevada: An Ethnobotany*. University of Utah Press, Salt Lake City.

Rhode, D. 2003. Coprolites from Hidden Cave, revisited: Evidence for site occupation history, diet and sex of occupants. *Journal of Archaeological Science* 30:909–922.

Rhode, D. 2008. Dietary plant use by middle Holocene foragers in the Bonneville Basin, western North America. *Before Farming* 2008(3), Article 2.

Rhode, D. 2009. Plant macrofossil assemblages from Alta Toquima. In D. H. Thomas and L. S. A. Pendleton, eds., *The Archaeology of Monitor Valley: 4. Alta Toquima and the Mt. Jefferson Tablelands Complex*. American Museum of Natural History Anthropological Papers, in press.

Rhode, D., and L. A. Louderback. 2007. Dietary plant use in the Bonneville Basin during the terminal Pleistocene/early Holocene transition. In K. E. Graf and D. N. Schmitt, eds., *Paleoindian or Paleoarchaic: Great Basin Human Ecology at the Pleistocene-Holocene Transition*, pp. 231–247. University of Utah Press, Salt Lake City.

Rhode, D., and D. B. Madsen. 1995. Late Wisconsin/early Holocene vegetation in the Bonneville Basin. *Quaternary Research* 44:246–256.

Rhode, D., and D. B. Madsen. 1998. Pine nut use in the early Holocene and beyond: The Danger Cave archaeological record. *Journal of Archaeological Science* 25:1199–1210.

Rhode, D., and D. H. Thomas. 1983. Flotation analysis of selected hearths. In D. H. Thomas, *The Archaeology of Monitor Valley. 2. Gatecliff Shelter*, pp. 151–157. American Museum of Natural History Anthropological Papers 59(1).

Rhode, D., K. D. Adams, and R. E. Elston. 2000. Geoarchaeology and Holocene landscape history of the Carson Desert, western Nevada. In D. R. Lageson, S. G. Peters, and M. M. Lahren, eds., *Great Basin and Sierra Nevada*, pp. 45–74. Geological Society of America Field Guide 2.

Rhode, D., D. B. Madsen, and K. T. Jones. 2006. Antiquity of early Holocene small-seed consumption and processing at Danger Cave. *Antiquity* 80:328–339.

Rhode, D., T. Goebel, K. E. Graf, B. S. Hockett, K. T. Jones, D. B. Madsen, C. G. Oviatt, and D. N. Schmitt. 2005. Latest Pleistocene–early Holocene human occupation and paleoenvironmental change in the Bonneville Basin, Utah-Nevada. In. J. Pederson and C. M. Dehler, eds., *Interior Western United States*, pp. 211–230. Geological Society of America Field Guide 6.

Richardson, D. M., and P. W. Rundel. 1998. Ecology and biogeography of *Pinus*: An introduction. In D. M. Richardson, ed., *Ecology and Biogeography of* Pinus, pp. 3–46. Cambridge University Press, Cambridge, UK.

Richmond, G. M. 1964. *Glaciation of Little Cottonwood and Bells Canyons, Wasatch Mountains, Utah*. U.S. Geological Survey Professional Paper 454-D.

Richmond, G. M. 1986. Stratigraphy and correlation of glacial deposits of the Rocky Mountains, the Colorado Plateau, and the ranges of the Great Basin. *Quaternary Science Reviews* 5:99–127.

Rickard, P. A. 1989. *A History of French Language*. Second edition. Unwin Hyman, London.

Rickart, E. A. 2001. Elevational diversity gradients, biogeography and the structure of montane mammal communities in the intermountain region of North America. *Global Ecology and Biogeography* 10:77–100.

Ricketts, T. H., E. Dinerstein, D. M. Olson, C. J. Loucks, W. Eichbaum, D. DellaSala, K. Kavanagh, P. Hedao, P. T. Hurley, K. M. Carney, R. Abell, and S. Walters. 1999. *Terrestrial Ecoregions of North America*. Island Press, Washington, DC.

Riddell, F. A. 1952. The recent occurrence of bison in northeastern California. *American Antiquity* 18, 168–169.

Riddell, F. A. 1960. *Honey Lake Paiute Ethnography*. Nevada State Museum Anthropological Papers 4.

Ridge, J. C., and F. D. Larsen. 1990. Re-evaluation of Antevs' New England varve chronology and new radiocarbon dates of sediments from glacial Lake Hitchcock. *Geological Society of America Bulletin* 102:889–899.

Riggs, A. C. 1992. Geohydrologic evidence for the development of Devils Hole, southern Nevada as an aquatic environment. *Proceedings of the Desert Fishes Council* 20–21:47–48.

Riggs, A. C., and J. E. Deacon. 2004. Connectivity in desert aquatic ecosystems: The Devils Hole story. In D. W. Sada and S. E. Sharpe, eds., *Conference Proceedings: Spring-fed Wetlands: Important Scientific and Cultural Resources of the Intermountain Region, May 7–9, 2002, Las Vegas, NV*. DHS Publication No. 41210. http://www.wetlands.dri.edu.

Roberts, S. M., and R. J. Spencer. 1998. A desert responds to Pleistocene climate change: Saline lacustrine sediments, Death Valley, California, USA. In A. S. Alsharhan, K. W. Glennie, G.

L. Whittle, and C. G. St. Kendall, eds., *Quaternary Deserts and Climatic Change*, pp. 357–370. Balkema, Rotterdam.

Robinson, G. S., L. P. Burney, and D. A. Burney. 2005. Landscape paleoecology and megafaunal extinction in southeastern New York. *Ecological Monographs* 75:295–315.

Rogers, G. F. 1982. *Then and Now: A Photographic History of Vegetation Change in the Central Great Basin Desert*. University of Utah Press, Salt Lake City.

Rolle, A. 1991. *John Charles Frémont: Character as Destiny*. University of Oklahoma Press, Norman.

Rominger, E. M., H. A. Whitlaw, D. L. Weybright, W. C. Dunn, and W. B. Ballard. 2004. The influence of mountain lion predation on bighorn sheep translocations. *Journal of Wildlife Management* 88:993–999.

Rosenbaum, J. G., and C. W. Heil, Jr. 2009. The glacial/deglacial history of sedimentation in Bear Lake, Utah and Idaho. In J. G. Rosenbaum and D. S. Kaufman, eds., *Paleoenvironments of Bear Lake, Utah and Idaho, and its Catchment*, pp. 247–261. Geological Society of America Special Paper 450.

Rosentreter, R., M. Bowker, and J. Belknap. 2007. *A Field Guide to Biological Soil Crusts*. U.S. Government Printing Office, Denver.

Roust, N. L. 1967. Preliminary examination of prehistoric human coprolites from four western Nevada caves. *University of California Archaeological Survey Reports* 70:49–88.

Ruhlen, M. 1991a. *A Guide to the World's Languages. Volume 1. Classification*. Stanford University Press, Stanford, CA.

Ruhlen, M. 1991b. The Amerind phylum and the prehistory of the New World. In S. M. Lamb and E. D. Mitchell, eds., *Sprung from some Common Source*, pp. 328–350. Stanford University Press, Stanford, CA.

Rush, F. E. 1972. Hydrologic reconnaissance of Big and Little Soda lakes, Churchill County, Nevada. In *Water for Nevada: Hydrologic Atlas*, Map L-5. Division of Water Resources, State of Nevada, Carson City.

Russell, I. C. 1885a. Existing glaciers of the United States. *Annual Report of the U.S. Geological Survey, 1883–1884*:303–355.

Russell, I. C. 1885b. *Geological History of Lake Lahontan, A Quaternary Lake of Northwestern Nevada*. U.S. Geological Survey Monograph 11.

Ruter, A., J. Arzt, S. Vavrus, R. A. Bryson, and J. E. Kutzbach. 2004. Climate and environment of the subtropical and tropical Americas (NH) in the mid-Holocene: Comparisons of observations with climate model simulations. *Quaternary Science Reviews* 23:663–679.

Ryser, F. A., Jr. 1985. *Birds of the Great Basin*. University of Nevada Press, Reno.

Sack, D. 1990. *Quaternary Geology of Tule Valley, West-Central Utah*. Utah Geological and Mineral Survey Map 124.

Sack, D. 1999. The composite nature of the Provo Level of Lake Bonneville, Great Basin, western North America. *Quaternary Research* 52:316–327.

Sack, D. 2002. Fluvial linkages in Lake Bonneville subbasin integration. In R. Hershler, D. B. Madsen, and D. R. Currey, eds., *Great Basin Aquatic Systems History*, pp. 129–144. Smithsonian Contributions to the Earth Sciences 33.

Salocks, C., and K. B. Kaley. 2003. *Ephedrine and pseudoephedrine*. Technical Support Document: Toxicology Clandestine Drug Labs/Methamphetamine, vol. 1 (13). Office of Environmental Health Hazard Assessment, Sacramento.

Salzer, M. W. 2000. Temperature variability and the northern Anasazi: Possible implications for regional abandonment. *Kiva* 65:295–318.

Sampson, C. G. 1985. *Nightfire Island: Late Holocene Lakemarsh Adaptation on the Western Edge of the Great Basin*. University of Oregon Anthropological Papers 33.

Saunders, J. J. 1996. North American Mammutidae. In J. Shoshani and P. Tassy, eds., *The Proboscidea: Evolution and Palaeoecology of Elephants and their Relatives*, pp. 271–279. Oxford University Press, Oxford.

Saysette, J. 1996. Preliminary report of molar and alveolus measurements as predictive indicators in late Pleistocene marmots. In R. E. Reynolds and J. Reynolds, eds., *Empty Basins, Vanished Lakes*, pp. 55–58. San Bernardino County Museum Association Quarterly 43(1).

Scharf, E. A. 1992. Archaeobotany of midway: Plant resource use at a high altitude site in the White Mountains of eastern California. M.A. thesis, University of Washington, Seattle.

Scharf, E. A. 2009. Foraging and prehistoric use of high elevations in the western Great Basin: Evidence from seed assemblages at Midway (CA-MNO-2196), California. *Journal of California and Great Basin Anthropology* 29:11–27.

Schlesinger, W. H., E. H. DeLucia, and D. W. Billings. 1989. Nutrient-use efficiency of woody plants on contrasting soils in the western Great Basin, Nevada. *Ecology* 70:105–113.

Schmitt, D. N. 1992. Faunal remains. In R. G. Elston and C. Raven, eds., *Archaeological Investigations at Tosawihi, A Great Basin Quarry. Part I: The Periphery*, pp. 303–328. Intermountain Research, Silver City, NV.

Schmitt, D. N., and K. D. Lupo. 1995. On mammalian taphonomy, taxonomic diversity, and measuring subsistence data in zooarchaeology. *American Antiquity* 60, 496–514.

Schmitt, D. N., and K. D. Lupo. 2005. The Camels Back Cave mammalian fauna. In D. N. Schmitt and D. B. Madsen, *Camels Back Cave*, pp. 136–176. University of Utah Anthropology Papers 125.

Schmitt, D. N., and D. B. Madsen. 2005. *Camels Back Cave*. University of Utah Anthropology Papers 125.

Schmitt, D. N., D. B. Madsen, and K. D. Lupo. 2002a. Small-mammal data and early and middle Holocene climates and biotic communities in the Bonneville Basin, USA. *Quaternary Research* 58, 255–260.

Schmitt, D. N., D. B. Madsen, and K. D. Lupo. 2002b. The worst of times, the best of times: Jackrabbit hunting by Middle Holocene human foragers in the Bonneville Basin of western North America. In M. Mondini, S. Muñoz, and S. Wickler, eds., *Colonisation, Migration, and Marginal Areas*, pp. 86–95. Oxbow Press, Oxford, UK.

Schmitt, D. N., D. B. Madsen, C. G. Oviatt, and R. Quist. 2007. Late Pleistocene/early Holocene geomorphology and human occupation of the Old River Bed Delta, western Utah. In K. E. Graf and D. N. Schmitt, eds., *Paleoindian or Paleoarchaic: Great Basin Human Ecology at the Pleistocene/Holocene Transition*, pp. 105–119. University of Utah Press, Salt Lake City.

Schneider von Deimling, T., A. Ganopolski, H. Held, and S. Rahmstorf. 2006. How cold was the Last Glacial Maximum? *Geophysical Research Letters* 33: L14709, doi:10.1029/2006GL026484.

Schneider von Deimling, T., H. Held, A. Ganopolski, and S. Rahmstorf. 2008. Are paleo-proxy data helpful for constraining future climate change? *PAGES News* 16(2):20–21.

Schott, M. J. 1997. Stones and shafts redux: The metric discrimination of chipped-stone dart and arrow points. *American Antiquity* 62:86–101.

Schramm, A., M. Stein, and S. L. Goldstein. 2000. Calibration of the ^{14}C time scale to >40 ka by ^{234}U-^{230}Th dating of Lake Lisan

Schramm, D. R. 1982. *Floristics and Vegetation of the Black Mountains, Death Valley National Monument, California*. Cooperative National Park Resources Studies Unit, University of Nevada, Las Vegas, Contribution CPSU/UNLV 012/13, Las Vegas, NV.

Schroedl, A., ed. 1995. *Open Site Archaeology in Little Boulder Basin: 1992 Data Recovery Excavations in the North Block Heap Leach Facility Area, North-Central Nevada. Report BLM1-2021(P)*. Manuscript on file, Bureau of Land Management, Reno, NV.

Schroth, A. B. 1994. The Pinto Point controversy in the western United States. Ph.D. dissertation, University of California, Riverside.

Schubert, B. W. 2010. Late Quaternary chronology and extinction of North American giant short-faced bears (*Arctodus*). *Quaternary International* 217:188–194.

Schultz, C. B., Howard, E. B. 1936. The fauna of Burnet Cave, Guadalupe Mountains, New Mexico. *Proceedings of the Academy of Natural Sciences of Philadelphia* 87:273–298.

Schultz, C. B., L. D. Martin, and L. G. Tanner. 1970. Mammalian distribution in the Great Plains and adjacent areas from 14,000 to 9,000 years ago. *American Quaternary Association Abstracts* 1970:119–120.

Schwarz, J. 1991. *A Water Odyssey: The Story of the Metropolitan Water District of Southern California*. Metropolitan Water District of Southern California, Los Angeles.

Schweikert, R. A., M. M. Lahren, R. Karlin, J. Howle, and K. Smith. 2000. Lake Tahoe active faults, landslides, and tsunamis. In D. R. Lageson, S. G. Peters, and M. M. Lahren, eds., *Great Basin and Sierra Nevada*, pp. 1–22. Geological Society of America Field Guide 2.

Scott, A. C., N. Pinter, M. E. Collinson, M. Hardiman, R. S. Anderson, A. P. R. Brain, S. Y. Smith, F. Marone, and M. Stampanoni. 2010. Fungus, not comet or catastrophe, accounts for carbonaceous spherules in the Younger Dryas "impact layer." *Geophysical Research Letters* 37, L14302.

Scott, B. R., ed. 1971. Average annual evaporation. In *Water for Nevada: Hydrologic Atlas*, Map S-5. Department of Water Resources, State of Nevada, Carson City.

Scott, W. E. 1988. Temporal relations of lacustrine and glacial events at Little Cottonwood and Bells canyons, Utah. In M. N. Machette, ed., *In the Footsteps of G. K. Gilbert: Lake Bonneville and Neotectonics of the Eastern Basin and Range Province*, pp. 78–81. Utah Geological and Mineral Survey Miscellaneous Publication 88-1.

Seck, S. M. 1980. The archaeology of Trego Hot Springs: 26Pe118. M.A. thesis, University of Nevada, Reno.

Service, E. R. 1962. *Primitive Social Organization*. Random House, New York.

Severinghaus, J. P., T. Sowers, E. J. Brook, R. B. Alley, and M. L. Bender. 1998. Timing of abrupt climate change at the end of the Younger Dryas interval from thermally fractionated gases in polar ice. *Nature* 391:141–146.

Shackleton, N. J., R. G. Fairbanks, T.-C. Chiu, and F. Parrenin. 2004. Absolute calibration of the Greenland time scale: Implications for Antarctic time scales and for $\Delta^{14}C$. *Quaternary Science Reviews* 23:1513–1522.

Shafer, S. L., P. J. Bartlein, and R. S. Thompson. 2001. Potential changes in the distributions of western North America tree and shrub taxa under future climatic scenarios. *Ecosystems* 4:200–215.

Sharp, N. D. 1992. Fremont hunters and farmers: Faunal resource exploitation at Nawthis Village, central Utah. Ph.D. dissertation, University of Washington, Seattle.

Sharp, R. P. 1938. Pleistocene glaciation in the Ruby–East Humboldt Range, northeastern Nevada. *Journal of Geomorphology* 1:298–323.

Shaul, D. L. 1986. Linguistic adaptation and the Great Basin. *American Antiquity* 51:415–416.

Shelley, P. 1976. A different view of crescentic artifacts: Preliminary report of opal phytolith deposition. Paper presented at the Northwest Anthropological Conference, Ellensburg, WA.

Shevenell, L. 1996. Statewide potential evapotranspiration maps for Nevada. *Nevada Bureau of Mines and Geology Report* 48.

Shields, W. F. 1978. *The Woodruff Bison Kill*. University of Utah Anthropological Papers 99:45–54.

Shields, W. F., and G. F. Dalley. 1978. *The Bear River No. 3 Site*. University of Utah Anthropological Papers 99.

Shindell, D. T., G. A. Schmidt, M. E. Mann, D. Rind, and A. Waple. 2001. Solar forcing of regional climate change during the Maunder Minimum. *Science* 294:2149–2151.

Shreve, F. 1942. The desert vegetation of North America. *Botanical Review* 8:195–246.

Shutler, D., Jr., M. E. Shutler, and J. S. Griffith. 1960. *Stuart Rockshelter: A Stratified Site in Southern Nevada*. Nevada State Museum Anthropological Papers 3.

Shutler, M. E., and R. Shutler, Jr. 1963. *Deer Creek Cave, Elko County, Nevada*. Nevada State Museum Anthropological Papers 11.

Sigler, J. W., and W. F. Sigler. 1994. Fishes of the Great Basin and the Colorado Plateau: Past and present forms. In K. T. Harper, L. L. St. Clair, K. H. Thorne, and W. M. Hess, eds., *Natural History of the Colorado Plateau and Great Basin*, pp. 163–208. University Press of Colorado, Niwot.

Sigler, W. F. 1962. *Bear Lake and Its Future*. Utah State University, Logan.

Sigler, W. F., and J. W. Sigler. 1987. *Fishes of the Great Basin*. University of Nevada Press, Reno.

Sigler, W. F., and J. W. Sigler. 1996. *Fishes of Utah: A Natural History*. University of Utah Press, Salt Lake City.

Silver, S., and W. R. Miller. 1997. *American Indian Languages: Cultural and Social Contexts*. University of Arizona Press, Tucson.

Simmons, W. S., R. Morales, V. Williams, and S. Camacho. 1997. Honey Lake Maidu ethnogeography of Lassen County, California. *Journal of California and Great Basin Anthropology* 19:2–31.

Simms, S. R. 1983a. Comments on Bettinger and Baumhoff's explanation for the "Numic Spread" in the Great Basin. *American Antiquity* 48:825–830.

Simms, S. R. 1983b. The effects of grinding stone reuse on the archaeological record of the eastern Great Basin. *Journal of California and Great Basin Anthropology* 5:98–102.

Simms, S. R. 1984. Aboriginal Great Basin foraging strategies: An evolutionary analysis. Ph.D. dissertation, University of Utah, Salt Lake City.

Simms, S. R. 1986. New evidence for Fremont adaptive diversity. *Utah Archaeology* 8:204–216.

Simms, S. R. 1987. *Behavioral Ecology and Hunter-Gatherer Foraging: An Example from the Great Basin*. BAR International Series 381.

Simms, S. R. 1990. Fremont transitions. *Utah Archaeology* 3:1–18.

Simms, S. R. 1999. Farmers, foragers, and adaptive diversity. In B. E. Hemphill, and C. S. Larsen, eds., *Prehistoric lifeways in the Great Basin Wetlands: Bioarchaeological Reconstruction and Interpretation*, pp. 21–54. University of Utah Press, Salt Lake City.

Simms, S. R. 2008. *Ancient Peoples of the Great Basin & Colorado Plateau*. Left Coast Press, Walnut Creek, CA.

Simms, S. R., and M. C. Isgreen. 1984. *Archaeological Excavations in the Sevier and Escalante Deserts, Western Utah*. University of Utah Archaeological Center Reports of Excavations 83-12.

Simms, S. R., and L. W. Lindsay. 1989. 42MD300, an early Holocene site in the Sevier Desert. *Utah Archaeology* 2:56–66.

Simms, S. R., and A. W. Raymond. 1999. The treatment of human remains from three Great Basin sites. In B. E. Hemphill, and C. S. Larsen, eds., *Prehistoric Lifeways in the Great Basin Wetlands: Bioarchaeological Reconstruction and Interpretation*, pp. 8–20. University of Utah Press, Salt Lake City.

Simms, S. R., C. J. Loveland, and M. E. Stuart. 1991. *Prehistoric Human Skeletal Remains and the Prehistory of the Great Salt Lake Wetlands*. Department of Sociology, Social Work, and Anthropology, Utah State University, Logan.

Simpson, J. H. 1983 [1876]. *Report of Explorations across the Great Basin of the Territory of Utah*. University of Nevada Press, Reno.

Simpson, R. D. 1989. *An Introduction to the Calico Early Man Site Lithic Assemblage*. San Bernardino County Museum Association Quarterly 36(3).

Simpson, R. D. 1999. *An Introduction to the Calico Early Man Site Lithic Assemblage: The 35th anniversary edition*. San Bernardino County Museum Association Quarterly 46(4).

Singer, V. J. 2004. Faunal assemblages of four early to mid-Holocene marsh-side sites in the Fort Rock Valley, south-central Oregon. In D. L. Jenkins, T. J. Connolly and C. M. Aikens, eds., *Early and Middle Holocene Archaeology of the Northern Great Basin*, pp. 167–185. University of Oregon Anthropological Papers 62.

Smales, T. J. 1972. Existing lakes and reservoirs. In *Water for Nevada: Hydrologic Atlas, Map L-1*. Division of Water Resources, State of Nevada, Carson City.

Smiley, T. L. 1977. Memorial to Ernst Valdemar Antevs, 1888–1974. *Geological Society of America Memorials* 6:1–7.

Smith, A. T. 1974a. The distribution and dispersal of pikas: Consequences of insular population structure. *Ecology* 55:1112–1119.

Smith, A. T. 1974b. The distribution and dispersal of pikas: Influences of behavior and climate. *Ecology* 55:1368–1376.

Smith, A. T., and Weston, M. L. 1990. *Ochotona princeps. Mammalian Species* 352.

Smith, D. G., J. Lorenz, B. K. Rolfs, R. L. Bettinger, B. Green, J. Eshleman, B. Schultz, and R. Malhi. 2000. Implications of the distribution of Albumin Naskapi and Albumin Mexico Mexico for New World prehistory. *American Journal of Physical Anthropology* 111:557–572.

Smith, E. A. 2004. Why do good hunters have higher reproductive success? *Human Nature* 15:343–364.

Smith, F. A., S. K. Lyons, S. K. M. Ernest, K. E. Jones, D. M. Kaufman, T. Dayan, P. A. Marquet, J. H. Brown, and J. P. Haskell. 2003. Body mass of late Quaternary animals. *Ecology* 84:3403; *Ecological Archives* E084-094.

Smith, G. 1988. The story behind the story of Eureka sand dunes: What happened when scientists jumped down from their pedestals into rough-and-tumble politics. In C. A. Hall Jr. and V. Doyle-Jones, eds., *Plant Biology of Eastern California*, pp. 195–298. Natural History of the White-Inyo Range Symposium Volume 2. White Mountain Research Station, University of California, Los Angeles.

Smith, G. I. 1979. *Subsurface Stratigraphy and Geochemistry of Late Quaternary Evaporites, Searles Lake, California*. U.S. Geological Survey Professional Paper 1043.

Smith, G. I. 1984. Paleohydrologic regimes in the southwestern Great Basin, 0–3.2 my ago, compared with other long records of "global" climate. *Quaternary Research* 22:1–17.

Smith, G. I., and F. A. Street-Perrott. 1983. Pluvial lakes of the western United States. In S. C. Porter, ed., *Late-Quaternary Environments of the United States, Volume 1*, pp. 190–212. University of Minnesota Press, Minneapolis.

Smith, G. I., and J. L. Bischoff, eds. 1997. *An 800,000-Year Paleoclimatic Record from Core OL-92, Owens Lake, Southeast California*. Geological Society of America Special Paper 317.

Smith, G. I., L. Benson, and D. R. Currey. 1989. *Quaternary Geology of the Great Basin: Field Trip Guidebook* T117. American Geophysical Union, Washington, DC.

Smith, G. I., V. J. Barczak, G. F. Moulton, and J. C. Liddicoat. 1983. *Core KM-3, a Surface-to-bedrock Record of Late Cenozoic Sedimentation in Searles Valley, California*. U.S. Geological Survey Professional Paper 1256.

Smith, G. M. 2006. Pre-Archaic technological organization, mobility, and settlement systems: A view from the Parman localities, Humboldt County, Nevada. M.A. thesis, University of Nevada, Reno.

Smith, G. M. 2007. Pre-Archaic mobility and technological activities at the Parman localities, Humboldt County, Nevada. In K. E. Graf and D. N. Schmitt, eds., *Paleoindian or Paleoarchaic: Great Basin Human Ecology at the Pleistocene-Holocene Transition*, pp. 139–155. University of Utah Press, Salt Lake City.

Smith, G. R. 1983. Paleontology of Hidden Cave: Fish. In D. H. Thomas, *The Archaeology of Monitor Valley. 2. Gatecliff Shelter*, pp. 171–178. American Museum of Natural History Anthropological Papers 59(1).

Smith, G. R., T. E. Dowling, K. W. Gobalet, T. Lugaski, D. K. Shiozawa, and R. P. Evans. 2002. Biogeography and timing of evolutionary events among Great Basin fishes. In R. Hershler, D. B. Madsen, and D. R. Currey, eds., *Great Basin Aquatic Systems History*, pp. 175–234. Smithsonian Contributions to the Earth Sciences 33.

Smith, S. D., and R. S. Nowak. 1990. Ecophysiology of plants in the intermountain lowlands. In C. B. Osmond, L. F. Pitelka, and G. M. Hidy, eds., *Plant Biology of the Basin and Range*, pp. 179–242. Springer-Verlag, Berlin.

Smith, S. J. 1994. Fremont settlement and subsistence practices in Skull Valley, northern Utah. *Utah Archaeology* 7:51–68.

Smith, S. J., and R. S. Anderson. 1992. Late Wisconsin paleoecologic record from Swamp Lake, Yosemite National Park, California. *Quaternary Research* 38:91–102.

Smith, W. K., and A. K. Knapp. 1990. Ecophysiology of high elevation forests. In C. B. Osmond, L. F. Pitelka, and G. M. Hidy, eds., *Plant Biology of the Basin and Range*, pp. 87–142. Springer-Verlag, Berlin.

Smoot, J. P. 2009. Late Quaternary sedimentary features of Bear Lake, Utah and Idaho. In J. G. Rosenbaum and D. S. Kaufman, eds., *Paleoenvironments of Bear Lake, Utah and Idaho, and its Catchment*, pp. 49–104. Geological Society of America Special Paper 450.

Smoot, J. P., and J. G. Rosenbaum. 2009. Sedimentary constraints on late Quaternary lake-level fluctuations at Bear Lake, Utah and Idaho. In J. G. Rosenbaum and D. S. Kaufman, eds., *Paleoenvironments of Bear Lake, Utah and Idaho, and its Catchment*, pp. 263–290. Geological Society of America Special Paper 450.

Snyder, C. T., G. Hardman, and F. F. Zdenek. 1964. *Pleistocene Lakes in the Great Basin*. U.S. Geological Survey Miscellaneous Geological Investigations Map I-416.

Snyder, N. F., and N. J. Schmitt. 2002. California Condor (*Gymnogyps californianus*). In A. Poole, ed., *The Birds of North America Online*. Cornell Lab of Ornithology, Ithaca. http://bna.birds.cornell.edu/bna/species/610.

Soltz, D. L., and R. J. Naiman. 1978. *The Natural History of Native Fishes in the Death Valley System*. Natural History Museum of Los Angeles County Science Series 30.

Southern Nevada Water Authority. 2009. *Water Resource Plan 09*. www.snwa.com/html/wr_resource_plan.html.

Southworth, J. 1987. *Death Valley in 1849: The Luck of the Gold Rush Emigrants*. Pegleg Books, Burbank, CA.

Spangler, J. D. 2000a. One-pot pithouses and Fremont paradoxes: Formative stage adaptations in the Tavaputs Plateau region of northeastern Utah. In D. B. Madsen and M. D. Metcalf, eds., *Intermountain Archaeology*, pp. 25–38. University of Utah Anthropological Papers 122.

Spangler, J. D. 2000b. Radiocarbon dates, acquired wisdom, and the search for temporal order in the Uinta Basin. In D. B. Madsen and M. D. Metcalf, eds., *Intermountain Archaeology*, pp. 48–100. University of Utah Anthropological Papers 122.

Spaulding, W. G. 1977. Late Quaternary vegetational change in the Sheep Range, southern Nevada. *Journal of the Arizona Academy of Science* 12(2):3–8.

Spaulding, W. G. 1980. *The Presettlement Vegetation of the California Desert*. Desert Planning Staff, Bureau of Land Management, Riverside, CA.

Spaulding, W. G. 1981. The late Quaternary vegetation of a southern Nevada mountain range. Ph.D. dissertation, University of Arizona, Tucson.

Spaulding, W. G. 1983. Late Wisconsin macrofossil records of desert vegetation in the American Southwest. *Quaternary Research* 19:256–264.

Spaulding, W. G. 1985. *Vegetation and Climates of the Last 45,000 years in the Vicinity of the Nevada Test Site, South-Central Nevada*. U.S. Geological Survey Professional Paper 1329.

Spaulding, W. G. 1990a. Comparison of pollen and macrofossil based reconstructions of late Quaternary vegetation in western North America. *Review of Palaeobotany and Palynology* 64:359–366.

Spaulding, W. G. 1990b. Vegetational and climatic development of the Mojave Desert: The Last Glacial Maximum to the present. In J. L. Betancourt, T. R. Van Devender, and P. S. Martin, eds., *Packrat Middens: The Last 40,000 Years of Biotic Change*, pp. 166–199. University of Arizona Press, Tucson.

Spaulding, W. G. 1990c. Vegetation dynamics during the last deglaciation, southeastern Great Basin, U.S.A. *Quaternary Research* 33:188–203.

Spaulding, W. G. 1991. A middle Holocene vegetation record from the Mojave Desert of North America and its paleoclimatic significance. *Quaternary Research* 35:427–437.

Spaulding, W. G. 1994. *Paleohydrologic Investigations in the Vicinity of Yucca Mountain: Late Quaternary Paleobotanical and Palynological Records*. NWPO-TR-022-94. Dames & Moore Inc., Las Vegas, NV.

Spaulding, W. G. 1995. Environmental change, ecosystem responses, and the late Quaternary development of the Mojave Desert. In D. W. Steadman and J. I. Mead, eds., *Late Quaternary Environments and Deep History: A Tribute to Paul S. Martin*, pp. 139–164. The Mammoth Site of Hot Springs, Inc., Scientific Papers 3.

Spaulding, W. G., and L. J. Graumlich. 1986. The last pluvial climatic episodes in the deserts of southwestern North America. *Nature* 320:441–444.

Spaulding, W. G., and T. R. Van Devender. 1980. Late Pleistocene montane conifers in southeastern Utah. In J. D. Jennings, *Cowboy Cave*, pp. 159–161. University of Utah Anthropological Papers 104.

Spaulding, W. G., E. B. Leopold, and T. R. Van Devender. 1983. Late Wisconsin paleoecology of the American Southwest. In S. C. Porter, ed., *Late Quaternary Environments of the United States, Volume 1, The Late Pleistocene*, pp. 259–293. University of Minnesota Press, Minneapolis.

Spaulding, W. G., J. L. Betancourt, L. K. Croft, and K. L. Cole. 1990. Packrat middens: Their composition and methods of analysis. In J. L. Betancourt, T. R. Van Devender, and P. S. Martin, eds., *Packrat Middens: The Last 40,000 Years of Biotic Change*, pp. 59–84. University of Arizona Press, Tucson.

Speller, C. F., B. M. Kemp, S. D. Wyatt, C. Monroe, W. D. Lipe, U. M. Arndt, and D. Y. Yang. 2010. Ancient mitochondrial DNA analysis reveals complexity of indigenous North American turkey domestication. *Proceedings of the National Academy of Sciences* 107:2807–2812.

Spencer, L., R. C. Hanes, C. S. Fowler and S. Jaynes. 1987. *The South Fork Shelter Revisited: Excavation at Upper Shelter, Elko County, Nevada*. Bureau of Land Management Cultural Resource Series 11, Reno, NV.

Spencer, R. J., M. J. Baedecker, H. P. Eugster, R. M. Forester, M. B. Goldhaber, B. F. Jones, K. Kelts, J. Mckenzie, D. B. Madsen, S. L. Rettig, M. Rubin, and C. J. Bowser. 1984. Great Salt Lake, and precursors, Utah: The last 30,000 years. *Contributions to Mineralogy and Petrology* 86:321–224.

Spiess, A. 1974. Faunal remains from Bronco Charlie Cave (26EK801), Elko County, Nevada. In L. A. Casjens, The prehistoric human ecology of southern Ruby Valley, Nevada, pp. 452–486. Ph.D. dissertation, Harvard University, Cambridge, MA.

Sridhar, V., D. B. Loope, J. B. Swinehart, J. A. Mason, R. J. Oglesby, and C. M. Rowe. 2006. Large wind shift on the Great Plains during the Medieval Warm Period. *Science* 313:345–347.

Stafford, T. W., Jr., P. E. Hare, L. Currie, A. J. T. Jull and D. J. Donahue. 1991. Accelerator radiocarbon dating at the molecular level. *Journal of Archaeological Science* 18:35–72.

Stahl, D. W., F. K. Fye, E. R. Cook, and R. D. Griffin. 2007. Tree-ring reconstructed megadroughts over North America since A.D. 1300. *Climatic Change* 83:133–149.

Stahle, D. W., F. K. Fye, E. R. Cook, and R. D. Griffin. 2007. Tree-ring reconstructed megadroughts over North America since A.D. 1300. *Climatic Change* 83:133–149.

Stanford, D., and B. Bradley. 2002. Ocean trails and prairie paths? Thoughts about Clovis origins. In N. G. Jablonski, ed., *The First Americans: The Pleistocene Colonization of the Americas*, pp. 255–272. Memoirs of the California Academy of Science 27.

Steadman, D. W. 1980. A review of the osteology and paleontology of turkeys (Aves: Meleagridinae). In K. E. Campbell, Jr., ed., *Papers in Avian Paleontology Honoring Hildegarde Howard*, pp. 131–208. Natural History Museum of Los Angeles County Contributions in Science 330.

Steadman, D. W. 2006. *Extinction and Biogeography of Tropical Pacific Birds*. University of Chicago Press, Chicago.

Steadman, D. W., and N. G. Miller. 1987. California Condor associated with spruce-jack pine woodland in the late Pleistocene of New York. *Quaternary Research* 28:415–426.

Steadman, D. W., T. W. Stafford, Jr., and R. E. Funk. 1997. Non-association of Paleoindians with AMS-dated Late Pleistocene mammals from the Dutchess Quarry Caves, New York. *Quaternary Research* 47:105–116.

Stearns, C. E. 1942. A fossil marmot from New Mexico and its climatic significance. *American Journal of Science* 240, 867–878.

Steele, D. G., and J. F. Powell. 2002. Facing the past: A view of the North American human fossil record. In N. G. Jablonski, ed., *The First Americans: The Pleistocene Colonization of the Americas*, pp. 93–122. Memoirs of the California Academy of Science 27.

Steffensen, J. P., K. K. Andersen, M. Bigler, H. B. Clausen, D. Dahl-Jensen, H. Fischer, K. Goto-Azuma, M. Hansson, S. J. Johnsen, J. Jouzel, V. Masson-Delmotte, T. Popp, S. O. Rasmussen, R. Rothlisberger, U. Ruth, B. Stauffer, M.-L. Siggaard-Andersen, A. E. Sveinbjörnsdóttir, A. Svensson, J. W. C. White. 2008. High-resolution Greenland ice core data show abrupt climate change happens in few years. *Science* 321:680–684.

Stern, T. H. 1966. *The Klamath Tribe: A People and Their Reservation*. University of Washington Press, Seattle.

Sternberg, C. H. 1909. *The Life of a Fossil Hunter*. Henry Holt, New York.

Stevens, G. C., and J. F. Fox. 1991. The causes of treeline. *Annual Review of Ecology and Systematics* 22:177–191.

Steward, J. H. 1933. Ethnography of the Owens Valley Paiute. *University of California Publications in American Archaeology and Ethnology* 33:233–350.

Steward, J. H. 1937a. Ancient Caves of the Great Salt Lake Region. Bulletin of American Ethnology 116.

Steward, J. H. 1937b. Linguistic distributions and political groups of the Great Basin Shoshoneans. *American Anthropologist* 39:625–634.

Steward, J. H. 1938. *Basin-Plateau Aboriginal Sociopolitical Groups*. Bureau of American Ethnology Bulletin 120.

Steward, J. H. 1941. *Culture Element Distributions*: xiii. *Nevada Shoshone*. Anthropological Records 4:2.

Steward, J. H. 1970. The foundation of Basin-Plateau Shoshonean Society. In E. H. Swanson, Jr., ed., *Languages and Cultures of Western North America: Essays in Honor of Sven S. Liljebad*, pp. 113–151. Idaho State University Press, Pocatello.

Stewart, O. C. 1941. *Culture Element Distributions: XIV. Northern Paiute*. Anthropological Records 4:3.

Stine, S. 1984. Late Holocene lake level fluctuations and island volcanism at Mono Lake, California. In *Geologic Guide to Aspen Valley, Mono Lake, Mono Craters, and Inyo Craters*, pp. 21–49. Friends of the Pleistocene, Pacific Cell (reprinted by Genny Smith Books, Palo Alto, CA).

Stine, S. 1990. Late Holocene fluctuations of Mono Lake, eastern California. *Palaeogeography, Palaeoclimatology, Palaeoecology* 78:333–381.

Stine, S. 1994. Extreme and persistent drought in California and Patagonia during mediaeval time. *Nature* 369:546–549.

Stine, S. 1998. Medieval climatic anomaly in the Americas. In A. S. Issar and N. Brown, eds., *Water, Environment and Society in Times of Climatic Change*, pp. 43–67. Kluwer Academic Publishers, Dordrecht.

Stock, C. S. 1920. Origin of the supposed human footprints of Carson City, Nevada. *Science* 51:514.

Stock, C. S. 1925. *Cenozoic Gravigrade Edentates of Western North America*. Carnegie Institute of Washington Publication 331.

Stock, C. S. 1931. Problems of antiquity presented in Gypsum Cave, Nevada. *Scientific Monthly* 32:22–32.

Stock, C. S. 1936a. A new mountain goat from the Quaternary of Smith Creek Cave, Nevada. *Bulletin of the Southern California Academy of Sciences* 35:149–153.

Stock, C. S. 1936b. Sloth tracks in the Carson prison. *Westways*, 1936 (July):26–27.

Stone, R. 2008. Have desert researchers discovered a hidden loop in the carbon cycle? *Science* 320:1409–1410.

Storey, A. A., J. M. Ramirez, D. Quiroz, D. V. Burley, D. J. Addison, R. Walter, A. J. Anderson, T. L. Hunt, J. S. Athens, L. Huynen, and E. A. Matisoo-Smith. 2007. Radiocarbon and DNA evidence for a pre-Columbian introduction of Polynesian chickens to Chile. *Proceedings of the National Academy of Sciences* 104: 10335–10339.

Strahan, R. 1983. *The Complete Book of Australian Mammals*. Angus & Robertson, London.

Straus, L. G., D. J. Meltzer, and T. Goebel. 2005. Ice Age Atlantis? Exploring the Solutrean-Clovis "connection." *World Archaeology* 37:507–532.

Stuart, A. J. 1991. Mammalian extinctions in the late Pleistocene of northern Eurasia and North America. *Biological Review* 66: 453–562.

Stuart, A. J. 1999. Late Pleistocene megafaunal extinctions: A European perspective. In R. D. E. MacPhee, ed., *Extinctions in Near Time*, pp. 257–269. Kluwer Academic/Plenum Publishers, New York.

Sturm, P. A. 1980. The Great Salt Lake brine system. In J. W. Gwynn, ed., *Great Salt Lake: A Scientific, Historical, and Economic Overview*, pp. 147–162. Utah Geological and Mineral Survey Bulletin 116.

Stutte, N. A. 2004. The Holocene history of bison in the Intermountain West: A synthesis of archaeological and paleontological records from eastern Oregon. M.A. thesis, Portland State University, Portland, OR.

Suárez, W. 2004. The identity of the fossil raptor of the genus *Amplibuteo* (Aves: Accipitridae) from the Quaternary of Cuba. *Caribbean Journal of Science* 40:120–125.

Suring, L. H., M. J. Wisdom, R. J. Tausch, R. F. Miller, M. M. Rowland, L. Schueck, and C. W. Meinke. 2005. Modeling threats to sagebrush and other shrubland communities. In M. J. Wisdom, M. M. Rowland, and L. H. Suring, eds., *Habitat Threats in Sagebrush Ecosystems: Methods of Regional Assessment and Applications in the Great Basin*, pp. 114–149. Alliance Communications Group, Lawrence, KS.

Surovell, T. A., and P. A. Brantingham. 2007. A note on the use of temporal frequency distributions in studies of prehistoric demography. *Journal of Archaeological Science* 24:1868–1877.

Surovell, T. A., V. T. Holliday, J. A. M. Gingerich, C. Ketron, C. V. Haynes, Jr., A. C. Goodyear, I. Hilman, D. P. Wagner, and P. Claeys. 2009. Non-reproducibility of Younger Dryas extraterrestrial results. *Proceedings of the National Academy of Sciences*, 106:18155–18158.

Sutton, M. Q. 1986. Warfare and expansion: An ethnohistoric perspective on the Numic spread. *Journal of California and Great Basin Anthropology* 8:65–82.

Sutton, M. Q. 1988. *Insects as Food: Aboriginal Entomophagy in the Great Basin*. Ballena Press Anthropological Papers 33.

Sutton, M. Q., M. E. Basgall, J. K. Gardner, and M. W. Allen. 2007. Advances in understanding Mojave Desert prehistory. In T. L. Jones and K. A. Klar, eds., *California Prehistory: Colonization, Culture, and Complexity*, pp. 229–245. AltaMira Press, Lanham, MD.

Swadesh, M. 1950. Salish internal relationships. *International Journal of American Linguistics* 16:157–167.

Swadesh, M. 1952. Lexico-statistic dating of prehistoric ethnic contacts. *Proceedings of the American Philosophical Society* 96:452–463.

Swadesh, M. 1955. Algunas fechas glotocronológicas importantes para la prehistoria nahua. *Revista Mexicana de Estudios Antropólogicos* 14:173–192.

Swadesh, M. 1959. Linguistics as an instrument of prehistory. *Southwestern Journal of Anthropology* 15:20–35.

Swanson, T. W., D. L. Elliot-Fisk, and R. J. Southard. 1993. Soil development parameters in the absence of a chronosequence in a glaciated basin of the White Mountains, California-Nevada. *Quaternary Research* 39:186–200.

Sweet, A. T., and R. G. McBeth. 1942. Soil Survey of the Klamath Reclamation Project. In L. S. Cressman, *Archaeological Researches in the Northern Great Basin*, pp. 145–147. Carnegie Institute of Washington Publication 538.

Sweitzer, R. A., S. H. Jenkins, and J. Berger. 1997. Near-extinction of porcupines by mountain lions and consequences of ecosystem change in the Great Basin desert. *Conservation Biology* 11:1407–1417.

Szabo, B. J., P. T. Kolesar, A. C. Riggs, I. J. Winograd, and K. R. Ludwig. 1994. Paleoclimatic inferences from a 120,000-yr calcite record of water-table fluctuations in Browns Room of Devils Hole, Nevada. *Quaternary Research* 41:59–69.

Tadlock, W. L. 1966. Certain crescentic stone objects as a time marker in the western United States. *American Antiquity* 31:662–675.

Talbot, R. K. 2000a. Fremont architecture. In J. C. Janetski, R. K. Talbot, D. E. Newman, L. D. Richens, and J. D. Wilde, *Clear Creek Canyon Archaeological Project: Results and Synthesis*, pp. 131–184. Brigham Young University Museum of Peoples and Cultures Occasional Papers 7.

Talbot, R. K. 2000b. Fremont settlement patterns and demography. In J. C. Janetski, R. K. Talbot, D. E. Newman, L. D. Richens, and J. D. Wilde, *Clear Creek Canyon Archaeological Project: Results and Synthesis*, pp. 201–230. Brigham Young University Museum of Peoples and Cultures Occasional Papers 7.

Talbot, R. K., and L. D. Richens. 1996. *Steinaker Gap: An Early Fremont Homestead*. Brigham Young University Museum of Peoples and Cultures Occasional Papers 4.

Talbot, R. K., and L. D. Richens. 2004. *Fremont Farming and Mobility on the Far Northern Colorado Plateau*. Brigham Young University Museum of Peoples and Cultures Occasional Papers 10.

Talbot, R. K., S. A. Baker, and J. A. Janetski. 2005. Project synthesis: Archaeology in Capitol Reef National Park. In J. C. Janetski, L. Kreutzer, R. K. Talbot, L. D. Richens, and S. A. Baker. 2005. *Life on the Edge: Archaeology in Capitol Reef National Park*, pp. 351–407. Museum of Peoples and Cultures, Brigham Young University Occasional Paper 11.

Talbot, R. K., L. D. Richens, J. D. Wilde, J. C. Janetski and D. E. Newman. 2000. *Excavations at Five Finger Ridge, Clear Creek Canyon, Central Utah*. Brigham Young University Museum of Peoples and Cultures Occasional Papers 5.

Tankersley, K. B. 1997. Sheriden: A Clovis cave site in eastern North America. *Geoarchaeology* 12:713–724.

Tankersley, K. B., B. G. Redmond, and T. E. Grove. 2001. Radiocarbon dates associated with a single-beveled bone projectile point from Sheriden Cave, Ohio. *Current Research in the Pleistocene* 18:62–64.

Tanner, W. W. 1978. Zoogeography of reptiles and amphibians in the intermountain region. In K. T. Harper and J. L. Reveal, eds., *Intermountain Biogeography: A Symposium*, pp. 43–53. Great Basin Naturalist Memoirs 2.

Taulman, J. F., and L. W. Robbins. 1996. Recent range expansion and distributional limits of the nine-banded armadillo (*Dasypus novemcinctus*) in the United States. *Journal of Biogeography* 23:635–648.

Tausch, R. J., N. E. West, and A. A. Nabi. 1981. Tree age and dominance patterns in Great Basin pinyon-juniper woodlands. *Journal of Range Management* 34:259–264.

Taylor, A. 2002. Results of a Great Basin fluted point survey: Chronological and functional relationships between fluted and stemmed points. Senior thesis, Hamilton College, Clinton, NY.

Taylor, A. 2003. Results of a Great Basin fluted-point survey. *Current Research in the Pleistocene* 20:77–79.

Taylor, D. C. 1954. *The Garrison Site: A Report of Archaeological Excavations in Snake Valley, Nevada-Utah*. University of Utah Anthropological Papers 16.

Taylor, K. C., P. A. Mayewski, R. B. Alley, E. J. Brook, A. J. Gow, P. M. Grootes, D. A. Meese, E. S. Saltzman, J. P. Severinghaus, M. S. Twickler, J. W. S. White, S. Whitlow, and G. A. Zielinski. 1997. The Holocene–Younger Dryas transition recorded at Summit, Greenland. *Science* 278:825–827.

Taylor, R. E., and L. A. Payen. 1979. The role of archaeometry in American archaeology: Approaches to the evaluation of the antiquity of *Homo sapiens* in California. *Advances in Archaeological Method and Theory* 2:239–283.

Tchakerian, V. P., and N. Lancaster. 2002. Late Quaternary arid/humid cycles in the Mojave Desert and western Great Basin of North America. *Quaternary Science Reviews* 21:799–810.

Tennyson, A., and P. Martinson. 2006. *Extinct Birds of New Zealand*. Te Papa Press, Wellington, New Zealand.

Terry, R. C. 2008. Raptors, rodents, and paleoecology: Recovering ecological baselines from Great Basin caves. Ph.D. dissertation, University of Chicago, Chicago.

Thomas, D. H. 1972a. Unmodified faunal remains from NV-Pe-104. In R. A. Cowan, *The Barrel Spring Site (NV-Pe-104): An Occupation-Quarry Site in Northwestern Nevada*, pp. 41–48. Archaeological Research Facility, University of California, Berkeley.

Thomas, D. H. 1972b. Western Shoshone ecology: Settlement patterns and beyond. In D. D. Fowler, ed., *Great Basin Cultural Ecology: A Symposium*, pp. 135–153. Desert Research Institute Publications in the Social Sciences 8.

Thomas, D. H. 1978. Arrowheads and atlatl darts: How the stones got the shaft. *American Antiquity* 43:461–472.

Thomas, D. H. 1981. How to classify the projectile points from Monitor Valley, Nevada. *Journal of California and Great Basin Anthropology* 3(1):7–43.

Thomas, D. H. 1982. The 1981 Alta Toquima Village Project: A Preliminary Report. *Desert Research Institute Social Sciences Center Technical Report* 27.

Thomas, D. H. 1983a. *The Archaeology of Monitor Valley. 1. Epistemology*. American Museum of Natural History Anthropological Papers 58(1).

Thomas, D. H. 1983b. *The Archaeology of Monitor Valley. 2. Gatecliff Shelter*. American Museum of Natural History Anthropological Papers 59(1).

Thomas, D. H., ed. 1985. *The Archaeology of Hidden Cave, Nevada*. Anthropological Papers of the American Museum of Natural History 61(1).

Thomas, D. H., ed. 1986. *A Great Basin Shoshonean Source Book*. Garland, New York.

Thomas, D. H. 1988. *The Archaeology of Monitor Valley*, Volume 3, *Survey and Additional Excavations*. American Museum of Natural History Anthropological Papers 66(2).

Thomas, D. H. 1994. Chronology and the Numic Expansion. In D. B. Madsen and D. Rhode, eds., *Across the West: Human Population Movement and the Expansion of the Numa*, pp. 56–61. University of Utah Press, Salt Lake City.

Thomas, D. H. 2009. Pattern recognition: The temporal structure of the Mt. Jefferson complex. In D. H. Thomas and L. S. A. Pendleton, *The Archaeology of Monitor Valley: 4. Alta Toquima and the Mt. Jefferson Tablelands Complex*. American Museum of Natural History Anthropological Papers, in press.

Thomas, D. H., and L. S. A. Pendleton. 2009. *The Archaeology of Monitor Valley: 4. Alta Toquima and the Mt. Jefferson Tablelands Complex*. American Museum of Natural History Anthropological Papers, in press.

Thomas, D. H., J. M. Adovasio, I. Taylor, and E. M. Hattori. 2008. Earlier than we thought: Textile dates from Gatecliff Shelter, Nye County, Nevada. Paper presented at the Great Basin Anthropological Conference, Portland, OR.

Thompson, G. A., and D. E. White. 1964. *Regional Geology of the Steamboat Springs Area, Washoe County, Nevada*. U.S. Geological Survey Professional Paper 458-A.

Thompson, R. S. 1978. Late Pleistocene and Holocene packrat middens from Smith Creek Canyon, White Pine County, Nevada. In D. R. Tuohy and D. L. Rendall, eds., *The Archaeology of Smith Creek Canyon, Eastern Nevada*, pp. 363–380. Nevada State Museum Anthropological Papers 17.

Thompson, R. S. 1983. Modern vegetation and climate. In D. H. Thomas, *The Archaeology of Monitor Valley, Volume 1, Epistemology*, pp. 99–106. American Museum of Natural History Anthropological Papers 58(1).

Thompson, R. S. 1984. Late Pleistocene and Holocene environments in the Great Basin. Ph.D. dissertation, University of Arizona, Tucson.

Thompson, R. S. 1985. Palynology and *Neotoma* middens. In B. F. Jacobs, P. L. Fall, and O. K. Davis, eds., *Late Quaternary Vegetation and Climates of the American Southwest*, pp. 89–112. American Association of Stratigraphic Palynologists Contributions Series 16.

Thompson, R. S. 1990. Late Quaternary vegetation and climate in the Great Basin. In J. L. Betancourt, T. R. Van Devender, and P. S. Martin, eds., *Packrat Middens: The Last 40,000 Years of Biotic Change*, pp. 200–239. University of Arizona Press, Tucson.

Thompson, R. S. 1992. Late Quaternary environments in Ruby Valley, Nevada. *Quaternary Research* 37:1–15.

Thompson, R. S., and E. M. Hattori. 1983. Paleobotany of Gatecliff Shelter: Packrat (*Neotoma*) middens from Gatecliff Shelter and Holocene migrations of woodland plants. In D. H. Thomas, *The Archaeology of Monitor Valley. 2. Gatecliff Shelter*, pp. 157–167. American Museum of Natural History Anthropological Papers 59(1).

Thompson, R. S., and R. R. Kautz. 1983. Paleobotany of Gatecliff Shelter: Pollen analysis. In D. H. Thomas, ed., *The Archaeology of Monitor Valley. 2. Gatecliff Shelter*, pp. 136–151. American Museum of Natural History Anthropological Papers 59(1).

Thompson, R. S., and J. I. Mead. 1982. Late Quaternary environments and biogeography in the Great Basin. *Quaternary Research* 17:39–35.

Thompson, R. S., K. H. Anderson, and P. J. Bartlein. 1999. *Quantitative Paleoclimatic Reconstructions from Late Pleistocene Plant Macrofossils of the Yucca Mountain Region*. U.S. Geological Survey Open-File Report 99–338.

Thompson, R. S., K. H. Anderson, and P. J. Bartlein. 2000. *Atlas of Relations Between Climatic Parameters and Distributions of Important Trees and Shrubs in North America*. U.S. Geological Survey Professional Paper 1650.

Thompson, R. S., L. Benson, and E. M. Hattori. 1986. A revised chronology for the last Pleistocene cycle in the central Lahontan Basin. *Quaternary Research* 25:1–9.

Thompson, R. S., L. J. Toolin, R. M. Forester, and R. J. Spencer. 1990. Accelerator-mass spectrometer (AMS) radiocarbon dating of Pleistocene lake sediments in the Great Basin. *Palaeogeography, Palaeoclimatology, Palaeoecology* 78:301–313.

Thompson, R. S., C. Whitlock, P. J. Bartlein, S. P. Harrison, and W. G. Spaulding. 1992. Climatic changes in the western United States since 18,000 yr B.P. In H. E. Wright, J. E. Kutzbach, T. Webb III, W. F. Ruddiman, F. A. Street-Perrott, and P. J. Bartlein, eds., *Global Climate since the Last Glacial Maximum*, pp. 468–513. University of Minnesota Press, Minneapolis.

Thompson, R. S., C. G. Oviatt, A. P. Roberts, J. Buchner, R. Kelsey, C. Bracht, R. M. Forester, and J. P. Bradbury. 1995. *Stratigraphy, Sedimentology, Paleontology, and Paleomagnetism of Pliocene–Early Pleistocene Lacustrine Deposits in Two Cores from Western Utah*. U.S. Geological Survey Open-File Report 95-1.

Titmus, G. L., and J. C. Woods. 1991. A closer look at margin "grinding" on Folsom and Clovis points. *Journal of California and Great Basin Anthropology* 13:194–203.

Towner, R. H. 1996. *The Archaeology of Navajo Origins*. University of Utah Press, Salt Lake City.

Townley, J. M. 1977. *Turn This Water into Gold: The Story of the Newlands Project*. Nevada Historical Society, Reno.

Townley, J. M. 1984. *The Lost Fremont Cannon Guidebook*. Jamison Station Press, Reno.

Trimble, S. H. 1989. *The Sagebrush Ocean: A Natural History of the Great Basin*. University of Nevada Press, Reno.

Trimble, S. H. 2008. *Bargaining for Eden: The Fight for the Last Open Spaces in America*. University of California Press, Berkeley.

Truett, J. 1996. Bison and elk in the American Southwest: In search of the pristine. *Environmental Management* 20:195–206.

Tueller, P. T., C. D. Beeson, R. J. Tausch, N. E. West, and R. H. Rea. 1979. *Pinyon-juniper Woodlands of the Great Basin: Distribution, Flora, Vegetal Cover*. USDA Forest Service Research Paper INT-229.

Tuohy, D. R. 1970. The Coleman Locality: A basalt quarry and workshop near Falcon Hill, Nevada. *Nevada State Museum Anthropological Papers* 15:143–205.

Tuohy, D. R. 1974. A comparative study of late Paleo-Indian manifestations in the western Great Basin. In R. G. Elston and L. Sabini, eds., *A Collection of Papers on Great Basin Archaeology*, pp. 91–116. Nevada Archaeological Survey Research Paper 5.

Tuohy, D. R. 1979. Kachina Cave. In D. R. Tuohy and D. L. Rendall, eds., *The Archaeology of Smith Creek Canyon, Eastern Nevada*, pp. 1–89. Nevada State Museum Anthropological Papers 17.

Tuohy, D. R. 1981. A brief history of the discovery and exploration of pebble mounds, boulder cairns, and other rock features at the Sadmat Site, Churchill County, Nevada. *Nevada Archaeologist* 3(1):4–15.

Tuohy, D. R. 1982. Another Great Basin atlatl with dart foreshafts and other artifacts: Implications and ramifications. *Journal of California and Great Basin Anthropology* 4(1):80–106.

Tuohy, D. R. 1988. Paleoindian and early Archaic cultural complexes from three central Nevada localities. In J. A. Willig, C. M. Aikens, and J. L. Fagan, eds., *Early Human Occupation in Far Western North America: The Clovis-Archaic Interface*, pp. 217–230. Nevada State Museum Anthropological Papers 21.

Tuohy, D. R., and A. J. Dansie. 1997. New information regarding early Holocene manifestations in the western Great Basin. *Nevada Historical Society Quarterly* 40:24–53.

Tuohy, D. R., and E. M. Hattori. 1996. Lovelock wickerware in the lower Truckee River basin. *Journal of California and Great Basin Anthropology* 18:284–296.

Tuohy, D. R., and T. N. Layton. 1977. Towards the establishment of a new series of Great Basin projectile points. *Nevada Archaeological Survey Reporter* 10(6):1–5.

Turk, L. J. 1973. *Hydrogeology of the Bonneville Salt Flats, Utah.* Utah Geological and Mineral Survey Water-Resources Bulletin 19.

Turnmire, K. L. 1987. An analysis of the mammalian fauna from Owl Cave One and Two, Snake Range, east-central Nevada. M.S. thesis, University of Maine, Orono.

Turvey, S. T., J. R. Oliver, Y. M. Narganes Storde, and P. Rye. 2007. Late Holocene extinction of Puerto Rican native land mammals. *Biology Letters* 3:193–196.

Tuskes, P. M., J. P. Tuttle, and M. C. Collins. 1996. *The Wild Silk Moths of North America: A Natural History of the Saturnidae of the United States and Canada.* Comstock Publishing Associates, Ithaca, NY.

Twain, M. 1981 [1872]. *Roughing It.* Penguin Books, New York.

Ugan, A. 2005. Climate, bone density, and resource depression: What is driving variation in large and small game in Fremont archaeofaunas? *Journal of Anthropological Archaeology* 24:227–251.

Unmack, P. J., and W. L. Minckley. 2008. The demise of desert springs in North America. In L. E. Stevens and V. J. Meretsky., eds., *Aridland Springs in North America: Ecology and Conservation*, pp. 11–34. University of Arizona Press, Tucson.

U.S. Department of Commerce 1983. *Climatic Atlas of the United States.* Washington, D.C.

Van Devender, T. R. 1977. Holocene woodlands in the southwestern deserts. *Science* 198:189–192.

Van Devender, T. R. 1987. Holocene vegetation and climate in the Puerto Blanco Mountains, southwestern Arizona. *Quaternary Research* 27:51–72.

Van Devender, T. R. 1988. Pollen in packrat (*Neotoma*) middens: Pollen transport and the relationship of pollen to vegetation: Discussion. *Palynology* 12:221–229.

Van Devender, T. R. 1990a. Late Quaternary vegetation and climate of the Chihuahuan Desert, United States and Mexico. In J. L. Betancourt, T. R. Van Devender, and P. S. Martin, eds., *Packrat Middens: The Last 40,000 Years of Biotic Change*, pp. 104–133. University of Arizona Press, Tucson.

Van Devender, T. R. 1990b. Late Quaternary vegetation and climate of the Sonoran Desert, United States and Mexico. In J. L. Betancourt, T. R. Van Devender, and P. S. Martin, eds., *Packrat Middens: The Last 40,000 Years of Biotic Change*, pp. 134–165. University of Arizona Press, Tucson.

Van Devender, T. R., and W. G. Spaulding. 1979. Development of vegetation and climate in the southwestern United States. *Science* 204:701–710.

Van Devender, T. R., Phillips, A. M., III, Mead, J. I. 1977. Late Pleistocene reptiles and small mammals from the lower Grand Canyon of Arizona. *Southwestern Naturalist* 22:49–66.

Van Devender, T. R., R. S. Thompson, and J. L. Betancourt. 1987. Vegetation history of the deserts of southwestern North America: The nature and timing of the Late Wisconsin–Holocene transition. In W. F. Ruddiman and H. E. Wright, Jr., eds., *North America and Adjacent Oceans during the Last Deglaciation*, pp. 323–352. The Geology of North America, Volume K-3. Geological Society of America, Boulder, CO.

Van Devender, T. R., P. S. Martin, R. S. Thompson, K. L. Cole, A. J. T. Jull, A. Long, L. J. Toolin, and D. J. Donahue. 1985. Fossil packrat middens and the tandem accelerator mass spectrometer. *Nature* 317:610–613.

Van Valkenburgh, B., and F. Hertel. 1998. The decline of North American predators during the late Pleistocene. In J. J. Saunders, B. W. Styles, and G. F. Baryshnokov, eds., *Quaternary Paleozoology in the Northern Hemisphere*, pp. 357–374. Illinois State Museum Scientific Papers 27.

Van Vuren, D., and M. P. Bray. 1985. The recent geographic distribution of *Bison bison* in Oregon. *The Murrelet* 66:56–58.

Van Vuren, D., and F. C. Deitz. 1993. Evidence of *Bison bison* in the Great Basin. *Great Basin Naturalist* 53:318–319.

Van Winkle, W. 1914. *Quality of the Surface Waters of Oregon.* U.S. Geological Survey Water-Supply Paper 363.

Vander Wall, S. B., and R. P. Balda. 1977. Coadaptations of the Clark's Nutcracker and the piñon pine for efficient seed harvest and dispersal. *Ecological Monographs* 47:89–111.

Vasek, F. C. 1966. The distribution and taxonomy of three western junipers. *Brittonia* 18:350–372.

Vasek, F. C., and R. F. Thorne. 1988. Transmontane coniferous vegetation. In M. G. Barbour and J. Major, eds., *Terrestrial Vegetation of California*, pp. 797–832. California Native Plant Society Publication 9.

Vaughan, S. J., and C. J. Warren. 1987. Toward a definition of Pinto points. *Journal of California and Great Basin Anthropology* 9:199–213.

Vaughan, T. A. 1990. Ecology of living packrats. In J. L. Betancourt, T. R. Van Devender, and P. S. Martin, eds., *Packrat Middens: The Last 40,000 Years of Biotic Change*, pp. 14–27. University of Arizona Press, Tucson.

Vereshchagin, N. K., and F. F. Baryshnikov. 1984. Quaternary mammalian extinctions in northern Eurasia. In P. S. Martin and R. G. Klein, eds., *Quaternary Extinctions: A Prehistoric Revolution*, pp. 483–516. University of Arizona Press, Tucson.

Verts, B. J., and L. N. Carraway. 1998. *Land Mammals of Oregon.* University of California Press, Berkeley.

Waitt, R. B., Jr. 1984. Periodic jökulhlaups from Pleistocene glacial Lake Missoula: New evidence from varved sediment in northern Idaho and Washington. *Quaternary Research* 22:46–58.

Waitt, R. B., Jr. 1985. Case for periodic, colossal jökulhlaups from Pleistocene Lake Missoula. *Geological Society of America Bulletin* 96:1271–1286.

Walker, D. N. 2002. Faunal remains from archaeological excavations at the Rock Springs site, southeast Idaho. In B. S. Arkush, *Archaeology of the Rock Springs Site: A Multi-component Bison Kill and Processing Camp in Curlew Valley, Southeastern Idaho*, pp. 99–174. Boise State University Monographs in Archaeology 1.

Walker, M., S. Johnsen, S. O. Rasmussen, T. Popp. J.-P. Steffensen, P. Gibbard, W. Hoek, J. Lowe, J. Andrews, S. Björck, L. C. Cwynar, K. Hughen, P. Kershaw, B. Kromer, T. Litt, D. J. Lowe, T. Nakagawa, R. Newnham, and J. Schwander. 2009. Formal definition and dating of the GSSP (Global Stratotype Section and Point) for the base of the Holocene using the Greenland NGRIP ice core, and selected auxiliary records. *Journal of Quaternary Science* 24:3–17.

Walker, R. W., R. E. Turley, Jr., and G. M. Leonard. 2008. *Massacre at Mountain Meadows.* Oxford University Press, New York.

Wallace, M. P., and S. A. Temple. 1987. Competitive interactions within and between species in a guild of avian scavengers. *The Auk* 104:290–295.

Wallace, R. E. 1984. Patterns and timing of late Quaternary faulting in the Great Basin Province and relation to some regional tectonic features. *Journal of Geophysical Research* 89:5763–5769.

Wang, Y., A. H. Jahren, and R. Amundson. 1997. Potential for ^{14}C dating of biogenic carbonate in hackberry (*Celtis*) endocarps. *Quaternary Research* 47:337–343.

Wanner, H., J. Beer, J. Bütikofer, T. J. Crowley, U. Cubasch, J. Flückiger, H. Goosse, M. Grosjean, F. Joos, J. O. Kaplan, M. Küttel, S. A. Müller, I. C. Prentice, O. Solomina, T. F. Stocker, P. Tarasov, M. Wagner, and M. Widmann. 2008. Mid- to late Holocene climate change: An overview. *Quaternary Science Reviews* 27:1791–1828.

Ward, P. D. 2000. *Rivers in Time: The Search for Clues to Earth's Mass Extinctions*. Columbia University Press, New York.

Warren, C. N. 1967. The San Dieguito Complex: A review and hypothesis. *American Antiquity* 32:168–185.

Warren, C. N. 1991. *Archaeological Investigations at Nelson Wash, Fort Irwin, California*. Fort Irwin Archaeological Project Research Report 23.

Warren, C. N., and J. De Costa. 1964. Dating Lake Mohave artifacts and beaches. *American Antiquity* 30:206–209.

Warren, C. N., and H. T. Ore. 1978. Approach and process of dating Lake Mohave artifacts. *Journal of California Anthropology* 5:179–187.

Warren, C. N., and C. Phagan. 1988. Fluted points in the Mojave Desert: Their technology and cultural context. In J. A. Willig, C. M. Aikens, and J. L. Fagan, eds., *Early Human Occupation in Far Western North America: The Clovis-Archaic Interface*, pp. 121–120. Nevada State Museum Anthropological Papers 21.

Warren, C. N., and A. J. Ranere. 1968. Outside Danger Cave: A view of early man in the Great Basin. In C. Irwin-Williams, ed., *Archaic Prehistory in the Western United States*, pp. 6–18. Eastern New Mexico University Contributions in Anthropology 1(4).

Waters, M. R. 1989. Late Quaternary lacustrine history and paleoclimatic significance of pluvial Lake Cochise, southeastern Arizona. *Quaternary Research* 32:1–11.

Waters, M. R., and T. W. Stafford, Jr. 2007a. Redefining the age of Clovis: Implications for the peopling of the Americas. *Science* 315:1122–1126.

Waters, M. R., and T. W. Stafford, Jr. 2007b. Response to comment on "Redefining the age of Clovis: Implications for the Peopling of the Americas". *Science* 317:320.

Wayne, W. J. 1984. Glacial chronology of the Ruby Mountains–East Humboldt Range, Nevada. *Quaternary Research* 21:286–303.

Webb, R. H., J. W. Steiger, and R. M. Turner. 1987. Dynamics of Mojave Desert shrub assemblages in the Panamint Mountains, California. *Ecology* 68:478–490.

Webb, S. D., C. A. Hemmings, and M. P. Muniz. 1998. New radiocarbon dates for Vero tapir and stout-legged llama from Florida. *Current Research in the Pleistocene* 15:127–128.

Webb, T., III, K. H. Anderson, P. J. Bartlein, and R. S. Webb. 1998. Late Quaternary climate change in eastern North America: A comparison of pollen-derived estimates with climate model results. *Quaternary Science Reviews* 17:587–606.

Wehausen, J. D. 1996. Effects of mountain lion predation on bighorn sheep in the Sierra Nevada and Granite Mountains of California. *Wildlife Society Bulletin* 24:471–479.

Weiss, N. T., and B. J. Verts. 1984. Habitat and distribution of pygmy rabbits (*Sylvilagus idahoensis*) in Oregon. *Great Basin Naturalist* 44:563–571.

Wells, P. V. 1976. Macrofossil analysis of wood rat (*Neotoma*) middens as a key to the Quaternary vegetational history of arid America. *Quaternary Research* 6:223–248.

Wells, P. V. 1983. Paleobiogeography of montane islands in the Great Basin since the last glaciopluvial. *Ecological Monographs* 53:341–382.

Wells, P. V., and C. D. Jorgensen. 1964. Pleistocene wood rat middens and climatic change in Mohave Desert: A record of juniper woodlands. *Science* 143:1171–1174.

Wells, P. V., and R. Berger. 1967. Late Pleistocene history of coniferous woodland in the Mojave Desert. *Science* 155:1640–1647.

Wells, P. V., and D. Woodcock. 1985. Full-glacial vegetation of Death Valley, California: Juniper woodland opening to yucca semidesert. *Madroño* 32:11–23.

Wells, S. G, W. J. Brown, Y. Enzel, R. Y. Anderson, and L. D. McFadden. 2003. Late Quaternary geology and paleohydrology of pluvial Lake Mojave, southern California. In Y. Enzel, S. G. Wells, and N. Lancaster, eds. *Paleoenvironments and Paleohydrology of the Mojave and Southern Great Basin Deserts*, pp. 79–114. Geological Society of America Special Paper 368.

Wells, S. G., R. Y. Anderson, L. D. McFadden, W. J. Brown, Y. Enzel, and J.-L. Miossec. 1989. *Late Quaternary Paleohydrology of the Eastern Mojave River Drainage, Southern California: Quantitative Assessment of the Late Quaternary Hydrologic Cycle in Large Arid Watersheds*. New Mexico Water Resources Research Institute Technical Report 242.

West, G. J., W. Woolfenden, J. A. Wanket, and R. S. Anderson. 2007. Late Pleistocene and Holocene environments. In T. L. Jones and K. A. Klar, eds., *California Prehistory: Colonization, Culture, and Complexity*, pp. 11–34. AltaMira Press, Lanham, MD.

West, N. E. 2000. Intermountain valleys and lower mountain slopes. In M. G. Barbour and W. D. Billings, eds., *North American Terrestrial Vegetation*, pp. 255–284. Cambridge University Press, Cambridge, UK.

West, N. E., R. J. Tausch, and P. T. Tueller. 1998. *A Management-Oriented Classification of Pinyon-Juniper Woodlands of the Great Basin*. USDA Forest Service Rocky Mountain Research Station General Technical Report RMRS-GTR-12.

West, N. E., R. J. Tausch, K. H. Rea, and P. T. Tueller. 1978. Phytogeographical variation within juniper-pinyon woodlands of the Great Basin. In K. T. Harper, and J. L. Reveal, eds., *Intermountain Biogeography: A Symposium*, pp. 119–136. Great Basin Naturalist Memoirs 2.

Wheeler, S. M. 1997. Cave burials near Fallon, Nevada. *Nevada Historical Society Quarterly* 40:15–23.

Wheeler, S. M., and G. N. Wheeler. 1969. Cave burials near Fallon, Nevada. *Nevada State Museum Anthropological Papers* 14:70–79.

White, S. R. 1991. Late Quaternary paleobotany at Mad Chipmunk Cave, Elko County, Nevada. M.S. thesis, Northern Arizona University, Flagstaff.

Whitebread, D. H. 1969. *Geologic Map of the Wheeler Peak and Garrison Quadrangles, Nevada and Utah*. U.S. Geological Survey Miscellaneous Investigations Map I-578.

Wigand, P. E. 1985. Diamond Pond, Harney County, Oregon: Man and marsh in the eastern Oregon Desert. Ph.D. dissertation, Washington State University, Pullman.

Wigand, P. E. 1987. Diamond Pond, Harney County, Oregon: Vegetation history and water table in the eastern Oregon desert. *Great Basin Naturalist* 47:427–458.

Wigand, P. E. 1990b. Vegetation history. In S. D. Livingston and F. D. Nials, eds., *Archaeological and paleoenvironmental investigations in the Ash Meadows National Wildlife Refuge, Nye County, Nevada*, pp. 15–48. Quaternary Sciences Center, Desert Research Institute, Technical Report 70.

Wigand, P. E. 1997a. A late-Holocene pollen record from Lower Pahranagat Lake, southern Nevada, USA: High resolution paleoclimatic records and analysis of environmental responses to climate change. In C. M. Isaacs and V. L. Tharp, eds., *Proceedings of the Thirteenth Annual Pacific Climate (PACLIM) Workshop, April 15–18, 1996*, pp. 63–77. Interagency Ecological Program, Technical Report 53, California Department of Water Resources, Sacramento.

Wigand, P. E. 1997b. Woodrat midden and pollen analysis. In P. E. Buck, B. Hockett, F. Nials, and P. E. Wigand, eds., *Prehistory and Paleoenvironment of Pintwater Cave, Nevada: Results of Field Work during the 1996 Season*. Project Number OS-005071, Nellis Air Force Base, Nevada.

Wigand, P. E. 2002. Prehistoric dynamics of piñon woodland in the northern Owens Valley region and beyond as revealed in macrofossils from ancient woodrat middens. In J. W. Eerkens and J. H. King, *Phase II Archaeological Investigations for the Sherwin Summit Rehabilitation Project, U.S. Highway 395, Inyo and Mono Counties, California.*, Appendix I. Far Western Anthropological Research Group, Inc., Davis, CA.

Wigand, P. E. 2003. Middle to late Holocene climate and vegetation dynamics. In J. S. Rosenthal and J. Eerkens, eds., *The Archaeology of Coso Basin: Test Excavations at 28 Sites Located in the North Ranges Complex, NAWS. China Lake*. Far Western Anthropological Research Group, Inc., Davis, CA.

Wigand, P. E. 2004. The modern environment. In Wegener, R. M., Altschul, J. H., Keller, A. H., and Stoll, A. Q. eds., *Distant Shores: Cultural Resources Survey at Honey Lake, Lassen County, California*, pp. 7–32. Technical Report 04-10, Statistical Research Inc., Redlands, CA.

Wigand, P. E., and P. J. Mehringer, Jr. 1985. Pollen and seed analyses. In D. H. Thomas, ed., *The Archaeology of Hidden Cave, Nevada*, pp. 108–124. Anthropological Papers of the American Museum of Natural History 61(1).

Wigand, P. E., and C. L. Nowak. 1992. Dynamics of northwest Nevada plant communities during the last 30,000 years. In C. A. Hall, Jr., V. Doyle-Jones, and B. Widawski, eds., *The History of Water: Eastern Sierra Nevada, Owens Valley, White-Inyo Mountains*, pp. 40–62. White Mountain Research Station Symposium Volume 4.

Wigand, P. E., and D. Rhode. 2002. Great Basin vegetational history and aquatic systems. In R. Hershler, D. B. Madsen, and D. R. Currey, eds., *Great Basin Aquatic Systems History*, pp. 309–368. Smithsonian Contributions to the Earth Sciences 33.

Wigand, P. E., M. L. Hemphill, S. E. Sharpe, and S. Patra. 1995. Great Basin semi-arid woodland dynamics during the Holocene. In W. J. Waugh, K. L. Petersen, P. E. Wigand, B. D. Louthan, and R. D. Walker, eds., *Climate change in the Four Corners and adjacent regions: Implications for environmental restoration and land-use planning*, pp. 51–70. U.S. Department of Energy CONF-94093250. National Technical Information Service, Springfield, VA.

Wilcox, B. A., D. D. Murphy, P. R. Ehrlich, and G. T. Austin. 1986. Insular biogeography of the montane butterfly faunas in the Great Basin: Comparison with birds and mammals. *Oecologia* 69:188–194.

Wilde, J. D. 1985. Prehistoric settlements in the northern Great Basin: Excavations and collections analysis in the Steens Mountain area, southeastern Oregon. Ph.D. dissertation, University of Oregon, Eugene.

Wilde, J. D. 1992. Finding a date: Some thoughts on radiocarbon dating and the Baker Fremont site in eastern Nevada. *Utah Archaeology* 5:39–54.

Wilde, J. D., and D. E. Newman. 1989. Late Archaic corn in the eastern Great Basin. *American Anthropologist* 91:712–720.

Wilde, J. D., and R. A. Soper. 1999. *Baker Village: Report of Excavations, 1990–1994*. Brigham Young University Museum of Peoples and Cultures Technical Series 99–12.

Wilde, J. D., and G. L. Tasa. 1991. A woman at the edge of agriculture: Skeletal remains from the Elsinore Burial Site, Sevier Valley, Utah. *Journal of California and Great Basin Anthropology* 13:60–76.

Wilde, J. D., D. E. Newman, and A. E. Godfrey. 1986. *The Late Archaic/Early Formative Transition in Central Utah: Pre-Fremont Corn from the Elsinore Burial, Site 42 Sv 2111, Sevier County, Utah*. Museum of Peoples and Cultures, Brigham Young University Technical Series 86–20.

Wilke, P. J., and H. W. Lawton. 1976. *The Expedition of Capt. J. W. Davidson from Fort Tejon to the Owens Valley in 1859*. Ballena Press Publications in Archaeology, Ethnology, and History 8.

Wilke, P. J., T. F. King, and R. Bettinger. 1974. A comparative study of late Paleo-Indian manifestations in the western Great Basin. *Nevada Archaeological Survey Research Paper* 5:80–90.

Willden, R. 1964. *Geology and Mineral Deposits of Humboldt County, Nevada*. Nevada Bureau of Mines Bulletin 59.

Williams, J. E. 2003. *Empetrichthys latos*: Pahrump Poolfish. Desert Fishes Council. www.desertfishes.org/na/goodeida/empetric/elatos__/elatos__.html.

Williams, J. W., B. N. Shuman, and T. Webb, III. 2001. Dissimilarity analyses of late-Quaternary vegetation and climate in eastern North America. *Ecology* 82:3346–3362.

Williams, T. R., and M. S. Bedinger. 1984. *Selected Geologic and Hydrologic Characteristics of the Basin and Range Province, Western United States: Pleistocene Lakes and Marshes*. U.S. Geological Survey Miscellaneous Investigations Map I-1522-D.

Willig, J. A. 1988. Paleo-Archaic adaptations and lakeside settlement patterns in the Northern Alkali Basin, Oregon. In J. A. Willig, C. M. Aikens, and J. L. Fagan, eds., *Early Human Occupation in Far Western North America: The Clovis-Archaic Interface*, pp. 417–482. Nevada State Museum Anthropological Papers 21.

Willig, J. A. 1989. Paleo-Archaic broad spectrum adaptations at the Pleistocene-Holocene boundary in far western North America. Ph.D. dissertation, University of Oregon, Eugene.

Wilson, D. E., and F. R. Cole. 2000. *Common Names of Mammals of the World*. Smithsonian Institution Press, Washington, DC.

Wilson, D. E., and D. M. Reeder. 2005. *Mammal Species of the World: A Taxonomic and Geographic Reference*. Third edition. Johns Hopkins University Press, Baltimore.

Wingard, G. F. 2001. *Carlon Village: Land, Water, Subsistence, and Sedentism in the Northern Great Basin*. University of Oregon Anthropological Papers 57.

Wingerson, L. 2009. High life in the high mountains. *American Archaeology* 13(4):12–18.

Winograd, I. J., and E. H. Roseboom, Jr. 2008. Yucca Mountain revisited. *Science* 320:1426–1427.

Winograd, I. J., T. B. Coplen, J. M. Landwehr, A. C. Riggs, K. R. Ludwig, B. J. Szabo, P. T. Kolesar, and K. M. Revesz. 1992. Continuous 500,000-year climate record from vein calcite in Devils Hole, Nevada. *Science* 258:255–260.

Wisdom, M. J., M. M. Rowland, and L. H. Suring, eds. 2005. *Habitat Threats in the Sagebrush Ecosystem: Methods of Regional Assessment and Applications in the Great Basin*. Alliance Communications Group, Lawrence, KS.

Wohlfahrt, G., L. F. Fenstermaker, and J. A. Arnone III. 2008. Large annual net ecosystem CO^2 uptake of a Mojave Desert ecosystem. *Global Change Biology* 14:1475–1487.

Wolverton, S. 2006. Natural-trap ursid mortality and the Kurtén Response. *Journal of Human Evolution* 50:540–551.

Woodcock, D. 1986. The late Pleistocene of Death Valley: A climatic reconstruction based on macrofossil data. *Palaeogeography, Palaeoclimatology, Palaeoecology* 57:272–283.

Woodhouse, C. A. 2004. A paleo perspective on hydroclimatic variability in the western United States. *Aquatic Sciences* 66:346–356.

Woolfenden, W. B. 1996a. Late Quaternary vegetation history of the southern Owens Valley region, Inyo County, California. Ph.D. dissertation, University of Arizona, Tucson.

Woolfenden, W. B. 1996b. Quaternary vegetation history. In *Sierra Nevada Ecosystem Project: Final Report to Congress*, Vol. 2, *Assessments and Scientific Basis for Management Options*, pp. 47–70. University of California Centers for Water and Wildland Resources, Davis.

Woolfenden, W. B. 2003. A 180,000-year pollen record from Owens Lake, CA: Terrestrial vegetation change on orbital scales. *Quaternary Research* 59:430–444.

Wormington, H. M., and D. Ellis, eds. 1967. *Pleistocene Studies in Southern Nevada*. Nevada State Museum Anthropological Papers 13.

Worthy, T. H., and Holdaway, R. N. 2002. *The Lost World of the Moa*. Indiana University Press, Bloomington.

Wyatt, S. 2004. Ancient transpacific voyaging to the New World via Pleistocene South Pacific islands. *Geoarchaeology* 19:511–529.

Yang, W., T. K. Lowenstein, H. R. Krouse, R. J. Spencer, and T.-L. Lu. 2005. A 200,000 δ^{18} record of closed-basin lacustrine calcite, Death Valley, California. *Chemical Geology* 216:99–111.

Yohe, R. M., II. 1998. The introduction of the bow and arrow and lithic resource use at Rose Spring (CA-INY-372). *Journal of California and Great Basin Anthropology* 20:26–52.

Yohe, R. M., II. 2000. Rosegate revisited: Rose Spring point temporal range in the southwestern Great Basin. In J. S. Schneider, R. M. Yohe II, and J. K. Gardner, eds., *Archaeological Passages: A Volume in Honor of Claude Nelson Warren*, pp. 213–224. Western Center for Archaeology & Paleontology Publications in Archaeology 1. Hemet, CA.

Young, D. A., and R. L. Bettinger. 1992. The Numic spread: A computer simulation. *American Antiquity* 57:85–99.

Youngman, P. M. 1993. The Pleistocene small carnivores of eastern Beringia. *The Canadian Field-Naturalist* 107:139–163.

Youngman, P. M., and F. W. Schueler. 1991. *Martes nobilis* is a synonym of *Martes americana*, not an extinct Pleistocene-Holocene species. *Journal of Mammalogy* 72:567–577.

Yuan, F., B. K. Linsley, S. P. Lund, and J. P. McGeehin. 2004. A 1200 year record of hydrologic activity in the Sierra Nevada from sediments in Walker Lake, Nevada. *Geochemistry Geophysics Geosystems* 5(3):1–13.

Zancanella, J. K. 1988. Early lowland prehistory in south-central Nevada. In J. A. Willig, C. M. Aikens, and J. L. Fagan, eds., *Early Human Occupation in Far Western North America: The Clovis-Archaic Interface*, pp. 251–272. Nevada State Museum Anthropological Papers 21.

Zdanowicz, C. M., G. A. Zielinski, and M. S. Germani. 1999. Mount Mazama eruption: Calendrical age verified and atmospheric impact assessed. *Geology* 27:621–624.

Zeanah, D. W., and R. G. Elston. 2001. Testing a simple hypothesis concerning the resilience of dart point styles to hafting element repair. *Journal of California and Great Basin Anthropology* 23:93–124.

Zeveloff, S. I., and F. R. Collett. 1988. *Mammals of the Intermountain West*. University of Utah Press, Salt Lake City. UT.

Zhang, P., H. Cheng, R. L. Edwards, F. Chen, Y. Wang, X. Yang, J. Liu, M. Tan, X. Wang, J. Liu, C. An, Z. Dai, J. Zhou, D. Zhang, J. Jia, L. Jin, and K. R. Johnson. 2008. A test of climate, sun, and culture relationships from an 1810-year Chinese cave record. *Science* 322:940–942.

Ziegler, A. C. 1963. Unmodified mammal and bird remains from Deer Creek Cave, Elko County, Nevada. In M. E. Shutler and R. Shutler, Jr., *Deer Creek Cave, Elko County, Nevada*, pp. 15–22. Nevada State Museum Anthropological Papers 11.

Zielinski, G. A., and W. D. McCoy. 1987. Paleoclimatic implications of the relationship between modern snowpack and late Pleistocene equilibrium-line altitudes in the mountains of the Great Basin, western U.S.A. *Arctic and Alpine Research* 19:127–134.

Zigmond, M. L. 1981. *Kawaiisu Ethnobotany*. University of Utah Press, Salt Lake City.

Zigmond, M. 1986. Kawaiisu. In W. L. d'Azevedo, ed., *Great Basin, Handbook of North American Indians, Volume 11*, pp. 389–411. Smithsonian Institution Press, Washington, DC.

Zimmerman, S. H., S. R. Hemming, D. V. Kent, and S. Y. Searle. 2006. Revised chronology for late Pleistocene Mono Lake sediments based on paleointensity correlation to the global reference curve. *Earth and Planetary Science Letters* 252:94–106.

Zreda, M. G., and F. M. Phillips. 1995. Insights into alpine moraine development from cosmogenic ^{36}Cl buildup dating. *Geomorphology* 14:149–156.

INDEX

Page numbers followed by *f* refer to figures and *t* refer to tables.

Ababneh, Linah, 324
accelerator mass spectrometry (AMS), 55
Adams, Ken, 108, 109, 130, 220, 265, 266–267
Adobe Valley, 115, 116
Adovasio, James, 59–60, 327, 329
Adrian Valley, 268
Afton Canyon, 119
agave, 162, 164
Agency Lake, 88
Ager, Tom, 49
agriculture, 276, 319, 329, 330–331
Aikens, C. Melvin, 233, 305, 311, 313
air temperature. *See* temperature
Alabama Hills, 149–153
Alaska
    Bering Land Bridge, 45–46, 48–51, 54
    Nenana complex, 78
    Pleistocene climate, 49–50
Alcorn, Gordon, 33
Alcorn, Ray, 181
Alkali Lake Basin, 221, 290, 299
alkali sacaton (*Sporobolus airoides*), 153
Allen, Bruce, 130
Allison, Ira, 61, 112, 113, 120, 173, 174
alluvial fans, 240
alpine tundra, 27*f*, 30, 31, 320
Alpine Tundra Zone, 32, 143
alpine villages, 27*f*, 36–37, 320–324, 326, 329, 330
Alta Toquima Village, 27*f*, 320–322, 326
Alta Toquima Wilderness area, 22
Altithermal, 244, 245, 251, 252–253
Alvord Desert, 87
Alyawara, 301–302
Amargosa Desert, 158–159, 162, 230
Amargosa Range, 12, 155*t*, 161
Amargosa River, 12, 116, 119–120, 208
amberat, 136
American "cheetah," (*Miracinonyx trumani*), 69
American Flamingo (*Phoenicopterus ruber*), 181
American lion (*Panthera leo*), 69, 176
American marten (*Martes americana*), 282
American Museum of Natural History, 147–148

American pikas (*Ochotona princeps*), 33, 139, 199–204
American River, 5
Ana River, 112, 113, 114
Anabernicula, 181–182
Anasazi, 332
Anderson, Diana, 118, 119
Anderson, Elaine, 282
Anderson, Patricia, 50
Anderson, Scott, 149, 150, 152, 159, 160, 229, 242
Angel Lake glacier, 122, 123–124
Antarctic ice sheet, 48
antelope, 286
antelope bitterbrush, 152
Antelope Cave, 183
Antevs, Ernst, 61, 128–130, 242–244, 252–253, 293, 304, 313
antilocaprids, 70
Anza, Juan Bautista de, 6
Apache, 34, 315
archaeological sites, 289–338
    dating issues, 173–176
    earliest in America, 54–63
    late Holocene, 313–333
    late Pleistocene/early Holocene, 289–302
    middle Holocene, 302–313
    pre-Clovis sites, 289
arctic fox (*Alopex lagopus*), 73
Arctodus, 180
Arizona
    bison, 278
    Clovis sites, 75–77, 83, 218
    Frémont's expedition, 6
    glaciation, 120
    Harrington's mountain goat, 79
    late Pleistocene/early Holocene climate, 240–241
    See also specific locations
Arizona singleleaf pinyon (subspecies *fallax*), 24, 25, 240
Arkush, Booke, 38
armadillos, 67
Arnow, Ted, 99
Arpin, Trina, 59
arrowweed (*Pluchea sericea*), 153

artiodactyls, 70–71, 235–238
Ash Meadows National Wildlife Refuge, 207
Ash Meadows poolfish (*Empetrichthys merriami*), 206
Asphalt Stork (*Ciconia maltha*), 181
Astor Pass, 173
Athapaskan-speaking people, 333
Atlantic Ocean, 53–54
atlatl, 299, 309–311, 322, 324, 329
Austin, George, 29
Australia, arrival of humans, 49
Aysees Peak, 136–137
Aztlan rabbit (*Aztlanologus agilis*), 70, 81

Bacon, Steven, 117, 127, 219
badger (*Taxidea taxus*), 342
Badwater Basin, 16, 17
Bailey, Robert, 32
Bailey's greasewood (*Sarcobatus baileyi*), 146, 148
Baker, Victor, 102
Baker Village, 330
Baltic rush (*Juncus balticus*), 142
Barlow, Renee, 330–331
Bartholomew, George, 236
Bartlein, Patrick, 226, 344
Basgall, Mark, 295, 296, 298, 299, 302, 312, 313
Basin and Range Province, 13
Basin Canyon, 162
Basin-Plateau Aboriginal Sociopolitical Groups (Steward), 35
basketry, 275, 300, 307, 313, 325, 326, 327, 328, 329, 331, 333
beaked sedge (*Carex rostrata*), 142
Bear Lake, 127, 209–210
Bear River, 3, 12, 91, 104–106, 127, 135
bears, 49, 69, 180, 342
Beaty, Matthew, 249
beavers, 69–70
beavertail pricklypear (*Opuntia basilaris*), 157
Beck, Charlotte, 220, 289, 291, 295, 296, 298, 299, 304, 312
Bedwell, Stephen F., 62, 221, 292, 301
beetles, 50
Beever, Erik, 342–343
Behle, William, 29

409

Beiswenger, Jane, 166
Belearic Islands goat (*Myotragus balearicus*), 81
Bell, Chris, 204
Bells Canyon, 125
Belnap, Jayne, 24
Benson, Larry, 106, 108, 109, 111, 113, 116, 117, 126–127, 128, 129, 130, 149, 248–249, 251, 265, 268, 277, 327
Benton, Thomas Hart, 3, 7
Berelekh site, 52–53
Bergen site, 304
Berger, Joel, 342
Beringia, 45–46, 48–54
Bering Land Bridge, 45–46, 48–51, 54
Bering Strait, 48
Betancourt, Julio, 228, 229
Bettinger, Bob, 310, 322, 324–326
Bevis, Ken, 120, 122, 124
Bidwell-Bartleson party, 12
big greasewood (*Sarcobatus vermiculatus*), 22, 143, 146, 148, 149, 166
big sagebrush (*Artemisia tridentata*), 18, 143, 148, 149, 152, 154, 155, 166, 167, 240, 344
Big Smoky Valley, 299
bigtooth maple (*Acer grandidentatum*), 32
Billings, Dwight, 17, 25, 26–27, 28–30, 31–32, 146, 148, 259
biogeography, 168–169, 205–206
biological soil crusts, 23–24
birds
  disjunct populations, 199
  extinctions, 71–72, 80, 82
  floristic Great Basin native species, 33
  Fossil Lake excavations, 173
  of late Pleistocene, 181–186
  pinyon pine seed dispersal role, 256–257
  of Rocky Mountain origins in Great Basin, 29
Bischoff, James, 116, 126–127
bison (*Bison bison*), 36, 180, 235, 268–278, 343
bitterbrush (*Purshia tridentata*), 150
black mats, 76, 83, 218, 240, 251, 261
Black Rock Desert, 12, 15*f*, 87, 106, 220, 275, 295
black sagebrush (*Artemisia nova*), 143, 154
Black Vulture (*Coragyps atratus*), 183
blackbrush (*Coleogyne ramosissima*), 150
Blackwater Draw site, 77, 78
Blackwelder, Eliot, 122
Blitz, John, 324
Blitzen River, 11
Blue Lake, 139, 140, 166, 223, 226, 249–250, 255
blue spruce (*Picea pungens*), 17, 32
bobcats (*Lynx rufus*), 342
Bodie Hills, 28, 256
Bonneville Basin
  early Holocene climate, 241
  early Holocene lakes and marshes, 219–220
  early Holocene vegetation, 222–225
  fluted point discoveries, 290
  glacial-lake relationship, 125–126
  late Holocene archaeological sites, 313
  late Holocene mammals, 261
  late Holocene vegetation, 260
  late Pleistocene vegetation, 139–141, 164–165
  middle Holocene climate, 250
  middle Holocene human population, 305–307
  middle Holocene mammal decline, 249–250
  pikas not found in, 205
  pluvial lakes, 94–105

Bonneville Estates Rockshelter, 238, 292, 301, 302, 303, 326
Bonneville Flood, 102, 120
Bonneville Salt Flats, 87, 99
Botta's gopher (*Thomomys bottae*), 195, 196, 197
bow and arrow, 59, 65, 309–310, 315, 322, 324–325, 329
Bradbury, Platt, 268
Bradley, Bruce, 53, 59
Briggs, Richard, 110
Bright, Robert, 166
bristlecone pine (*Pinus longaeva*), 25, 27, 30, 143, 154, 156, 161, 164–165, 167, 168, 224, 264
brittlebush (*Encelia farinosa*), 153, 156, 159
broadleaf cattail (*Typha latifolia*), 142
Bronco Charlie Cave, 282
Brooks Range, 50
Broughton, Jack, 102, 103, 183, 187, 210, 236–238
Brown, Jim, 205–206, 342
Brubaker, Linda, 50
Brussard, Peter, 259, 343
Bryan, Alan, 253, 299
Buck, Paul, 296
Buckland, William, 63
bud sagebrush (*Picrothamnus desertorum*), 148, 149, 167
Budinger, Fred, 58
Buenaventura River, 6, 7, 8
Bull Lake glaciation, 125
bull trout (*Salvelinus confluentus*), 210
bulrush, 140, 142, 146, 314, 321
Bureau of Land Management, 175–176
Burgett, Ruth, 327, 332
burials, 300, 330
Burr, David, 7
bushy-tailed woodrat (*Neotoma cinerea*), 136, 191*t*, 192–195, 195*f*, 206, 232–233, 235, 241, 249, 264–265
Butler, Bob, 258
butterflies, 29
Bynon, Theresa, 316
Byrne, Roger, 248

calendar years, vs. radiocarbon years, 55–56, 347–350
Calico Hills, 57–58
Calico site, 289
California
  Bear Flag Revolt (1846), 3
  Frémont's expedition, 3, 5, 6, 7
  Holocene bison, 274–275
  hydrographic Great Basin, 11
  Walker's expedition, 8
  See also specific locations
California Condor (*Gymnogyps californianus*), 181, 182–183, 184, 186
California singleleaf pinyon (*Pinus californiarum*), 24, 25
camels, 49, 70, 109, 174, 175, 179
Camels Back Cave, 192, 195, 199, 205, 232, 233, 238, 241, 249, 303, 329
Campbell, Elizabeth, 292–294
Campbell, Kenneth, 181, 184
Campbell, William, 292–294
Campito Mountain, 244, 245*f*, 261
Cane, Scott, 302
Cannon, Bill, 253
Cannon, Mike, 78
capybaras, 70
carbon-14, 55
caribou, 49, 57, 72
Carlin's Cave, 224

carnivores, 69, 179, 186–187, 204, 257, 342
Carson, Kit, 3, 6, 7
Carson Desert, 145–149, 159, 166, 248, 265–268
Carson Lake, 8, 11, 146, 248
Carson Pass, 5
Carson Range, 28, 30, 32, 204
Carson River, 11, 106, 108*t*, 145
Carson Sink, 11, 109, 127, 145–149, 265–268
Carson Valley, 5
Carter, Deron, 120
Cascade Range, 5, 7, 8, 11, 13, 17, 90, 92, 113, 120, 124, 134, 296
Cathedral Cave, 204
cattail, 140, 142, 146, 218, 314, 315
cattle egrets (*Bubulcus ibis*), 163, 199, 257
ceramics, 307, 324–325, 328, 329, 333
Chamberlain, C. Page, 186
Charlet, David, 24, 26, 142
cheatgrass (*Bromus tectorum*), 23
Chemult Lake, 94
Chewaucan River, 112–113
Chihuahuan Desert, 18, 21*f*, 228, 230
China Lake, 115, 116, 117, 290, 299
chipmunks, 59, 204
Chippewa, 342
chisel-toothed kangaroo rat (*Dipodomys microps*), 23, 187, 189*f*
cicada burrows, 218
cingulates, 67–69
cinquefoil, 139, 140, 142
cirque lakes, 123
Clark, William, 7
Clark Mountain, 161
Clark's Condor (*Breagyps clarki*), 182, 184, 185
Clark's Nutcracker (*Nucifraga columbiana*), 256, 257
Clear Creek Canyon, 292, 330
Clements, Frederick E., 163
climate
  in Bonneville Basin, 126, 139–140
  in Death Valley, 157
  of early Holocene, 226–227, 237, 239–242
  glaciation impact, 50, 127–130
  lakes and, 88–93, 127–130
  of late Holocene, 260–265, 282
  of late Pleistocene, 124, 127–130
  of middle Holocene, 242–253
  in Mojave Desert, 162–163, 239–241
  in Ruby Valley, 141, 144
  See also precipitation; temperature
climate change
  future predictions, 343
  during late Holocene, 260–265
  Pleistocene extinctions role, 82–83
  Younger Dryas, 104–105, 110, 117
Clovis culture, 53–54, 57, 59, 61, 62, 63, 75–83, 176, 218, 289, 290, 291, 292, 299
coastal migration route, 51–52
Colby site, 77
Cole, Ken, 25, 228, 258
collared peccary (*Tayassu tajacu*), 70
Colorado, floristic Great Basin, 17, 19
Colorado pinyon (*Pinus edulis*), 24, 25, 36
Colorado Plateau, 8, 13, 17, 71, 168, 276, 328
Colorado River, 6, 7, 8, 11, 115, 208
Coltrain, Joan, 330
Columbia Plateau, 13, 342
Columbia River, 4–5, 7, 8, 9, 11, 102
Columbian mammoth (*Mammuthus columbi*), 71, 72*f*, 180–181
Comanche, 34
Common Black Hawk (*Buteogallus anthracinus*), 184

condors, 181, 182–186
Confusion Range, 164–165
conifers, 168–169, 223–225. *See also specific species*
Connley Caves, 221, 292, 312, 313, 314
Connolly, Tom, 296
Conservation Blueprint (The Nature Conservancy), 32
Cope's Flamingo (*Phoenicopterus copei*), 181
coprolites, 63, 176, 301, 303, 314, 315
Cordilleran Ice Sheet, 46, 47, 48, 49, 51
corn, 276, 319, 329, 330–331
cosmopolitan bulrush (*Schoenoplectus maritimus*), 321
Coso Basin, 296
Cottonball Marsh pupfish (*Cyprinodon milleri*), 208
cottonwoods, 146
Cougar Mountain points, 294
cougars, 69, 342
coyote (*Canis latrans*), 176, 342
Coyote Basin, 120
creosote bush (*Larrea tridentata*), 18, 153, 155, 156–157, 158, 159, 160, 161, 162, 228–230, 232, 251–252, 344
Cressman, Luther S., 61–63, 305, 313, 327
Crested Caracara (*Polyborus plancus*), 181, 184
"Cro-Magnon Man," 52
Crooked Creek, 120
Crookshanks, Chris, 27
Crowley, Tom, 265
cryptobiotic soil crusts, 24
Crystal Ball Cave, 174–175, 184, 204
Cuba, ground sloths of, 80
cultural Great Basin, 33–40
culture area groups, 33
Cummins, Kathleen, 126–127
curl-leaf mountain mahogany (*Cercocarpus ledifolius*), 26
currant (*Ribes montigenum*), 139, 140, 226
Currey, Don, 101, 222–223, 250
Cutler Dam cycle, 105
Cutting, Lisa, 88

The Dalles, 5
Danger Cave, 62, 187, 192, 195, 224, 233–235, 236, 238, 255, 256–257, 290–291, 302–303, 305, 313
Dansie, Amy, 301
dark kangaroo mice (*Microdipodops megacephalus*), 198–199
Darwin Pass, 108, 109, 111
Davis, Jonathan, 106, 108, 113, 248, 249
Davis, Owen, 105, 152, 160, 166, 229, 240–241, 242
Davis, William, 290
Deacon, James, 207
Dead Horse Lake, 226
Death Valley, 16, 17, 20f, 34, 114–116, 118–120, 135, 153–159
Deep Creek Range, 16, 31, 195
deer, 36, 59, 70, 74, 82, 180, 235, 341–342
deforestation, 80
De Geer, Gerard, 242
DeLucia, E. H., 29
deMay, Ida, 61
Dent site, 77
Desatoya Range, 33
desert almond (*Prunus fasciculata*), 164
desert holly (*Atriplex hymenelytra*), 153, 155
desert peach (*Prunus andersonii*), 167
Desert Series points, 309, 310, 311, 333
desert snowberry (*Symphoricarpos longiflorus*), 162

desert spruce (*Peucephyllum schottii*), 252
desert woodrat (*Neotoma lepida*), 136
deuterium/hydrogen ratios, 264
Devils Hole, 206–209
Devils Hole pupfish (*Cyprinodon diabolis*), 33, 207
dhole (*Cuon alpinus*), 69, 176
Diamond, Jared, 80
Diamond Pond, 247, 260–261, 264
diet, human, 36–38, 300–302, 314–315, 326, 330–331
Dietz Basin, 221, 226
Dillehay, Tom, 60–61
diminutive pronghorn (*Capromeryx minor*), 70
dire wolf (*Canis dirus*), 69, 176
disjunct distributions, 168–169, 198–199
ditchgrass (*Ruppia*), 143
Dixie Valley earthquake, 40
DNA analysis, 54, 63, 72, 242, 331–332
Dome Canyon, 135–136
Domebo site, 77
Domínguez, Francisco, 6
Donner-Reed party, 12, 94–99
Douglas, Charles, 301
Douglas-fir (*Pseudotsuga menziesii*), 17, 32, 167, 168, 169
Drake Peak, 120
dromedary (*Camelus dromedarius*), 70
drought, 261–262, 264, 277, 331
Droz, Michael, 253
Dry Cave, New Mexico, 192
Dry Creek Cave, 282
Dry Creek glaciation, 125
ducks, 181–182
Dugas, Dan, 120, 221, 247
Dugway Proving Ground, 219, 296, 298
Duke, Daron, 297
Dyke, Arthur, 47

early Holocene, 217–242
  archaeological sites, 289–302
  climate, 226–227, 237, 239–242
  shallow lakes and marshes, 217–222
  vegetation, 222–232
earth, axis of, 239
earthquakes, 40
East Humboldt Range, 122, 123
East Walker River, 5, 11
Easter Island, 53
eastern chipmunk (*Tamias striatus*), 59
Ebbett's Pass, 7
Echelle, Anthony, 208
ecosystem-based Great Basin, 32
Edgar, Heather, 300
Eerkens, Jelmer, 296, 310, 325
Eetza Mountain, 146–148
Egan Range, 16
Eiselt, Sunday, 327
ELA (equilibrium-line altitude), 124
Eleana Range, 162
Elias, Scott, 49, 50
elk (*Cervus elaphus*), 180, 235, 278–282
elk-moose (*Cervalces scotti*), 70
Elko Series points, 309, 310–312
Elliot-Fisk, Deborah, 124, 151–152, 253
Elston, Bob, 109, 248, 249, 266
Emerson Pass threshold, 109, 110
empetrichthyids, 206
Emslie, Steve, 183, 184
Engelmann spruce (*Picea engelmannii*), 17, 27, 30, 32, 139, 143, 165, 180
English sparrow (*Passer domesticus*), 199
Ephedra, 23, 138
Epstein, Samuel, 264

equilibrium-line altitude (ELA), 124
Errant Eagle (*Neogyps errans*), 184
Escalante, Francisco, 6
Eskimos, 46, 49
Estancia Basin, 130
ethnographic Great Basin, 33–40
Eurasian extinctions, 73–74, 82, 83
Eureka Valley, 157–158, 229, 230, 252
Europeans, arrival in North America, 38, 341
evaporation rates, 92, 93–94, 99, 114, 116–117, 128–129, 226
extant muskox (*Ovibos moschatus*), 71, 72
extinctions
  of birds, 71–72, 181–186
  causes of, 79–83, 180
  in Eurasia, 73–74
  of mammals, 67–74, 176–181
  timing of, 72–73
extraterrestrial impact events, as extinction cause, 83

Fagan, John, 312
Faith, Tyler, 72
Fallen Leaf Lake, 263
fandango, 37
fecal pellets, 237–238
Fedje, Daryl, 52
Feng, Xiahong, 264
Ferguson, Wes, 161
festival gatherings, 37
Finley, Alexander, 7
fire control practices, 26
Firestone, R. B., 83
firs (*Abies*), 17, 27, 30, 32, 143, 154, 161, 164, 168, 169
fish, 33, 102–103, 206–210, 248
Fish Lake, 123, 245–247, 260, 314
Five Finger Ridge, 330
Five Mile Flat, 297
Five Points site, 296
Fivemile Point, 61, 62f
Fladmark, Knut, 52
flamingos, 181
flat-headed peccary (*Platygonus compressus*), 70
flesher tool, 57
Floating Island, 99
floods, 102, 120, 127, 130, 144, 195
floristic Great Basin
  definition of, 17–22, 40
  early Holocene, 219–221, 241
  late Pleistocene vegetation, 164–168
  middle Holocene pinyon pine arrival, 253–258
  Monitor Valley to Mount Jefferson, 22–31
  vegetation zones, 31–32
  vertebrates of, 32–33
fluted points, 289–292, 299–302
Folsom, 79, 289
Font, Pedro, 6
Forester, Rick, 143
Fort Irwin Military Reservation, 291, 299
Fort Rock Basin, 21f, 61, 112, 120, 221, 253, 297, 301, 304, 313
Fort Rock Cave, 61, 62, 289
Fort Vancouver, 4, 7
Forty Mile Desert, 145
Fortymile Canyon, 231
Fossil Lake, 173–174, 181, 184
Four Corners, 318
fourwing saltbush (*Atriplex canescens*), 18, 22, 146, 148, 153
Fowler, Kay, 38, 300, 319
fox, 73

Fox-Dobbs, Kena, 183, 184, 186
foxtail pine (*Pinus balfouriana*), 168, 263
Fragile Eagle (*Buteogallus fragilis*), 181, 184
Francaviglia, Richard, 7
Freidel, Dorothy, 114
Frémont, John C., 3–9, 18, 135
Fremont cottonwoods (*Populus fremontii*), 146
Fremont culture, 276–277, 327, 328–332
Frenchman Flat area, 137
fungi, 73, 81–82
Furnace Creek, 114, 159

Gale, Hoyt S., 243
gambel oak (*Quercus gambelii*), 32
Gandy Mountain, 174
Garcés, Francisco, 6
Gatecliff Series points, 308, 309, 310
Gatecliff Shelter, 187, 204, 235, 255–256, 258–259, 312, 314, 327, 343
Gehr, Keith, 221
gender roles, 36, 322
Gerisch, Herbert, 174
gestation periods, 236
giant beaver (*Castoroides ohioensis*), 70
giant deer (*Megaloceros giganteus*), 74, 82
giant ground sloth (*Eremotherium laurillardi*), 69
giant sequoia (*Sequoiadendron*), 152
giant short-faced bear (*Arctodus simus*), 49, 69, 180
Gilbert shoreline, 104–105, 140, 223
Gillette, Dave, 180–181
glaciation
 and Bering Land Bridge, 48–51
 climate impact, 50, 127–130
 equilibrium-line altitude calculation, 124
 and lake expansion, 124–127
 Little Ice Age, 264–265
 net budget calculation, 124
 Pleistocene, 45–48, 120–130
 Wisconsin era, 46–48, 50, 51, 123–124, 126
 Younger Dryas, 104–105, 110, 117
Glass Creek, 263
Gleason, Henry A., 163
global warming, 343–344
glottochronology, 315–319
glyptodonts, 67–68
Goddard, Ives, 34
Godey, Alexander, 6
Godsey, Holly, 102
Goebel, Ted, 54, 78, 238, 292
Goldberg, Paul, 59
golden-mantled ground squirrels (*Spermophilus lateralis*), 203
gomphotheres, 71
Goodwin, Thomas, 204
Goose Lake, 88, 94, 120
gophers, 196–197, 205–206
Goshute Mountains, 139, 165
Goss, James, 317–318
Graf, Kelly, 297
Graham, Russ, 82
Grand Canyon, 71, 79, 157, 183, 228
grand fir (*Abies grandis*), 17
Granite Canyon, 165
Granite Mountains, 229, 251
Grapevine Mountains, 154–155
grasses, 26, 139–140, 225, 246–247, 260, 261, 264
Graumlich, Lisa, 239, 240, 263, 264
gravel fans, 153
Gray, Stephen, 264
gray wolf (*Canis lupus*), 176
Grays Lake, 166

greasewood, 22, 143, 144, 146, 148, 149, 166, 225, 247, 248
Great Basin
 archaeological sites, 289–338
 current and future prospects, 341–344
 ecosystem-based definitions, 32
 ethnographic definition, 33–40
 floristic definition, 17–33, 40
 Frémont's expedition and naming of, 3–9
 during Holocene (See Holocene)
 hydrographic definition, 11–12, 40
 lakes, 87–120
 mapping of, 7
 physiographic definition, 12–17, 40
 during Pleistocene (See Pleistocene)
 precipitation, 88–93
 total area in acres, 91f
 vegetation, 135–171
Great Basin Desert, 18, 21f
Great Basin pocket mice (*Perognathus parvus*), 233, 234f, 249
Great Basin Shrub Steppe, 32
Great Basin singleleaf pinyon (*Pinus monophylla monophylla*), 24, 25, 257–258
Great Basin wildrye (*Leymus cinereus*), 143
Great Salt Lake
 within Bonneville Basin, 29
 Domínguez/Escalante expedition, 7
 fish, 209–210
 Fremont culture, 330
 Frémont's expedition, 3–4, 6
 hydrographic Great Basin, 12
 late Holocene, 261, 262
 level fluctuations, 87, 99, 127, 264
 middle Holocene, 250
 pollen analysis, 166
 salinity, 99
 Smith's expedition, 7
 sources for, 91
 Walker's expedition, 8
Great Salt Lake Desert, 87, 99, 262
greater scaup (*Aythya marila*), 33
green algae, 143
green rabbitbrush (*Ericameria teretifolia*), 150
Green River, 6
Greenland ice sheet, 48
grinding stones, 302–303, 312–313
Grosscup, Gordon, 147
Gruhn, Ruth, 253
Gugadja, 302
Guilday, John, 59, 60
Guthrie, Dale, 71, 82
Guthrie, R. D., 50
Gypsum Cave, 173, 183

hackberry, 222, 231–232, 239–240
Hale, Kenneth, 317
Hall, E. R., 195–196, 278–279, 282
Hall, Matt, 299
Hamrick, J. L., 168
Hansen, Henry P., 137, 243
Hansen, Richard, 228
Harding, Sidney T., 262
hardstem bulrush (*Schoenoplectus acutus*), 142
Harlan's ground sloth (*Paramylodon harlani*), 69
Harney Basin, 220, 222, 247, 274
Harney Lake, 11
Harper, Kim, 28
Harrington, C. R., 57
Harrington's mountain goat (*Oreamnos harringtoni*), 71, 79, 82, 176
Hart Mountain, 120
Hart, William, 106

Harvey, Adrian, 240
Haskett points, 294
Hattori, Gene, 255, 295, 300, 327, 332
Haury, Emil, 75, 76
Hawai'i, 53
Hawk eagles, 184
Haynes, C. Vance, 57, 58, 59, 77, 82, 83, 217, 231
Haynes, Gregory, 296
Haystack Mountain, 150, 151f
heather vole (*Phenacomys intermedius*), 72, 204, 258
Heaton, Tim, 174–175, 184
Heizer, Robert, 305, 307–308, 313
helmeted muskox (*Bootherium bombifrons*), 71, 179
Helzer, Margaret, 304
Hendrys Creek Canyon, 27
Henwood site, 291, 301
Hersh, Larry, 260
Hershler, Robert, 208
Hertel, Fritz, 183, 184, 185
Heusser, Calvin, 52
Hickerson, Harold, 342
Hidden Cave, 146–148, 166, 225, 241, 248, 261, 282, 313, 314
Hiebert, Ronald, 168
High Rock Creek, 5
Hildebrandt, William, 312
Hill, Jane, 318–319, 326, 332
Hockett, Bryan, 175–176, 195, 238, 241, 301
Hogup Cave, 233, 238, 305, 311, 313, 329
Hokan, 35
Holliday, Vance, 83
Holmer, Rick, 310–311
Holmgren, Noel, 17–18
Holocene, 217–286
 archaeological sites, 289–302, 302–313
 climate and climate change, 226–227, 237, 239–242, 242–253, 260–265, 282
 definition of, 46, 259–260
 mammals, 249, 258–259, 268–278, 268–282
 pinyon pine arrival, 253–258
 shallow lakes and marshes, 217–222
 vegetation, 222–232
 Walker Lake-Carson Sink conundrum, 265–268
Homestead Cave, 102, 103, 187–190, 192, 193, 196, 199, 205–206, 210, 222, 232, 233, 235–236, 237, 238, 241, 249, 262, 264–265, 343
Homo erectus, 49
Honey Lake Basin, 311, 312
honey mesquite (*Prosopis glandulosa*), 153
Hooke, Roger leB, 118
Hopi, 34
horsebrush (*Tetradymia*), 139
horses, 49, 70, 72, 174, 175, 176, 179
Horse Thief Hills, 158
Hostetler, Steve, 126
Hot Springs Mountains, 149
Houghton, John, 90
house sparrow (*Passer domesticus*), 199
houses
 early Holocene, 296
 late Holocene, 313–314, 320–324, 328, 329, 330
 middle Holocene, 303–304
 Monte Verde, Chile, 60
Howard, Alan, 128
Howard, Hildegarde, 181, 183, 184
Hubbs, Carl L., 206
Huckell, Bruce, 82

Huckleberry, Gary, 220
Hughes, Malcolm, 263
humans and human artifacts
    African origins, 45
    Clovis culture, 53–54, 57, 59, 63, 75–79, 81–82, 176, 289
    diet, 36–38, 300–302, 314–315, 326, 330–331
    early migrations, 45–46, 48–54
    extinction role, 79–82
    Fremont culture, 276–277, 327, 328–332
    gender roles, 36, 322
    hunting, 37, 78, 79–80, 236, 292, 301, 322, 324, 342
    impact on lakes, 87–88
    languages, 34–35, 315–319, 327–328
    late Holocene, 313–333
    late Pleistocene/early Holocene, 289–302
    middle Holocene, 302–313
    population density, 304–307, 311–313, 314
    pre-Clovis sites, 54–63, 289
    See also houses
Humboldt Lake, 6, 8, 12, 92, 106, 127, 145, 248, 302
Humboldt River, 6, 8, 12, 25, 106, 108*t*, 145, 261–262, 292, 302
Humboldt Sink, 8, 12, 233
Hunt, George B., 153
hunting, 37, 78, 79–80, 236, 292, 301, 322, 324, 342
Huntington Mammoth site, 180–181
hydrographic Great Basin, 11–12, 40

Ibex Pass, 229
Ice Age. *See* Pleistocene
ice-free corridor, 51
Idaho
    ecosystem-based Great Basin, 32
    floristic Great Basin, 17–18, 19
    hydrographic Great Basin, 11
    physiographic Great Basin, 13
    See also specific locations
incense cedar (*Calocedrus decurrens*), 17, 28
Incredible Teratorn (*Aiolornis incredibilis*), 183–184
Indian ricegrass (*Achnatherum hymenoides*), 146
Innuitian Ice Sheet, 46, 47
Intermountain Brown Ware, 325
internal drainage, 8–9, 11–12
International Commission on Stratigraphy, 47
iodinebush (*Allenrolfea occidentalis*), 148
Irving, W. N., 57
Isgreen, Marilyn, 329
island colonization, 80–81
Isleta Cave, 190
isotope ratios, 116–117
isotropic signature, 106

jackrabbits (*Lepus californicus*), 37
Jackson, Donald, 8
jaguars (*Panthera onca*), 69, 176
Jahren, Hope, 239
Jakes Valley, 299
Janetski, Joel, 39, 329, 330, 333
Jass, Chris, 204
Jayko, Angela, 118
Jefferson, George, 181
Jefferson's ground sloth (*Megalonyx jeffersoni*), 69, 179
Jefferson's mammoth (*Mammuthus jeffersoni*), 71
jeffrey pine (*Pinus jeffreyi*), 17, 28, 32

Jenkins, Dennis, 62–63, 221
Jennings, Jesse, 62, 291, 302, 305
Jennings, Steven, 151–152, 253
jet streams, 128, 130
Johnson, Catherine, 278
Johnston, W. A., 51
jointfir, 23, 155, 159, 232
Jolie, Ed, 327, 332
Jones, Terry, 264
Jones, Tom, 220, 289, 291, 295, 296, 297, 298, 299, 304
Jordan River, 12, 91
Jorgensen, Clive D., 136–137
Joshua tree (*Yucca brevifolia*), 18, 21*f*, 25, 149, 150, 156, 159, 344
junipers, 24–27, 139, 144, 149, 150, 151, 152, 153–154, 156, 158–160, 161, 162, 164–165, 167, 168, 224, 227, 261, 263, 264

Kaestle, Frederika, 332
kangaroo mice, 198–199
kangaroo rats, 22–23, 33, 187, 189*f*, 241, 249
Kautz, Bob, 256, 259
Kawaiisu language, 34, 317
Kearny, Stephen Watts, 6
Kelly, Bob, 266, 324
Kelly, Isabel, 275
Kelson, K. R., 278–279
Kenagy, Jim, 23, 236
Kern River, 6
Kiahtipes, Chris, 250
Kimmswick site, 77
King, Guy, 266, 267, 268
King, Tom, 228, 230–231
King's Dog site, 274–275, 303–304, 313–314
Klamath Lake, 5, 6
Klamath Marsh, 5, 6, 7
Klamath people, 327
Klamath River, 5, 7
Kleppe, John, 263
Koehler, Peter, 149, 150, 151, 159, 160, 230, 231, 251, 252, 261
Kramer Cave, 313
Kroeber, Alfred, 317, 318
krummholz, 30, 244
Ku, Teh-Lung, 118
Kutzbach, John, 128

La Brea tar pits, 69, 70, 181, 183, 184
lagomorphs, 70
Lahontan Basin
    early Holocene, 220, 221
    early Holocene vegetation, 225–226
    human occupation, 332
    late Holocene, 313, 327
    late Pleistocene, 128, 167
    middle Holocene, 247–249
Lahontan Trough, 29, 145–149
Lajoie, Ken, 125
Lake Abert, 5, 112, 114, 243, 244, 252–253, 314, 326, 327
Lake Adobe, 115, 116
Lake Alvord, 120
Lake Bonneville, 94–105, 111, 126, 128, 130, 139–140, 165–166, 209–210, 219
Lake Chewaucan, 111–114
Lake Cloverdale, 130
Lake Cochise, 130
Lake Dumont, 119
Lake Estancia, 130
Lake Franklin, 141, 143–144
Lake Gale, 118
Lake Gunnison, 104, 219
Lake Hubbs, 220

Lake Lahontan, 106–111, 130, 145, 148, 159, 225, 305
Lake Manix, 119, 181
Lake Manly, 115, 118–119, 128, 130
Lake Mojave, 119–120, 130, 218–219, 222, 261, 292–294
Lake Russell, 116, 125, 127, 220
Lake San Agustin, 130
lakes, 87–120
    climate impact, 88–93, 127–130
    and glacier expansion, 124–127
    late Holocene, 260–261
    major cycles, 105
    pluvial lakes, 93–120, 127–130
    total area of, 91*f*, 242
    valley bottom lakes, 88, 89–90*t*
    variation in characteristics, 87–88
    See also specific lakes
Lake Tahoe, 5, 11, 35, 248–249, 261, 302
Lake Tecopa, 119
Lake Tonopah, 290, 299
Lakeside Cave, 313
LaMarche, Val, 244, 264
Lamb, Hubert H., 262
Lamb, Sydney, 316–317
Lamoille Canyon, 122–123
land-speed records, 87
Lange/Ferguson site, 77
languages, 34–35, 315–319, 327–328
Lanner, Ron, 256
Lao, Yong, 108
large-headed llama (*Hemiauchenia macrocephala*), 70
Larrea divaricata, 228
Larrucea, Eveline, 259, 343
Larsen, Vonn, 292
Las Vegas Valley, 217–218, 222, 231, 240
Las Vegas Wash, 217
Last Chance Range, 157–158, 230, 231
Last Glacial Maximum, 47–48, 51, 52, 128
Last Supper Cave, 296, 301
late Holocene, 259–282
    bison, 268–278
    climate change, 260–265, 282
    definition of, 259–260
    elk, 278–282
    martens, 282
    Walker Lake-Carson Sink conundrum, 265–268
late Pleistocene. *See* Pleistocene
Laurentide Ice Sheet, 46, 47–48, 50, 51, 83
Lava Creek cycle, 105, 106
Lawlor, Tim, 205
Lead Lake, 262, 267
Leakey, Louis, 57–58
least chipmunks (*Tamias minimus*), 204
least weasel (*Mustela nivalis*), 204
Lees, Robert, 316
Lehner, Ed, 75–77
Lehner, Lyn, 75
Leonard Rockshelter, 248
Leonard, Zenas, 8
Licciardi, Joseph, 114
Lillquist, Karl, 144
Lily Lake, 226, 249
limber pine (*Pinus flexilis*), 25, 27, 29, 30, 31–32, 139–140, 143, 154, 161, 162, 164–165, 167, 223–224, 241
Limber Pine-Bristlecone Zone, 31–32
Lin, Jo, 118
Lind Coulee points, 294
Lindström, Susan, 248
lions, 69, 176
Lips, Elliot, 125

Little Cottonwood Canyon, 125, 222–223
Little Ice Age, 264–265
Little Lake, 251
little sagebrush (*Artemisia arbuscula*), 22
Little Skull Mountain, 229, 231
Little Valley cycle, 105, 106
livestock, 26
Livingston, Stephanie, 233
llamas, 70, 175, 176
Lodgepole pine (*Pinus contorta*), 28
loess, 50
Logan River, 12
Lomolino, Mark, 205
long-nosed peccary (*Mylohyus nasutus*), 70
Long Valley, 299
Loope, Lloyd, 142
Los Angeles, water diversion
    projects, 88, 116
Louderback, Lisbeth, 140, 223, 249, 301
Lovelock Cave, 233, 312, 313, 314, 327
Lower Klamath Lake, 88, 341–342
Lower Pahranagat Lake, 262, 264
Lowie, Robert, 37
Lucerne Valley, 227, 228, 229, 230, 231
Lundelius, Ernie, 82
Lupo, Karen, 195, 199, 249, 276

Mack, Richard, 178–179
macrofossils, 137, 138, 152, 160, 167,
    180, 228–229, 230, 231, 247,
    253–255, 259, 261, 319
Madagascar, large vertebrates of, 81
Madsen, David, 103, 104, 125, 139, 140,
    165, 166, 167, 180–181, 219, 222–223,
    224, 249, 250, 255, 257, 262, 296, 297,
    302–303, 305, 313, 315, 329, 330
Maguari stork (*Ciconia maguari*), 181
Maher, Louis, 138
mahogany, 26
Mahogany Flat, 153–154
Maidu, 35, 327
maize, 276, 319, 329, 330–331
Malheur Basin, 220–221, 274
Malheur Gap, 220, 221
Malheur Lake, 11, 120, 127, 220, 302
Malheur National Wildlife Refuge,
    33, 220, 247, 343
Malheur River, 120
mammals
    current situation, 341–343
    early Holocene, 232–238
    Great Basin native species, 33
    late Holocene, 268–282
    late Pleistocene, 67–74, 176–181, 186–206
    middle Holocene, 249, 258–259
    See also specific species
Mammoth Steppe, 50
mammoths, 48, 49, 71, 72, 74, 75, 76, 82,
    179, 180–181
Mamontovaya Kurya, 52
Manly, William Lewis, 135
maple, 32
Marble Mountains, 229
marmots, 190–192, 193f, 205, 232, 233, 241
Marr, John, 30
marshes, 217–222, 264
martens, 204, 282
Martin, Jim, 173–174
Martin, Paul, 79–81, 137, 228, 252
Mary's Lake, 6
mastodon (*Mammut americanum*),
    71, 179, 180–181
Matsubara, Yo, 128
Matthes, François, 264

McCullough Range, 252
McGuire, Kelly, 175
McLane, Alvin, 16
Mead, Jim, 79, 166, 176, 204, 278
Meadow Valley Wash, 167, 255
Meadowcroft Rockshelter, 53, 58–60, 79
Medieval climate anomaly,
    262–264, 277, 331
Meek, Norman, 119
Mehringer, Peter, 123, 146, 148, 159,
    161, 162, 166, 218, 225, 231, 245–247,
    251–252, 253, 256, 260, 261, 264,
    289, 315
Melish, John, 7
Meltzer, David, 54, 60, 78, 83
Menges, Christopher, 119
Menlo Baths site, 303
Mensing, Scott, 149, 153, 248, 262
Merced River, 8
Mercury Ridge, 227
Merriam, C. Hart, 278
Merriam's kangaroo rat
    (*Dipodomys merriami*), 33
Merriam's teratorn (*Teratornis merriami*),
    183, 184
Merrill, William, 319
mesquite, 153
Miami site, 77
microbiotic soil crusts, 24
microphytic soil crusts, 24
middens. *See* packrat middens
middle Holocene, 242–259
    archaeological sites, 302–313
    climate, 242–253
    mammals, 249, 258–259
    pinyon pine arrival, 253–258
Millar, Constance, 168, 244, 263
Miller, Alden H., 184
Miller, Becky, 210
Miller, Jerry, 248, 261–262
Miller, R. F., 142
Miller, Richard, 26
Miller, Robert Rush, 206, 207
Miller, Wade, 174, 180
Millikin, Randall, 312
Minckley, Tom, 226, 237, 241
Mineral Hill Cave, 175–176, 184
Missoula floods, 102
mitochondrial DNA (mtDNA), 331–332
moas, 80
Modoc Lake, 94
Modoc people, 303–304
Moffit, James, 341–342
Mojave Basin, 264
Mojave Desert, 21f
    Calico Hills site, 57–58
    early Holocene climate, 239–241
    early Holocene lakes and
        marshes, 217–219
    early Holocene vegetation, 227–232
    expeditions to, 6, 7
    floristic Great Basin, 18, 22
    fluted point discoveries, 291
    late Holocene, 261
    late Pleistocene climate, 162–163
    late Pleistocene vegetation, 159–164, 167
    middle Holocene climate, 251–252
    middle Holocene human population, 307
    See also Death Valley
Mojave River, 6, 12, 13, 119, 120, 218
Mojave seablite (*Sueada moquinii*), 153
Monitor Lake, 22, 23, 31
Monitor Range, 33, 37
Monitor Valley, 22–31

Mono Basin, 125
Mono Lake, 87, 88, 113–114, 115,
    116, 152, 262, 267–268
Mono language, 34
monsoons, 239–240, 252, 277
montane mammals, 205–206
Monte Verde, Chile, 60–61, 79, 81, 176
Montenegro, Álvaro, 45
Monterey, 6–7
moose, 70
moraines, 122
Mormon tea (*Ephedra viridis*),
    23, 154, 159–160
Mormons, 3, 6
Mosquito Willie's spring, 250, 255
Mount Jefferson, 22–31, 320
Mount Mazama, 61, 62, 113, 245, 253
Mount McKinley, 49
Mount Moriah, 30
Mount St. Helens, 113
mountain deer (*Navahoceros fricki*), 70
mountain goats, 71, 79, 82, 176
mountain hemlock (*Tsuga mertensiana*), 32
mountain lion (*Felis concolor*), 342
Mountain Meadows, 6, 8–9
mountain ranges, in physiographic
    Great Basin with summits about
    10,000 feet, 14t, 16. *See also specific ranges*
mountain sheep (*Ovis canadensis*), 36, 71,
    180, 235, 341, 342
Mt. Washington, 30
Mud Lake Slough, 108, 109
Mule deer (*Odocoileus hemionus*), 235
Murchison, Stuart, 250
Murie, Olaus, 278–282
Murphy, Dennis, 29
Murray Springs, 77, 83
muskoxen, 71, 72, 74, 82, 179

Na-Dene, 34
Nahuatl language, 34
Narrative (Leonard), 8
narrow-headed vole (*Microtus gregalis*), 73
Native Americans
    Frémont expedition's encounters
        with, 5, 6
    Great Basin defined by, 33–40
    mitochondrial DNA lineages, 331–332
    oral histories, 268, 274
    See also specific groups
The Nature Conservancy, 32
Navajo, 34, 315
Neanderthals, 52
Nebraska Sand Hills, 264
Negrini, Robert, 113–114
Nelson Basin, 229
Nenana complex, 78
net budget calculation, 124
Netleaf hackberry (*Celtis reticulata*),
    222, 231–232, 239–240
Nevada
    birds, 33
    expeditions, 5, 6, 7, 8
    Holocene bison, 275–276
    hydrographic Great Basin, 11
    mountains of physiographic Great
        Basin, 16
    valleys of physiographic Great Basin,
        16–17
    water projects, 341
    See also specific locations
Nevada jointfir (*Ephedra nevadensis*),
    23, 155, 159, 232
New Zealand, 53, 80

Newberry Cave, 313
Newberry Crater, 61, 62
Newman, Deborah, 180–181, 329
nine-banded armadillo (*Dasypus novemcinctus*), 67
noble marten (*Martes nobilis*), 282
Nogués-Bravo, David, 83
northern bog lemmings (*Synaptomys borealis*), 204
northern pamapathere (*Holmesina septentrionalis*), 67
northern pocket gophers (*Thomomys talpoides*), 195–197, 205–206
northern shrike (*Lanius excubitor*), 33
Northern Side-notched points, 312–313
Northern Uto-Aztecan language, 318
Norway rat (*Rattus norvegicus*), 199
Nowak, Cheryl, 149, 167, 226, 257
nuclear families, 35, 37, 325–326
Numic expansion, 314–320, 325–328, 332–333
Numic languages, 34–35
Nuttall's saltbush (*Atriplex nuttallii*), 143

oak-hickory forests, 163
oaks, 32
obsidian, 296–298
ocean spray (*Holodiscus*), 164
O'Connell, Jim, 274–275, 301–302, 303, 313–314
O'Connor, Jim, 102
Oetting, Albert, 297, 327
Ogden River, 12
Old Crow Basin, 57
Old Humboldt site, 292, 296, 301
Old River Bed, 103, 104, 219–220, 222, 240, 249, 296, 297, 298
Olivella shells, 304
Omernik, James, 32
oral histories, 268, 274
Ord's kangaroo rat (*Dipodomys ordii*), 23, 187, 189f, 241, 249
Oregon
  birds, 33
  floristic Great Basin, 19
  Frémont's expedition, 4, 5
  Great Basin definitions, 13
  hydrographic Great Basin, 11
  Paisley Caves, 61–63, 74, 81, 176, 275, 289
  See also specific locations
Oregon Trail, 6
Organ Pipe National Monument, 229
Orme, Amalie, 117, 219
Orme, Antony, 117, 219
O'Rourke, Dennis, 332
Osborn, Gerald, 120, 122, 124
ostracodes, 143
overkill extinction hypothesis, 79–82
Oviatt, Jack, 101, 104, 105, 219, 264
Owens-Death Valley system, 208
Owens Lake, 18, 88, 115, 116–118, 126–127, 130, 149, 152–153, 217–219, 222, 243, 250–251, 261, 263, 302
Owens River, 12, 34, 115
Owens Valley, 8, 18, 22, 36, 38–39, 149–153, 159, 326
Owens Valley earthquake, 40
Owl Canyon, 162, 229
owl pellets, 186
owls, 33
Owyhee Desert, 19
oxygen isotope values, 116–117, 248
Pacific Ocean, 53

Pacific rats (*Rattus exulans*), 80
packrat middens, 136–141, 146, 149, 151–152, 155–158, 159–162, 164–168, 224, 228–230, 231, 235, 247, 253–256
Pahrump poolfish (*Empetrichthys latos*), 206
Paillet, Fred, 128
Painted Hills, 226
Paisley Caves, Oregon, 61–63, 74, 81, 176, 275, 289
Paiute, 38–39, 317
paleomagnetism, 113–114
palynology, 137
pampatheres, 67
Panamint language, 34
Panamint Range, 17, 26, 114, 153–156
Panamint Valley lakes, 115, 118
Pancake Range, 16
Pandora moth (*Coloradia pandora*), 38, 39f
Papoose Lake, 135
Paquay, François, 83
Parmalee, Paul, 59, 60
Parman points, 294
Parry's saltbush (*Atriplex parryi*), 149
passenger pigeon (*Ectopistes migratorius*), 59
Patterson Lake, 226
Patterson, Robert, 174
Paulina Lake site, 296
Paulina Marsh, 221, 222, 304
Paviotso language, 34
Payen, Louis, 58
peccaries, 70
Pedler, David, 59
Penutian, 35–36, 327, 332
perissodactyls, 70
Perry Aiken glaciation, 124
Peterson, Ken, 277
Phillips, Fred, 124, 127
photographic studies, 260
physiographic Great Basin, 12–17
Picacho Peak, 228
Pie Creek Shelter, 313
pikas (*Ochotona princeps*), 33, 139, 199–204, 205, 232, 235, 258, 342–343
Pilosa, 67, 69, 79, 80, 82
Pilot Range, 97
Pine Creek Canyon, 27, 30
Pine Forest Range, 120
pine marten (*Martes americana*), 204
Pine Nut Range, 28
Pinedale glaciation, 124
pines (*Pinus*), 17, 24–27, 28, 29, 30, 31–32, 36–37, 138, 139–140, 141, 143, 144, 148–154, 155, 156, 158, 159, 160, 161, 162, 164–165, 167, 168, 169, 223–224, 225, 240, 241, 253–258, 263, 264, 314
Pinson, Arianne, 223, 238
Pinto points, 312
Pintwater Cave, 195, 196, 229–230, 232, 233, 239–240, 241
Pinyon Jay (*Gymnorhinus cyanocephalus*), 256
pinyon-juniper woodland, 24–27, 149, 153–154, 160–162, 168, 258–259
Pinyon-Juniper Zone, 31, 36
pinyon nuts, 36–37, 256–257, 303, 325
pinyon pines, 24–27, 36–37, 141, 149, 150–153, 153–154, 155, 158, 159, 160, 161, 162, 167, 168, 169, 224, 240, 253–258, 257–258, 314
Pippin, Lonnie, 325
Pister, Edwin, 207

Pit River, 94, 120
pithouses, 330
plant macrofossils. *See* macrofossils
Pleasant Valley earthquake, 40
Pleistocene
  archaeological sites, 54–63, 289–302
  Bering Land Bridge, 45–46, 48–51, 54
  climate, 124, 127–130
  definition of, 46
  extinctions at end of, 67–74, 79–83
  glaciation in, 45–48, 120–130
  humans during, 74–83, 289–302
  lakes, 93–120
  stages of, 47
  vegetation, 135–171
  vertebrates, 173–213
Plew, Mark, 278
pluvial lakes, 93–120, 127–130
pocket mice, 233, 234f, 249
Point of Rocks, 229, 230, 231
Pokes Point cycle, 105, 106
pollen analysis, 50, 61, 137–138, 140, 144, 146–148, 149, 152–153, 166–167, 223, 229, 230–231, 248
Polynesians, 53
ponderosa pine (*Pinus ponderosa*), 17, 32
Pony Express, 135
population density, 304–307, 312–313, 314
Porter, Stephen, 124
potassium–argon (K/Ar) dating, 113
pottery, 307, 324–325, 328, 329, 333
precipitation
  Bering Land Bridge, 49–50
  in Carson Desert, 146
  in Death Valley, 114
  drainage patterns, 11
  of early Holocene, 226, 239–240
  future predictions, 343
  lake level impact, 127–128
  of late Pleistocene, 88–93
  of middle Holocene, 252
  in Mojave Desert, 162, 163, 240–241
  in Ruby Valley, 141
Preuss, Charles, 5, 6, 8
pricklypear (*Opuntia*), 25, 157
proboscideans, 71
projectile points, 76–77, 79, 289–302, 307–313
Promontory Culture, 333
pronghorn (*Antilocapra americana*), 36, 37, 38, 70, 180, 235, 342
Proto-Uto-Aztecan peoples, 319
Provo River, 12
Provo Shoreline, 101
Pueblo people, 264, 332
Puerto Rico, large rodents of, 80
pupfish, 33, 206–209
Putnam, William, 125
pygmy rabbit (*Brachylagus idahoensis*), 33, 187–190, 191t, 232, 249, 258–259, 343
Pyramid Lake, 5, 6, 11, 87, 92, 108, 109, 110, 220, 248, 249, 261, 262, 263, 302

Quade, Jay, 76, 83, 106, 217–218, 219, 231, 251
quaking aspen (*Populus tremuloides*), 180
Quaternary period, 46
Quinn River, 11–12, 106, 108t

rabbitbrush, 22, 143, 149, 154, 155
rabbits, 33, 70, 81, 187–190, 191t, 232, 249, 258–259, 343

radiocarbon dating, 48–49, 50, 54–56, 57, 58, 59, 61, 62, 79, 82, 104, 109, 111, 113, 114, 116, 123, 124, 125, 137, 138, 139, 143, 144, 173, 175, 176, 181, 183, 187, 219, 221, 229, 233, 243, 244, 245, 247, 248–249, 253, 262, 263, 266, 291, 292, 293, 294, 299, 304–307, 311, 313, 318, 320–321, 323, 325
Railroad Valley, 299
Railroad Valley springfish (*Crenichthys nevadae*), 206
Rampart Cave, 228
Rancho La Brea, 69, 70, 181, 183, 184
Raspberry Square phase, 127
rattlesnakes (*Crotalus viridis*), 23
Rea, John, 316
red fir (*Abies magnifica*), 17, 32
Red Rock Pass, 101–102
Redwine, Joanna, 119
Reese River valley, 326
Reheis, Mareth, 108, 116, 119
reindeer, 74, 82
Remote Oceania, 53
Report of the Exploring Expedition to Oregon and North California in the Years 1843–1844 (Frémont), 3, 5, 6, 7, 8
Reveal, James, 29, 256
Reynolds, Robert, 204
Rhode, David, 103, 109, 139, 140, 149, 164, 165, 166, 167, 219, 223, 224, 225, 244, 249, 255, 257, 292, 301, 302–303, 314, 321
Richens, Lane, 330
Rickart, Eric, 195, 205
Riggs, Alan, 207
rivers and streams, 91–92. *See also specific rivers and streams*
rock flour, 126–127
Rocky Mountain juniper (*Juniperus scopulorum*), 25, 149, 167
Rocky Mountains, 8, 28–29, 124, 125, 126, 168
rodents, extinctions at end of Pleistocene, 69–70. *See also specific species*
Rodgers, Garry, 259–260
Rosegate Series points, 309–310
Rosenbaum, Joseph, 127
Roughing It (Twain), 11
Rouse, Norman, 147
rubber rabbitbrush (*Ericameria nauseosa*), 143
Ruby Crest Trail, 123
Ruby Lake, 141, 143, 144, 264
Ruby Lake National Wildlife Refuge, 33
Ruby Marshes, 245, 302, 314
Ruby Mountains, 25, 31, 92, 122, 123, 142, 143
Ruby Valley, 141–145, 159, 166, 220, 222, 241, 260
Ryser, Fred, 181

saber-toothed cat (*Smilodon fatalis*), 69
Sack, Dorothy, 101, 166
Sage grouse (*Centrocercus urophasianus*), 37
sage vole (*Lemmiscus curtatus*), 204–205, 233, 234f
sagebrush (*Artemisia*), 17, 18, 22, 139, 140, 143, 144, 148–149, 152, 154, 155, 159, 162, 166, 167, 187, 225, 230–232, 240, 246–247, 249, 261, 344
sagebrush-grass steppe, 19, 159
Sagebrush-Grass Zone, 26, 31, 143, 145, 225, 259
sagebrush steppe, 144, 153, 166, 167, 225, 232, 241, 260

saguaro cactus, 344
saiga (*Saiga tatarica*), 71
Salt Creek Marsh pupfish (*Cyprinodon salinus*), 208
Salt Creek pupfish (*Cyprinodon salinus*), 33
salt lakes, 11
saltbush (*Atriplex*), 17, 18, 22, 106, 138, 140, 143, 144, 146, 148, 149, 153, 162, 167, 225, 247, 248
saltgrass (*Distichlis spicata*), 22, 142, 153
Salton Basin, 11
Salton Sea, 11
San Bernardino Valley, 7
San Buenaventura River, 6
Sandy Valley, 231
San Pedro Riparian National Conservation Area, 75
Santa Rosa Range, 120
Sarcobatus, 146
Saunders, Jeff, 71
Scharf, Elizabeth, 322–323, 326
Schell Creek Range, 16, 224, 255, 256
Schlesinger, W. H., 29
Schmitt, Dave, 104, 195, 199, 219, 249, 276, 329
Schroth, Adela, 312
scimitar cat (*Homotherium serum*), 69
Scott, Andrew, 83
screwbean mesquite (*Prosopis pubescens*), 153
Scrub Jay (*Aphelocoma coerulescens*), 256
sea crossings, 49, 53
sea level, 48
Searles Lake, 115, 116, 117–118, 130
seasonal extremes, 239
sediment core analysis, 106, 111, 140, 143, 146, 225, 245, 248, 249
seeds, 301–302, 302–303, 325, 326
sequoia, 152
Service, Elman R., 38
Sevier Basin, 103, 104, 219
Sevier Desert, 299, 329–330
Sevier Lake, 12, 88, 264
Sevier River, 6, 7, 12
Sevier River Valley, 329
shadscale (*Atriplex confertifolia*), 18, 22, 140, 143, 146, 149, 152, 156, 158, 159, 164, 167–168, 232, 261
Shadscale Zone, 31, 149–150, 167–168
Shafer, David, 240–241
Shafer, Sarah, 344
Sharp, Robert, 122
Shasta ground sloth (*Nothrotheriops shastensis*), 69, 79, 82, 179, 228
Sheaman site, 77
Sheep Range, 162, 164, 228, 244, 261
shelducks, 182
Shelley, Phil, 295
Sheppard, John, 251
short-faced skunk (*Brachyprotoma*), 69
Shoshone language, 34, 317
Shoshone Mountains, 33, 120
Shoshone people, 39
Shott, Michael, 309
Shreve, Forrest, 17, 19
shrubby cinquefoil (*Dasiphora fruticosa*), 139, 140, 142
shrubby common juniper, 164–165
shrub-ox (*Eucerathrium*), 71, 175, 178
Siberia
  Bering Land Bridge, 45–46, 48–51, 54
  human occupation of, 52–53
Sierra Nevada
  early Holocene climate, 242
  expeditions, 5, 6, 7, 8

floral and fauna compared to Great Basin, 28–29
floristic Great Basin, 17
glaciation, 47, 124, 127, 264
hydrographic Great Basin, 11
physiographic Great Basin, 13
Silurian Valley, 119, 120, 251
silver buffaloberry (*Shepherdia argentea*), 24
Silver Island Range, 224, 257
Silver Lake, 119
Silver Lake points, 293–294
silverweed cinquefoil (*Argentina anserina*), 142
Simms, Steve, 302, 326–327, 329, 330
Simpson, George Gaylord, 208
Simpson, James H., 135–136
Simpson, Ruth D., 57–58
Simpson's glyptodont (*Glypotherium floridanum*), 67–69
Simpson's pediocactus (*Pediocactus simpsonii*), 321
singleleaf pinyon (*Pinus monophylla*), 24, 25, 36, 149, 150–151, 155, 158, 159, 161, 162, 167, 168, 169, 240, 253–258
Sioux, 342
Skeleton Hills, 158–159, 228
Skull Creek Dunes site, 311
skunks, 69
sloths, 67, 69, 79, 80, 82, 179, 228
Smith Creek Canyon, 27, 224
Smith Creek Cave, 173, 176, 182, 183, 184, 197, 204, 282
Smith, Elmer, 290–291
Smith, Geoff, 297
Smith, George, 116, 117, 251
Smith, Gerald, 208, 248
Smith, Jedediah, 7, 11
Smith, Travis, 27
Smoke Creek Desert, 167, 220
smokebush (*Psorothamnus polydenius*), 146
Snake Creek Burial Cave, 204, 282
Snake Range, 16, 24, 27, 30, 164–165, 195, 197, 204, 224, 255
Snake River, 4, 8, 18, 19
Snake River Plain, 278, 342
snakes, 23, 33
Snake Valley, 165, 167, 174–175, 204
snowberry, 139, 159, 162
Snowbird Bog, 222–223, 250
snowy owl (*Nyctea scandiaca*), 33
Soda Lake, 119
Soldier Meadow, 5
Solutrean culture, 53–54, 59
Sonoran Desert, 18, 21f, 228, 230, 239, 240–241, 344
South America, Polynesian voyages to, 53
South Fork Shelter, 313
South Pass, 3
southern flying squirrel (*Glaucomys volans*), 59
southern pampathere (*Pampatherium*), 67
Spaulding, Geof, 136, 155, 156, 157–158, 159, 160, 161, 162, 163, 164, 165, 167, 228, 229, 230, 231, 239, 240, 252
spectacled bear (*Tremarctos ornatus*), 69
Specter Range, 158
spikerush (*Eleocharis*), 321
spineless horsebrush (*Tetradymia canescens*), 152
Spirit Cave burials, 300, 301
Sporomiella, 73, 82
spotted skunk (*Spilogale*), 69
Spring Range, 161
spring-snails, 208

spruce, 17, 27, 30, 32, 139, 143, 165, 180
squirrels, 203, 204
Stanbury Oscillation, 101
Stanford, Dennis, 53, 59
Stanislaus River, 7
Stansbury cliffrose (*Purshia stansburiana*), 154, 159
Stansbury Mountains, 195
Steens Mountain, 120, 122, 123, 245–247, 253, 327–328
Stein, John, 277
Steinaker Gap, 329
Steller's Jay (*Cyanocitta stelleri*), 256
stemmed points, 292–302
Stephens, Doyle, 99
stepping-stone hypothesis, 53
Steward, Julian, 35–39, 333
Stewart, Omer, 38
Stillwater Marsh, 266–267, 322, 331–332
Stillwater Range, 240
Stine, Scott, 262, 263
Stock, Chester, 61
storks, 181
stout-legged llama (*Palaeolama mirifica*), 70
stranded shorelines, 52
Straus, Lawrence, 54
strontium isotopes, 106
Stuart Rockshelter, 313
subalpine fir (*Abies lasiocarpa*), 27, 30, 32
sugar pine (*Pinus lambertiana*), 17
Sulphur Spring Range, 175
Summer Lake, 112–114, 243, 244, 252–253
Sundell, Taya, 278
Sunshine Locality, 291, 295, 296, 298
Surovell, Todd, 72, 83
Surprise Valley, 274–275, 303–304, 313, 326
Susan River, 106, 108t
Sutter's Fort, 5
Sutton, Mark, 307
Swadesh, Morris, 316, 317
Swan Lake, 166
Sweitzer, Richard, 342
Szabo, Barney, 207

Tabeau, Jean Baptiste, 6
taiga vole (*Microtus xanthognathus*), 72
Talbot, Richard, 330
tamarisks (*Tamarix*), 153, 155
Tambora, eruption of, 265
tapirs, 70, 176
Tasmanian devil (*Sarcophilus harrisii*), 199
Tausch, Robin, 26
Taylor, Alan, 249
Taylor, Amanda, 292
Tecopa Basin, 119
teddybear cholla (*Cylindropuntia bigelovii*), 232
Tehachapi Mountains, 6
Telescope Peak, 17
temperature
    in Death Valley, 114, 157
    of early Holocene, 226, 239
    and evaporation rates, 92
    future predictions, 343
    of late Pleistocene, 128–129
    of middle Holocene, 243–244
    in Mojave Desert, 162, 163
tephras, 113–114
teratorns, 183–186
Terry, Rebecca, 187, 262
thermohaline circulation, 105
thermoluminescence (TL) dating, 113
Thomas, Dave, 36–37, 147–148, 255, 258, 308, 309, 310, 312, 320–322, 323, 324

Thompson, Bob, 105, 123, 142, 143, 144, 160, 162–163, 164, 165, 166, 167, 220, 224–225, 231, 240, 241, 242, 245, 249, 255, 256, 259, 260, 264, 344
Thompson, John, 178–179
Timpanogot, 39
Tinajas Atlas Mountains, 228
Tioga glaciation, 124, 126
Toiyabe Range, 27, 30, 31, 33
Tonni, Eduardo, 184
tools
    archaeological site dating, 56
    Clovis culture, 53–54, 75–79
    flesher tool at Old Crow Basin, 57
    Meadowcroft Rockshelter artifacts, 59
    Monte Verde, Chile, 60
    projectile points, 76–77, 79, 289–302, 307–313
    Solutrean vs. Clovis artifacts, 53–54
Top of the Terrace, 139–140, 165
Toquima Range, 22, 23–24, 26, 27, 29, 30, 31, 33, 196, 197, 204, 258, 314
Torrey saltbush (*Atriplex torreyi*), 146, 148
Townsend, Gail, 248
Townsend's ground squirrels (*Spermophilus townsendii*), 204
transatlantic human migration route, 53–54
transpacific human migration route, 53
treelines, 29–30, 32, 244
tree-ring analysis, 55, 161, 242, 244, 262, 263, 264, 265
Trego Hot Springs tephra, 113
Trinity River, 7
trout, 210
Truckee River, 11, 106–107, 108t, 145, 195, 248
Tübatulabal Indians, 8, 317
Tule Lake, 88
Tule Springs, 162, 183, 231, 252, 289
Tule Valley, 165–166
tundra, 31, 49, 50–51
Turkey Vulture (*Cathartes aura*), 182, 184–185
Twain, Mark, 11
Twin Peaks, 165
Two Goblin site, 150
two-needle pinyon (*Pinus edulis*), 24

Uinta Mountains, 11, 91, 126, 127, 195, 204
Umpqua River, 7
ungulates, 70–71
University of Arizona, Laboratory of Tree-Ring Research, 244
University of Oregon, 253
Upper Dollar Lake, 123, 143
Upper Klamath Lake, 88
Upper Sagebrush-Grass Zone, 31, 225
uranium-234, 58
uranium-thorium (U-Th) dating, 58
U.S. Army Corps of Topographical Engineers, 135
U.S. Bureau of Topographical Engineers, 3
U.S. Fish and Wildlife Service, 207
U.S. Forest Service, 149
U.S. Geological Survey, 17, 88, 110–111, 116, 117, 143, 146, 149, 153, 166
Ushki-1/-5, 53
Utah
    floristic Great Basin, 17, 19
    hydrographic Great Basin, 11
    See also specific locations
Utah agave (*Agave utahensis*), 162, 164
Utah junipers (*Juniperus osteosperma*), 24, 25, 139, 149, 150, 151, 152, 153–154, 156, 158–160, 161, 162, 167, 168, 224, 227

Utah Lake, 6–7, 12, 39, 87, 209
Ute language, 34, 317
U-Th (uranium-thorium) dating, 58
Uto-Aztecan language family, 34, 317–319

valley-bottom vegetation, 166–167
valley-bottom elevations, 18f
valley-bottom lakes, 89–90t
valleys, of physiographic Great Basin, 16–17
Van Devender, Tom, 167, 228, 229, 239
Van Valkenburgh, Blaire, 183, 184
Van Winkle, Walton, 243, 244
Vasek, Frank, 154
vegetation, 135–171
    Bering Land Bridge, 49, 50–51
    Clements-Gleason plant community debate, 163–164
    of early Holocene, 222–232
    in floristic Great Basin, 31–32, 164–168
    future predictions, 344
    Great Basin zones, 31–32, 143, 144–145
    in Mojave Desert, 159–164, 167
    in northwestern Bonneville Basin, 139–141
    in Ruby Valley, 141–145
vertebrates
    of floristic Great Basin, 32–33
    of late Pleistocene, 173–213
    See also specific species
villages, 27f, 36–37, 320–324, 326, 329, 330
Virgin River, 6, 217
Virginia Range, 25, 28, 247
volcanic activity, 263, 265
volcanic ash deposits, 61, 62, 113, 246, 247
Volcanic Tablelands, 151–152
voles, 72, 73, 204–205, 233, 234f, 258
vultures, 181, 183, 184–186

Wah Wah Mountains, 164
Wahoe language, 35
Walker, Joseph, 6, 8–9
Walker Lake, 7, 87, 91, 110–111, 221, 263, 265–268
Walker Pass, 8
Walker River, 11, 91, 106, 107–108, 111, 265–268
Ward, Peter, 80
Warner Mountains, 28, 226
Warner Valley, 275
Warren, Claude, 291, 312
Wasatch Range, 13, 17, 32, 90, 91, 125, 204
Wasden site, 258
Washoe pine (*Pinus washoensis*), 28
Wassuk Range, 28, 29, 120
water table, 142–143
Wayne, William, 122, 123–124
weasels, 204
Weber River, 12, 91
Wehausen, John, 342
Wells, Philip V., 136–137, 155, 156, 159, 164, 165, 167, 224, 228
Wells, Stephen, 119, 240
Wesnousky, Steve, 109
West Humboldt Range, 146, 248
West, Neil, 25, 26
West Walker River, 5, 11, 262–263
Western Black Vulture (*Coragyps occidentalis*), 181, 183, 184
western ground snake (*Sonora semianulata*), 33
western harvest mouse (*Reithrodontomys megalotis*), 233, 249
western juniper (*Juniperus occidentalis*), 25–26, 154, 167, 263
western migration, 12, 94–99, 145

western patch-nosed snake (*Salvadora hexalepis*), 33
western white pine (*Pinus monticola*), 17, 32
Wetmore, Alexander, 61
Wheeler, Georgia N., 147, 300
Wheeler Peak, 16
Wheeler, Sydney, 147, 300
Whipple yucca (*Hesperoyucca whipplei*), 156, 157, 159
whitebark pine (*Pinus albicaulis*), 27, 30, 143, 149
white bursage or burrobush (*Ambrosia dumosa*), 153, 155, 156, 159, 161, 162, 227–228, 232, 252
white fir (*Abies concolor*), 17, 32, 143, 154, 161, 164, 168, 169
White Mountain Peak, 16
White Mountains, 16, 27, 28, 29, 31, 33, 124, 261, 263, 264, 302, 322–324, 326, 341
White River springfish (*Crenichthys baileyi*), 206
white-footed woodrat (*Neotoma fuscipes*), 136
white-tailed deer (*Odocoileus virginianus*), 59, 235
Whitewing Mountain, 263
Whitlock, Cathy, 226
Wigand, Peter, 24, 146, 148, 149, 151, 159, 160, 161, 166, 167, 225, 229, 231–232, 240, 247, 256, 260–261, 262, 315
wild potato (*Solanum maglia*), 60
Wilde, Jim, 311, 329
wilderness areas, 344
Wildhorse Lake, 123, 245–247, 260
Willamette Valley, 4
Willard Peak, 30
Williams, Howel, 61
Windust points, 294
Wingate Pass, 118
Winnemucca Lake, 11, 88, 109, 110, 167, 220
Winograd, Isaac, 207
Winter Lake, 112, 114
winter villages, 36–37
winterfat (*Krascheninnikovia lanata*), 143, 146
Wisconsin glaciation, 46–48, 50, 51, 123–124, 126
wolves, 69, 176, 342
Wono tephra, 113
Wood Stork (*Mycteria americana*), 181
Woodcock, Deborah, 155, 156, 157, 163, 228
woodrats, 136–141, 191t, 192–195, 195f, 206, 232–233, 235, 241, 249, 264–265. *See also* packrat middens

Woods' rose (*Rosa woodsii*), 226
Woolfenden, Wally, 149, 152–153
woolly mammoth (*Mammuthus primigenius*), 71
wooly rhinoceros (*Coelodonta antiquitatus*), 49
World Wildlife Fund, 32
Wyoming
 floristic Great Basin, 17, 18
 hydrographic Great Basin, 11
 See also specific locations

Yana RHS site, 52
yellow rabbitbrush (*Chrysothamnus viscidiflorus*), 22, 154, 155
yellow-bellied marmots (*Marmota flaviventris*), 190–192, 193f, 205, 233, 241
Yosemite Valley, 8
Young, Craig, 297
Younger Dryas, 104–105, 110, 117
Yuan, Fasong, 263
Yucca, 18, 21f, 25, 149, 150, 156, 157, 159, 344
Yucca Mountain, 111, 162, 230
Yukon, 46, 48, 50, 51, 57, 69, 71, 78, 176, 282

Zenda threshold, 101, 102
Zreda, Marek, 124
ZX Lake, 112, 113, 114

Indexer: Stephanie Reymann
Composition: Michael Bass Associates
Text: ITC Stone Serif Med
Display: Akzidenz Grotesk
Printer and Binder: Thomson-Shore